THE ROUTLEDGE HANDBOOK OF GRASSROOTS CLIMATE ACTIVISM

The Routledge Handbook of Grassroots Climate Activism introduces contemporary forms of grassroots climate activism from around the world through the lenses of a variety of academic disciplines, methodologies, and perspectives. Focusing on bottom-up case studies, it showcases innovative and creative approaches, as well as the knowledge of those working towards swift decarbonisation, just transitions, and climate justice.

Grassroots climate activism presents a rich body of material to be studied not only by anthropologists, sociologists, geographers, and political scientists but also by scholars in the humanities and the creative arts. This timely handbook explores climate activism across six continents, and it provides perspectives from climate activists themselves. The authors interrogate a range of key questions: what forms of mobilisation, organisation, and practice constitute grassroots climate activism, and how have these changed over the last decade? What are the boundaries of the climate movement and how does it interact with, or differ from, other social movements? How do activists engage with the moral dimensions of the climate crisis? How do grassroots engagements with climate struggles give shape to plural, site-specific, but nonetheless interconnected, forms of climate activism? What tools do climate activists use to create functioning and effective local, national, and transnational networks? How has climate activism been impacted by the Covid-19 pandemic? What is the relationship between critical scholarship and climate activism? What methodologies are particularly effective for studying climate activism, and why?

This handbook aims to inspire others to devote more attention to grassroots climate activism. It brings together established and up-and-coming scholars, scholar-activists, and practitioners who present novel, cutting-edge research and new findings exploring current developments in different parts of the world. This book will be of particular interest to students and scholars of climate activism, climate solutions, climate and society, human-environmental crises, grassroots activism, and social movements. It will also be of interest to practitioners involved in climate action and to all those who are ready to launch their own grassroots initiatives, or support one of the many already underway.

Sabine von Mering is Director of the Center for German and European Studies, Professor of German and Women's, Gender, and Sexuality Studies, and a core faculty member in the Environmental Studies Program at Brandeis University in Waltham, Massachusetts, USA.

Thomas E. Bell has a PhD in social anthropology from the University of Kent, UK, where he is an associate lecturer at the Durrell Institute of Conservation and Ecology.

Alexandre da Silva Faustino has a PhD in urban geography completed at the School of Global, Urban, and Social Studies, RMIT University, Melbourne, Australia, and is a member of The Alliance for Praxis Research.

Wendy Steele is Professor of Sustainability and Critical Urban Governance in the Centre for Urban Research at RMIT University, Australia and co-Chair of Future Earth Australia (FEA).

Ann Ward has a PhD in sociology from Brandeis University, USA and works at the Office of Sustainability at Tufts University, USA.

Mariana Arjona Soberón is a PhD candidate at the Rachel Carson Center and the Institute of Social and Cultural Anthropology at the Ludwig-Maximilian-University of Munich, Germany.

"This book, brimming with ideas, offers activists, practitioners and academics a much needed critical climate praxis"

Julian Agyeman, Professor, Tufts University

"Just what we have needed: a global, diverse, and accessible must-have collection for anyone who studies, teaches or is active in the climate movement."

Juliet B. Schor, Professor of Sociology, Boston College, UK

"An incredible resource for activists and activists-to-be."

Luisa Neubauer, climate activist, co-founder of the Fridays for Future youth climate movement in Germany, and co-author of *Beginning to End the Climate Crisis. A History of Our Future* (2023)

"In Brazil, across the Latin American continent and globally, grassroots climate activism movements are fighting for justice. They fight to overcome historical inequalities. They fight for processes of decoloniality. They fight for racialized and urbanized communities to be heard, and their rights respected, protected and fulfilled. Therefore, this handbook makes a significant academic and political contribution to shed light and resituate the knowledge production process itself. And to understand the achievements of activist movements, opening new possibilities for thinking and acting."

Renata Bovo Peres, Professor of Environmental Sciences and Urban Planning at Federal University of Sao Carlos (UFSCar), Brazil

"This multidisciplinary approach to the study of grassroots climate activism is full of hopeful case studies of what we can do to address the wicked problem of climate change."

Annette Kehnel, Chair of Medieval History, University of Mannheim, Germany, author of *The Green Ages. Medieval Innovations in Sustainability*

"Each of us can make a difference in the climate movement. This book shows us how."

Betty Lai, Associate Professor in Counseling Psychology at Boston College and author of *The Public Scholar* (forthcoming from Princeton University Press)

"A must-read primer for anyone seeking to navigate the ever-increasing multiplicity of diverse and emergent forms of climate activism. A remarkable achievement."

Professor Rob Fish, Centre for Environmental Policy, Imperial College London, UK

THE ROUTLEDGE HANDBOOK OF GRASSROOTS CLIMATE ACTIVISM

Edited by Sabine von Mering, Thomas E. Bell,
Alexandre da Silva Faustino, Wendy Steele,
Ann Ward, and Mariana Arjona Soberón

Routledge
Taylor & Francis Group

LONDON AND NEW YORK

Designed cover image: Mirror Shield Project, by Cannupa Hanska Luger.
Photograph courtesy of the artist and Garth Greenan Gallery, New York City.

First published 2025
by Routledge
4 Park Square, Milton Park, Abingdon, Oxon OX14 4RN

and by Routledge
605 Third Avenue, New York, NY 10158

Routledge is an imprint of the Taylor & Francis Group, an informa business

British Library Cataloguing-in-Publication Data
A catalogue record for this book is available from the British Library

Library of Congress Cataloging-in-Publication Data
Names: Von Mering, Sabine, editor.
Title: The Routledge handbook of grassroots climate activism / edited by Sabine von Mering,
Thomas E. Bell, Alexandre da Silva Faustino, Wendy Steele, Ann Ward, and Mariana Arjona Soberón.
Other titles: Handbook of grassroots climate activism
Description: Abingdon, Oxon ; New York, NY : Routledge, 2025. | Series: Routledge international
handbooks | Includes bibliographical references and index.
Identifiers: LCCN 2024031832 | ISBN 9781032500232 (hardback) | ISBN 9781032500263 (paperback) |
ISBN 9781003396567 (ebook)
Subjects: LCSH: Environmentalism—Citizen participation. | Climatic changes—Prevention—Citizen
participation. | Climate justice—Citizen participation.
Classification: LCC GE195 .R6785 2025 | DDC 363.7—dc23/eng/20240922
LC record available at https://lccn.loc.gov/2024031832

ISBN: 978-1-032-50023-2 (hbk)
ISBN: 978-1-032-50026-3 (pbk)
ISBN: 978-1-003-39656-7 (ebk)

DOI: 10.4324/9781003396567

Typeset in Sabon LT Pro
by Apex CoVantage, LLC

CONTENTS

Contents

FIGURES

TABLES

NOTES ON CONTRIBUTORS

Co-editors

Sabine von Mering is Director of the Center for German and European Studies, Professor of German and Women's, Gender, and Sexuality Studies, and a core faculty member in the Environmental Studies Program at Brandeis University in Waltham, Massachusetts, USA.

Thomas E. Bell has a PhD in social anthropology from the University of Kent, UK where he is an Associate Lecturer at the Durrell Institute of Conservation and Ecology.

Alexandre da Silva Faustino has a PhD in urban geography completed at the School of Global, Urban, and Social Studies, RMIT University, Melbourne, Australia and is a member of The Alliance for Praxis Research.

Wendy Steele is Research Director at RMIT Europe, Spain and Professor of sustainability and critical urban governance at the Centre for Urban Research at RMIT University, Australia.

Ann Ward has a PhD in sociology from Brandeis University, USA and works at the Office of Sustainability at Tufts University, USA.

Mariana Arjona Soberón is a PhD candidate at the Rachel Carson Center and the Institute of Social and Cultural Anthropology at the Ludwig-Maximilian University of Munich, Germany.

Contributing Authors

Maximilian Becker is an economist and economic lawyer. For many years he has been an active member of the climate justice movement in Germany, since 2016 with the anti-coal movement *Ende Gelände*.

Judith Bessant is a professor at RMIT University, Melbourne, Australia. She was awarded a Member of the Order of Australia in 2017 for "Significant service to education as a social scientist, advocate and academic in youth studies research." She writes and publishes extensively in the fields of sociology, politics, youth studies, policy, media-technology studies, and history.

Benjamin Bowman is Senior Lecturer in the Department of Sociology at Manchester Metropolitan University. He specialises in collaborative research with young people at the Manchester Centre for Youth Studies, an interdisciplinary centre for research with a specialism in youth participatory action research (YPAR).

Caelli Jo Brooker is Lecturer in Indigenous Studies at the Wollotuka Institute for Indigenous Education and Research at The University of Newcastle, Australia. She is a multidisciplinary academic and creative professional working across the creative and cultural industries, tertiary education, and the galleries, libraries, archives and museums sector.

Paloma Bugedo is a visual creative from between the Andes and the Pacific. Throughout her career she has been interested in activating public spaces, promoting active mobility, and engaging with communities through collaborative and creative ecological art practices.

Elena Apostoli Cappello is a social anthropologist. She is currently a researcher at Université Libre de Bruxelles (Sasha) and an associate researcher at the Laboratoire d'Anthropologie Politique in Paris (EHESS–CNRS). She analyses, with an ethnographic anchoring, emerging political issues in Europe, the processes of subjectivation in contexts of activism, and public participation, especially in eco-socialist and environmentalist movements.

Dr. Michelle Catanzaro is Senior Lecturer in Visual Communication Design, School of Humanities and Communication Arts, Western Sydney University, Parramatta, NSW, Australia. Michelle's current research focuses on understanding how young people draw on creativity and visual language to express their cultural, political, and social ideals.

Andrea Catellani is Professor of Communication at the Catholic University of Louvain (UCLouvain, Belgium). He has published various scientific articles and books, notably on environmental communication and rhetoric, CSR, the semiotic approach to organisations, ethics in communication, and the relationship between religion and communication. He was Lead Project Investigator of the research project "Overcoming Obstacles and Disincentives to Climate Change Mitigation" (funded by JPI Climate, 2020–2024).

Linda J. Chase, composer and professor, integrates sustainability education into the musical arts curriculum at Berklee College of Music and New England Conservatory. Her recent cross-genre oratorio, based on *Laudato Si'*, is a musical call to action. Highlighting compassion, *For Our Common Home – Resounding Ecojustice* addresses the environmental crisis and the need to heal our relationship with Earth and each other.

Innocent Chirisa is Vice Chancellor of the Zimbabwe Ezekiel Guti University (ZEGU) in Bindura, Zimbabwe and a Research Fellow with the Department of Urban and Regional Planning, University of the Free State, Bloemfontein, South Africa. As a full Professor of Environmental Planning, his research has focused on urban and regional planning, resilience, and environmental systems dynamics with a proclivity to land use, housing, ecology, public health, water, and energy.

Sylvia Cifuentes is Assistant Professor of Environmental and Social Equity and Justice at Mount Holyoke College. As an interdisciplinary scholar, she investigates the connections among Indigenous politics and global environmental challenges, including climate change, deforestation, and "Smart Earth" technologies with a geographical focus on Amazonia.

Her research has been published in the *Journal of Latin American Geography*, *Digital Geography and Society*, and *Perspectives on Global Development and Technology*, among other publications. She has collaborated with several Indigenous and environmental organisations as part of her research and prior to academia.

Philippa Collin is Professor and Co-Director of the Young and Resilient Research Centre at Western Sydney University. She researches political participation and citizenship in digital society with a focus on the implications for health, wellbeing, and democracy.

Louise-Amélie Cougnon is Head of Research at the Media Innovation & Intelligibility Lab (UCLouvain). Today, her main work is focused on the development of an important project of *Digital Corpora Monitoring* (including Twitter, Instagram, Twitch, YouTube, and TikTok) that would act as an historical archive of national interest, notably related to crisis communication (climate, COVID-19, wars . . .) and empowerment needs (children, seniors, NEETs, migrants, alternative organisations . . .) among the population.

Joy Egbe is a social entrepreneur, renewable energy researcher, sustainability consultant, and climate activist with many years of experience in advocacy, policy, and entrepreneurship. She is the co-founder of Newdigit Technologies and worked as Chief Impact Officer with a vision to eradicate energy poverty in Nigeria.

Reba Elliott has 15 years of leadership experience, with a focus on organisational development and partner engagement for faith-based environmental NGOs. Global projects she has led have engaged 1 billion people, including the campaign for *The Letter*, a documentary film on frontline leaders and Pope Francis; a global sustainability platform with guidance on high-impact actions; and strategic communications to engage new audiences.

Joe Ellis is the Sigrid Rausing postdoctoral fellow at the Mongolia and Inner Asia Studies Unit, Affiliated Lecturer in the Department of Social Anthropology, and Fellow and Director of Studies of St Catharine's College at the University of Cambridge in Cambridge, UK. His research is founded upon long-term ethnographic fieldwork in Mongolia, particularly in the ger districts of Ulaanbaatar and the more remote Khovd province in the far west of the country.

Ellen Field is an assistant professor in the Faculty of Education at Lakehead University in Orillia, Ontario, Canada. Her research interests are in policy and practice of climate-change education in the Canadian K–12 system. She teaches environmental education (B.Ed) and climate-change education (M.Ed) in the Faculty of Education and has engaged with hundreds of in-service teachers in professional development workshops.

Hannah Fitchett is a PhD candidate in social anthropology at the University of St Andrews in Scotland. Her PhD research explores the occurrence of ethical change within the transnational Extinction Rebellion (XR) movement, focusing specifically on how XR's "regenerative culture" approach manifests and compares in XR London and XR Madrid.

Adam Fleischmann is an anthropologist, writer, and teacher with a PhD in anthropology from McGill University in Tiohtià:ke/Mooniyang/Montréal, Québec. His research and teaching practices bridge anthropological and related approaches to the study of global climate change, science and technology, social movements, the environment, institutions, energy, expertise, knowledge, and ethics. He is Postdoctoral Fellow in Urban Sustainability at the University of Ottawa in Ottawa, Ontario, Canada.

Andy Gheorghiu is a freelance anti-gas campaigner and consultant for climate and environmental protection. He is a co-founder of the German Climate Alliance Against LNG and a member of the Break Free From Plastic movement and Saving Okavango's Unique Life (SOUL) Alliance.

Keerti Gopal is a New York City-based climate journalist covering grassroots activism in the climate movement. Previously, she conducted research on youth climate activism in the United States and, as a Fulbright-National Geographic Storytelling Fellow in Taiwan. While she was a student at Northwestern University, Keerti also organised for climate and economic justice with the Sunrise Movement.

Faith Gordon is Associate Professor and Deputy Associate Dean of Research at the ANU College of Law, Australian National University. She is Director of the Interdisciplinary International Youth Justice Network and Co-Founder of the Australian and New Zealand Society of Criminology's Thematic Group on children, young people, and the criminal justice system.

Emilia Groupp is a PhD candidate in the Department of Anthropology at Stanford University, working at the intersection of environmental anthropology and science and technology studies. Her current research focuses on renewable energy development, grassroots political organising, and the cultivation of "climate resilient" landscapes.

Bowen Gu is a postdoctoral research fellow at the Boston University Global Development Policy Center. Her research concerns the political ecology of coal and transition minerals in China and "Global China" and explores the intersection of environmental governance, energy transition, and climate justice.

Nícolas Guerra-Tão has a PhD in art, design, and architecture from Monash University, and is a researcher at both Monash and RMIT Universities, based in Naarm|Melbourne, Australia. He is the co-Founder of The Alliance for Praxis Research, a collective of young researchers and artists nurturing spaces and encounters for social change and knowledge sharing.

Colin Harris is the executive director for Take Me Outside, a national non-profit organisation committed to outdoor learning. He is also a PhD student at the Werklund School of Education at the University of Calgary, Canada.

Ashley Hemmers, NPM, MPA (Master Public Administration, Non-Profit Management) is an enrolled member of the Fort Mojave Indian Tribe, a Federally Recognised Indian Tribe, located in California, Arizona, and Nevada. She is the Tribal Administrator for her Nation specialising in multi-state cross-jurisdictional development and management of Tribal economies and government with emphasis in Tribal enterprising, sovereign fiscal and capital wealth strategies, Nevada gaming, and public service operations.

Fiona Hillary is a Naarm/Melbourne based artist/academic working in the public realm. Her passion lies in site specific fieldwork practices and the human/non-human relationships that reveal themselves across time. Exploring scale through publicly shared moments of awe and wonder, working with site, neon, sound, human, and non-human companion species, her work focuses on temporary, fleeting encounters in and of the everyday.

Daniel Hofinger has been a climate justice organiser in the Rhenish Coal Fields for over a decade. He was involved in the Climate Camps of the 2010s and is a co-founding member of the grassroots climate justice group ausgeCO$_2$hlt. He supported the Hambach Forest occupation and co-founded Ende Gelände ("Here and no further"), an organisation dedicated to empowering thousands of people in mass actions of civil disobedience.

Byambabaatar Ichinkhorloo is a senior lecturer in the Department of Anthropology and Archaeology and an Academic Secretary at the Institute for Mongolian Studies, National University of Mongolia. He was also a director of the IISNC under the auspices of UNESCO in 2021–2023 and a senior lecturer at the Department of Social Anthropology and Cultural Studies at the University of Zurich in 2019–2021 while being a co-investigator of the "Gobi Framework" research project at the University of Oxford.

Stewart Jackson is Senior Lecturer in the Department of Government and International Relations at the University of Sydney. His research focuses on green politics in Australia and the Asia Pacific, specialising in party development. These interests extend to green political theory, environmental feminism, and the intersection of social movements and parliamentary politics – particularly the role street protest plays in fostering political identities.

Thomas Karakadzai is Projects Officer at Dialogue on Shelter Trust, an affiliate of Slum Dwellers International (SDI) and a temporary full-time lecturer at the Department of Community and Social Development, University of Zimbabwe. He has five years' experience working with informal settlements in Zimbabwe. His research interests include but are not limited to human settlements (informal and planned), climate change, resilience, disaster management, and environmental planning and policy.

Thomas A. King is Associate Professor of English at Brandeis University in Waltham, Massachusetts, USA. He researches and teaches in theatre and performance studies, queer studies, and seventeenth- and eighteenth-century British literary and cultural studies.

Juan Liu holds a PhD in development studies. She is Associate Professor of political ecology and agrarian studies at the College of Humanities and Development Studies, China Agricultural University (COHD-CAU), Researcher at the Institute of Environmental Science and Technology, Universitat Autònoma de Barcelona (ICTA-UAB), and Associate Editor of *World Development*.

Gregory Lowan-Trudeau is an associate professor with a focus on environmental education and communication in the Werklund School of Education at the University of Calgary, Canada.

Lea Main-Klingst is a legal expert for public international law and fundamental rights with ClientEarth, a legal environmental charity. She holds an LL.M. in human rights from University College London and an LL.M. in international environmental law from the George Washington University Law School.

Prashanti Mayfield is an urban geographer from so-called Australia and Post-Doctoral Research Fellow at UiT – The Arctic University of Norway. She is currently a contributor to the collaborative project Urban Transformations in a Warming Arctic and her current research is based in Nuuk, the capital city of Kalaallit Nunaat (Greenland), looking at how the city is positioning itself as a "capital of the Arctic," considering how the city is being developed – and for whom.

Rumbidzai Irene Mpahlo is the Africa Climate Justice (ACJC) coordinator based in Zimbabwe. She currently works with 27 climate justice-oriented organisations and movements from different parts of the African region to demand climate justice for all.

Sally Neas is a 4-H youth development program advisor with a specialty in environmental education in the Cooperative Extension San Mateo-San Francisco Counties of the University of California Agriculture and Natural Resources. Sally has a PhD in geography from UC Davis and her research focuses on youth climate activism and environmental education.

Chloe Nguyen is a secondary school student and coordinator of Climate Education Reform BC who is passionate about environmental activism, education, and empowering youth to take action towards a better future.

Rebecca Olive is Senior Research Fellow in the Centre for Urban Research at RMIT University. Her work explores the role of lifestyle sport and leisure in human-environmental health and wellbeing, with a focus on ocean sports, such as swimming and surfing. Drawing on feminist and cultural studies theories and ethnographic approaches, she is interested in how everyday physical activities shape our relationships of care for multispecies places and communities.

Mark Ortiz is currently Presidential Postdoctoral Fellow and Assistant Professor of Geography at the Pennsylvania State University. His work focuses on transnational youth climate activism, intergenerational climate justice, digital storytelling, and youth politics.

Gabriella Sesti Osséo is a postdoctoral researcher at the Department of Social, Political, and Cognitive Sciences, University of Siena (Italy). She obtained a PhD in political science and international and European studies at Roma Tre University in 2023. She has contributed to editorial and educational projects at the Italian Institute for Philosophical Studies and Rosa Luxemburg Stiftung. Her research interests include social movement and youth studies, critical theory, and party politics.

Prof. Dr. Hermann E. Ott is a lawyer and consultant on international environmental law and policy. Prior to this he was Managing Director of the Germany Office for ClientEarth. He holds a law degree and a PhD in international environmental law from the Free University of Berlin and is a professor at the University of Sustainable Development Eberswalde (HNEE).

Joseph Emile Honour Percival is a postdoctoral researcher at the University of Hawaii, where his research focuses on analysing carbon sequestration trade-offs in the context of land-use decisions. His work contributes to understanding how land cover and land-use decisions impact climate-change resilience, offering insights for developing more sustainable and adaptive land management strategies.

Aishwarya Puttur is an undergraduate student at UBC, climate justice organiser, and the education and outreach coordinator for Banking on a Better Future, a youth-led organisation focused on institutional fossil fuel divestment.

Alexander Repenning is a sociologist and writer engaged in climate justice with a background in social science (BA), economics (MA), and human rights (MA). He is co-author of the book *Beginning to End the Climate Crisis. A History of Our Future* and is currently working on strategic capacity building and the development of educational formats for activists with organisations such as the Albert Einstein Institution and the International Institute on Nonviolent Action (NovAct).

Damien Rudd is undertaking a practice-led PhD at RMIT University in Melbourne, within the Aegis Research Network. His creative practice aims to imagine nonhuman entities as agential actors in speculative climate fiction. Rudd is part of the Degrowth Belgium network, writes on the culture of climate activism, and works as a climate-change communication consultant.

Yuliya Samofalova is a lecturer in Information and Communication at the University of Lille, France and an affiliated researcher at the Catholic University of Louvain, Belgium. Her interdisciplinary research interests include social media communication, data science, opinion leadership, communication campaigns about the climate crisis, and the transition to a sustainable future.

Lotte Schack is a PhD student in sociology at the Department of Sociology and Work Science, University of Gothenburg, Denmark. She holds a BSc and an MSc in social anthropology from the University of Copenhagen. Her current research explores visions of a fossil-free and just future within the Swedish climate movement.

Madison Shakespeare, a proud Australian First Nations saltwater woman born on and of Gadigal and Bidjigal Country, is a multidisciplinary lawyer, filmmaker, artist, writer, and scholar with extensive academic experience. With jurisdiction and cultural custodianship over the Indigenous Australian Studies Major (IASM) at Western Sydney University for the past eight years, Madison has innovatively employed Indigenous pedagogical approaches to deliver a curriculum focused on truth-telling, which is essential for First Nations Communities asserting their rights over sovereign knowledge and practices. Similarly decolonising are her research methodologies which privilege Indigenous episteme for knowledge inquiry. Aspects of Madison's contributions were recognised and showcased internationally at the United Nations COP28UAE in 2023.

Tahere Talaina Si'isi'ialafia-Mau is a sociology lecturer at the National University of Samoa. She is also currently a doctoral candidate in sociology at the University of Otago in Dunedin, New Zealand. Her research focuses on Pacific youth advocacy and activism, exploring their perspectives, motivations, and outcomes. Outside of her occupation and PhD, she is Chairwoman for the Pacific Youth Council, a regional youth-led NGO that advocates for the interests and development of youth in the Pacific.

Katrina Simon's teaching and research focus on the ways memory is embedded in urban landscapes, particularly in the design and history of cemeteries and memorials and in landscapes that have been altered through events such as abandonment, flooding, and earthquakes.

Annika Skoglund holds expertise in sustainability, climate change, and technology-based entrepreneurship, areas she has researched based in Sweden and the United Kingdom. Through ethnographic long-term collaborations with industry partners, Skoglund advances interdisciplinary perspectives within industrial engineering and management.

Sarah Stapleton is an associate professor with a focus on environmental and science education at the College of Education at the University of Oregon in Eugene, Oregon, USA.

Naomi Stead is Director of the Design and Creative Practice Enabling Capability Platform at RMIT, where she works with researchers across the creative fields to engage in interdisciplinary research leading to social and environmental benefit.

Dr. Maria Antonia Tigre is Director of Global Climate Change Litigation at the Sabin Center for Climate Change Law at Columbia Law School in New York, NY, USA, bringing extensive expertise to her role in managing the Global Climate Litigation Database.

Georgina Treloar is a PhD candidate in environmental social science at the University of Kent in Canterbury, UK. Her thesis is an analysis of Extinction Rebellion's nascent mobilisation from the social movement framing perspective. Georgina has previously been involved with formal politics as an elected district councillor and parliamentary candidate for the Green Party.

Adriana Voss-Andreae, MD PhD, has a background in food justice policy and was founding Executive Director of 350PDX, a grassroots climate justice organisation based in Portland, OR, USA. She helped grow the organisation to thousands of activists and a series of major climate justice action and policy wins that have become national models. She currently serves as founding President of Ariel's ARC, an organisation working at the intersection of climate and food justice.

Mattias Wahlström is Associate Professor of sociology in the Department of Sociology and Work Science, University of Gothenburg, where he is also Assistant Head of Department with responsibility for research and postgraduate studies.

Fleur Watson is Associate Professor, School of Architecture and Design at RMIT University and founder of independent curatorial practice *Something Together*. She is an experienced researcher within an "integrated scholarship" framework with a focus on curating contemporary art and design within a cross-disciplinary context.

Rob Watts is Professor of Social Policy at RMIT and a Fellow of the Australian Academy of Social Sciences. Rob is an internationally recognised sociologist of youth, political dissent, and policy; he has produced a substantial and distinguished body of high-quality scholarship and research since the 1980s.

Linda Williams is Emeritus Professor of cultural history at RMIT University where she leads the Aegis Research Network. She has a particular research focus on cultural and environmental history and has published widely in the field of the environmental humanities. She currently leads a research project assessing emotional responses to visual images of extinction in Australasia funded by the Australian Research Council <LP 220100191>, Saffron Aid, and Greenpeace Australia Pacific.

Kit Wise explores the application of interdisciplinary curriculum and pedagogy within tertiary contexts. In 2014, Wise was the recipient of a major research grant from the Australian Government Office of Learning and Teaching that addressed interdisciplinary assessment design in the Humanities and Creative Arts. He has published extensively on arts education, including co-editing *Transformative Pedagogies and the Environment*, 2018.

Joseph Worthy Jr. is a community organiser, facilitator, and scholar activist with over 15 years experience organising frontline communities in the United States around mass incarceration, education, police brutality, and women's rights. He has studied under and been mentored by Rev. James Lawson, the theoretician behind the American Civil Rights Movement; Marshall Ganz of the Farm Workers Movement; and Gene Sharp, founder of the Albert Einstein Institution.

Haneen Ali Zeglam is a non-resident scholar at the Middle East Institute's North Africa and Sahel Program and a PhD student in University College London's (UCL) Political Science Department. Her research focuses primarily on political attitudes and behaviour across North Africa, but her interests span many of the other political, social, and economic trends across the Maghreb and the Sahel.

FOREWORD

I've been working on climate since 1988, when I was writing what is generally regarded as the first book on this greatest of crises. That means I've gotten to watch rise of this global activist movement, which is one of the two happy developments in what's been a fairly dismal three and a half decades. By now I think it's safe to say it's as large as any movement has ever been, which makes sense given the size of the problem.

But it's worth pointing out how odd it is that we even need climate activism. After all, key to the story of global warming is that it came with a timely warning: beginning in the late 1980s scientists told us in detail and with consensus that we faced a serious problem. They detailed, accurately, how hot it would get and what we needed to do to prevent that warming. And they were believed – in those years the Republican president of the United States, George W. Bush, said "we will fight the greenhouse effect with the White House effect." It didn't feel like we'd need to mobilise.

And then the fossil fuel industry got its venal act together, building an architecture of deceit, denial, and disinformation that told a very big lie – that climate change was not a serious problem – which cost us the years when we could have been making progress. It's in opposition to their obstruction that grassroots climate activists have organised. And they have taken on every aspect of the fight – blocking pipelines, divesting stock, heading to court, organising boycotts, throwing soup on paintings. This book makes clear the vast and creative array of tactics that have grown up over time – and it also makes clear how much of that opposition has come not from old-line environmental groups but from the front-line communities, often people of colour, Indigenous, and poor, who have borne the brunt of the impacts so far. I know many of the movements described here – my admiration for the people who built them is boundless.

I said before that this was one of two happy developments. The other is the rapid decline in the price of renewable energy, to the point where we now live on a planet where the cheapest way to produce power is to point a sheet of glass at the sun. This opens up possibilities for new kinds of activism: along with anger and fear and guilt and sadness, we can add a certain kind of adventure and romance to the roster of emotions we might call on, as we try to quickly wean the world off combustion and rely instead on the large ball of burning gas hanging in the sky. To make that transition fast enough to catch up with the physics of climate change will require every bit of courage and cleverness that activists can muster.

For all the good work described here, we're not going to stop global warming – though we may still stop it short of the worst cases. But we're also going to have to take care of those who are already suffering: activism turning also to mutual aid, opposition morphing into community building. Thank heaven we have lots of people who care and who know how to act.

Bill McKibben,
founder of 350.org and Third Act

ACKNOWLEDGEMENTS

This volume is the result of beautiful cooperation among many dedicated people from around the world. As editors we would like to express our gratitude to our contributing authors for trusting us with their work and for responding so kindly and patiently to our suggestions and comments throughout this multi-year process. We know how much time and unpaid labour went into realising this project, and we are deeply grateful.

A special thank you to Hannah Rich, our editor at Routledge, Lauren Powell, and Jayachandran Rajendiran, and to Preston Merrill, Brandeis University class of 2025, who, thanks to a grant from the Hearst Foundation, served as our Undergraduate Research Assistant in 2022–2023. Thomas A. King co-authored the introduction to Part II – a special thanks to him. Thank you also to Poppy Bardwell for helping with copy editing and proofreading.

The seedling idea for this volume was planted on 14 December 2021 when Sabine hosted Ann, Mariana, and Tom to present their research about the climate movement at a Center for German and European Studies (CGES) Online webinar entitled *Fighting for a Liveable Future: The Climate Movement in Germany, Europe, and the Americas*, at Brandeis University. Ann, who was completing her doctorate at Brandeis at the time, had previously served as a teaching assistant for Sabine's course *Human/Nature: European Perspectives on Climate Change* in Spring 2020, and Tom and Sabine had first met at the Harvard Co-op Bookstore in February 2018 when Tom was preparing for his fieldwork. Mariana joined from across the pond from the Rachel Carson Center in Munich, which was the result of a connection created with the help of Kiley Kost at Carleton College, with whom Sabine was then working on the Climate Emergency and Technology (CLEAT) committee for the German Studies Association (thanks, Kiley!). After Tom helped Sabine clean out her house for a year-long sabbatical in January of 2022, they were able to finalise the initial proposal in person, when Sabine visited Brighton, UK in May 2022.

It was thanks to the encouragement from Emily Ross, our first editor at Routledge, that the initial proposal for a much smaller multidisciplinary volume on the climate movement grew into the full-fledged handbook you are now reading. In response to helpful input from reviewers at Routledge, the initial group of editors was looking to develop and deepen the disciplinary and geographical scope of work and welcomed Wendy and Alex from RMIT University (Australia) to the team in October 2022. Wendy's transdisciplinary research centres around wild cities, planetary civics, and socially innovative climate action through an exploration of quiet activism and the extraordinary ways in which ordinary people take action on climate change at the local scale. As a founding member of the Alliance for Praxis Research, Alex is an activist and scholar working with communities and grassroots initiatives in both Brazilian and Australian contexts. Despite the 16,932 kilometres and, more importantly, 14–16 hours of time difference between Boston and Melbourne depending on the time of year, the team was able to work closely together on a number of rounds of revisions of the 40 chapters and joyfully celebrate the arrival of Mariana's Luca, as well as Ann's, Alex's, and Tom's completed dissertations, in the course of the 19 months of working together.

We humbly acknowledge that we are by no means representative of the vast diversity of scholars and practitioners in this field and that our positionalities and encounters come with various levels of unearned privilege that allow us to engage in this work. While we have made strides to ensure that our contributors represent as many different voices as possible and are proud to have achieved not only gender parity but a very healthy representation of female authors among our contributors, we would have loved to have done more to ensure maximum inclusivity in all other respects as well. We know that our volume is unconventional in many ways – we are proudly "undisciplined" and hope that this handbook will encourage others to follow in our footsteps. We dedicate this work to all the climate activists around the world who are working tirelessly to protect a liveable future for all creatures, communities, and places on this beautiful planet.

1

GRASSROOTS CLIMATE ACTIVISM ACROSS SIX CONTINENTS

Sabine von Mering, Thomas E. Bell, Alexandre da Silva Faustino, Wendy Steele, Ann Ward, and Mariana Arjona Soberón

Responding to the "unprecedented, catastrophic flooding" that devastated local communities in the town of Lismore in 2022 in the biggest modern flood disaster in Australian history,[1] Nicole Rogers writes that

> [a]ll academics, and not only the academics of a university already experiencing the ravages of climate change, have an overriding obligation to be climate activists. There is no possible justification for neutrality, dissembling or prevarication in light of what we know, and have experienced, at 1.1 degrees of global warming.
>
> *(Rogers, 2023, p. 133)*

Animated by her impassioned appeal, this "undisciplined" handbook sets out to examine grassroots climate activism as a crucial feature of an epoch of perilous human–environmental crises. It starts from three primary questions: what does grassroots climate activism – or taking action on climate change – look like in different parts of the world and in different political, economic, cultural, and institutional contexts? How can it be studied in "undisciplined" ways – without the restraints of disciplinary boundaries, and by giving space to nuance, dilemma, and critical analysis – while remaining sensitive to the phenomenal urgency of climate breakdown? And in what ways can a handbook such as this serve as a prompt for others to get involved in tackling the climate crisis, in the circumstances of their own lives, and using the repertoires of action available to them?

Grassroots Climate Activism(s): Ground-Up and Urgent

As attested by the chapters in this handbook, what it means to be a climate activist – someone motivated to act "[in relation to the] difference . . . between what is and what ought to be" (Heywood, 2018, p. 4) – is culturally variable, as are the repertoires of "critique, invention, and creative practice" (Dave, 2012, p. 3) that constitute climate activism. This handbook is necessarily planetary-scale in outlook, and it treats grassroots climate activism as a dynamic term that incorporates a diversity of forms, as well as scales and place-based

DOI: 10.4324/9781003396567-1

modes of action. In this spirit of openness and inclusivity, we provide (an account of) critical analyses from a variety of academic disciplines, as well as creative and practice-based perspectives. The handbook not only employs a variety of methodologies but includes a range of texts based on different knowledges and approaches to transformation with a much longer history than modern academe – from storytelling and listening to civil disobedience.

Indeed, this is at the heart of what is suggested by our use of "grassroots" activism: a practice that is rooted in the ground, it grows up from the local soil through and among people acting alongside each other (Mihaylov and Perkins, 2015), and it evokes a sense of both regenerative flourishing and solidarity with more-than-human environments and communities (Crowley et al., 2021). It is also at the grassroots level that we are most likely to discover different ways of knowing that deepen, broaden, and complement but, at times, bring into question established ways of thinking in academia, policy, and think tanks. As demonstrated by many chapters in this volume and shown by Apostolopoulou et al.'s (2022) work on radical social innovations propelled by community-led or social movements, activist groups and communities often create and deploy grassroots modes of organisation and action to counter traditional structures of governance – those institution-alised through the nation-state, intergovernmental bodies, markets, and corporations – that are often insufficient, negligent, and exclusionary by design (Graeber, 2007, 2009; Juris and Khasnabish, 2013; Shukaitis et al., 2007; Swyngedouw, 2004). Grassroots activism, then, plays a pivotal role in contesting hegemonic power arrangements and the structural injustices that are foundational to the climate crisis (Klein, 2014). It is a mode of organisation and action that can respond in a diversity of creative ways to the void created by governmental inaction and address the inherent contradictions and encroachments of capitalist enclosure, dispossession, and accumulation (Mascarenhas-Swan, 2017; Steele et al., 2021). And by creating opportunities for resistance, solidarity, adaptation, and increased resilience and regeneration (Moore and Russell, 2011), it remains at the forefront of advocating for just and democratic socio-ecological solutions, often led by marginalised communities rooted in particular environments (Fairchild and Weinrub, 2017; Faustino, 2023; MGJEP, 2017; Tokar and Gilbertson, 2020).

Our use of grassroots signals how this handbook centres on climate activists themselves, on the people who are driving change in their local communities, inside their professions or disciplines, on different continents, and within very different political, cultural, and eco-logical contexts. Through its focus on grassroots modes of participation and action, the handbook presents a fresh look at climate activism from *within*, and it provides accounts of climate activism *as it is practised* in different parts of the world. Indeed, what constitutes collective action in a given cultural and political context is a matter for investigation, as are the ways in which individuals connect their activism to broader struggles for transforma-tion. Directed at policymakers, grassroots climate activism may target those responsible for greenhouse gas emissions (fossil fuel executives; banks; insurances; etc.) through traditional forms of political advocacy. But it can be enacted as a climate activist-author's blank page, a climate artist's performance or curatorial practice, a teacher's decision to innovate the curriculum, or as a research method of a scholar-activist aiming to raise awareness, educate, and inspire change.

Just as grass is rooted in the specific conditions of the soil from where it spreads, grass-roots climate activism is bottom-up activism that is conditioned by – but has the potential to spread far beyond – its original locale. In the chapters of this handbook, we find grass-roots climate activists who are deeply committed to their local context but who equally understand that addressing a planetary-scale phenomenon is a task that requires different levels of cooperation; climate activism produces not only translocal relations but increasingly more radical forms of planetary governance beyond the sphere of the nation state. As this collection of chapters demonstrates, the limitations of nation states in defining and addressing global climate agendas is a recurrent factor that animates the responses of grassroots activists.

After a growing number of scientists began raising the alarm about global warming in the late 1980s, the United Nations convened the "Earth Summit" in Rio de Janeiro, Brazil, in 1992 and founded the United Nations Framework Convention on Climate Change (UNFCC). The UNFCC held its first "Conference of the Parties" (COP1) in Berlin in March/April 1995, and its establishment led to the Kyoto Protocol in 1997 (UN Climate Change, n.d.) – a binding treaty aimed at reducing carbon emissions worldwide. However, President George W. Bush pulled the United States, the largest per capita emitter, out of the Kyoto Protocol in 2001. It took until 2015 (COP21) for the Paris Agreement to be signed, through which 195 Parties to the UNFCC (194 states and the European Union) have agreed to limit global warming to 2 degrees Celsius (2.7 F) and ideally to 1.5 degrees (United Nations, n.d). Sluggish negotiations at the COPs, hijacked repeatedly by fossil fuel interests (Gills and Morgan, 2020), contrast with ever-more-dire scientific warnings (McGuire, 2022; Richardson et al., 2023). At the same time, an ever-increasing cascade of extreme weather events and growing calls for global climate justice in support of vulnerable populations have brought waves of urgency to the climate movement in recent years (Sultana, 2022), which in turn inspires and demands that climate activists both expand and deepen their own rhizomatic activities (Deleuze and Guattari, 2004) and co-create responses to taking urgent action on climate change.

A Brief (and Selective) Overview of a Shifting Terrain

Since the publication of Dietz and Garrelts' (2014) *Routledge Handbook of the Climate Change Movement*, grassroots climate activism has evolved rapidly in many parts of the world, and the climate movement has developed considerably over the past few years. As introduced later in the chapter, it is within this context that this handbook has been produced. That being said, it is necessary to begin by noting that, contrary to the widely held public perception in the Global North that climate activism is a rather recent phenomenon, climate activists in the Global South have been demanding more attention to the issue of climate change for many decades (Bond, 2016; Burman, 2017; Cifuentes, Chapter 3 in this handbook; Dwivedi, 2001). While many environmental activists Ken Saro-Wiwa in Nigeria and Bertha Caceres in Honduras, have paid for their activism with their lives, that is not keeping others from continuing to follow bravely in their footsteps.

Similarly, it must be acknowledged that the climate movement in Global North countries is composed of diverse elements, with varying historical roots (Dietz and Garrelts, 2014, Nulman 2022). Historical analyses have tended to spotlight two dimensions of the movement (e.g.: Garrelts and Dietz, 2014; De Lucia, 2014; Dietz, 2014; della Porta and Parks, 2014): a (traditionally more white and middle-class) movement for climate mitigation that, in the main, developed out of the ambition to combat global warming through "taking action" on carbon emissions, and a climate justice movement that is rooted in organising for environmental justice, racial justice, Indigenous rights, and so on.[2] Closely related to the environmental justice movement (see Schlosberg and Collins, 2014), Indigenous struggles against settler-colonialism (see Estes, 2019), and the alterglobalisation (global justice) movement (Moore and Russell, 2011, p. 21), the climate justice movement spotlights an important consideration to take into account when contextualising grassroots climate activism: the climate movement has a multiplicity of histories. The story we tell here is merely *one* version that reveals key themes that speak to this handbook.

The climate justice movement has been at the forefront of emphasising how climate change is symptomatic of industrial capitalism, systems of oppression, and histories of slavery and colonialism (Bahvani et al., 2019; Bedall and Görg, 2014; De Lucia, 2014; Klein, 2014; Tokar, 2014a, 2019, 2020). By way of example, frontline organising across the United States led by groups within the Climate Justice Alliance (Mascarenhas-Swan, 2017), which was founded in 2013 (CJA, n.d.), is underpinned by a conceptualisation of the "Just Transition [as] a vision-led, unifying and place-based set of principles, processes, and practises that build economic and political power to shift from an extractive economy to a regenerative economy" (CJA, 2019, p. 1). This is encapsulated by the renowned movement zine *From Banks and Tanks to Cooperation and Caring: A Strategic Framework for a Just Transition* (MGJEP, 2017). Predicated on repairing planetary relations,[3] it recognises that engendering structural transformation "requires stopping the bad while at the same time building the new" (CJA, 2019, p. 3). In other words, resistance to an extractive and violent political economy is fused with organising for – and investing in – "visionary and oppositional" solutions (MGJEP, 2017, p. 21) that, for example, "[d]rive racial justice and social equity" (*ibid*, pp. 16–17), promote "social and ecological well-being" (*ibid*, p. 30), and engender economic localisation, "ecological restoration" (*ibid*, p. 15), and "deep democracy" (*ibid*, p. 18).

The Just Transition Framework signals how the distinctions – and tensions – between the aforementioned dimensions of the climate movement must be understood in the context of the environmental justice movement's direct challenges to environmentalism (Pezzulo, 2022; Sandler and Pezzullo, 2007). Crucially, these include the criticisms of majority-white environmental organisations that emerged from the United States in the 1990s (e.g. Bullard, 1992; Moore and other signatories, 1990; Wright et al., 1994; see: Dowie, 1995, Chapter Six; Pezzullo and Sandler, 2007). In such critiques, racism was linked not only to the exclusion of environmental justice from "prevalent terms and conceptual frames" of prominent environmental organisations (Pezzullo and Sandler, 2007, p. 9) but their "inaction around addressing environmental problems in . . . communities [of colour]" (Moore and other signatories, 1990, p. 3), as well as "ignorance, ambivalence, and complicity" in relation to the perpetuation of environmental injustices (Pezzullo and Sandler, 2007, p. 4).

Moving forward to 2010, frontline climate justice organisations rooted in the environmental justice movement wrote a powerful public response to an "open letter" by 1Sky (Bailey et al., 2010), a climate advocacy organisation that was incorporated into 350.org in 2011 (Tokar, 2014b, p. 253). Following the failure of the US Congress to deliver climate legislation in 2009, 1Sky's letter called for a "redoubl[ing] of efforts to focus on grassroots movement-building" (Bailey et al., 2010). In reply, *Grassroots Organizing Cools the Planet: Letter From the Grassroots to 1Sky* detailed successful – but by implication sidelined – grassroots resistance to fossil fuel infrastructure already being led by frontline communities: "directly impacted communities who have been able to collectively name the ways they are burdened and are organising for action together" (Moore and Russell, 2011, p. 13). It also outlined an "organizing strategy for climate justice" based upon "building power from the bottom-up", generating cross-movement solidarity, "investing in grassroots action at frontline struggles", and "prioritizing local organizing to build the resilient communities, economic alternatives, and political infrastructure that we need" (MGJEP and other signatories, 2010). This went hand-in-hand with critiques of climate advocates' technocratic focus on "carbon targets . . .[that] reinforce the 'carbon fundamentalism' frame that hides the root causes of climate change", as well as the ways in which "advocacy work, however well intentioned, migrated towards false solutions that hurt communities and compromised on key issues such as carbon markets and giveaways to polluters" (MGJEP and other signatories, 2010; see: Amorelli et al., 2021; CJN!, 2012).

In recent years, there has been a growing convergence within the climate movement of the Global North around the concept of climate justice (Cassegård and Thörn, 2017; della Porta and Parks, 2014). This is highlighted by the development of 350.org, an international climate movement organisation that is principally located in the United States, with chapters across the country. Founded in 2008 by Bill McKibben and others, 350.org's name "is a reference to 350 parts per million (ppm) – the concentration of carbon dioxide (CO_2) in the atmosphere considered the safe limit to avoid the worst impacts of climate change" (350. org, 2024a). With an aim of building a worldwide grassroots climate movement "united around the imperative of reducing atmospheric carbon" (Tokar, 2014b, p. 253), 350.org's early work characteristically focused on spearheading mobilisations for climate action at the state and federal levels and in the international arena. Despite 350.org's growing ability to mobilise, its focus on advocacy around emissions and clean energy policy came with criticisms that the organisation was "politically cautious" and relatively uncombative (*ibid*, p. 253), especially when compared to "the more radical, direct-action oriented elements of the movement" (della Porta and Parks, 2014, p. 23).

But the tenor of 350.org's organising shifted from 2011 (Tokar, 2014b, p. 253), and it became increasingly "system-critical in aspiration and orientation" (De Lucia, 2014, p. 69; see Foran, Gross and Hornick, 2019). In the 2010s, 350.org established itself as a critical player in struggles against fossil fuel extraction and the institutional apparatus of the fossil fuel industry (350.org, 2024b), as shown by its role in mobilising opposition to the Keystone XL tar sands pipeline during the Obama era, its efforts to foreground the subsidies and assets underpinning fossil fuel corporations (McKibben, 2012), and its influence on fossil fuel divestment. At the time of writing, 350.org positions itself as "an international movement of ordinary people working to end the age of fossil fuels and build a world of community-centred renewable energy for all", and it "[calls] for a deeper transition, one which places energy justice at the beating heart of its values" (350.org, 2024a). As such, its values frame "the fight against climate change [as] a fight for justice" that involves "listening to the communities who are getting hit hardest, amplifying the voices that are being silenced, and following the leadership of the people on the frontlines of the crisis" (*ibid*).

That there has been a continued "discursive shift" towards climate justice (della Porta and Parks, 2014, p. 21) is also demonstrated by the emergence of the Green New Deal as a dominant framing of the U.S. climate movement, during 2018–2019 (Bell, 2024; Klein, 2019). At its heart, the Green New Deal connects decarbonisation and rapidly tackling carbon emissions to equitable political-economic transformation: major government investment in public infrastructure and climate resilience; job creation and economic redistribution; and struggles for racial and environmental justice (Aronoff et al., 2019; Chomsky and Pollin, 2020; Pettifor, 2019; Tienharra and Robinson, 2022b; see Tienharra and Robinson, 2022a). The Green New Deal's rapid rise to prominence in the United States resulted from the youth-led Sunrise Movement (Prakash and Girgenti, 2020), who sought to compel the Democratic Party establishment to adopt the Green New Deal as an overarching policy framework. In effect, Sunrise established itself as an "activist arm" of the progressive wing of the Democratic Party (Girgenti and Shahid, 2020; Rojas and Shahid, 2020), which was strengthened after the Democrats flipped the House in the midterm elections during Donald Trump's presidency in November 2018. A non-binding congressional resolution for a Green New Deal, introduced by newly elected movement-ally Rep. Ocasio-Cortez (2019) and longstanding Massachusetts Senator Ed Markey (2019) in February 2019, came off the back of Sunrise's late 2018 civil disobedience in Washington D.C., which famously included a "sit-in" in Speaker Nancy Pelosi's office. Similarly, Sunrise's "Green New Deal Tour" in 2019 involved Sen. Ed Markey, Rep. Alexandria Ocasio Cortez, Sen. Elizabeth Warren, and Sen. Bernie Sanders, and they agitated for the Green New Deal to be part of the televised debates during the Democratic Primaries for the 2020 Presidential election.

The Sunrise Movement's organising for a Green New Deal in 2018–2019 took place in the context of the IPCC's (2018a) discourse-shifting *Special Report on Global Warming of 1.5* (SR1.5), released in October 2018. Inspiring the refrain "we have twelve years", popularised by Sunrise (Prakash, 2020, p. xi), SR1.5 argues that limiting global warming to 1.5 degrees Celsius would require net carbon dioxide emissions to be reduced by around 45% by 2030, from 2010 levels; net-zero would have to be reached by around 2050 (IPCC, 2018a, p. 12). And in the opening sentence of its press release, the IPCC declared that "[l]imiting global warming to 1.5°C [2.7°F] would require rapid, far-reaching and unprecedented changes in all aspects of society" (2018b, p. 1). Crucially, then, SR1.5 showed that "there is absolutely no path to staying below 1.5 or 2 degrees of warming that does not involve a society-wide plan for decarbonization in the next decade" (Wallace-Wells, 2020, p. 6). It also made this argument in light of the catastrophic differences between the impacts of 1.5 degrees and 2 degrees:

> by 2100, global sea level rise would be 10cm lower with global warming of 1.5°C compared with 2°C. The likelihood of an Artic Ocean free of sea ice in summer would be once per century with global warming of 1.5°C, compared with at least once per decade with 2°C. Coral reefs would decline by 70–90 percent with global warming of 1.5°C, whereas virtually all (>99 percent) would be lost with 2°C.
>
> *(IPCC, 2018b, p. 1)*

In short, "[t]he differences between 1.5 degrees and 2 degrees may not seem like much, but with climate change every tiny differential comes with an enormous, if not unprecedented, amount of human suffering" (Wallace-Wells, 2020, p. 7).

As put by David Wallace-Wells, SR 1.5 "was followed by a surge of political mobilization around climate change, with activists and disengaged citizens alike terrified by how short the runway seemed: twelve years to halve emissions" (2020, p. 5). A few months previously, in August 2018, Swedish teen Greta Thunberg decided to strike for climate in front of the Swedish parliament; she quickly became a media sensation, and hundreds of thousands of young people were inspired to organise what became known as the Fridays For Future school strikes, first in Europe, and then around the world. While the school strikes were heavily influenced by the actions and words of Thunberg, the proliferation of places where Fridays for Future chapters emerged began to change what had at the beginning been a relatively centralised discourse (see Chapters 6, 7, 10, 15, 16, 17, 18, 21, 29, 32). Also in this period, others joined or initiated local chapters of Extinction Rebellion (see Chapters 19, 34, and 38), which shot to prominence after occupations of central London in

2018 and 2019, and 350.org, as well as a continuously expanding number of other climate organisations, including the Sunrise Movement in the U.S. (see Chapters 7 and 15).

On 20 September 2019, millions were brought to the streets for the first-ever "Global Climate Strike" (Von Storch et al., 2021). A few months later, the COVID-19 pandemic demonstrated the fragility of the current planetary systems that keep our lives and routines moving forward. The pandemic also forced climate activists everywhere to abandon forms of activism that relied on the occupation of public space, due to the healthcare and safety restrictions that were in place in many places around the world. This altered reality, which regulated the movement of bodies in public space, forced activists to reimagine what climate activism could look like. Although activists use digital tools extensively (Youmans and York, 2012), digital landscapes acquired renewed importance during the pandemic, as they offered an alternative space to meet and act collectively. Activists had to innovate ways to strategise and become more adept at networking across distances (Christou et al., 2023; Wahlström and Schack, Chapter 16 in this handbook), while other groups opted for more aggressive tactics (Delina, 2022). During this time some activists felt that the global pandemic led to greater global connectedness, though campaign coordination across long distances continues to pose challenges (Fisher and Nasrin, 2021; Samofalova *et al.*, Chapter 17 in this handbook). The effects of the pandemic are still being felt; the diversity of tactics that emerged in this time has changed the activist landscape as we know it, and it will continue to influence repertoires of activist practice in years to come.

Situating This Handbook: Taking an "Undisciplined" Approach

By providing a snapshot of the shifting terrain outlined earlier through the lens of grassroots climate activism, this handbook adds to a proliferation of scholarship on, for example, the climate movement (Cassegård et al., 2017; Dietz and Garrelts, 2014), climate justice (Bahvani et al., 2019; Jafry, 2018; Tokar and Gilbertson, 2020), "energy democracy" (Fairchild and Weinrub, 2017), and contemporary environmentalism (Grasso and Giugni, 2022). But it is distinctive in a number of ways. Other handbooks have been written primarily by social scientists, with political scientists predominantly in the lead. This handbook intentionally combines social science research with perspectives and diverse methods from the humanities and the creative arts, with some contributors drawing upon ethnographic research, including data derived from participant observation, digital methods, and action research to provide in-depth cases of climate activism in specific locales. As shown by the expansive range of the handbook's chapters, this editorial approach has served to broaden our view of the practices and modes of participation that comprise climate activism in different contexts. We hope that the handbook can encourage engagement and continued knowledge exchange among anthropologists, geographers, sociologists, political scientists, and urban studies scholars, as well as scholars in the humanities, the creative arts, and beyond.

Indeed, it is our view that the climate crisis – and the study of grassroots climate activism – *necessitate* that scholars are not confined to traditional disciplinary silos: they overwhelm the boundaries of our academic disciplines. As the human-environmental crises of the twenty-first century play out with increasing severity, new forms of climate activism are likely to be triggered. What we are seeing – and what we hope to establish with this handbook – is that as the climate crisis worsens, more people (will) define themselves as climate activists wherever they are, often in non-traditional ways: as scholars, artists, youth activists, entrepreneurs, and more. Transdisciplinary scholarship can help us to access, document, and analyse a wealth of climate activism that is otherwise at risk of being over-looked at best – or actively ignored at worst.

Through its spotlighting of a range of disciplinary perspectives, this handbook also makes the case that studying grassroots climate activism requires the deployment of a diverse and innovative range of methodological tools: ethnographic methods, such as participant obser-vation, multiple forms of interviews, and digital ethnography (see Chapters 3, 4, 5, 18, and 28); qualitative and quantitative survey analysis (see Chapters 15 and 16); hermeneutics and content analyses of primary sources, social media posts, and movement publications (see Chapters 6, 17, 21, and 29); visual data analysis using software such as Visme (Chap-ter 17), and mapping (see Chapters 27 and 28); and arts-based lenses (see Chapters in Part II and Chapter 33). In other words, this handbook is receptive to methodological hetero-geneity, and we view this as a crucial part of studying a phenomenon that is by its nature multifaceted and highly dynamic.

As a result of the handbook's "undisciplined" openness to a range of disciplinary and methodological perspectives, its chapters interrogate a broad array of themes that can serve as prompts for further critical, engaged, and comparative investigation. These include forms of political mobilisation, organisation, and practice that constitute grassroots climate activism, and how they have changed (see Chapters in Part III); the boundaries of grassroots climate activism in different contexts and how it is defined and demarcated, both in practice and by actors themselves (see Chapters 22 and 25); how climate activists understand and derive meaning from their activism in different contexts (see Chapters 5, 20 and 36); activist engagements with the moral dimensions of the climate crisis, such as with respect to respon-sibility and justice (see Chapter 4); etc. A number of chapters examine the tools activists use to create functioning and effective (transnational) networks (see Chapters 26, 30, and 38); while others are concerned with the ways in which climate activism has been impacted by (other) global disruptions caused by the cascading social, ecological, and political crises of the twenty-first century (see Chapters 30 and 32). The handbook's chapters also highlight a variety of visions of societal transformation articulated by climate activists in different contexts, and the collection brings to light the forms of politics, socio-economic futures, and alternative ways of living, working, and art-making that such visions generate (see Chapters 31, 33, and 34).

Crucially, this handbook poses questions about the relationship between critical schol-arship and activism (see Chapters 35 and 39), in view of the injustices that come with the devastations caused by climate change and out of which the climate crisis manifests. The handbook also asks how we can most effectively provide the leaders of the next generation with a rich and transdisciplinary set of tools that serve to make sense of a phenomenon that will no doubt continue to change, perhaps in unexpected ways, as the climate crisis continues to unfold (see Chapter 38). As Adam Fleischmann discusses in Chapter 36: "[cli-mate change] is a problem whose underlying cause is not . . . fossil fuels,. . . but a history of exploitation and of violence toward particular human and non-human worlds . . .[I]t is an issue of knowledge and relations, ethical and political". With this in mind, this handbook emphasises the efforts of grassroots activists who are confronting the climate issue from a variety of perspectives, including but not restricted to their roles within the academy. Therefore, we invited a wide range of practitioners, artists, and scholar activists to join us on this journey and provide us with their most promising ideas and wisdom.

By including contributors with a variety of different backgrounds and skills, we not only aim to bridge the gap between academia and "the field" but, more importantly, embrace different forms of knowledge creation and show how crucial – and indeed enriching – this is for tackling the climate crisis. Almost a quarter of the chapters are written by "practitioners"' rather than "academics" – climate lawyers, writers, social workers, and organisers – and many of our contributing authors bring years of practical experience to this work, providing not only a uniquely close-up look at the phenomenon but also turning the scholarship itself into a form of climate activism. As discussed by Hannah Fitchett (Chapter 38), this positionality of course poses challenges, but, as attested by this handbook, it also produces scholarship that is highly accessible, meaningfully engaged, and responsive to the phenomenal gravity of the climate crisis.

Contrary to the doom and gloom that characterises most treatments of the climate crisis, this handbook is buoyed by the wealth of creativity and stamina of grassroots climate activists and activist-scholars, whose inspiring determination to work towards a better future, we hope, will serve readers as a source of inspiration (Solnit and Young Lutunatabua, 2023; Neubauer and Repenning, 2023). There are many ways to support effective decarbonisation, just transitions, and different paths towards climate justice, as the chapters collectively emphasise. Through this handbook, we aim to provide both an introduction to and an overview of these efforts, and we seek to inspire others to join us in this work. This is an integral part of a handbook that brings together established scholars and early career researchers, who have authored key chapters; three of the five editors (Ward, Faustino, and Bell) have recently completed their PhDs, and Arjona Soberón is a PhD candidate. This has been a strength of the editorial process that we celebrate, and it highlights how research on grassroots responses to the climate crisis is being led by scholars at all stages of their academic journeys.

To counter the underrepresentation of women, Indigenous voices, and BIPOC scholars and their knowledge, we have taken special care to invite a broadly diverse group of contributors. It has been our goal to be as inclusive as possible, but we also know that the very nature of scholarship often perpetuates inequities between those with access to institutions of higher education and those without. The commitment to inclusivity is therefore not a badge of honour we aspire to broadcast but an ongoing pursuit, which involves continuous, critical self-reflection, including about our positions within institutions and disciplines that are entangled within the climate crisis and the structural conditions out of which it has arisen (Rogers, 2023).

Brief Overview of the Handbook

This handbook is divided into seven parts. Each part begins with a brief introduction and summary of the chapters contained therein. The arc of the parts follows from our decision to centre frontline communities. We therefore begin not with a traditional methodology and theory section; instead, "Part I: Accounts from the Ground: The Fight for Climate Justice" is devoted to (Indigenous) people and youth at the frontlines in the United States, Brazil, the Pacific Islands, Australia, Taiwan, and Canada. "Part II: Climate Artivism" highlights the broad interdisciplinarity of our approach by featuring a number of studies devoted to the role of the arts in climate activism. "Part III: Diverse Modes of Organising Different Constituencies" takes a closer look at the various groups that are organising at the grassroots level, highlighting elders, youth, and faith-based communities. "Part IV: Civil Disobedience, Litigation, and the Suppression of Climate Activism", introduces chapters studying the breaking, making, and shaping of laws in various contexts around the world.

"Part V: Grassroots Critical Perspectives on Climate Actions" presents chapters focused on disagreements among those who are taking action on climate. "Part VI: Fighting Against False Solutions and Prefiguring Alternatives" is devoted to the campaigns against the false solution of fossil gas and the many initiatives, from local "climatepreneurs" in Sub-Saharan Africa (Egbe, in Chapter 32) to films devoted to climate solutions to insider activism, that aim to prefigure the future climate activists envision. Finally, in "Part VII: Epistemological Questions and New Praxis", we return to the conditions, challenges, and opportunities for researching and teaching the phenomenon of grassroots climate activism. As a final outlook, Chapter 40 addresses key themes related to grassroots climate activism based on the work presented in this handbook and potential future developments.

We hope that, by bringing together and making more visible just a small part of the diverse and critical work of grassroots activists and scholar-activists in different parts of the world and in different political, economic, cultural, and institutional contexts, this handbook will serve as a prompt for others to get involved in tackling the climate crisis, within the circumstances of their own lives, using all of the repertoires of action available to them.

Notes

1 See The Visual Stories Team and Gilmor (2022).
2 The master's thesis of one of Bell's interlocutor's has greatly influenced his understandings of the climate movement's contrasting elements, movement histories, and growing alignment around climate justice. While he would like to uplift this work, he has made the difficult decision not to cite the dissertation here, as part of efforts to help preserve the anonymity of both his interlocutor and the climate movement organisation involved in his research.
3 Thanks to Miguel Alexiedes for this phrasing.

References

350.org. (2024a). *About 350*. https://350.org/about/ [Accessed 5 June 2023].

350.org. (2024b). *2009–2019: 350.Org Celebrates a Decade of Action*. https://350.org/10-years/ [Accessed 5 May 2024].

Amorelli, L., Gibson, D. and Gilbertson, T. (eds.) (2021). *Hoodwinked in the Hothouse: Resist False Solutions to Climate Change*. climatefalsesolutions.org. https://climatefalsesolutions.org/wp-content/uploads/HOODWINKED_ThirdEdition_On-Screen_version.pdf [Accessed 9 February 2021].

Apostolopoulou, E., et al. (2022). Radical social innovations and the spatialities of grassroots activism: Navigating pathways for tackling inequality and reinventing the commons. *Journal of Political Ecology*, 29, 143–188. https://doi.org/10.2458/jpe.2292

Aronoff, K., et al. (2019). *A Planet to Win: Why We Need a Green New Deal*. London and New York: Verso.

Bailey, J., et al. (2010). An Open Letter from 1Sky to All People and Organizations Working to Combat Global Warming. *HuffPost*, 6 August. https://www.huffpost.com/entry/an-open-letter-from-1sky_b_673798 [Accessed 2 September 2021].

Bedall, P. and Görg, C. (2014). Antagonistic standpoints: The climate justice coalition viewed in light of a theory of societal relationships with nature. In M. Dietz and H. Garrelts (eds.) *Routledge Handbook of the Climate Change Movement* (pp. 44–65). Abingdon and New York: Routledge.

Bell, T. E. (2024). *Taking Responsibility for the Climate Crisis in Massachusetts: Climate Activism, Justice, and Organizing to Construct a Green New Deal*. PhD Thesis. Canterbury, University of Kent. http://doi.org/10.22024/UniKent/01.02.105510

Bhavnani, K.-K., Foran, J., Kurian, P.A., Munshi, D. (eds.) (2019). *Climate Futures: Reimagining Global Climate Justice*. London: Zed Books.

Bond, P. (2016). Who wins from "climate apartheid"? African climate justice narratives about the Paris COP21. *New Politics*, 15(4), 83–90.

Bullard, R. D. (1992). Anatomy of environmental racism and the environmental justice movement. In R. D. Bullard (ed.) *Confronting Environmental Racism: Voices from the Grassroots* (pp. 15–39). Boston, MA: South End Press.

Burman, A. (2017). The political ontology of climate change: Moral meteorology, climate justice, and the coloniality of reality in the Bolivian Andes. *Journal of Political Ecology*, 24(1), 921–930. https://doi.org/10.2458/v24i1.20974

Cassegård, C., Soneryd, L., Thörn, H. and Wettergren, Å. (eds.) (2017). *Climate Action in a Globalizing World: Comparative Perspectives on Environmental Movements in the Global North*. Abingdon and New York: Routledge.

Cassegård, C. and Thörn, H. (2017). Climate justice, equity and movement mobilization. In C. Cassegård, et al. (eds.) *Climate Action in a Globalizing World: Comparative Perspectives on Environmental Movements in the Global North* (pp. 33–56). Abingdon and New York: Routledge.

Chomsky, N. and Pollin, R. (2020). *Climate Crisis and the Global Green New Deal*. London and New York: Verso Books.

Christou, G., Theodorou, E. and Spyrou, S. (2023). "The slow pandemic": Youth's climate activism and the stakes for youth movements under Covid-19. *Children's Geographies*, 21(2), 191–204. https://doi.org/10.1080/14733285.2022.2026885

CJA (2019). *Just Transition Principles*. Climate Justice Alliance. https://climatejusticealliance.org/wp-content/uploads/2019/11/CJA_JustTransition_highres.pdf [Accessed 3 September 2021].

CJA (n.d.). *About*. Climate Justice Alliance. https://climatejusticealliance.org/about/ [Accessed 5 May 2024].

CJN! (2012). *CJ in the USA: Root-Cause Remedies, Rights, Reparations and Representation*. Climate Justice Now! https://www.climate-justice-now.org/cj-in-the-usa-root-cause-remedies-rights-reparations-and-representation/ [Accessed 14 October 2021].

Crowley, D., Marat-Mendes, T., Falanga, R., Henfrey, T. and Penha-Lopes, G. (2021). Towards a necessary regenerative urban planning. *Cidades*, Spring. http://journals.openedition.org/cidades/3384

Dave, N. N. (2012). *Queer Activism in India: A Story in the Anthropology of Ethics*. Durham and London: Duke University Press.

Deleuze, G. and Guattari, F. (2004). *A Thousand Plateaus: Capitalism and Schizophrenia*. Translation by Brian Massumi. London and New York: Continuum.

Delina, L. L. (2022). Moving people from the balcony to the trenches: Time to adopt "climatage" in climate activism? *Energy Research & Social Science*, 90, 102586. https://doi.org/10.1016/j.erss.2022.102586

della Porta, D. and Parks, L. (2014). Framing processes in the climate movement: From climate change to climate justice. In M. Dietz and H. Garrelts (eds.) *Routledge Handbook of the Climate Change Movement* (pp. 19–30). Abingdon and New York: Routledge.

De Lucia, V. (2014). The climate justice movement and the hegemonic discourse of technology. In M. Dietz and H. Garrelts (eds.) *Routledge Handbook of the Climate Change Movement* (pp. 66–83). Abingdon and New York: Routledge.

Dietz, M. (2014). Debates and conflicts in the climate movement. In M. Dietz and H. Garrelts (eds.) *Routledge Handbook of the Climate Change Movement* (pp. 292–307). Abingdon and New York: Routledge.

Dietz, M. and Garrelts, H. (eds.) (2014). *Routledge Handbook of the Climate Change Movement*. Abingdon and New York: Routledge.

Dowie, M. (1995). *Losing Ground: American Environmentalism at the Close of the Twentieth Century*. Cambridge: The MIT Press.

Dwivedi, R. (2001). Environmental movements in the Global South. *International Sociology*, 16(1), 11–31. https://doi.org/10.1177/0268580901016001003

Estes, N. (2019). *Our History Is the Future: Standing Rock Versus the Dakota Access Pipeline, and the Long Tradition of Indigenous Resistance*. London and New York: Verso Books.

Fairchild, D. and Weinrub, A. (eds.) (2017). *Energy Democracy: Advancing Equity in Clean Energy Solutions*. Washington, Covelo and London: Island Press.

Faustino, A. da S. (2023). *Dissolving the Concrete: Reconfiguring Urban Waterscapes Through Grassroots Activism in São Paulo, Brazil*. PhD thesis. Melbourne, Australia, RMIT University. https://researchrepository.rmit.edu.au/esploro/outputs/9922282713301341

Fisher, D. R. and Nasrin, S. (2021). Climate activism and its effects. *WIREs Climate Change*, 12, e683. https://doi.org/10.1002/wcc.683

Foran, J., Grosse, C., Hornik, B. (2019). Frontlines, Intersections, and Creativity: The Growth of the North American Climate Justice Movement. In: Bhavnani, K.-K., Foran, J., Kurian, P.A., Munshi, D. (eds.). *Climate Futures: Reimagining Global Climate Justice*. London: Zed Books, pp. 222–231.

Garrelts, H. and Dietz, M. (2014). Introduction: Contours of the transnational climate movement – conception and contents of the handbook. In M. Dietz and H. Garrelts (eds.) *Routledge Handbook of the Climate Change Movement* (pp. 1–15). Abingdon and New York: Routledge.

Gills, B. and Morgan, J. (2020). Global climate emergency: After COP24, climate science, urgency, and the threat to humanity. *Globalizations*, 17(6), 885–902. https://doi.org/10.1080/14747731.2019.1669915

Girgenti, G. and Shahid, W. (2020). The next era of American politics. In V. Prakash and G. Girgenti (eds.) *Winning the Green New Deal: Why We Must, How We Can* (pp. 212–240). New York: Simon & Schuster.

Graeber, D. (2007). *Possibilities: Essays on Hierarchy, Rebellion and Desire*. Oakland and Edinburgh: AK Press.

Graeber, D. (2009). *Direct Action: An Ethnography*. Oakland and Edinburgh: AK Press.

Grasso, M. and Giugni, M. (eds.) (2022). *The Routledge Handbook of Environmental Movements*. Abingdon and New York: Routledge.

Heywood, P. (2018). *After Difference: Queer Activism in Italy and Anthropological Theory*. New York and Oxford: Berghahn Books.

IPCC (2018a). *Global Warming of 1.5°C: IPCC Special Report on Impacts of Global Warming of 1.5°C above Pre-Industrial Levels in Context of Strengthening Response to Climate Change, Sustainable Development, and Efforts to Eradicate Poverty* [V. Masson-Delmotte, P. Zhai, H.-O. Pörtner, D. Roberts, J. Skea, P. R. Shukla, A. Pirani, W. Moufouma-Okia, C. Péan, R. Pidcock, S. Connors, J. B. R. Matthews, Y. Chen, X. Zhou, M. I. Gomis, E. Lonnoy, T. Maycock, M. Tignor, and T. Waterfield (eds.)]. The Intergovernmental Panel on Climate Change. Cambridge and New York: Cambridge University Press. https://doi.org/10.1017/9781009157940

IPCC (2018b). *IPCC Press Release: Summary for Policymakers of IPCC Special Report on Global Warming of 1.5oC Approved by Governments*. The Intergovernmental Panel on Climate Change. https://www.ipcc.ch/site/assets/uploads/sites/2/2019/05/pr_181008_P48_spm_en.pdf [Accessed 12 February 2023].

Jafry, T. (ed.) (2018). *Routledge Handbook of Climate Justice*. Abingdon and New York: Routledge.

Juris, J. S. and Khasnabish, A. (eds.) (2013). *Insurgent Encounters: Transnational Activism, Ethnography, and the Political*. Durham and London: Duke University Press.

Klein, N. (2014). *This Changes Everything: Capitalism vs. the Climate*. London: Penguin Books.

Klein, N. (2019). *On Fire: The (Burning) Case for a Green New Deal*. New York: Simon & Schuster.

Markey, E. J. (2019). *A Resolution Recognizing the Duty of the Federal Government to Create a Green New Deal*. S.Res.59. 116th Congress (2019–2020). https://www.congress.gov/bill/116th-congress/senate-resolution/59/text [Accessed 30 May 2023].

Mascarenhas-Swan, M. (2017). The case for a just transition. In D. Fairchild and A. Weinrub (eds.) *Energy Democracy: Advancing Equity in Clean Energy Solutions* (pp. 37–56). Washington, Covelo and London: Island Press.

McGuire, B. (2022). *Hothouse Earth: An Inhabitant's Guide*. London: Icon Books.

McKibben, B. (2012). Global Warming's Terrifying New Math. *Rolling Stone*, 19 July. https://www.rollingstone.com/politics/politics-news/global-warmings-terrifying-new-math-188550/ [Accessed 5 May 2024].

MGJEP (2017). *From Banks and Tanks to Cooperation and Caring: A Strategic Framework for a Just Transition*. Movement Generation Justice & Ecology Project. https://movementgeneration.org/wp-content/uploads/2016/11/JT_booklet_English_SPREADs_web.pdf [Accessed 3 September 2021].

MGJEP and other signatories (2010). *Grassroots Organizing Cools the Planet: Letter from the Grassroots to 1Sky*. Movement Generation Justice & Ecology Project. https://movementgeneration.org/grassroots-organizing-cools-the-planet-letter-from-the-grassroots-to-1sky/ [Accessed 27 June 2023].

Mihaylov, N. and Perkins, D. (2015). Local environmental grassroots activism: Contributions from environmental psychology, sociology and politics. *Behavioral Sciences*, 5(1), 121–153. https://doi.org/10.3390/bs5010121

Moore, H. and Russell, J. K. (2011). *Organizing Cools the Planet: Tools and Reflections to Navigate the Climate Crisis*. Pamphlet Series. Number 0011. Oakland: PM Press.

Moore, R. and other signatories (1990). *Southwest Organizing Project Letter – March 15, 1990.* Southwest Organizing Project, Harwood Training and Service Center, New Mexico 87102. https://www.ejnet.org/ej/swop.pdf [Accessed 3 September 2021].

Neubauer, L. and Repenning, A. (2023). *Beginning to End the Climate Crisis: A History of Our Future.* Translated by S. von Mering. Waltham: Brandeis University Press.

Nulman, E. (2022). Climate change movements in the Global North. In: Grasso, M. and Giugni, M. (eds). *The Routledge Handbook of Environmental Movements.* Abingdon and New York: Routledge, pp. 185–198.

Pettifor, A. (2019). *The Case for the Green New Deal.* London and New York: Verso Books.

Pezzullo, P. C. and Sandler, R. (2007). Introduction: Revisiting the environmental justice challenge to environmentalism. In R. Sandler and P. C. Pezzullo (eds.) *Environmental Justice and Environmentalism: The Social Justice Challenge to the Environmental Movement* (pp. 1–24). Cambridge, MA: MIT Press.

Pezzullo, P. C. (2022). Environmental justice and climate justice. In: Grasso, M. and Giugni, M. (eds). *The Routledge Handbook of Environmental Movements.* Abingdon and New York: Routledge, pp. 229–244.

Prakash, V. (2020). Introduction: The adults in the room. In V. Prakash and G. Girgenti (eds.) *Winning the Green New Deal: Why We Must, How We Can* (pp. vii–xx). New York: Simon & Schuster.

Prakash, V. and Girgenti, G. (eds.) (2020). *Winning the Green New Deal: Why We Must, How We Can.* New York: Simon & Schuster.

Rep. Ocasio-Cortez, A. (2019). *Recognizing the Duty of the Federal Government to Create a Green New Deal.* H.Res.109. 116th Congress (2019–2020). https://www.congress.gov/bill/116th-congress/house-resolution/109/text [Accessed 30 May 2023].

Richardson, K., Steffen, W., Lucht, W., Bendtsen, J., Cornell, S. E., Donges, J. F., Drüke, M., Fetzer, I., Bala, G., von Bloh, W., Feulner, G., Fiedler, S., Gerten, D., Gleeson, T., Hofmann, M., Huiskamp, W., Kummu, M., Mohan, C., Nogués-Bravo, D.,. . . Rockström, J. (2023). Earth beyond six of nine planetary boundaries. *Science Advances,* 9(37). https://doi.org/10.1126/sciadv.adh2458

Rogers, N. (2023). Stepping out of the ivory tower into fire and flood: Climate activism and the academy. *Alternative Law Journal,* 48(2), 133–139. https://doi.org/10.1177/1037969X231161741

Rojas, A. and Shahid, W. (2020). From protest to primaries: The movement in the democratic party. In V. Prakash and G. Girgenti (eds.) *Winning the Green New Deal: Why We Must, How We Can* (pp. 241–261). New York: Simon & Schuster.

Sandler, R. and Pezzullo, P. C. (eds.) (2007). *Environmental Justice and Environmentalism: The Social Justice Challenge to the Environmental Movement.* Cambridge, MA: MIT Press.

Schlosberg, D. and Collins, L. B. (2014). From environmental to climate justice: Climate change and the discourse of environmental justice. *Wiley Interdisciplinary Reviews: Climate Change,* 5(3), 359–374. https://doi.org/10.1002/wcc.275

Shukaitis, S., Graeber, D. and Biddle, E. (eds.) (2007). *Constituent Imagination: Militant Investigations//Collective Theorization.* Oakland and Edinburgh: AK Press.

Solnit, R. and Young Lutunatabua, T. (eds.) (2023). *Not Too Late: Changing the Climate Story from Despair to Possibility.* Chicago: Haymarket Books.

Steele, W., Hillier, J., MacCallum, D., Byrne, J. and Houston, D. (2021). *Quiet Activism: Climate Action at the Local Scale.* New York and London: Palgrave Macmillan.

Sultana, F. (2022). Critical climate justice. *The Geographical Journal,* 188(1), 118–124. https://doi.org/10.1111/geoj.12417

Swyngedouw, E. (2004). *Social Power and the Urbanization of Water: Flows of Power.* Oxford: Oxford University Press.

Tienharra, K. and Robinson, J. (eds.) (2022a). *Routledge Handbook of the Green New Deal.* Abingdon and New York: Routledge.

Tienharra, K. and Robinson, J. (2022b). Introduction. In K. Tienharra and J. Robinson (eds.) *Routledge Handbook of the Green New Deal* (pp. 1–20). Abingdon and New York: Routledge.

Tokar, B. (2014a). *Toward Climate Justice: Perspectives on the Climate Crisis and Social Change* (2nd ed.). Porsgrunn: New Compass Press.

Tokar, B. (2014b). Organization profile – 350.org. In M. Dietz and H. Garrelts (eds.) *Routledge Handbook of the Climate Change Movement* (pp. 252–254). Abingdon and New York: Routledge.

Tokar, B. (2019). On the evolution and continuing development of the climate justice movement. In T. Jafry (ed.) *Routledge Handbook of Climate Justice* (pp. 13–25). Abingdon and New York: Routledge.

Tokar, B. (2020). Climate justice and community renewal: An introduction. In B. Tokar and T. Gilbertson (eds.) *Climate Justice and Community Renewal: Resistance and Grassroots Solutions* (pp. 1–16). Abingdon and New York: Routledge.

Tokar, B. and Gilbertson, T. (eds.) (2020). *Climate Justice and Community Renewal: Resistance and Grassroots Solutions*. Abingdon and New York: Routledge.

UN Climate Change (n.d.). *What Is the Kyoto Protocol?* https://unfccc.int/kyoto_protocol [Accessed 3 Sep 2024].

United Nations (n.d.). *The Paris Agreement.* https://www.un.org/en/climatechange/paris-agreement [Accessed 3 Sep 2024].

The Visual Stories Team and Gilmore, H. (2022). Anatomy of the Lismore disaser. *The Age*, 30 June. https://www.theage.com.au/interactive/2022/lismore-flooding/ [Accessed 6 May 2024].

Von Storch, L., Ley, L. and Sun, J. (2021). New climate change activism: Before and after the Covid-19 pandemic. *Social Anthropology*, 29(1), 205–209. https://doi.org/10.1111/1469-8676.13005

Wallace-Wells, D. (2020). The crisis here and now. In V. Prakash and G. Girgenti (eds.) *Winning the Green New Deal: Why We Must, How We Can* (pp. 3–11). New York: Simon & Schuster.

Wright, B. H., Bryant, P. and Bullard, R. D. (1994). Coping with poisons in Cancer Alley. In R. D. Bullard (ed.) *Unequal Protection: Environmental Justice and Communities of Color* (pp. 110–129). San Francisco: Sierra Club Books.

Youmans, W. L. and York, J. C. (2012). Social media and the activist toolkit: User agreements, corporate interests, and the information infrastructure of modern social movements. *Journal of Communication*, 62, 315–329. https://doi.org/10.1111/j.1460-2466.2012.01636.x

PART I

Accounts from the Ground
The Fight for Climate Justice

**Introduction to Part I: Accounts from the Ground:
The Fight for Climate Justice**

In the rapidly shifting landscapes of climate change, the voices and perspectives from grassroots movements have become crucial to the ongoing struggle for climate justice. While large-scale summits and international agreements may capture the headlines (Tavares et al., 2020), the real work of change is most often happening at the community level, in the form of activism, resistance, and advocacy. The entanglement between effective climate activist engagements with collective, localised, and emplaced experiences should not be seen as a surprise. The ubiquity of the climate crisis across all continents and ecosystems leads to more diverse forms of climate activist organisation and resistance, in response to the multifaceted disruptions in communities, forests, cities, villages, rivers, and deserts (Stanley et al., 2021). This intimate entanglement of grounded experiences and collective mobilisations denotes a key trace of the grassroots activism that shapes and transforms the climate agenda across multiple territories.

This opening section of the handbook brings together six chapters that illustrate just a small fraction of the depth and diversity of contexts, motivations, scales, and strategies that emerge from climate activism rooted on the ground. Such an emphasis is not only one of focus but also of perspective, as the authors of these chapters are often protagonists in the grassroots organisations or activist contexts they are analysing and discussing. This is reflected in their disciplinary positions and methodological approaches, with a particular prominence of ethnographically oriented and critical action research accounts, which bring rich perspectives and ways of storytelling to this work.

Climate-induced disruptions and challenges affect each territory in distinct ways, as complex and dynamic biophysical processes are enmeshed in the asymmetries and distortions of a global political economy of accumulation often rooted in localised systems of dispossession and extraction (O'Hara, 2009). These structural asymmetries mark the ubiquity of the climate crisis, which includes many expressions of injustice: uneven impacts, political marginalisation, institutional silences, and epistemic violences, to name a few (Sultana, 2022). Such factors of injustice are another core element of many of the grassroots engagements

DOI: 10.4324/9781003396567-2

intersecting the global climate movement, as organised groups urgently seek alternatives to processes that deeply threaten their territories.

In Chapter 2, "NoDAPL and the continued fight for Tribal Self-determination by Federally Recognized Indian Tribes in the United States," Ashley Hemmers uses her personal experience as a Mojave leader who accompanied a group of Elders to support fellow Indigenous activists and community leaders at Standing Rock in the fight against the North Dakota Access Pipeline to highlight the United States' ongoing treaty violations and discusses how the struggle against NoDaPL exemplifies the layered intersectional complexity that exists in pursuit of Tribal self-determination to protect the rights of Native people and their territories. The author walks readers through the interjurisdictional complexity Tribal governments face when working to protect traditional territories, and how Tribal self-determination, free prior and informed consent, and Tribal governance contribute to the protection of Tribal sovereignty. She also provides a realistic perspective of where these systems fall short, as in the case of proposed large-scale energy developments like the Dakota Access Pipeline, which seeks to extract resources from Tribal homelands.

In Chapter 3, "Indigenous Climate Justice and Epistemic Politics in Amazonia," Sylvia Cifuentes brings us into the plural and conflictive context of Amazonia, where Indigenous communities are at the forefront of climate activism. This chapter delves into the epistemic dimensions of climate justice, focusing on the ways Indigenous people in this region define and experience the concept. Ancestral knowledge, deeply embedded in the territory, plays a pivotal role in Indigenous strategies for climate action as it relationally implicates the natural and supernatural world in climate politics. The author explores the intersection of Indigenous leadership and climate activism, acknowledging the diversity within Indigenous cultures and the importance of not essentialising ancestral knowledge. Cifuentes explores the ways climate justice is not a static concept for Indigenous Amazonian communities. As she argues, it is deeply rooted in place-based experiences and shaped by the plurinational nature of the Indigenous movement.

In Chapter 4, "Brazilian Grassroots Climate Activism in Unequal Waterscapes," Alexandre da Silva Faustino, Nicolas Guerra-Tão and Wendy Steele shift our focus to the urban peripheries of São Paulo, Brazil, where grassroots climate activism is implicated in the insurgent struggles of communities living in precarious urbanisation. The authors employ ethnographic and digital methods to investigate how flooding, water struggles, and urban injustice are entwined with climate activism in these communities. In these urban peripheries, grassroots actors work to co-produce alternatives and spaces of refuge in uneven waterscapes through solidarity, collectivism, and connection to place. These strategies are key to countering the tenets of the dominant pulse of urbanisation in São Paulo, which is rooted in dispossession, displacement, and commodification. The chapter highlights the challenges faced by marginalised populations under climate emergency and the creative strategies employed by grassroots activists to resist violence, reimagine places, and create systems of community and horizontal organisation.

In Chapter 5, "Pacific Youth and Climate Activism: Safeguarding Our Fa'asinomaga", Tahere Talaina Siisiialafia-Mau and Joseph Emile Honour Percival explore the vulnerabilities of Indigenous populations in the Pacific Island Countries and Territories (PICTs), and their engagement with climate activism as a response. Through in-depth interviews across 15 PICTs, the authors uncover the significant role played by Pacific youth in global climate activism, particularly at events like the Conference of the Parties (COP) summits. The concept of *fa'asinomaga*, denoting a strong collective sense of identity and heritage across

Pacific Island cultures, is central to understanding Pacific climate activism. The chapter demonstrates how fighting climate change is not only about resisting the loss of lands and devastation of traditional territories; it includes deep concern for protecting cultural identity and heritage. This deep-rooted activism is driven by stories of hope, resilience, and solidarity.

In Chapter 6, "Realising Common Ground: Custodianship of Country and Youth Climate Action," Madison Shakespeare, Michelle Catanzaro, and Caelli Jo Brooker focus on the intersections of youth and Indigenous perspectives in climate activism, particularly in the School Strikes for Climate movement in Australia. The authors use an Indigenous-led methodology to assess climate activism, emphasising relationality, custodianship, and ways of knowing. The analysis of visual expressions in the strikes sheds light on the different belief systems, priorities, and motivations carried by groups participating in the movement. Indigenous perspectives, although still limited in Australian climate-change protests, are pivotal to bridging grassroots activism with broader land rights and environmental justice struggles. These perspectives foreground critical dimensions, such as time/urgency, biodiversity crisis, and the need for collectivism. Overall, this chapter underscores the importance of understanding the politics of place in grassroots activism and how visual representations in protests reflect the diversity of voices in the movement.

The section closes with Chapter 7, "Action, Inertia, and the Stories that Get us Moving," written by climate storyteller Keerti Gopal. A journalist and former member of the Sunrise Movement in the United States. Gopal presents the results of her interviews with four youth climate activists from California, New Jersey, and Taiwan, highlighting what motivated them to join the movement, how they sustain movement participation, and how they now, in turn, activate others in a generation coming of age at a time of multiple crises. Gopal concludes the chapter by considering the value of stories for collective action, particularly concerning how stories can combat climate *inaction*.

Overall, this first part of the handbook – "Accounts from the Ground: The Fight for Climate Justice" – explores grassroots efforts and unique contributions to the broader climate justice movement. Together, these chapters provide a multifaceted perspective on grassroots climate activism, demonstrating the diversity and resilience of climate justice movements around the world. From the Amazonian Indigenous communities to the urban peripheries of São Paulo, from Indigenous homelands in North America to the Pacific Islands, these stories showcase the power of grassroots activism to drive change and challenge entrenched systems of power. As the climate crisis continues to escalate, these grassroots accounts offer hope and inspiration, reminding us that the fight for climate justice is ultimately a fight for a more just and equitable world.

References

O'Hara, P. A. (2009). Political economy of climate change, ecological destruction and uneven development. *Ecological Economics*, 69(2), 223–234. https://doi.org/10.1016/j.ecolecon.2009.09.015.

Stanley, S. K., Hogg, T. L., Leviston, Z. and Walker, I. (2021). From anger to action: Differential impacts of eco-anxiety, eco-depression, and eco-anger on climate action and wellbeing. *The Journal of Climate Change and Health*, 1, 100003. https://doi.org/10.1016/j.joclim.2021.100003.

Sultana, F. (2022). Critical climate justice. *The Geographical Journal*, 188(1), 118–124. https://doi.org/10.1111/geoj.12417.

Tavares, A. O., Areia, N. P., Mellett, S., James, J., Intrigliolo, D. S., Couldrick, L. B. and Berthoumieu, J.-F. (2020). The European media portrayal of climate change: Implications for the social mobilization towards climate action. *Sustainability*, 12(20), 8300. https://doi.org/10.3390/su12208300.

2

NODAPL AND THE CONTINUED FIGHT FOR TRIBAL SELF-DETERMINATION BY FEDERALLY RECOGNIZED INDIAN TRIBES IN THE UNITED STATES

Ashley Hemmers

One of the most memorable moments of my adult life was a morning spent with my grandmother in North Dakota – interestingly enough, considering I am a Mojave woman from Tribal homelands over 1,200 miles away located in the tristate area of Nevada, California, and Arizona.

The morning was not particularly unusual in any way. I woke up, prayed, said good morning to my grandmother, and then began to get ready for the day. I had a lot of work ahead of me that day and my grandmother offered to braid my hair. I did not think anything of it. Having my hair braided is so normal to me that it is one of my earliest memories. But on this day, instead of sitting still in a chair, I was on the foot of a minivan while she was sitting in the backseat. We talked as we usually did and for a moment I forgot where we were. We were parked in Sacred Stone Camp with members of our Tribe who had answered the call to join the members of the Standing Rock Sioux Tribe in their efforts to prevent the Dakota Access Pipeline from destroying their ancestral homelands.

The camp was quiet. My capacity during our time at Sacred Stone Camp was to provide operational and logistical support to ensure that our Tribal Chairman and Tribal Council members who had travelled with our group had up-to-date information, to ensure that we had adequate supplies so that the 54 Tribal members we had brought from our reservation to help hold ground did not hinder existing resources, and perhaps most importantly, to ensure that the contingency of Tribal Elders, my grandmother being one of them, were supported and safe.

Our rented minivan became my operational headquarters and was parked along a dirt road inside the camp where we had set up our tents. While my grandmother was braiding my hair and we shared a laugh to ourselves, a woman walking the road approached us smiling and said, "This is beautiful, you are lucky to have this blessing from an Elder, it is

DOI: 10.4324/9781003396567-3

healing for my land." We both smiled at her and she came closer so that I could take her hands, and then she said, "thank you."

I have been holding the frontlines in my own community alongside my grandmother since a very young age. In fact, that is one of the reasons my Tribe felt so compelled to travel to Sacred Stone Camp. For the Mojave, our Tribe led the successful 120-day occupation of Ward Valley that prevented nuclear waste from being dumped in unlined trenches, risking irreparable ramifications to our water supply in our ancestral homelands of the Mojave Desert in southern California. Our Tribe is uniquely familiar with the complexity of engagement during times of occupation, the strain these efforts have on their associated Tribal communities, and the folks that these movements invite. So, we answered the call in hopes that our Tribal leaders could share this knowledge with the Tribal leaders at Standing Rock, that the fresh arms and legs we brought could help with things around the camp, that the presence of our Elders who had won our fight could boost morale, and to share our songs in prayer.

When I heard "thank you" from the woman who saw hope from my grandmother braiding my hair, it hit me like a bolt of lightning. This was not just an environmental justice issue, neither was it just a social justice issue – Native people want to live in peace on their homelands without threat from harm or destruction, and while we have managed to live under threat, we want to share moments with our families without it, even if that's just simply braiding our hair.

Calling Out the Black Snake, #NoDAPL

The Dakota Access Pipeline was a project proposed by Energy Transfer Partners LLC that would span four states and 1,172 miles to connect the Bakken Oil Field crude oil to an oil field tank farm in the state of Illinois. Once constructed, the proposed pipeline would lead to a flow of approximately 470,000 barrels of oil per day. The No Dakota Access Pipeline (#NoDAPL) resistance was initiated in 2016 by a group of women from the Standing Rock Sioux Tribe (SRST). The women set up Sacred Stone Camp to monitor the construction of the Dakota Access Pipeline while the Standing Rock Sioux Tribe issued dissent for the project through the government-to-government consultation process. The Dakota Access Pipeline was formally introduced to the Tribe by Energy Transfer Partners (ETP) during a 2014 meeting with the SRST Tribal Council during which ETP informed the Tribe of their intention to build a crude oil route less than a mile from the present reservation boundary that would cross under the Missouri River at Lake Oahe. The proposed ETP crude oil route would fall within boundaries established in the 1851 Treaty of Fort Laramie and the Treaty of 1868. In that meeting, SRST informed ETP of a 2012 SRST Tribal Council resolution that opposed all pipelines within treaty boundaries.

Tribal Self-determination

Federally Recognized Indian Tribes share a unique government-to-government relationship with the United States federal government. This relationship originates from the inherent sovereignty Tribal Nations possess over themselves and their ancestral homelands. The Indian Reorganization Act of 1934 affirms Tribal sovereignty through the federal recognition process.

Officially passed by the United States Congress as the Wheeler-Howard Act on June 18, 1934 (United States Congress, n.d.), the act established a process that restored portions of

Tribal lands to Tribal ownership if Tribes adopted federal governing style constitutions and governing councils. Tribes that adopted these governing styles were given federal recognition which vest the governing of Tribal lands to the governing council elected by the Tribes' members. The Standing Rock Sioux Tribal Council is this governing body for the Tribe and their Tribal homelands.

The Indian Reorganization Act's purpose of restoration is in many ways misleading in that it came during an era of active termination policy. During the termination era, Tribes were displaced from their ancestral homelands, relocated to undesirable areas or encampments, all while their land was stolen and sold. Some Tribes, particularly in the plains, were engaged by officials from the new U.S. federal government to establish treaties to provide settlers reprieve from the chaos of encroachment. The Fort Laramie Treaty of 1851 and later 1868 (National Park Service, n.d.) was one of these attempts. The treaties promised to protect Indian resources and Tribal hunting grounds from depredations by white settlers moving West in exchange for the right of way for these new U.S. citizens to cross Tribal homelands when making their way West. The Standing Rock Sioux Tribal Council is the present governing body of one of the Tribes represented in the Fort Laramie Treaties. The Fort Laramie Treaties were a painful exercise in the exploitation of concession for the right of way to an infantile government who had no systematic framework to control the behaviour of its citizens. It is important to understand that treaties established with Indian Tribes are still present agreements even though the United States has historically failed and continues to fall short of meeting many of the concessions it made in these agreements.

The 1851 and 1868 treaties established boundaries between the Tribes represented in these treaties and the United States. In an effort to protect Tribal homelands from resource extraction and harmful impacts, Tribal councils often pass resolutions within the governing framework established under the Wheeler-Howard Act to formalise Tribal position. The Standing Rock Sioux Tribe's passing of the 2012 Tribal Council Resolution opposing all pipelines within established treaty boundaries was yet another action to remind the United States of its treaty responsibility to protect Indian resources from depredation, this time from their U.S. oil industry, for use of right of way.

Implications of State Jurisdiction in Public Law 280 States

Tribal homelands in the United States, often rich with minerals or sheer land mass, have been the continual targets for resource extraction by large-scale energy developments. These enterprises help to mobilise localities through an enticement of the possibility of high paying jobs that could potentially generate revenue for adjacent cities or townships to garner public support for the need for the project.

In 1953, the United States Congress passed Public Law 83–280 granting certain states criminal jurisdiction over American Indians on reservations and to allow civil litigation that had come under Tribal or federal court jurisdiction to be handled by state courts (U.S. Department of Interior, n.d.). North Dakota is one of the states that elected to assume full or partial jurisdiction over American Indians on reservations subject to Tribal consent. It is important to note that, while the law is currently still enacted, the act was passed without the consultation of Tribes and federal jurisdiction to engage with Tribes has since been restored by later acts of Congress, court decisions, and state actions due to Article 1, Section 8 of the U.S. Constitution that rests the authority to engage with Tribes at the federal level.

While the U.S. Constitution clearly defines Tribal and federal relations, the history of the United States and U.S. statehood is a constant tension in this relationship. The United States is a republic currently comprised of 50 states and various territories. While the U.S. Constitution provides the framework for the union, the implementation of that framework is built upon additional jurisdictional layers of local governance. These layers complicate Tribal and federal relations because the federal government during its establishment had to and arguably still does deploy loose authority over its asserted territory largely in part to its sheer size and diversity. Localities then became members of a state and built their state governing systems to meet their constituents' needs. Native voices were often left out of these conversations for various reasons. First, these new statesmen encroached upon Native lands, often settling in areas already agreed to as Indian Territory by their federal officials. These new settlers responded better to physical barriers of ownership like lands that are fortified, walled, or designated by other structures. When new settlers saw open land, their lack of knowledge of land history gave them the impression then that continues to be misconceived now, like in the case of DAPL, that land is available for their taking. This tension increased us-versus-them mentalities that still overshadow relationship building and partnership opportunities between Tribes and states today. Second, when a dispute occurred, the federal government did not deploy timely interventions to address the issues. This lack of federal engagement bottlenecked resolution of disputes, creating deeper tensions between the two governments. Historically, these tensions never fully recover even once a resolution has been identified.

This backdrop of interjurisdictional state tension is important to understanding the environment at the #NoDAPL standoff. The Tribe was at a point where it had to remind the federal government of its treaty obligations while voicing opposition to a pipeline route that had secured a large contingency of non-Tribal local state support.

The Virality of Social Media

One thing that no one could account for at the time was the enormous reach social media would have in making localised Indigenous issues globally known. The hashtag #NoDAPL began a social media campaign that helped the conditions at Standing Rock become part of the national and international dialogue in a way that no one could have predicted or had witnessed since the Indian Reorganization Act and the global call to address the horrendous conditions Native Americans endured from the United States. Young SRST members were providing first-hand accounts of the threat to their community in their own words.

The pipeline route was making headway towards reality and SRST youth began to run for the water. They documented these actions on Facebook Live and other social media platforms and soon their networks reached other networks, and all of Indian Country and beyond learned about the attempt to put a pipeline through treaty territory near the Tribe's water supply after a failed attempt by the company to put the pipeline near Bismarck.

The Dakota Access Pipeline LLC engaged local state authorities to protect their equipment and remove individuals, mostly women and Native Youth, from blocking the path of the pipeline. These individuals cited protection under the 1978 Native American Religious Freedom Act for the right to protect their lands and lifeways. The SRST Tribal Chairman Dave Archambault and SRST Council member Dana Yellow Fat were arrested by state law enforcement on August 12, 2016, charged with disorderly conduct, and booked into the Morton County jail (Donovan, 2016). Witnesses recorded the state arrest of the Tribal chairman and council member on treaty territory, and the video stream went viral.

The video of state intervention on a federal issue resonated with many Natives across the United States and globally. There is an ongoing frustration with the lack of enforcement from the federal government to ensure states within their union are acting in accordance with the agreements made with Indian Tribes. Unfortunately, because the federal government has not streamlined an approach to creating capacity at the state level, it is easy for states to interject prematurely and complicate an already complex situation.

In hopes of protecting their land, Native youth and others gave first-hand accounts of their fear on social media. They began asking for prayers and help because of which thousands of activists, Indigenous and non-Indigenous, began making their way to sacred stone camp on the banks of the Missouri river, treaty territory of the Standing Rock Sioux Tribe, to support the Tribe in protest of the Dakota Access pipeline. The protest at Standing Rock became a global rallying cry for Indigenous rights and climate activists. For the Tribe, it was about their right to live in peace on their land without fear from threat, land already protected under treaty.

Many Tribes also answered the call to come to Standing Rock. Tribes from all over the country brought resources, posted their Tribal flags, and sat in prayer for the Standing Rock Sioux people. As someone who has had to occupy my own Tribal homeland so that my own people would not be poisoned by lazy business practices implemented to increase profit margins at the expense of the land and all who rely upon it – occupations are a strange place. They are at once a place of deep reflection, prayer, connection, fear, courage, and stress.

More than 200 Tribes came to Standing Rock hailing from across the United States, Canada, and other Indigenous communities around the globe. Federally Recognized Indian Tribes offered support by passing Tribal Resolutions in protest of the Dakota Access Pipeline calling for the United States to honour the treaties. Tribal flags were raised in the entry to Sacred Stone Camp to illustrate the Nations represented in protest. When a member of a new Nation came to camp, they were able to represent themselves on behalf of their Tribe to the Standing Rock Sioux Tribal Leadership, Tribal Elders, and people. Once getting there, many would share songs, prayer, stories, and resources from their Tribe.

On our second day with the Standing Rock Sioux Tribe, the men from our group ran with feather staffs our people had made for the Standing Rock Sioux people in prayer. Behind the runners, more of our men walked singing our traditional Tribal bird songs. This group of runners and singers started at the frontline and carried those prayers to the entrance of Sacred Stone Camp where the rest of our group was waiting. The combined group danced behind the singers as we walked down the hill lined with Tribal flags to an area at the bottom of the camp entrance. Our Tribal Chairman introduced who we were to the SRST representatives, introduced members of our Council and our Elders, and shared more of our songs in prayer. We did this to let the SRST people know that we would be willing to stand beside them in civil protest on behalf of their people in a good way. As a Tribe who has led a successful occupation, we understand the pressures that Tribes can face in being vilified by authorities as disruptors when, in actuality, when it comes down to its most simplistic form, occupations are birthed out of provocation because of an external disruption preventing the ability of Tribal people to live in established peace.

An ongoing tactic to suppress peaceful protests is to label peaceful protestors as miscreants, criminals, violent, or agitators. Standing Rock was no different. Local media outlets and state authorities claimed that the individuals at Standing Rock were violent agitators to justify their heavy force. Throughout the protest, it was widely documented by the media and official reports that local authorities used rubber bullets, pepper spray, tear gas, tasers, sound weapons, and other "less than lethal" methods on protestors (Wong, 2016). These tactics got worse as the encampment continued into colder weather. It was not uncommon for authorities to use water cannons in the early fall when temperatures in North Dakota began to drop. One of the last violent encounters between local authorities and peaceful protestors resulted in a climate activist sustaining injury, by her family's account, from a police concussion grenade. The woman's father contended that his daughter was hit by an exploding concussion grenade thrown by police (Wong, 2016). The victim experienced life-threatening injuries and was eventually flown out to a Minneapolis hospital due to the likely amputation of one of her arms from severe tissue damage. Local authorities maintain that the injury was caused by an explosion caused by protestors. The FBI conducted a multi-year investigation but have not released their findings.

Accountability for injury or destabilisation often falls to the Tribes in hopes that these pressures will deter Tribes from continuing pursuit. Individuals come to protests for many different reasons. For me, it was to share hope. For others it was to fight for what was right. It's hard to trust personal agendas during times of occupation. Trust deteriorates further when there are multiple instances, particularly in the plains, of a historical record of payment to agitators to disrupt situations so that Tribes would be admonished. This history gave grave concern that all visitors coming to Standing Rock did not have the Tribe's best interest at heart and even greater concern that the Tribe would not be able to manage the individuals at the camps.

U.S. District Judge Daniel Hovland issued a seven-page decision dissolving the restraining order against Tribal Chairman Dave Archambault and other Tribal officials over the Dakota Access Pipeline that even went so far as to cite protestors as "hooligans," condemning out-of-state visitors for in his view "preempting the interests of the Tribe by engaging in mindless and senseless criminal mayhem," (ICT Staff, 2016). While the decision was favourable to the Tribe and identified the "legitimate interests of Native Americans of the Standing Rock Sioux Tribe who are actually impacted by the pipeline" (ICT Staff, 2016), it also included a purview perpetuated by the impact of Public Law 280 that there were severe losses by DAPL and the state of North Dakota condemning the protest altogether.

On our last night in Sacred Stone Camp, I headed ten miles to Prairie Knights Quik Mart with my grandmother and a member of my staff to prepare for the trip back home the following day. One broad misconception is that Native Americans in prayerful protest to protect lands should not be able to use gas or other technology. This train of thought is not only offensive; it is unrealistic at best. Indigenous ecological knowledge is built upon complex systems of relational use of the land. However, consumptive extraction for profit when better stewardship options are available is less relational use and more convenience. The Dakota Access Pipeline LLC knew the pipeline was not a good idea when proposing it. The City of Bismarck knew it was not a good idea when they said "not in our neighbourhood." The federal government knew there was no justification for permit under guidance from federal law, specifically the National Environmental Policy Act that it used to deny Dakota Access Pipeline LLC the required permit to continue. Native peoples knew that even under these truths, visible protection was necessary.

We left the camp in the evening; in our homelands in the desert we enjoy longer days, but we learned at Standing Rock the days were shorter. There were volunteers posted at the entrance and we told them where we were headed and that we would be back. The volunteer looked inside the minivan to find two women and an Elder and asked if we would be willing to give someone a ride. I looked at my grandma, who said "okay." The volunteer called to someone sitting in a chair; when he got up it was a tall young man, who, we found out on our drive, was just 16. The volunteer told the young man "These Aunties will give you a ride home" and gave him a side hug. When the young man got in the car, he let us know that he lived with his grandma in Tribal housing. The drive to his grandma's house was no more than five minutes away from the camp on the route to the Quik Mart. In conversation he let us know he was in high school and I urged him to consider applying for college; I let him know that there's always a way to pay for it, at which he lit up. He thanked us for coming; he said that a lot of his friends and younger people knew how important it was but that it had been hard for his community. We told him doing the right thing can feel hard but it is the best way through. As I watched him walk into the square HUD home so close to Sacred Stone Camp, something inside me broke. I could not help but think about why it was okay to protect households in Bismarck but so difficult to protect this young man and his family, even with the law on their side.

Free, Prior, and Informed Consent

An undeniable aspect of Tribal Sovereignty is the ability for free, prior, and informed consent for the Tribe, their territories, and membership. Free, prior, and informed consent (FPIC) is recognised by the United Nations Declaration of the Rights of Indigenous Peoples (UNDRIP) as the right for universal self-determination. FPIC is the right for Indigenous peoples to "withhold/withdraw consent, at any point, regarding projects impacting their territories." (Working Group of the Memorandum of Understanding Regarding Interagency Coordination & Collaboration for the Protection of Tribal Treaty & Reserved Rights, 2022)

The #NoDAPL protest illustrates a call for SRST's right to free, prior, and informed consent regarding projects that impact their treaty-protected ancestral homelands. SRST Tribal Leadership were invited to the United Nations because of the wide disregard of their FPIC in relation to the pipeline. The Department of the Interior through the Bureau of Indian Affairs established a multi-agency working group that identified best practices to increase consultation due to the quagmire that occurred at Standing Rock. In this report, the U.S. government identified the shortcomings and failure of governance to protect Tribal FPIC through government-to-government consultation. The U.S. can answer the call to their responsibility in the agreements it made with Indian Tribes by streamlining consultation that is inclusive of free, prior, and informed consent.

Tribal Sovereignty

The #NoDAPL movement amplified the complexity of Tribal Sovereignty in a world where there are few leaders and even fewer people, specifically in the United States, that are familiar with U.S. Federal Indian Policy, the unique government-to-government relationship that is shared between Federally Recognized Indian Tribes and the U.S. federal government, and the binding agreements that have forged it.

The standoff at Standing Rock is another example in the well-documented historic record of threats to ancestral Tribal Homelands by the United States. On 24 January 2017, on his second day in office, newly elected President Donald Trump signed a Presidential Executive Order entitled "Regarding the Construction of the Dakota Access Pipeline" that authorised the construction of both the Dakota Access Pipeline and the Keystone XL Pipeline (UC Berkeley, 2021). In fact, the intervention of an elected U.S. President usurping authority over the treatment of Treaty lands signalled to Indigenous advocates and scholars to prep for a coming rise in set-backs of Indian Policy to levels unseen since President Jackson. Indeed, President Trump overruled the finding by the Army Corps of Engineers to not issue the permit under the National Environmental Protection Act and took action against federal law to allow DAPL.

In 1975, the U.S. Government passed the Indian Self-Determination and Education Assistance Act, Public Law 93–638. The Act allows for autonomy of Federally Recognized Indian Tribes to govern their territories and members. It also assures that Indian Tribes have paramount involvement in the direction of services provided by the federal government to better target the delivery of services to the needs and desires of Tribal communities. This Act further cements the government-to-government relationship so that Tribal and federal leaders have a framework for consultation regarding issues that impact Tribes and their homelands. This law set the groundwork for present day consultation policy between Tribal leaders and the U.S. Executive Administration. In 2000, President Clinton issued Executive Order 13175 which until 2017 had been issued by every bi-partisan U.S. President following him until Donald Trump.

Executive Order 13175 states that the United States "recognizes the right of Indian Tribes to self-government; and supports Tribal sovereignty and self-determination" (Federal Registrar, 2000). The order outlines that the U.S. Executive Branch will work under the law of the United States, in accordance with treaties, statutes, Executive Orders, and judicial decisions that recognise the right of Indian Tribes to self-govern. It reminds elected officials that the U.S. Executive Administration is working with domestic dependent nations, and that Indian Tribes exercise inherent sovereign powers over their members and territory. The United States continues to work with Indian Tribes on a government-to-government basis to address issues concerning Indian Tribal self-government, Tribal trust resources, and Indian Tribal treaty and other rights. However, on more occasions than not, despite precedence and federal law, the United States fails to meet the obligations of their agreements.

Pass Down the Hot Sauce

I vividly remember the sense of relief I felt landing safely in Las Vegas with the group of Elders I was responsible for during our trip to Standing Rock. I took them to dinner before we made our drive back to the reservation about an hour South. They had chosen to eat at a BBQ restaurant founded by a Native American Pitmaster. An interesting thing happens when you spend a lot of time in public service; you somehow learn how to smile and pretend things are okay so that you can help others. As we were sitting at the dinner table waiting for our food, one of my Elders read me like a book in spite of my best efforts to look positive. When I noticed his glare, I smiled and he said "our people are not going anywhere, now pass down the hot sauce." Truth is, I needed that reminder. When living under external threat the feelings of helplessness, the questioning of what can be done and whether it would be enough, are constant. His reminder helped me switch back from individual expectation and ego to communal responsibility and connection. While the events that happened at Standing Rock are felt directly by the people of Standing Rock, their willingness to share their story and welcome Tribes and allies all while heeding pressures only they will know is the larger call to action. It's not just another history of failure and disregard of federal law by a sitting U.S. President but a model of courage. A reminder of the shared duty we have to pass down the truth.

References

Donovan, L. (2016). Standing Rock Sioux chairman Dave Archambault arrested at Dakota access. *The Bismarck Tribune*, 12 August. https://bismarcktribune.com/news/state-and-regional/standing-rock-sioux-chairman-dave-archambault-arrested-at-dakota-access/article_fb12da36-3e84-5694-8eba-000013930cb2.html [Accessed 11 March 2024].

Federal Registrar (2000). *Consultation and Coordination with Indian Trivial Governments.* https://www.federalregister.gov/documents/2000/11/09/00-29003/consultation-and-coordination-with-indian-tribal-governments [Accessed 5 May 2024].

ICT Staff (2016). Judge dissolves restraining order against standing Rock Sioux leaders over DAPL. *Indian Country Today.* https://ictnews.org/archive/judge-dissolves-restraining-order-against-standing-rock-sioux-leaders-over-dapl [Accessed 11 March 2024].

National Park Service (n.d.). *Fort Laramie Treaty of 1851.* https://www.nps.gov/articles/000/horse-creek-treaty.htm [Accessed 11 March 2024].

UC Berkeley (2021). *In Berkeley Library Standing Rock and the Dakota Access Pipeline: Native American Perspectives.* https://guides.lib.berkeley.edu/DAPL [Accessed 11 March 2024].

United States Congress (n.d.). *National Archives.* https://www.archives.gov/research/native-americans/indian-reorganization-act [Accessed 11 March 2024].

U.S. Department of Interior. (n.d.). What is public law 280 and where does it apply? *Bureau of Indian Affairs.* https://www.bia.gov/faqs/what-public-law-280-and-where-does-it-apply [Accessed 11 March 2024].

Wong, J. C. (2016). *The Guardian*, 23 November. https://www.theguardian.com/us-news/2016/nov/23/standing-rock-dakota-access-pipeline-sophia-wilansky-injury [Accessed 11 March 2024].

Working Group of the Memorandum of Understanding Regarding Interagency Coordination & Collaboration for the Protection of Tribal Treaty & Reserved Rights. (2022). *Bureau of Indian Affairs.* https://www.bia.gov/sites/default/files/dup/inline-files/best_practices_guide.pdf [Accessed 11 March 2024].

3

INDIGENOUS CLIMATE JUSTICE AND EPISTEMIC POLITICS IN AMAZONIA

Sylvia Cifuentes

Introduction

Globally, the role of Indigenous peoples in climate activism is critical and increasingly visible. In the United States, the No Dakota Access Pipeline (No DAPL) movement in Standing Rock, North Dakota drew the attention and solidarity of Indigenous peoples and allies from all over the world (Estes, 2019; Wehelie, 2016). In Amazonia, the Waorani leader Nemonte Nenquimo was recognised as one of 2020's most influential people in the world by *Time Magazine* for leading her peoples' "historic legal victory against the Ecuadorian government, protecting half-a-million acres of primary rainforest from oil drilling" (Amazon Frontlines, 2020). Indigenous Peoples[1] have been at the frontlines of efforts to resist, and eventually dismantle, the fossil fuel power structure – which is one of the forms and definitions of climate justice (c.f., Jafry et al., 2019). Such widely publicised cases of Indigenous resistance are thus demonstrating that climate justice is unthinkable without them.

Yet, Indigenous resistance to climate change – and climate justice – go well beyond such episodes of mobilisation against fossil fuel developments. This chapter analyses the meanings of Indigenous climate justice in Amazonia, focusing on the epistemic dimensions of territorial defence. Present in the discourses and actions of leaders from organisations across Amazonia, territorial defence can be understood as a struggle to not only secure legal rights to Indigenous territories, but rather as an all-encompassing political goal that involves self-determination, human rights, anti-extractivism, food sovereignty, and climate action and justice, among other goals (Cifuentes, 2020, 2021). For many Indigenous scholars, confronting climate change requires a distinct formulation of Indigenous climate justice (ICJ), which engages with "Indigenous philosophies, ontologies, and epistemologies . . . [and] conceptions of what constitutes justice" (McGregor et al., 2020, p. 35). Scholars and activists further highlight the importance of Indigenous knowledges (IK) to respond to climate change, increasing the calls to include them in global climate governance (e.g., Jasanoff and Martello, 2004). But they have yet to fully engage with global and plurinational Indigenous political strategies for climate change and how they integrate – and are inspired by – epistemologies or ways of knowing. Additionally, the literature on Indigenous environmental and climate justice largely responds to a North American context or remains mostly theoretical (e.g., Gilio-Whitaker, 2019; Whyte, 2020).

DOI: 10.4324/9781003396567-4

Ancestral knowledges (AK)[2] are simultaneously an element that territorial defence seeks to safeguard and one of the mechanisms that guide strategies for territorial defence and climate action. This is because AK are inextricably linked to *integral territorial ontologies* – by which territories are indivisible lifeworlds or systems that "contain forests, biodiversity (or animals/plants), humans, sacred sites,. . . underground resources, supernatural beings, and other elements," and they comprise worlds "above and below," which transcend the tangible one (Cifuentes, 2021, p. 143). Indigenous territories further encompass multiple relationships among human and more-than-human beings which explain forest vitality according to Amazonian Indigenous leaders (Cifuentes, 2021).

The agency of natural and supernatural beings also influences the creation of knowledge and political strategies. However, AK in climate action involves different types of expertise, which international climate activist campaigns can make invisible. These dynamics are evident in the politics of the Coordinator of Indigenous Organizations in the Amazon Basin (COICA) and its Amazon Indigenous initiative to Reduce Deforestation (RIA) and in the School of Political Training of the Organization of Indigenous Peoples of the Colombian Amazon (henceforth OPIAC School).

The roles of these organisations and initiatives are particularly notable in Amazonia, as it stores an important amount of the planet's carbon stocks, while also accounting for around 27% of all emissions attributed to deforestation (Hecht, 2011). Indigenous territories, about a third of Amazonian lands,[3] have higher proportions of primary forest cover and carbon storage, with lower rates of deforestation (Blackman and Veit, 2018; RAISG, 2020). Indigenous territories are increasingly at the centre of climate mitigation debates, which motivates leaders to create initiatives that build on the recognition that Indigenous peoples and their knowledges are fundamental in keeping forests standing. In that regard, leaders such as Robinson Lopez[4] (a Colombian Inga leader) highlighted the importance of AK for climate action, while also explaining some of the motivations behind designing climate initiatives:

> We want to strengthen [ancestral] knowledges, they are disappearing . . . many Indigenous peoples do not have their wise people [anymore], and . . . such knowledges . . . allow us to keep conserving forests and maintaining culture . . . they are the essence of the life of the peoples.

Situating Ancestral Knowledges

This chapter focuses on the question of how Indigenous organisations and leaders engage with and integrate AK in climate action, as part of a multi-sited and qualitative project about Amazonian Indigenous climate politics, involving various primary and secondary data sources. My methodology applies a political ecology of scale perspective, which sees scale as socially, politically, and biogeographically defined (Neumann, 2009) and Indigenous methodologies, which incorporate a decolonising lens and centre Indigenous voices and epistemologies through conversational and open-ended methods (Kovach, 2010; Smith, 2013).

Fourteen months of collaborative fieldwork and an additional period of document collection and analysis – between 2017 and 2021 – led to over 45 open-ended interviews, participant observation, and a review of secondary sources (e.g., OPIAC School manuals and RIA materials). I carried out participant observation while volunteering with the organisations at the centre of this study.

First, COICA is a Pan-Amazonian organisation headquartered in Quito, Ecuador. Founded 1984, it is, according to its leaders, the largest grassroots organisation in the world. This is because, in most Amazon Basin countries, political representatives are selected starting at the grassroots or community scale. These leaders subsequently select others at the sub-national and national scales. Every 4 years, the leaders of national organisations meet in a congress to select one leader from each country to be part of COICA's Directive Council. Therefore, COICA's mandate and political structure originates in over five thousand communities across Amazonia, and leaders must ideally represent organisations at the different scales before joining COICA – so they are aware of the common struggles that communities have across the Basin.

COICA is a transnational and plurinational organisation. The leaders of the nine national organisations – from Colombia, Ecuador, Peru, Bolivia, Brazil, Venezuela, Guyana, Suriname, and French Guiana – are also part of its Coordinating Council, further representing over five hundred Indigenous Peoples. Among such peoples are those of COICA leaders: Curripaco, Shuar, Tacana, Asháninka, Manchineri, Inga, Patamona, and several others. In my interviews, leaders shared that it is often difficult to balance the interests of such a diversity of Amazonian peoples and communities, as (broadly speaking) some of them seek to – or are willing to – collaborate with NGOs or government agencies, while others do not want to have anything to do with them – not to mention those that choose to remain uncontacted altogether. However, by defining strategic priorities every 4 years, COICA's Councils seek to address some of the issues that they identify as common to these communities.

For COICA leaders, the struggle that unites Amazonian communities is territorial defence. COICA's actions are thus oriented towards "promoting, protecting, and securing Indigenous peoples and territories, through the defence of their lifeways" (COICA, n.d.). In identifying climate action as one of its strategic priorities, COICA further demonstrates that it has embraced it, both in response to an international interest in working with Amazonian organisations and to leaders' interests in simultaneously addressing other issues that territorial defence encompasses.

In that context, COICA began conceptualising the Amazon Indigenous Initiative to Reduce Emissions from Deforestation (RIA) in 2012 as an alternative to the international REDD+ mechanism. This climate mitigation strategy would represent a "Holistic Governance of Territories for a Full Life" and an Indigenous vision for climate action that could resonate with communities and grassroot organisations across Amazonia. RIA thus intends to support communities' Life Plans while drawing from AK, respecting Indigenous development preferences, titling territories, and "valuing forests as human-nature integrating systems" (Unkuch, 2014, p. 20).[5]

Second, created by OPIAC, COICA's Colombian member, OPIAC School similarly aims to support Amazonian Indigenous peoples in "their defence of life, autonomy, and territories, by training their own leaders integrally" (OPIAC, n.d.). This means that it works to promote and recover knowledge systems that are "propios" to Amazonia – meaning that they are characteristic or unique of its peoples – while fostering a sense of belonging (OPIAC, n.d.). By foregrounding AK's role in keeping deforestation low, OPIAC secured a project funded by the Norwegian Agency for Development Cooperation's Climate and Forest Initiative, to launch the school in 2016. Consequently, its Territory and Biodiversity Program integrates several interconnected modules about climate change and territorial defence, providing training for youth leaders that represent Indigenous Peoples and grassroots organisations across the Colombian Amazon.

My collaborative fieldwork had the purpose of supporting the organisations' territorial defence and climate initiatives. With COICA, documents were created including project proposals and reports or research about women's leadership for its Women's Council. With OPIAC School, we co-taught a module that focused on REDD+, RIA, and global climate politics. Therefore, the political ecology of scale perspective was suitable for identifying how the proposals from Amazonian Indigenous organisations take place on the ground and to what extent they support the concerns of, for instance, students that represent grass-roots organisations. The literature about Indigenous and decolonial methodologies also supported reflections about how to engage with Indigenous organisations more respectfully and how my knowledge about mechanisms such as REDD+ could support their goals.

However, these methodological approaches usually focus on an engagement with Indigenous communities at the local scale only, rather than with transnational Indigenous organisations, spaces, and politics. Moreover, because this research engages with plurinational organisations, it is not possible to adapt it to one single peoples or epistemology (as Kovach, 2010 suggests). To address those challenges, methods such as participant observation in decision-making spaces or where Indigenous leaders meet and discuss were purposefully incorporated. This included assemblies, demonstrations, the meetings of COICA's Councils; *knowledge mingas* – in other words, collaborative workshops; *guayusadas* – planning spaces in political meetings that occur at dawn and incorporate the herb guayusa – among others.

I acknowledge my own positionality in this collaborative research as a South American mestiza scholar. As such, I am familiar with Indigenous thought and lifeways from personal and professional experiences. But this positionality may also limit my comprehension of Indigenous epistemologies.

Centring Indigenous Definitions of Justice

Emerging scholarship about Indigenous definitions of justice shows how Indigenous resistance to climate change and colonialism go well beyond the episodes of mobilisation that the media has popularised. Indigenous scholars have pointed out that the No DAPL movement in Standing Rock is part of a long history of Indigenous resistance and a fight for environmental justice that can be traced back to colonisation (Estes, 2019; Gilio-Whitaker, 2019). For Potawatomi scholar Kyle Whyte, current "anthropogenic climate change is an intensification of environmental change imposed on Indigenous peoples by colonialism" (2017, p. 153). Hence, the threats, vulnerability, and displacement that climate change causes are only intensified forms of the challenging conditions that Indigenous Peoples have had to face – and have resisted and survived – for centuries (Whyte, 2017; Wilson, 2014). Indigenous scholars thus recognise that colonialism – through capitalism and industrialisation – has induced changes that transformed the ecological conditions supporting Indigenous Peoples' lives and livelihoods.

The roots of current environmental and climate injustices, therefore, are a simultaneous environmental disruption and cultural genocide (Gilio-Whitaker, 2019). For McGregor and colleagues (2020), recognising this – and the urgency of responding to the ecological crisis – necessitates a distinct formulation of Indigenous climate justice (ICJ). For Native American and allied scholars, climate justice must thus involve *restorative* processes that transform "the conditions that perpetuate violence, domination and denial of rights" (Whyte, 2018, p. 280). Such restorative goals involve not only Indigenous territories, self-government, and economic and bodily security (Whyte, 2018) but also healing, horizontality, and mutual respect in human and more-than-human relationships (Gilio-Whitaker, 2019). This further entails the recognition of Indigenous identity, self-determination, and, in forested lands like Amazonia, forest governance (Whyte, 2020). All in all, an ICJ approach must incorporate Indigenous conceptions of what constitutes justice, involving "Indigenous philosophies, ontologies, and epistemologies" and Indigenous-determined futures (McGregor et al., 2020, p. 35).

Consequently, Indigenous epistemologies are central for ICJ, also in relation but going beyond other denominations of justice including restorative (e.g., Whyte, 2018) recognition (c.f., Fraser, 2020), or transformative (Newell et al., 2021). For similar reasons, scholars including Kyle Whyte and Michi Saagiig Nishnaabeg scholar Leanne Betasamosake Simpson advocate for a renewal of Indigenous knowledges to inform self-determined climate-change planning. But while engaging with IK as part of an ICJ is important for Indigenous Peoples around the world, from the Amazon and Congo Basins to Indonesia and Mesoamerica (see Global Alliance, n.d.), scholarship on Indigenous definitions of environmental and climate justice has largely responded to North American context and experiences, with a few exceptions (e.g., Nuñez, 2018; Osborne et al., 2024; Sauls, 2020). This chapter therefore contributes to addressing that gap.

Additionally, when examining climate governance, discussions about Indigenous knowledges tend to remain abstract rather than to elaborate on their role in practical alternatives (e.g., Jasanoff and Martello, 2004). Or they emphasise communities' interactions with nature, rather than political uses of knowledge (c.f., Forsyth and Walker, 2008). Furthermore, scholars have recognised that international climate governance and current climate actions are not including Indigenous voices and their knowledges, pointing to a coloniality of knowledge (Nuñez, 2018) within those spaces. This echoes analyses of a "coloniality of justice" within the environmental justice scholarship, which often sees communities in the Global South as subjects of study only, rather than as "knowledge-holders capable of reimagining the meaning of EJ and its underlying concepts" (Alvarez and Coolsaet, 2020, p. 63).

For decolonial theorists, the way colonial institutions have suppressed and even killed multiple knowledges is also a form of injustice – i.e., epistemicide (c.f., Escobar, 2007b). This underlies current problems with climate mechanisms like REDD+, which, scholars argue, tend to prioritise certain forms of "expert" or technocratic knowledge – other than neglecting and often perpetuating structural inequality (Thompson et al., 2011; Cifuentes, 2017). Even with participatory mechanisms and safeguards in place, REDD+ national strategies often exclude Indigenous knowledges from their design, implementation, and evaluation (Cifuentes, 2017; Osborne et al., 2024) and fail to incorporate a "vision of Indigenous knowledge, rights, and lifeways" (Powless, 2012, p. 412). But analyses of how specific alternatives to REDD+ actualise aspects of justice and embrace AK remain scant.

Furthermore, Indigenous leaders' references to more-than-human agency resonate with scholarship about posthumanism and multispecies justice. Posthumanism critiques how modernity establishes separations between humans and nature, due to its anthropocentric character based on human exceptionalism (e.g., Braun et al., 2010; Haraway, 2008). Decolonial scholars similarly question the modern/colonial assumption that humans are the "foundation for all knowledge . . . of the world, separate from the natural" (Escobar, 2007b). Nonetheless, posthumanism can also "reproduce colonial ways of knowing and being by enacting universalizing claims" (Sundberg, 2014, p. 33), while failing to credit Indigenous thinkers who have long referred to complex relationships of humans and other beings (Todd, 2016).

Thus, this chapter speaks to multispecies justice scholarship in considering the forms of knowledge and communication of non-human beings (Celermajer et al., 2020) and in decentring the human in explaining imaginaries of more just climate futures (c.f., Tschakert et al., 2021). It also speaks to Latin American decolonial theorists who refer to cognitive justice as a concept that can guide the creation of more just relationships between different types of knowledge (e.g., Escobar, 2007a). In the case of global environmental change, cognitive justice would entail foregrounding a plurality of knowledges – and their bearers – as they can unveil different problem definitions and solutions that can have tangible impacts on the environment and people's lives.

However, I highlight the perspectives of Indigenous leaders from COICA and OPIAC School about how, as part of integral territorial ontologies, AK emerge from the relationships among Indigenous Peoples and more-than-human beings. The following section thus details how Indigenous leaders conceive and understand AK and climate activism in their own terms, giving new meanings to ICJ.

Ancestral Knowledges as Part of Integral Territorial Ontologies and Climate Initiatives

The importance that Amazonian Indigenous leaders give to AK within Indigenous climate initiatives is evident in several ways. RIA's materials argue that Indigenous Peoples have unique knowledges about forests as recognized by UN agreements, and point to the use of AK as a significant commonality among Amazonian peoples, despite the diversity of their cultures and "Indigenous economies" (COICA, 2018). The role of these knowledges within RIA is central both in keeping the forest standing and as part of alternatives to deforestation (COICA, 2018). Thus, for COICA, climate solutions should "include . . . ancestral wisdom, innovation, and practices" (COICA, 2010).

But beyond a mere inclusion of these knowledges in international climate governance, for Indigenous leaders, AK should be at the basis of self-determined climate initiatives because they are inextricably linked to Indigenous territories. As Robinson explains,

> The territory for Indigenous Peoples is . . . where we unfold our traditional and spiritual knowledge systems, our own (forms of) government . . . our history, our cosmogony, our culture . . . These systems [guide] climate change actions.
> *(Personal communication, August 2019)*

More specifically, this inextricable relationship between AK and territories is present in instruments – or Indigenous technologies – that organise territorial activities and that leaders identify as having potential for guiding climate-change mitigation and adaptation strategies. For instance, ecological calendars show how natural cycles influence the relationships among different elements of the territory (see Figure 3.1). Created by communities' wise people, these calendars portray yearly cycles and indicate when to hunt, collect fruits, or plant. They also integrate spheres like productive or ceremonial activities with the cycles and biogeographic determinants of the territory, since "it is impossible to disconnect (them)" (OPIAC, 2019c, p. 1). Ecological calendars are a central part of the training of youth leaders of the OPIAC School, since they can also be adapted to climatic changes by modifying the cycles that they portray (OPIAC, 2019c).

Figure 3.1 Ecological calendar, Andoque peoples. Elaborated by 2019 OPIAC School students.
Source: Cifuentes, reproduced with permission.

Another Indigenous technology where AK are inextricably linked to Amazonian territories are *chacras* or Indigenous agricultural parcels/systems in Ecuador, Peru, and Colombia (where the term is *chagra* or *conuco*) – and across Amazonia, according to COICA leaders. OPIAC School dedicates an entire module to chacras as knowledge systems, explaining how these are scenarios of resistance that protect AK and its connection to territorial use and ordering. As such, it encourages students to create models of how a chagra of their peoples would look (see Figure 3.2).

For OPIAC School and COICA leaders, chacras are technologies for climate-change mitigation and adaptation simultaneously, while involving the agency of more-than human beings. Leaders see them as alternatives to the monocultures that cause deforestation and that can confront climate vulnerability – some RIA-associated projects have thus focused on recuperating ancestral crops in the Peruvian Amazon. OPIAC School further mentions that they are:

Filled with symbolic content where plants are considered people . . . [with] social relationships . . . that are like those that regulate the human world."

(OPIAC, 2019b, p. 6, citing Tropembos Internacional)

Figure 3.2 OPIAC School student presenting his chagra model.
Source: Cifuentes, reproduced with permission

These examples therefore demonstrate how AK are part of *integral territorial ontologies*, and how multiple knowledges can inform climate action. But AK are also part of what territorial defence and climate action seek to safeguard. For Alonso, a Tacana and Bolivian leader, climate initiatives like RIA can also "reinforce [part of] what is being lost in terms of AK, to tell the youth: this is the richness of a territory, which our elders can share" (personal communication, January 2019). Leaders across different scales of political organisation thus highlighted how, very often, elders pass away or are no longer able or willing to transmit their knowledges to younger generations.

According to OPIAC School, some of the weaknesses that are present in the leadership of the Colombian Amazon thus relate to its distancing and superficial engagement with "the Amazonian knowledge system that is inscribed in the peoples' worldviews, from and about their territory" (OPIAC, 2019a, p. 5). For the school, there is a lack of understanding of how the ancestral wisdom of those who have cultural legitimacy – "due to having received and inherited their knowledge through ancestral rituals and 'remedies', and many years of practice" (idem) – can inform political leadership today, particularly that which takes place in spheres that involve dealing with non-Indigenous persons. This includes spaces for climate-change action or biodiversity conservation but also negotiations – and resistance to – extractive projects and fossil fuel extraction in Indigenous territories. Thus, OPIAC school seeks to address issues including a "lack of a sense of belonging and a weakness in the collective self-esteem" (idem) by teaching AK to the youth political leaders of grassroots organisations. It does this by encouraging them to seek out their elders, *pajés*, and wise people while completing projects as part of the school's training, while completing research projects after attending the school, and as a continuous practice when returning to their territories and/or taking on different leadership roles.

Tatiana and Marta (Peruvian leaders, Asháninka and Shipibo youth, respectively) similarly noted that the peoples lose out when elders cannot transmit their knowledges, but that AK has been key in the preparation of youth leaders since colonial times (personal communications, February and April 2019). For them and other leaders, there may be different reasons that explain this loss, including an increasing interest of the younger generations in "Western" or "modern" life and artefacts and decreased interest in community life. But OPIAC School points to broader and more structural factors, as it aims to "strengthen Indigenous cultures in view of colonising education processes" (OPIAC, 2021).

That echoes scholarship suggesting a continuous undermining of efforts to strengthen IK systems by colonial infrastructures (Simpson, 2004). Contributions about intercultural education in Peru, Bolivia, and Ecuador, for instance, refer to the "centuries of official prohibition and social denigration [of Indigenous languages] dating from the imposition of colonial rule by Spaniards in the 16th century" (Hornberger, 2000, p. 174). They explain how this has only begun to change since the last quarter of the twentieth century, thanks to Indigenous mobilisations that have demanded change in many areas including education, as part of a rejection to assimilationist policies promoted by states that have "long modelled themselves after the myth of the linguistically and culturally homogeneous nation" (idem, p. 177).

Decolonial scholars further discuss how modernity, as imposed by coloniality, has established ideologies like secularisation or of (Western) science as the only valid form of knowledge (Escobar, 2007b) which have similarly excluded AK or have labelled it as "superstition" or "beliefs" (Lyons et al., 2017, p. 35). Conversely, anthropologists have noted how cultural changes brought by Christianity in Amazonia can have "negative impacts on biodiversity patterns and species abundances . . . [by] weaken[ing] community and regional-level governance systems needed for effective management of resources and interaction with government authorities and other outside interests" (Luzar and Fragoso, 2013, p. 309) – even if some forms of Christianity can also coexist with Indigenous belief systems (Capiberibe, 2023). While an in-depth analysis of how education, secularism, or religion impact ancestral knowledges for political leadership goes beyond the scope of this research, when volunteering with the OPIAC School, I similarly witnessed how two students argued that Christianity had "enlightened" them to realise that the cosmovision of their peoples was just a superstition. Thus, it is worth reiterating that OPIAC School – and RIA to an extent – seek to counter these different influences by highlighting how AK are profoundly connected to territorial governance systems and contemporary decision-making.

Ancestral Knowledges as a Guide for
Territorial Defence and Climate Politics

Some struggles that are taking place . . . [are] absolutely directed from the traditional spiritual knowledge systems.

(Robinson López)

In my interviews, leaders also demonstrated how AK are at the basis of and guide political strategies for climate action and territorial defence. For instance, ceremonies with sacred plants can be considered as political spaces of planning and visualising these strategies. Robinson illustrated this as follows:

When we are in the ambi wasi,[6] during yage[7] ceremonies, we contact all living and non-living beings, those who are underground, those who are in the centre, those who are in the cosmos . . . through them, it is possible to make visible the problems that are coming, that we are living, the threats to the leaders for defending the territory, but also those which can affect Earth as a whole [such as climate change].
(Personal communication, August 2019).

Furthermore, leaders referred to how wise people – and some plants – can guide leaders in determining how to move forward and how to identify the right path to do what is best for the struggle. Shuar youth leaders in Ecuador narrated how their family members put chilli in their eyes with the purpose of envisioning this path and their role in territorial defence (personal observation, January 2019). In guayusadas,[8] tobacco is also inhaled to "clear the mind, and to have visions that show where to go, how to maintain the struggle for collective rights and the territory, and how to obtain results to take back to the communities and base (grassroots) organisations" (personal observation, February 2019). Sofia (a Shuar Ecuadorian leader) also talked about how plants guide leaders:

[From the time of our] grandparents, we maintain [the tradition] of drinking guayusa, natema, malecos, tobacco, to have visions . . . to know where [to go], to see what can come after, to see the future . . . this tradition for decision-making is maintained until now, so our leaders . . . do these rituals to have more strength, to be able to move the organisation, and our nationalities, forward.
(personal communication, April 2019)

Even more, for Enrique, a Machinieri and Brazilian COICA leader, ceremonies with pajés (shamans) and the use of plants can even intervene in the context of international environmental governance. He explained how:

Knowledges are the essence of the struggle . . . in 2007, I was in my community and I drank ayahuasca, and it took me in a dream, it took me far away, to a valley that I had never seen in my life . . . and later, in 2007, I went to Bonn, to the biodiversity convention . . . [there] I saw a valley and remembered that it was there where ayahuasca was pointing me to.
(personal communication, April 2019)

Thus, for Enrique, when there are a lot of important political matters about to happen, leaders meet with their pajé to see if the struggle will be positive, and ayahuasca (or another channel or plant) shows how the struggle will or can be. From then on, he said, you can create strategies for territorial defence.

Additionally, for Enrique, understanding the pajé's visions is necessary as it is through them that it is possible to connect with the forest and with the spirits of the animals and to know how to heal. He explained that "knowledge is thus fundamental to understand the Earth because it allows us to enter another world, the spiritual world . . . which is connected to the territories" (personal communication, April 2019). But while several leaders highlight the roles of AK, plants, and pajés in their political roles in international meetings, others also mention how this is connected to their longstanding struggle for the defence of the territories – which the "international community" often reads as a struggle for climate justice.

Robinson also saw AK guiding resistance mingas,[9] mobilisations that are meant to protest government actions, most notably the approval of projects of oil and mining extraction – and others that Indigenous communities and/or organisations may consider adverse for Indigenous rights. This leader defined these spaces as

> Setting(s) for thought (pensamiento) for sharing, for resistance, [places] to exchange knowledges, that evidence the problems that Indigenous Peoples live, and where we make demands to the government . . . as well as propositions to safeguard our rights.
> *(Personal communication, August 2019)*

Consequently, AK are an active part of Indigenous advocacy for territorial and collective rights and are thus at the basis of political strategies for territorial defence and climate action. Even though they are "ancestral", leaders engage with them frequently in "modern" politics to articulate their demands for more rights. As this and the previous subsections have shown, for Indigenous leaders in Amazonia, the agency of more-than-human beings such as plants or spiritual beings influences knowledge creation and the political strategies that are fundamental in the context of climate change. This is in addition to how the political discourse of Amazonian Indigenous leaders is increasingly centred on AK as a central aspect in keeping forests standing and climate-change mitigation.

Ancestral Knowledges, Expertise, and International Climate Activism

The previous discussion further suggests that climate initiatives have become a venue for Indigenous organisations to reach broader, global audiences while still strengthening their longstanding political goal of territorial defence. Most notably for this chapter, the latter includes safeguarding AK, securing territorial rights, and rejecting extractive projects as well. But while international environmental non-government organisations (IENGOs) and other actors identify Indigenous leaders as climate activists because of such goals, their role is not solely – or even primarily – that of an activist but rather to respond to "las bases" – which is to the grassroots – or the communities and organisations at different political scales that have elected them to safeguard their rights. However, this is often unknown or ignored in international climate activism.

Consequently, a final aspect to consider when discussing an ICJ and epistemic politics in Amazonia concerns how different actors – for example elders, leaders, cultivators, among many others – hold different types of expertise that are relevant for climate action. An assumption behind the OPIAC School is that all Indigenous Peoples (and persons) can engage with AK from their various roles and experiences. That is why, as mentioned earlier, the school encourages students to approach the "Amazonian knowledge system" in their own communities and cultural contexts so they can be good leaders. This involves connecting with those who have *cultural legitimacy* as they have "inherited" knowledges ancestrally and through specific rituals and (human and more-than-human) laws (OPIAC,

2019a, my emphasis). The latter include the elders, pajés/shamans, and traditional authorities who hold specific kinds of knowledge for the survival of their communities and peoples. OPIAC School explains how in many communities, for instance, elders oversee the ecological calendars' regulations and ordering. Similarly, in the knowledge systems of chacras, women are generally the knowledge holders, as they know about seeds, techniques of planting, or tending to the soil, among other aspects (OPIAC, 2019b).

However, that elaborated acknowledgment of the expertise of different knowledge holders does not travel easily across scales and publics. COICA leaders have also been part of an international campaign (and project) called "Guardians of the Forests" which seeks to raise awareness about the role of Indigenous peoples in fighting climate change. Thus, this campaign should ideally garner support for initiatives like RIA. However, the narratives within this campaign talk about "incorporating AK in climate strategies," often without further explanations. The same happens with knowledge holders and expertise. In the campaign, the "Guardians" are mainly the leaders of the organisations that are involved, including COICA.

However, political leaders do not necessarily engage with AK in their roles as project creators and negotiators, neither are they necessarily or always at the frontlines of, for instance, rejecting extractive projects. They are generally not the ones putting knowledges in practice by, for example, creating diverse and nutritious chacras. Political leaders are usually different from traditional communal authorities or shamans (with notable exceptions, like some OPIAC School's students who are also training to become pajés or Davi Kopenawa – see Kopenawa and Albert, 2013). There is of course some overlap and changing roles, but the Guardians of the Forest campaign (and narrative) makes these distinct roles invisible, often referring to undifferentiated Indigenous persons and to AK in general terms.

While this type of campaign seeks to reach to the largest, more international public possible by simplifying narratives, this also has drawbacks. Forsyth and Walker (2008) have warned against making inevitable links between ethnicity and environmental outcomes. For Mihnea Tănăsescu, guardianship is a deeply rooted western notion that can obscure the meanings of ethics of care like the Māori kaitiakitanga – where "trustees" are specifically nonhuman and which is also about "managing people" (2020, p. 446). Similarly, erasing differences with respect to AK and expertise runs the risk of essentialising Indigenous Peoples. It can make AK seem like an immutable body of knowledge, when it has changed and adapted to different conditions over time – including to environmental changes – and while facing different threats.

Therefore, while leaders can and often do have important roles in, for instance, strengthening knowledge systems and facilitating the transmission of knowledges to younger generations – for example, through initiatives like OPIAC School – making these different types of expertise and specific knowledges invisible could take the attention away from those roles, responsibilities, and possibilities.

Conclusion: What Does an Indigenous (and) Epistemic Climate Justice Entail in Amazonia?

Here, I have referred to climate justice as territorial defence in Amazonia, as Indigenous-led climate initiatives illustrate, drawing from and expanding discussions about the need for a distinct understanding of ICJ. In addition to climate initiatives, territorial defence encompasses multiple other political practices – for example, actions rejecting oil and mining projects and safeguarding AK to respond to climate change – that share the same purpose of defending the territories as lifeworlds. These practices are unfolding at all scales of Indigenous political organisation and, directly or indirectly, confront the threats of climate change and other factors to territorial integrity.

As such, it is not sufficient to simply acknowledge Indigenous leaders as climate activists since their different roles include but also differ and go beyond that. Neither is it sufficient to seek to achieve a form of cognitive justice by simply recognising the existence of multiple knowledges that can inform climate action – or even including IK in policies that are designed far from their territories, such as the UN climate treaties. As the scholarship notes, there are multiple definitions and meanings of both "justice" and "the environment" for a diversity of peoples (Álvarez and Coolsaet, 2020), and this is a central aspect to consider when thinking about what ICJ means for specific Indigenous Peoples and in considering specific dimensions such as the epistemic.

Instead, this chapter has demonstrated that there are many ways in which scholars and activists can and must understand justice in epistemic terms. First, as RIA and the OPIAC School have demonstrated, multiple knowledges may indeed inform climate-change action, as illustrated by Indigenous technologies like ecological calendars and chacras – which address climate-change mitigation and adaptation in agriculture. While these technologies can have thousands of variations considering the number of Indigenous Peoples in Amazonia, many leaders across the Basin highlight them as important for climate initiatives.

Second, epistemic justice in Indigenous climate politics also involves recognising that there are different knowledge holders including elders, wise people, pajés/shamans, cultivators, leaders, among many others. These knowledge holders have specific kinds of knowledge that are relevant for territorial defence and climate action, including – but not limited to – agricultural, spiritual, political, and ecological. They also take on different roles and responsibilities within Indigenous communities and organisations. Recognising and valuing this can prevent international climate activism from one important problem that can emerge when talking about Indigenous Peoples, AK, and climate change: essentialising Indigenous individuals while considering AK as a static body of knowledge. This can also prevent viewing "the role allotted to colonised peoples in this dark fairytale [of the 'Anthropocene' or 'Capitalocene,' etc.] (a)s one of aggrieved victim or heroic activist" (Curley and Smith, 2023, p. 15). This chapter has shown how Indigenous leaders are not simply heroic activists but people who actively choose to participate in climate politics to further their political goals – often in response to demands from the grassroots, although not without challenges and contradictions.

Third, epistemic justice also entails recognising the different relationships among knowledges and "nature" – or rather, the territories. This chapter has shown how AK are an inextricable part of integral territorial ontologies, where territories are lifeworlds conformed by multiple relationships among human and more-than-human beings. As such, the agency of more-than-human beings is important in instances of knowledge creation, as is again illustrated by ecological calendars and chacras. However, this agency exists in relationship to that of leaders, elders, shamans, or cultivators that engage with these beings for knowledge creation.

Fourth, an Indigenous epistemic climate justice entails making the multiple relationships among knowledges and climate politics visible. Safeguarding, transmitting, and questioning the institutions that undermine the maintenance and renewal of AK are a part of territorial defence that cannot be separated from struggles to, for instance, reject extractive projects. But I have also shown how AK are fundamental for Indigenous leaders to guide their strategies for territorial defence and climate action. This often involves their use in ceremonies, and the agency of more-than-human beings such as plants like ayahuasca, guayusa, or chilli, which can show the path that leaders should follow to strengthen the struggle.

These findings thus expand discussions about environmental knowledges, politics, and expertise. They also show how grassroots climate activism in the Global South often engages with the more longstanding – and self-determined – forms of politics of historically marginalised groups of people – who may not define or understand their actions as activism even though their purposes are often similar. They further illuminate how grassroots organisations and AK can simultaneously be global (and plurinational) and place-based, and the various roles that Indigenous ways of knowing have in visualising climate transformations.

Notes

1 Most Indigenous leaders across Amazonia prefer the term "peoples" when referring to their nation/ nationality, ethnic group, identity, and/or sense of belonging. Terms like "tribes" can be considered inappropriate due to their colonial legacies.
2 Indigenous leaders employ "ancestral" to refer to those knowledges and practices that emerge from the historical relationships of Indigenous Peoples and territories. AK are passed from generation to generation and are part of different areas of life including medicine, nourishment, and forest vitality. Here, I use AK and IK interchangeably.
3 This estimate includes lands that countries have officially recognised as Indigenous, those that are in process, and those which still lack recognition (RAISG, 2020).
4 Due to IRB protocols, I use a pseudonym (a protected first name) when quoting most of my interviewees. I use Robinson López's full name as a posthumous tribute, as his interview was one of the most inspiring for my work and he was a public figure when he unfortunately passed away.
5 Life Plans seek to capture Indigenous communities' own, diverse visions of how to achieve a better life "through elements ranging from opposition to extractive industries and strengthening Indigenous cultures to promoting income-generating endeavours" (Cifuentes, 2021).
6 A communal central house where gatherings take place.
7 *Yagé* is another name for ayahuasca.
8 See previous description/definition above.
9 Mingas are gatherings of cooperative work or voluntary communal labour, held in various Indigenous and sometimes mestize communities.

References

Álvarez, L. and Coolsaet, B. (2020). Decolonizing environmental justice studies: A Latin American perspective. *Capitalism Nature Socialism*, 31(2), 50–69.

Amazon Frontlines (2020). *Indigenous Amazonian Leader Nemonte Nenquimo Is Named TIME 100 Most Influential People in the World*. https://amazonfrontlines.org/chronicles/indigenous-amazonian-leader-nemonte-nenquimo-is-named-time-100-most-influential-people-in-the-world/ [Accessed 2 August 2022].

Blackman, A. and Veit, P. (2018). Amazon indigenous communities cut forest carbon emissions. *Ecological Economics*, 153, 56–67.

Braun, B., Whatmore, S. J. and Stengers, I. (2010). *Political Matter: Technoscience, Democracy, and Public Life*. Minneapolis: University of Minnesota Press.

Capiberibe, A. (2023). Indigenous evangelicals, adventists, and catholics: Intersections of Christianity and Shamanism on the Brazil/French Guiana Amazonian Border. In *Indigenous Churches: Anthropology of Christianity in Lowland South America* (pp. 109–130). Palgrave Macmillan.

Celermajer, D., Schlosberg, D., Rickards, L., Stewart-Harawira, M., Thaler, M., Tschakert, P., Verlie, B. and Winter, C. (2020). Multispecies justice: Theories, challenges, and a research agenda for environmental politics. *Environmental Politics*, 30(1–2), 119–140.

Cifuentes, S. (2017). Global forest governance, poverty, and expert knowledge in Ecuador: The case of UN-REDD. *Perspectives on Global Development and Technology*, 16(1–3), 315–328.

Cifuentes, S. (2020). Territory, autonomy and rights: Indigenous politics and COVID-19 in the Amazon basin. *Environment and Planning D: Society and Space (Digital Magazine)*. https://www.societyandspace.org/articles/territory-autonomy-and-rights

Cifuentes, S. (2021). Rethinking climate governance: Amazonian indigenous climate politics and integral territorial ontologies. *Journal of Latin American Geography*, 17(4).

COICA (2010). *Posición Política de COICA sobre REDD+*.

COICA (2018). *Mandato de Macapá*. IV Cumbre Amazónica.

COICA (n.d.). *¿Qué es la COICA?* https://coica.org.ec/que-es-la-coica/ [Accessed 24 May 2021].

Curley, A. and Smith, S. (2023). The cene scene: Who gets to theorize global time and how do we center indigenous and black futurities? *Environment and Planning E: Nature and Space*, 25148486231173865.

Escobar, A. (2007a). Actors, networks, and new knowledge producers: Social movements and the paradigmatic transition in the sciences. In Boaventura Santos (ed.) *Cognitive Justice in a Global World: Prudent Knowledges for a Decent Life* (pp. 273–294). Lexington Books.

Escobar, A. (2007b). Worlds and knowledges otherwise: The Latin American modernity/coloniality research program. *Cultural Studies*, 21(2–3), 179–210.

Estes, N. (2019). *Our History Is the Future: Standing Rock versus the Dakota Access Pipeline, and the Long Tradition of Indigenous Resistance*. Verso.

Forsyth, T. and Walker, A. (2008). *Forest Guardians, Forest Destroyers: The Politics of Environmental Knowledge in Northern Thailand*. University of Washington Press.

Fraser, N. (2000). From redistribution to recognition?: Dilemmas of justice in a 'postsocialist' age. In *The New Social Theory Reader* (pp. 188–196). Abingdon: Routledge.

Gilio-Whitaker, D. (2019). *As Long as Grass Grows: The Indigenous Fight for Environmental Justice, from Colonization to Standing Rock*. Beacon Press.

Global Alliance (n.d.). *Homepage*. https://globalalliance.me/ [Accessed 15 March 2022].

Haraway, D. J. (2008). *When Species Meet*. Minneapolis: University of Minnesota Press.

Hecht, S. B. (2011). From eco-catastrophe to zero deforestation? Interdisciplinarities, politics, environmentalisms, and reduced clearing in Amazonia. *Environmental Conservation*, 39(1), 4–19.

Hornberger, N. H. (2000). Bilingual education policy and practice in the Andes: Ideological paradox and intercultural possibility. *Anthropology & Education Quarterly*, 31(2), 173–201.

Jafry, T., Helwig, K. and Mikulewicz, M. (ed.) (2019). *Routledge Handbook of Climate Justice*. Routledge.

Jasanoff, S. and Martello, M. L. (2004). *Earthly politics: Local and Global in Environmental Governance*. Cambridge, MA: MIT Press.

Kopenawa, D. and Albert, B. (2013) *The Falling Sky: Words of a Yanomami Shaman*. Cambridge, MA: Harvard University Press.

Kovach, M. (2010). *Indigenous Methodologies: Characteristics, Conversations, and Contexts*. University of Toronto Press.

Luzar, J. B. and Fragoso, J. M. (2013). Shamanism, Christianity and culture change in Amazonia. *Human Ecology*, 41, 299–311.

Lyons, K., Parreñas, S. and Tamarkin, N. (2017). Engagements with decolonization and decoloniality in and at the interfaces of STS. *Catalyst: Feminism, Theory, Technoscience* 3(1), 1–47.

McGregor, D., Whitaker, S. and Sritharan, M. (2020). Indigenous environmental justice and sustainability. *Current Opinion in Environmental Sustainability*, 43, 35–40.

Neumann, R. P. (2009). Political ecology: Theorizing scale. *Progress in Human Geography*, 33(3), 398–406.

Newell, P., Srivastava, S., . . . & Price, R. (2021). Toward transformative climate justice: An emerging research agenda. *Wiley Interdisciplinary Reviews: Climate Change*, 12(6), e733.

Nuñez, A. J. (2018). Mother earth and climate justice. In T. Jafry (ed.) *Routledge Handbook of Climate Justice* (pp. 420–430). Routledge.

OPIAC (2021). *Historia*. OPIAC's website. https://opiac.org.co/opiac/historia [Accessed 17 June 2021].

OPIAC (n.d.). *OPIAC Escuela de Formación Política*. https://www.opiacescuela-copaiba.info/ [Accessed 15 June 2021].

OPIAC School of Political Training (2019a). *Programa de Territorio y Biodiversidad.* Módulo 1 – Cosmovisión y Territorio.

OPIAC School of Political Training (2019b). *Programa de Territorio y Biodiversidad.* Modulo 3 – Chagra o Conuco Amazónico.

OPIAC School of Political Training (2019c). *Programa de Territorio y Biodiversidad.* Modulo 5— Conocimiento, Uso y Manejo de la Naturaleza.

Osborne, T., Cifuentes, S., Dev, L., Howard, S., Marchi, E., Withey, L. and Santos Rocha da Silva, M. (2024). Climate justice, forests, and Indigenous peoples: Toward an alternative to REDD+ for the Amazon. *Climatic Change*, 177(8), 1–28.

Powless, B. (2012). An indigenous movement to confront climate change. *Globalizations*, 9, 411–424.

RAISG. (2020). *Amazonía bajo presión* (1st ed.). São Paulo: ISA – Instituto Socioambiental.

Sauls, L. A. (2020). Becoming fundable? Converting climate justice claims into climate finance in Mesoamerica's forests. *Climatic Change*, 161(2), 307–325.

Simpson, L. R. (2004). Anticolonial strategies for the recovery and maintenance of Indigenous knowledge. *American Indian Quarterly*, 373–384.

Smith, L. T. (2013). *Decolonizing Methodologies: Research and Indigenous Peoples.* Zed Books.

Sundberg, J. (2014). Decolonizing posthumanist geographies. *Cultural Geographies*, 21(1), 33–47.

Tănăsescu, M. (2020). Rights of nature, legal personality, and indigenous philosophies. *Transnational Environmental Law*, 9(3), 429–453.

Thompson, M. C., Baruah, M. and Carr, E. R. (2011). Seeing REDD+ as a project of environmental governance. *Environmental Science & Policy*, 14(2), 100–110.

Todd, Z. (2016). An indigenous feminist's take on the ontological turn: 'Ontology' is just another word for colonialism. *Journal of Historical Sociology*, 29(1), 4–22.

Tschakert, P., Schlosberg, D., Celermajer, D., Rickards, L., Winter, C., Thaler, M., Stewart-Harawira, M. and Verlie, B. (2021). Multispecies justice: Climate-just futures with, for and beyond humans. *Wiley Interdisciplinary Reviews: Climate Change*, 12(2), e699.

Unkuch, K. (2014). *REDD+ Indígena Amazónico-RIA.* Quito, Ecuador: COICA.

Wehelie, B. (2016). Sacred ground: Inside the Dakota pipeline protests. *CNN* [online]. https://www.cnn.com/interactive/2016/12/us/dapl-protestscnnphotos/ [Accessed 2 August 2022].

Whyte, K. (2017). Indigenous climate change studies: Indigenizing futures, decolonizing the anthropocene. *English Language Notes*, 55(1), 153–162.

Whyte, K. (2018). On resilient parasitisms, or why I'm skeptical of Indigenous/settler reconciliation. *Journal of Global Ethics*, 14(2), 277–289.

Whyte, K. (2020). Too late for indigenous climate justice: Ecological and relational tipping points. *Wiley Interdisciplinary Reviews: Climate Change*, 11(1), e603.

Wilson, N. J. (2014). The politics of adaptation: Subsistence livelihoods and vulnerability to climate change in the Koyukon Athabascan village of Ruby, Alaska. *Human Ecology*, 42(1), 87–101.

4

BRAZILIAN GRASSROOTS CLIMATE ACTIVISM IN UNEQUAL WATERSCAPES

Alexandre da Silva Faustino,
Nícolas Guerra-Tão, and Wendy Steele

Introduction

The politics of urban water governance and climate activism are contextualised in this chapter through the lived struggles of riverine communities within an urbanised territory of the Global South periphery of capitalism (Rolnik, 1997; Maricato, 1996) – *São Paulo*, Brazil. In *Metrópole na Pariferia do Capitalismo* (1996), Ermínia Maricato refers to these peripheries as the urban geographies commodified by the political economy of global capitalism. In such territories of accumulation by violent dispossession, the long cycles of colonialism and slavery structured highly unequal societies marked by dependent favour, privilege, and arbitrariness, reflected in profound contradictions of the dominant governance system that deepen the social and legal detachment from the real city. Economic growth led to deeper social exclusion, enacting a modernisation linked with exclusion, a late industrialisation of low wages, and a modern development of backwardness. Particularly at the periphery of capitalism, configured in the planetary landscapes of extraction produced by the imperialism and colonialism of Global North states and nations (Mezzadra, 2023; Brenner and Schmid, 2011; Arboleda, 2016), neoliberalism – the late stage of this process – went wild long ago (Angotti, 2013; Parnell and Robinson, 2012). The implications of decades of socio-ecological devastation can be seen in cities like São Paulo, Mexico City, and Bogota.

In these 'wild capitalist' contexts, legislation, market, and real-estate profits are purposefully entangled in a contradictory unity (Franco, 1979) that creates built environments whose illegal production is informally consented to by a state that does not secure formal access to land and city (Maricato, 2003). Insurgency originating from such contradictions and against imposed exclusions, has emerged within marginalised communities struggling for spatial justice. Through activism they establish 'occupations' on the land as both a world-making process and political strategy to disrupt the hegemonic model of private property primacy that keeps reproducing social inequality (Caulkins, 2022; Ferrara et al., 2019; Maricato, 2009). In that struggle, while state and capital constantly work to erase occupations from urban land highly valued by real estate, a blind eye is cast when occupations take place in more neglected territories, such as environmentally protected areas (Valencio et al., 2009).

DOI: 10.4324/9781003396567-5

Amidst this political-economic backdrop, Latin American cities are confronting the escalating threat of climate change, exacerbating their vulnerability to extreme weather events. The most common events associated with disasters in Latin America are heat waves, droughts, heavy storms, and landslides, all becoming more intense and frequent due to climate change (UN-Habitat, 2018; Lizarralde et al., 2021). Between 2000 and 2019, the region has reported 1,205 catastrophes affecting more than 152 million people, where floods feature as the most frequent type of disaster (Jacobi, 2023; OCHAN, 2020). Latin America is the most urbanised region in the world, with 81% of its population living in cities, projected to reach 90% by 2050 (UN-Habitat, 2018). The sheer mobilisation of materials, infrastructure, and energy to produce such densification and urban growth represents billions of tons of carbon – about 40% of global emissions come from cities (Rosenzweig et al., 2011) – which poses a serious challenge for keeping targets of global temperature increase below 2 degrees Celsius by 2100 (PBMC, 2016).

Brazilian cities are threatened by changes of planetary hydrological cycles driven by global warming, exacerbating floods, landslides, heat waves, shifting availability of freshwater, increased food insecurity, sea level rise, and coastal erosion (PBMC, 2016; Calvin et al., 2023). Urbanised territories in Brazil are highly vulnerable to such events, especially given the structural precarity and informality that is a trait of its urban production, whose socio-nature is rooted in the historical colonial projects of appropriation of land and resources (Lampis et al., 2020). The collapse of critical infrastructure such as water, sanitation, energy, and transportation are some of the major threats, given the networked nature of these systems and their different levels of vulnerability associated with precarity and lack of adaptability (PBMC, 2016).

As an urban epicentre with 21 million inhabitants (IBGE, 2022), São Paulo reflects in its massive metropolitan scale many of the contradictions, inequalities, and conflicts that mark global mega-cities in a climate, biodiversity, and humanitarian crisis. São Paulo is a place that has experienced both catastrophic floods and drought (Cruxên, 2016; Henrique and Tschakert, 2019a; Millington, 2016), and global projections on climate change highlight floods as key threats to be intensified in scale, frequency, and volume (Calvin et al., 2023). The impact of these disasters on already marginalised and vulnerable communities has been compounded by a lack of universal sanitation, increasingly speculative urban development, and the political and spatial marginalisation of many of its citizens, including those living in favelas and Indigenous communities (Cruxên, 2016; Henrique and Tschakert, 2019a; Cohen, 2016; Fracalanza and Campos, 2006; Henrique and Tschakert, 2019b; Martins, 2011).

Our focus in this chapter is on the grassroots strategies of riverine communities that seek to change and disrupt the unequal structures of power relations shaping São Paulo's water systems in climate change, which creates the context for their political and social struggle. The lived experience of the climate emergency is more severe for marginalised and vulnerable urban populations. Extreme climate events intensified by disruptions of planetary systems further impact communities and landscapes already marked by socio-economic precarity. Structural and political inequalities disproportionally affect oppressed and marginalised populations (e.g. poor, black, migrants, women, etc.). Their experience provides highly valuable insights for the climate emergency debate as they illustrate what the struggle with planetary disruptions looks like while also offering opportunities to theorise about and potentially disseminate the processes of socio-ecological change they can enact. The everyday lives of riverine communities in this context are marked by the social and environmental injustices of São Paulo's uneven development whose suffering is barely recognised by the public urban authorities.

We use riverine (*ribeirinho*) to refer to the communities and neighbourhoods of the informal settlements along Tietê flood plains in the east zone (ZL) of São Paulo. Riverine is an accepted concept in Brazilian anthropology, although more commonly applied to Indigenous and traditional communities whose lives are deeply shaped by the ecology of large river systems (Schwartzman et al., 2013). However, it was also used in previous works to discuss communities in the same circumstances of urbanisation as those examined here (Henrique and Tschakert, 2019b, 2019a). While typically used in English as an adjective, we will 'contaminate' (Tsing, 2015) the Anglo-Saxon vocabulary by adopting the Portuguese use of riverine as a noun, naming the people who reside in territories highly influenced by the river dynamics.

The flooding struggles of riverine communities in São Paulo's eastern periphery have been documented in the media and literature (Cohen, 2016; Henrique and Tschakert, 2019a; Millington, 2018a) and this underpinned our intention to study their grassroots actions in context. Research fieldwork developed during the COVID-19 pandemic (2019 to 2022) included the use of online interviews, focus groups, co-mapping exercises, and social media as a central strategy to explore the context and actions of mobilised communities in the urban peripheries of São Paulo. The exploratory phases of the research in local news and social media had already indicated the scale and significance of the water crisis events and the effects on riverine communities.

Through participation in the activist movement (Brazilian) *Observatório Nacional dos Direitos à Água e Saneamento* (National Observatory for Rights to Water and Sanitation – ONDAS), we were able to contact Tamires and Gabriel, activists of the *Movimento dos Atingidos por Barragens* (MAB – Movement of the Affected by Dams), who later facilitated wider introductions to the community – people who are simultaneously residents, leaders, and activists – and discussions with figures like Souza and Ancelmo. Tamires and Gabriel work as local coordinators of MAB in São Paulo's peripheries of the East and North regions, while also being residents of the Eastern districts. Their life experiences have shaped their personal histories with the precarity of urban water infrastructures and were critical to the research.[1]

In the first part of the chapter, we outline the conceptual lens of the waterscape and empirical approach (Faustino, 2023), before highlighting key findings related to the co-production of grassroots politics in riverine communities. This is a population that has historically experienced the most intense facets of crisis in water infrastructures (sewage, water supply and drainage), further troubled by the increasing vulnerability of these infrastructures in the context of climate emergency (e.g. less reliable, less resilient). This is also a community with a strong history of grassroots activism, as such a mode of mobilisation has been essential to their establishment, survival, resistance, and flourishing. The chapter concludes by emphasising the strategies the riverine communities found to mobilise and fight for their needs as they open new fronts of struggle and find places of refuge, offering contributions to the debate on grassroots climate activism.

Disrupting Unequal Waterscapes

We are interested in the waterscape as a concept that draws particular attention to the lived and material experience of communities within the context of complex pressures promoting urban and climate change, such as conditions of access and use, conflicting demands, climate change, and dynamics of land use. Karpouzoglou and Vij (2017) argue that a distinctive trait of the waterscape is that it opens the empirical scope to rich ethnographies and detailed case-studies, bringing attention to the asymmetric and uneven dimensions of power relations at play in place-based contexts. As a conceptual lens for activist practices, the waterscape addresses 'the production and constitution of waterscapes, the processes and technologies that are harnessed to produce them, and the ideologies and agendas that drive such harnessing' (Acharya, 2015, p. 378).

A significant character of the waterscape vision is to promote resistance and critical responses to dominant water discourses which creates space for alternative understandings and practices (Karpouzoglou and Vij, 2017). As exposed by Swyngedouw (2004), the focus on the involvement of local people in the governance of urban water is crucial to challenge the systematically uneven water power that constitutes urban mega-city regions such as São Paulo. He argues that grassroots organisations are cornerstones for the emancipation of urban life, and their voice and participation on water governance must be raised: a process that usually follows sustained activist action, championed by activist leadership in negotiating power and the implementation of grassroots projects (Swyngedouw, 2004).

An important theme of waterscapes scholarship is advancing discourses and practices of social emancipation, a commitment that brings emphasis to the social struggles always present over water and through activism (Acharya, 2015). As a stance against positions that accept inequalities of access to water or seek to rely on technology as the main way out of a crisis, the waterscape is envisaged as an emancipatory project for all, and has at its centre the politicisation of the understandings and decisions regarding water (Loftus, 2009). The aim is to disrupt the dominant narratives driving the water crisis, which in the latest contexts of capitalist reestablishment and growing planetary crisis have mobilised water privatisation strategies worldwide and in multiple scales (Loftus, 2009; Swyngedouw, 2005; Bakker, 2010). We outline five key themes that support this analysis:

1. **materiality of socio-natural hybrids**: the expression of physical aspects of a geography, the contextual backdrop for the framework orienting the representation of places, including the ways and shapes that water assumes and the significant urban hybrids
2. **water fluidity**: the expression of water dynamics and the variations of the ways water manifests in space
3. **social actors**: how the agents, groups, institutions, classes, and clusters are placed
4. **relations**: the interaction between the elements of the waterscape in terms of the power they hold and the relationships that are shaped as a result
5. **injustices**: the exclusions and inequities that result from power asymmetries that can be settled and/or revealed by the materiality of the geography, by the variability and occurrence of the dynamics, by the materiality of the urban hybrid, by the organisation of society, and ultimately by the arrangement of the relations

A central idea of the representation of waterscapes for this research is to capture – or even map – the reshaping of materiality and relations in ways that are challenging the injustices and asymmetries. Figure 4.1 presents a conceptual illustration of these categories representing general situations without referring to any real place.

When using such a framework to analyse the historical and material development of São Paulo entangled in the transformation of its main river, *Tietê*, a highly unequal

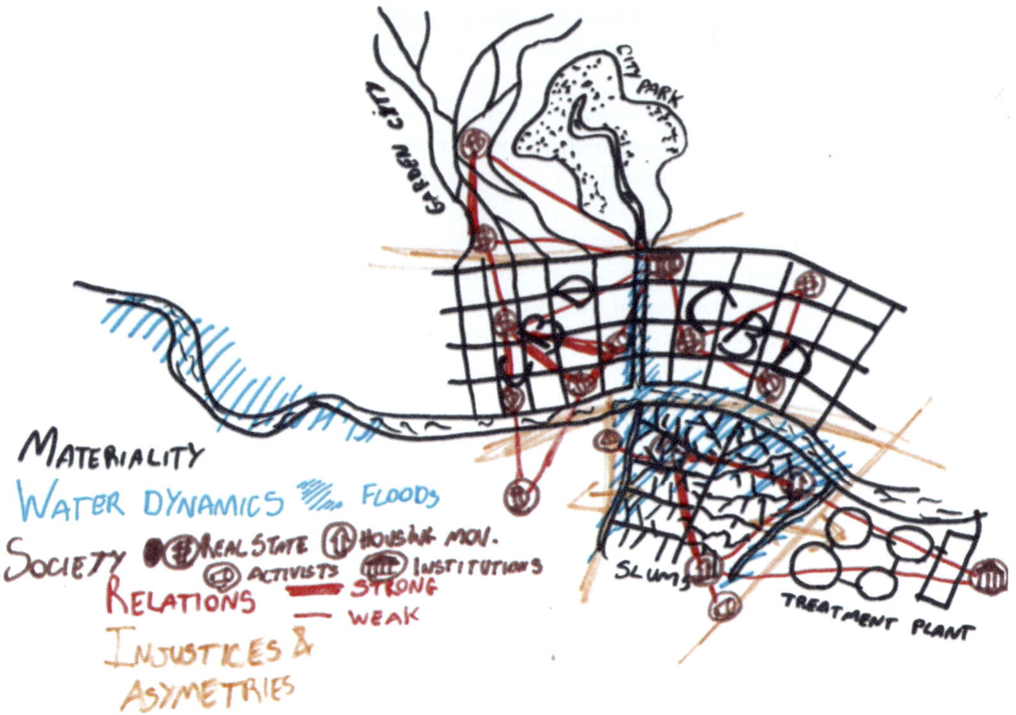

Figure 4.1   An illustrative example of waterscape visual representation.

Source: Faustino (2023). Reproduced with permission.

waterscape takes shape. The river's transformations reflect the extractive development and values associated with colonial occupation, resource extraction, and settlement development (Vilardaga, 2020). Over time, as the population grew and industrial production increased, the river became increasingly polluted with waste and sewage (Fracalanza and Campos, 2006). The exploitation of the river's resources to support the urbanisation process, including alluvial gold mines and extraction of clay and sand for construction, further impacted its ecosystem (Nóbrega, 1978; Pinheiro, 2008). Power dynamics inscribed in labour and land ownership have shaped the urban and rural spaces surrounding the river (São Paulo, 2013).

Indigenous dispossession and slavery played a significant role in the early settlements, and the history of informal urban settlements in São Paulo is connected to the socio-spatial position assigned to the black population (Oliveira and Fischer, 2017). Structural racism and exclusionary land policies further marginalised these communities and led to the establishment of favelas and informal urban territories (Carril, 2003). Unequal urbanisation, characterised by the concentration of wealth in central areas and emergence of favelas in the periphery and other spaces rejected by the market, is a historical characteristic of Brazilian urban production (Maricato, 2003). The recent rise of ultra-conservatism in the shape of extreme right movements like Bolsonarism (Santos, 2020) has exacerbated socio-spatial inequalities and led to civil unrest in São Paulo and the rest of Brazil.

Riverine Communities of São Paulo's Eastern Periphery

In the context of São Paulo's urban waters, two issues have recurrent importance in the context of climate change: flooding and water shortage. The Tietê river, the city's most important river, is situated on a flood plain that is the lowest altitude of São Paulo metropolis and an open flat region that decelerates water as it flows through the river. The disruptive flooding events impact occupants spread out alongside the urban waterways whose informality was partially consolidated by traditional planning more intensively and persistently. The floods of the Tietê river impact dwellers of informal neighbourhoods in São Paulo's east zone located along the riverbanks and close to its meanders not only because of the proximity to the river but also due to lack of drainage infrastructures and proper maintenance of the existing ones. União de Vila Nova, Jardim Lapena, and Vila Nair are some of these communities dealing with the most frequent and persistent flood conditions in the east (Figure 4.2).

These riverine communities live under constant threat of displacement as few residents have legal tenure of land to secure residence. The displacement pressure is also reinforced by discourses of environmental protection that wrap evictions as part of the risk control strategies of floodplain management (Millington, 2016). A series of state actions between the 1990s and 2000s sought to remove riverine neighbourhoods in the east under the pretext of environmental risk, criminalising land occupations in floodplain areas, and employing militarised vocabulary and tactics (Barbosa, 2015). Between the abandonment and oppression from the state, the neighbourhood developed largely based on 'collective or

Figure 4.2 Indicative delimitation of União de Vila Nova and Jardim Lapenna/Vila Nair communities and the main waterways present in their waterscape.

Source: Faustino (2023). Reproduced with permission.

auto-construction', where the implementation of infrastructures of housing, sidewalks, roads, and sewerage was made possible through the community's own efforts (Millington, 2016; Kowarick, 1979). There is a clash between the definition of environmental protection zones (floodplain protection overlay) and the continued extension of informal urbanisation in these areas (Coelho and Borelli, 2003; Martins, 2011). Beyond exposing the contradictions and limits of the urban expansion dominant rationale, these dwellings/communities, as many similar informal peripheral ones, live under the constant struggle of lack of access to potable water, as the provision of water supply system and infrastructures takes its most negligent expression in these contexts.

A recent initiative articulated by *Movimento dos Atingidos por Barragens* – Movement of those Affected by Dams (MAB) is building a narrative around the inequalities created by the retention of water at Penha Dam, an in-line detention[2] of Tietê's waters that protects central areas from flooding but makes these neighbourhoods upstream more vulnerable to it. The community seeks to bring visibility to this matter and the struggles of decades that the population has been suffering. Through their actions they have been calling out the state authorities for their responsibility to provide support and aid to the families. The organised groups are proposing an action plan to be followed by the government but until now have only started some dialogue with the local councils.

This condition of accentuated water detention creates the longest periods of water stagnation during flood episodes in São Paulo. The environmental attributes of the area are formally protected by the environmental instrument of Area of Environmental Protection (AEP) overlay – *Área de Proteção Ambiental* (*APA*), which created the Tietê AEP (São Paulo, 2013). Tiete river flow is tamed by four dams from its origin until its passage through São Paulo's city centre. The fourth interruption of its flow is located at Penha district, part of São Paulo's east zone. This marks a clear divide between the river and its surrounding characteristics from upstream and downstream Penha Dam. Before reaching this point, the river channel follows a meandric journey along its banks surrounded by extensive vegetated areas. Just after crossing Penha Dam, Tietê meets an imprisoned concrete channel of linear geometry that gives place to marginal highways over its banks.

While Penha Dam retains water to safeguard the most valuable areas and infrastructures of the city centre from Tietê's floods, it redirects water to the areas upstream that will accommodate the flood. The neighbourhoods located close to the riverbanks before Penha Dam go underwater for the protection of the rest of the city (Figure 4.3). Some of the neighbourhoods most affected by Tietê floods include Jardim Pantanal, Jardim La Pena, Vila Seabra, Vila Itaim, and União de Vila Nova. As the dam constantly operates to regulate the river flow, some neighbourhoods have reported flooding conditions even during days without rainfall.

The water struggles of these riverine communities speak directly about the conditions (spatial, material, social, political) of informal settlements in São Paulo. A critical analysis of such unequal waterscapes creates a useful basis for in-depth discussion of the role of infrastructures in urban climate justice. Such examination must consider the internal contradictions and how river, city, and drainage infrastructures are implicated in each other; the environmental and social divides created by the dam; the precarity of sanitation, either absent or incomplete; and the claims of the affected population, including their requests for infrastructural solutions. The presence of infrastructural interventions in such neighbourhoods is intense. There have been 119 interventions in the region that were directed to reduce the problems of flooding episodes. But most of these works ended up silting the riverbeds more than before. At the same time, precarity marks the formal implementation

Figure 4.3 Registers of a flooding episode on União de Vila Nova in February 2023 and some of the material losses residents experienced in its aftermath.

Source: MAB São Paulo (2023). Reproduced with permission.

of infrastructures for basic sanitation, as a single sewer serves – insufficiently – three entire neighbourhoods.

The role of critical infrastructure is strong evidence to draw from this case. The region is marked by a series of interventions designed to alleviate the city drainage and flooding problems, but most of them ended up worsening the disaster episodes for local communities. This includes the localised negative impacts of large infrastructure projects protecting central and more valued areas of the city as well as poorly executed works intended to improve drainage conditions locally, with many episodes mentioned by locals during the interviews. The whole region is surrounded and crossed by infrastructure serving systems beyond sanitation, like transport, energy distribution, and industry. The communities live with the problems produced by a whole settlement sitting on lands of an abandoned chemical industry, without full knowledge of the systemic and intergenerational nature of the risks their exposure to this type of contamination holds.

In one interview one of the participants mentioned how water provision is interrupted every night in most of the communities, a practice that became normalised after the water crisis of 2015–2016. This indicates a critical failure of the dominant water system, an analysis well explored by the scholarship on São Paulo's water governance (Jacobi et al., 2015; Empinotti et al., 2019; Fracalanza and Freire, 2015; Millington, 2018b). Many of the drainage problems come from channels and drains left without decluttering for years, which end up not being able to conduct water. Some of the pressure communities put on local governments is to clean these channels and stormwater drains to alleviate problems of micro drainage, and they have used recent flooding episodes to amplify that pressure through social media and rallies in the streets.

There is a strong spirit of rebellion among the local population, and *Movimento dos Atingidos por Barragens* (MAB) coordinators working with the communities see the situation as a timebomb. Some of the communities' leaders have already spoken about occupying local government buildings and offices to demand effective responses. To further complicate their history of abandonment, these people are facing the deepest neglect from their local government, which is not acknowledging the episodes of flooding affecting them. The local mayor alleges the images registered by residents during flood episodes are fake, contending there is no more flooding happening in the area. The city council is deeply negligent in processing the official requests for action presented by MAB – especially during the years of the COVID-19 pandemic – and the local mayor had threatened Tamires, one of the MAB coordinators in the region, by being aggressive and abusive with her during their phone calls. These activists are working under the hardest conditions possible, on top of the physical and emotional stress unavoidable from such work, exacerbated by the neglect of the council.

A second core narrative is the struggle of these communities to be seen, to have their existence and basic human rights acknowledged. They already exist on the margins, impacted by systematic failures of society, under conditions that deny them their basic human rights. And governments have been negligent not only by not addressing the matter but by not even acknowledging its existence. Instead, the dominant discourse always blames the victims of these systems of exclusion for the hardships they face. They are deemed guilty of having to live in precarious territories because they did not work hard enough or did not seek more job opportunities.

During one of the interviews, Tamires questioned 'how can someone go looking for jobs when they have to wash their clothes in creeks filled with sewage?[3]' A large portion of the public and most members of the government executive also blame the population impacted by floods for causing them by disposing rubbish and litter in the streets and thus blocking stormwater gullies and drains. This same discursive mechanism was used during water supply crises, when most of the public debate focused on the population's waste of water, diverting critiques and attention away from the failure to invest in new infrastructure on the part of São Paulo's State Water Agency (SABEP) and also from the high rate of water waste on the part of industry and agriculture.

The invisibility of these riverine communities is reinforced by wider São Paulo's ignorance about their existence or struggles, something the activists mentioned repeatedly in our interviews. They mentioned how a lot of what they do is to work on increasing the visibility of their struggles and the lack of assistance. Many times, instead of resolutions, what they get from local government is even further levels of rejection and invisibility. This is dramatically illustrated by the concealment of a community living on an island in the middle of the river. To move through and use the services in the neighbourhood, they had to cross the Tietê River by floating on a truck's roof, using a vacant property on the riverbank to access the river's waters. The city council's response was to build a wall around that property to discourage anyone from relocating to the island but did not offer any other support or relocation. The invisibility therefore extends to the lack of effort from authorities to understand the situation and evaluate the people's needs.

Many of these informal riverine settlements are not mapped officially, and authorities have not registered them in socio-demographic studies about the population. Instead, a local feminist grassroots group composed of União de Vila Nova residents is making an effort to conduct a socio-demographic study of the area.

Mobilising Grassroots Activism

Drawing on the framework put forth by Apostolopoulou et al. (2022) regarding radical social innovations driven by community-led and/or social movements, we posit that grassroots activism plays a crucial role in challenging dimensions of the dominant system (in this case, characterised as neoliberal, colonialist, extractive, and degenerative). Grassroots activism stays in the forefront of advocating for just and democratic hydrosocial solutions that prioritise overlooked or marginalised communities and regions. By delving into grassroots engagements, it is possible to uncover opportunities for resistance, adaptation, and increased resilience in the realm of water justice and sovereignty. These grassroots initiatives occupy the void left by the inaction of authorities (the state) and confront the inherent contradictions and encroachments of capitalist enclosure, dispossession, and accumulation.

Grounded in the waterscape, grassroots activists are deeply embedded in the social, economic, cultural, and geographical aspects that drive their organised response for social change. These factors shape their livelihoods, struggles, and agendas, while also serving as the basis for envisioning alternative actions and prefigurative futures. Even when physically absent, the conditions of the waterscapes and the presence of water continue to influence and flow with these actors.

The grassroots activist work we discuss in this chapter traces the strategies, tactics and achievements embedded in the relationship between MAB and Tietê urban riverines as they organise against water injustice. MAB contributions to this context highlight the transformative power of networked organisations, whose legacy is mobilised in their engagement methodology, one that seeks emancipation by articulating struggles in multiple territories. Riverines' work illustrates the localised experience of struggle with the uneven production of dwellings but also the strength and strategies of grassroots activism to respond to and reshape these challenges.

MAB arose in the late 1980s from local and regional experiences of popular organisation of communities and families affected by social, environmental, and economic harms driven by implementation of hydropower dams, fighting as well for environmental justice to ecosystems affected by such infrastructure projects. It soon became a national organisation, with representatives and organised groups in most states of the country, with national, regional, and local managing groups, but with more prominent action in the Southeast region. MAB (2019) defines itself as:

> a national, autonomous, mass, fighting movement, with regional faces, without distinction of skin colour, gender, sexual orientation, religion, political party, or level of education. We are an organisation with collective participation and leadership at all levels. Our goal is to organise those affected by dams (before, during or after the construction of the projects).[4]

As a movement, MAB carries a vast expertise of activism and community organisation, to enact their ethos of collectivism, which largely builds from decades of confrontation to dominant top-down structures of power in water and energy politics. The grassroots organisation of eastern riverines and their modes of activism for urban justice reflect an assemblage of groups weaving collaborations in solidarity networks. MAB activists have been bringing a range of important contributions to the organisation and support of families, while centring on the riverines' own tradition of collectivism and organisation to fight for human rights and justice.

Regional MAB coordinators of São Paulo's East zone like Tamires work with the riverine communities, seeking to bring visibility to the population's decade-long struggle. Through their actions, they have been calling out the state authorities to alert them to their responsibility to provide support and aid to the families. The organised groups are proposing an action plan to be followed by the government, but at the time of this research fieldwork they had only started some dialogues with local councils.

Through our interviews it became clear that a prominent characteristic of people from these neighbourhoods is their sense of a collective. However, as each collective and group builds their own agenda for action, conflicts between local leaders are a common reality. When MAB started to work with local leaders and the collective organised groups in the area it sought to unify them under the cause of the dam's impacts and housing conditions. These local groups, mostly neighbourhood associations, have been fighting for more than 40 years for proper housing, transport, sanitation, and better living conditions in general. According to Tamires, MAB's understanding of these communities' organisational process was that União de Vila Nova presented a more cohesive and unified agenda, aligning leaders with common causes. These characteristics of cohesion and coordination are referred to as a strong differential of União de Vila Nova. Anselmo also highlighted that União de Vila Nova only became the community and territory that it is today (referring to the positive transformations such as reduction of violence to zero and partial urbanisation of the territory) because of the dedicated work of leaders such as Souza and Raquel, facilitating dialogue among all residents to create a sense of identity with their place based on affection and on nurturing a spirit of collective work from the very beginning of the community.

As an activist leader Souza comments on the uniqueness of how UVN people have achieved what they did, compared to other neighbourhoods coming from a similar history after 40 or 50 years of profound housing struggle. He said that in the beginning, the leaders would fight on their own for improvements and take credit for bringing about some changes. Souza's vision is that this approach has a limit, that it ultimately does not meet collective interests, and that the struggle/resistance must instead be based on the unification of ideas and strengths. For that reason, they created the *Vila Nova Fórum*, to unify ideas under a common roof for the whole community. After implementing that approach, they achieved more results, for example, the establishment of day-care centres in the neighbourhood.

Souza sees collaboration as a crucial way to ground activists' actions as legitimate and in the interest of the residents, not coming from a political figure that could imprint their label onto the achieved outcomes. He says, 'Here the movement is the union of the residents. This is what has strengthened our neighbourhood'. Souza highlights as some achievements the creation of schools with nutritional monitoring of children, a school focused on the arts[5] (NUA – *Instituto Nova União da Arte*; New Art Union Institute), a women-led movement that grows plants locally and cooks them, providing catering services for events[6] (*Mulheres do GAU*; Women of GAU), waste recycling cooperatives, and the House of Memory[7] (*Casa da Memória*), among others.

Explaining the work MAB carries out with the communities, another one of the Grassroots community leaders, Gabriel, shared that their aim is to offer education and empowerment to the people. Going beyond mere emergency relief (welfarism), MAB uses mutual aid actions as well to train people and mobilise them in the processes of planning and distribution of donations. They organise the population in grassroots groups, which must appoint two coordinators, preferably one woman and one young person, to reduce gender exclusion and increase intergenerational work. This organisational and empowering methodology followed by MAB is the same in every territory where they are present, mobilising affects and empathy to unify various communities under the same banner, which made the movement stronger, more legitimate, and national in their reach:

> There was a leader of ours who said something like "MAB allows those affected from the south of the country to feel the pain of those affected from Rondônia [state of Brazil's north region]". And [having] this sensitivity means [for the activist in the movement] also knowing the organisation that the militant[8] (activist) belongs to (Gabriel).[9]

Activist leader Tamires mentioned that when MAB begins engaging with a community, their first step is establishing conversations to understand what to expect from a collective engagement with the movement:

> So we always have this concern to say "Look, you are the MAB, the MAB are you here, no one is promising anything, we promise the construction of a collective agenda where everyone has the same direction, the same goal: the fight against floods, the fight for decent housing, improving the region". Because these are regions extremely lacking in health, education, and culture [referring to formal infrastructures providing such services].[10]

Tamires emphasised the crucial role of organisation in the politicisation of the community, from understanding their struggles to defining their agenda and how it could bring about change:

> Organisation is key, it is critical that people are on site fighting for something, that they know what they are doing. It's not just reaching the door of city hall and saying, "Oh I came." Okay, but what did you come for? What's your fight? What are you doing here? What is it that you want? And what do you desire for society? So, I think above anything [we need] popular organisation and popular education.[11]

These are just some of the grassroots organisational strategies used by these communities which reveal how most of their actions and achievements revolve around collectivism, essential to the achievement of higher degrees of legitimation and to articulate a stronger transformative capacity for the community. It is also following the principle that multiple partnerships emerge in a network of collaborators that support the riverine communities' practices of local governance in this unequal waterscape. At the same time, the collective arenas are the spaces where their agendas are proposed, disputed, and shaped as urban waterscapes.

Successful Strategies and Demands

An overall area of demand for these grassroots community activists is the improvement of infrastructural conditions and the caretaking of the waterways in the area as the underpinning conditions required to access better living conditions. They claim improvements could be achieved with at least some maintenance of the existing infrastructure. Many of the drainage problems come from channels and drains which end up too obstructed to conduct water. The demand for infrastructural improvements can also be seen in the projects designed for the Jardim Lapenna Neighbourhood Plan, which calls for streets to receive permeable paving and the development of micro drainage devices (Fundação Tide Setubal, 2019). The plan projects the relocation of the local UBS (Unit of Basic Health) to an area not affected by floods, which often compromises the facility's operation.

One strategy MAB encourages is the political formation of youth in the community, bringing young people from Vila Nova into communication courses from MAB. They want the youth to learn how to fight for representation and their neighbourhood. Tamires highlighted that MAB seeks to be a movement led by the population, and they aim to foster a popular struggle, emancipate the population, and raise people's awareness that they need to become active subjects of their own history and, consequently, continuously contribute to the fight. According to Tamires, the definition of MAB's agenda is an ongoing collective and adaptable process that allows each region to consider its own emerging issues and priorities. For example, at Vila do Aimore (an adjacent neighbourhood to União de Vila Nova), they need people to be resettled, whilst at União de Vila Nova they need the waterways to be cleaned and cared for to prevent floods. While sharing about how the MAB coordinator defines agendas with community members, Tamires expressed how laborious it can be to manage the many different demands that are constantly emerging and synthesise them in a single agenda:

> So we progress talking, in dialogue with the population and seeing what are the main problems. And then we're trying to build a collective bridge, because it's no use . . . we coming as MAB and saying "the agenda is this, this and this", no. Every day something different comes up. Every day that passes I am more shocked by how much trouble you have in these areas. So, when you think that the agenda is defined, [you realise that] it's not only that; the agenda [actually] has thousands of other things happening and we continue adapting as we go fighting for the rights of those affected.[12]

MAB coordinators shared some of the issues that make up the MAB agenda with the riverine communities affected by floods in Tietê floodplains between 2019 and 2021:

- Resolution of housing problems, involving (1) the resettlement of people in extreme cases, and (2) land title regularisation of properties of informal status so people can have better agency to enact formal citizenship and land rights. For example, the lack of land title is also a barrier to submitting some requests to the local government to spur them to action
- Cleaning and desilting of the waterways
- Environmental education
- Basic food provision, to reduce hunger and to support people in being healthy and active in the movement
- Resolving administrative jurisdiction fights between local governments (São Paulo and Guarulhos), which is an obstacle to many of the other issues such as land title regularisation and requests for cleaning waterways
- Reparation for material losses, as the continuous loss of essential goods can be a huge toll on people living for more than 40 years with the presence of floods
- Including these issues as a public health agenda as well (clean rivers, regularised creeks)

- As a main agenda of MAB, establishing legislative acts for the rights of people affected by dams. After 40 years of MAB mobilising in policymaking across all levels of government, this goal has been recently achieved in November 2023 by having a National Policy of Rights of Populations Affected by Dams (Brasil, 2023) approved by National Congress and sanctioned by President Luiz Inácio Lula da Silva.

Tamires emphasises how inaccessibility to housing triggers new illegal occupations in the face of a lack of policies and programs to shelter people in well-qualified dwellings:

The issue is housing, because people live in an extremely decadent situation, on stilts within the largest city in Brazil. It's an absurd thing. And then they [authorities] say they don't have any housing programs, either by the city council, or the state government. In that case, people are going to start to occupy [occupy vacant land and build improvised dwellings]. And it will be drastic, because in addition to not having a housing plan, they do not support the population, especially during the time of the pandemic.[13]

To draw attention to these issues, activist collectives encourage more residents to join the movement and to pressure local governments into action. One of MAB's strategies is to organise rallies on the main streets in the region. Tamires shared that she was happy with the outcome of a recent rally on March 8, 2021, emerging from the feminist fights promoted by MAB to empower local women leaders (Figure 4.4). Amidst the highly

Figure 4.4 Activists of MAB marching in the protest 'Women fighting for Life' on 8 March 2021.
Source: MAB São Paulo (2021). Reproduced with permission.

challenging times of COVID-19 pandemic and Bolsonaro's (then president) extremely negligent and genocidal health policies (Rosario, 2020), the positive response these activists received from the public to their demands for vaccines and emergency aid was encouraging for their mobilisation:

> We had the rally on the 8th of March too, which was very good. We had the participation of women, it was really beautiful. The women, the kids, they went out with us. We had a very good reception. I, in particular, was afraid because we were bringing the banner of 'Bolsonaro Out', 'Vaccine Now' and 'Emergency Aid', and we had a very good reception. We blocked Avenida Doutor Assis Ribeiro, which is a strategic avenue here in the East Zone, which connects with highways. A lot of cars passed, a lot of trucks, and where we marched, people shouted 'That's right, vaccine now, emergency aid!'. So it was a very gratifying action for us. It was really good.[14]

Conclusions

Our research findings underscore the significance of urban riverine grassroots activism and community mobilisation, particularly exemplified by organisations like Movimento dos Atingidos por Barragens (MAB). These grassroots efforts have emerged from years of advocacy for improved housing, transportation, sanitation, and overall living conditions, reflecting communities with a robust sense of cohesion and political coordination. Their advocacy extends to demands for infrastructural enhancements, the preservation of waterways, safeguarding water sources, and promoting environmental awareness. All of these issues are contextualised in a wider arena of water governance – in this case entangled in the production of and response to flooding events in urbanised contexts – and are intricately connected to the socio-political and environmental challenges posed by climate change. Such conflicting, fluid, and still creative struggles inscribe the grounded practices examined through the chapter within the radical landscapes of grassroots climate activism. Ultimately, the study highlights the pivotal role of collective action and the organisational capacity of grassroots activism in addressing the multifaceted challenges faced by these marginalised communities in shaping their urban waterscapes (Faustino, 2023).

Importantly this chapter does not aim to provide a comprehensive overview or fully represent the plurality of activist organisations or social movements operating in various ways, contexts, and realities across Latin America or Brazil. Rather, it offers insight into the nature of grassroots activism embedded in São Paulo's eastern waterscape, which could be further enriched by additional perspectives. For instance, while beyond the scope of this chapter, it's noteworthy that we observe the emergence of a climate activism movement closely intertwined with Indigenous peoples' struggle for sovereignty over land, forests, water, and culture (Faustino, 2023; Phillips, 2023). Attention should be directed towards these fronts of resistance.

Through this chapter, we have outlined the existence of unequal urban waterscapes in the peripheries of São Paulo and the socio-ecological inequities present in these areas. We highlighted how the dominant ideology disconnects nature, including water, from humanity and views it as a resource for exploitation. This logic contributes to social oppression and reinforces the control of the dominant class over land and resources. Dispossession, displacement, and commodification are enacted to maintain this control, leading to the marginalisation of communities and their relationship with land and water. This depiction of inequalities and injustices affects groups exploited and many times forgotten in the Global South periphery of capitalism. However, through solidarity networks, collective efforts, and alternative forms of caring for themselves and their communities, they are able to challenge these issues.

The riverine communities highlight the collective networks and solidarity among community groups as critical for enacting care interventions. Grassroots activism plays a crucial role in providing solutions and aid and battling political barriers to support vulnerable people. The narratives of riverine communities show how they organise collectively to care for their territories and people, despite the neglect from government and authorities. Principles such as solidarity, collectivism, and connection to place inform their grassroots practices and efforts to reshape the modern waterscape. Examples include community groups working to preserve public spaces, provide emergency relief, conduct surveys, and engage in advocacy efforts.

In the context of climate change, where institutional responses often prove inadequate to addressing the crisis, it becomes imperative to comprehend the emergence, governance mechanisms, and strategies of grassroots activism – which tackles the challenges at the frontline. This understanding serves as a critical tool in preparing to confront present and future demands effectively. Grassroots activism emerges and evolves from the encounter of people's experiences with other long-established social movements such as the housing movements, labour unions, and arrival of wider social-environmental movements like MAB. Therefore, activist movements are not closed in themselves; they are organic, permeable, and ever-evolving. Their governance relies on groups of individuals organised by practices of engagement with their community and territory. This entails the definition of leadership and hence shapes local politics, creating a routine of neighbours' meetings to discuss relevant issues and ways to pursue action, meetings that might use cultural or community centres. Their strategies involve the definition of targeted political agendas, pressure over wider stakeholders, protests, and the offer of support and emergency aid to communities. It also encompasses the use of digital systems to share communication, stories, interventions, and protests, allowing a reach beyond the individuals closely involved in the groups. This presence becomes an activist platform for visibility – and more importantly solidarity in the face of a cascading crisis. In climate change, grassroots activism is the amalgamation of efforts to transform community realities, driven by senses of need and responsibility, navigating the organisation of actions towards the achievement of shared and commonly defined goals.

Through this narrative, we evidenced the legacy of grassroots strategies that have been practised by marginalised communities responding to the challenges and needs of their territories, here seen as unequal waterscapes. It circles issues communities in many other contexts are already facing, such as fighting against authoritarian action, negotiating eviction processes unleashed by interventions related to the climate crisis, and continuously creating spaces of refuge and mobilisation. The modes of knowing and addressing challenges articulated by such a plural ecosystem of resistances emerge from grassroots engagements because they are responding to how their immediate territories have been affected by disruptions of weather patterns and by the modern drivers of climate change. It can be expected that such an intimate relationship with place, nature, and Planet is going to orient many of the strategies adopted by the growing climate movement in São Paulo, Brazil, and other sites of resistance and regeneration across the globe.

Notes

1 Qualitative data was provided by participants mostly through interviews and workshops. Quotes from such material will be included in the chapter translated to English, but a footnote will be added with the original words in Portuguese.

2 Detention refers to when water is temporarily stored on-site by slowing down or restricting the drainage of a stream or river. This effect can be engineered by implementation infrastructures such as dams, with intention to reduce flooding downstream from the dam, but the accumulated water volume can affect the areas upstream of the intervention.

3 Translation from Portuguese original: 'Então assim me fala como uma pessoa que . . . lava a roupa dentro do córrego, a roupa fedendo a esgoto, tem condições de arranjar emprego?'

4 Translation from Portuguese original: 'O MAB é definido como um movimento de caráter nacional, autônomo, de massa, de luta, com rostos regionais, sem distinção de cor da pele, gênero, orientação sexual, partido político ou nível de instrução. Somos uma organização de participação e liderança coletiva em todos os níveis. Nosso objetivo é de organizar os afetados pelas barragens (antes, durante ou depois da construção dos projetos)'.

5 https://www.novauniaodaarte.org

6 https://www.instagram.com/mulheresdogau/?hl=en

7 https://www.facebook.com/casadamemoriauvn

8 Gabriel uses *militant* to refer to the activist participating in the movement because it is a word widely used in Brazilian activism culture, even when it does not involve the use of methods based in confrontation and violent engagement. Here it carries a meaning closer to *activist* in English.

9 Translation from Portuguese original: 'Tinha uma liderança nossa que ele falava assim 'o MAB ele permite que um atingindo do sul do país sinta a dor do atingido de Rondônia'. E essa sensibilidade significa que não só da sensibilidade, mas também saber da organisação e da pertença que aquele militante tem'.

10 Translation from Portuguese original: 'Então a gente tem sempre essa preocupação de falar 'ó, vocês, vocês são o MAB, o MAB são vocês aqui, ninguém está prometendo nada, a gente promete a construção de uma pauta coletiva que todo mundo tenha o mesmo, a mesma direção, o mesmo objetivo: combate às enchentes, a luta por uma moradia digna, melhorar a região'. Porque são regiões extremamente carentes de cultura de saúde e de educação'.

11 Translation from Portuguese original: 'Organização é fundamental, é fundamental que as pessoas estejam no local lutando por algo que elas saibam o que elas estão fazendo. Não é simplesmente chegar na porta da prefeitura e falar 'Ah eu vim'. Tá, mas você veio o quê? Qual a sua luta? O que você veio fazer aqui? O que você quer e o que você almeja para a sociedade? Então acho que acima de qualquer coisa organização popular e a educação popular'.

12 Translation from Portuguese original: 'Então assim a gente vai conversando, dialogando com a população e vendo quais são os principais problemas. E aí a gente vai tentando construir uma ponte coletiva, porque não adianta a gente chegar e falar, a gente enquanto MAB chegar e falar "a pauta é essa, essa e essa", não. Todo dia aparece uma coisa diferente. Cada dia que passa eu fico mais chocada com o tanto de problema que tem nas regiões. Então assim quando você pensar a pauta que está redonda isso, não é só isso, a pauta tem tem milhares de outras coisas acontecendo e a gente vai adaptando e vamo na luta pelos direitos dos atingidos'.

13 Translation from Portuguese original: 'A questão é moradia, porque as pessoas vivem em uma situação extremamente decadente, de palafitas dentro da maior cidade do Brasil. É uma coisa absurda. E aí eles falam não têm programa nenhum de moradia, nem por parte da prefeitura, nem do governo. Então, assim o pessoal vai começar a ocupar. E vai ser drástico, porque além de não ter projeto de moradia, eles não dão suporte para a população, principalmente na época da pandemia'.

14 Translation from Portuguese original: 'A gente teve o ato do 8 de Março também, que foi muito bom. A gente teve a participação das mulheres assim, foi muito bonito mesmo. As mulheres, crianças foram com a gente pra rua. A gente teve uma recepção muito boa. Eu estava, eu, particularmente, fiquei com medo porque a gente estava levantando na faixa Fora Bolsonaro, Vacina já e Auxilio emergencial, e a gente teve uma recepção muito boa. A gente parou a Avenida Doutor Assis Ribeiro, que é uma avenida estratégica aqui na Zona Leste, que ela se conecta com rodovias. Então tem muita, passa muito carro, muito caminhão, e assim onde a gente passava, o pessoal gritava 'É isso mesmo, vacina já, auxilio emergencial'. Então foi uma ação muito gratificante pra gente. Foi muito boa mesmo'.

References

Acharya, A. (2015). The cultural politics of waterscapes. In Raymond L. Bryant (ed.) *The International Handbook of Political Ecology* (pp. 373–386). Cheltenham: Edward Elgar Publishing.

Angotti, T. (2013). Urban Latin America. *Latin American Perspectives*, 40(2), 5–20.

Apostolopoulou, E., et al. (2022). Radical social innovations and the spatialities of grassroots activism: Navigating pathways for tackling inequality and reinventing the commons. *Journal of Political Ecology*, 29, 143–188.

Arboleda, M. (2016). In the nature of the non-city: Expanded infrastructural networks and the political ecology of planetary urbanisation. *Antipode*, 48(2), 233–251.

Bakker, K. (2010). Commons versus commodities: Political ecologies of water privatization. In *Global Political Ecology* (pp. 347–370). London: Routledge.

Barbosa, L. da S. (2015). *A Identidade do Movimento por Urbanização e Legalização do Pantanal da Zona Leste – SP: Esperança e Desalento na Luta Contra o Deslocamento Populacional*. Sao Paulo: Scortecci Editora.

Brasil (2023). *Política Nacional de Direitos das Populações Atingidas por Barragens* (PNAB). Berlin: Jovis.

Brenner, N. and Schmid, C. (2011). Planetary urbanisation. In M. Gandy (ed.) *Urban constellations* (pp. 10–13). Berlin: Jovis.

Calvin, K., et al. (2023). IPCC, 2023: Climate change 2023: Synthesis report. In Core Writing Team, H. Lee and J. Romero (eds.) *Contribution of Working Groups I, II and III to the Sixth Assessment Report of the Intergovernmental Panel on Climate Change*. Geneva, Switzerland: IPCC.

Carril, L. de F. B. (2003). *Quilombo, favela e periferia: a longa busca da cidadania*. Universidade de São Paulo.

Caulkins, M. W. (2022). Insurgent property relations and the spatio-legal work of housing the urban poor: The case of the *Prestes Maia* occupation in São Paulo. *Geografiska Annaler: Series B, Human Geography*, 104(2), 93–111.

Coelho, M. A. and Borelli, D. L. (2003). A disputa pela água em São Paulo: Entrevista com Gerôncio Albuquerque Rocha. *Estudos Avançados*, 17(47), 153–165.

Cohen, D. A. (2016). The rationed city: The politics of water, housing, and land use in drought-parched São Paulo. *Public Culture*, 28(2), 261–289.

Cruxên, I. A. (2016). *Fluid Dynamics: Politics and Social Struggle in São Paulo's Water Crisis (2014–2015)*. Master Thesis. Massachusetts Institute of Technology.

Empinotti, V. L., et al. (2019). Governance and water security: The role of the water institutional framework in the 2013–15 water crisis in São Paulo, Brazil. *Geoforum* [Online], 98, 9846–9854. https://doi.org/10.1016/j.geoforum.2018.09.022

Faustino, A. da S. (2023). *Dissolving the Concrete: Reconfiguring Urban Waterscapes Through Grassroots Activism in São Paulo, Brazil*. Doctor of Philosophy (PhD) Thesis. Melbourne: RMIT University.

Ferrara, L. N., et al. (2019). Espoliação urbana e insurgência: conflitos e contradições sobre produção imobiliária e moradia a partir de ocupações recentes em São Paulo. *Cadernos Metrópole*, 21(46), 807–830.

Fracalanza, A. P. and Campos, V. N. de O. (2006). Produção social do espaço urbano e conflitos pela água na região metropolitana de São Paulo. *São Paulo Em Perspectiva*, 20(2), 32–45.

Fracalanza, A. P. and Freire, T. M. (2015). Crise da água na Região Metropolitana de São Paulo: a injustiça ambiental e a privatização de um bem comum. *GEOUSP: Espaço e Tempo (Online)*, 19(3), 464.

Franco, M. S. de C. (1979). As idéias estão no lugar. In *Cadernos de Dabate – História do Brasil* (pp. 61–64). São Paulo: Brasiliense.

Fundação Tide Setubal (2019). *Plano de Bairro Jardim Lapenna: a rota para um território de direitos*. São Paulo: Fundação Tide Setubal. https://fundacaotidesetubal.org.br/publicacoes/plano-de-bairro-jardim-lapenna/#boletim-modal [Accessed 8 September 2024].

Henrique, K. P. and Tschakert, P. (2019a). Contested grounds: Adaptation to flooding and the politics of (in)visibility in São Paulo's eastern periphery. *Geoforum*, 104(May), 181–192 [online]. https://doi.org/10.1016/j.geoforum.2019.04.026

Henrique, K. P. and Tschakert, P. (2019b). Taming São Paulo's floods: Dominant discourses, exclusionary practices, and the complicity of the media. *Global Environmental Change*, [Online] 58(June), 101940. https://doi.org/10.1016/j.gloenvcha.2019.101940

IBGE, I. B. de G. e E. (2022). *IBGE Cidades – Panorama Sao Paulo* [online]. https://cidades.ibge.gov.br/brasil/sp/sao-paulo/panorama [Accessed 8 September 2022].

Jacobi, P. R. (2023). Desafios da governança ambiental urbana face à emergência climática. *Revista Cadernos de Pós-Graduação em Arquitetura*, 22(3).

Jacobi, P. R., et al. (2015). Crise hídrica na Macrometrópole Paulista e respostas da sociedade civil. *Estudos Avançados*, 29(84), 27–42. https://www.scielo.br/j/ea/a/V6K8tDqY3sSqgFGSWGqDVJh/

Karpouzoglou, T. and Vij, S. (2017). Waterscape: A perspective for understanding the contested geography of water. *Wiley Interdisciplinary Reviews: Water*, 4(3).

Kowarick, L. (1979). *A espoliação urbana*. Rio de Janeiro: Paz e Terra.

Lampis, A., et al. (2020). A produção de riscos e desastres na América Latina em um contexto de emergência climática. *O Social em Questão*, 23(48), 75–96.

Lizarralde, G., et al. (2021). Does climate change cause disasters? How citizens, academics, and leaders explain climate-related risk and disasters in Latin America and the Caribbean. *International Journal of Disaster Risk Reduction*, 58102173.

Loftus, A. (2009). Rethinking political ecologies of water. *Third World Quarterly*, 30(5), 953–968.

MAB (2019). *Quem somos*. https://mab.org.br/quem-somos/ [Accessed 3 November 2023].

MAB (2021). *MAB Sao Paulo*. https://www.facebook.com/mabsaopaulo.

Maricato, E. (1996). *Metrópole na periferia do capitalismo: ilegalidade, desigualdade e violência*. São Paulo: HUCITEC.

Maricato, E. (2003). Metrópole, legislação e desigualdade. *Revista Estudos Avançados*, 17, 151–167.

Maricato, E. (2009). Fighting for just cities in capitalism's periphery. In Peter Marcuse et al. (eds.) *Searching for the Just City: Debates in Urban Theory and Practice* (pp. 194–213). New York: Routledge.

Martins, M. L. R. (2011). São Paulo, centro e periferia: A retórica ambiental e os limites da política urbana. *Estudos Avancados*, 25(71), 59–72.

Mezzadra, S. (2023). Exploring the landscapes of extraction: Colonial continuities, postcolonial assemblages of power, anti-colonial struggles. In Naoki Sakai, et al. (eds.) *Knowledge Production and Epistemic Decolonization at the End of Pax Americana* (1st ed., p. 23). Routledge. https://www.taylorfrancis.com/chapters/edit/10.4324/9781003036661-4/exploring-landscapes-extraction-sandro-mezzadra [Accessed 21 March 2024].

Millington, N. (2016). *Impermeable Assemblages: Flooding, Urban Infrastructure, and Stormwater Politics in São Paulo, Brazil*. Doctoral Dissertation Thesis. The University of Kentucky. https://uknowledge.uky.edu/geography_etds/49/

Millington, N. (2018a). Linear Parks and the political ecologies of permeability: Environmental displacement in São Paulo, Brazil. *International Journal of Urban and Regional Research*, 42(5), 864–881.

Millington, N. (2018b). Producing water scarcity in São Paulo, Brazil: The 2014–2015 water crisis and the binding politics of infrastructure. *Political Geography*, 65(May 2017), 26–34 [online]. https://doi.org/10.1016/j.polgeo.2018.04.007

Nóbrega, H. M. (1978). *História do Rio Tietê* (Vol. 8). São Paulo: Governo do Estado de São Paulo.

OCHAN (2020). *Natural Disasters in Latin America and the Caribbean 2000–2019*. https://reliefweb.int/report/world/natural-disasters-latin-america-and-caribbean-2000-2019 [Accessed 18 March 2024].

Oliveira, N. A. S. de and Fischer, L. R. da C. (2017). Segregação espacial urbana e os efeitos da lei de terras de 1850. *Revista Brasileira de História do Direito*, 3(1), 36–54.

Parnell, S. and Robinson, J. (2012). (Re)theorizing cities from the Global South: Looking beyond neoliberalism. *Urban Geography*, 33(4), 593–617.

PBMC (2016). *Mudanças Climáticas e Cidades*. http://www.pbmc.coppe.ufrj.br/index.php/pt/publicacoes/relatorios-especiais-pbmc/item/relatorio-especial-mudancas-climaticas-e-cidades?category_id=19 [Accessed 18 March 2024].

Phillips, T. (2023). Indigenous activists gather in Brazil to discuss future of the Amazon. *The Guardian – Amazon Rainforest*, 7 August. https://www.theguardian.com/environment/2023/aug/06/indigenous-activists-gather-in-brazil-to-discuss-future-of-the-amazon [Accessed 26 March 2024].

Pinheiro, J. E. de M. (2008). Ciclo do ouro de Guarulhos. In Elmi E. H. Omar (ed.) *Guarulhos tem história*. Guarulhos: Ananda. https://www.scribd.com/doc/63984107/Guarulhos-tem-Historia [Accessed 28 June 2023].

Rolnik, R. (1997). *A cidade e a lei: legislação, política urbana e territórios na cidade de São Paulo [The City and the Law: Legislation, Urban Policy and Territories in the City of São Paulo]*. São Paulo: Studio Nobel.

Rosario, L. (2020). The genocidic necropolitics of bolsonaro in pandemic times and the ultra-neoliberal project. *Revista Interdisciplinar em Cultura e Sociedade (RICS)*, 6(2), 28–49 [online]. https://oglobo.globo.com/sociedade/educacao/negros-sao-maioria-pela-primeira-vez-nas-universidades-

Rosenzweig, C., et al. (2011). Urban climate change in context. Climate change and cities. In Cynthia Rosenzweig et al. (eds.) *Climate Change and Cities: First Assessment Report of the Urban Climate Change Research Network* (pp. 3–11). Cambridge: Cambridge University Press.

Santos, G. V. dos (2020). Governo Bolsonaro: o retorno da velha política genocida indígena. *Revista da ANPEGE*, 16(29), 420–451 [online]. https://ojs.ufgd.edu.br/index.php/anpege/article/view/12527

São Paulo (2013). *Plano de Manejo da Área de Proteção Ambiental Várzea do Rio Tietê*. São Paulo: Fundação Florestal. https://fflorestal.sp.gov.br/wp-content/uploads/2022/12/diagnostico-1-2.pdf [Accessed 8 September 2024].

Schwartzman, S., et al. (2013). The natural and social history of the indigenous lands and protected areas corridor of the Xingu River basin. *Philosophical Transactions of the Royal Society B: Biological Sciences*, 368(1619), 20120164.

Swyngedouw, E. (2004). *Social Power and the Urbanization of Water: Flows of Power*. Oxford: Oxford University Press.

Swyngedouw, E. (2005). Dispossessing H 2 O: The contested terrain of water privatization. *Capitalism Nature Socialism*, 16(1), 81–98.

Tsing, A. L. (2015). *The Mushroom as the End of the World*. Princeton: Princeton University Press.

UN-Habitat (2018). *Working for a Better Urban Future. Annual Progress Report*. Kenya: UN-Habitat. https://unhabitat.org/sites/default/files/documents/2019-05/annual_progress_report_2018.pdf [Accessed 8 September 2024].

Valencio, N., et al. (2009). *Sociologia Dos Desastres. Construção, Interfaces E Perspectivas No Brasil*. São Carlos: RiMa.

Vilardaga, J. C. (2020). No fluxo do Anhembi-tietê: o rio e a colonização da capitania de São Vicente nos séculos XVI e XVII. *Nuevo Mundo Mundos Nuevos*. https://doi.org/10.4000/nuevomundo.82993

5

PACIFIC YOUTH AND CLIMATE ACTIVISM

Safeguarding Our Fa'asinomaga

*Tahere Talaina Si'isi'ialafia-Mau and
Joseph Emile Honour Percival*

Ukōt Bōkā Eo! – Turn the Tides

Climate change presents an existential threat to the Pacific Island Countries and Territories (PICTs), exacerbating their vulnerability and testing their resilience despite their minimal contributions to global greenhouse gas emissions. Many PICTs face escalating impacts such as rising sea levels, coastal flooding, ocean acidification, and increasingly frequent and intense extreme weather events, causing severe damage to infrastructure and natural ecosystems (Hoegh-Guldberg et al., 2007; Yamano et al., 2007; Campbell and Barnett, 2010; Knutson et al., 2010). The looming threat of forced relocation due to these changes is a growing concern, particularly for low-lying PICTs such as Tokelau, with a population of just over 1,400, roughly 10 square kilometres in total land area, and a highest point above sea-level of 5 metres. Thus, this urgent and multidimensional crisis not only jeopardises the livelihoods of island communities and the unique biodiversity of their ecosystems but also undermines the cultural heritage of PICTs, threatening their ways of life (McNamara et al., 2021).

In the face of these existential threats, PICTs have actively developed and implemented a variety of adaptation and mitigation measures to increase resilience and decrease vulnerability. National development strategies and policies, such as Fiji's comprehensive National Development Plan, are progressively incorporating climate-change considerations to promote sustainable development, resilience, and a transition to low-carbon economies (Republic of Fiji, 2017). To improve infrastructure resilience against extreme weather, PICTs are required to invest in sea walls, community relocations, and integrated water resource management to alleviate the issues associated with water scarcity (Campbell and Barnett, 2010). Alongside these efforts, PICTs actively participate in global climate policy discussions, with coalitions such as the Alliance of Small Island States (AOSIS) leading the charge in advocating for their rights and interests (Ashe et al., 1999). Through key international platforms such as the United Nations Framework Convention on Climate Change (UNFCCC), Conference of the Parties (COP) and regional forums like the Pacific Islands Forum (PIF), PICTs push for ambitious global action on greenhouse gas mitigation and recognition of climate-change-induced loss and damage.

DOI: 10.4324/9781003396567-6

Yet, despite these measures and the proactive actions by PICT governments, a strong sense of injustice pervades Pacific Island communities. The disparity between their minimal contribution to the climate crisis and the disproportionate impact they endure underscores a grave inequity. Many Pacific Islanders feel that the current efforts to curb climate change are insufficient to address the magnitude of the crisis faced (Sefeti, 2023). This feeling of injustice has not only been a catalyst for an emerging climate activism movement but also marks the beginning of a new chapter in their response to the crisis. Amid this backdrop, climate activism has become a vital avenue for change, also driven by the rise of a robust youth-led movement (McNamara and Farbotko, 2017).

Recognising the disproportionate impact of these climatic shifts on their communities, youth in the Pacific have mobilised into action. Harnessing both traditional and digital avenues, Pacific youth activists are engaging in climate meetings and protests and directing their concerns primarily towards major intergovernmental organisations involved in UN meetings and countries that bear a greater responsibility for the climate crisis (McNamara and Farbotko, 2017; Kelly, 2021; Anthony, 2023; Guardian Staff, 2023). Their concerted efforts effectively illuminate the urgency and human face of climate change. As modern-day warriors, they push back against the climate injustice that necessitates building sea walls, relocating their communities, and adapting to a changing environment. Their role is pivotal in contesting global power dynamics, as they demand equitable representation in meaningful climate dialogues and decisions. This can be seen, for example, in the recent strong call to regional governments demanding a Fossil Fuel Free Pacific and a swift scale-up of community-centred renewable energy programs (Anthony, 2023), the subsequent recognition of such actions by the Pacific Island Forum (PIF) (Pacific Island Forum, 2023) as well as in the formation of the PIF Secretary General's Young Climate Leaders Alliance. This alliance underscores a transformative approach to youth inclusion and engagement in climate negotiations and action in the Pacific. It invites and engages Pacific youth to contribute to climate-change projects and advocacy leading up to COP every year and is the first of its kind in its aim to deliver opportunities for intergenerational advocacy and dialogue for Pacific responses to the climate crisis (Pacific Island Forum, no date).

This chapter gives a comprehensive view of climate activism in the Pacific Islands. It highlights the experiences of Pacific youth in climate activism and focuses on their underlying motivations. Our goal is to shed light on what it means to be a Pacific climate activist and to understand what factors shape their activism in their pursuit of climate justice. The title, "Safeguarding our Fa'asinomaga" is derived from the analysis of 44 in-depth interviews with Pacific youths from 13 PICTs. The data revealed a range of motivations, with a fundamental emphasis on protecting their *fa'asinomaga* – a Samoan concept that encapsulates a shared sense of identity, heritage, and sense of belonging that are ingrained in many Pacific cultures. Utilising Pierre Bourdieu's concept of habitus, our study interprets the alignment of Pacific youth's perspectives, actions, and motivations in climate activism with their rooted *fa'asinomaga*. In this light, a *fa'asinomaga-framed habitus* was derived to articulate the very essence of Pacific youths' motivations for climate activism, one that we believe sets their realities apart from Western and Eurocentric notions of climate activism. This perspective reveals their fight against climate change as more than a reaction to physical challenges; it is an integrated struggle deeply intertwined with the preservation of their cultural identity, their heritage, and the existential threat to their way of life.

Research Design and Methodological Framework

Our primary data source consists of 44 in-depth, open-ended interviews with Pacific youth, aged 18–35, from the following 13 PICTs: Federated States of Micronesia, Fiji, French Polynesia, Kiribati, Republic of the Marshall Islands, Niue, Palau, Papua New Guinea, Samoa, Solomon Islands, Tonga, Tuvalu, Hawaii. These interviews were designed to elicit rich, descriptive narratives that provide insight into the participants' backgrounds, motivations, and perspectives on advocacy and activism in general. Each was broken down into three stages and the following focus topics:

1. **exploring perceptions and experiences:** the interview questions in this stage were designed to explore how Pacific youth perceive and experience advocacy and activism, delving into personal stories, organisational involvement, challenges, and motivations
2. **understanding factors and motivations:** these questions sought to understand the underlying factors and motivations behind youth activism, including cultural influences, barriers faced, and the impacts of these activities
3. **documenting outcomes and determinants:** the final stage aimed to document the outcomes of advocacy and activism and to determine what factors contribute to successful youth advocacy in the Pacific context

These interviews were transcribed and we then engaged in initial data coding for habitus elements as examined through the lens of Pierre Bourdieu's concept of habitus (Bourdieu, 1990). These interviews took on average between 3 and 4 hours each and all interviews and interview transcriptions were conducted by T.T.S. Mau during her doctoral work at the University of Otago. Data coding involved breaking the transcripts into manageable segments and then extracting key concepts and keywords that corresponded to: (1) detailed instances of ingrained habits, (2) specific perceptions that were shaped by the participants' background, and (3) clear dispositions that were influenced by past experiences. In instances where quotes from interviewees are used in this chapter and the respondents wished to remain anonymous, we have assigned the pseudonym "Te'o" to represent their contributions.

As authors of this chapter, we bring our own unique experiences and viewpoints to the study of climate activism in the Pacific. T.T.S. Mau, from Samoa, is deeply engaged in the field of youth advocacy and youth citizen engagement and brings a strong insider's perspective to this study. She holds a leadership role in the Pacific Youth Council and her extensive experience in the field offers rich insight. J.E.H. Percival, also from Samoa, is a tropical forest ecologist by training and his research focuses on understanding climate-change dynamics and developing sustainable practices in the Pacific region. Both contributed equally to this chapter and together, our joint perspectives provide a comprehensive view of the complex dynamics in climate activism in the Pacific.

Climate Activism in the Pacific: Reshaping Narratives and Forging Change

We in Oceania are also creating a new regional identity, part of which is our common recognition of the fact that we are not just small island states, but also large oceanic states, our boundaries defined by the ocean that surrounds us. This perspective, which derives from our ancestral understanding of our sea of islands, leads us to reassert our dignity and pride as a people, confident in our ability to make our own contribution to the world.

(Hau'ofa, 1994)

In the Pacific, the perception of its island nations and their peoples regarding their identity holds significant implications in their approach to policymaking, social action, and overall regional collaboration. The refinement and creation of their narratives are integral to the way Pacific Islanders articulate their role in the fight against climate change. Despite the acknowledged smallness and remoteness of the islands, Hau'ofa's (1994) idea of the Pacific as "a sea of islands" emphasises the belief that their ancestors envisioned the world as a vast sea interspersed with interconnected islands, which in turn fostered a collective identity steeped in oceanic immensity. Within this complex network of interconnections, societies and cultures traversed the oceans, resulting in dynamic communities of exchange where wealth, skills, and arts flourished (Hau'ofa, 1994). Stories of resilience, cultural heritage, and interconnectivity are embedded within the social, environmental, and political framework of the Pacific region. These narratives highlight the unique nature of Pacific climate activism today, which draws its strength from ancestral wisdom and Indigenous perspectives. The resurrection and dissemination of these narratives underscore the Pacific Islanders' dignity, pride, and capacity to make a significant contribution to the global struggle against climate change.

As we explore the development of climate activism in PICTs, we find that this self-perception and positioning has been instrumental in shaping political dynamics that have influenced the strategies and pathways employed by activists. Climate activism in the Pacific materialised most pronouncedly during the early stages of global climate policy dialogues, most notably with the formation of AOSIS in the lead-up to the Second World Climate Conference in 1990 (Carter, 2015). The establishment of AOSIS is significant because it not only became one of the most influential vehicles for PICTs to address climate concerns in the global policy arena but also because it established itself as one of the first examples of Pacific regionalism in climate policy discussions (Morgan, 2022). Connell (2013) and Hau'ofa (1994) stress that a shared, regional identity strengthens the collective influence and ability of Pacific nations to navigate global challenges, and the formation of AOSIS has reflected how these island nations have coalesced to amplify their voices in global environmental discussions (Morgan, 2022). By presenting themselves as a unified front, active participation by AOSIS in the initial Climate Conference of 1990 and subsequent meetings, culminating in and beyond the 1992 Rio Summit, marked critical turning points in the dynamic journey of PICTs, wherein the Pacific's voice demanded global recognition (Shibuya, 1996; Ashe et al., 1999; Kirsch, 2020).

As such, many have argued that regionalism has played a vital role in amplifying the influence of PICTs in international climate discussions (Tarte, 2014; Williams and McDuie-Ra, 2018; Keen and Connell, 2019; Morgan, 2022). This unified approach has enabled these small island nations to address environmental challenges more effectively on both regional and global scales. To illustrate, their participation at annual COP meetings has been central for advocating stronger global action on emissions reductions and adaptation efforts in PICTs (Carter, 2015; Ourbak and Magnan, 2018). The "High Ambition Coalition", led by the Marshall Islands, for example, played a crucial role in securing ambitious temperature targets in the Paris Agreement, encapsulated in the slogan "1.5 to stay alive". The "Kainaki II Declaration for Urgent Climate Change Action Now" by regional governments is another recent and significant example, highlighting the region's stance on the climate crisis, which is to collectively urge all parties to the Paris Agreement to rapidly phase out fossil fuel subsidies, enhance global climate finance, and strengthen measures for loss and damage associated with climate impacts, particularly emphasising the acute threats posed to the Pacific region (Pacific Island Forum, 2019).

Influenced by this rising tide of Pacific regionalism in global climate politics, activists in the region are uniting their efforts for greater impact. By aligning with Pacific regionalism, they are advocating for stronger climate action on regional and global stages. Initiatives such as the PIF Secretary General's Young Climate Leaders Alliance and Pacific Islands Students Fighting Climate Change highlight this shift towards a united, strategic approach, leveraging collective strength to influence climate policy changes.

Pacific Youth and Climate Action

Pacific youth are the generation that has the most to lose; but that also means that they have the biggest stake in ensuring our survival. As Pacific youth, our very existence and every part of our being is at stake. If we allow our economies to be run in this disgusting capitalist world, then we lose all the fundamentals that make us Pacific Islanders. It's because of capitalism that we are confronted with many injustices, especially with climate change. I cannot just sit and let it continue!

(E. Gibson, personal communication, 2021)

This sentiment expressed by Ernest, one of the activists interviewed in this study, epitomises the collective struggle and mobilisation of Pacific youth in response to climate change. Groups and alliances with significant youth participation are central in driving the Pacific region's response to climate change (Steiner, 2015; McNamara and Farbotko, 2017). Notable among these are the Pacific Climate Warriors – aligned with the global network 350. org – and the Pacific Islands Climate Action Network (PICAN), both characterised by their strong youth involvement. Additionally, the Pacific Islands Students Fighting Climate Change (PISFCC), grassroots initiatives like Jo-Jikum in the Marshall Islands, and numerous National and Provincial Youth Councils across the PICTs play pivotal roles in the climate-change movement.

Pacific youth activists skilfully weave traditional knowledge, cultural sentiments, and practices into their climate activism. Various grassroots projects like Pacific Climate Tales, Mana Moana – Pasifika Voices, and The Islands are Calling, employ art, storytelling, and song to powerfully convey the climate challenges faced by the Pacific community. This activism also involves community-based dialogues and consultations that integrate their causes into cultural norms and ensuring respect for customary practices and community ways of life (Enari and Viliamu Jameson, 2021). Originating from community-level engagement and driven by local concerns, these expressions and approaches creatively combine elements of Pacific cultural heritage with modern advocacy methods. As such, Pacific Islanders perform their climate activism while maintaining a deep respect for their traditions and a commitment to maintaining them. Rather than relying on traditionally mainstream methods like public demonstrations, rallies, and protests, activism manifests in more culturally resonant and community-focused ways. This form of activism not only addresses regional challenges but also strengthens the cultural identity of Pacific Island communities.

In an example of acknowledging the importance of cultural norms in Pacific activism, one of our interviewed activists from Kiribati explained the concept of *katei* (or *katei ni Kiribati*, often translated as "the Kiribati way") and how it affects his participation in activism in Kiribati. He describes how Kiribati people believe in being accepted by both the goddess of the island and its people as it ensures protection by the island's spirit and successful work execution. He explained that *katei* is a central construct within Kiribati families symbolising their standing, personal honour, and mannerism. To navigate the cultural protocols and engage effectively within any local community or organisation, one must understand *katei*. This suggests consulting local governments and elders when undertaking activism work, as they can provide essential introductions and connections to the community. *Katei* must also be reciprocated, exemplifying a core value for Kiribati people. This respectful mannerism fosters invitations to homes and evokes protective sentiments within the community.

There are several examples of how climate activism manifests itself in the Pacific and is broadcast to the world. A recent example of this was observed at COP27 in 2022, where the PISFCC played a central role in successfully advancing a UN resolution to hold polluting countries legally accountable for inaction on climate change. This resolution was born from a grassroots movement in the Pacific that eventually made a direct call for the International Court of Justice to clarify nations' obligations in tackling the climate crisis, with particular attention to the damage endured by small island states (Guardian Staff, 2023). The success of PISFCC's campaign at COP27 highlights both the effectiveness of the Pacific model as well as the strength of Pacific regionalism in their approach. By uniting the voices of various Pacific nations and drawing on the collective power of regional solidarity, PISFCC effectively amplified their message and influence on the global stage. This regional unity, spearheaded by Vanuatu and supported by other Pacific Island nations, was crucial in driving the campaign's success and demonstrates PISFCC's evolving strategy that leveraged both local and regional cooperation for impactful climate advocacy.

Another prominent example is the Pacific Climate Warriors' 2014 canoe blockade at Newcastle Coal Port in Australia. This initiative, originating from the grassroots and uniting activists from 12 PICTs, used canoes crafted in traditional style for a non-violent protest against climate-change impacts. This peaceful demonstration not only highlighted the vulnerability of their islands but also effectively brought significant international attention to the urgent need for climate action (McNamara and Farbotko, 2017). The demonstration showcased the impact of integrating traditional cultural practices with contemporary forms of activism, a method rooted in the community-based values and experiences of Pacific Islanders. This combination has helped to position the Pacific not just as a frontline but also as a central leader in global climate resistance, demonstrating how grassroots movements can catalyse significant global awareness and action (McNamara and Gibson, 2009; Farbotko and McMichael, 2019).

Consequently, Pacific youth activists are not just responding to climate change; they are actively reshaping the global narrative around it (McNamara and Farbotko, 2017; Fair, 2020). Pacific youth, by stepping into roles as transformative agents, are rejecting the portrayal of their communities as passive victims of climate change. Groups like the Pacific Climate Warriors exemplify this shift, embracing their agency to challenge preconceived narratives of vulnerability (McNamara and Farbotko, 2017; Fair, 2020, p. 20). Their unifying declaration, "We are not drowning. We are fighting", embodies their tenacity and resilience (Fair, 2020). As portrayed by Fair (2020), the Pacific Climate Warriors exemplify the strength and interconnectedness of their Oceanic identities, thereby challenging the pervasive narrative of "drowning islands". This proactive stance in Pacific climate advocacy disassembles the global narratives that depict Pacific islands solely as victims.

Thus, the collective spirit of Pacific climate activists is transforming their approach to climate challenges. By uniting under a shared Oceanic identity, they are shifting the narrative from victimhood to proactive leadership, effectively reshaping the global dialogue on climate action and emphasising the Pacific's role in advocating for sustainable solutions.

Pacific Youth Perspectives on Climate Activism:
A Mirror of Identity, Heritage, and Community

Engaging as a youth in advocacy and activism is hard, painful, and exhausting; but our role is important in ensuring that our communities thrive, and the vision of youth being fully recognised and appreciated in that role is within our grasp, and that makes me excited!
(E. Gibson, personal communication, 2021).

This chapter has so far explored the realities of climate change as experienced by Pacific Islanders, underscored the central role of Pacific regionalism in shaping climate discourse, and illuminated the influence of youth in this arena. We have also examined various manifestations of activism within the Pacific context. Building on this foundation, our next focus is to gain a deeper understanding of how Pacific youth perceive activism. Drawing from the interviews with Pacific youth in our study, two underlying aspects play a significant role in how Pacific youth engage. The first is *how they perceive themselves as climate activists*, and the second is the *context in which they live and engage as activists*. The subsequent discussion unpacks each in detail, beginning with the former.

The responses from Pacific youth reveal a deep-seated narrative of stewardship in their climate activism. They see themselves not just as residents but as vigilant custodians of their environment, tasked with safeguarding it for future generations. This stewardship, deeply entwined with their cultural identity, is pivotal to their perception of self within their climate activism efforts. Kathy, a young Marshallese woman who was interviewed, articulates this sentiment:

Knowing who you are is not enough! Really understanding your identity encompasses everything that makes you who you are – kinship, culture, faith, community, values, beliefs. And this gives me the confidence I need to do my work as a climate activist. More importantly, when I am out there representing my island home, I understand the value of who I am as a young Marshallese.

(K. Jetnil-Kijiner, personal communication, 2021)

This sense of stewardship arises not just from their cultural identity but also from lived experiences shaped by a close relationship with their environment. As such, Pacific youth climate activism emphasises the lived experiences and narratives of the climate crisis, which stands as a vital counterpoint to the Western-dominated narrative that often prioritises scientific discourse (Farbotko, 2010). These activists point out that while in many Western societies climate change is understood predominantly through scientific data, in the Pacific, it is a reality felt in the day-to-day lives of its inhabitants. As another youth activist from the Federated States of Micronesia succinctly put it, "Climate change is no longer a distant, abstract concept; it's a pressing reality for families who are striving to survive with the limited resources they possess" (Heidi, personal communication, 2021). Another recounts:

Having lived in Fiji for many years, I subsequently moved to Kiribati. Here, I witnessed first-hand the urgent effects of climate change on low-lying atolls such as Kiribati, perched merely 3 metres above sea level. The consequences of climate change were palpable, from flying into the ever-narrowing landscapes to confronting the reality of using brackish water for bathing which often led to skin conditions like scabies. Everything that was once taken for granted while residing in Fiji now carried a tinge of discomfort.

(Te'o, personal communication, 2021)

Amid these experiences, representation becomes central for many Pacific youth, especially as they engage in global spaces. Many feel that this is not just about systemic change; it is about empowering their communities and defending their rights. Representation is about honouring their ancestors and carrying forward their community's hopes. One activist said:

I feel that there is an ever-present dichotomy in the treatment of the Pacific Islanders. On one hand, they are paraded like spectacles, perhaps akin to exhibits in a zoo, used to demonstrate purported efforts of involvement with the Pacific. On the other hand, they are dismissed entirely, overlooked so thoroughly it's as if not even a pretence of remembrance is offered. This recurring marginalisation fuels my sentiments of defiance. Each day is devoted to counteracting harmful and inaccurate narratives.

(G. Jiva, personal communication, 2021)

These sentiments, echoed by many Pacific youth activists, highlight a common challenge: navigating the complexities of global perception and representation. As such, it goes back to the idea of counteracting misleading narratives (McNamara and Farbotko, 2017) and advocating for accurate, respectful representation that speaks to the resilience and deep commitment of Pacific youth to their communities. This commitment runs deep. For Pacific youth, climate activism is an extension of their cultural heritage and a platform to ensure their voices and experiences are acknowledged and respected in both the local and global arena.

Milan, another activist from the Marshall Islands who was interviewed, shares her perspective on what activism means in the Pacific context:

Being an activist in the Pacific sense, we bring our stories of how climate change is impacting our islands in the Pacific. We do so in a way that is peaceful, showcasing our culture, history, and stories through songs, dance, and chants. These are the mediums that already exist and are embedded in our culture. They provide a different perspective to outsiders and are authentic to who we are as Pacific Islanders.

Milan's reflections illustrate how Pacific youth perceive themselves in climate activism: as custodians of both their environment and heritage, underlining their resilience and dedication to driving meaningful change.

Having explored how Pacific youth perceive themselves as activists, our focus shifts to the context in which they engage in activism. This context is defined by the enduring cultural structures and traditional practices that remain deeply ingrained within family, community, and faith (Teariki and Leau, 2023). Many interviewees frequently mentioned elements such as culture, kinship, faith, and community, highlighting the communal ethos that dispositions Pacific youth in their climate activism. Understanding these elements and their placement within Pacific societies adds significant depth to comprehending the perspectives and experiences of Pacific youth.

In Pacific societies, kinship, culture, faith, and community are distinct yet intertwined aspects defining social relations (Keesing, 1989; Teariki and Leau, 2023). Kinship networks establish familial roles within the broader community context. In many Pacific societies, the extended family forms the primary family structure, with shared family names, chief titles, inheritance rights, and familial rituals as key markers (Lilomaiava-Doktor, 2009). For example, in Samoan society, the *matai* (chief) title system is a crucial kinship marker, central to the basic political and social unit. Selected family members, holding these chiefly titles, become trustees of family inheritance, such as customary land, and bear decision-making authority in village governance. Therefore, the right to reside on and use customary land is directly tied to one's kinship relations (Meleisea, 1991; So'o, 2008).

However, within Pacific societal structures, the role of youth faces complex challenges, especially given the power dynamics between older and younger generations. Despite recognising the importance of intergenerational efforts, Pacific youth frequently confront issues like tokenism, resentment, and misunderstanding. Kinship and cultural traditions, emphasising respect for elders and inherited leadership, often place children and youth in a subordinate status. This position does not signify their insignificance but indicates a period of waiting before they assume decision-making and leadership roles. Such cultural norms can inadvertently erect barriers to their active engagement in crucial discussions, including climate action. Therefore, while cultural values do not completely prevent young leadership, they can lead to their involvement being more symbolic, reflecting a paternalistic approach to youth participation.

Like kinship, community and faith are integral to the lives of Pacific youth, embodying shared realities and strong influences. Their concept of community is often characterised by shared values and responsibilities, a deep sense of belonging, and a collective way of life. For example, the *fa'asamoa* (the Samoan way) illustrates this communal ethos through practices like resource sharing, hospitality, and community involvement. Likewise, faith plays a crucial role in many Pacific communities, influencing Indigenous life profoundly. Although the experience of faith varies across cultures, a common thread is a reciprocal commitment to both the community and the environment, shaped by spiritual beliefs (Teariki and Leau, 2023).

Understanding these elements and how they are situated in the lives of Pacific youth adds value to perceiving the role of youth in climate activism, the challenges youth face, the opportunities that exist for youth, and the motives that drive their activism. From this, we find that climate activism in the Pacific is deeply entwined with two fundamental elements: the first is that activism serves primarily as an affirmation of identity and heritage, and the second is that there is a strong inclination towards community-centric action. The emphasis on these core aspects of their activism sheds light on the distinctive perspective of Pacific climate activism within the wider global dialogue on climate change that focuses on changing narratives, cultural preservation, community resilience, and intergenerational justice.

Activism as an Affirmation of Identity and Connection to Heritage

Pacific youth climate activists intimately link their activism with their cultural identity and heritage. For them, activism is not simply a form of advocacy or protest; instead, it serves as an embodiment of who they are and their roots. Their advocacy goes beyond raising awareness or catalysing systemic changes in issues like climate change. Instead, it is a conduit for empowering their communities, addressing pressing social and environmental dilemmas, and defending their rights. These facets of activism are deeply intertwined with their cultural identity and heritage. The perspectives that have been shared thus far illuminate the critical role that culture and spirituality play in shaping climate activism among Pacific youth. These elements are not just backdrop or context; they are the driving force of their activism. The activists see themselves as custodians of their cultures and narrators of their community's stories, representing their people's concerns on a global stage. Informed by a deep sense of duty to serve their communities, preserve their cultural traditions, and safeguard future generations, they view their activism as an indelible component of their cultural identity.

Community-centric and Collaborative Activism

The second defining feature of Pacific climate activism in youth is its profound embeddedness within the community. This relationship, deeply anchored in a cultural ethos that values extended family and communal ties, critically influences their activism. Each aspect of their activism is interlinked, collectively demonstrating a robust, community-centred, and collaborative approach to their advocacy efforts. Despite diverse inspirations, all interviewees were united in their central goal to uplift and advocate for their communities. Their activism, firmly anchored in social, political, and environmental change, unfailingly honours their unique cultural contexts.

Safeguarding Fa'asinomaga: The Heart of Climate Activism in the Pacific

The emphasis on cultural identity and heritage gives Pacific youth a distinctive character in climate action, differentiating them from their Western counterparts and realities. It is not just an element added; it is inherent. In being cognisant of this, we draw on Pierre Bourdieu's social theory to provide a more nuanced understanding of the interplay between *fa'asinomaga* and Pacific youth motivations. His concept of the "habitus" is a fitting lens. According to Bourdieu, habitus refers to a system of internalised dispositions and tendencies that guide an individual's perceptions, thoughts, and actions within their social world (Bourdieu, 1990). It is rooted in the interaction between one's personal history and present social conditions, shaping the response to the environment, including the participation in social action and activism (Reay, 2004). Building on this, we introduce the concept of a *fa'asinomaga*-framed habitus, a culturally specific adaptation of Bourdieu's habitus that captures the essence of Pacific cultural, societal, and climatic realities influencing youth motivations towards climate action. Their diverse experiences and realities, while individually unique, are bound together by a sense of cultural identity and heritage to their islands. The Samoan concept of *fa'asinomaga* was adopted to encapsulate that shared sense of belonging, identity, and heritage, ingrained in many Pacific cultures. Although the term *fa'asinomaga* can simply be used to mean identity and heritage, the essence of its meaning is much deeper. The following quote is a Samoan conceptualisation that conveys that deep essence of its meaning:

> The Samoan person does not exist as an individual. There is myself and yourself. Through you, my being is contextually meaningful and whole. Through myself, you are given primacy in light of our collective identity and places of belonging, our genealogical lineage, and our roles, responsibilities and heritage.
>
> *(Tamasese et al., 1997)*

Considering this conceptualisation, the safeguarding of our *fa'asinomaga* is profoundly shaped by cultural signals that encompass ancestry, language, lineage, inheritance, heritage, and social responsibilities. In this light, safeguarding their *fa'asinomaga* is what connects the motivations of these Pacific youth in climate activism. Understanding the motivations of Pacific youth rooted in *fa'asinomaga* provides a lens through which we can reshape the narration of climate action in the Pacific.

The *fa'asinomaga*-framed habitus positions us to understand how Pacific youth's structured experiences (family, culture, community, faith, etc.) and their individual agency (subjective experiences, conscious choices) meld together to guide their motivation in climate activism. This intertwining collectively constructs the Pacific identity and motivates us as Pacific Islanders to protect our *fa'asinomaga*. The *fa'asinomaga*-framed habitus underlines the importance of recognising that our past informs the present, and together they shape our future – a timeless continuum that serves as a source of resilience, motivation, and continuity for the generations to come. It underscores the dynamic and reciprocal nature reflected in the experiences of Pacific youth in climate action.

The experiences of Pacific youth in climate activism highlight the critical role *fa'asinomaga* plays in grounding and shaping identities and call into vivid focus the devastating effects of displacement when this connection is severed. Their narratives underscore the urgency of continuing to protect *fa'asinomaga* to sustain individual and collective identities and as a bulwark for preserving cultural heritage. Thus, the safeguarding of *fa'asinomaga* is inexorably entwined with the resilience of Pacific societies in the face of climate change.

In conclusion, the narrative of Pacific youth activism, rooted in the significance of Indigenous knowledge, shines a light on a story of hope, resilience, and solidarity. Despite the geographical perception of being small island states, these Pacific Island nations embrace an identity as vast oceanic states, their boundaries defined by the expansive ocean that surrounds them (Hau'ofa, 1994; Fair, 2020). This perspective, derived from the ancestral understanding of their "sea of islands", echoes the thinking of Hau'ofa, who emphasised the expansive nature of their cultural and ecological identities (Hau'ofa, 1994). Their voices, enriched by timeless stories, songs, and wisdom, resonate powerfully on the global stage. The profound wisdom of these voices is beautifully encapsulated in an old Samoan song from the village of Asau in Savaii. Its verse, *"E lē la'a le uto i le maene pe sopoa'e le tai i le 'ele'ele"*, translated as "the floater (uto) cannot change places with the sinker (maene), nor can the sea (tai) walk on the land (eleele)", serves as a poignant reminder of the irreversible changes inflicted by exploitative human behaviour, echoing the challenges faced in our current era of anthropogenic climate change. This new generation of climate warriors, determined to protect their islands and their heritage, offers a compelling call to action for the global community to recognise and respond to their plight. They fight not only for a sustainable future but also for the preservation of their *fa'asinomaga* – their diverse cultures, identities, and ways of life, reinforcing the fact that the battle against climate change is as much a cultural endeavour as it is environmental. Their assertion of their expansive identity serves to instil dignity, pride, and confidence and underscores their contributions to the global discourse on climate change.

References

Anthony, K. (2023). Pacific youth activists present Forum with climate demands. *Radio New Zealand*, 31 October. https://www.rnz.co.nz/international/pacific-news/501365/pacific-youth-activists-present-forum-with-climate-demands

Ashe, J. W., Van Lierop, R. and Cherian, A. (1999). The role of the Alliance of Small Island States (AOSIS) in the negotiation of the United Nations Framework Convention on Climate Change (UNFCCC). *Natural Resources Forum*, 23(3), 209–220. https://doi.org/10.1111/j.1477-8947.1999.tb00910.x

Bourdieu, P. (1990). *The Logic of Practice*. Stanford, CA: Stanford University Press.

Campbell, J. and Barnett, J. (2010). *Climate Change and Small Island States: Power, Knowledge and the South Pacific* (1st ed.). London: Routledge.

Carter, G. (2015). Establishing a pacific voice in the climate change negotiations. In G. Fry and S. Tarte (eds.) *The New Pacific Diplomacy* (1st ed.). ANU Press. https://doi.org/10.22459/NPD.12.2015.17

Connell, J. (2013). *Islands at Risk: Environments, Economies and Contemporary Change*. Edward Elgar Publishing.

Enari, D. and Viliamu Jameson, L. (2021). Climate justice: A Pacific Island perspective. *Australian Journal of Human Rights*, 27(1), 149–160. https://doi.org/10.1080/1323238X.2021.1950905

Fair, H. (2020). Their sea of islands? Pacific climate warriors, oceanic identities, and world enlargement. *The Contemporary Pacific*, 32(2), 341–369. https://doi.org/10.1353/cp.2020.0033

Farbotko, C. (2010). Wishful sinking: Disappearing islands, climate refugees and cosmopolitan experimentation. *Asia Pacific Viewpoint*, 51(1), 47–60. https://doi.org/10.1111/j.1467-8373.2010.001413.x

Farbotko, C. and McMichael, C. (2019). Voluntary immobility and existential security in a changing climate in the Pacific. *Asia Pacific Viewpoint*, 60(2), 148–162. https://doi.org/10.1111/apv.12231

Guardian Staff (2023). "Beginning of a new era": Pacific islanders hail UN vote on climate Justice. *The Guardian*, 30 March. https://www.theguardian.com/world/2023/mar/30/un-vote-on-climate-justice-pacific-island-change-crisis-united-nations-vanuatu

Hau'ofa, E. (1994). Our sea of islands. *The Contemporary Pacific*, 6(1), 148–161.

Hoegh-Guldberg, O., et al. (2007). Coral reefs under rapid climate change and ocean acidification. *Science*, 318(5857), 1737–1742. https://doi.org/10.1126/science.1152509

Keen, M. and Connell, J. (2019). Regionalism and resilience? Meeting urban challenges in pacific island states. *Urban Policy and Research*, 37(3), 324–337. https://doi.org/10.1080/08111146.2019.1626710

Keesing, R. M. (1989). Creating the past: Custom and identity in the contemporary pacific. *The Contemporary Pacific*, 1(12), 19–42.

Kelly, J. (2021). My day at COP26: "I told world leaders: We're not drowning, we're fighting". *BBC*, 1 November. https://www.bbc.com/news/science-environment-59121480

Kirsch, S. (2020). Why pacific islanders stopped worrying about the apocalypse and started fighting climate change. *American Anthropologist*, 122(4), 827–839. https://doi.org/10.1111/aman.13471

Knutson, T. R., et al. (2010). Tropical cyclones and climate change. *Nature Geoscience*, 3(3), 157–163. https://doi.org/10.1038/ngeo779

Lilomaiava-Doktor, S. (2009). Beyond "migration": Samoan population movement (Malaga) and the geography of social space (Vā). *The Contemporary Pacific*, 21(1), 1–32. https://doi.org/10.1353/cp.0.0035

McNamara, K. E. and Farbotko, C. (2017). Resisting a "doomed" fate: An analysis of the pacific climate warriors. *Australian Geographer*, 48(1), 17–26. https://doi.org/10.1080/00049182.2016.1266631

McNamara, K. E. and Gibson, C. (2009). "We do not want to leave our land": Pacific ambassadors at the United Nations resist the category of "climate refugees". *Geoforum*, 40(3), 475–483. https://doi.org/10.1016/j.geoforum.2009.03.006

McNamara, K. E., Westoby, R. and Chandra, A. (2021). Exploring climate-driven non-economic loss and damage in the Pacific Islands. *Current Opinion in Environmental Sustainability*, 50, 1–11. https://doi.org/10.1016/j.cosust.2020.07.004

Meleisea, M. (ed.) (1991). *Lagaga: A Short History of Western Samoa*. Repr. by Suva: Fiji Times Ltd. Suva: Institute of Pacific Studies, University of the South Pacific.

Morgan, W. (2022). Large ocean states: Pacific regionalism and climate security in a new era of geostrategic competition. *East Asia*, 39(1), 45–62. https://doi.org/10.1007/s12140-021-09377-8

Ourbak, T. and Magnan, A. K. (2018). The Paris agreement and climate change negotiations: Small islands, big players. *Regional Environmental Change*, 18(8), 2201–2207. https://doi.org/10.1007/s10113-017-1247-9

Pacific Island Forum (2019). *Kainaki II Declaration for Urgent Climate Action Now*. https://www.forumsec.org/2020/11/11/kainaki/

Pacific Island Forum (2023). *Fifty-Second Pacific Islands Forum*. Forum Communiqué. Rarotonga, Cook Islands: Pacific Island Forum. https://www.forumsec.org/wp-content/uploads/2023/11/FINAL-52nd-PIF-Communique-9-November-2023-1.pdf

Pacific Island Forum (n.d.). *PIF-SG's Young Climate Leaders Alliance*. https://www.forumsec.org/pacific-islands-forum-secretary-generals-young-climate-leaders-alliance/

Reay, D. (2004). "It's all becoming a habitus": Beyond the habitual use of habitus in educational research. *British Journal of Sociology of Education*, 25(4), 431–444. https://doi.org/10.1080/0142569042000236934

Fiji Ministry of Economy. (2017). *5 Year and 20 Year National Development Plan: Transforming Fiji*. https://www.fiji.gov.fj/getattachment/15b0ba03-825e-47f7-bf69-094ad33004dd/5-Year-20-Year-NATIONAL-DEVELOPMENT-PLAN.aspx

Sefeti, S. (2023). "Not conducive to our survival": Pacific islands on the climate frontline respond to Cop28 deal. *The Guardian*, 19 December. https://www.theguardian.com/environment/2023/dec/20/not-conducive-to-our-survival-pacific-islands-on-the-climate-frontline-respond-to-cop28-deal

Shibuya, E. (1996). "Roaring mice against the tide": The South Pacific Islands and agenda-building on global warming. *Pacific Affairs* [Preprint]. https://doi.org/10.2307/2761186.

So'o, A. (2008). *Democracy and Custom in Sāmoa: An Uneasy Alliance*. Suva, Fiji: IPS Publications, University of the South Pacific.

Steiner, C. E. (2015). A sea of warriors: Performing an identity of resilience and empowerment in the face of climate change in the pacific. *The Contemporary Pacific*, 27(1), 147–180. https://doi.org/10.1353/cp.2015.0002

Tamasese, K., Peteru, C. and Waldegrave, C. (1997). *Ole Taeao Afua, the New Morning: A Qualitative Investigation into Samoan Perspectives on Mental Health and Culturally Appropriate Services*. A Research Project Carried Out by the Family Centre. http://familycentre.org.nz/wp-content/uploads/2019/04/Ole-Taofua.pdf

Tarte, S. (2014). Regionalism and changing regional order in the Pacific Islands. *Asia & the Pacific Policy Studies*, 1(2), 312–324. https://doi.org/10.1002/app5.27

Teariki, M. A. and Leau, E. (2023). Understanding Pacific worldviews: Principles and connections for research. *Kōtuitui: New Zealand Journal of Social Sciences*. https://doi.org/10.1080/1177083X.2023.2292268

Williams, M. and McDuie-Ra, D. (2018). Organising a regional response to climate change in the Pacific. In *Combatting Climate Change in the Pacific: The Role of Regional Organizations* (pp. 13–38). Springer.

Yamano, H., et al. (2007). Atoll island vulnerability to flooding and inundation revealed by historical reconstruction: Fongafale Islet, Funafuti Atoll, Tuvalu. *Global and Planetary Change*, 57(3–4), 407–416. https://doi.org/10.1016/j.gloplacha.2007.02.007

6

REALISING COMMON GROUND

Custodianship of Country and Youth Climate Action

Madison Shakespeare, Michelle Catanzaro and Caelli Jo Brooker

Introduction

In this chapter we consider the ways visual expressions of climate justice are evident in grassroots climate action in Australia. We contend that climate justice must manifest First Nations' sovereign rights of custodianship, implicit to which are decolonising approaches to care of Country[1] and climate wellbeing. This involves returning the land to First Nations Traditional Custodians who have looked after and cared for Country for tens of thousands of years.

Young people have been organising, advocating and protesting in coordinated efforts since the late 1990s in protests that advocated for climate health, most recently in their millions (Hilder and Collin, 2022). In Australia the School Strike for Climate (SS4C) movement has led protests seeking immediate intervention from politicians, corporations and others who can enact change to redress the extensive detriment that continues to impact on Australia's natural environment. As part of ongoing grassroots cultural activism in Australia, these youth School Strike for Climate protests have seen over 80,000 people attend in Sydney and an estimated 100,000 in Melbourne, making these protests the largest since those held against the Iraq war in 2003 (Henriques-Gomes et al., 2019), which reveals a marked increase in activism in Australia since 2017 (Thackeray et al., 2020).

It is remarkable to note that these strikes are youth-led. Images of youth participants are documented by multiple media platforms along with their personal expressions of protest in the form of placards carried at the strikes containing messages, text and images that communicate their climate concerns as youth advocates for climate justice. The primary call to arms from student attendees is to confront climate inaction, but there are various visual and textual associations that reveal or represent clear and emergent themes in the visual expressions included in the strikes.

This chapter asserts that youth perspectives and First Nations standpoints are essential to caring for climate. In this chapter we assess the (in)visibility of Indigenous representation, activism and awareness at Australian youth climate action events to provide insight into the ways First Nations sovereign rights are included and represented in youth protest activist narratives.

Broadly, at each climate strike, the collective aim of participants is to prompt others to act now to 'stop global warming' and 'save our planet'. However, individuals at the strikes have

DOI: 10.4324/9781003396567-7

different belief systems, priorities and motivations in the fight for climate action, and these differences are made visible in the placards (Catanzaro, 2024). For many students, the protests are a unique space for their political voice, and their messages delivered through signs and placards carry diverse images, words and metaphors (Catanzaro and Collin, 2023). Previous consideration of the thematic and semiotic implications of student narratives on climate activist placards (see Catanzaro and Collin, 2023) informed understandings of the power of design and communication to shift student voices beyond the constraints of classroom settings. Catanzaro and Collin's (2023) preliminary analysis also illuminated how the possible coexistence of different individual student perspectives can evolve congruent collective activist voices that hold potential for enacting real change. For this chapter, the images from Catanzaro and Collin's (2023) data set have been combined with the growing archive of placard photographs from 2022 and 2023 climate strikes for analysis as they pertain to a type of visibility for young people whose voices are often marginalised and diminished in the fight for climate justice.

There are multifaceted visual expressions found in the placards at the strikes reflecting Australian Aboriginal perspectives, including references to the Aboriginal flag and the application of its colours (red, black and yellow) alongside textual references to Country (expressed through Aboriginal place names), as well as resilience and decolonisation. These forms of visual communication are depicted by a range of activists that include students, Australian First Nations Peoples, Indigenous organisations and non-Indigenous allies.

As part of a broader project on youth climate activism (in which one of the authors is a chief investigator: see acknowledgements) five SS4C protests on Gadigal Country (Sydney) from 2019–2023 have been photographed by researchers and student researchers. For this chapter, the photographic documentation of 850 placards has been reviewed and analysed to identify the prevalence and recurrence of Indigenous Australian visual references found within the strikes, and we expand on this methodological approach later. In doing this, we sought to identify and explore the extent of First Nations climate perspectives found in the protest signage through an Indigenous worldview lens (Redvers et al., 2022). What we discovered was that, despite the visibility, inclusion and influence of Indigenous organisations and peoples within the SS4C movement, there are still somewhat limited representations of explicitly Indigenous perspectives evident in the visual data. However, the data does demonstrate that the inclusion of Indigenous perspectives at the strikes has been increasing in number since the documentation of the SS4C protests began in September 2019. While the structure of the strikes actively includes First Nations perspectives – organisers privileging First Nations voices as keynote speakers to welcome activists to Country and to talk about specific place-based violence to Country, values and insights (Bessant et al., 2023), for example – the observed deference to Indigenous knowledge is not as readily evidenced in the visual messaging found in the student placards.

Our work joins scholars in the call for 'theoretical and methodological innovations' (Bowman, 2019) that provide more nuanced, bottom-up insights into young people's political participation (Bessant et al., 2023; Wall, 2021). The authors saw deep value in focusing on and exploring the powerful and educative visual references that did emerge at the climate strikes surrounding Australian First Nations representations. As a result, this chapter examines how Indigenous perspectives, voices and references can be positioned and situated in Australian climate action contexts.

To gain a holistic understanding of the data set, we harnessed an Indigenous worldview approach focused on expansive notions of Country (Yap and Yu, 2016) through which to understand and interpret visual and written placards. This Country-centred approach acted as a 'common ground' element (Wilson, 2008) that revealed complementary and connected themes within the broader data set. There were powerful correlations of Indigenous thinking and youth perspectives articulated through the visual examples, and it is these overlapping concerns that have shaped the understandings explored in this chapter.

Finding Common Ground

In providing a background to a discussion of climate justice in Australia, it is important to acknowledge that Australia is home to the longest continuing cultures of First Nations Peoples in the world (Marshall, 2019) and to honour Aboriginal and Torres Strait Islander Peoples' sovereign knowledges, practices and rights as First Peoples. First Nations Custodianship of Country has continued across at least 65,000 years, and the diverse knowledges and practices that surround the care of Country are an essential spiritual responsibility evoked through relational connectedness with the natural environment (Redvers et al., 2022).

Whilst climate health is an imminent contemporary crisis, understanding the political history in Australia can inform on ways in which Indigenous-led climate activist mobilisation can achieve meaningful outcomes. To fully understand this, readers need to engage with perspectives on the history of invasion in Australia in 1788 (Mencevska, 2019; Watson, 2018). This history provides insight into the ways continuing colonising ideologies and practices are contrary to the health and sustainability of Australia's natural environment and undermine fundamental outcomes sought by climate activists. Our research is grounded in this understanding; however, there is no scope to break this down in detail within this chapter. Instead, we focus on contemporary manifestations of activism that respond to and subvert dominant settler narratives through contemporary climate strike action.

In an Australian context, First Nations people have a history of activism, fighting for land rights and sovereignty. Markedly, the Aboriginal Tent Embassy in Canberra, which began in 1972, is now the longest running protest for Indigenous land rights in the world. This is but one moment in history where Australian First Nations peoples have engaged in protest action to bring visibility to injustice. As explored through the discussion of Indigenous connection to Country later, land rights for Aboriginal and Torres Strait Islander peoples are inextricably connected to climate justice.

In Figure 6.1, we can see one of the placards held by the 1972 protesters, which states 'Land Rights Now'. This same message can be found in the placard on the right at a June

Figure 6.1   Image on the left: aboriginal Embassy, 27 January 1972. Image on the right: June 2023. School Strike for Climate on Gadigal land.

Source: Mitchell Library, State Library of New South Wales and Courtesy SEARCH Foundation.

2023 SS4C protest. This connects the persistent and continuing need for political action from First Nations peoples across multiple generations.

Many other examples such as the Mabo judgement for sovereignty (Smyth, 1995), the Aboriginal Tent Embassy (Watson, 2009) and the suburb of Redfern (Robinson, 1994) spotlight Aboriginal peoples' continued struggles for sovereign rights, through representation and visibility in Australia against the continued negligence of successive Australian Governments. Chicka Dixon, who was a critical leader in the Aboriginal Tent Embassy movement, strategically involved the media in First Nations activism, stating, 'I wanted to put our plight into the eyes of the world' (Cowan, 2006, p. 4).

Motivated by prior First Nations activism that harnessed global media visibility of First Nations perspectives and considerations, this chapter explores the representations of Aboriginal perspectives that include realising sovereign rights through Custodianship of Country in the SS4C protest placards to inform on the extent of (in)visibility of Australian First Nations' cultural and political considerations. The increased public awareness of the climate crisis in Australia since 2018 (Thackeray et al., 2020) and the implicit nexus that exists between First Nations' care of Country and whole-population activism that seeks climate wellbeing evokes common ground between First Nations peoples and non-Indigenous allies. An assessment of visual narratives can illuminate whether this common ground is understood by young people, as raising awareness of shared interest is a powerful strategy through which to enact meaningful social change around First Nations' cultural care of Country.

In doing this, we consider the lack of formal visibility of both youth voices and that of First Nations Australians and contemplate how climate activism can be a meeting point between the shared investment of First Nations people and young people in Australia. We see this represented in the ways students have encompassed intersectional approaches to climate justice:

> Despite their marginal status in Australian institutional politics, university students and their young peers have led the establishment of national environment initiatives, networks and organisations such as Students of Sustainability in 1991, the Australian Student Environment Network in 1997 (www.asen.org.au), the Australian Youth Climate Coalition in 2006 (www.aycc.org.au) (AYCC) and the Indigenous youth climate network Seed Mob, in 2014 (www.seedmob.org.au) (Partridge, 2008; Collin, 2015) In Australia, the AYCC has 120,000 members and a further 100,000 supporters (AYCC, 2019). Yet, little scholarly attention has been given to the nature and role of such organisations for climate activism, including the SS4C.
>
> *(Hilder and Collin, 2022, p. 794)*

Advancing the work of the aforementioned scholars, we observe that even less attention has been given to the representation of Indigenous perspectives within the movement. This chapter renders this visible by examining the intersection between youth voices for climate action and Australian Indigenous perspectives. Figure 6.2 demonstrates this coalescence between the two perspectives. The leading message in this placard is that the land which the protest is held on (and all land in Australia) 'always was and always will be' Aboriginal land. Whilst this message leads front and centre in the placard, an alliance between the School Strike 4 Climate movement (SS4C) and First Nations Australians is identified. This demonstrates an understanding of the inextricable relationship between what this student understands and believes is critical within the context of the climate strikes and the role that colonisation and dispossession of stolen land plays in climate degradation.

School Strike For Climate, Sydney: 24 September, 2022

Figure 6.2   'Always was and always will be' Aboriginal land – 24 September 2022.

Source: Authors' own image, taken as part of the ARC Discovery Project.

What is particularly insightful in this placard is that, whilst there is an apparent absence of written and visual narrative specific to 'saving the planet', there is an implicit broader overarching reliance on Indigenous principles of Custodianship of Country for planet care. The clarity of messaging in this sign is formidable, embodying a climate justice ethos and demonstrating the potentiality of a powerful union between First Nations people and the SS4C movement. In continuing this idea, we explore how visual expressions of Australian Aboriginal perspectives facilitate deeper contemplation of the ways visual strategies of cultural empowerment and pride can be realised through self-representation and how visible signs of solidarity and allyship can be formative in building relationships premised on reconciliation between First Nations and non-Indigenous allies.

Methodological Approaches

Our methodology contemplates how visual and/or written inferences on climate activist placards involve First Nations considerations as strategic approaches towards a shared care of Country through which to redress climate harm. Essential to our approach is identifying common ground considerations (Wilson, 2008) that unite diverse communities of non-Indigenous youth climate activists in spaces evoked through visual and written climate narratives and action. Harnessing this grassroots approach evokes opportunities to identify, construe and understand why First Nations Custodianship of Country should be embedded in climate politics. Beneficially, this can develop opportunities for ongoing yarning focused on First Nations informed political activism (Carlson and Frazer, 2018), facilitate First Nations leadership in natural resource management (Muller et al., 2019) and promote dialogues focused on perpetuity of a felt, shared, healthy, natural environment (Latulippe and Klenk, 2020) that benefits all humans.

Decolonising Climate Activism

Aligning with Richards' 'emergent method' (2014) where the analysis begins with no preconceived categorisations and instead allowing the findings to emerge, our approach privileges Indigenous epistemologies that impose a relational worldview approach through which to consider written and visual data captured at School Strike for Climate protests on Gadigal Country (Sydney) across 2019, 2022 and 2023. Whilst this data set has been used in conjunction with a number of other findings surrounding youth climate action in Australia (Bessant et al., 2023; Catanzaro and Collin, 2023; Hilder and Collin, 2022), in this case a Gadigal First Nations Scholar, Madison Shakespeare, explored the images, free of any previous thematic coding. Shakespeare's role as a Gadigal Traditional Owner is implicit to this analysis as all the protests from which data is considered were held on her nation Country. As a culturally ethical approach, this methodology connects place and people of the protest and informs on intersections that may arise between climate activist narratives and specific First Nations knowledges and practices involved in Custodianship of Country.

> For the Indigenous world, Western conceptions of space, of arrangements and display, of the relationship between people and the landscape, of culture as an object of study, have meant that not only has the Indigenous world been represented in particular ways back to the West, but the Indigenous world view, the land and the people, have been radically transformed in the spatial image of the West. In other words, Indigenous space has been colonized.
>
> *(Smith, 2021, p. 51)*

Imposing an Australian First Nations' lens through which to construe and interpret how visual and written representations may involve Indigenous ontologies and axiologies privileges Indigenous ways of knowing through which to engage in research practice (Rigney, 1999). This decolonising approach to research is a strengths-based way to perceive non-Indigenous allies' understanding of First Nations perception of Custodianship of Country, an element essential to meaningful care and protection of Country and climate sustainability. As a methodological approach, this develops opportunities for research practice that are culturally relevant, appropriate and safe.

Data Set, Data Collection and Analysis

The analysis discussed in this chapter is drawn from photographs of 850 placards taken at strikes on Gadigal land (Sydney, New South Wales) across 2019–2023. In this chapter, we ground this analysis in the decolonised methodology that positions and privileges the First Nations perspectives prevalent in the visual activism of the protests. This analysis and approach bring visibility to the voices and intentions of First Nations perspectives present at the strikes and highlight the importance of enacting Indigenous methodologies to understand and conceptualise First Nations cultural activism and care for Country.

The photographs were assessed according to key First Nations cultural thematic considerations; the selection criteria included:

1. Australian Indigenous notions of Country including visual and or linguistic representations of Indigenous worldviews

2. representations of Indigenous knowledges and cultural practices
3. basic acknowledgement of Australian First Nations and or Custodianship of Country

Following this, the themes were explored by the authors who drew on multidisciplinary backgrounds that include visual communication design, global Indigenous health, geography, law, creative arts, film making, humanities and education. Implementing this Indigenous paradigm facilitated transdisciplinary approaches where the focus of inquiry is united through worldview approaches, whilst still meaningfully engaging the skills and specialist knowledge of each researcher. The results of this analysis are discussed in the following sections.

Indigenous Perspectives and Grassroots Approaches

Eurocentric notions of ownership and property rights to land starkly contrast First Nations' custodianship perspectives. Care of Country has evolved and developed through deep connection and relational oneness with Country over tens of thousands of years, and these knowledges and practices realise communal (Yunkaporta and McGinty, 2009), relational (Wilson, 2003) and environmental worldviews (Hewitt, 2000) that are founded on whole-of-life spiritual considerations (Gee et al., 2014). Land is cared for, rather than apportioned and owned for the benefit of individuals realising fiscal wealth. When considering Australian First Nations peoples' connection with the natural environment, Dwyer (2012) understands this to be a reciprocal relationship, and whilst care of Country (custodianship) realises reciprocity for both, Australian Indigenous ontologies inherently construe custodianship as a cultural responsibility that is directed by spiritual mandate.

Relationality between Australian First Nations and Country prioritises the health and wellbeing of Country and is more expansive than an interrelationship between humans and nature. Fundamental to these ontologies is 'Country', perceived as more than just a term that represents aspects of the natural environment and is encoded with cultural connotations and spiritual Lore responsibilities and obligations:

> My Ancestors . . . accumulated networks of meaningful, deep, fluid, intimate collective and individual relationships of trust. In times of hardship, we did not rely to any great degree on accumulated capital or individualism but on the strength of relationships with others.
>
> *(Simpson, 2017, p. 77)*

This relational connectedness that involves relationships with others is a powerful potential source of insight and connection for grassroots climate activism.

Articulating climate care around narratives that are construed through Indigenous worldview approaches can create beneficial yarning spaces of climate activism (Brooker et al., 2023). Harnessing First Nations-led yarning about climate care evokes outcomes of reciprocity for non-Indigenous allies by providing deeper insight into the value of Indigenous ways of knowing, doing and being surrounding custodianship of our shared natural environment. In the spirit of reconciliation, it can also create opportunities for truth telling (Council, 2017) about the distressing detriment still experienced by Australian First Nations peoples through dispossession from Country, kin and community as a result of invasion in 1788 and colonisation thereafter.

Recent publications have clearly identified the undeniable nexus between climate and human psychological distress (Latkin et al., 2022). Whilst human emotional responses evoked through climate distress include anger, fear and guilt, there is still an underlying hope that change can be enacted, and these responses are key motivators in climate activism (Kleres and Wettergren, 2017).

Visual Themes/Discussions

After analysing the placards from the Gadigal-based strikes from 2019–2023, we found that initially the visual data itself seemed to be missing obvious references to Indigenous perspectives and there appeared to be a paucity of First Nations viewpoints being expressed. This lack of visual representation was surprising as the School Strike 4 Climate protests actively prioritise First Nations voices, with First Nations speakers welcoming activists to Country, alongside yarning about specific place-based and Country-based violence, values and insights. The authors can hypothesise a number of different reasons as to why more Indigenous perspectives are lacking in the visual activism at the strikes.

First, the Indigenous population in New South Wales is 3.4% (Australian Bureau of Statistics, 2021), with the majority of Aboriginal and Torres Strait Islander people in NSW living outside of city centres. The Gadigal-based strike is held in Sydney, in an urban city centre which is not an accessible site for all to visit, and hence the strikes themselves are representative of a certain type of activist. Aware of this, as a means to counter our potential bias, the data set was cross compared to include 114 photographs in the SS4C gallery that can be found on Flickr (https://www.flickr.com/photos/ss4c/albums/72157710891050723/page2). These images represent strikes at various sites across Australia in 2019 (our largest comparable data set) and this gallery includes more documented Indigenous representations. We discuss why this may be the case in the 'Place-based Specificity' section to follow.

Second, despite a smaller percentage of direct Indigenous visual representation in the signage, our findings were able to connect references to climate justice, decolonisation, protesting inaction, nature, care, concern and earth-centred expressions to Indigenous worldviews. We explore this further in the following analysis. We argue that First Nations-led custodianship of the shared natural environment offers positive long-term solutions to climate change within Australia and globally. This idea was prevalent in a small number of placards, as seen in Figure 6.3.

In Figure 6.3, we see three students grouped together; one sign is asking for 100% eco-energy and another states she 'shouldn't have to miss school for this'. In the middle, the student sign states, 'Resource First Nations-led Solutions'. Broadly, all of these signs can be categorised or understood under the theme of education, with the left and middle placard offering solutions-driven responses to the climate crisis and the student on the right begrudging the fact that she is missing school as a result of the uneducated leaders of the Country who are refusing to listen to the overwhelming evidence of the climate crisis, in this case the young people at the strikes.

Diverse Indigenous-led practices offer solutions that address climate concerns (see cultural burning in Williamson, 2022, and water management in Goff, 2020). In the middle placard from Figure 6.3, the First Nations Solution sign visually incorporates a symbol of the earth inside the 'O' of the word 'solutions'. Conceptually, this is a profound visual symbol that communicates relationality between local and global contexts, the need to

Figure 6.3   First Nations solutions acknowledged – March 2023.

Source: Authors' own image, taken as part of the ARC Discovery Project.

acknowledge and resource Indigenous-led approaches and the global implications of Indigenous knowledges lying within solutions that are led by First Nations peoples.

Whilst, in this example, the reference to First Nations perspectives and knowledge is very clear, in other placards, the references may be more subtle. In the following sections we explore how climate activism can evoke strong covalence between Indigenous Custodianship of Country and non-Indigenous youth allyship through shared intentions focused on the common goal of protecting the planet (Brooker et al., 2023). This potential relationship acknowledges the importance of Indigenous knowledges, place-based rights and sovereignties and how they can be respected and embedded in grassroots climate activism for successful and specific action.

Custodianship of Country

Understanding Country and climate through First Nations perspectives informs on sovereign knowledges and practices that successfully ensure a healthy environment for all to share. An Indigenous relational worldview mandates Custodianship of Country as an essential spiritual responsibility evoked through relational connectedness with the natural environment (Redvers et al., 2022). Australian First Nations knowledge systems are kincentric and involve 'Raw Law' (Watson, 2016), where 'the law is filled with the spirit of creation' (Watson, 1998, p. 1). Significantly, Dudgeon highlights that this law 'comes from the land, is the land'. Whilst Australian First Nations worldviews are diverse (Wilson, 2009), in Aboriginal contexts, Custodianship of Country is implicit to realising cultural and spiritual law (lore) obligations and responsibilities (Redvers et al., 2022).

Custodianship of Country not only ensures the need for protection and care of Country but advocates for immediate positive change and climate justice (Bawaka Country et al., 2022). This results in positive action towards larger reconciliation goals to manifest genuine change for all Australians. It can be conceptualised that this connection to and privilege of Country and

Figure 6.4   'Respect and Protect Sacred Land' September 2019.
Source: Authors' own image, taken as part of the ARC Discovery Project.

community embodies the essence of grassroots climate justice. Whilst the term 'climate justice' was not introduced until 1999 (see Bruno et al., 1999), the Aboriginal Tent embassy occupation in 1972 was an act for climate justice. Climate justice values Indigenous and intersectional voices, Indigenous rights, sovereignty and sustainable practices (Climate Justice Now, 2007, 2008).

> Over the past decade, climate justice has come to encompass several distinct but complementary currents from various parts of the world. In the Global South, demands for climate justice unite an impressive diversity of Indigenous and other land-based people's movements.
>
> *(Tokar, 2018, p. 17)*

This united need to respect and protect our land emerges in the placard signage at the SS4C protests as seen in Figure 6.4. This placard on the left uses text and symbols to connect notions of respect, protection and sacred land with symbols of nature (flowers) that swirl around the lettering, referencing and furthering an understanding of connection and reciprocity between humans and non-humans that is in alignment with Indigenous worldviews. The placard on the right is more explicit and has filled the shape of Australia with the Indigenous flag. In doing this, the sign holder is drawing a nexus between climate and First Nations culture and identity.

Thematic Reflections

Assessing the images revealed visual and or written depictions of climate that have been expressed in the following sections:

1.   *Global Climate Health*

Within the placards, the most frequently recurring symbol is **Planet Earth**. This signifies the fact that young people acknowledge that climate change is a global issue that affects us all. It was typical to find visual and written representations that evidenced climate activist perceptions and understanding that their specific experiences in Australia are implicitly linked to global climate health.

We conceptualise that the multiple representations of the earth can also serve as a point of communication exchange between non-Indigenous climate activists and First Nations Custodians of Country through which to engage in deeper understanding about care and protection of shared Country. This correlation evokes common ground understanding through which to inform and understand Indigenous connectedness to Country and Custodianship of Country.

In Figure 6.5, we see an example placard that ties youth concerns for their future with the planet. We see this in recurring images in the strikes such as 'there is no PLANet B', 'Bail out the Planet' and 'Earth used to be cool'. These climate concerns are focussed on the global arena (Earth) but do not strictly offer a grassroots representation in the sense that it is not always connected to action or place and not locally tangible. What these placards demonstrate is that the issue is big and the consequences are even bigger. There are so many multifaceted issues at stake in the climate crisis that, whilst this generalised 'global' approach may create a sense of global community and awareness, it does little to offer solutions or make climate issues tangible or easy to tackle.

Figure 6.5 Global Climate Health perspectives, September 2019.

Source: Authors' own image, taken as part of the ARC Discovery Project.

2. *'Mother Earth'*

Multiple representations of Earth assign it a gender as a woman, a mother and female. In this way, the visual representation aligns with notions of female leadership, wisdom and care. This is a commonality found in Australian Aboriginal culture where women practise collective leadership that has roots in Indigenous approaches like the 'upside-down hierarchy' (see Sveiby, 2011).

Climate activist perspectives often embrace responsibility and relationships (as highlighted in placards presenting 'Mother Earth'); the identity and explicit role of First Nations people as part of Country itself is a key aspect of Aboriginal ontologies (Wildcat, 2013) and a worldview that conceives of 'more-than-human' relationships (Bawaka Country et al., 2022). This 'ethics of care' and deference to county and otherhood is reflected in the signage in multiple ways:

a. Calls to Save Mother Earth

Figure 6.6   Protecting Global 'Mother Earth' perspectives, September 2019.

Source: Authors' own image, taken as part of the ARC Discovery Project.

b. Appeals to Respect Mother Earth

Figure 6.7   Respect Mother Earth, September 2022.

Source: Authors' own image, taken as part of the ARC Discovery Project.

c. Calls for Better Treatment of Mother Earth

Figure 6.8   Treatment of Mother Earth perspectives, September 2019.

Source: Authors' own image, taken as part of the ARC Discovery Project.

An Indigenous relational worldview conceptualises Country as kin, and accordingly care of Country and climate is implicit to the physical, mental and cultural health of the people involved in looking after it (Garnett et al., 2009) and critical to the spiritual development and holistic well-being of Aboriginal children (Salmon et al., 2018). Country is seen 'as a mother and a source of identity and spirituality, with identity, cultural practices, systems of authority and social control, traditions and concepts of spirituality all being tied to the land' (Ganesharajah, 2009, p. iv).

Globally, Indigenous peoples realise holistic wellbeing through their relationality with their natural environment; it is an 'enduring relationship between populations, their territories, and the natural environment' (Shakespeare, 2021, p. 2).

Fundamental to First Nations Australians is the knowledge gained through experience that if you are custodian and look after Country, then it will look after you (Griffiths and Kinnane, 2010). This adage reflects more than a common-sense reciprocity realised when humans dwell in clean natural environments; it reflects a relational connectedness with Country through which First Nations peoples attain holistic physical, social, emotional, individual, familial and communal wellbeing (Morrissey, 2015). Ngarrindjeri Elder Tom Trevorrow from Camp Coorong spoke of the effects that changes to Country have on health:

> The land and waters is a living body. We the Ngarrindjeri people are a part of its existence. The land and waters must be healthy for the Ngarrindjeri people to be healthy. We are hurting for our Country. The Land is dying, the River is dying, the Kurangk (Coorong) is dying and the Murray Mouth is closing. What does the future hold for us?
>
> *(Walker, 2002, p. iii)*

Care for the earth is a shared plight among all activists, regardless of age. The climate debate is intergenerational, and whilst there are increasingly more and more adults present at the strikes, it is the young people that are instigating, organising and leading the charge (Bessant et al., 2023). Consequently, the intergenerational nature of the strikes is apparent across the people at the strikes but also in their visual messaging.

Often adults are seen carrying signs that use humour, empathy and other visual communication approaches to direct audiences to listen to and consider children/kids. This repositions young people as not only the people who will inhabit the future earth but also as the authorities (Figure 6.9).

Other signs state 'we made this mess, now it is time to clean it up' alongside handwritten promises to children, grandchildren and future generations that we can fix things if we enact change. Climate-change arguments focused on the perpetuity of future generations are not exclusive to Australia. In Switzerland, climate activists have designed arguments around a 'grandchildren frame' that strategically promotes messages of altruism that appeal to the population to realise their moral responsibilities to ensure perpetuity of future generations through a healthy planet (Keller and Bornemann, 2021).

3. *Place-based Specificity*

Exceptions to the broader calls to climate action presented in the posters include activists who are clearly protesting against geographically specific environmental harm that ensues from place-based extractive enterprise. Examples of this can be seen in subgroups of climate activists who clearly articulate their protests against coal seam gas, mining and fracking (Figure 6.10). At times, the representation of mother and mining the earth come together as seen in Figure 6.11.

Figure 6.9 'Climate Justice for All – Listen to the Kids', September 2019.
Source: Authors' own image, taken as part of the ARC Discovery Project.

Figure 6.10 Place-based perspectives – 'What I stand on is what I stand for'. September 2019.
Source: Authors' own image, taken as part of the ARC Discovery Project.

Figure 6.11 Fracking Mother Nature, September 2019.

Source: Authors' own image, taken as part of the ARC Discovery Project.

These placards represent the exercise of democratic rights in contemporary Australian society through opinions voiced against contemporaneous commercial enterprises and an explicit protest against the loss of sovereign and human rights (Assembly, 2007).

Place-based and issue-based visuals evident in the protests referenced key political aspects of the climate crisis with reference to place but also connected these to the health and strength of Indigenous peoples.

References to specific countries and Aboriginal Nations – Pilliga, the Burrup Peninsula and the Gomeroi peoples, for example, highlight specific climate-related issues and pressure points of the larger movement that are place-based. The inclusion of the phrase seen in Figure 6.12 of 'Gomeroi Strong' also points to larger conceptualisations of Country, Aboriginal nationhood and the broader health and wellbeing of the Aboriginal community. These references are strongly connected to Indigenous knowledges, rights and sovereignties – evoking powerful issues and the nexus between Indigenous perspectives and grassroots activism. At times, the links to grassroots activism are broader sweeping and refer to the broader concepts underpinning activism such as resisting colonialism and neoliberalism, as seen in Figure 6.13.

Figure 6.12 Specific environmental issues and place-based perspectives. June 2023.
Source: Authors' own image, taken as part of the ARC Discovery Project.

Figure 6.13 Resist colonialism, September 2019.
Source: Authors' own image, taken as part of the ARC Discovery Project.

As a part of this analysis process, we also examined the official SS4C gallery (2019) that housed images of student strikers from all over Australia. While there is no room to discuss this in detail here, it is of interest to note that, within this curated gallery, there was an increased representation of Aboriginal people on the front lines of the climate protests. This was specifically prevalent in the regional areas of Australia and this phenomenon warrants further research and investigation.

4. *Collective Themes and Broader Messaging*

a. Time

There were connective points that emerged between the thematic analysis conducted by Catanzaro and Collin (2023) and the findings in this study. One similarity was the recurrent representation of time. Many signs use capital letters and exclamation marks to reinforce the urgency of the climate crisis. Symbols associated with death and destruction reinforce a sense of alarm and panic and demand an immediate call to action. In Figure 6.14, we see an image that connects three Australian birds – often impacted by the climate crisis, seemingly lifeless, alongside an empty hourglass timer and a roaring red background to emphasise the interrelationship between increasing temperatures, natural species and time running out. In referencing time and visually depicting and positioning nature in danger, these placards communicate a shared sense of urgency seen across many of the visuals.

Figure 6.14   Time and urgency perspectives, September 2019.

Source: Authors' own image, taken as part of the ARC Discovery Project.

b. Australian Flora and Fauna

A number of posters feature depictions of native animals specific to Australia (Figures 6.14 and 6.15). Whilst carried by a number of different protesters and often combined with political critique and humour, Australian youth participants in the strikes demonstrate pride in the animals that are special to them, anthropomorphising them and differentiating their national identity through these depictions. Importantly, these placards also evidence a shared sense of protection, particularly for those beings that cannot protect themselves as essential participants in – and integral parts of – the larger sense of Country.

c. Collectivism and Community

Youth activists are often committed voices in centring Indigenous perspectives for climate justice and calling for urgent climate action, framing the importance of climate justice and the fact that this does not exist without justice for First Nations people. This sentiment is expressed on the Australian Youth Climate Coalitions website, which states:

> We see the climate crisis as an issue of social and environmental injustice. Climate change affects everyone, but not equally. . . . This is particularly true for Aboriginal and Torres Strait Islander people, who are already experiencing the impacts of climate change and fossil fuel extraction on community, culture and country. It is also true for Indigenous peoples the world over, people in the global south, people of colour, people with a disability, poor communities, workers, fossil fuel communities, young people and future generations.[2]

Figure 6.15   Australian Flora and Fauna references, September 2019.

Source: Authors' own image, taken as part of the ARC Discovery Project.

Figure 6.16   Community references, 2019 and 2023.

Source: Authors' own image, taken as part of the ARC Discovery Project.

Within the signage there is a collective call to arms evidenced in representations of 'we' and 'our' that reflect the fact that one person cannot do this alone. This is done by individuals in references to the collective power of the people and the use of the royal 'we', such as in Figure 6.16 where we see 'With Earth, we rise' amongst other signs such as 'The fires are raging and so are we' and 'like the sea, we rise'. This solidarity can also be found in more formal banners that bring together and unite organisations fighting for a common cause such as the banner on the right in Figure 6.16 that validates each organisation and groups in a linear way that does not exert power but collectivity. This banner led the march and was held by children alongside young and older adults. Calling for solidarity across communities brings together Indigenous and non-Indigenous allies and evokes a mutual sensibility and an elevated sense of collective responsibility.

Conclusion

In examining visual expressions present in youth climate protest, this research highlights the role of visual communication in activism and public discourse and the potential benefits of the overlapping dynamics between youth perspectives in Aboriginal activism and climate-change protest.

Adopting a decolonising First Nations approach creates culturally relevant and meaningful ways through which to consider Aboriginal representation in the climate movement. This methodology not only brings visibility to the voices and intentions of First Nations people present at the strikes but also spotlights the importance of enacting Indigenous methodologies to understand and conceptualise First Nations cultural activism and custodianship of care for Country. In turn, this facilitates deeper understanding about the beneficial nexus between the Australian First Nations Custodians of Country and non-Indigenous allies and how this can be harnessed as a powerful element of allyship through which to redress climate distress and promote meaningful care of Country. Insight can also be gained into communal care of the environment for all communities, life forms and environmental sustainability for the future.

If approaches to climate activism, education, grassroots action and planning were premised on acknowledgement of Country, such as land, water and air as entities, this would be a conceptual shift that moved closer to a productive understanding and incorporation of Indigenous perspectives and worldviews. As premised earlier in the chapter, our observations lead us to believe that the grassroots climate movement is beginning to acknowledge and incorporate this into their messaging more and more as the strikes and the knowledge surrounding them is evolving, as exemplified in Figure 6.17.

Figure 6.17   Climate Justice, 14 September 2023.
Source: Authors' own image, taken as part of the ARC Discovery Project.

Recognising this mutuality and relationship might also shift the ethical basis and motivation of development, policy and planning in ways that respect the needs and demands of overlapping sectors of Indigenous youth and between youth perspectives in Aboriginal activism and climate-change protest. In considering the implications of this research, we conclude that deeper learning and acknowledgement of Indigenous place-based rights, sovereignties and knowledges needs to be embedded in grassroots activism for successful and specific action. This can support a rich and renewed understanding of individual and collective power and responsibility in terms of autonomy, sovereignty and effective activism by engaging a Country-centred approach that transforms care and knowledge into intention and action.

Acknowledgements

We would like to thank and acknowledge all the First Nations people, young people and allies who have taken part in the Gadigal-based climate strikes and thank them for their contribution to climate action and our research more broadly. We would like to acknowledge the efforts of students and co-researchers in the collection of strike data.

The data analysed has been collected as a part of the below ARC project. Catanzaro's contribution is funded by the Australian Research Council Discovery Program Grants (DP230101704). New Possibilities: Young People and Democratic Renewal. Project team: Philippa Collin, Western Sydney University (project lead); Judith Bessant and Rob Watts, RMIT; Stewart Jackson, USyd and Faith Gordon, ANU. Post doc: Luigi Di Martino, Student Researcher: Dinusha Soo. We acknowledge their contribution to the intellectual work of the project which has informed this chapter.

Notes

1 The notion of "Country" with a capital "C" denotes "land to which Indigenous people have a unique spiritual and cultural connection" (MacLean et al., 2017, p. 309).
2 https://www.aycc.org.au/climate_justice.

References

Assembly, U. G. (2007). United Nations declaration on the rights of indigenous peoples. *UN Wash*, 12, 1–18.

Australian Bureau of Statistics (2021). https://www.abs.gov.au/statistics/people/aboriginal-and-torres-strait-islander-peoples/census-population-and-housing-counts-aboriginal-and-torres-strait-islander-australians/latest-release

AYCC (Australian Youth Climate Coalition) (2019). *Annual Report*. Melbourne: AYCC. https://acncpubfilesprodstorage.blob.core.windows.net/public/68b770e1-39af-e811-a95e-000d3ad24c60-aa000e0b-8ae7-4150-bc92-6e20c010e50f-Financial%20Report-334ca9c4-6dba-ea11-a812-000d3ad1cc03-AYCC_2019_Financial_Statements.pdf

Bawaka Country including Burarrwanga, L., Ganambarr, R., Ganambarr-Stubbs, M., Ganambarr, B., Maymuru, D., . . . and Daley, L. (2022). Gapu, water, creates knowledge and is a life force to be respected. *PLOS Water*, 1(4), e0000020.

Bessant, J., Collin, P. and Watts, R. (2023). 'Blah, Blah, Blah . . . [not] business as usual': Politics through the lens of young female climate leaders. *Australian Journal of Political Science*, 1–17.

Bowman, B. (2019). Imagining future worlds alongside young climate activists: A new framework for research. *Fennia*, 197(2), 295–305.

Brooker, C. J., Catanzaro, M. and Shakespeare, M. (2023). A creative workshop model for understanding climate activism as custodianship of country. In *Indigenizing Education for Climate Action: Strategies, Case Studies and Testimonios*. Space4Innovation.

Bruno, K., Karliner, J. and Brotsky, C. (1999). *Greenhouse Gangsters vs. Climate Justice*. San Francisco: CorpWatch.

Carlson, B. and Frazer, R. (2018). Yarning circles and social media activism. *Media International Australia*, 169(1), 43–53.

Catanzaro, M. (2024). Understanding young people's visual politics: Making young people's participation (more) visible. In J. Bessant, P. Collins and P. O'Keefe (eds.) *Research Handbook for the Sociology of Youth*. Edward Elgar Publishers.

Catanzaro, M. and Collin, P. (2023). Kids communicating climate change: Learning from the visual language of the SchoolStrike4Climate protests. *Educational Review*, 75(1), 9–32.

Climate Justice Now (2007). *What's Missing from the Climate Talks? Justice!* Press Release. Via Durban Group for Climate Justice email list 14 December.

Climate Justice Now (2008). *Principles of Unity*. Via Climate Justice Now email list 12 May .

Collin, P. (2015). *Young Citizens and Political Participation in a Digital Society. Addressing the Democratic Disconnect*. Basingstoke: Palgrave Macmillan.

Council, R. (2017). *Uluru Statement from the Heart*. National First Nations Constitutional Convention. https://ulurustatement.org/the-statement/view-the-statement/ [Accessed 29 August 2024].

Cowan, G. (2006). *Nomadic Resistance: Tent Embassies and Collapsible Architecture: Illegal Architecture and Protest*. Masters Dissertation, The University of Adelaide. Kooriweb.org. https://www.kooriweb.org/foley/resources/history/newstuff2016/new%20folder/Nomadic%20Resistance.pdf

Dwyer, A. (2012). Pukarrikarta-jangka muwarr–Stories about caring for Karajarri country. *The Australian Community Psychologist*, 24(1).

Ganesharajah, C. (2009). *Indigenous Health and Wellbeing: The Importance of Country*. Native Title Research Report No. 1 (2009), Native Title Research Unit, Australian Institute of Aboriginal and Torres Strait Islander Studies (AIATSIS).

Garnett, S., Sithole, B., Whitehead, P., Burgess, C., Johnston, F. and Lea, T. (2009). Healthy country, healthy people: Policy implications of links between Indigenous human health and environmental condition in tropical Australia. *Australian Journal of Public Administration*, 68(1), 53–66.

Gee, G., Dudgeon, P., Schultz, C., Hart, A. and Kelly, K. (2014). Aboriginal and Torres Strait Islander social and emotional wellbeing. In P. Dudgeon, H. Milroy and R. Walker (eds.) *Working Together: Aboriginal and Torres Strait Islander Mental Health and Wellbeing Principles and Practice* (2nd ed., pp. 55–68). Canberra: Department of the Prime Minister and Cabinet.

Goff, S. (2020). Visionary evaluation: Approaching Aboriginal ontological equity in water management evaluation. *Evaluation and Program Planning*, 79, 101776.

Griffiths, S. and Kinnane, S. (2010). *Kimberley Aboriginal Caring for Country Plan: Healthy Country, Healthy People: Right People, Right Country, Right Way*. Kimberley Language Resource Centre.

Henriques-Gomes, L., Davidson, H., Cox, L., Smee, B. and Zhou, N. (2019). Hundreds of thousands attend school climate strike rallies across Australia. *The Guardian*, 20 September. https://www.theguardian.com/environment/2019/sep/20/hundreds-of-thousands-attend-school-climate-strike-rallies-across-australia

Hewitt, D. (2000). A clash of worldviews: Experiences from teaching Aboriginal students. *Theory into Practice*, 39(2), 111–117.

Hilder, C. and Collin, P. (2022). The role of youth-led activist organisations for contemporary climate activism: The case of the Australian Youth Climate Coalition. *Journal of Youth Studies*, 25(6), 793–811.

Keller, S. and Bornemann, B. (2021). New climate activism between politics and law: Analyzing the strategy of the KlimaSeniorinnen Schweiz. *Politics and Governance*, 9(2), 124–134.

Kleres, J. and Wettergren, Å. (2017). Fear, hope, anger, and guilt in climate activism. *Social Movement Studies*, 16(5), 507–519.

Latkin, C., Dayton, L., Scherkoske, M., Countess, K. and Thrul, J. (2022). What predicts climate change activism? An examination of how depressive symptoms, climate change distress, and social norms are associated with climate change activism. *The Journal of Climate Change and Health*, 8, 100146.

Latulippe, N. and Klenk, N. (2020). Making room and moving over: Knowledge co-production, Indigenous knowledge sovereignty and the politics of global environmental change decision-making. *Current Opinion in Environmental Sustainability*, 42, 7–14.

Marshall, V. (2019). Removing the veil from the 'Rights of Nature': The dichotomy between First Nations customary rights and environmental legal personhood. *Australian Feminist Law Journal*, 45(2), 233–248.

Mencevska, I. (2019). Truth telling in Australia's historical narrative. *NEW: Emerging Scholars in Australian Indigenous Studies*, 5(1).

Morrissey, P. (2015). Bill Neidjie's story about feeling: Notes on its themes and philosophy. *Journal of the Association for the Study of Australian Literature*, 15(2). https://openjournals.library.sydney.edu.au/index.php/JASAL/article/view/9943

Muller, S., Hemming, S. and Rigney, D. (2019). Indigenous sovereignties: Relational ontologies and environmental management. *Geographical Research*, 57(4), 399–410.

Partridge, E. (2008). From ambivalence to activism young people's environmental views and actions. *Youth Studies Australia*, 27(2), 18–25. https://research.fit.edu/media/site-specific/researchfit-edu/coast-climate-adaptation-library/climate-communications/youth-climate-ampsocial-media/Partridge.-2008.-Youth-Environmental-Views–Actions.pdf

Redvers, N., Celidwen, Y., Schultz, C., Horn, O., Githaiga, C., Vera, M., . . . and Rojas, J. N. (2022). The determinants of planetary health: An Indigenous consensus perspective. *The Lancet Planetary Health*, 6(2), e156–e163.

Richards, C. (2014). Critical thinking for an emergent, relevant, and productive 'knowledge-building' approach to research inquiry and academic writing. *Argumentation, Rhetoric, Debate and Pedagogy*, 79–93.

Rigney, L. I. (1999). Internationalization of an Indigenous anticolonial cultural critique of research methodologies: A guide to Indigenist research methodology and its principles. *Wicazo sa Review*, 14(2), 109–121.

Robinson, S. (1994). The Aboriginal Embassy: An account of the protests of 1972. *Aboriginal History*, 49–63.

Salmon, M., Doery, K., Dance, P., Chapman, J., Gilbert, R., Williams, R. and Lovett, R. (2018). *Defining the Indefinable: Descriptors of Aboriginal and Torres Strait Islander Peoples' Cultures and Their Links to Health and Wellbeing*. Canberra: Aboriginal and Torres Strait Islander Health Team, Research School of Population Health, The Australian National University. https://openresearch-repository.anu.edu.au/server/api/core/bitstreams/f3e9d39d-19fa-4f13-8e19-7d4e1dd48295/content

Shakespeare, M., Fisher, M., Mackean, T. and Wilson, R. (2021). Theories of Indigenous and non-Indigenous wellbeing in Australian health policies. *Health Promotion International*, 36, 669–679. doi: 10.1093/heapro/daaa097. PMID: 32968777

Simpson, L. B. (2017). *As We Have Always Done: Indigenous Freedom Through Radical Resistance*. University of Minnesota Press.

Smith, L. T. (2021). *Decolonizing Methodologies: Research and Indigenous Peoples*. Bloomsbury Publishing.

Smyth, D. (1995). Caring for sea country – accommodating Indigenous peoples' interests in marine protected areas. In *Marine Protected Areas: Principles and Techniques for Management* (pp. 149–173). Dordrecht: Springer Netherlands.

Sveiby, K. E. (2011). Collective leadership with power symmetry: Lessons from Aboriginal prehistory. *Leadership*, 7(4), 385–414.

Thackeray, S. J., Robinson, S. A., Smith, P., Bruno, R., Kirschbaum, M., Bernacchi, C., Byrne, M., Cheung, W., Cotrufo, M., Francesca, Gienapp, P., Harley, S., Janssens, I., Jones, T. H., Kobayashi1, K., Luo, Y., Penuelas, J., Sage, R., Suggett, D. J., Way, D. and Long, S. (2020). Civil disobedience movements such as School Strike for the Climate are raising public awareness of the climate change emergency. *Global Change Biology, Online First* 1–3. https://ro.uow.edu.au/smhpapers1/1130

Tokar, B. (2018). On the evolution and continuing development of the climate justice movement. In *Routledge Handbook of Climate Justice* (pp. 13–25). https://doi.org/10.4324/9781315537689

Walker, D. (2002). What is possible: Hydrology and morphology. In N. Goodwin and S. Bennett (eds.) *The Murray Mouth, Exploring the Implications of Closure or Restricted Flow, Murray Darling Basin Commission and the Department of Water*. Adelaide, South Australia: Land and Biodiversity Conservation, July.

Wall, J. (2021). *Ethics in Light of Childhood*. Washington, DC: Georgetown University Press.

Watson, I. (1998). Naked peoples: Rules and regulations. *Law Text Culture*, 4(1).

Watson, I. (2009). Sovereign spaces, caring for country, and the homeless position of Aboriginal peoples. *South Atlantic Quarterly*, 108(1), 27–51.

Watson, I. (2016). *Aboriginal Peoples, Colonialism and International Law: Raw Law*. Routledge.

Watson, I. (2018). Aboriginal recognition: Treaties and colonial constitutions we have been here forever. *Bond Law Review*, 30, 7.

Wildcat, D. R. (2013). Introduction: Climate change and Indigenous peoples of the USA. *Climatic Change*, 120(3), 509– 515.

Williamson, B. (2022). Cultural burning and public forests: Convergences and divergences between Aboriginal groups and forest management in south-eastern Australia. *Australian Forestry*, 85(1), 1–5.

Wilson, S. (2003). Progressing toward an Indigenous research paradigm in Canada and Australia. *Canadian Journal of Native Education*, 27(2), 161.

Wilson, S. (2008). *Research Is Ceremony. Indigenous Research Methods*. Winnipeg: Fernwood.

Yap, M. and Yu, E. (2016). Operationalising the capability approach: Developing culturally relevant indicators of Indigenous wellbeing – an Australian example. *Oxford Development Studies*, 44(3), 315–331. http://doi.org/10.1080/13600818.2016.1178223

Yunkaporta, T. and McGinty, S. (2009). Reclaiming aboriginal knowledge at the cultural interface. *The Australian Educational Researcher*, 36(2), 55–72.

7
ACTION, INERTIA, AND THE STORIES THAT GET US MOVING

Keerti Gopal

In July of 2019, I found myself in the middle of a crowd of thousands, spread out across the dry, sun-soaked grass of Cass Park in Detroit, Michigan. All around me, people were holding signs and posters with slogans like "We Demand Union Jobs" and "We Demand Breathable Air" or "Make Detroit the Engine of the Green New Deal." I was holding a huge green fist that had been cut out of plywood and mounted onto a thick wooden stake. The crowd was singing a Pete Seeger song with the lyrics swapped out:

> *Which side are you on now, which side are you on?*
> *Storms surge and fires burn but you don't hear the call.*
> *'Cause fossil fuels keep paying you,*
> *Does it weigh on you at all?*

Singing, chanting, and holding signs, we tumbled out of Cass Park and onto a street I didn't recognise, and the crowd marched through the city, a teeming, shifting mass of bodies walking as one. We made our way to the Fox Theater – where the candidates for the Democratic Party's presidential primary were gathered for the first of two primary debates – and stopped there. From where I stood in the crowd, I could see television reporters sitting around a table outside the theatre facing big black cameras. Later, we'd learn they were talking about us.

The march was organised by Frontline Detroit, a coalition of progressive groups including environmental justice organisations, labour unions, advocates for racial justice, and the Sunrise Movement, a national, youth-led climate justice organisation fighting for a Green New Deal. I'd first begun researching Sunrise earlier that summer, as part of a project exploring youth motivations for activism, but by the time I travelled to Detroit, I'd joined the movement.

That weekend, hundreds of Sunrisers – as the organisation called its members – had travelled to Detroit from all over the Midwest for what Sunrise was calling a summit. For a weekend, we gathered in a borrowed theatre building for a crash course in organising: during the day, we did workshops on Sunrise's theory of change, learned history lessons about the civil rights movement, learned chants and songs, and practised storytelling or talking to

DOI: 10.4324/9781003396567-8

people with opposing politics. At night, local high schoolers got picked up by their parents and the rest of us covered the floor of the auditorium in sleeping bags and with portable fans to keep out the summer's heat. During lunch, we took breaks – some people gathered around a piano and sang, some played hacky sack in the parking lot, some sat in circles on the carpet talking about the state of the climate crisis.

At 20 years old, that weekend was a turning point for me. I'd never before been in a room with so many young people who looked at politics as a matter of life or death. Sunrise defines "youth" as anyone under 35, but almost everyone at the summit was at least 10 years younger than that, and a solid contingent weren't even old enough to vote. The leaders of the trainings were mostly in their late teens and early twenties; some had taken a semester off college several years ago and never gone back. The dominant consensus in the auditorium was that the scale of the climate crisis is so big that nothing else could be as important as fighting this fight. Why go to school and get a degree if the world is set to burn before you get old enough to use it?

I'd gotten involved with Sunrise that year, attempting to channel my growing climate desperation by conducting narrative-based interview research with youth climate organisers in Chicago as part of an undergraduate summer research project. As I listened to their stories – their fears, hopes, anxieties, and triumphs – I was energised. Surrounded by stories of action, I couldn't help but feel like real change was possible. Soon enough, I'd joined the youth-led climate and economic justice collective, the Sunrise Movement, becoming a youth organiser myself.

Social Movement research has found that peer narratives are crucial in motivating youth movement participation. This was true for me, and since 2019, I've continued collecting stories of youth mobilisation: first in Chicago, then across the US, and then in Taiwan. In this chapter, I'll share a few of these stories, the stories of activists like Isaac, from California, who was activated by the feeling of effecting political change. Or Liang-Yi, who fell into activism and then dedicated over a decade to building the organising capacity of Taiwan's youth. These stories are only a few among the millions of youth activists who continue to mobilise worldwide in the face of climate change, and they are not a representative sample. But the act of storytelling has always been central in movement organising. Stories like these are how we believe in our own capacity, how we get inspired to build power. Stories like these are how we combat the inertia of apathy and get moving.[1]

Isaac: Maybe Good Things Can Happen

Growing up in Northern California, Isaac Larkin remembers being aware of climate change but not seeing avenues to address the crisis beyond the confines of conservation and green capitalism. His parents were involved in disaster preparedness, and he first got Red Cross certified in high school as part of an emergency response initiative his mom organised for community response to natural disasters like forest fires and earthquakes. But as he learned more and more about the scope and speed of the climate crisis, he felt increasingly hopeless about the possibility of society achieving any kind of real change.

"My initial response for a couple of years was just to get increasingly despairing about the future of the world," he said. His mind sprang to doomsday prepping: he began thinking about the skills he might need to learn to sustain himself in the event of full-on civilisational collapse. It was a solitary reaction, focused on himself as an individual.

"I was just learning these things in my isolated lab and thinking of ways I would prepare myself for the future," Isaac said.

Then, in 2018, Isaac's perspective began to shift. He remembered watching the primary campaign of Alexandria Ocasio Cortez, the rising star progressive running for Congress in New York City. From her campaign, Isaac learned about the Green New Deal, Ocasio-Cortez and Ed Markey's proposal for a New Deal-style economic stimulus that would push for green jobs to fuel the energy transition. On an impulse, the day of Ocasio-Cortez's election, Isaac decided to phone bank for the first time. He downloaded an app that provided names and numbers of potential voters and made a few calls.

"It's not like I was very helpful, but I had sort of taken a step towards actually getting involved in something." When AOC won, it upended Isaac's whole outlook. He didn't feel hopeless anymore.

"I was like, 'oh, maybe the bad future is not inevitable, maybe good things can happen.'"

That was Isaac's activation. After that, he started phone banking for more and more candidates. He started learning about the concept of the Green New Deal, a means of tackling climate change while addressing historical social and racial injustices and building a new economic order. He joined the Sunrise Movement, and it gave him momentum to keep taking action.

"It is astonishing how invigorating activist organizing is in terms of like making you feel hopeful and feel like you have power and can engage in a society and change it for the better," he said.

Liang-Yi: Crazy People

Liang-Yi Chang never intended to be a climate activist. His interest in climate change came more from curiosity. On an exchange program in Sweden during college, he became aware of a growing global movement for climate action, and he was curious to see if any organisations were working on the issue back home, in Taiwan.

That initial curiosity sparked what has become a career dedicated to fighting the climate crisis. He founded the Taiwan Youth Climate Coalition, mobilising hundreds of young people in his country, and has expanded his work to collaborate and build coalitions with youth groups across Asia, now working for 350.org. Over the past decade, Chang has been involved in local organising, urban planning consultancy, and global UN sustainability initiatives. He's done research in Antarctica, been detained for tracking air pollution in Vietnam, and led training sessions for climate organisers around the world.

When Liang-Yi first tried to organise his peers around climate change, he found that there wasn't a lot of interest, but as the political organising power of Taiwan's young people grew, so did his movement.[2]

Chang, who was 34 at the time of our interview, is older than most of the other activists I've interviewed, and he said that he's trying to move on from the youth branding and pass the torch to younger organisers. Still, his decade of climate organising experience is an important perspective, especially as youth movements are often subject to high rates of turnover. I asked Chang a few times what's kept him in the movement for so long. His first answer was simple.

"At the beginning it was curiosity, like why don't people care," he said. "Now, I just love to work with crazy people."

A couple weeks later, I asked Liang-Yi again what kept him in the movement, pressing for more of an explanation. How did he go from being a third-year biology major who just wanted to travel around Europe, to dedicating more than a decade of his life to climate justice, with no signs of slowing down? Chang walks into rooms with a smile; he laughs easily and often, showing an impressive lack of cynicism or fatigue for someone who's spent his entire adulthood fighting for urgency against an existential crisis in a slow-moving world. At the time of our interview, I was about the same age as Chang was when he started TWYCC, and to me, his continued idealism was powerfully inspiring. How does he keep up the energy? How does he stay so hopeful, optimistic, and poised for change?

Again, Chang's answer was the same.

"I want to work with crazy people," he said, smiling. "Did I tell you last time? They are crazy enough, they want to change the whole world."[3]

Han-Wei: Still Some Hope[4]

Han-Wei Chang doesn't know if she'll ever want kids. When she thinks about the future and the possibility of starting a family, her mind jumps to climate change. Sitting in a cafe in New Taipei City, Han-Wei shook her head and looked away from the table in contemplation.

"How dare I?" she asked rhetorically at the prospect of bringing a child into a world of widespread destruction and dwindling resources.

"I'm afraid to have a baby," she continued. "I think it's not fair for the newborn child, because they have to face the impact of climate change for a longer period. I don't know if I want to see my child face those disasters."

When we met in 2021, Han-Wei was co-director of the Taiwan Youth Climate Coalition (TWYCC), Taiwan's largest youth-led non-profit, dedicated to fighting climate change. She grew up in Taichung, a city in the central south with a history of activism against air pollution. Han-Wei said environmentalism and air quality were mainstream conversations, but she rarely heard any mentions of climate change. Han-Wei remembers her consciousness around climate growing during university, as she started watching documentaries and following global news with growing trepidation. But these feelings weren't shared by those around her, and at first, her parents weren't pleased with her new direction. She said they didn't understand the importance of climate change or her work; it didn't track onto a recognisable path for success.

"They don't know because there are no examples," she said. "They just don't know the importance of my job."

That all changed last spring, when record droughts caused a months-long water shortage in Han-Wei's hometown and forced the climate crisis centre-stage. Han-Wei remembers her parents calling from Taichung to let her know that they were down to five days of water access per week, relying for the remaining two days on whatever water they had been able to save. The stress of the drought was exacerbated by the pandemic.

"They were strong, they tried to adapt to the situation, but it's quite [inconvenient]," Han-Wei told me. "We want to clean everything to not get the virus, but they also have no water . . . they told me that it was very depressing."

2020 marked the first[5] year since 1964 where Taiwan experienced zero typhoons, and the lack of rainfall hit the nation hard, particularly in the central region where Taichung is located. Residents like Han-Wei's parents had to ration water used for drinking, cooking, and hygiene. The drought also caused major distress to agriculture and production industries and impacted global supply chains.[6] Scientists predict[7] that with increased global warming, typhoons will decrease in frequency but increase in severity. That's bad news:

fewer typhoons mean less rainfall and more ubiquitous drought, while more severe typhoons might resemble Typhoon Morakot, the devastating 2009 storm that killed almost 700 people and caused staggering destruction across Taiwan's southern and central regions.[8]

Last year's drought was a wake-up call to citizens like Han-Wei's parents, who began to realise that the climate crisis isn't abstract or distant: it's about a pressing need for local adaptation and security at home. Now, they've started taking an interest in her work, asking her about how climate change might impact their country.

"They will ask me, like, '[how] will sea level rising affect Taiwan?" she said. "They are aware that the sea level rising is caused from climate change . . . and based on this knowledge, they ask the question."

Han-Wei's parents were experiencing a reckoning, forced to understand the link between the broad, amorphous climate crisis and their lives. The truth is, environmental movements have not, historically, been all that great at telling people why they should care about climate change. Images of polar bears and melting glaciers have painted the crisis as something abstract and distant, and although in the past decade there's been a shift away from that kind of rhetoric and a mounting awareness of the link between climate change and dangers like wildfires, hurricanes, typhoons, floods, and droughts, for many people it can still be hard to see the link between the global crisis and day to day life.

"I have hope and faith in the Taiwanese people," Han-Wei told me. "Ten years ago there is [nobody] talking about climate change . . . now our president talks about climate change on her Facebook or [in speeches]."

To young people in the anxiety stage, trying to transform their distress into action, Han-Wei's advice is first to get educated and then to find your personal niche in the movement.

"The youth can find their passion first," she said. "Maybe they love communicating with people . . . maybe they love coding . . . find your passion first and . . . see how your strength and your passion can help to solve the problems from climate change."

Turning anxiety into action isn't easy, but Chang knows it's possible. She's done it herself.

"I think some Taiwanese youths are now facing the climate depression," Chang said. "They are worried about their future under the impact of climate change and that makes them maybe anxious, or sad, or disappointed. But I really want to, if I can, encourage them that we already have some change [in] public awareness . . . don't give up, there's still some hope there."

Anjali Mitra: Moral Authority

By the time she was an 18-year-old high school senior, Anjali was already organising her peers across the United States, taking on leadership for Sunrise Movement's national middle- and high-school student network. But as a younger kid, Anjali didn't feel a sense of urgency about climate action. The threat of climate change hovered over her childhood, but like many members of her generation, she felt it was somewhat divorced from her daily life. She knew about global warming and polar bears and melting ice caps, but it all felt so far away. Sad – but not necessarily urgent.

That changed in 2012, when Hurricane Sandy hit the northeast. Anjali's grandparents lived in New Jersey, in a little house in the middle of the woods, and when the storm hit they were cut off from the rest of their community. They had no water, no heat, and the trees surrounding their home collapsed, cutting off their ability to leave for two weeks. Sandy brought cold weather and unexpected snow, and Anjali's grandparents would turn on their car intermittently just to charge their phones and get some warmth, rationing the vehicle's gas, not knowing how long they'd be stranded. Every day, they'd go out into the woods and collect firewood for the one room in their house that they could keep warm. It scared Anjali to see the impacts of climate change so close to home.

"It made me realize that climate change wasn't just like, polar bears in the Arctic," she said. "It was people and communities right here, right now, and it was only getting worse."

Anjali started doing more research on the climate crisis and began to grow her understanding of how fossil fuel extraction and corporate greed had knowingly enabled the crisis to grow while throwing money at politicians to stagnate action and solutions.

"Fossil fuel executives [and] politicians in power were the people who are making this possible and who were harming communities like my grandparents' for power and for money," Anjali said. "That fear that I had changed to anger, and into politicisation. I was like, I can't just sit here and watch this happen and watch people get hurt while other people benefit off of that suffering."

Anjali started getting involved in a local climate club that took part in some political organising alongside service actions like park clean-ups. It was a good start, but she knew she wanted to do something with a larger-scale impact, something that would address the systemic problems she'd been learning about: that's when she found the Sunrise Movement.

Reflecting on the challenges facing young organisers, especially those below voting age, Anjali noted how intimidating it can be to enter into political activism as a kid, when adults are constantly underestimating your intelligence and capacity for creating change. Young people are underestimated, but they have more than enough reasons to be energised and engaged.

Anjali said she feels her generation in the United States is growing up in an era of political extremes, with growing suppression of movements and free thought fighting against growing politicisation and understanding of the importance of civic action and protest. Anjali cited the election of Donald Trump and the uprisings in support of Black Lives Matter as opposing forces on the spectrum of mobilisation. It's a politicised moment, Anjali said. But it's also a turning point, where the decisions we make, particularly in regard to climate change, will have permanent repercussions down the line.

Middle and high schoolers who can't vote understand that this is a pivotal moment in history, Anjali said.

"Honestly, because of that, they have more of a moral authority, if you will, to speak out to fight for what they believe in," she said. "Because if the people in power won't do it for them and the people who vote won't do it for them, who will?"

Anjali observed a lack of engagement beyond voting from adults. People would vote and feel like their civic duty was done.

"That is not true," she said. "That is the bare minimum, and maybe not even, and middle and high schoolers are showing that more direct action and showing up for what you believe in is more powerful than just dropping your ballot in a box."

Anjali believes that the generation of non-voting-age young people that's coming into social activism and organising is going to be a mass wave for progressive action, especially in the name of addressing climate change. It scares older conservatives, she said.

As an organiser with Sunrise, Anjali has put together direct actions that have garnered outsized responses from law enforcement, things like 12 teenagers standing and chanting on a bridge with banners and getting surrounded by four or five police cars.

"That's more powerful," Anjali said. "The fact that people are afraid of middle and high schoolers, and middle and high schoolers have this much power."

There's a moral authority to the argument of the youth fighting for climate action. We didn't create this crisis, and it'll impact our lives most acutely and most completely. Youth can also be a powerful force of energy.

"There's much more of a belief in change and hope that sometimes people forget as they get older," she reflected.

What's the Point of Storytelling?

Recently, I've been asking myself this question a lot. The word "storytelling" comes up often, in so many contexts – from corporate conferences to movement spaces and everywhere in between. It's starting to feel like just another buzzword. I sometimes wonder if it's lost its meaning.

So, what is storytelling? Simply put, storytelling is just a means of sharing: an account of events, real or imaginary. There are university classes, professional development workshops, and internet courses on storytelling. Storytelling competitions, YouTube tutorials, textbooks, and how-to-guides on the art of storytelling. And though it's true that honing your craft as a storyteller can make you more effective at sharing truths and compelling your audience – after all, everything can get better with practice – storytelling is, really, just a very basic form of sharing. From the walls of Lascaux to our social media today, it's perhaps the most basic form of human communication: we share accounts of our lives and our dreams to see and to be seen, to expand our understanding of ourselves, or of the world, or of one another.

In the context of the climate movement, sharing our stories – of why we care or how we stay mobilised – is only one piece of the equation to counteract the inertia of apathy. The real question is, what will you do with the stories you hear? Will you listen from afar, nod your head in agreement, shake your head in dismay, and continue on with your day? Will you read theories of activism and changemaking in your university library and write essays about movements past? Or will you listen to these stories with your head and your heart and let them move you, not only internally but also externally, let them propel you to action?

Social movement research has found time and time again that the most effective motivators of movement participation have to do with proximity – often, having a personal stake in a movement's outcome and having social relationships with others who are mobilised is a key factor in compelling participation. Distance is the enemy of action, but being in community with people who are already activated can help us do the same. Both apathy and action can hold a kind of inertia: once you get moving it's easier to stay in motion.

So, tell your story to combat your own inertia, or someone else's. And take the stories you hear to heart; let them move you to action. Put yourself in proximity to people who are steeped in the fight that moves you and listen to the stories they tell you. We must document the events of the past and present, and we must imagine the futures we want to build. But ultimately, it's not just the stories we hear but what we do with them that matters.

Notes

1 This research has been supported by the Northwestern University Undergraduate Research Office and American Studies Department, the US Fulbright Program, and the National Geographic Society. Interviews and quotations have been edited minimally for length and clarity.

  Also, portions and/or versions of some of the text in this chapter, written by Keerti Gopal, have been published non-exclusively in the *Northwestern Undergraduate Research Journal*, on the National Geographic Society's *Fieldnotes* platform, and on *Medium*.

2 See Daybreak Project. (2018). https://daybreak.newbloommag.net/.
3 Gopal, K. (2022a). *How to Build a Movement, with Liang-Yi Chang, Medium.* https://keertigopal. medium.com/how-to-build-a-movement-with-liang-yi-chang-ac88d1f63091. A version of Liang-Yi's story was published on Gopal's Medium page.
4 Gopal, K. (2022b). *Turning Anxiety into Action, with Han-Wei Chang, Medium.* https://keerti-gopal.medium.com/turning-anxiety-into-action-with-han-wei-chang-2145155195c9. A version of Han-Wei's story was published on Gopal's Medium page.
5 Chien, A. C., Ives, M. and C, B. H. (2021). Taiwan Prays for Rain and Scrambles to Save Water. *The New York Times.* https://www.nytimes.com/2021/05/28/world/asia/taiwan-drought.html.
6 Barbiroglio, E. (2022). No water no microchips: What is happening in Taiwan? *Forbes.* https://www. forbes.com/sites/emanuelabarbiroglio/2021/05/31/no-water-no-microchips-what-is-happening-in-taiwan/?sh=52437d7d22af.
7 Tzu-ti, H. (2021). Taiwan to see less rain, more droughts and violent typhoons: Taiwan news: 2021-11-15 11:04:00. *Taiwan News.* https://www.taiwannews.com.tw/en/news/4345496.
8 Wu, C.-C. (2013). *Typhoon Morakot: Key Findings from the Journal Tao for Improving Prediction of Extreme Rains at Landfall, AMETSOC.* https://journals.ametsoc.org/view/journals/bams/94/2/bams-d-11-00155.1.xml.

References

Barbiroglio, E. (2022). No water no microchips: What is happening in Taiwan? *Forbes.* https://www. forbes.com/sites/emanuelabarbiroglio/2021/05/31/no-water-no-microchips-what-is-happening-in-taiwan/?sh=52437d7d22af [Accessed 25 May 2022].

Chien, A. C., Ives, M. and C, B. H. (2021). Taiwan prays for rain and scrambles to Save Water. *The New York Times.* https://www.nytimes.com/2021/05/28/world/asia/taiwan-drought.html [Accessed 25 May 2023].

Daybreak Project (2018). An introduction to the sunflower movement. https://daybreak.newbloom-mag.net/2017/07/22/what-was-the-sunflower-movement/ [Accessed 20 February 2022].

Gopal, K. (2022a). *How to Build a Movement, with Liang-Yi Chang, Medium.* https://keertigopal. medium.com/how-to-build-a-movement-with-liang-yi-chang-ac88d1f63091

Gopal, K. (2022b). *Turning Anxiety into Action, with Han-Wei Chang, Medium.* https://keertigopal. medium.com/turning-anxiety-into-action-with-han-wei-chang-2145155195c9

Tzu-ti, H. (2021). *Taiwan to See Less Rain, More Droughts and Violent Typhoons: Taiwan News: 2021-11-15 11:04:00, Taiwan News.* https://www.taiwannews.com.tw/en/news/4345496 [Accessed 18 February 2022].

Wu, C.-C. (2013). *Typhoon Morakot: Key Findings from the Journal Tao for Improving Prediction of Extreme Rains at Landfall, AMETSOC.* https://journals.ametsoc.org/view/journals/bams/94/2/bams-d-11-00155.1.xml [Accessed 18 February 2022].

PART II

Climate Artivism

Introduction to Part II: Climate Artivism

When U.N. Secretary-General António Guterres described the second Working Group's contribution to the IPCC Sixth Assessment Report ("Climate Change, 2022: Impacts, Adaptation and Vulnerability") as "an atlas of human suffering and a damning indictment of failed climate leadership" (Borenstein, 2022), he was recognising what climate and environmental justice activists had long stated: that the climate emergency is already underway – and not in the future; that it is causing great harm to humans and the ecosystems of which they are (only) a part and on which they depend; and that these harms fall disproportionately on poor and vulnerable populations and communities of colour (see also IPCC, 2022, pp. 3, 12). As author and activist Naomi Klein has put it in her critique of "the logic of indiscriminate economic growth" (2014, p. 86):

> [W]e have an economic system that fetishizes GDP growth above all else, regardless of the human or ecological consequences, while failing to place value on those things that most of us cherish above all – a decent standard of living, a measure of future security, and our relationships with one another.
>
> *(2014, p. 88)*

Telling (performing, dancing, singing, drawing) stories, creating space for reflection and debate, and visualising data are important ways we make sense of this situation and discover new ways to respond and return to being, guided by our shared humanity. (see also Boyd et al., 2016)

Insisting on the necessity of "engaged optimism," in which imagining and rehearsing may reveal actionable solutions, U.K.-based transitional arts practitioner and theorist Lucy Neal writes: "Transitional arts practice proposes how [such] 'macro' challenges [as energy, finance, food, community resilience, climate change, and wildness] are transformed creatively by a daily 'micro' personal practice" (Neal, 2015, p. 7, 14). This would be "an art that could be practised by everyone, inseparable from daily life" (4). The act of improvising solutions in democratic and participatory fashions can help us better see not only the feedback loops from which we must divest but those in which we must invest; that is, we can

DOI: 10.4324/9781003396567-9

see divestment and investment as realistic and necessary responses to the lived experience of climate change. Collaboratively imagining, designing, and rehearsing transition at the level of one's town, Neal proposes, is a way of "playing for time," creatively finding solutions for degrowth and reinvestment in real-time and in local space, solutions that may also buy time for mitigating the most dire scenarios projected by climate models. This section introduces a number of investigations of artists, writers, and performers as they engage in climate activism.

It begins with Chapter 8, in which Rebecca Olive, Fiona Hillary, Wendy Steele, Kit Wise, Alexandre da Silva Faustino, Nicolas Guerra-Tão and Paloma Bugedo draw inspiration from the grassroots climate activist collective *Gunditjmara First Nations People's Ocean Defenders* to engage with the nature of human-water entanglements in Naarm/Melbourne. By centring First Nations stories, colonial legacies, ghost rivers, ecological insights, transport lines, counter maps, underwater recordings, poetic reflections, artworks, and installations as a creative response to the climate crisis at the intersection of oceans research, the authors bring to light the role of performance and art in the creative practices of everyday climate activism in coastal settings. The authors explore the ordinary and profound impacts that everyday interactions with oceans can have on people's lives, as well as the response-ability we have for finding meaningful ways of supporting and safeguarding an ecological near-future.

In Chapter 9, "A Contemplative Pedagogy of Listening," Linda J. Chase discusses how her ecomusicology students at Berklee College of Music and New England Conservatory in Massachusetts, USA, engage with climate-change issues through listening practices. She asks how musical artists can articulate a vision of hope that leads to action by focusing on *how* musicians can become leaders in social and environmental transformation. She reveals connections among listening, artmaking, and activism in her work.

In Chapter 10, "The Sonic Work of Transnational Youth Climate Movements," Mark Ortiz looks at a very different function of music in climate activism. Ortiz illustrates that transnational youth climate networks have generated powerful sonic and musical cultures, which have helped to mobilise youth publics around the world, by analysing what he thinks of as the sonic work of transnational youth climate movements. By sonic work, Ortiz indicates a range of sound expressions which, while inclusive of music, also reach beyond that category in important ways. This includes exploring how activist movements navigate the interplay of sound and silence in protest actions, how words are said in loud voices and hushed tones in spaces known for their monotony, and how youth networks irreverently remix pop culture music in new multimedia formats to communicate the complexity of ecological emotions.

In Chapter 11, "Performing Transformative Climate Justice," Thomas A. King explores the capacity of theatre and performance, ranging from protest performance to scripted plays and interactive installations, to imagine alternative ways of being in relation to others and to our planet. To offer critique, to enable hope and inspire joy, and to offer an embodied ethics of care, compassion, and repair in a world that may be increasingly uninhabitable.

In Chapter 12, "Turning a Leaf: Fiction and Nonfiction as Grassroots Climate Activism," Sabine von Mering provides an overview of the large number of English language publications devoted to addressing the climate crisis. Works of cli-fi exist alongside a growing body of nonfiction works by authors like Bill McKibben and Naomi Klein, David Wallace Wells, and many others. Von Mering asks whether – and if so how – the written word can serve as a form of grassroots climate activism, based on examples from Afrofuturism to children's literature and a discussion of the role the written word may play in raising the alarm about climate change and mobilising people to take action. The chapter introduces efforts to anthologise cli-fi, as well as other initiatives by writers to confront the climate crisis through collective organising – such as the Climate Fiction Writers League founded by Lauren James in 2020.

The section closes with Chapter 13, "Wild Hope and the Curatorial Activism," where Fleur Watson, Wendy Steele, Naomi Stead, and Katrina Simon draw attention to creative practices and the role of the Exhibition "curatorium" as an arts-based grassroots activist response to the climate crisis. They describe the curation of research-led exhibitions within the context of large festival programs as creating opportunities for charged public encounters. As a form of grassroots activism, the authors argue that progressive curatorial practices offer an important experimental space for exploring, testing, and mediating speculative design ideas at the radical interface of performing "research-in-action" and civic engagement. The role of the curator as Translator, Activist, and Dramaturge frames the chapter, with critical insights drawn from the large public exhibition and programme *Wild Hope: Conversations for a Planetary Commons*. Activist ideas and practices can help explore and open up opportunities for immersion, imagination, reflection, and discussion – and in doing so can help to reveal alternative places and spaces of wild hope in dark times.

Overall, the chapters in this part provide just a glimpse into the vast production of climate art and artivism that has proliferated especially – but not only – in countries that bear the bulk of the responsibility for the crisis. Whether art, music, performance, or text – the artists, curators, musicians, writers, and filmmakers are united in their efforts to counter ignorance, apathy, and active attempts to silence local communities' and scientists' urgent calls for change.[1]

Note

1 Thanks to Thomas A. King for generously and substantially contributing to this introduction.

References

Borenstein, S. (2022). UN climate report: Atlas of human suffering, worse, bigger. *AP*, 28 February. https://apnews.com/article/climate-science-europe-united-nations-weather-8d5e277660f7125ffda b7a833d9856a3

Boyd, A. and Mitchell, D.O. (2016). *Beautiful Trouble: A Toolbox for Revolution*. New York: OR Books.

IPCC. (2022). *Sixth Assessment Report: Climate Change 2022: Impacts, Adaptation, and Vulnerability*. https://www.ipcc.ch/report/ar6/wg2/ [Accessed 6 May 2024].

Klein, N. (2014). *This Changes Everything. Capitalism vs. The Climate*. New York, London, Sydney and New Delhi: Simon & Schuster.

Neal, L. (2015). *Playing for Time: Making Art as If the World Mattered*. London: Oberon.

8

CLIMATE ACTIVISM AND ATTUNEMENT THROUGH CREATIVE PRACTICES

*Rebecca Olive, Fiona Hillary, Wendy Steele,
Kit Wise, Alexandre da Silva Faustino, Nícolas Guerra-Tão,
and Paloma Bugedo*

Introduction: Sea Country

The *Southern Ocean Protection Embassy Collective: Gunditjmara First Nations People's Ocean Defenders* are Indigenous grassroots climate activists fighting for the ancient oceans they see as their kin. In the words of Vicki's daughter Yaaran Couzens-Bundle, 'The saltwater and the freshwater, they're living entities, they're alive, they breathe – the waves going in and out and the tides and the swells, that's the ocean breathing'.[1] They are the custodians and caretakers of their Sea Country: the Southern Right Whale songlines and all Southern ocean life that is under attack from deep sea mining, gas drilling, sonic blasting, plastics, pollution, urbanisation, and disease. In response they run protests, write submissions, organise rallies, and build alliances with other climate activist and community groups with a 'powerful love, true respect and responsibility . . . not for the spotlights, or financial gain, just true Blak resistance for Sacred Country!'[2] Through dance, film, music, and art performing song stone and whalebone heart, they honour their ancient connections as part of the still-unfolding saltwater Earth story.

These practices are expressions of culture and offer ways of developing, nurturing, and honouring connections to Country; place, ecologies, ancestors, spirits, lore, kin, and community. Linked to deep cultural traditions and connections they are from Country and part of Country in return. For newcomer settlers and other non-Gunditjmara people, establishing meaningful connections to Gunditjmara places and communities requires building knowledges and relationalities, which can take time and requires deep ethical and reflexive consideration.

In this chapter, we explore practices that aim to facilitate place-based interactions with ocean and coastal communities that can result in deeper knowledges of and a sense of connection to ecologies, people, and histories. These interactions are aimed at encouraging everyday climate activisms based in ethics and relationalities. We refer to these practices in terms of *attunement* (see Tsing, 2021) to indicate the forms of everyday knowledge, sensibilities, and relationships that can emerge from coming to know the complexity of place and the responsibilities we might have in terms of caring for place. At the same time, we

DOI: 10.4324/9781003396567-10

use *attunement* to recognise the limits of what these knowledges and relationships confer on settlers. Most significantly, that they do not confer new levels of authority for settlers over Indigenous and First Nations people, and neither do they replicate the cultural or genealogical depth of Indigenous and First Nations relationships in settler-colonial contexts like Australia (Lucashenko, 2006). This does not mean that attunements to place for non-Indigenous people are without significance. Indeed, we believe they are essential to more ethical and caring place relations, but we emphasise that understandings of the complex cultural, ancestral, and spiritual histories and politics of places must be central to how we develop our relationships to oceanic and other terrestrial places.

To explore these ideas, this chapter draws inspiration from grassroots climate activist collectives like the Gunditjmara First Nations People's Ocean Defenders to critically engage with the nature of human–water urban entanglements. Their outrage is clear as articulated by Vicki Couzens, Vice-Chancellor's Indigenous Research Fellow at RMIT University and Gunditjmara Keeray Whurrong Elder[3]

> ***Wantayngeenkopa leekanyoong ngootook?? What's the matter within you?*** *What's the matter with you that you destroy future generations' right to live?? Mine and future generations have inherited a world of pollution, disease and mass extinction!*

As uninvited guests living and working in Naarm/Melbourne in the country known as Australia, we stand in solidarity with Vicki, Yaaran on Southern land and sea country that was never ceded. We are an Oceans Research and Climate Action (ORCA) network that came together in response to the need for greater action on climate change in the context of oceans and coastal areas. We are particularly interested in diverse ways of engaging *with*, as opposed to against, the more-than-human urban ocean community. Our transdisciplinary focus is the intersection of oceans research, community activism and engagement, and the powerful role of performance and art in the creative attunement of everyday climate activist practices in coastal settings.

In our discussion of attunements in practice, we will be drawing on practical examples of human–water entanglements in Naarm/Melbourne. The examples we offer are place- and arts-based activities with Nairm/Port Phillip Bay, which is an important – but often overlooked – part of Naarm/Melbourne. As well as facilitating creative interactions, these activities centred First Nations stories, colonial legacies, ghost rivers, ecological insights, transport lines, counter maps, underwater recordings, poetic reflections, materialities, artworks, and installations as a creative response to the climate crisis. In our work, the focus was on everyday coastal practices and the role of creative practices as an intimate and sensory form of climate activism. Through an emphasis on *attunement*, we are drawn to explore both the ordinary and profound impacts that everyday interactions with oceans can have on our lives, the response-ability we have for finding meaningful ways of supporting and safeguarding an ecological near-future, and the ethics with which these must be considered.

Creative Climate Activism and Arts-based Practices

The Oceans Research Climate Action (ORCA) collective is the result of a series of collaborative encounters, iterations across time, immersed in watery thinking and its planetariness, among researchers, industry partners, and audiences in Naarm/Melbourne. The first public iteration of the collective was an interactive workshop around 'Urban Undercurrents: The

Hidden Infrastructure of Wild Cities' which featured in both the Festival of Endangered Urbanism of 2021 – a national series of events hosted by the University of Sydney – and the internationally recognised Melbourne Design Week of 2022 in Victoria. The workshop involved a live encounter with various modes of creative infrastructural and more-than-human entanglements at Collingwood Yards, one of Melbourne's not-for-profit independent arts precincts and a co-created video/sound work and spoken word piece performed online during the Covid lockdown.[4]

This live/digital performance was inspired by the spaces in and between the city and its environs, inviting audiences into an experience of the city's metaphoric and literal subterranean through the lens of dirt theory[5] (see Lagerspetz, 2018; Steele et al., 2020). As artistic intervention, speculative proposition, and community discussion – this experimental panel focused on the hidden infrastructures and wild undercurrents of the city from an inter and transdisciplinary perspective, bringing into view the subversive, dirty, multi-sensory, lived nature of the city – and the many ways we are all emplaced within it (see Figure 8.1).

This was designed to be 'a panel that roamed' – tracing the human and more-than-human entanglements of what lies beneath the city and its culture/natures. Key provocations and questions raised by the panel include: what happens when climate activism goes underground? How do government and industry agendas reverberate beneath our feet? What are the wild undercurrents and hidden infrastructures coursing all around? How do we encounter feral ecologies, contaminated creativity, and stray ethics? How can our sensory imagination act as a precursor to critical praxis? And how do we create space for conversations and rituals that create the conditions for regeneration and 'wild' life? To really shift the sustainability of city-regions requires a thoroughly problematic glimpse of

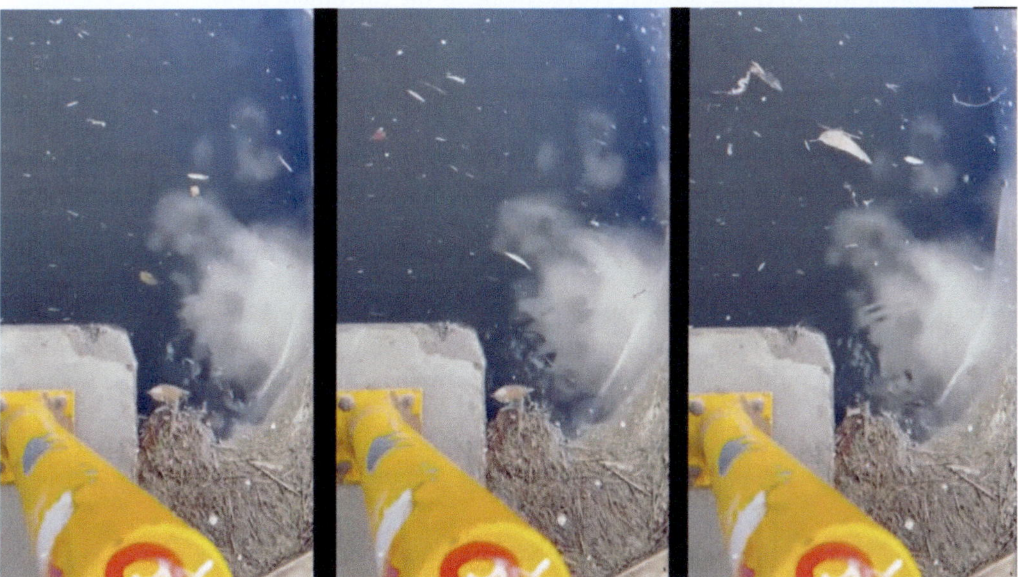

Figure 8.1 Still from 'Urban Undercurrents: The Hidden Infrastructure of Wild Cities' (2021).[6]

Source: Urban Undercurrents group (2021). Reproduced with permission.

our urban habitats and fetishes and greater recognition that nature is woven through our intimate lives in ways that are obvious and immersive, implicit yet hidden in the context of a climate crisis.

The activities of the Oceans Research and Climate Action (ORCA) collective built on the thematic of urban undercurrents but more literally in relation to oceans and coastal edge communities, with a strong focus on the everyday activist activities that bring people to the water or which offer immersive artistic forms of radical engagement through civic performativity. In keeping with our iterative process, we invited a group of transdisciplinary researchers to join in a creative practice workshop. Using seaweed collected from their last coastal visit, we undertook Cyanotype printing and haptic processes to 'think with' algae as we shared our ideas for a larger, public participatory experience and engagement (see Figure 8.2). The workshop drew inspiration from the techniques employed by Anna Atkins, renowned botanist and artist noted for creating the first photographic book documenting British algae.[7] Through engaging with creative artistic methodologies, we explored approaches to climate activism that specifically seek to make the links among the climate emergency and oceanic ecologies, ecosystems, and communities that are increasingly threatened and vulnerable.

The formation and dispersal of temporary communities occurs as an iterative process through this approach. As we reach out to industry partners, invite fellow researchers to collaborate with us, and engage with the broader public, we begin to trace how we move through the world and our watery immersions. One such context is Nairm – or Port Phillip Bay,

> the largest enclosed seawater embayment in the Southern Hemisphere, more like an enclosed 'sea' than what is commonly thought of as a bay. The bay is over 1950 km² in area and stretches 58 km from north to south and 41 km from east to west at its widest point. Its coastline is 264 km long, and the water catchment feeding the bay is 9790 km², with over 3.2 million people living in the catchment area.[8]

Noticing how people attune to and in Nairm, the Port Phillip Ferry caught our attention as a means to cross the bay, to connect with broader communities and immerse ourselves in different perspectives. The first field trip in early 2023 allowed us to imagine how we could engage commuters with an assemblage of works to explore varied encounters with the bay. We visited the Geelong-based contemporary arts space *Platform Arts* and immersed ourselves literally at Eastern Beach, Geelong's public swimming beach. With an expanded group we planned another field trip, engaged in partnership conversations with Port Phillip Ferries and Platform Arts, and co-imagined a performative crossing where the Ferry was enlivened through drawing with floating philosophy, sound works experienced in silent disco headsets, storytelling shared through online platforms and recorded on postcards (made with the cyanotypes from our previous gathering), and salty meditations. A shared meal and a situating entanglement of oceanic encounters transpired as a co-created film and sound work offered an ambient backdrop. Participants' attunements were captured on the cyanotype postcards to inform future iterations (see Figure 8.3).

Feeling, Thinking, and Remembering with the Ocean

Through shared reflective practice we sought to register some of the ideas, memories, feelings, and references about home, belonging, attunement, and embodiment that were resurfaced by facilitating and experiencing the diverse possibilities of engagement with

Figure 8.2 Cyanotype workshop outputs drying.
Source: Nicolas Guerra-Tão' (2023). Reproduced with permission.

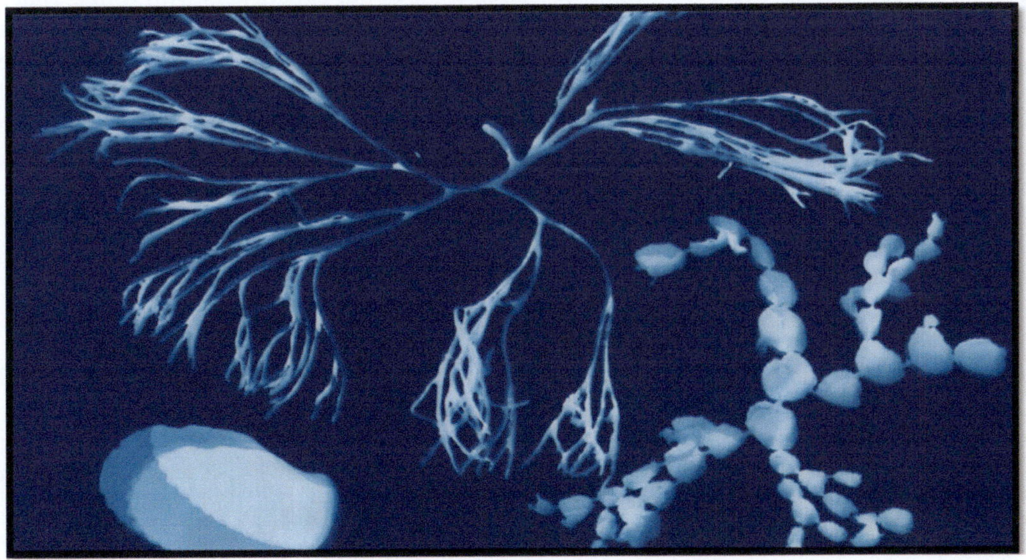

Figure 8.3 Cyanotype postcards developed by ORCA collective.
Source: Fiona Hillary (2023). Reproduced with permission.

ocean and water in this journey with Nairm bay. The following vignettes from Nícolas, Alexandre, and Paloma are fragments from their fieldnotes exercise 'written in flesh' in which they explored creative and affective writing while returning to Nairm from Geelong.

Nícolas

I always enjoy being part of experiences with this crew! The only certainty that we have is that of flexibility, that we'll be free to collaborate however we like and that people are free to enjoy this encounter on their own terms. As expected, there was a lot to be surprised by. It was like playing a game, choose your own adventure! You could be with Alex, meditating in the freezing cold and wild winds outside – feeling moist in the air, the dryness on your lips, touching the very little salts accumulating in the ferry – feeling the duality of absence and abundance of water surrounding (inside) us. You could be rambling around, listening to the soundscapes designed by Jordan. I felt the vulnerability and exposure of floating on an infinite liquid surface. How many kilos does a ferry have? As the sounds guided me, I was being pushed by the waves, in and out of the water. Catching a breath and diving again.

I remembered lyrics of a song, *Oceano* (Ocean), from Brazilian artist MC Tha: *Flutuo, nesse vazio/Mas a tristeza? A tristeza é oceano raso* (I float, on this emptiness/But sadness? Sadness is shallow ocean). The song talks about depression and how the feeling is, according to my interpretation, similar to being floating in the immensity of the ocean. It looks like it's never-ending. But unlike the ocean, sadness is shallow. If you try to reach down you will find the floor, you will ground yourself and see that sadness is actually a 'shallow ocean'.

At Platform Arts Geelong, Wendy prompted us with a beautiful ritual – letting your body react to the sounds of water. After that, we shared some of our experiences with water. That's a difficult one. I was raised in a house in front of a river. I would wake up every day to that landscape. During school holidays, I would stay home alone most of the time, as the house was far from the city and public transport. I was basically barking at the end of the school holidays – I had very little contact with humans and lots of dogs. Recently I've been doing therapy, after much resistance. I have difficulty expressing exactly what I feel to my therapist, so we've been doing this exercise: I close my eyes and connect with conflicting feelings. Anxiety, my dear little companion. Then, I look for a safe space in my mind. Where am I?

I see the water, I see the river. I'm home.

Alexandre

When I started gathering the materials for our activity, I noticed there was no crystallised salt on the handlebars of the ferry, maybe someone cleaned them, maybe the humidity in the air and cloudy sky didn't allow the water to evaporate. The freedom of the ocean surprised me. With no salt to scrape, Nick suggested that I return to a water meditation we facilitated in a previous experience. I was concerned that participants would resist the exposure to the weather, but giving them the option to go back inside the ferry didn't make them abandon the experience. I was happy to see that something was drawing them to stay feeling the bay. A few asked about the use of salt. For me, it is a clear reminder, noticeable in our everyday lives, of the deep entanglements we have with the ocean. It's also used in many rituals, creating boundaries of protection, and preparing baths to cleanse your body from energies you shouldn't be carrying. After every swim in the ocean, I do feel it took something from me, I walk out of it lighter, calmer.

Just before we left Platform Geelong, Paloma shared the thought of expatriates' relationship to water and the ocean in Nairm. Our original homes are kept beyond the end of the horizon, from here our Latin America stays where the sun rises. For some of our ancestors, it was where the sun set as they left their homelands, sometimes through violent capture and forced displacement. I had the words of Brazilian singer Luedji Luna in my head speaking about crossing the sea guided by the sun while carrying a handbag, a prayer, and a goodbye: *Atravessei o mar/um Sol da América do Sul me guia. / Levo uma mala de mão, dentro uma oração, um adeus* (I crossed the sea/A South American sun guides me/I bring a hand-bag, inside it [there's] a prayer, a goodbye).

So I wonder, where is home?

Paloma

The beautiful ocean-oriented meditation led by Alex made me reconnect with my own story: *home*. This (re)connection with water made me think of its capacity to travel through time and embody the memories from tasting my own salty lips: *home*. Cahuil, where the ocean carves its way through meandering river shores, shaped into pools, tamed to harvest ocean salt. It's become such a biodiversity hotspot in the Pacific (far from peaceful) that again contrasts with how salt mining, high up in the Andes responds to logics of disconnection and exploitation. *Ay Chile. Cuánto me dueles y aún así te llevo en el corazón* (Oh Chile. How bad you hurt me and still I carry you in my heart). A journey to the memories of home is also a dislocation . . . diasporas that play around time and space, lost in the horizon.

The horizon of the ocean

How many layers does it hold?
How many contradictions can it hold?
Peace and fear . . .
A bath, a lab, a game, an aim . . .
An escape, a route home . . .
A vessel of memories.
Flowing stillness and (eternal) reminder
of a fleeting present, easily washed out
by a wave, an adventure, a challenge,
a war.

The ocean is a place that makes linkages unravel.
A shared coastal culture to relate to.
A wordless common language.
An opportunity to reconnect.

(poem by Paloma Bugedo)

Attuning to Our Entanglements: More-than Human Ecologies

The creative arts-based practices of the Oceans Research and Climate Action (ORCA) collective aim to facilitate ecological exchanges that encourage new possibilities for imagining ways of living with bodies of water, to create a fluid archive of human, more-than-human, and oceanic entanglements. This is an expression of grassroots activism that interacts with oceans as a planetary commons in a climate crisis. By making ourselves more aware and even vulnerable to the sensory experience of human–ocean entanglements, diverse media can help to re-imagine new relationships and possibilities with multispecies places. The work of the ORCA collective and other climate activist groups like this invites audiences to think about their relationships with the ocean in new, multi-sensory ways. Creative activist practices can help communities explore oceanic knowing as a form of kinship, through sound, film, sensory activities, and virtual reality within the context of – and as a response to – the climate emergency.

These grassroots activist practices are underpinned by the challenge and opportunity of bridging different cosmologies – ways of knowing and being that are informed by and support Indigenous knowledge systems. As Indigenous scholar Tyson Yunkaporta's describes

> In Aboriginal Australia our Elders tell us stories, ancient narratives to show us that 'if you don't move with the land, the land will move you'. There is nothing permanent about settlements and the civilisations that spawned them . . . unsustainable settlements don't last long. . . . An unsettling thought.[9]

In his work he outlines four key areas relevant to grassroots climate activism: *respect* (values and boundaries/spirit); *connection* (relationships and equity/heart); *reflection* (collective knowledge/head); and *direction* (action and negotiation/hands).

The work of the ORCA collective is also influenced by an emphasis on relationality and ethics. For example, from within the Western philosophical context, Rosi Braidotti's posthumanist work draws on Deleuze and Guattari's notion of assemblages to suggest that we need 'to rethink subjectivity as a collective assemblage that encompasses human and nonhuman actors, technological mediation, animals, plants, and the planet as a whole' (Braidotti, 2017, p. 9). Such assemblages can be viewed close up or from a distance, and we have focused on site specificity in the kinds of activities we have developed in the hope of activating more ecological ways of thinking amongst temporary participant communities. Of course, these communities may reform to attend other activities together, but it is never assured. Understanding the human and non-human entanglement of a site establishes a contextual model of shared possibilities.

We are interested in framing how ecological ways of thinking emerge in practice (including creative practice) through everyday acts of 'attuning' to places. Tsing's use of 'attunement' emerges from her ethnographic work using 'the arts of noticing' to follow communities that develop around sites where highly prized matsutake mushrooms grow (Tsing, 2015). Matsutake mushrooms thrive in environments that have suffered human intervention – for example, forestry – and the mushroom foraging communities offer an example of 'living together' and 'living in common' with a changing and damaged planet (see Tsing et al., 2017). Through engagement with the matsutake mushrooms and the environments in which they grow, the mushroom gathering community learns how a damaged environment remains abundant in terms that emerge from our responses to the ruins. As Tsing describes of her own everyday relationship to matsutake mushrooms,

> *Attunement* is a form of love or, at least, commitment, responsibility, care, attention. Learning to smell the matsutake mushroom gives me this moment of attunement, where it enters my lifeworld, now we inhabit together and live with each other in a way that we were not before. When the roots of trees are attuning themselves through their own chemical sensitivities, a kind of smelling with the fungi and forming these mutualisms with each other, this, too, is a form of living together. It doesn't mean that it's happily ever after, and that everything supports each other, but you see this kind of living in common. Such a huge part of human civilisational practices of the last couple of centuries has been about not paying attention. Haunting and more-than-secular ways of observing can be the means to bring us outside of our training in particular civilisational practices to notice things that we wouldn't have noticed before.
>
> *(Tsing in conversation with Sarah Shin 2021)*

For Tsing the project is learning to live together in the world as it is, not in a romanticised past. Tsing posits that: 'staying alive – for every species – requires liveable collaborations. Collaboration means working across difference, which leads to contaminations. Without collaboration, we all die' (Tsing, 2015, p. 28). In this sense we are all 'contaminated by our encounters; they change who we are as we make way for others' (p. 27). This echoes Neimanis' use of attunement in her discussion of human–water relations in which we, ourselves, are bodies of water:

> Attuning ourselves to these hydro-logics and adopting an attitude of humility and curiosity towards water, is also the demand of this figuration. As bodies of water, we need to cultivate the ability to respond, rather than master.
>
> *(Neimani, 2013, p. 37)*

For both Tsing and Neimanis, attunement is an ethics of relating to environments and ecologies that requires us to both be open with our existing knowledges and to bring an openness to the possibilities of other worldviews, knowledges, and relationships. It is a way to develop an ethos of knowing-with and alongside, instead of an emphasis on colonial mastery (Neimanis, 2013, p. 37). In a settler-colonial and multicultural society like Australia, finding ways to decentralise authority away from positivist epistemologies and towards forms of knowing with attunements are essential.

When attuning to environments through creative practices, we are able to observe what Deborah Bird Rose (2017, p. 51) called the 'shimmer of the biosphere/of life', a concept that is translated from the word Bir'yun in the cosmology of the Yolgnu people's language. Transdisciplinary and relational attuning to everyday life can enable us to understand the relational multiplicities of the world.

In this framework, creative practice yields attentive registers where lines of flight can offer new readings of human/ocean entanglements and conjure new relationalities that manifest stories for our times. This is the hope behind the work we do collectively and in relation to the living waters that shape our urban homes – be they rivers, seas, or oceans.

Everyday Oceans, Climate Change, and Activist Practices

Human relationships to coasts and oceans are often romanticised through the sublime, yet most of our encounters with these environments are through everyday practices like swimming, dog walking, beach combing, surfing, diving, sailing, and fishing. These practices are part of community living and as part of critically reflective praxis can help to build meaningful relationships to places, ecologies in a changing climate crisis. Undertaken with a regenerative and political sensibility it is through these activities that people can experience – with a view to better supporting and understanding – the need to protect coasts and oceans in ways that create 'an acute embodied sense of connectivity' (Evers, 2009, p. 900) that is more-than-human.

We swim alongside the dolphins and fish and tread carefully to avoid the patterns of balled up sand excavated by little crabs building their homes. We come to know the rhythms of whale and fish migrations and the seasons of storms, sunshine, and winds. We learn how shifting sandbars impact the breaking waves and observe skittish shorebirds nesting and caring for their chicks. We watch sand dunes erode under ever-rising tides and collect bright fragments of plastic along with shells and driftwood. These lived experiences can have impacts on how people understand the effects of the changing climate and how they are directly impacting the places we know and love. Recognising Indigenous sovereignty and custodianship is critical.

Satchell (2008) describes the process of developing individual/communal 'sustainability-based ethics' as ecological sensibilities, which refers to the development of an eco-centric perspective that 'transcends the narrow sense of self and human superiority' (p. 110). Ecological sensibilities operate through the developments of relationships and a sense of connection, 'whereby emotive responses to environmental conditions are attributed a deterministic ethical power: if we feel, we will act in a productive way' (Potter, 2005, p. 2). That is, human relationships to place, culture, and experience manifest in ethical decisions about lifestyle choices and responsibilities relating to the environment. With a sense of personal connection and relationality that is so key to the development of ecological sensibilities, our everyday experiences become powerful pathways to activating ethics and practices of coastal care. Of course, this is not without its tensions, a key one being the sense of ownership and entitlement that a sustained relationship to place can bring.

These relationships are evident in what Kim Satchell describes as 'possibilities for an "eco-politics" deriving from a close analysis of the vicissitudes of belonging, the contested spaces of everyday life and the performance of care and creativity in an alternative teleology of emplaced encounters and returns' (Olive, 2015, p. 3). Satchell's work emerges from his own everyday practices of beach walking, rockpool gazing, and photography at the beach close to his home. These multisensory, ecological practices become part of how we understand and experience intimacy with the world and can change how we feel a part of it (see Olive, 2015). They offer experiential, affective engagements that challenge enduring human/nature binaries in Western and European worldviews and to 'disrupt this deep historical dualism by resituating humans in ecological terms at the same time as it resituates non-humans in ethical and cultural terms' (Bird Rose, 2015, p. 3).

Of course, entanglements with and re-situation into ecologies are about more than the joyful encounters and relationships. At the same time as we walk and paddle alongside sharks and fish, we are also alongside sewage, plastics, and chemicals. We experience environmental impacts in ways that have consequences; coastal change is in our communities and in our bodies, not abstract and happening somewhere else. We observe changes in biodiversity, experience increased severity of storms and erosion, and absorb the pollution in the water, just as we leave forms of pollution behind – microplastics, sunscreen, pharmaceuticals (Olive, 2023). In this way, everyday practices act to resituate people in ecologies, and remind us how we are entangled with the fate of the coasts and oceans we love. Activities like beach clean-ups certainly remind us of such tensions (see for example Wyles et al., 2017).

This shift in perspective – to feeling ourselves as part of ecologies – results in powerful forms of environmental activism that are based in a deep sense of care for places and communities. This is clear in emerging forms of community activism such as the Southern Ocean Protection Embassy Collective that we discussed at the opening of this chapter. Groups like Surfrider and Surfers for Climate in Australia and Right to Roam and Surfers Against Sewage in the UK are all reactions to environmental challenges observed by recreational walkers, fishers, surfers, swimmers, and hikers in their everyday relationships to oceans and rivers. These groups are also navigating how climate activism remains tied up in cultural politics over who has authority to speak for the coasts and oceans we care so deeply about.

Previous research has shown that coastal management, science, and media have been dominated by particular groups, with fewer opportunities for contributions by First Nations people, women, people of colour, and people with disabilities. Yet the politics of climate change – and rising and polluted oceans – are not separate from the ongoing effects of colonisation and imperialism and the marginalisation of diverse groups and world-views (Bennett, 2019). Approaches to these complex problems require ecological ways of thinking, which help account for *human* diversity and systems, as much as they do for non-human animals, plants, and geographies. This is the power and everyday nature of grassroots climate activism and attunement through creative practices that we have highlighted in this chapter. Where art as activism draws on key contemporary art methodologies of attuning to site – being present and in situ across time – it can be harnessed to reveal creative nature–culture convergences (Latour, 2014).

Conclusion: Beyond Climate Literacy

Landscapes enact more-than-human rhythms. To follow these rhythms, we need new histories and descriptions, crossing the sciences and humanities. As artists, we conjure magical figures, weave speculative fictions, animate feral and partial connections. We necessarily stumble. And try again. With every mark, difference haunts and struggles to appear anew.
(Tsing et al., 2017, p. 12)

As climate activists, our approach is that attunements to ecologies are achieved through repeat visits to a place or site, by being present and in situ, by noticing the movement of the world informing our relationship with the place, and by having our attention drawn to the multiplicity of ecological entanglements always in motion. Creative practices and time-based sensory activities have a clear role in activating people to think in new ways with places and ecologies and to challenge established, structured, instrumental learning activities. Painting, drawing, dancing, walking, foraging, collecting, sculpting, recording, listening, tasting, and more require the art of noticing or attuning. Experiencing places and ecologies in such ways can draw our attention to forms, qualities, and relationalities that we then re/produce through our own interpretations. Such processes can produce a range of outcomes, both personal and collective, that can hold space for ongoing creative inquiry as they are shared with others in new forms of noticing. This includes the formation and dispersal of temporary communities and collective acts that are both personal and intimate, as well as collaborative and connecting. New perspectives and conversations are enabled, which in turn can lead to new values and practices.

Arts of attunement can also result in new ethical relationships to caring for places. The kind of learning we are discussing in this chapter is derived from a pedagogy of creative experiences and entanglements with the coasts and oceans, rather than acquiring scientific knowledges. Although such specific knowledges can also be an outcome of creative attunements to places, this research is focused on how ocean encounters and knowledge sharing through creative practices facilitate meaning-making and relationship-building opportunities. They are based in experience, affect, and emotion; things felt, seen, tasted, heard, and shared. In order to access diverse understandings of oceans, attunements require relationships of trust, dialogue, and learning between people which reject any one claim to authority over knowledge (Neilson and Seixas, 2022). The grassroots activism and attunement we are describing is not designed to emphasise human/Western exceptionalism but to activate recognition that people should not remain at the centre of how we think about the ocean and collective community wellbeing. This is not a rejection of Western, science-based knowledge but an opening up to complex, diverse, and multiple ways of understanding multispecies, polluted, warming ocean ecologies, which is more aligned with Indigenous and First Nations worldviews, amongst others.

As we recognised in the introduction, it is important, however, that we do not overstate what kinds of knowledges and authorities can be accessed by participants through attunement activities. This is especially important in settler-colonial contexts where narratives of deep connections to places and ecologies are claimed as a result of developing intimacies with ecologies (Booth, 2020). While many settler 'locals' do possess significant and intergenerational knowledges of places, these cannot be equated with the genealogical and cultural knowledges and custodianship of traditional owners (Schenirer, 2021). Still, settler knowledges developed through personal, family, and community histories and experiences with coasts and oceans are key everyday practices through which participants build cumulative knowledge through observation, experience, and conversations (Lucashenko, 2006). As grassroots climate activist settler participants, we have much to offer in our understandings of human-ocean ecologies and kin-based entanglements, but this must be done with recognition of colonial injustice and First Nations peoples' sovereignty at the heart of activist practices.

The participatory climate activist attunements we have described in this chapter aim to facilitate forms of intimacy and vulnerability with ecologies that can enhance individual and collective climate action in everyday life, including away from the water. As the climate crisis deepens, we believe we need to encourage attunement to places through activities that meet people where they are and ensure that relationships are rooted in relationalities that remain accessible, relational, potential, non-prescriptive, and affective.

Notes

1 Djab warrung Whale, Eel and Platypus Songline story. Credit: Yaraan Couzens-Bundle.
2 Accessed on https://www.facebook.com/SouthernOceanProtectionEmbassyCollective, November 14, 2023 post 9.51 am
3 See https://drillwatch.org.au/
4 The interdisciplinary team included Wendy Steele (Urban Research), David Carlin (Creative Writing), Fiona Hillary (Art in Public Space), Jordan Lacey (Soundscape Design), Aviva Reed (Visual Ecologist), Liam Fenaughty (*KERB Journal of Landscape Architecture*) and Ana Lara Heyns, Alexandre da Silva Faustino, and Rachel Iampolski and Nicolas Guerra-Tão (The Alliance for Praxis Research).
5 See Lagerspetz (2018), Steele et al. (2020).
6 This recording can be accessed on https://www.festivalofurbanism.com/events/fou2021/urban-undercurrents-the-hidden-infrastructure-of-wild-cities
7 https://www.nhm.ac.uk/discover/anna-atkins-cyanotypes-the-first-book-of-photographs.html
8 Port Phillip Bay, The City of Melbourne – https://www.emelbourne.net.au/biogs/EM01165b.htm)
9 Yunkaporta, T. (2019). *Sand Talk: How Indigenous Thinking Can Save the World*. Melbourne: Text Publishing, pp. 2–3.

References

Bennett, N. J. (2019). Marine social science for the peopled seas. *Coastal Management*, 47(2), 244–252.

Bird Rose, D. (2015). The ecological humanities. In K. Gibson, D. B. Rose and R. Fincher (eds.) *Manifesto for Living in the Anthropocene* (pp. 1–5). Punctum Books.

Bird Rose, D. (2017). Shimmer: When all you love is being trashed. In A. Tsing, H. Swanson, E. Gan and N. Bubandt (eds.) *Arts of Living on a Damaged Planet* (pp. G51–G61). Minneapolis: University of Minnesota Press.

Booth, D. (2020). Nature sports: Ontology, embodied being, politics. *Annals of Leisure Research*, 23(1), 19–33.

Braidotti, R. (2017). Posthuman critical theory. *Journal of Posthuman Studies*, 1(1), 9–25. https://doi.org/10.5325/jpoststud.1.1.0009

Evers, C. (2009). 'The point': Surfing, geography and a sensual life of men and masculinity on the Gold Coast, Australia. *Social and Cultural Geography*, 10(18), 893–908.

Lagerspetz, O. (2018). *A Philosophy of Dirt*. London: Reaktion Books.

Latour, B. (2014). Agency at the time of the anthropocene. *New Literary History*, 45(1), 1–18.

Lucashenko, M. (2006). All my relations: Being and belonging in Byron Shire. In Anna Haebich and Baden Offord (eds.) *Landscapes of Exile: Once Perilous, Now Safe* (pp. 61–67). Bern: Peter Lang.

Neilson, A. L. and Seixas, E. C. (2022). Ocean literacies: Splashing around on the beach or venturing into the dark abyssal sea. In *Sustainable Policies and Practices in Energy, Environment and Health Research* (pp. 61–77). Cham: Springer.

Neimanis, A. (2013). Feminist subjectivity, watered. *Feminist Review*, 103, 23–41.

Olive, R. (2015). Surfing, localism, place-based pedagogies, and ecological sensibilities in Australia. In B. Humberstone, H. Prince and K. A. Henderson (eds.) *Routledge International Handbook of Outdoor Studies* (pp. 501–510). Abingdon, Oxon: Routledge.

Olive, R. (2023). Swimming and surfing in ocean ecologies: Encounter and vulnerability in nature-based sport and physical activity. *Leisure Studies*, 42(5), 679–692.

Potter, E. (2005). Ecological consciousness in Australian literature: Outside the limits of environmental crisis. *Hawke Research Institute for Sustainable Societies, Working Paper Series*, No. 29. http://www.unisa.edu.au/Documents/EASS/HRI/working-papers/wp29.pdf

Satchell, K. (2007). Shacked: The ecology of surfing and the surfing of ecology. In *Online Proceedings of, 'Sustaining Culture', Annual Conference of the Cultural Studies Association of Australia*. University of South Australia, Adelaide, December 6–8.

Satchell, K. (2008). Reveries of the Solitary Islands: From sensuous geography to ecological sensibility. In A. Haebich and B. Offord (eds.) *Landscapes of Exile: Once Perilous, Now Safe*. Bern: Peter Lang.

Schenirer, S. (2021). *Indigenous Ecological Knowledges, Interviewed for Saltwater Library (Podcast)*. https://movingoceans.com/saltwater-library-podcast/01-indigenous-ecological-knowledges/ [Accessed 30 August 2022].

Steele, W., Davison, A. and Reed, A. (2020). Imagining the dirty green city. *Australian Geographer*, 51(2), 239–256.

Tsing, A. (2015). *The Mushroom at the End of the World: On the Possibility of Life in Capitalist Ruins* (Pilot project. eBook available to selected US libraries only). Princeton: Princeton University Press. https://doi.org/10.1515/9781400873548

Tsing, A. (2021). EXTRACT| carrier bag fiction: Sarah Shin in conversation with Anna Tsing, Tank TV. *Tank Magazine*. https://magazine.tank.tv/tank/2021/06/carrier-bag-anna-tsing [Accessed 10 November 2023].

Tsing, A., Swanson, H. A., Gan, E. and Bubandt, N. (eds.) (2017). *Arts of Living on a Damaged Planet: Ghosts of the Anthropocene*. Minneapolis: University of Minnesota Press.

Wyles, K. J., Pahl, S., Carroll, L. and Thompson, R. C. (2017). Can beach cleans do more than clean-up litter? Comparing beach cleans to other coastal activities. *Environment & Behaviours*, 49(5), 509–535. http://doi.org/10.1177/0013916516649412

Yunkaporta, T. (2019). *Sand Talk: How Indigenous Thinking Can Save the World*. Text Publishing.

9
A CONTEMPLATIVE PEDAGOGY OF LISTENING

Linda J. Chase

Prelude

How can musical artists articulate a vision of hope that leads to action? Musicians are already listening, but can contemplative listening increase the capacity to hear the falling world?[1] *A Contemplative Pedagogy of Listening* addresses climate justice issues in ways that are relevant and applicable to musical artists. Although listening applies to all of the arts, this work focuses primarily on music and sound. This arts-based methodology cultivates compassion, reignites relationship with the natural world, and bridges the arts and ecology. Based in praxis and influenced by the capacity to advance climate justice through aesthetic education (Greene, 2001), this pedagogy explores listening practices that nurture transformation. Grounded in the need for reconnection with the earth and each other, these listening practices artfully motivate, build solidarity, and develop courage to challenge systems perpetuating climate collapse. Activism in the classroom guides emerging artists to discover how they can participate in the climate movement.

How does climate activism relate to the musical and career choices of an artist? At this precarious time on the planet, artists have great potential to inspire action. Emphasising environmental studies and sustainability education in arts curriculum directs students to discover how their voices can contribute to building an equitable and sustainable world. Confronting realities of the climate emergency through artistic practice allows musicians to respond using their musical language. Socially engaged artist Pablo Helguera describes this interdisciplinary process.

> Socially engaged art functions by attaching itself to subjects and problems that normally belong to other disciplines, moving them temporarily into a space of ambiguity. It is this temporary snatching away of subjects into the realm of art making that brings new insights to a particular problem or condition and in turn makes it visible to other disciplines.
>
> *(Helguera, 2011, p. 5)*

Moving into "a space of ambiguity" leads to reinterpretation. Scholar/artist James Rolling proposes that artistic practices of reinterpretation become "reflexive tools for inquiry"

DOI: 10.4324/9781003396567-11

(Rolling, 2018, pp. 496–497). When a concept is made visible (or audible) the artist extends the discourse (Rolling, 2018). With arts-based research (ABR) as a foundation, this pedagogy develops multidisciplinary thinking and shapes modes of understanding using creative practices. A process-oriented research, arts-based methods reframe questions through artmaking.

Creating pieces that grapple with climate issues encourages musicians to engage *through* the process of making sound. This process extends perception and deepens understanding of climate consequences and solutions.

Examples:

1 Accompany listening to climate news reports with your instrument. Improvise to images of climate disaster. Portray the acoustic environment. Process the questions.
2 Improvise urban and forest soundscapes. Invite places into your awareness through sonic vibration. Create pieces about protests. Acknowledge disconnection.
3 Listen to personal stories of climate impact and restoration. Retell them musically.
4 Create a collage of climate-change articles with text and photos. Use it as a graphic score.

This chapter describes ways that my Ecomusicology classes engage with climate issues through listening practices that empower action. We examine performance arts that are influenced by nature, analyse songs voicing resistance, and practise contemplative listening. We discuss how climate change and biodiversity loss are reported in the news and join organisations that are pushing back against injustice. Students create pieces inspired by more-than-human beings, write community activist songs, make sound documents, and create listening meditations. Highlighting solutions, students collaborate with environmental and activist organisations.

Ecosystems (like art) rely on connection. Ecology investigates interrelationships within an environment. Fields of ecomusicology make connections between the natural environment and musical understanding. The *Grove Dictionary of American Music* offers an extensive definition of ecomusicology, including this excerpt: "[Ecomusicology is] the study of music, culture, and nature in all the complexities of those terms. Ecomusicology considers musical and sonic issues, both textual and performative, related to ecology and the natural environment" (Allen, 2013). The interdisciplinary fields of ecomusicology attract musicologists and ethnomusicologists but also musicians, activists, and scientists (Allen, 2023). My perspective as a composer/improviser brings arts-based inquiry into ecomusicology and also draws from acoustic ecology and sustainability education.

Without a universal definition of music, one must remember how perceptions of sound vary according to culture, experience, and place. For some, the process of listening transforms sound into music. Appreciating the musicality of nature brings attention to relationships with more-than-human beings and illuminates interdependence between species. *A Contemplative Pedagogy of Listening* explores various approaches to listening and artmaking with activism. Listening practices bridge artistry with action because engaging with the planetary crisis through artistic forms deepens understanding and encourages creative approaches to change. Compassionate listening nurtures the ability to not only sympathise but to suffer with, opening the heart to make a difference in the kind of art we make and the kind of world we create.

Reflection Confronts Distraction

Noise and distraction masking the problems
Questions of justice unravelling.[2]

As the pace of life intensifies, the need to reflect increases. Rather than reacting to information overload with a shield of detachment, devoting time for mindful listening resists a culture infatuated with screens of distraction. Contemplative listening provides space to reflect and translate reflections to the world, cultivating artistic *and* ecologically conscious lives.

In the midst of multiple environmental and social catastrophes facing humanity, it is critical to address how the climate crisis triggers anxiety and feelings of helplessness which hinder the ability to respond. How do we confront climate breakdown without mental breakdown? We mourn for what has been lost, but how do we move forward? How do we respond in a time of such uncertainty? Ecophilosopher and activist Joanna Macy guides people to "honor our pain for the world." She instructs that "Intellectual awareness by itself is not enough. We need to digest the bad news. This is what rouses us to respond" (Macy and Johnstone, 2022, p. 69).

Students feel the burden that their generation will have to bear and are hungry to engage in honest dialogue. When educators provide a place to face fears, students become empowered to transition from individual to collective, from despair to hope, from inaction to action. Contemplative listening clears a path to express grief and anger artfully. Music provides comfort but also offers a way to process the uncertainties of our future, helping us cope with climate trauma. When we feel like screaming, music agitates, expressing dissonance or vocalising contradictions.

In order to sustain our spirits and energise activism we must also remember that celebration challenges despair. Faith-based community activists advocating creation justice embrace rituals of celebration and gratitude. In the ecology encyclical, *Laudato Si'*, Pope Francis writes, "May our struggles and concern for our planet never take away the joy of our hope . . . let us sing as we go" (Francis, 2015). Music invites celebration. A brass band draws attention to and accompanies a park clean-up; a climate change lecture shares the stage with a string quartet. When Louis Armstrong sings "What a Wonderful World" (Thiele and Weiss, 1967), he reminds us to savour the beauty of the green trees and blooming roses.

Students respond by writing a planetary concern on a 3 × 5 card. On the other side, they write a celebration of nature. Cards are exchanged and students engage in call and response musical conversations practising compassionate listening. This practice allows them to process concerns through musical ways of knowing.

Wisdom of Places

Writer in Residence for the Chickasaw Nation, Linda Hogan, tells us, "There is a language . . . that rises from the land itself." Drawing from Indigenous knowledge, she describes a way of being in the world that recognises relationships with the "larger-than-human" community. "There is a way that nature speaks, that land speaks. Most of the time we are simply not patient enough, quiet enough, to pay attention to the story, to be attentive" (Hogan, 2004). The land speaks but much of what is heard does not fit into the objective, mechanical point of view found in Western ways of thinking (Deloria and Wildcat, 2001). Belief systems from cultures that represent humans as interdependent with other life forms differ from prevailing ideas of the industrialised West that promote the individual as the centre of a technology-and consumer-dependent lifestyle. Tewa author and professor Greg Cajete instructs, "As a world searches for how to address current environmental, social, and cultural crises, it makes sense to look to cultures that inherently respect – in thought and practice – a sacred sensibility for our relationships with each other and the Earth" (Cajete, 2015, p. 211).

Hearing landscape begins with listening to other-than-human species, but the opportunity to cultivate relationships with non-humans in undeveloped places has diminished for many people. Worldwide, the majority of people reside in cities without regular access to rural or wilderness areas, and in the US, most people work inside. For an indoor species, music about nature reminds us of our longing for connection.

Slow Down to Listen to the Land

Landscape music contributes to a pedagogy of listening by fostering a sense of place. Musical representations of nature transport us into imagined environments. Anthropologist Mark Pedelty suggests that "Place is re-created in some small part through music" (Pedelty, 2016, p. 193). Music about nature creates a space to build associations and emotional connections that initiate relationship with place from afar and trigger a sense of nostalgia even during the first encounter. While Artist in Residence at Grand Canyon National Park, I composed *Grand Canyon Sketches* for voice and strings. Several months following the Boston premiere, one of the violinists visited the Canyon. She didn't want to take in the view until she could listen at the same time. Before that moment of awe upon first sight, she put our recording on her headphones. Opening her eyes, she listened.

To see beyond eternity
and reach into the sky
longing just to hold you
never let them take away
the beauty of your sighs.[3]

Throughout history and across geography, composers from many cultures have described environments by musically portraying the feeling of being there. The Japanese *shakuhachi* repertoire features pieces such as "Cranes on a Nest." Navajo-Ute R. Carlos Nakai evokes the Southwest with "Whippoorwill" on cedar flute. In the Western art music tradition, Florence Price's "Mississippi River Suite" incorporates African-American elements with European musical traditions (Blumhofer, 2018). Charles Ives paraphrases hymns and marches to recreate "Places in New England." Composers mimic sounds of birds as heard in Olivier Messiaen's ornithological compositions. Composer/zoomusicologist Emily Doolittle incorporates electronics and transcriptions of whale songs in "Social Sounds of Whales at Night." Relationships between place and sound are described in ethnomusicologist Stephen Feld's research about sonic ways of knowing place (acoustemology). He reports that the Kaluli people from Papua, New Guinea improvise duets with cicadas and birds and sing with waterfalls (Feld, 1982). "Kaluli not only take inspiration from, listen to the forest, but become part of it" (Feld, 1984, p. 395). Another example of acoustemology is found with the Baka Pygmy people in central Africa who practise *liquindi* – or water drumming. While standing in a lake or river, women and girls strike the surface of the water with their hands, producing a percussive effect. Singing accompanies their collaborative improvisations. These interactions facilitate becoming part of place (Nettl et al., 2001).

Experimental music incorporates recordings of non-human voices to facilitate interspecies understanding. "Syrinx" by Pamela Z for solo voice, electronics, and sampled birdsong offers an insightful way to perceive non-human language. Pamela Z electronically slows and lowers the pitches of birdsong. Using a gesture-based MIDI controller, she manipulates a birdsong sample, shifting it down in pitch until it is in a range where she can sing it. Then she speeds up her voice recording until she sounds like a bird, demonstrating language similarities without saying a word.

The main purpose in asking students to create pieces informed by being in nature is to cultivate relationship with place. Scholar/activist Manulani Meyer informs, "It does not matter if you view land as Mother or sky as Father. . . . What matters is that you have some kind of relationship with this place we live and breathe in" (Meyer, 2013, p. 99). Students discuss how culture informs worldviews and how ongoing colonisation exploits lands and severs relationships with the natural world, community, and culture (Wickham, 2012, pp. 188–199). Students contemplate how they as culture-makers might cultivate ecological relationships.

Students respond by observing mimetic and associative musical techniques, spending time in nature, and then composing pieces that describe, express, or converse with more-than-human beings. Topics have included: transcribed birdcall for clarinet and viola, imagined sounds of nature's cry, recreated backyard sounds, and improvisational interactions with pigeons. When students share their love for places, they renew connections. This process influences composer, performer, and the audience.

A Failure of Culture

"In short, the environmental crisis is a failure of culture" (Allen, 2011). This statement by musicologist Aaron Allen fuelled my exploration into the fields of ecomusicology. Pedelty tackles questions of artmaking. "How could the music we love ignore the greatest crisis we face?" (Pedelty, 2012, p. 6). Reflecting on the consequences of my culture's belief systems and the significance of the artist's capacity to shape public opinion challenged me to realise *my* responsibility as an artist.

How has culture failed? Although hits like "Mercy, Mercy Me", "Big Yellow Taxi" and "Earth Song" come to mind, cultural response to the planetary crisis has been minimal. Despite decades of warnings by climate scientists, mainstream culture (movies, music, TV, social media) has failed by not *centring* attention on the crisis. Culture remains compliant when programming repertoire to satisfy donors rather than highlighting compelling pieces about the planetary emergency. Popular culture fails when producing spectacle for consumption that supplies escape without providing critical reflection. Media fails when conforming to economic systems rather than using art to emphasise issues with a compassionate voice. Live performance electrifies crowds, but what are we energised about? What happens when that energy is harnessed for positive change?

Climate change is a symptom of an economic system that prioritises profit at all costs and normalises exploitative and extractive practices as "business as usual." Emerging musicians who are already convinced "that's just the way it is" may not realise that choosing to engage in business as usual means perpetuating systems that propel us further into climate breakdown.

But it's complicated. Capitalism sells the cultural products that artists need to survive and most artists do not have celebrity platforms or resources. How do these economic realities influence decisions made by artists and how do creative responses lead to healing the planet?

There are clear relationships between lifestyles supported by exploitation and values that are taught in schools. Author, environmentalist, and educator David W. Orr advocates for teaching ecoliteracy across the curriculum in higher education. In his book, *Earth in Mind*, Orr explains that "The crisis we face is one of mind, perception and values" (Orr, 2004). Education must participate in a transition to a more "empathic civilization" along with political changes, social movements, and activism (Orr, 2018). Brazilian educator Paolo Freire instructs in *Pedagogy of the Oppressed* (1970) that if people continue to be educated to fit into the world as it is, there will be no understanding of processes required to change the world in pursuit of a fuller experience of humanity (Freire, 1970). When conventional education teaches students how to succeed in the current system rather than how to decolonise knowledge, we miss an opportunity to expand environmental consciousness and develop stewardship. When music entrepreneurship promotes how to make the most money possible without teaching ethical ways to participate in building a just and sustainable world, the potential of artistic thinkers is misdirected. Activism in arts education requires a creative approach to social change, challenging artists to question what kind of world we want to live in.

Unjustified dominion destroys the bond we seek
Respect the laws of nature, take only what you need.[4]

Culture endorses ways of thinking. Helguera observes that it is not so much the facts as the *perception* of the facts that makes a difference in people's reactions (Helguera, 2011). Artists shape belief systems and can initiate conversations that direct us to envision the environment as being more than just a resource for consumption. Nature is not a commodity and music doesn't have to be either. Shifting priorities away from profit directs artists to redefine success and imagine a better world while deepening the purpose of their musical calling. Environmentalist David W. Orr proposes that the world does not need more successful people but rather courageous people willing to work for change.

> The plain fact is that the planet does not need more successful people. But it does desperately need more peacemakers, healers, restorers, storytellers, and lovers of every kind. It needs people who live well in their places. It needs people of moral courage willing to join the fight to make the world habitable and humane. And these qualities have little to do with success as we have defined it.
>
> *(Orr, 2004)*

In the music industry success means touring (flying) around the world with celebrity status. Performing at large sold-out venues defines success. The bigger the better. But success can mean community engagement and regional recognition. Success can mean healing the earth. When musicians bring stories of resilience onto the stage and promote ethical response, cultural activism extends out into the world.

> As our field of action extends, via our students and research, from educational institutions into the public sphere of planning, management, policy, and even cultural thought, ecomusicology becomes activism.
>
> *(Allen, 2014)*

Is the environmental crisis fundamentally a crisis of culture? If so, this poses a huge challenge to musical artists today. Ignoring problems results in perpetuating them. Rather than assuming human dominance, musicians are listening to the earth and responding. Musical artists inspire change by highlighting solutions, telling the stories of those who are affected by climate disaster and visualising a better world. Music creates a teachable moment and the audience has come to listen.

Sounding Resilience

Art helps us imagine how things could be different. Aesthetics lift us up and help us live compassionately by offering positive and challenging images. Scholar/activist bell hooks writes, "The function of art is to do more than tell it like it is – it's to imagine what is possible" (hooks, 1994). In practices of socially engaged art, when artists work with community, the art itself educates and becomes a form of activism. Political critique voiced through the arts amplifies the call to action. Political theatre groups like Bread and Puppet Theatre in Vermont bring art to the streets. Their manifesto, "WHY CHEAP ART?" proclaims:

ART IS NOT BUSINESS! ART IS FOOD. You can't EAT it BUT it FEEDS you.
ART has to be CHEAP & available to EVERYBODY (Bread and Puppet, 1984).

Social scientist Courtney Brown points out in *Politics in Music* (2008) that throughout history musicians have been contributing to social and political movements. Raising voices through song nurtures solidarity and collective identity (Brown, 2008). In *Civil Rights Music: The Soundtracks of the Civil Rights Movement*, Reiland Rabaka relates that music's purpose "was much more than merely music" (Rabaka, 2016). When activist Fannie Lou Hamer sang "This Little Light of Mine," the power of song rallied people to face voter suppression, violence, and struggles to come. Freedom singer Rutha Mae Harris remembers, "Music was an anchor. It kept us from being afraid" (Deggans, 2018). Sounding resilience is activism.

Activists have critiqued cultural attitudes limiting the definition of environment as undeveloped wilderness, where climate change wears the face of a polar bear. All environments will be affected by climate change, but when disaster strikes, the ability to relocate or rebuild is not equal. The "most severe harms from climate change fall disproportionately upon underserved communities who are least able to prepare for, and recover from heat waves, poor air quality, flooding, and other impacts" (EPA, 2021). It is critical to focus on racism and classism in environmental policy and advocate for climate justice in all communities.

Yale University Dean and hip-hop artist Thomas RaShad Easley discusses how the environmental movement has failed to connect with urban communities and people of colour but explains that music can be an entry point by showing how environmental issues are relevant. When hip hop communicates the realities of environmental justice, it resonates with those most affected. "By confronting issues from contaminated water to climate change, hip hop music can help bridge [the environmental movement] divide and bring home the realities of environmental injustice" (Easley, 2020).

Musicians bring the dialogue to popular spaces. Sam Cooke, Nina Simone, Pete Seeger, Marvin Gaye, Bernice Reagan, Mahalia Jackson, Mavis Staples, and Joni Mitchell have led the way to challenge the status quo. Popular culture is "where the learning is," observes bell hooks. (hooks, 1997, p. 2). High-profile artists can use their platforms to raise awareness. In *Rape of the World*, Tracy Chapman challenges us to testify that we've witnessed crimes.

Once we are aware of these crimes (strip mining, clear cutting, poisoning, bombing, global warming), she tells us not to stand aside. "Where is the Love?" (Black Eyed Peas, 2003) confronts disinformation and negative images in the media. Instead of spreading animosity, the song urges us to spread love. Tom Paxton mocks climate denial with satire, joking that to address climate change we just need more sunscreen or a wide brimmed hat in "What's so bad about that?" (Paxton, 2018). The music industry organisation Music Declares Emergency offers a "Climate Pack" for performing artists. Although the "NO MUSIC ON A DEAD PLANET" T-shirts are catchy, we must delve deeper to figure out *how* music motivates the collective action needed to tackle the climate emergency.[5]

Noriko Manabe discusses the role of music and political resistance in Japan's antinuclear movement in *The Revolution Will Not Be Televised: Protest Music After Fukushima* (Manabe, 2015). Considering cultural views on the reluctance of media to discuss antinuclear positions, Manabe observes that, compared to major-label musicians, independent musicians have more freedom of expression through online communication and demonstrations.

Grassroots groups outside of the commercial music industry are leading by example with significant work. Encouraging critical listening and creating interdisciplinary, transformative, and transgressive art, artists challenge conventional ideology with new narratives and collaboration. Musicians are traveling by train, and producers are being selective about performance spaces and merchandise. More and more artists are amplifying activism through cultural expression.

Bridging science with art, the San Francisco-based Climate Music Project joins scientists with musicians to engage with the audience. Guided by scientific data, composers write multi-genre music to bring climate science into new spaces. For example, using pitch and rhythm to represent temperature and carbon pollution levels, with data graphs projected behind the stage. After the concert, scientists answer questions and have noticed how music seems to increase audience members' motivation to take action (YCC, 2019).

Musicians are expanding the confines of performance and redefining collaboration between nature and culture, helping us listen more deeply. Promoting the necessity of renewable energy, UK-based Orchestra for the Earth presents place-based concerts such as Mozart's "Serenade for Winds" performed next to the first commercial wind farm in the UK. The orchestra partners with environmental groups, and ticket sales support tree planting. Conductor John Warner points out that concerts "allow us to see that we are not alone in worrying about the planet or in our efforts to protect it" (Head, 2022). German-based orchestra Orchester des Wandels confronts the climate crisis through creative concert programming. On display at their performance of Strauss' "Alpine Symphony" are photographs of glaciers taken when the piece was composed alongside recent photos revealing climate-related melting. This group also organised flash mob performances of "Ode to Joy" to bring attention to the UN's 17 goals for sustainable development. Music gets our attention.

Artist, activist, and founder of the Center for Cultural Power Favianna Rodriguez supports culture makers to engage with climate justice issues. Rodriguez explains the urgent need for artists to tell compelling stories that show us what the world could look like.

> We are not seeing diverse stories about climate because the folks who control the cultural engine in the United States . . . are overwhelmingly White men. And they exist in a status quo that has entrenched their power and dramatically narrowed our collective gaze.
>
> *(Rodriguez, 2021)*

Stories need to be heard. Journalist Devi Lockwood believes "Listening is a form of activism" (*MIT Technology Review*, 2021a). She bicycled around the world to hear stories by people whose lives have been drastically altered by climate change. Her book *1,001 Voices of Climate Change* (Lockwood, 2021b) shares stories that personalise abstract climate science. Artists are speaking out for change through storytelling. Climate activist and musician Tanya Kalmanovitch created the *Tar Sands Songbook*, a solo performance combining violin, field recordings, and storytelling to investigate our "unseen relationships to oil." Born near the Athabasca Oil Sands, Kalmanovitch gathered voices of oil patch workers, family members, activists, and First Nations elders to offer audiences a personal "meditation on our complicity in extractivist culture" (Kalmanovitch, n.d.).

Composer Judith Shatin articulates her call to action in "Ice Becomes Water" for string orchestra, electronics, and field recordings of melting glaciers. When glaciers melt, the air bubbles make popping sounds when they meet the sea. The faster that chunks of glaciers melt, the more sound they make, suggesting that acoustics could help track the speed of glacier melt. Musicologist Denise Von Glahn observes how this work "might speak to large numbers of listeners far beyond the typical audience of electroacoustic music and . . . connect with others' understanding of how our world works" (Von Glahn, 2023, p. 180). Listening to the sound of climate change through music presents research in a captivating way and reaches beyond the scientific community. Hearing the complexity and the loss invites emotional response.

Another important work, "Requiem for Our Earth" by Cecilia Damström for women's voices, uses documentary texts related to climate change along with electronics, video projections, and choreography. It incorporates voices including climate activist Greta Thunberg and pre-recorded sounds such as fire, explosions, logging, and melting ice. The mixed-media techniques amplify the intensity and add to the sensual experience. Musicologists Torvininen and Välimäki explain that this piece musically represents "the very process of atmospheric changes in the context of environmental crisis" (Torvinen, and Välimäki, 2023, p. 98).

Changes in climate lead to migration, unravelling
Masking the problems, concealing the symptoms.[6]

Climate change, homeland security, and border policies are interrelated. Climate drought escalates migration, forcing people to travel to the border only to arrive at a highly militarised wall. But music fights back. Northern Arizona University professor Robert Neustedt, along with local and well-known musicians, produced *Border Songs,* an album addressing border policy consequences. Proceeds went to No More Deaths, an organisation that places jugs of water in the desert for migrants trying to cross the border. In addition to the humanitarian aid, the album generated discussion about the global border crises.

My most recent musical call to action, an oratorio for ecojustice, highlights compassion and solidarity. Based on *Laudato Si'* (Pope Francis, 2015), "For Our Common Home: Resounding Ecojustice" (Chase, 2021) lifts up humanity's urgent need to heal our relationship with Earth and each other.

Earth was here before us
A gift for us to keep
To share the fruit with those in need

Protect and oversee
Unjustified dominion
Destroys the bond we seek
Respect the laws of nature
Take only what you need[7]

Confronting our planetary crisis through system change will require massive non-violent protest and artful expressions of love and rage. Drastic lifestyle changes are needed along with the commitment to vote, attend community meetings, and take to the streets. And sing in solidarity. Pete Seeger endorses singing for social change. "[When] a crowd joins in on a chorus as though to raise the ceiling a few feet higher, then they also know there is hope for the world." (Seeger, 2007). Composer John Luther Adams considers the responsibility of the artist.

Our survival as a species depends on a fundamental change of our way of being in the world. If my music can inspire people to listen more deeply to this miraculous world we inhabit, then I will have done what I can as a composer to help us navigate this perilous era of our own creation.

(Adams, 2015)

Students respond by writing songs that are intended to be sung collectively and empower communities. Informed by music from social movements, their songs adapt to issues in different communities. For many students, this is the first time composing for group participation. They report that their songs didn't seem complete until the class sang them together. Composers working with electronic music craft pieces that incorporate clips of recordings from IPCC reports or climate activist speeches. Demonstrating the value of working together, one student observed, "Being in service to each other is at the heart of activism."

Environmental Listening

There are many kinds of listening. Listening to recordings of music is customary in music conservatories. Discussing musical details creates an environment conducive to sharing ideas. Like listening to music, paying attention to any sound expands consciousness. Listening to non-human sounds plants seeds of awareness. Robin Wall Kimmerer tells us in *Braiding Sweetgrass* that "Listening in wild places, we are audience to conversations in a language not our own" (Kimmerer, 2013, p. 48). Denise Von Glahn explores how listening to nature cultivates relationship with the environment and makes connections between music, nature, and place in *Music and the Skillful Listener: American women compose the natural world*.

We live in a sounding world, and musicians are especially aware of that reality; our job is to attend to sound. Paying attention increases the value of everything we hear and reminds us of our place within this shared vibrating organism.

(Von Glahn, 2013)

Listening is a way of knowing. Deep Listening® meditations, whole body listening, and nature listening practices are foundational in this pedagogy. Composer, electronic music pioneer, and humanitarian Pauline Oliveros developed Deep Listening® which she describes as "learning to expand the perception of sounds to include the whole space/time continuum of sound." Deep Listening® "taps into the beyond, the in-between, to an unseen or a spiritual state of communication" (Oliveros, 2005, p. xxiii). Consider her piece "Environmental Dialogue."

Each person finds a place and becomes aware of the field of sounds from the environment. Gradually they begin to reinforce the pitch of any one sound source that has drawn their attention. The sound source can be reinforced with an instrument, vocally or mentally.

(Oliveros, 2005)

Deep Listening® incorporates sonic meditations, movement practices, listening in dreams, and listening to daily life. According to Oliveros, inclusive listening expands environmental consciousness. All sounds compose the soundscape, although "we may not be of it, we are in it" (Oliveros, 2005, p. 18).

Along the bike path behind my house, I observe that the barking dog and the skateboarder inhabit my soundscape. We occupy a sound environment, and for a moment our sounds become a shared world. Our bodies are resonant chambers. Unknowingly, we connect with each other through sound. We are in it together.

The entire body senses sound. Perceiving the tangibility of sound develops deep awareness because touch, like sound, senses vibrations. Removing the protective veil by going outside exposes our senses to intermittent birdsong, unexpected rain, or rumbling thunder. Deep Listening® incorporates whole body listening utilising movement practices by artist/dancer Heloise Gold. Her walking meditations awaken kinesthetic listening. In one movement, participants voice connection with each step. "I caress the earth with my steps: the earth in turn caresses me" (Gold, 2008, p. 8).

As a result of Deep Listening®, I've discovered a rare and most precious sound. While walking my dog on a freezing, windy afternoon, a mysterious sound drew me to the edge of the pond near my house. The pond had frozen, thawed, then partially froze again. That afternoon the wind blew broken ice pieces along the shore creating the most beautiful ice-wind-chimes. Although I've walked along the shore many winters, I had never noticed this before. Deep Listening® invited me outside, opened my ears and my eyes.

Listen
and I will speak
said the tree.

Listening to the natural world was central to Canadian composer R. Murray Schafer. He composed pieces to be performed outside and incorporated music of nature in his compositions. Founder of Soundscape Studies, Schafer also developed innovative listening and sounding exercises. *A Sound Education: 100 Exercises in Listening and Sound Making* describes practices to teach different ways to listen. This work is about paying attention to listening. For training the perception of moving sounds he suggests:

Find a portable sound and move around the room with it while the group, with eyes closed points to it . . . now find a contrasting sound and move about in different directions while the group points to both sounds.

(Schafer, 1992, p. 20)

Some nature listening practices acknowledge spiritual purpose. According to Shinto, all sounds have spiritual power. Scholar and *shakuhachi* player Koji Matsunobu describes *otodamaho*, a Japanese meditative practice focused on the embodiment and appreciation of natural sounds. To achieve a state of unity, practitioners immerse themselves in a sound such as a waterfall or the rain. *Otodamaho* purifies the mind and body through listening (Matsunobu, 2007a).

Students respond by paying attention to sounds in their neighbourhoods and practising listening meditations. These actions inform the creation of their own practices intended to enhance or redefine listening. We examine how these practices might nurture empathy, address climate anxiety, or motivate action.

Sonic Journalism

Listening to the acoustic environment and incorporating recorded sounds into composition directs attention to the impact that humans impose on the planet. Field recordings contribute to this pedagogical practice by offering ecological insights through sound. Soundscape artists reflect on environmental issues through field recordings, spoken narrative, and/or music. A significant voice in soundscape studies is heard from composer and sound ecologist Hildegard Westerkamp. Her field recordings take place in urban and wilderness areas to report on issues through sound. Listening requires presence in the moment. "To be in the present as a listener is a revolutionary act" (Westerkamp, 2017). Westerkamp's sound document, "Under the Flight Path," (1981) leads listeners to a neighbourhood near the Vancouver International Airport where a citizen's group was trying to stop an expansion. The piece begins with loud aeroplane noise, making it difficult to hear the interview with a resident describing life under the flightpath. The conversation continues, but another aeroplane interrupts.

Sound artist Peter Cusack explains that in sonic journalism "field recordings are allowed adequate space and time to be heard in their own right, when the focus is on their original factual and emotional content" (Cusack, 2009). Cusack brings attention to places facing major human-inflicted damages such as oil fields, landfills, and the Chernobyl Exclusion Zone. His recording, "Sounds from Dangerous Places," stimulates conversation about destructive impacts on communities and ecosystems.

Appreciating sound also leads to conservation. The Japanese Ministry of Environment hosted a soundscape project called the *100 Soundscapes of Japan*. People nominated soundscapes such as waves on pebbles, ice floes, and the Hiroshima peace bell. These places are now protected. Furthermore, field recordings are addressing dying forests. Due to the drought in the Western US, beetle infestations have killed millions of trees. Composer David Dunn wondered if sounds were associated with this activity, so he invented a device to listen to the interior of the trees. Dunn explains:

> After making hundreds of hours of recordings inside hundreds of trees, I made a large sound composition that represented the incredible diversity of sounds made by bark beetles and their changing responses to the life cycle of tree hosts that they invade.
>
> *(Dunn, 2017)*

Scientists joined Dunn to stop bark beetles from tunnelling through the drought-stressed trees. They collaboratively created a sonic device that disrupts insect behaviours such as feeding, communication, and reproduction. The sounds from this device cause beetles to halt their basic functions, demonstrating how listening to trees leads to creative solutions (Rapaport, 2017).

Students respond by creating sound documents that report on a climate issue. To centre environmental listening as the creative process, students inquire: what can sounds teach us? Many students base their projects on urban noise. They observe whether streets sound wet or dry, listen to the scurrying of squirrel paws on tree trunks, and capture the feeling of being in places. Commenting on energy use, one outstanding piece was based on the rhythms of a malfunctioning radiator. Another effective project recorded the sound of filling up a gas tank. Not having a car, students walked to a gas station and asked a truck owner if they could record the sound of filling up at the pump. It took a long time.

Noise

Describe sounds with words
Rustles, bangs, swishes, thuds, squawks.

In this loud, fragmented world, blaring sounds generated by cities, traffic, and industrial sites negatively impact all beings. When the acoustic environment is disrupted, the ability for animals to communicate, find food, or locate mates is affected. Bird populations decline when exposed to the continuous noise of oil and gas operations. Whales in the Arctic are forced to retreat to areas without adequate breathing holes because seismic surveys and shipping noise masks their ability to communicate (Westdal, 2013).

Listening practices in urban settings give students entry points to empathise with species impacted by noise pollution. Environmental listening alters the way humans perceive the natural world because listening nurtures empathy. One method to increase sound awareness is through soundwalking.[8] In a soundwalk, the group leader directs attention to sounds in the environment and chooses which sounds are worthy of our attention. Participants listen for pitch, dynamics, rhythm, density, activity, and rest. Students discuss sonic perceptions, observe soundmarks unique to place, and notice sounds that evoke feeling or provoke judgement. All places have unique sounds.

One of my classes included both classical musicians training for orchestral work and improvising musicians looking for something else. For our first walk we visited a community garden. Students contributed to the soundscape with their instruments. Birdsong contrasted with air conditioners and sirens. We compared the sound of our footsteps on the dirt path, the sidewalk, and the crunchy fallen leaves.

In urban environments, human sounds block out the sounds of other species. The scope, degree, and volume of constant urban noise demonstrate how much humans tune out. Information overload and too much noise causes people to habitually tune out, but reactivating our sense of hearing can expand our capacity for relationship with the environment. Paying attention requires intention. What is the effect of *not* listening? Students were assigned to wear earplugs for a day, and although many found it to be a calming experience, others found it disorienting. Some discovered more space for their own thoughts, relieved of feeling constantly bombarded by external sounds. These sounding and listening activities have produced insightful reflections. Students observed that social media causes them to be more tuned out in their daily lives. They expressed that multi-tasking teaches us to tune out and noticed that their attention spans are diminishing. They reported that the modern world is overwhelming and that people don't tune into their surroundings because it is exhausting.

Students respond by making sound maps. After tuning in to the city, students choose a place to sketch a map showing where sounds originate and overlap. Considering foreground, background, tempo, and dynamics, these sound maps become improvisational structures or notated scores. Connections between noise and climate change are drawn.

Applied Listening

Listening practices reveal the world through sound, but what are the next steps? An essential ingredient in grassroots activism is believing that we can make a difference and realising that we have the tools to act. *A Contemplative Pedagogy of Listening* demonstrates to emerging artists that they already possess the skills needed to act, empowering them to engage with the larger community. These listening practices guide students to play their part in the climate movement as they articulate the complexities of increasing climate instability.

As Freire suggests, rather than teaching students to fit into the existing system, we must practise an education that models humane and sustainable ways of being. Facing climate issues through music connects artistic ways of knowing with scientific knowledge and encourages a creative approach to process and respond. The goal of the final project is to collaborate with an environmental organisation and artistically express how culture means caring for all inhabitants on Earth. To write the story we want to live in, made desirable through art. There are thousands of organisations developing climate solutions, but each person must figure out where and how to plug in. Hearing landscape, nurturing relationships with more-than-humans, creating sound documents, practising sound meditations, and examining the music of social movements prepares students for their final project and beyond.

Students respond with projects intending to artfully motivate, express solidarity, and ethically challenge destructive systems. Groups brainstorm how to artistically address the climate emergency. They join organisations, attend trainings, and highlight solutions such as: ocean farming, stopping pipelines, and non-violent protest. Projects have included recording music for educational videos, writing activist songbooks, organising climate strikes, presenting benefit concerts, and collaborating with Extinction Rebellion.

Performance is the Window and Reflection

Lift me up so I can raise my voice and join in the song.
Wake me up, there's many ways to make a world where all can belong.
Shake me up so I can feel the life all around.
Take away what isn't real so I can hear the sound.[9]

Melodies move us and lyrics tell stories, but even vibration communicates. Sufi teacher and musician Hazrat Inayat Khan explores spirituality, harmonies of nature, and vibrations through music. "All things and beings in the universe are connected with each other, visibly or invisibly, and through vibrations a communication is established between them" (Khan, 1963). Khan writes that music *tunes us* to be in harmony with life, bridging formlessness with form. Bridging formlessness with form transpires when ideas and emotions translate into music. Listening tunes musicians to interact together through sound. Individual vibrations create the whole through collective music-making. Ethnomusicologist Jeff Todd Titon observes:

> Music is the corporeal expression of a sound connection as a vibrational exchange. It is also an information exchange, a behavioral exchange, a social exchange, an economic exchange and an ecological exchange.
>
> *(Titon, 2020, p. 274)*

Titon's ideas about vibrations, copresence, and cultural sustainability in *Toward a Sound Ecology* (Titon, 2020) have led me to a new understanding of the power in live music performance. He explains that "sounds travel through a medium such as air or water or solids, and they pick up something of the character of that medium" (Titon, 2015, p. 30). Titon reminds us of Henry David Thoreau's observations hearing church bells from the distant forest, and how differently they sounded nearby. Continuing to discuss Thoreau, Titon writes, "the vibrations coming through the forest assume to some extent the echo of the wood and the leaves and the earth and the air before they reach the ear" (Titon, 2015, p. 30). This leads me to imagine that in a concert hall, musical vibrations "assume to some extent the echo" of each person in that room creating connections between all who are present. Vibrational exchange transforms the audience, who transform the sound.

Music connects us. But isn't being a musician doing enough good for the world? Students are attending music conservatories to become professional musicians, not activists. How do we balance artistry with activism? There will always be compromises, inconsistencies, and contradictions, but balance can be found. We can write compelling lyrics and also participate in community. We can elevate the energy in protests with flutes and drums and also play Bach. We can organise benefit concerts and also play shows that pay the rent. Some seek the spotlight (soloist) but others prefer to speak as a group (string section or chorus), demonstrating multiple forms of participation. There are many paths. Activism doesn't have to mean all or nothing at all.

> There's no single answer that will solve all of our future problems. There's no magic bullet. Instead there are thousands of answers – at least. You can be one of them if you choose to be.
>
> *(Butler, 2000)*

Performance is a window inviting us to venture outside but at the same time reveals our reflection. Reaching beyond lyrics, titles, and program notes, performers relate to the audience with a presence of sharing. This moment raises awareness and transports listeners to places they can only imagine and in that moment, share consolation and inspiration.

Investigating climate-change issues through our common sonic language extends an invitation to join the multi-species conversation. Listening to the trees and joining shouts for justice awakens compassion to embrace all communities. Expanding the boundaries of art with the conviction that music can transform collective consciousness is activism. When activated listening promotes social and environmental consciousness through grassroots action, it brings activism to venues of all kinds and the influence of music is intensified – especially when everyone is singing. The vibrational interconnection of sounding together *feels* empowering. And we need that now.

Coda

Open the window. Go outside. Listen to a park, empty lot, or beach. Create a piece to be performed in the fullness of sounds, smells, and textures. Sharpen your sense of place and listen to the hum. Reveal the dialogue that you hear.[10]

Notes

1 "To hear the falling world" from a Jane Hirshfield poem, used with permission.
2 Lyrics from: "For Our Common Home – Resounding Ecojustice" (2021) by Linda J. Chase An oratorio based on *Laudato Si'* (Pope Francis, 2015) More information can be found here: https://www.lindajchase.com/commonhome
3 From "Grand Canyon Sketches," by Linda J. Chase written in 2012 during her Artist Residency at Grand Canyon National Park. https://soundcloud.com/lchase-1/beyond-eternity
4 Lyrics from: "For Our Common Home – Resounding Ecojustice" (2021) by Linda J. Chase An oratorio based on *Laudato Si'* (Pope Francis, 2015)
5 More songs are found in *Music for Our Emergency*. Currin focuses on music as a process to digest climate despair. "These songs simply exist for the people who need them, their writers included. They're not expressions of what they know or what they want you to do but only how they feel" (Currin, 2019).
6 Lyrics from: "For Our Common Home – Resounding Ecojustice" (2021) by Linda J. Chase An oratorio based on *Laudato Si'* (Pope Francis, 2015)
7 Lyrics from: "For Our Common Home – Resounding Ecojustice" (2021) by Linda J. Chase An oratorio based on *Laudato Si'* (Pope Francis, 2015)
8 For more on soundwalks see The World Soundscape Project https://www.sfu.ca/~truax/wsp.html.
9 Lyrics from "The City is Burning" (2017) by Linda J. Chase. https://soundcloud.com/lchase-1/lift-me-up
10 Closing songs from "For Our Common Home – Resounding Ecojustice" by Linda J. Chase https://soundcloud.com/lchase-1/harmonize-with-creation.

References

Adams, J. L. (2015). Making music in the Anthropocene. *Slate.* http://www.slate.com/articles/arts/culturebox/2015/02/john_luther_adams_grammy_winn er_for_become_ocean_discusses_politics_and.html

Allen, A. (2011). Prospects and problems for ecomusicology in confronting a crisis of culture. *Journal of the American Musicological Society*, 64(2), 414–424.

Allen, A. (2013). Ecomusicology. In *The Grove Dictionary of American Music* (2nd ed.). Oxford: Oxford University Press.

Allen, A. and Titon, J. (eds.) (2023). *Sounds, Ecologies, Musics* (p. 2). New York: Oxford University Press.

Allen, A., Titon, J. and Von Glahn, D. (2014). Sustainability and sound: Ecomusicology inside and outside the academy. *Music and Politics*, 8(2). http://quod.lib.umich.edu/m/mp/9460447.0008.20 5?view=text;rgn=main

Black Eyed Peas (2003). Where is the love. *Elephunk*. UNK Recordings Inc.

Blumhofer, J. (2018). *Rethinking the Repertoire #22 – Florence Price's "Mississippi River Suite" the Arts Fuse*, 6 June. https://artsfuse.org/171130/rethinking-the-repertoire-22-florence-prices-mississippi-river-suite/

Bread & Puppet (1984). *Why Cheap Art Manifesto*. Bread and Puppet. https://1284bf722a.nxcli.net/cheap-art/why-cheap-art-manifesto?_ga=2.265995919.1029407119.1705848081-540546362.1705848081

Brown, C. (2008). *Politics in Music: Music and Political Transformation from Beethoven to Hiphop.* N.p: Farsight Press.

Butler, O. (2000, May). *A Few Rules for Predicting the Future*. Common Good Collective. https://commongood.cc/reader/a-few-rules-for-predicting-the-future-by-octavia-e-butler/

Cajete, G. (2015). *Indigenous Community: Rekindling the Teachings of the Seventh Fire*. St. Paul, MN: Living Justice Press.

Chapman, T. (1995). *The Rape of the World*. https://www.youtube.com/watch?v=BUkKl3FDlyw

Chase, L. (2021). *For Our Common Home – Resounding Ecojustice*. Retrieved From https://www.lindajchase.com/commonhome

Currin, G. (2019). *Music for Our Emergency*. NPR, 5 December, Music Features. https://www.npr.org/2019/12/05/784818349/songs-our-emergency-how-music-approaching-climate-change-crisis

Cusack, P. (2009). *Sounds from Dangerous Places* [Compact disc]. https://www.sounds-from-dangerous-places.org/sonic-journalism

Deggans, E. (2018). *This Little Light of Mine' Shines on, a Timeless Tool of Resistance*. NPR, 6 August, American Anthem. https://www.npr.org/2018/08/06/630051651/american-anthem-this-little-light-of-mine-resistance

Deloria, V. and Wildcat, D. (2001). *Power and Place: Indian Education in America*. Golden, CO: Fulcrum Publishing.

Dunn, D. (2017). *Music Professor Receives Patent to Help Fighting Bark Beetles Ravaging Western Forests*. UC Santa Cruz, 24 February. https://www.universityofcalifornia.edu/news/music-professor-receives-patent-help-fight-bark-beetles-ravaging-western-forests

Easley, T. (2020). *How Hip Hope Can Bring Green Issues to Communities of Color*. Yale Environment 360, 3 March. https://e360.yale.edu/features/how-hip-hop-can-bring-green-issues-to-communities-of-color

EPA (2021). *EPA Report Shows Disproportionate Impacts of Climate Change on Socially Vulnerable Populations in the United States*. US Environmental Protection Agency, September. https://www.epa.gov/newsreleases/epa-report-shows-disproportionate-impacts-climate-change-socially-vulnerable

Feld, S. (1982). *Sound and Sentiment: Birds, Weeping, Poetics, and Song in Kaluli Expression*. Philadelphia, PA: University of Pennsylvania Press.

Feld, S. (1984). Sound Structure as Social Structure. *Ethnomusicology*, 28(3), 383–409. https://doi.org/10.2307/851232

Francis (2015). *IX 244 Laudato Si' On Care For our Common Home*. https://www.vatican.va/content/francesco/en/encyclicals/documents/papa-francesco_20150524_enciclica-laudato-si.html

Freire, P. (1970). *Pedagogy of the Oppressed*. New York: Bloomsbury Publishing.

Gold, H. (2008). *Deeply Listening Body*. Kingston, NY: Deep Listening Publications.

Greene, M. (2001). *Variations on a Blue Guitar*. New York: Lincoln Center Institute.

Head, F. (2022). *Music, Musicians, and Climate Change, an Interview with John Warner, Founder of the Orchestra for the Earth*. Interlude, 20 June. https://interlude.hk/music-musicians-and-climate-change-an-interview-with-john-warner-founder-of-the-orchestra-for-the-earth/

Helguera, P. (2011). *Education for Socially Engaged Art*. New York: Jorge Pinto Books.

Hogan, L. (2004). Linda Hogan. In D. Jensen (ed.) *Listening to the Land: Conversations about Nature, Culture, and Eros* (p. 124). White River Junction, VT: Chelsea Green Publishing.

hooks, b. (1994). *Outlaw Culture: Resisting representations* (1st ed.). New York: Routledge

hooks, b. (1997). *Bell Hooks: Cultural Criticism and Transformation*. Media Foundation Transcript. http://www.mediaed.org/transcripts/Bell-Hooks-Transcript.pdf

Kalmonovitch, T. (n.d.) Retrieved from Tar Sands Songbook. http://tanya-kalmanovitch-neng.squarespace.com/

Khan, H. I. (1963). *The Music of Life*. New Lebanon, NY: Omega Publications.

Kimmerer, R. W. (2013). *Braiding Sweetgrass: Indigenous Wisdom, Scientific Knowledge, and the Teachings of Plants*. Minneapolis, MN: Milkweed Editions.

Lockwood, D. (2021a). Another tool in the fight against climate change: Storytelling. *MIT Technology Review*, December. https://www.technologyreview.com/2021/12/23/1041331/climate-change-storytelling/

Lockwood, D. (2021b). *1,001 Voices of Climate Change*. New York: Simon & Schuster.

Macy, J. and Johnstone, C. (2022). *Active Hope: How to Face the Mess We're in with Unexpected Resilience & Creative Power*. Novato, CA: New World Library.

Manabe, N. (2015). *The Revolution Will Not Be Televised: Protest Music After Fukushima*. Oxford: Oxford University Press.

Matsunobu, K. (2007a). Spiritual arts and education of less is more: Japanese perspectives, western possibilities. *Journal of the Association of Curriculum Studies*, 5(1).

Meyer, M. (2013). Holographic epistemology: Native Common Sense. *China Media Research*. https://education.illinois.edu/docs/default-source/default-document- library/hereca256a3980b76a29a33dff4b008a8698.pdf?sfvrsn=0

Nettl, B., Capwell, C., Bohlman, P., Wong, I. and Turino, T. (eds.) (2001). *Excursions in World Music*. Upper Saddle River, NY: Prentice Hall.

Oliveros, P. (2005). *Deep Listening: A Composer's Sound Practice*. Lincoln, NE: iUniverse.

Orr, D. (2004). *Earth in Mind: On Education, Environment, and the Human Prospect*. Washington, DC: Island Press.

Orr, D. (2018). *Dangerous Years: Climate Change, the Long Emergency, and the Way Forward*. New Haven, CT: Yale University Press.

Paxton, T. (2018). *What's So Bad About That*. Bandcamp. https://tompaxton.bandcamp.com/track/whats-so-bad-about-that

Pedelty, M. (2012). *Ecomusicology: Rock, Folk, and the Environment*. Philadelphia: Temple University Press.

Pedelty, M. (2016). *A Song to Save the Salish Sea: Musical Performance as Environmental Activism*. Bloomington, IN: Indiana University Press.

Rabaka, R. (2016). *Civil Rights Music: The Soundtracks of the Civil Rights Movement*. Lanham, MD: Lexington Books.

Rapaport, S. (2017). Music professor receives patent to help fighting bark beetles ravaging western forests. *UC Santa Cruz*, 24 February. https://www.universityofcalifornia.edu/news/music-professor-receives-patent-help-fight-bark-beetles-ravaging-western-forests

Rodriguez, F. (2021). Harnessing cultural power. In A. E. Johnson and K. Wilkinson (eds.) *All We Can Save: Truth Courage, and Solutions for the Climate Crisis* (p. 122). New York: One World Trade.

Rolling, J. H. (2018). Arts-based research in education. In P. Leavy (ed.) *Handbook of Arts- Based Research*. New York: Guilford Press.

Schafer, R. M. (1992). *A Sound Education*. Tokyo: Shunjusha.

Seeger, P. (2007). *Pete Seeger: The Power of Song* (Director J. Brown) [film].

Thiele, B. and Weiss, G. D. (1967). What a wonderful world [Lyrics]. *ABC Records*. ABILENE MUSIC INC.

Titon, J. T. (2015). Exhibiting Music in a Sound Community. *Ethnologies*, 37(1), 23–41. https://doi.org/10.7202/1039654ar

Titon, J. T. (2020). *Towards a Sound Ecology: New and Selected Essays* (p. 274). Bloomington, IN: Indiana University Press.

Torvinen, J. and Välimäki, S. (2023). Music, ecology and atmosphere: Environmental feelings and sociocultural crisis in contemporary finnish classical music. In Aaron S. Allen and Jeff Todd Titon (eds.) *Sounds, Ecologies, Musics* (pp. 91–98). New York: Oxford University Press.

Von Glahn, D. (2013). *Music and the Skillful Listener: American Women Compose the Natural World*. Bloomington, IN: Indiana University Press.

Von Glahn, D. (2023). Relational capacities, musical ecologies: Judith Shatin's ice becomes water. In A. S. Allen and J. T. Titon (eds.) *Sounds, Ecologies, Musics* (p. 180). New York: Oxford University Press.

Westdal, K. (2013). *The Unseen Threat of Noise in Our Oceans* [Video file]. https://www.youtube.com/watch?v=bo_GZ3q_Ays

Westerkamp, H. (1981). *Under the Flightpath*. https://www.hildegardwesterkamp.ca/sound/docs/flightpath/.

Westerkamp, H. (2017). How opening our ears can open our minds: Hildegard Westerkamp. *CBC Radio*, 2 February. https://www.cbc.ca/radio/ideas/how-opening-our-ears-can-open-our-minds-hildegard-westerkamp-1.3962163

Wickham, M. (2012). Initiating the process of youth decolonization: Reclaiming our right to know and act on our experiences. In Waziyatawin and M. Yellow Bird (eds.) *For Indigenous Minds Only: A Decolonization Handbook*. Santa Fe, NM: SAR Press.

YCC Team (2019). *Climate Music Project Turns Data into Concerts*. Yale Climate Connections, February. https://yaleclimateconnections.org/2019/02/climate-music-project-turns-data-into-concerts/

10
THE SONIC WORK OF TRANSNATIONAL YOUTH CLIMATE MOVEMENTS

Mark Ortiz

Diminishment and Domination: Sonic Indices of Ecological Loss

The sonic indices of climate change are multiplying. Parables of environmental breakdown have long centred sonic loss as a key indicator. Rachel Carson famously forecast the loss of birdsong in *Silent Spring*, and writers like Julian Aguon (2022) are documenting such losses in their lived realities. Climate change is fundamentally affecting the flow of sound in the deep oceans; indeed it may be fundamentally "changing Earth's natural acoustic fabric" (Sueur et al., 2019, p. 971).

What Krause (1987) and other soundscape ecologists (see Pijanowski et al., 2011) refer to as "biophony" – the suite of biologically produced sounds that animate ecosystems – is replete with forms of "echolocation." Some species use echolocation in a literal sense to establish relational coordinates in space, producing sound and relying on its echoes to situate themselves (Jones, 2005). Beyond its biological application, I would contend that most beings echolocate in a variety of ways to feel situated in time and space. As writer Carina del Valle Schorske (2020), quoting Ralph Ellison, reminds us, "one of the chief values of living with music lies in its power to give us an orientation in time." Much of the music on earth, much of its sound, is "social music," to borrow a phrase from the composer and trumpeter Miles Davis who regularly used the phrase as a substitute for the term "jazz" (Hart, 2016). And, in an era of climate change when soundscape loss is trending towards the "dystopic and pathological," such cultural echolocation may be in jeopardy (Winters, 2022).

Many artists and activists have worked to draw attention to the diminished soundscapes of climate change. Climate change and its associated losses have largely been limned in the elegiac mood. In 2016, Greenpeace released a video featuring the composer and pianist Ludovico Einuadi playing a plaintive "Elegy for the Arctic" in front of a glacier near Svalbard, Norway (Einaudi, 2016). The camera alternates in its panning between Einaudi to aerial and wide shots of the landscape and more focused visions of the calving of the surrounding glaciers. This type of sonic expression is often intended to send a message about the urgency of climate change and the scale of natural loss.

The dominance of this type of sonic representation of climate change as the "end of nature," as Bill McKibben framed it long ago, carries cultural and emotional resonance in

DOI: 10.4324/9781003396567-12

the West (McKibben, 1989). But sound is not innocent, and far from an idyllic return to nature, sonic work must contend with the "audibilities of colonialism and extractivism" (Clark, 2021) and projecting emptiness onto ice-worlds as Einaudi does, along with the sonic vocabularies of global loss, may sediment rather than destabilise colonial power relations in sound (see Dodds and Rose Smith, 2022). What Kanngieser (2023) calls "sonic colonialities" characterises much of the sonic production responding to climate change from historically dominant subject positions. Thus, it is critical to turn critical and reflexive attention to how "listening and sound methods are inherently determined by sonic colonialities" and to examine the "natures enacted through sonic geographical practices" (Kanngieser, 2023, p. 2). Sonic cultures lamenting the loss of Euro-American conceptualisations of nature may in fact obscure the ways that activists and organisers are experimenting with sound in efforts to trouble the nature/culture binary and resist the "sonic colonialities" inherent therein (Kanngieser, 2023).

In the context of prevailing discourses about soundscapes of diminution, my long-term work with transnational youth climate movements attuned me to different soundscapes of climate change. Children and young people represent a unique sonic category. As Nolas (2021) points out, children and young people are regularly denied access to political agency and influence and particular sonic hegemonies form part of the logic of this restriction. The "babble of voices" and the "idiomatic" modes of expression characteristic of children and young peoples' communication do not fit easily into the sonic rationalities of adult-dominated politics (Nolas, 2021, p. 326). Yet youth climate activists have formed transnational communities in recent years working through the channels of social media and expressing themselves in voices, words, and sounds that are registering and resonating with other young people (Nolas, 2021).

I argue that listening and paying attention to the sonic work of transnational youth climate movements reveals their efforts to create soundscapes of relation across generations and across geographic contexts. In this sonic work, youth climate movements offer a mode of resistance to sonics of loss, soundscapes of carelessness, and the apocalyptic overtones of climate politics as usual. I use the phrase "sonic work" in a gesture to sound studies, thinking broadly about soundscapes and the social-spatial relations therein (Gallagher et al., 2016). Thus, I pay attention not only to the music that youth climate movements create – though this is important – but also more broadly to the ways that youth movements employ and navigate sound, silence, noise, remixing, singing, and other forms of sonic creation to modulate the geopolitics of climate change. Looking at the range of sonic work of youth climate movements, I follow Tausig (2018, p. 26) in analysing the "diverse ways that political dissent is made audible."

There is a fragmentary pluralism in the sonic work of youth climate movements that mirrors their diffuse, networked structures and creates profound moments of connection and resonance in sonic relation. In the place of a singular anthem, we have a variety of sonic work happening across youth movements. From multimedia applications of popular music in TikTok and Instagram videos to acapella updates of old movement songbooks to the use of live music in building support for climate activism, the sonic work of movements is diverse and varied and it is not strictly relegated to what we might consider music. In the following sections, I look at various examples of how transnational youth movements have engaged in sonic work, drawing on analysis of key multimedia and digital texts. The digital approach to data collection and analysis modelled here is a critical method for archiving and understanding transnational youth activism in a social media age.

The Sonic Work of Transnational Youth Climate Movements

Echolocation: Re-sounding Intergenerational Sonic Traditions

The US-based Sunrise Movement has a Soundcloud account featuring acapella recordings of their versions of popular movement songs like "Rise Up" and "Which Side Are You On" cleverly updated for the climate-change era (Szostak, 2020). In a profile of the movement for the New Yorker, Emily Witt (2018) notes the musicality of the Sunrise Movement which uses singing in both its internal culture and outward-facing protests. Sunrise has displayed a profound sensitivity to sonic textures from its inception, and several of the organisation's protest actions have been modelled on the sonic work of previous movements. Following the pre-Civil War abolitionist group the Wide Awakes, for example, during the summer of 2020 Sunrise Movement hubs coordinated a series of "wide awake actions," standing outside of the doors of political figures and shouting and singing throughout the night with the hope of conveying that "We are wide awake and we'll ensure that the architects of this death economy will be too" (Sunrise Movement, 2020).

For the Sunrise Movement, this return to a movement tactic and sound economy over a century old is a form of what I call "echolocation," a way of situating a sonic-political project in the present moment in relation to its echoes of the past and its silences. Employing similar metaphors of sleep and wakefulness that would come to define the public image and legacy of the Wide Awakes, Sunrise Movement activists have pointed out that adult politicians are "asleep at the wheel" (Sunrise Movement, 2019). Members of the Sunrise Movement see their mission in part as mobilising young people around the country to awaken a political class that structurally silences the voices of young people much like the Wide Awakes who, Grinspan (2009, p. 359) argues, "amplified the murmurings of a generation." The process of this awakening, for Sunrise as for the Wide Awakes, is equal parts loud, obnoxious, festive, carnivalesque, joyful, and rageful.

As a further testament to how the movement deliberately situates itself in intergenerational sonic traditions, Sunrise Movement has developed a "Sunrise Songbook," noting that singing "helps us connect with one another while offering reflective, uplifting and comforting shared moments" (Sunrise Movement Songbook, 2022, p. 5). In public actions, the authors write "singing can do anything from inspiring courage to acknowledging pain, marking a transition, or boosting the energy of the crowd" (Ibid.). Further, "singing together also allows us to envision new futures even while they feel out of reach, and it honours the musical history of racial, labour, environmental and other justice movements that the Sunrise Movement aligns with" (Ibid.). The Songbook is a critical resource for movement activists and features historical context for each song included as well as detailed discussions of issues of cultural appropriation which recognise histories of co-optation of movement music. The notes included with each song help to conceptualise the ways in which sonic work travels, at times in ways which are complicated by histories of racial and class injustice. Such a project speaks to the wish on the part of Sunrise Movement to situate itself, through sonic work, in long traditions of labour, racial justice, and other activism and to be intersectional not only in its contemporary work but in the tracing of its sonic legacies.

While the sonics of loss characteristic of colonial encounters with climate crisis leave little space for sonics of hope, recreation, and re-enlivenment, efforts towards echolocation such as the Songbook work to create sonics of relation across time. As another counterpoint to these elegies of loss, I am reminded of the long-time phrase of Pacific climate organisers: "We are not drowning – we are fighting!" (Wiseman, 2022). I am reminded of how these words have echoed across digital space over the years of global climate justice organising, how recently they rang again in the COP 26 climate negotiations when Samoan youth climate organiser Brianna Fruean with 350 Pacific described how Pacific youth climate organisers used this phrase as a "warrior cry." Fruean's speech at COP 26 makes an important case about words and their echoes that "even stones decay, but words remain" (Doha Debates, 2021). This is the sonic work of echolocation. Re-sounding words across generations of organising, ensuring that those words do not decay but instead continue to ground climate organising and ensure that more words and actions continue, in Jetñil-Kijiner's (2014) poetic words, "blooming from teenage fingertips." These words find their way into song, into protest chants, into speeches, into writings, poetry, and digital content. These words have enlivened hope and work actively to script different futures that disrupt climate geopolitical narratives of the Pacific Islands and other places as inevitable tales of loss.

Sonically Weaving Generational and Transnational Ethics:
Remix, Re-Enliven, Amplify

Goffe (2020) illustrates how the DJ functions as a "time machine," able to bend time and temporality and act as a portal into different future possibilities. Stevenson (2014, p. 167) postulates that the importance of song resides in the capacity to "call into being" a different "ethical universe," a space of indeterminate ethical trajectories and obligations that accommodates insoluble contradictions. Morality that moves beyond the insensitivities of our moment resides in learning to listen differently (Hache et al., 2010; see also Chase, this volume).

If we think of the soundscapes that typically characterise climate politics in the rooms of high policy, a few things come to mind. Emotional sonic outbursts and expressions are generally marginalised; song is almost absent, save for the sonic work of activists and Indigenous leaders who work to reclaim or enliven such spaces differently. Youth voice has been a central feature and concern of institutional climate politics since at least 2009, when the YOUNGO youth constituency to the U.N. climate negotiations was formed. But youth voice, as understood and mobilised by adultist institutions, fundamentally traffics in adultist categories, in using languages and grammars that are legible and often in sonically holding back. We know from listening to youth climate organisers like Greta Thunberg that much of what they hear at these conferences sounds like "blah blah blah" and the words of leaders ring like "empty promises" (Carrington, 2021). But what do "empty promises" sound like, and what types of sonics would characterise a climate politics which centres intergenerational and transnational justice?

Thinking intergenerational and transnational ethics through sonic work asks us to reconsider what we pay attention to when we analyse the politics of climate. In 2009, at the anticipated Copenhagen climate conference, youth organisers coordinated a "thunderstorm" action in the conference halls. The "thunderstorm" of the action consisted of a transnational group of youth organisers creating a sonic environment meant to mimic the sound of a thunderstorm. Advocates simulated the pattering of rainfall, the pitch of intense wind and cracking of thunder, by snapping, clapping against their bodies with open palms, exhaling loudly, and replicating the noises of rain droplets with their mouths. This thunderstorm bookended the action which featured speeches of youth climate activists from the Pacific (Rainforest Action Network, 2009).

Creating a sonic intervention like the one detailed earlier contributes to drawing attention to and amplifying voices which are generally lost in the political outcomes of these conventions and opens the possibility of audiences in that space listening differently and engaging differently as a result. It allows us to remember the Copenhagen conference not solely as a monumental diplomatic failure constrained by the *realpolitik* of the moment but as a gathering space for a growing international youth climate movement. Similarly, what might it mean to remember the Doha climate conference in 2012 – not through its political failures and the politicking of powerful men – but through the sonic work of giggling with a transnational group of youth activists, the sound of the "icebreakers" where activists introduce themselves, the sounds of "doubling over with laughter at 1 in the morning" as youth advocate Nikki Hodgson (2013) describes in recollections of the conference? These are a far cry from the range of sonics typical of the climate negotiations.

Part of what sonic work may accomplish involves undoing the dominant sonics of power and hegemony in adult-dominated spaces. For Kadich (2021, p. 1313), young people in Sarajevo, Bosnia and Herzegovina find in the transnational art form of hip hop a "grammar for unpolitics," a way of undoing hegemonic political categories and trajectories and differently establishing the terms and scales of youth politics. Music and sonic work hold the potential to temporarily fracture both the present of climate injustice and apocalyptic overtones of the futures being written by adult-dominated institutions and their soundscapes. For youth climate organisers, I suggest that praxes of remixing, amplifying, and re-enlivening help to compose a planetary ethics and politics which subverts the dominant geopolitics of climate change. What digital-media savvy youth climate movements present us are richly referential, often nonsensical, multimedia creations that weave together pop songs about love and romance, soaring ballads which dramatise intimate loss, and other tunes spliced together with climate multimedia of various stripes. The emotional landscape is broad. Humour, satire, grief, sadness, joy, and hope all exist alongside despair, exhaustion, and hopelessness. These multimedia creations and their sonic dimensions permit space to reckon with and rebuild an emotionally nuanced climate politics and ethics.

For example, one Instagram reel by the Fridays for Future (FFF) youth climate network features a short introduction with a narrator mouthing the words "who got you smiling like that" dubbed with an audio clip. The video then cuts to scenes of youth climate strikers with megaphones and signs in Sierra Leone, the Philippines, Sweden, Brazil, and Italy. The entire video is set to the tune of A-Wall's song "Loverboy." The song is upbeat and zippy, memorialising in sound the feelings of being mesmerised and in love. The Instagram caption reads:

> Fridays for Future is about striking, fighting relentlessly for climate justice, trying to disrupt a system that for so long has been exploiting the planet and its people. But it is also about joy, about love for others, about sharing moments we'll never forget with people we never thought we would have met if it wasn't for activism. It's about finding a second family and being surrounded by the warmth and strength of those who, just like you, believe that another world is possible.
>
> *(Fridays for Future, 2021)*

Much of the recent digital work of transnational climate movements, such as Fridays for Future Digital, has focused on "amplifying" youth activists from what the movement refers to as most affected peoples and areas (MAPA). In 2020, the Fridays for Future network called on Sir David Attenborough to, instead of retiring his Instagram account, "pass the mic" to youth activists and frontline environmental defenders to move against the pervasive silencing of activists from most impacted communities (Fridays for Future, n.d.). In a piece describing their efforts to amplify activists from MAPA, FFF organisers write:

> We should be able to uplift the voices of those who have been unheard; to stop and listen, reconstruct our opinions, we must question ourselves: are we opening a platform for MAPA? . . . Supporting MAPA and fighting for an intersectional movement means hearing those who have been unheard and empowering those who have not yet dared to raise their voices.
>
> *(Reyes and Calderón, 2021)*

These strategic efforts at amplification are usefully thought of in terms of sound, of rebalancing the mix of signal and noise in relation to climate-change justice to ensure that signals which must be heard are featured more prominently in the blend. These efforts also interrogate whose voices are structurally amplified and draw attention to voices that are structurally diminished.

Redirecting Popular Culture: Fandom, Internet Celebrity, and Live Aid for the Climate

Pedelty (2016) makes the case that environmentalist music operates at local and regional scales given the decentralised nature of the movement and that environmental themes do not fit easily into the genre conventions of popular music. But youth climate movements are changing this calculus with ambivalent results. Contemporary iterations of youth-led climate movements are translocal and transnational in character and their musical expressions seem to cross boundaries and bring different scales into focus. In a world in which certain youth climate activists are beginning to experience growing levels of internet celebrity, what is popular and what takes climate as its topic are no longer separate.

"No K-Pop on a Dead Planet" reads a protest sign at the center of the webpage of K-Pop 4 Planet (KPop4Planet, n.d.). K-Pop for Planet is a "platform and community dedicated to mobilizing global K-pop fans to fight the climate crisis" (Suen, 2022). The platform relies on the networking power of social media as well as K-pop fan interest in supporting socially and environmentally conscious initiatives. The largely youth-led group uses tactics like "Twitter storms" – tweeting at influential figures – to press institutions involved in the K-Pop music industry to go green and adopt sustainability practices, and they also publicly protest by dancing to K-Pop songs (see Suen, 2022).

What are the stakes of taking the climate fight to the domain of pop culture? Looking to the music of K-Pop group BTS, Ju Oak Kim (2021, p. 1072) argues that BTS represents a "counter- hegemonic culture in the network society" which centres "non-western/peripheral views, experiences and methodologies." Fans of BTS are able to interact with the group on social media, and BTS represents far more than those members of the band or the music itself – it is a relational sphere of fandom, a networked space in which otherwise marginal actors are able to enter a participatory and interactive process of deliberation and shared interest (Ibid.). Networks like Fridays for Future and the organisers of K-Pop on a Dead Planet recognise the political, economic, and cultural power carried by fandom and youth digital cultures and seek to redirect these forces to environmental ends.

While many adults may think of this style of protest as unserious or trivial, there is deep and thoughtful strategy grounding the political analysis and selection of tactics by these groups. A podcast by Fridays for Future Digital elaborates the strategy behind K-Pop for Planet and also makes the point that K-Pop lyricism itself touches on themes of generation, youth, politics, and economy, distinguishing artists from the genre as potential allies in climate campaigns (Voice of the Youth, 2022). The discussion notes the online power of the K-Pop fan community, which logged nearly eight billion tweets during 2021 (Ibid.). The modes of activism that K-Pop for Planet and Fridays for Future enact also trace complex political-economic linkages and help their audiences visualise the impacts that polluting companies have on distant places and people (Ibid.).

Increasingly, young popular musicians are also turning to climate change as a key theme in their work and climate activism as a site of influence. The popular musician Billie Eilish hosted a six-day climate awareness event entitled "Overheated" during one of her tours which convened youth climate activists from around the world (Kaufman, 2022). She also released a documentary film centring the voices of a collective of youth climate activists, fashion designers, and others who are working on climate change in the popular sphere. At the beginning of her documentary *Overheated*, (2021) Eilish tells viewers "I definitely suffer from climate anxiety . . . and I know that so many people my age do too." The documentary offers viewers perspectives from activists, musicians, and writers from Cuba and Uganda as well as a section entitled "Letters from the Frontline" featuring perspectives from the Sami region, the Waorani people in the Amazon, and Burundi (Ibid.). Like the work of Fridays for Future, this project amplifies the work of frontline climate justice activists by leveraging the popular standing of Eilish to communicate climate justice to broader audiences. *Overheated* connects the personal experiences of youth climate advocates with the planetary transformations of climate change.

The Climate Live Concerts, started by Fridays for Future organisers, represent another example of redirecting popular culture and engaging fandom to spur climate action. Started after a youth organiser heard Queen guitarist Brian May's call for a "Live Aid for the climate," the Climate Live concerts would follow this mould while also capitalising on the transnational networking capabilities of youth climate networks (Aubrey, 2021). Here we see another example of contemporary youth movements drawing on and re-sounding an intergenerational tradition of humanitarian or justice-centred concert-throwing (Voice of the Youth, 2021). The tagline of the events – "Can You Hear Us Yet?" – also serves as a trenchant critique of adultist policies and institutions, playing on the presumed diminutive character of youth voice as it is registered in spaces of policy (Climate Live, n.d.).

The stated objectives of the Climate Live events were threefold: to "enlarge the global movement by engaging with a new audience through music"; to "raise awareness of the challenges faced today by people on the frontlines of ecological breakdown, and the predictions of scientists for the future"; and to "pressure world leaders (political economic and cultural) to take action to combat the climate crisis, with a focus on COP 28" (Climate-Live, n.d.). The inspiration and organisation of Climate Live reflects a unique example of intergenerational collaboration in sonic work, with humanitarian events past forming the template for a new mode of live crowd engagement.

But Climate Live was also something that Live Aid could not be – a networked live music event that was interconnected across transnational networks on platforms like Twitter, Instagram, and TikTok. "Guerrilla gigs" (e.g. Climate Live Global, a), full concerts, and individual performances (see Climate Live Global, b), would weave together different languages, instruments, musical styles, and sonic textures, together imploring institutions: "Can you hear us yet?" The backdrops were varied, from Swedish-Sámi musician Maxida Märak's performance in front of the Swedish parliamentary building (Climate Live Global, a) to musician Thaline Karajá's performance overlooking a forest landscape (Climate Live Global, b). If, as Goffe (2020) writes, "Sun Ra's time machine is the Moog synthesizer" then I would suggest that youth climate advocates use the visual and connective capabilities of social media as openings onto other places, times, and emotional registers. Different musicians in different contexts bring with them different audiences and publics who can then plug into a transnational platform and movement and mobilise in new ways.

Conclusion

This chapter has identified and elaborated three aspects of the "sonic work" that youth climate movements perform: (1) to re-sound intergenerational traditions through sonic work; (2) to weave intergenerational and transnational ethics and to amplify the work of advocates in MAPA contexts; and (3) to redirect popular culture and capitalise on the fandom of youth cultures. Yet, following Stuart Hall, Ju Oak Kim (2021) reminds us that practices of "cultural resistance" contain elements of disruption to dominant formations as well as continuations, intensifications, and evolutions of existing inequities. While the musicality and sonic work of transnational youth climate movements disrupts certain aspects of mainstream, adultist climate politics, in other ways the auditory cultures of youth climate movements may serve to reinforce dominant power relations in ways in which it is important for youth advocates, scholars, and others to reflect on. As we work to weave sonic cultures of connection across generations and continents, it is important to consider where we are hearing and sounding from, in time, in space, and in the context of asymmetrical power relations and the climatic and ecological manifestations of empire.

References

Aguon, J. (2022). On Guam there is no birdsong, you cannot imagine the trauma of a silent island. *The Guardian*, 31 October. https://www.theguardian.com/world/2022/nov/01/on-guam-there-is-no-birdsong-you-cannot-imagine-the-trauma-of-a-silent-island

Aubrey, E. (2021). Watch Declan McKenna perform 'British Bomb's outside parliament for 'climate live' event. *NME*, 26 April. https://www.nme.com/news/music/watch-declan-mckenna-perform-british-bombs-outside-parliament-for-climate-live-event-2928119

Carrington, D. (2021). 'Blah, blah, blah': Greta Thunberg lambasts leaders over climate crisis. *The Guardian*, 28 September. https://www.theguardian.com/environment/2021/sep/28/blah-greta-thunberg-leaders-climate-crisis-co2-emissions

Clark, E. H. (2021). Introduction: The audibilities of colonialism and extractivism. *The World of Music*, 10(2), 5–21.

Climate Live. (n.d.). *Climate Live: Can You Hear Us Yet?* https://climatelive.org/

Climate Live Global(a). [@climateliveglobal]. (2021). @maxidamarak In all times, they have tried to kill us Indigenous people. I represent them now. We try not to live for profit, but in a cycle. When everything collapses, we must learn to live with nature. This is for my Indigenous people. *Video File & Social Media Post*, 5 May. https://www.instagram.com/p/COgJ3M2HgGA/

Climate Live Global(b). [@climateliveglobal]. (2021). Thaline Karajá for Climate Live. Thaline is an Indigenous singer, songwriter, and environmentalist from Brazil. She is best known for her participation in The Voice Brazil in 2020. We are endlessly happy that she is part of Climate Live, not only because of her incredible talent but also because Indigenous voices are fundamental in the fight for climate justice. Go follow her! @thaline_maxim_karaja. #ClimateLive2020 #CanYouHearUsYet. *Video File & Social Media Post*, 6 June. https://www.instagram.com/reel/CPyhExYHcmH/

Del Valle Schorske, C. (2020). The world according to Bad Bunny. *The New York Times Magazine*, 11 October. https://www.nytimes.com/interactive/2020/10/07/magazine/bad-bunny.html

Dodds, K. and Smith, J. R. (2022). Against decline? The geographies and temporalities of the Arctic cryosphere. *Geographical Journal*. https://doi.org/10.1111/geoj.12481

Doha Debates. (2021). Climate activists Brianna Fruean's full speech at COP26 [Video]. *YouTube*, 3 November. https://www.youtube.com/watch?v=2ahG5gur7m0

Einaudi, L. (2016). Ludovico Einaudi – "Elegy for the arctic" – official live (Greenpeace) [Video]. *YouTube*, 20 June. https://www.youtube.com/watch?v=2DLnhdnSUVs

Fridays for Future [@fridaysforfuture]. (2021). *Fridays for Future Is about Striking, Fighting Relentlessly for Climate Justice, Trying to Disrupt a System That for So Long Has Been Exploiting the Planet and Its People*, 14 September. https://www.instagram.com/reel/CTztCu8jnCr/

Fridays for Future (n.d.). *Please #PassTheMic: Pass the Mic Sir David*. https://fridaysforfuture.org/pass-the-mic-sir-david/

Gallagher, M., Kanngieser, A. and Prior, J. (2016). Listening geographies: Landscape, affect, and geo-technologies. *Progress in Human Geography*, 41(5), 618–637.

Goffe, T. L. (2020). The DJ is a time machine. *Public Books*. https://www.publicbooks.org/the-dj-is-a-time-machine/

Grinspan, J. (2009). "Young men for war": The Wide Awakes and Lincoln's 1860 presidential campaign. *The Journal of American History*, 96(2), 357–378.

Hache, E., Latour, B. and Camiller, P. (2010). Morality or moralism?: An exercise in sensitization. *Common Knowledge*, 16(2), 311–330. https://doi.org/10.1215/0961754X-2009-109

Hart, R. (2016). Inside Miles Davis' prince obsession, as detailed by Davis' family and collaborators. *Pitchfork*, 22 April. https://pitchfork.com/thepitch/1114-inside-miles-davis-prince-obsession-as-detailed-by-davis-family-and-collaborators/

Hodgson, N. (2013). Notes from the international youth climate movement, Doha. *Matador Network*, 21 January. https://matadornetwork.com/change/notes-from-the-international-youth-climate-movement-doha/

Jetñil-Kijiner, K. (2014). Dear Matafele Peinam. *Marshallese Arts Project, The University of Edinburgh*. https://www.map.llc.ed.ac.uk/creative-writing/dear-matafele-peinem/

Jones, G. (2005). Echolocation. *Current Biology*, 15(13), 484–488.

Kadich, D. (2021). Young people, hip-hop, and the making of a "grammar for unpolitics" in Sarajevo, Bosnia, and Herzegovina. *Geopolitics*, 26(5), 1307–1330.

Kanngieser, A. M. (2023). Sonic colonialities: Listening, disposession, and the (re)making of anglo-European nature. *Transactions of the Institute of British Geographers*, 1–13.

Kaufman, G. (2022). *Billie Eilish and Finneas to Kick-Off London 'Overheated' Climate Event*, 3 May. https://www.billboard.com/music/pop/billie-eilish-finneas-london-overheated-climate-event-1235066074/

Kim, J. O. (2021). BTS as method: A counter-hegemonic culture in the network society. *Media, Culture, and Society*, 43(6), 1061–1077.

KPop4Planet. (n.d.). *KPop4Planet*. https://www.kpop4planet.com/

Krause, B. (1987). Bioacoustics, habitat ambience in ecological balance. *Whole Earth Review*, 57, 14–18.

McKibben, W. (1989). The end of nature. *The New Yorker*, 3 September. https://www.newyorker.com/magazine/1989/09/11/the-end-of-nature

Nolas, S. M. (2021). Childhood publics in search of an audience: Reflections on the childrens' environmental movement. *Children's Geographies*, 19(3), 324–331.

Overheated (2021). Overheated: The documentary [Video]. *YouTube*. https://www.youtube.com/watch?v=4suoAkkZy7c&t=212s

Pedelty, M. (2016). *A Song to Save the Salish Sea: Musical Performance as Environmental Activism*. Bloomington: Indiana University Press.

Pijanowski, B. C., Farina, A., Gage, S. H., Dumyahn, S. L. and Krause, B. L. (2011). What is soundscape ecology? An introduction and overview of an emerging new science. *Landscape Ecology*, 26, 1213–1232.

Rainforest Action Network. (2009). Youth create a thunderstorm in the UN at Copenhagen climate talks [Video]. *YouTube*, 10 December. https://www.youtube.com/watch?v=URsnZqXfs2Q&t=22s

Reyes, M. and Calderón, A. (2021, March 15). What is MAPA and why should we pay attention to it. *Newsletter for Future Edition No. 1*. https://fridaysforfuture.org/newsletter/edition-no-1-what-is-mapa-and-why-should-we-pay-attention-to-it/

Stevenson, L. (2014). *Life Beside Itself: Imagining Care in the Canadian Arctic*. Berkeley: University of California Press.

Suen, Z. (2022). Can K-Pop fans save the planet? *Atmos*, 24 May. https://atmos.earth/kpop-fans-blackpink-climate-change/

Sueur, J., Krause, B. and Farina, A. (2019). Climate change is breaking earth's beat. *Trends in Ecology and Evolution*, 34(11), 971–973.

Sunrise Movement (@sunrisemvmt) (2019). Our generation is taking this crisis very seriously, but the adults in the room seem to be either asleep at the wheel or selling us out to oil and gas donors. *Twitter Post*, 25 August. https://twitter.com/sunrisemvmt/status/1165710857239547904?ref_src=twsrc%5Etfw.

Sunrise Movement. (2020). Summer 2020 – Wide awake actions. *Sunrisemovement.org*. https://www.sunrisemovement.org/actions/wide-awake/

Sunrise Movement (2022). *Sunrise Movement Songbook*. https://drive.google.com/file/d/120wccqh6jdV-Cb1gCZgxtuE7ELnYPUXL/view.

Szostak, M. (2020). Sunrise movement advocates, empowers through protest music. *Tufts Daily*, 9 November. https://tuftsdaily.com/arts/2020/11/09/sunrise-movement-advocates-empowers-through-protest-music/

Tausig, B. (2018). Sound and movement: Vernaculars of Sonic Dissent. *Social Text*, 36(3), 25–45.

Voice of the Youth – The FFFD Podcast. (2021, February). Ep. 6 – Music, musicians, and the climate movement [Podcast]. *Spotify*. https://open.spotify.com/episode/412xFIBs79f7U32HtDcBbg

Voice of the Youth – The FFFD Podcast. (2022, July). Ep 13 – Kpop fandoms mobilizing climate action [Podcast]. *Spotify*. https://open.spotify.com/episode/22ZdD2wxSN2JzvVzSYonnC

Winters, J. (2022). How climate change is muting nature's symphony. *Grist*. https://grist.org/culture/nature-sounds-bird-insect-silence-climate-change/

Wiseman, J. (2022). Narrative matters: Fighting not drowning – facing a harsh climate future with wisdom, hope, and courage. *Child and Adolescent Mental Health*, 27(1), 84–86.

Witt, E. (2018). The optimistic activists for a Green New Deal: Inside the youth-led singing Sunrise Movement. *The New Yorker*, 23 December. https://www.newyorker.com/news/news-desk/the-optimistic-activists-for-a-green-new-deal-inside-the-youth-led-singing-sunrise-movement

11
PERFORMING TRANSFORMATIVE CLIMATE JUSTICE

Thomas A. King

The first Working Group's contribution to the Intergovernmental Panel on Climate Change (IPCC) Sixth Assessment Report, a "Summary for Policy Makers" (August 2021), offered policymakers three scenarios:

> Compared to 1850–1900, global surface temperature averaged over 2081–2100 is very likely to be higher by 1.0C to 1.8C under the very low GHG emissions scenario considered (SSP1–1.9), by 2.1C to 3.5C in the intermediate GHG emissions scenario (SSP2–4.5) and by 3.3C to 5.7C under the very high GHG emissions scenario (SSP5–8.5).
>
> *(Section B.1.1)*

The message was clear: which scenario will we choose? To suggest that we have a choice among scenarios is to imagine what performance theorist Diana Taylor has called the *performatic* capacity of social actors (Taylor, 2016), their creative ability to (re-)construct the real in mundane ways, both by taking embodied action here and now and through everyday acts of imagining, remembering, and rehearsing, prerequisite to bringing new capacities for action into being.[1] This chapter will explore the capacity of theatre and performance, ranging from scripted plays to interactive installations, to engage and support transformative climate justice.[2] Reusing and revising what performance studies scholar Margaret Thompson Drewal has called "acquired, in-body techniques" ("embodied skills and techniques" applied "to the task of taking action"), theatre and performance artists build capacity for imagining alternative ways of being in relation to others and to our planet; responding to asymmetrically distributed impacts and harms; enabling hope and inspiring joy; and enacting an embodied ethics of care, compassion, and repair in a world that may be increasingly uninhabitable (Drewal, 1991, p. 1).

While the use of arts and creativity to capture media attention for protests and occupations is best known, artist-activists – or "artivists" (Taylor, 2016, p. 147) – may be involved in community engagement and decision-making processes from the ground up. Helping to

DOI: 10.4324/9781003396567-13

facilitate conversations with frontline communities and to visibilise both the experience of harms and knowledge of solutions held by those communities, artivists democratise access to techniques and media giving aesthetic form to experience and knowledge.[3] Artivism provides cognitive, sensory, emotional, and kinaesthetic resources to support everyday transformation, movement building, and regenerative desire. Artivists uncover, create, illustrate, perform, and share stories about the harm of commitments to material instead of moral growth and where positive investments may be made. Artivists help us recognise and commit to reducing the current and future harms of climate change while also investing in fundamental rights to cultural heritage, beauty, and the full embodiment of our possibilities as human beings – understood not as the rights of individuals but of collectives (Tom Goldtooth, qtd. in Moore and Russell, 2011, p. 20). Transforming passive observers into spect-actors, artivists invite us to become who we must be to steward a just future for all human and nonhuman life.[4]

A game-changing recent study led by Timothy M. Lenton has argued that the ethical costs of climate change be expressed "in terms of numbers of people left outside the 'human climate niche' – defined as the historically highly conserved distribution of relative human population density with respect to mean annual temperature." Lenton et al find that "climate change has already put ~9% of people (>600 million) outside this niche" and that "current policies [that is, the voluntary Nationally Determined Contributions (NDCs) agreed on by nation-states participating in the Paris Agreement] leading to around 2.7 °C global warming could leave one-third (22–39%) of people outside the niche" by 2100 (2023, p. 1/23). Rethinking our response to climate change in terms of harm reduction might motivate both individual and collective accountability. Individual, because, as Lenton et al. assert, "The lifetime emissions of ~3.5 global average citizens today (or ~1.2 average US citizens) expose one future person to unprecedented heat by end-of-century. That person comes from a place where emissions today are around half of the global average" (2023, p. 6/23).

But, more crucially, collective: artivists centre collective knowledge and collective rights, for, as Lenton et al. point out, "[r]educing global warming from 2.7 to 1.5 °C results in a ~5-fold decrease in the population exposed to unprecedented heat (mean annual temperature ≥29 °C)" (2023, p. 1/23). If transformative climate justice calls us to act ethically in our lives, this disparity between the accountability of average (and, even more so, extreme) consumers and the distribution of harm also demands that we act on behalf of others, to work collectively to repair past harms and ensure the foundations of future thriving. The urgent political question – what will we do to prevent global warming beyond 1.5 °C? becomes the ethical question – who do we need to become to recognise our interdependence, affirm our mutual precariousness and possibility, and promote thriving – of the world's populations today, across generations, and across species?

Artivists insist that there is no choice between zeroing emissions of greenhouse gases (most urgently, CO_2 and methane) and dismantling extractive capitalism; white supremacy, anti-Blackness, and settler colonialism; heteropatriarchy; lack of access to jobs, housing, and health care; and income inequality. The sense of urgency can itself replicate some of the injustices of racialised, extractive, fossil-fuel capitalism – instating hierarchies of expertise, seeking top-down "solutions" at the expense of bottom-up participatory decision-making, emphasising product over process. Artivism slows us down, facilitates active listening and conversation, and reminds us that, given the duration and spatial expanse of harm, avoiding further harm ought to be the premise, the driver, of our shared work.

Several perspectives inform the choice of case studies for this chapter. First, performers and other artivists grapple with the gap between collective knowledge and personally held belief or feelings on the one hand and individual and collective action on the other. Second, the IPCC's Sixth Assessment Report acknowledges that (predominantly AngloEuropean) rational systems of evidence must be in dialogue with other, especially Indigenous, knowledge systems. Knowledge motivates action, bringing about a change in habitus, when it is fully embodied (head, heart, and muscle memory) and fosters relationships – between human and nonhuman beings, across and within generations, and among voice, bodily action, and place. Third, the capacity of artivists to appeal to and generate a "we" – the collective body or commons appealed to in performance and for which the performers or the audience (or both) stands in – must be accountable to the asymmetrical distribution of climate impacts and climate harms.

Finally, this chapter will take up problems of scale and representation. In its centring of the human body in expressive and communicative action with others, in its idealisation of rehumanisation and plenitude of being, (formal) theatre and performance has historically assumed – indeed, relied on and reinforced – the legibility of the human performer and of some ostensibly universal humanness, for example, storytelling, to create empathy with those whose lives are most precarious. In the works considered here, artivists have explored how the human and nonhuman elements of a performance work give form and utterance to nonhuman animals; bodies of water and ice; once-in-a-thousand-years disasters; or the time and space of flows, melts, currents, and global conveyor belts. They have aimed to use the resources of performance to connect the unrepresentable scale of global climate change to the all-too-human scales (and his/stories) of agriculture, exploration, colonialism, belief in progress, extraction, and consumption that constitute the capitalocene.

Theatre and Performance as Direct Action

Transformative climate justice consists of performances in both everyday spacetimes and spacetimes set apart or "framed" as special, ritual, or aesthetic opportunities for imagining, reflecting, judging, and engaging.[5] It is not just a vision but a set of practices. But the "work" of theatre and performance (in all the senses to be developed here) is not only to be evaluated in terms of outcomes or impacts, reductively considered (say, persuading legislators to pass laws) or the persuasiveness of its message. As modes of shared doing, grassroots performances such as bike parties, street occupations, and sit ins and stand outs do more than support actions directed at targets beyond themselves. They are also vehicles of transformation, training individuals to rehearse alternative futures in the here-and-now, to engage across their disparate identifications and interests, and to move in a shared direction determined by commitments to solidarity.

The theory of change informing theatre and performance's contributions to transformative climate justice, as transitional arts practitioner Lucy Neal (UK) writes, shifts focus from investment in legislative wins or technological fixes (e.g., geoengineering) to changing the underlying way of knowing, perceiving, and questioning. Transitional arts practice shifts focus as well from the juridical or governmental level to the daily, local, embodied (and "rehearsable") level. Aware that "[t]ransition is inevitable" but "[j]ustice is not" (Climate Justice Alliance), theatre and performance artists may, for example, bring residents together in conversation, creating opportunities for discovering, *across and respecting their distinct identities and interests*, vectors of solidarity. Creative techniques can be used as icebreakers, methods for creating and holding space, invitations to share stories and listen actively to other perspectives, requests to sit with uncomfortable moments, and methods for designing direct actions such as a protest, sit in or stand out, occupation, or rally. In its "Theory of Change," the Center for Story-Based Strategy observes that storytelling provides "a connective tissue for movement-building – more individuals and groups consider who else has a shared narrative with their work and increasingly see their work connected to others" (Center for Story-Based Strategy, n.d.).

Sharing stories toward solidarity does not imply shared identities or a romantic notion of community consensus. Facilitating difference and dissent, and recognising conflict, it nevertheless invests in discovering and building capacities for moving together toward a shared goal.[6] As Neal puts it, theatre and performance, along with other creative techniques, have "a tremendous 'power to convene', to get all manner of people in a room together, dreaming, planning and doing," allowing for the unexpected to happen (2015, p. 31).

More than just sending messages about the sustainability of cycling, Bike Parties in Cambridge (UK), San José, San Francisco, the Bay Area (California, USA), and other locations create "a performative way of relating to the world" (Terry and Todd, 2013, n.p.), allowing discoveries to be made in the very flow of performing. That is, new knowledge and new skills are found in the moment of applying "embodied skills and techniques to the task of taking action," transmitted by embodied means from one social actor to another and adding recursively to the know-how capable of being employed, reflected on, and debated in other times and spaces. Biking through the centres of urban, capitalist growth can provide a new perspective on the familiar and mundane, for example, by changing the spect-actor's perspective (looking at the city, and the ideology of incessant capitalist growth from a different angle, from a bicycle seat), by changing a spatial relationship (an unfamiliar route across the city), or by changing a temporal relationship (seeing the city from the speed of a bike rather than that of a car). Biking is a "performance of the local," bringing awareness of location and locality (proximity to what is nearby or even at hand) into the action of moving together in space and also by remapping and restorying the neighbourhood boundaries of the city, of the translocal (Terry and Todd, 2013, n.p.). It thereby also both practices and visibilises movement in all its politicised senses: the right to and liberatory capacities of mobility and access; on-the-ground political organising; putting an agenda into motion to create a win.

Dance scholar Susan Leigh Foster says of activist "techniques of the body":

Over the time that they are practiced, they acquire increasing influence over corporeal and also individual identity. Not a script that the protestor learns to execute, these are, rather, actions that both require and provide strong commitment and, once practiced, slowly change the world in which they occur.

(2003, p. 408)

Artivists can transform mundane daily activities such as biking, gardening, and walking into such a "performative relation to the world," one which (1) requires performers to get to know their materials (say, a bike, a ball of yarn, a body of water, a path, earth, voice); (2) can become a regular practice, creating kinaesthetic memory; (3) is portable across institutional spatial and temporal boundaries (what one discovers walking off the path can be reactivated on it); (4) can gather other participants together in the flow of action; and (5) facilitates doing as ways of knowing and transmitting knowledge and of analysis, critique, and revision. We might recall here the occupation method of amplifying a voice across space – the human microphone, a performance technique both solving an immediate problem and building organisational skills and embodied knowledge that were portable to other actions. Irreducible to a message, performance may be a "convivial technology" facilitating the "social-ecological transformation" of our current commitments to growth at all costs, alongside "tool-lending libraries, repair cafés, do-it-yourself spaces, and some ecologically and non-commercially oriented hacker spaces, maker spaces, or fab labs" (Schmelzer et al., 2022, section 5.3).

Performance Constitutes the Duality of Indicative and Subjunctive, What Is and What May Be

Performance is not opposed to the real. Performance is that which is always actually occurring (rather than the ideal, official, normative, expected, anticipated . . .); it is in the *indicative* mood. Performance is embodied in the deictic relations of here-and-now and "I" and "you," that is, performance is always actually occurring via bodies in relation and in motion alongside or relative to each other, such that the elemental unit of a performance is not the (sovereign, intending) "I" but the *relationship* through which "I" and "you" are mutually constituted as agents vis-à-vis one another. Thus, performance is always situated in a specific time and space (one could repeat the same behaviour in another time or space, and it would be a different performance). But at the same time, performance enfolds the *subjunctive* mood into the here-and-now; it tries out or rehearses, here and now, what might or ought to be (the mood of futurity, possibility, and becoming) (Turner, 1982, pp. 34, 82–84). "Possibility is as wide/as the space/we create/to hold it," assert Climbing PoeTree (Alixa Garcia and Naima Penniman, USA) in their 2014 performance poetry piece, "When the Last Tree Stands Alone," using voice, gesture, and gaze to make us feel possibility's expanse.

Paradigmatic examples are *scenarios* and *simulations*, which create efficacy, that is, real effects in the actually existing world. As Taylor explores, scenarios and simulations are rehearsals for future action, involving the prescribed and normative on the one hand and the imaginative and improvisational on the other (2016, pp. 140–141). Sculptor Mary Mattingly's (USA) large-scale simulation *Swale* (New York, 2016–ongoing) creates a fully participatory, floating edible landscape on a repurposed barge anchored at various New York City harbours. Created in response to New York laws prohibiting foraging on public lands – a sort of ongoing enclosure act limiting use of the commons and investing in GDP growth at the expense of sustenance and sharing economies – *Swale* poaches on marine law. Its edible landscape is both public and outside NY public land law. Enabling visitors to learn about and pick edible and medical plants and to explore the project's experimental regeneration of its resources, including its waste, *Swale* facilitates rethinking normative, everyday habits constructed by the enclosure of common lands, the capitalist divisions of public and private space, and fossil-fuel-driven ideologies of growth disincentivising sustenance and shared economies. Working like a research lab, *Swale* allows participants to try out, make, and transmit knowledge that can be portable to and actionable in other spacetimes as well, creating new norms for action. *Swale* expands the field of performance to include non-human and non-organic objects and entities (for example, New York harbour, wildlife, inorganic substances such as toxins, machines for filtering and recycling water or waste) that interact agentively, creating real effects that condition future performances, despite lacking "intentionality" or "consciousness" as defined by Western objectivism.

For its climate emergency simulation *3rd Ring Out* (2010–2011), Metis (U.K.) repurposed shipping containers – the detritus of a fossil capitalism driven by neoliberal policies that, as author and activist Naomi Klein has explained, enable

> the mass export of products across vast distances (relentlessly burning carbon all the way), and the import of a uniquely wasteful model of production, consumption, and agriculture to every corner of the world (also based on the profligate burning of fossil fuels).
>
> *(2014, p. 20)*

Inside a shipping container rigged with communicative and digital mapping technologies, *3rd Ring Out* positioned 12 participants as having "responsibility for a 'sector' of the UK, as de facto emergency planners" in a highly sensory rehearsal of a climate emergency specific to their location. Working with and on performance compulsion and at the swift speed of emergency, Metis performers required spect-actors to vote, privately, on actions to take, for example, whether to allow in climate refugees. As in democracy, the percentage of (privately cast) votes determined the (publicly shared) trajectory of the simulation, that is, which future would be rehearsed, while opportunities for conversation brought home that democratic participation is characterised more by dissent than idealised notions of community and consensus.

3rd Ring Out responded to and engaged, in embodied fashion, a research question: would participants collectively, because democratically, choose a scenario that protects investments in infrastructure, economic growth, private property, and policing against extreme weather events or one that safeguards people and cultural knowledge and heritage? Metis performers proposed *rehearsing* as a vehicle for the creative discovery of possible futures (possible scenarios to be implemented) (Metis, "Third Ring Out"). What was rehearsed in *3rd Ring Out* produced both muscle memory and propositional knowledge that might be brought to bear on current debates about climate adaptation and resilience, for example, about which version of the future should inform urban engineering and immigration policy.[7]

In addition to these modes of action, theatre and performance also contribute to justice in ways specific to its modes of articulation and aesthetic aims. Aesthetic form, rigour, and pleasure are not secondary but essential to the efficacy of theatre and performance promoting transformative climate justice. Theatre and performance as modes of *art* acknowledge and promote:

- the right to pleasure and beauty (the aesthetic)
- the right to full embodiment

- the right to imagination, the capacity to imagine oneself and one's world differently
- the right to recognition and the capacity to respond to another's demand for recognition (empathy)
- the right to cultural and historical memory and to heritage
- the right to the articulation and representation of events and experiences
- the right to space, time, and resources for enabling the creativity and the full becoming of oneself and others

Figure 11.1 *Mirror Shield Project*, Cannupa Hanska Luger, 2016–present. Image is of *Water Serpent Action*, a site-specific performance organised by Luger in collaboration with Rory Wakemup on 18 November 2016 at Oceti Sakowin Camp, Standing Rock, North Dakota.

Source: Image courtesy of the artist.

Indeed, aesthetic objects themselves exist as crucial vessels of practice, as in Cannupa Hanska Luger's (enrolled member of the Three Affiliated Tribes of Fort Berthold, USA) participatory art making. His *Mirror Shield Project* (2016) supplied both defensive force and aesthetic beauty to Water Protectors resisting the Dakota Access Pipeline near the Standing Rock Reservation (North Dakota, USA), where Luger was born. Provided by Luger with video instructions (*How To Build Mirror Shields For Water Protectors*), allied artivists constructed over 1,000 16" x 48" mirror shields of Masonite board and adhesive foil and shipped them to Water Protectors standing against pipeline security forces, the police, and the National Guard. There, the mirror shields were also used in a collaborative performance with Anishinaabe artist Rory Erler Wakemup, the *Water Serpent Action* (2016), in which hundreds of Water Protectors walked the Oceti Sakowin camp site, embodying the spiralling movements of the serpent and aiming their mirror shields at police surveillance planes flying overhead (see Figure 11.1).[8]

In their subsequent role as circulating and displayed objects, the shields serve as memorial/witness of their participatory making (of the coming together of a "we" and the embodied transmission of embodied technique) and as inspiration for future actions. Aesthetically complex and pleasing, taking up and persisting in (objective) time and space (for example, as a collectible object in museums), the mirror shields continue to summon new uses. Enfolding what performance studies theorist and ethnographer Dwight Conquergood called the three "A's" of performance, Luger's objects have a triple status, as aesthetic *accomplishment* (delight in the making and the made), embodied *analysis* (here, both research and the archive of that research, for example, of Indigenous memory, desire, and technique, able to be drawn on for future practice), and *activism* (convening a "we" capable of acting, providing material defence, and visibilising resilience) (Conquergood, 2002, p. 152).

Earlier, this chapter pointed to the potential of theatre and performance, not to build consensus but to make space and time for dissent, as key to participatory democracy, while fostering solidarity across distinct identities and interests. Likewise, performance has been a key mode for exploring the open-ended potential of desire and identity. Certain strands of feminist/queer ethics, for example, propose that we are called to act now on behalf of a future that we must imagine but can never fully know in advance.[9] What ethical responsibility do we have to a future world that we can anticipate but never fully know? David P. Terry and Anne Marie Todd describe Bike Party as not a performance for an audience but "an embodied articulation of diverse co-performers around the co-incidence of bicycling," which "forges an as-yet-to-be-fully-defined community around multiple intersecting ways of riding bikes)" (2013, n.p.). Perhaps preferable to "community" here would be an open-ended solidarity-in-action, one that must remain open to critique and revision in the very action of generating motion.

Performance as Action Matter

Creating performance for – or as – activism (artivism) involves relating the formal *vehicles* of each action (for example, plot, noise, spectacle, movement, use of aesthetic media) to the legibility of *aims* (a message directed at a target or a transformation in the participants themselves); *location* and *duration* (space and time as both contextual markers and markers of commitment, endurance, and resilience, as in a blockade, standout, or sit in, or building and occupying treehouses to protect forests); and what it produces, achieves, or brings into existence (its *efficacy*). Such events centre the performing body and bodily practices as what applied performance scholar James Thompson has called "action matter" and Foster the "body as articulate matter" (Thompson, 2012, p. 28; Foster, 2003, p. 395). Embodied social actors, differentially bearing the bodily traces of heteropatriarchal, settler colonial, racist, ableist, and extractivist histories and asymmetrically situated as able to thrive or to be disqualified from thriving, enact physical interference in the habitus of fossil capitalism. The human (animal) body can be at once the *vehicle* of the performance (as "action matter"), *aim* (as "target," as in the embodied production and transmission of know-how or the rehearsal of techniques for doing), *marker of location and duration* (taken into the body as history, memory, trauma, joy, and other emotions), and *efficacy* (the rehearsal of a future action strengthens the muscle memory of resistance or effects a transformation of the habits of the body).

Outdoor dance-circus company Motionhouse's *Cascade* (U.K., 2010) relocated home and domesticity to a public plaza, performing the flooding of a home and its occupants as a demand on public space and common time, interrupting the institutional uses of a plaza – typically in the U.K. and North America a space to cross through efficiently on one's way to a destination where something is to be done or effected, while taking in symbols of governmental and corporate power. Dance is particularly effective here; embodied and emotional rather than rational, repetitive and ornamented rather than efficient, as much or more medium than message, *Cascade* disrupted the normative and disciplinary habitus of the bodies collected in institutionalised spaces.

In The Yes Men's "RefuGreenErgy" (Brussels, 2017), likewise performed in (so-called) public spaces (spaces organised and maintained by corporate, governmental, police, and military power), unwitting passers-by witnessed the rollout of a new green tech start-up using bicycling refugees to generate electricity, "giving the refugees 24-hour periods of amnesty and 1.60 euros per day, and Belgians 'green electricity' and a 'guilt-free way to help others.'"[10] Lending "realness" to this spectacle of cycling refugees – artivists from We Are All Refugees (WAAR; Belgium), working in collaboration with refugee advocacy organisations – was The Yes Men's trademark savvy use of media legitimation and branding (based on the Situationist International technique of "culture jamming"). A website, promotional video, and social media accounts magically converted the horrified response of "this can't be real" to the realness of the commodity form. Visibilising the dehumanisation of refugees – along with access to fresh water and food, one of the key flashpoints for territorial and geopolitical conflicts of the ongoing climate emergency – and calling attention to the elite's habitual performance of looking away, "RefuGreenErgy" ridiculed corporate greenwashing of the continued expropriation of human labour and human thriving, in the Global South and the less resourced regions of the Global North, to feed the allegedly "greener" but still energy intensive, consumerist lifestyles of the privileged. The aim of many such performances and performatic installations is overturning what Taylor calls "percepticide," an attitude in which bystanders "conclude that their survival depends on not understanding or recognizing what is actually happening before their eyes" (2016, pp. 75, 77), providing instead what Julia Handschuh names a "perception score" (2013).

Regina José Galindo's (Guatemala) *Tierra* (France, 2013) enacts the intersecting violence against Indigenous women's bodies and the earth. Standing naked and unmoving on the soil, Galindo stands witness to an industrial excavator, moving repetitively around her, with each jab and dig removing more and more of the earth until she is left standing on a pillar of earth the size of her footprint. Galindo's body art/earth art, like that of her predecessor Ana Mendieta, neither makes nor is reducible to an explicit political statement; performing climate justice may not necessarily be propositional in form. Here, the artist's personal and distinct body, resonating on behalf of but never identical to or reducible to the political body (polis, community, or the counterhegemonic body of feminist or Indigenous women), measures environmental degradation/violence in terms of her own capacity to withstand, to refuse to move, to persist. A 90-minute performance without an audience but made available as a 35-minute film, Galindo's *Tierra* also raises questions such as, what is our position as spectators of the artist's body? How are we called on to respond? Who do "we" need to become to witness this work?

The Paradox of Scale and the Problem of Representation

Climate change exceeds the boundaries of any one individual's sensory awareness, experience, or capacity for acting intentionally, even as (and because) fossil capitalism shapes that everyday awareness and capacity for action. This is another way of saying that climate change is performative: it is the cumulative and recursive (looping back on itself) effect of systemic relations that produce human peoples and populations as their collective subject. As Metis observes of the inhuman scale of the climate emergency,

> The link between an apparently ordinary act of eating meat or driving a car, and floods on the other side of the world, is not only difficult to trace scientifically, it is difficult to represent through theatre: a medium that invites close scrutiny of the observable effects of individual instances of human behaviour on other human beings.
>
> *(n.d., n.p.)*

Beyond communicating scientific data, then, performers aim to embody, materially here and now, the vast spatial and temporal scale of climate change, the immensity of (differentially distributed) human accountability, and the scale of harm to human and nonhuman life (compare Lavery, 2018a). This is the first scalar problem of performing climate justice. The time and space of climate change exceeds one's awareness, knowledge, and intention but not one's culpability.

A second difficulty is that of representation. How can the playwright or performer represent, using the medium of human (animal) bodies, the *nonhuman* consequences of climate change and the permeability and relationality of human and nonhuman entities?

The gap between individual experience and the impersonality of infrastructure is the subject of Kamil Haque's (Singapore/U.S.) five-minute play *The Olde Woman Who Lives in a Shoe*, produced for the 2019 cycle of Climate Change Theatre Action (CCTA), an international collaboration among playwrights and producers/performers to stage, in a variety of public spaces and venues, five-minute plays responding to the climate emergency. (Playwright Chantal Bilodeau [Canada/USA], founding artistic director of the Arts and Climate Initiative, leads the initiative.) Each short, specific story can provide a vehicle for the collective or communal expression of emotion and an opportunity for cautionary and critical thinking. Reflecting on the losses sustained in Hurricane Katrina in the U.S. in 2005, in part because of the failure of the levees (Pruitt, 2020), Haque's play calls on audience members to witness the grief of a mother (designated "the Olde Woman") for her drowned child. Only the child's shoe has been found. In it the mother invests her words, her emotions, her life. As a kind of refrain in this spoken word piece, Haque, who is founder of the Haque Centre of Acting and Creativity, Singapore's first professional acting studio in English, presents the Olde Woman as pledging allegiance to "the Almighty/The Government, that is," exchanging her labour for the promise of a "right to life, liberty, and The Pursuit of Happyness [sic]" (Haque, 2020, p. 177). Has the Olde Woman been correct to position the US government in this way? Were Hurricane Katrina and the collapsed levees also "act[s] of God, that is, the Government"? Haque's short play, like most theatre and performance, does not answer the questions it poses – this is work for the audience to do and their work is part of the ongoing performance event. If we are moved to ask how we are to invest trust in the adaptation and resilience strategies of an often invisible, removed, and even disinterested higher power, then we may be moved to collective action: dissent, mutual aid, benchmarking new governmental promises, and repairing community ties.

As the title of David Finnigan's (2020, Australia) solo play asserts, *There are a Lot of Stories You Can Tell about Humanity* (CCTA, 2019). Finnigan situates a story about humans (human animals) within two twinned stories of inhuman scale – evolution and the future/beyond of the climate emergency. Our smallness is part of those stories, and yet telling the story of that smallness places us, as if discovering we are actors in – and at home in – the vastly impersonal unfolding of world systems. If Finnigan's play can be said to re-center humanness, through its capacity for storying its world, Caridad Svich's [U.S.A.] *Letter from the Ocean*, from the same round of CCTA commissions, shares the agency of storytelling with the ocean. In the here-and-now of performance time, a letter arrives from the future. Sent by the Ocean, it is addressed to the "Dear person who went away to the hut in the woods when all the world was on fire" and humans were refugees. In this shared story of the future, what has survived of humanness are words, music, and the hut: a place for helping refugees in flight from a world on fire but also a gathering place for discussing civil liberties, for reading poetry, and for storytelling (after the day's work has been done). Here is a basic motif of humanist literature: human travellers – in this case refugees – gather to tell stories. These refugee stories are archived in small, makeshift books written by hand, stored on a shelf in the hut. "If you were to run your hand along the shelf," says the speaker, "you could hear all the voices." Speaking of her rage but also of her love and the need for kindness, the Ocean suggests that, because humans have both reason and imagination, they can learn and create beauty. Storytelling is a gift of love from and with the Ocean, then, a way of recalling relationship and no longer that which defines the human.

Envoicing the More-Than-Human Natural World

Indigenous and ecoperformance scholars and practitioners extend this relationality to forms of agency other than the human and social. "How are our understandings of power and subjectivities changed," suggests Handschuh,

> if, instead of being defined by a single call-and-response of one human to another, we instead turn toward the soil, a river, or a storm? Understanding our identities as

constituted by both human and environmental relations thus requires that we give credence to the landscapes we inhabit.

(2013, n.p.)

An adequate response to climate change would demand reconsidering not just what we must do but who "we" must become and become-with.

In Bilodeau's five-minute play *Mother* (CCTA, 2015, Figure 11.2), the speaker imagines a time before the rationalist separation of human (animals) from Earth:

From dusk to dawn and dawn to dusk
I held you close and we lost ourselves into each other's mystery
That was before you started calling me Mother

Does the ability to say "Mother" (Nature, Earth) already imply separation and distance? To say "Mother" is to acknowledge that we have already turned away, to accept the capitalist tenet that separateness is necessary for human thriving:

Look at me
I wish you would look at me.

The play asks a question of us: What would looking at Mother require? Bring about? It is not a question that can be answered but one that is to be practised: rehearsed by

Figure 11.2 *Mother* by Chantal Bilodeau presented by the Box Collective as part of A Swan Song: Voices on Climate Change, a 2015 Climate Change Theatre Action event, New York, NY, 14 November 2015. Performed by Esther Sophia Artner. Directed by Sara Fay George.

Source: Photo by Julia Deffet. Image courtesy of Chantal Bilodeau.

spectators becoming spect-actors and continually revised in response to the changing eco-systems of which our everyday performances constitute one part.

Many of the plays commissioned by CCTA centre us human (animals) as one narrative thread in the story of ecosystems and invite us to speak for the fuller story, on behalf of the ecosystems of which we are (only) a part. CCTA playwrights accordingly explore different aesthetic strategies for giving "voice" to nonhuman beings. Should the playwright/performer resist or risk anthropomorphism? Who – what sort of embodied humans marked by which settler colonial, racialised, ableist, and heteropatriarchal histories of subjugation or privilege – should personate nature's voice? Such questions point to the paradox of (hegemonic) human consciousness, both a part of and apart from the natural world, both embedded in ecosystems beyond our full awareness and knowledge and granted an exceptional normative agency: protection of the rights of nature. (See also Donald, 2018, p. 24.)

Some theatre and performance artists advocate tactically embracing anthropomorphism and reworking the "pathetic fallacy" toward reparative ends (Donald, 2018, p. 35). Phrasing the problem this way, however, can still centre white, hegemonic rationalism and objectivism, erasing cultural/spiritual and subaltern understandings of the vitalism of nonhuman nature, including Indigenous ways of knowing and relating to the world. By contrast, scholars Kyle Powys Whyte and Chris Cuomo suggest that both the human and nonhuman have *reciprocal* responsibilities toward one another, such that stewardship involves "acknowledgment of one's place in a web of interdependent relationships that create moral responsibilities, and it recognises that there are methods and forms of expertise involved in carrying out such responsibilities" (2016, p. 238). Such responsibilities extend to nonhuman entities like water as well:

> The responsibilities that maintain and strengthen those relationships are between and among active agents, and they are reciprocal, not one-directional. . . . As responsible agents, a range of human and nonhuman entities, understood as relatives of one another, have caretaking roles within their communities and networks.
>
> *(p. 239)*

In such terms, our responsibility *qua* humans might include envoicing, representing, and giving personhood to nonhuman nature (Bilodeau and The Arctic Cycle, 2021). To "represent" nature here is that performance in which one body possessing the legal right of personhood (or, in jurisprudence, to have legal standing) is authorised to speak "on behalf of" or "for the benefit of" those they represent. The Global Alliance for the Rights of Nature (GARN) has this understanding of representation in mind when it asserts, "[W]e – the people – have the legal authority and responsibility to enforce these rights on behalf of ecosystems. The ecosystem itself can be named as the injured party, with its own legal standing rights, in cases alleging rights violations." "I am standing here for a body that cannot stand here," declares the speaker of Hanna Cormick's (Finland/Australia) *Canary* (CCTA, 2019), setting up analogies among the canary in the coal mine, the human (animal) suffering from Multiple Chemical Sensitivity (the playwright), and the earth – and also the actor representing (speaking on behalf of/speaking in the voice of) the playwright's experience of disability:

> My body asks her body what it feels like
> For her body to be so damaged by all the tiny personal choices of our day
> Her body tells my body that she feels like the Earth

So this body here also stands here for the Earth that cannot stand here
And that cannot speak to us
Except through the envelope of silent corpses.

(2020, p. 103)

In Mindi Dickstein's (USA) *Starving to Death in Midtown,* (CCTA online, 2020), a human animal stands for a bee who, standing there in front of us, claims legal standing: "I am the last bee standing. The last of my colony, here to testify. To make this last plea for my people." Representation as standing for – standing for the rights of, speaking on behalf of, occupying time and space, taking a stand – is irreducible to that more familiar theatrical art of the symbolic representation of social reality. As theatre theorist Carl Lavery suggests, to aim at representation must necessarily, and productively, fail (2018a); such representation is *pathetic* given the inhuman scale of climate change. "Do I have to paint you a picture?" the bee scoffs.

"Okay, we're not actually people," Dickstein's bee allows, "but I'm going to speak in terms you can understand." If bees are not human but could be a "people," what rights and responsibilities might follow? (Not all humans [animals] enjoy the full rights of personhood, including, at a minimum, what the bees lack, the right not to be killed or expropriated for the thriving of others.) Jhan Hochman finds an ethical conundrum here. How do we represent (speak on behalf of) nature without appropriating nature to ourselves as knowable to, useful to, or like us? How can we "grant [natural] being and entities unromanticized difference, an autonomy apart from humans, a kind of privacy and regard heretofore granted almost exclusively only to those considered human[?]" Hochman asks (2000, p. 192).

In an account of the performance *Water Borrow* (Canada, 2013), created and performed with Nick Millar as part of the *Guddling About* series, which deserves to be quoted at length, Minty Donald has reflected on performing the task of asking to borrow water from the tributaries of the Bow River for use in later performances:

> While it has become evident to me that I am undergoing a performance of sham courtesy, which affords the water no agency in acquiescing to or refusing my request, it is only through the actual utterance of the question [the request to "borrow" water] that the force of this realisation strikes me. In voicing my request, I also become aware that unsurprisingly, given my cultural background, it has never before occurred to me to ask for permission to 'borrow' water. As is common across affluent Western societies, I take for granted my dominion over water – my right to help myself to water for drinking, cooking and washing and, by extension, my implication in its use for more contentious purposes such as industrial-scale agriculture and fracking.

(2018, p. 34)

To perform this request is to hold space and time for a change in subjectivity, to invite becoming-with.

Calling for a shift from an economic model of adaptation to "a humanist view of adaptation," ethicists Allen Thompson and Jeremy Bendik-Keymer raise the key question, who do "we" humans (animals) need to *become* today (2012, p. 6)? If humans (animals) are called on to take up the role of "adaptively manag[ing] the basic ecological conditions of the global biosphere," they observe, then "[a]t present, it appears that we – humans – are not ready to adopt this role and its attendant responsibilities. *Who we are* today is not ready for this and *who we have been* got us into this mess" (p. 15). Adaptation to build infrastructural resilience, to continue our current ideologies of human thriving, must become adapting *ourselves*, individually and collectively.

Collaboratively creating an urban community garden, for example, may not just be about beautifying a vacant lot or making fresh produce available in a food desert (although these are urgent and important). It may be about creating new forms of urban subjectivity. As Barbara Willard observes of such a garden in Fuller Park, Chicago,

> The neighbors engaging in this reclamation often have little power to change the physical conditions (let alone the socioeconomic conditions) of their neighborhood; but, the transformation of vacant lots allows them to enact civic agency and actively participate in rewriting the future of their community.
>
> *(2013)*

The performatic transformation of embodied skill and know-how, as much as the creation of a new urban spacetime, is the project's efficacy. Embodied practice "without a discourse" (yet) *may* also be practice "without a subject" (yet) and thus to some extent a departure from or temporary suspension of the forms of subjectivities we take up, negotiate, and/ or resist in other aspects of our lives. As a shared withdrawal of time, space, and energy from other activities, say, habitual enactments of the dominant cultural commitments to growth, the *collective* (this is crucial) action of imagining, relating, exploring dependencies and vulnerability, and caring does more than propose a future; it creates here and now a spacetime alternative to the hegemonic spacetime of fossil capitalism. Brought to the level of discursive consciousness and thus to the level of a proposition or demand, we might organise around the infrastructures needed for supporting and sustaining these alternative spacetimes.

Notes

1 Performance studies explores both formal performance such as theatre or installation art and "performance in everyday life," that is, how we transmit knowledge, pass on and transform heritage and culture, and (re)produce the building blocks of social reality in the routines of our everyday behaviour. Our everyday routines can reproduce or resist dominant assumptions and expectations for behaviour (social scripts), and each repetition of a routine provides opportunities for reflection, critique, and transformation of that routine, creating openings for change in the ongoing flow of social performance.

2 Following Kinol et al. (2023), this chapter adopts the phrase "transformative climate justice," defined as "societal transformation to simultaneously address worsening economic and health inequities and growing climate vulnerabilities," for example, "water access, food production, physical and mental health, and physical and economic infrastructure, particularly for vulnerable communities around the world" (pp. 2–3/29). Transformative climate justice shifts the story of climate action from the binarism of mitigation or adaptation to addressing the root causes of the climate emergency, including the ideology of "growthism" and racialised, heteropatriarchal, and colonial economies of extraction and consumption. "[B]roaden[ing] responsibility to redress the legacy of injustice and exploitation resulting from systemic practices, policies, and priorities that perpetuate inequities in climate vulnerabilities – locally, regionally, and globally," transformative climate justice entails "investing in social innovation, social infrastructure, and social justice" (p. 5/29). Centering and growing the knowledge and solutions proposed by frontline and

Indigenous communities is key, as is building political will for payment of climate debt from the Global North to the Global South, reparations for land theft and stolen labor, and redistribution of technologies to the Global South enabling low-carbon economic development. Theatre and performance artists working in community might aim, for example, to build audiences for frontline knowledge; address climate emotions; dismantle human exceptionalism; advocate for investing in the infrastructure necessary for building and sustaining thriving, including the right to cultural heritage, growing community, and repairing the commons; and develop creative methods for responding to conflict resulting from climate migration and welcoming immigrants.

3 As organisers Hilary Moore and Joshua Kahn Russell observe, "The communities most directly impacted on the frontlines are not only dealing with the brunt of the problem, but are also best equipped with the knowledge and skills to chart the way forward" (2011, p. 12).

4 Contemporary, socially engaged theatre and performance strives to transform spectators into spect-actors, "people capable of acting and interrupting the performance or changing the roles they've been assigned" (Taylor, 2016, p. 80, drawing on Boal, 1985).

5 Taken from postmodern geography, "spacetime" names a switch from an objectivist and rational to an embodied and processual account of reality. Following Henri Lefebvre (*The Production of Space*, 1991), urban studies scholar Helen Liggett argues that space is not a universal, not an "empty container" or "abstract category" external to the self but "an active component of constructing, maintaining, and challenging social order." Not the object-like shapes on a map, space is "process," for example, mapping, that is "continually being produced." In short: "Actions, relations producing space, are what space is" (Liggett, 1995, p. 255). So too time, like space, is dominantly conceived as an empty container or abstract category, objective and "outside" of us. The term "spacetime" recognises that time is entailed in (taken up in, implied by) space, once the latter is conceived as a process and a production.

6 It is crucial both to consider the limits of theatre and performance artists' capacities to build community and to be cautious of the limitations of the idea of community itself. Transformations of subjective experience (experiences of joy, for example or embodied skill) may not always work toward progressive ends; these may leave intact systems of consumerism and risk commodification in turn. More immediately dangerous, they may promote majoritarian, populist, and nationalist ideologies or enact surveillance of difference and dissent. Emphasising human community, moreover, risks centering the human and human consciousness. Nevertheless, the response to climate change cannot be successful without thoughtful and critical investments in place-, land-, and community-based and engaged practice.

7 Sociologist Anthony Giddens calls embodied knowledge, not necessarily put into words, "practical consciousness" (1984, p. 90). "Discursive consciousness" names social actors' capacity to put what they know and do into words, reflecting on these in the medium of propositional language. Practical and discursive consciousness are recursively engaged with each other, so that, for example, critical reflection might add to the "know-how" with which we go about our daily routines, just as embodied knowledge gained in the flow of everyday performance (for example, whether a gendered article of clothing "fits" our embodied sense of self or playing with a tool or artistic medium "discovers" a new potential) can be "brought to the level of words" and shared and debated discursively.

8 In a press release, Luger states that the shields were "inspired by images of women holding mirrors up to riot police in the Ukraine, so that the police could see themselves" and by the legendary Greek Perseus's use of a mirror shield to slay Medusa. A contemporary reuse of older technologies, the mirror shields aid in "slaying modern day monsters such as extractive industry and capitalism" (courtesy of the artist).

9 Consider Elizabeth Freeman: "sexual dissidents must create continuing queer lifeworlds while not being witness to this future or able to guarantee its form in advance, on the wager that there will be more queers to inhabit such worlds: we are 'bound' to queer successors whom we might not recognize" (2005, p. 61).

10 According to their website, US-based Jacques Servin (pseud. Andy Bichlbaum) and Igor Vamos (pseud. Mike Bonanno) founded The Yes Men as an "anti-corporate sabotage group," performing as "yes men" to infiltrate corporate brands. Dedicated to "having fun" in the process, The Yes Men, now an international network of collaborators and trainers, create acts of "laughtivism" (https://theyesmen.org/about#faqit186, accessed 27 June 2023).

References

Besel, R. D. and Blau, J. A. (eds.) (2013). *Performance on Behalf of the Environment*. Lanham, MD: Lexington Books [eBook]. https://ebookcentral.proquest.com/lib/brandeis-ebooks/detail. action?docID=1574393 [Accessed 28 June 2023].

Bilodeau, C. (2015). Mother. *Script*. https://www.cbilodeau.com/one-acts; performed version available at: https://youtu.be/GN-ktVId_OY [Accessed 29 August 2023].

Bilodeau, C. and The Arctic Cycle (2021). Indigenous sovereignty. Webinar, from the Series *Artists Envisioning a Green New Deal*, 5 May. https://artsandclimate.org/indigenous-sovereignty [Accessed 29 June 2023].

Bilodeau, C. and Peterson, T. (eds.) (2020). *Lighting the Way: An Anthology of Short Plays About the Climate Crisis*. US: The Centre for Sustainable Practice in the Arts.

Boal, A. (1985). *Theatre of the Oppressed*. Translated by Charles A. and Maria-Odilia Leal McBride. New York: Theatre Communications Group.

Center for Story-Based Strategy. (n.d.). *Theory of Change*. https://static1.squarespace.com/static/59b848d980bd5ee35b495f6e/t/59d91e6029f187b71a7bcceb/1507401313613/css_theory_of_change_final.pdf [Accessed 28 June 2023].

Climate Justice Alliance. (n.d.). *Just Transition: A Framework for Change*. https://climatejusticealliance.org/just-transition/ [Accessed 28 June 2023].

Climbing PoeTree. (2014). When the last tree stands alone. *Pachamama Alliance*. https://youtu.be/6vhjV6kWCnM [Accessed 28 June 2023].

Conquergood, D. (2002). Performance studies: Interventions and radical research. *The Drama Review (TDR)*, 46(2) (Summer), 145–156.

Cormick, H. (2020). Canary. In C. Bilodeau and T. Peterson (eds.) *Lighting the Way* (pp. 99–105).

Dickstein, M. (2020). *Starving to Death in Midtown*. Performed Version. https://youtu.be/mLuuC0TVyQk [Accessed 29 June 2023].

Donald, M. (2018). The performance 'apparatus': Performance and its documentation as ecological practice. *Green Letters: Studies in Ecocriticism*, 20(3) (November 2016), 251–269. Reprint in *Performance and Ecology: What Can Theatre Do?* Edited by Lavery (pp. 23–41). https://guddlingabout.com/portfolio/performance-apparatus-essay/ [Accessed 29 June 2023].

Drewal, M. T. (1991). The state of research on performance in Africa. *African Studies Review*, 34(3) (December), 1–64.

Finnigan, D. (2020). There are a lot of stories you can tell about humanity. In C. Bilodeau and T. Peterson (eds.) *Lighting the Way* (pp. 151–153). Performed Version. https://youtu.be/jqrrlbAR_f8 [Accessed 29 June 2023].

Foster, S. L. (2003). Choreographies of protest. *Theatre Journal*, 55(3) (October), 395–412.

Freeman, E. (2005). Time binds, or, erotohistoriography. *Social Text*, 23(3–4) (Fall-Winter), 57–68.

Galindo, R. J. (2013). *Tierra*. Documentation. https://youtu.be/eSRqMMieSIA [Accessed 28 June 2023].

Giddens, A. (1984). *The Constitution of Society: Outline of the Theory of Structuration*. Berkeley: University of California Press.

The Global Alliance for the Rights of Nature (GARN). (n.d.). *What Are the Rights of Nature?* [online]. https://www.garn.org/rights-of-nature/ [Accessed 29 June 2023].

Handschuh, J. (2013). "On finding ways of being": Kinesthetic empathy in dance and ecology. In R. D. Besel and J. A. Blau (eds.) *Performance on Behalf of the Environment* (chapter 8).

Haque, K. (2020). The old woman who lives in a shoe. In C. Bilodeau and T. Peterson (eds.) *Lighting the Way* (pp. 177–179). Performed Version. https://youtu.be/1fVOkFAbRWE [Accessed 29 June 2023].

Hochman, J. (2000). Green cultural studies. In L. Coupe (ed.) *The Green Studies Reader: From Romanticism to Ecocriticism* (pp. 187–192). London: Routledge.

IPCC. (2021). Summary for policymakers. In V. Masson-Delmotte et al. (eds.) *Climate Change 2021: The Physical Science Basis. Contribution of Working Group I to the Sixth Assessment Report of the Intergovernmental Panel on Climate Change* (pp. 3–32). Cambridge: Cambridge University Press. https://www.ipcc.ch/report/ar6/wg1/chapter/summary-for-policymakers/ [Accessed 28 June 2023].

Kinol, A., et al. (2023). Climate justice in higher education: A proposed paradigm shift towards a transformative role for colleges and universities. *Climatic Change*, 176(15). https://doi.org/10.1007/s10584-023-03486-4 [Accessed 28 June 2023].

Klein, N. (2014). *This Changes Everything: Capitalism vs. the Climate*. New York: Simon & Schuster.

Lavery, C. (2018a). Theatre and Time Ecology: Deceleration in *Stifters Dinge* and *L'Effet de Serge*. *Green Letters: Studies in Ecocriticism*, 20(3) (November 2016), 304–323. Reprint in *Performance and Ecology: What Can Theatre Do?*, ed. Lavery, pp. 76–95.

Lavery, C. (ed.) (2018b). *Performance and Ecology: What Can Theatre Do?* London and New York: Routledge.

Lenton, T. M., et al. (2023). Quantifying the human cost of global warming. *Nature Sustainability*, 22 May.

Liggett, H. (1995). City sights/sites of memories and dreams. In H. Liggett and D. C. Perry (eds.) *Spatial Practices: Critical Explorations in Social Spatial Theory* (pp. 243–273). Thousand Oaks, CA: Sage.

Luger, C. H. (2016). *Mirror Shield Project*. Documentation. https://www.cannupahanska.com/social-engagement/mirror-shield-project [Accessed 28 June 2023].

Mattingly, M. (2016-present). *Swale*. Documentation. https://marymattingly.com/blogs/portfolio/swale-2016-ongoing [Accessed 28 June 2023].

Metis. (2010–11). *3rd Ring Out*. Documentation. https://metisarts.co.uk/projects/3rd-ring-out [Accessed 28 June 2023].

Metis. (n.d.). *Third Ring Out: The Research Context*. https://metisarts.co.uk/research/3rd-ring-out-research [Accessed 26 June 2023].

Moore, H. and Russell, J. K. (2011). *Organizing Cools the Planet: Tools and Reflections to Navigate the Climate Crisis*. PM Press Pamphlet Series no. 0011. Oakland, CA: PM Press. https://organizing coolstheplanet.wordpress.com/ [Accessed 28 June 2023].

Motionhouse. (2010). *Cascade*. Documentation. https://www.motionhouse.co.uk/production/cascade/ [Accessed 28 June 2023].

Neal, L. (2015). *Playing for Time: Making Art as If the World Mattered*. London: Oberon.

Pruitt, S. (2020). How levee failures made hurricane Katrina a bigger disaster. *History*. https://www.history.com/news/hurricane-katrina-levee-failures [Accessed 29 June 2023].

Schmelzer, M., Vetter, A. and Vansintjan, A. (2022). *The Future Is Degrowth: A Guide to a World Beyond Capitalism*. New York: Verso [eBook]. https://ebookcentral.proquest.com/lib/brandeis-ebooks/detail.action?docID=7012344 [Accessed 29 June 2023].

Svich, C. (2020). Letter from the ocean. In C. Bilodeau and T. Peterson (eds.) *Lighting the Way* (pp. 313–316). Performed Version. https://youtu.be/wNtokd5ZaHI [Accessed 29 June 2023].

Taylor, D. (2016). *Performance*. Translated by Abigail Levine; adapted into English by Diana Taylor. Durham and London: Duke University Press.

Terry, D. P. and Todd, A. M. (2013). It's a party, not a protest: Environmental community, co-incident performance, and the San José Bike Party. In R. D. Besel and J. A. Blau (eds.) *Performance on Behalf of the Environment* (Chapter 2).

Thompson, A. and Bendik-Keymer, J. (2012). Introduction: Adapting humanity. In A. Thompson and J. Bendik-Keymer (eds.) *Ethical Adaptation to Climate Change: Human Virtues of the Future* (pp. 1–24). Cambridge, MA: MIT Press [eBook]. ProQuest Ebook Central. https://ebookcentral. proquest.com/lib/brandeis-ebooks/detail.action?docID=3339406 [Accessed 29 June 2023].

Thompson, J. (2012). *Applied Theatre: Bewilderment and Beyond*. Oxford: Peter Lang.

Turner, V. (1982). *From Ritual to Theatre: The Human Seriousness of Play*. New York: PAJ Publications.

Whyte, K. P. and Cuomo, C. (2016). Ethics of caring in environmental ethics: Indigenous and feminist philosophies. In S. M. Gardiner and A. Thompson (eds.) *The Oxford Handbook of Environmental Ethics* (pp. 234–247) [eBook]. https://doi.org/10.1093/oxfordhb/9780199941339.013.22 [Accessed 29 June 2023].

Willard, B. (2013). Reinhabiting the land: From vacant lot to garden plot. In R. D. Besel and J. A. Blau (eds.) *Performance on Behalf of the Environment* (Chapter 6).

The Yes Men. (2017). *RefuGreenErgy*. Documentation. https://theyesmen.org/project/refugreenergy [Accessed 27 June 2023].

12

TURNING A LEAF

Fiction and Nonfiction as Grassroots Climate Activism

Sabine von Mering

Introduction

As global warming and its consequences are beginning to force societies everywhere to undergo dramatic transformations whether by design or by disaster, a special responsibility falls to authors of fiction and nonfiction. As readers turn to texts about the climate crisis for information, reflection, consolation and inspiration, these may also serve as seeds of grassroots climate activism. Indeed, some of these seeds are beginning to bear fruit: "Studies are beginning to show that environmental literature can change its readers' perceptions . . . and that the arts . . . and imagination . . . can develop the type of thinking needed to create societal transformations." (Martinez, 2024, p. 39) After much polemicising about the conspicuous absence of serious climate writing (Ghosh, 2017), the number of publications about the climate crisis and what to do about it has exploded in recent decades. This chapter provides an overview of the existing literature and identifies characteristics of climate writing we can categorise as a form of grassroots climate activism.[1]

Practising the Futures We Long For

In 2021, the editors of *Grist*'s online "Fix Solutions Lab" compiled a "climate-fiction glossary" introducing a list of climate fiction and cli-fi genres, from Afrofuturism to Hopepunk and Solarpunk. As the words suggest, genres like Hopepunk and Solarpunk "embrace optimism and celebrate a clean, green, just society."[2] The *Grist* editors interviewed American fiction writer Sarena Ulibarri, an editor of several anthologies of solarpunk, who describes her work: "Making or remaking your own clothing, growing your own food, and right-to-repair are ways that solarpunks reclaim their basic needs from corporate control." The editors contend, though, that not all solarpunk presents idyllic utopian societies in which sustainability and egalitarianism are the norm but that it may also involve making the best of challenging circumstances (like a changing climate) and times of conflict. Solarpunk, they say, is still punk, still a form of rebellion – in this case a rebellion against the fossil fuel industry and the inevitable climate apocalypse: "When the status quo is despair and defeat," Ulibarri says, "then hope becomes an act of resistance."[3] In "Climate Fiction

DOI: 10.4324/9781003396567-14

Glossary: From Afrofuturism to Ecotopia," Claire Elise Thompson suggests that "Fiction is how we imagine a different future . . . even in times when that future feels unreachable" and invites her readers to "explore the power of climate fiction to help far-off visions become realities, and experience what happens when you live in someone else's imagined future."[4] Maddie Stone asks, "can climate fiction deliver climate justice?" Other articles focus on green burials, stories "of Abundance, Adaptation, and Progress," and "Plastic-eating Organisms" but also on intersectionality and queer issues in climate fiction. The *Grist* editors also confidently provide a "Definitive Climate Fiction Reading List" which includes "20 wildly imaginative books that will shape the way you think about our planet's future – and humanity's place in it."[5]

They also discuss the purpose of this new genre of writing: "At its best, fiction inspires people and makes them think. Sometimes it even spurs them to action." The list of books discussed includes anthologies as well as monographs, the large majority of which are recent publications by young American authors like Sam J. Miller, Rebecca Roanhorse, Egyptian-Canadian Omar El Akkad, and several volumes by Arizona State University's Center for Climate and Imagination (Eschrich and Miller, 2018). Octavia E. Butler is celebrated as a pioneer: "Octavia E. Butler warned us about the societal impacts of climate change more than a decade before *An Inconvenient Truth* introduced the concept to mainstream audiences."[6]

Contrary to Solarpunk and Hopepunk, the genre of Afrofuturism is more multifaceted. Not solely devoted to works about the climate crisis,[7] it is an artistic movement that spans literature, music, and other art forms, incorporating Black history and culture into futuristic texts, films, etc. (Womack, 2013). In *Imagining the Future of Climate Change: World Making through Science Fiction and Activism* (2018), Shelley Streeby presents an in-depth reading of a number of Afrofuturist texts and points to the direct connection to climate activism: "people of color and Indigenous people use science fiction and other speculative genres to remember the past and imagine futures that help us think critically about the present and connect climate change to social movements" (Streeby, 2018, p. 5). To her, the leadership of Indigenous writers stands out: "Indigenous futurisms are at the forefront of efforts to imagine a future of climate change other than that envisioned by the fossil fuel industry, and they take many different cultural forms" (Streeby, 2018, p. 28). Streeby highlights Walidah Imarisha and adrienne maree brown's collection of short stories *Octavia's Brood: Science Fiction Stories from Social Justice Movements* (2015) and explains how the authors characterise activists and speculative fiction writers as equally engaged in envisioning other worlds. In the spirit of Octavia E. Butler, such writings represent a form of "decolonization of the imagination" and thus pose a direct threat to the status quo and should be seen as "the most dangerous and subversive form there is" (Imarisha and brown, 2015, p. 4).

Forms of Afrofuturism are explored by authors in African countries as well. The "Share Africa Climate Fiction Award" established in Botswana in 2022, for example, recently shortlisted South African writer Vuyokzi Ngemntu for her story *The Serpent's Handmaiden* in which she describes a dystopian future with a hopeful ending (Ngemntu, 2022). Today's African climate fiction, like African climate activism in the continent (see chapters in Part III and IV), is primarily concerned with adaptation and resilience or, as Nigerian author and editor Oghenechovwe Donald Ekpeki describes it in an interview, "real-time, in the now. Who has no home to live in? (Who has) no path to go to school or work as their roads are literally broken by flood waters? Who has no way to eat, no future?" (Kayawe, 2023). For the same reason, African writers also focus on climate migration, as for example Samuel Kôláwolé in *The Road to the Salt Sea* (2024).

Climate writing that aims to spur readers into action can take very different forms. Streeby mentions the *New York City Stands with Standing Rock Collective Syllabus*[8] that came out of the #NoDAPL fight against the North Dakota Access Pipeline as an example of a productive transfer of ideas into action. Another example of Indigenous climate leadership is the Mǎori website *Anamata Future News*, many episodes of which are devoted to the themes of "climate change, resource extraction, and renewable energy." Streeby also discusses how adrienne maree brown uses Octavia E. Butler's *Parable of the Sower* to inform her own activist work, emphasising the need for an intersectional approach to climate justice:

> brown builds on Butler's work to imagine the future of climate change . . . brown's approach is intersectional: she understands climate change cannot be isolated from other economic, social scientific and technological problems. In other words: confronting climate change necessarily involves confronting other inequalities. This clustering together of issues is apparent in the many different kinds of movements with which brown has worked. The insistence on not isolating climate change problems from larger economic, racial, and social problems and conflicts over colonialism is one of the biggest differences between mainstream environmental movements and movements that enjoy significant leadership from Indigenous people and people of color.
>
> *(Streeby, 2018, p. 118)*

In another interview from 2013, brown articulates how Butler's work inspires her activism as a fictional space that allows readers to imagine and thus try out what life could be like in a different future: "the realm of science and speculative fiction could be a great place to intentionally practice the futures we long for" (Streeby, 2018, p. 119).

More recent scholarship on climate writing embraces intersectional approaches similar to brown's and Streeby's. This is also the result of a learning curve, as Adam Trexler concludes in his "Epilogue: What Has Changed Since *Anthropocene Fictions?*" in Debra J. Rosenthal and Jason De La Mara Molesky's co-edited volume *Cli-Fi and Class: Socioeconomic Justice in Contemporary American Climate Fiction* (2023): "apocalyptic threats have spectacularly failed to marshal adequate emissions reductions; this rhetorical move must be judged an abysmal failure" (Trexler, 2023, p. 239). Trexler concedes that his own scholarship in *Anthropocene Fictions: The Novel in a Time of Climate Change* (Trexler, 2015) lacked the important focus on class provided by Rosenthal and Molesky's collection, which obliges us to consider that climate change is experienced vastly differently by different people already now: "climate change already shaped supply-side facets like industrial regulation, global trade, fisheries, agricultural production, and labour, as well as demand-side considerations like how we eat, where we live, how we move, and even what options we have in a crisis" (Trexler, 2023, p. 239). Trexler therefore sees most climate fiction's shortcomings in large part in this lack of attention to race and class:

> climate fiction's strong reliance on white characters evades consideration of poor and minority experiences of contemporary climate discourse and Anthropocene futurity. It would seem that the totalizing, planetary discourse of climate science and emissions

rhetoric, a supposed system on behalf of "all," was too easily and too often mobilized by supremacist worldviews. So much climate fiction privileges a technocratic, hierarchical mode of emissions governance that would dictate access to energy and consumption, with little interrogation of its own values and a contempt for positions that would challenge it.

(Trexler, 2023, p. 241)

Like for Streeby, the white supremacist perspective is confronted in Trexler's view above all by Indigenous authors: "Indigenous technologies of adaptation, survival, and narrative decenter white dreams of technocratic, planetary harmony" (Trexler, 2023, p. 241). Black and Indigenous writers, Trexler concludes, are not only making demands of the polluting industries but also of the white climate activist movement:

Black and Indigenous climate narratives can expressly refuse assimilation, challenging climate politics to model nonunitary, inclusive, and locally controlled modes of enacting emissions reductions and impact amelioration. Without such explicit, conscious perspective-building, climate criticism seems doomed to continue in the inadequacy that has defined so much activism around global heating.

(Trexler, 2023, p. 242)

Speculative fiction, though focused on an unknown future, allows us not only to envision that future but also to envision ourselves and our societies as doing things differently in that future. In a subchapter to "Educational Imaginaries for a Child-Climate Future" called "Cruel Optimism and Speculative Fiction," Emily Ashton writes: "Speculative fiction that engages with the climate crisis is often criticised because it is future-oriented so it encourages passiveness now. It can be escapist instead of realist – entertainment rather than activism" (Ashton, 2022). On the contrary, though, Ashton believes that speculative fiction is necessary and urgent as it acknowledges that "only transformative change will do, change akin to the end of the world as we know it" (Ashton, 2022). It thus can "produce meaning and material with which to build (and destroy) what we call 'the real world'" (Ashton, 2022)."

Destroying a destructive world and dealing with the terror of that destruction is what often motivates writers to write. In a short video for @HotMessPBS, host Miriam Nielsen interviews Lindsay Ellis of *It's Lit* and Amy Brady, author (with Tajja Isen) of the essay collection *The World as We Knew It: Dispatches from a Changing Climate* (2022) about the question "Can these Books Save the Planet? The Rise of Climate Fiction" (Nielsen, 2019)[9] Brady muses that "most climate fiction writers are creating their novels, because they have anxiety about what's happening in the world right now," (though Ellis also wonders whether some writers may not simply wish to capitalise on a "hot topic"). The three women provide a brief overview of existing climate fiction and films and conclude that although cli-fi may also make people more anxious rather than active, the books are likely to make people talk about the problem, which they agree is an important first step.

We Are Water Protectors: Picture Books

First steps may also motivate children's picture book authors, who have produced beautiful pleas for action in recent years. In their Caldecott-Medal-winning 2020 picture book *We are Water Protectors*, Native American author/illustrator team Carole Lindstrom and

Michaela Goade describe a young Ojibwe girl who encourages her young readers to become "earth stewards" like herself – the book even includes an "Earth Steward and Water Protector Pledge" for the reader to sign and date on the last page. Goade's beautiful illustrations of the threat/prophecy of the "black snake" pipeline that mobilises the young girl to organise her community to protest for a liveable climate is one of several recently published American picture books with an explicit climate activist agenda. Others include author Zoë Tucker and illustrator Zoe Persico's *Greta and the Giants*, inspired by *Greta Thunberg's Stand to Change the World* (2019*)* which ends with a list of "things you can do," Andrew Joyner's *Stand Up! Speak Up! A Story inspired by the Climate Change Revolution* (Schwartz and Wade, 2020), and author Gabi Snyder and illustrator Sarah Walsh's *Count on Us. Climate Activists from One to a Billion* (Barefoot Books, 2022). Author Christina Soontornvat's and illustrator Rahele Jomepour Bell's *To Change a Planet* (New York, Scholastic, 2022) also emphasise to their young readers that when "one person and one person become many, they can change a planet." All these children's books carry similar messages aimed at encouraging young readers to discover their voices and use them together by mobilising a cheerful chorus of colourful people-power for the greater good.

Clifi from Doomsday Clock Narratives to a Ministry for the Future

Literary works written for adults tend to employ a variety of methods to "crack the world open to alternative ways of thinking and being . . . to get in the way of business as usual" as Caroline Levine describes it in *The Activist Humanist: Form and Method in the Climate Crisis* (Levine, 2023). But the large majority of existing texts prioritises depicting the danger of human action/inaction through the description of collapsing order, climate chaos, and the resulting dystopian futures – what Gregers Andersen calls the "unhomeliness" (Andersen, 2020, p. 38) of the climate-changed planet in what Dominika Oramus calls "Doomsday Clock Narratives." (Oramus, 2023) This is true for many of the most well-known works – from early examples like John Wyndham's *The Kraken Wakes* (1953) – an early "climate fiction masterpiece (Seidel, 2022), J. G. Ballard's *The Drowned World* (1962) and *The Burning World* (1964), or Ursula Le Guin's *The New Atlantis* (1975) to twenty-first-century novels like Cormac McCarthy's *The Road* (2006), Paolo Bacigalupi's *The Windup Girl* (2009), Barbara Kingsolver's *Flight Behavior* (2012), Maggie Gee's *The Ice People* (2012), Emmi Itäranta's *Memory of Water* (2014), Jeff Vandermeer's *Annihilation* (2014), and Bacigalupi's *The Water Knife* (2015), as well as Stephen Markley's *The Deluge* (2023), among many others.

A smaller set of texts, which includes Octavia E. Butler's *The Parable of the Sower*, Margaret Atwood's trilogy including *Oryx and Crake* (2003), *The Year of the Flood* (2009), and *Maddaddam* (2004), or Richard Powers' *The Overstory* (2018) and Kim Stanley Robinson's *New York 2140* (2018) and *Ministry for the Future* (2021), do devote at least some attention to grassroots efforts to disrupt and combat the status quo. Thus, the protagonist in Octavia E. Butler's *The Parable of the Sower* persists in pursuing her dream of her utopian "Earthseed" community, the "gardeners" in Atwood's trilogy are driven by a commitment to each other's survival, and Robinson's *Ministry for the Future*, though it opens with an apocalyptic scene in which a large city in India is gripped by a major heatwave which crashes the electricity generation leading millions to die of heat stroke within hours, eventually focuses on the survivor of this catastrophe who joins a clandestine operation to re-invent the global economy. In Robinson's novel, with the help of a combination of activists engaging in civil disobedience, ingenuity, technological innovation, and sacrifice, humanity manages to pull back from the brink. Richard Powers also devotes a significant portion of *The Overstory* to describing the efforts of a small group of people deciding to sacrifice for the protection of trees. In most of these examples, the climate activists ultimately fail at scaling up their operations in part due to ineffective movement-building. And the survival of a few is typically accompanied by the death of millions.

Nonfi from *The End of Nature* to *All We Can Save*

For a long time, non-fiction texts explicitly intended to raise awareness about the climate crisis prioritised depicting the danger of inaction as well.[10] Examples of this approach include Bill McKibben, from early warnings in *The End of Nature* (1989) to the more recent *Falter: Has the Human Game Begun to Play Itself Out?* (2020) or David Wallace-Wells in *The Uninhabitable Earth. Life After Warming* (2019). Others highlight the injustices of climate change, like the UK's Jeremy Williams and Shola Mos-Shogbamimu's *Climate Change is Racist. Race, Privilege, and the Struggle for Climate Justice* (2021), Christina Gerhardt's *Sea Change. An Atlas of Islands in a Rising Ocean* (2023) or Rebecca Kormos' *Intertwined. Women, Nature, and Climate Justice* (2024).

Most of Bill McKibben's writing is devoted to spurring people into action, however. He has not only co-founded the climate movement 350.org and, more recently Third Act, but his published writing, too, is often directly aimed at mobilising people into action – from his seminal Rolling Stone article "Global Warming's Terrifying New Math" (2012) which helped launch the fossil fuel divestment movement to his latest substack newsletter *The Crucial Years*. Canadian author Naomi Klein, too, has devoted part of her efforts to spurring action. In *This Changes Everything* (2014) she describes the encouraging movement of "blockadia" against new fossil fuel infrastructure. One woman who has devoted her life to inspiring people to fight for the climate is Mary Robinson, the former President of Ireland and UN Special Envoy on Climate Change. In *Climate Justice. Hope, Resilience, and the Fight for a Sustainable Future* (2018), Robinson portrays a number of successful activists from around the world in order to inspire others to emulate them. Wen Stephenson is an author-activist who invites his readers to embrace arguments for direct action and civil disobedience in light of climate disruption in *What We are Fighting For Now Is Each Other. Dispatches from the Frontlines of Climate Justice* (2016).

A number of powerful works that prioritise imagining protagonists who forge a counter-movement were penned by women, including Indigenous writers and people of colour – in fiction, most prominently Octavia E. Butler in *Parable of the Sower* (1993) and *Parable of the Talents* (1998), as well as Leslie Marmon Silko in *Almanac of the Dead* (1999). In a growing number of nonfiction texts, urgency and warning are increasingly mixed with encouragement and with the explicit intention to activate readers. This includes Indigenous authors like Robin Wall Kimmerer in *Braiding Sweetgrass. Indigenous Wisdom, Scientific Knowledge, and the Teachings of Plants* (2013), Winona LaDuke in *To Be a Water Protector: The Rise of the Wiindigoo Slayers* (2020), and Vanessa Machado De Oliveira in *Hospicing Modernity: Facing Humanity's Wrongs and the Implications for Social Activism* (2021), as well as Rebecca Solnit in *Hope in the Dark. Untold Histories. Wild Possibilities* (2020), Joanna Macy in *Active Hope: How to Face the Mess We're in with Unexpected Resilience and Creative Power* (Macy, 2022), and Ayana E. Johnson and Katherine K. Wilkinson in *All We Can Save. Truth, Courage, and Solutions for the Climate Crisis* (2020), which started out as a nonfiction anthology and has evolved into a full-fledged organisation, the

All We Can Save Project.[11] In her latest book *What If We Get It Right? Visions of Climate Futures* (2024), Johnson presents positive visions of attainable climate futures in interviews with thought-leaders in the climate space from Brian Donahue to Rhiana Gunn-Wright.

Writing About the Power of Just Doing Stuff

After earlier works like Colin Beavan's *No Impact Man. The Adventures of a Guilty Liberal Who Attempts to Save the Planet, and the Discoveries he Makes about Himself and Our Way of Life in the Process* (2010) – later made into a movie – or Paul Hawken's *Project Drawdown* (2017) and Kate Aronoff et al.'s *A Planet to Win. Why We Need a Green New Deal* (2019), the past decade saw a great number of nonfiction books dispense practical advice on what to do about climate change – from Bill Gates' *How to Avoid a Climate Disaster. The Solutions We Have and the Breakthroughs We Need* (2021) and Mike Berners-Lee's *There is No Planet B* (2021) to Peter Fiekowsky and Carol Douglis' *Climate Restoration: The Only Future that Will Sustain the Human Race* (2022) to Swedish climate activist Greta Thunberg's edited volume *The Climate Book* (2022) with contributions by over 100 leading scientists, policymakers, and activists. (see also Kalmus, 2017; Hayhoe, 2021; Garcia, 2022; Göpel, 2023; Gunter, 2023; Roop, 2023; Wilson-Powell, 2023; White, 2024). A few are explicitly about mobilising people into collective action, such as Rob Hopkins' *The Power of Just Doing Stuff. How Local Action Can Change the World* (2013) and Luisa Neubauer and Alexander Repenning's *Beginning to End the Climate Crisis. A History of Our Future* (2023).

Embracing the Collective

McKibben's example notwithstanding, perhaps writers' propensity to seek solitude is the reason why so few fictional texts successfully describe grassroots climate activist mobilisation. Contrary to the quiet solitary act of writing, grassroots climate activism is often characterised by noisy and conflict-prone social engagement and collective organising. Still, writers have accepted the responsibility for the collective in a number of ways:

- in the narrative itself at the level of the story, as in *The Overstory*
- at the level of the publisher – such as *Grist*'s decision to profile a large collection of climate writing
- at the level of scholarly engagement, such as the work of the critic Dan Bloom who calls Cli-Fi an "urgent genre . . . a route to wake people up via storytelling" and chronicled hundreds of works of climate fiction in his *Cli-Fi Report*[12]

There is also an additional way, as many especially younger authors of fiction engage in forms of grassroots climate organising in their role as writers. The *Climate Fiction Writers League*, for example, is a gathering of writers who "believe in the necessity of climate action, immediately and absolutely" and are convinced that "fiction is one of the best ways to inspire passion, empathy and action in readers." The group aims to write works which "raise awareness of climate change, and encourage action at the individual, corporate and government levels." It sees itself in the lineage of the Women Writers Suffrage League from 1908, repeating their mantra: "A body of writers working for a common cause cannot fail to influence public opinion" (Liptak, 2022). As founder Lauren James told interviewer

Andrew Liptak: "The comparison between the suffragettes and modern Extinction Rebellion activists is something I'd been thinking about a lot while writing my own climate fiction novel." She explains,

> Immediately, I wanted to join a similar movement for climate fiction writers – but I couldn't find any when I googled it. In fact, I couldn't really find any comprehensive resources about climate fiction at all. It seemed like something that would be really useful to a lot of people, so I decided to set one up.[13]

As of this writing, the league has grown to over 200 members, with a popular Substack channel and a newsletter.[14]

Rob Hopkins and the Trouble with Our Imagination

Writers must believe in the power of the imagination. Our inability to truly imagine a better future, however, may well be part of the problem, as the UK's Rob Hopkins suggests. The founder of the *Transition Town Network*[15] that originated in Totnes, UK and has mushroomed into a worldwide network of neighbourhood-focused climate activism and resilience is also the author of a number of books on climate action. One of his more recent nonfiction books, *From What Is to What if*, (Hopkins, 2019) is all about the power of the imagination. At the outset, Hopkins emphasises that in order to create a different future we must first be able to imagine it: "to tell stories about it, to long for its realisation. If we can imagine it, describe it, dream about it, it is so much more likely that we will put our energy and determination into making it reality." (Hopkins, 2019, p. 9). His book is based on interviews he conducted with over 100 climate activists, artists, and thinkers. Like Amitav Ghosh in *The Great Derangement. Climate Change and the Unthinkable* (2017), Hopkins sees our inability to address the climate crisis as "an incredible failure of imagination" (Hopkins, 2019, p. 9) and deplores that this is the case at the very moment when imagination is so sorely needed:

> Nobody seems able to explain why our imaginations are failing us so spectacularly. Why are we so incapable of coming together to create, sustain and carry out a vision in which we capably address global crises and enjoy our lives more in the process? It seems as though we are becoming less imaginative at the very time in history when we need to be at our most imaginative. Our imagination muscle should be taut and well exercised; instead it is flaccid and untoned. . . . As we know from the field of positive psychology, imagining a certain outcome can increase the likelihood of its coming to pass.
>
> *(Hopkins, 2019, p. 11)*

The author explores a number of possible answers to that question. For one, he believes "we don't give imagination enough time"(Hopkins, 2019, p. 13). In his view, children should be given more free, unstructured time to be able to practise playing and improvising, because "imagination is the only thing we have that is – or could be – radical enough to get us through, provided it is accompanied, of course, by bravery, and by action"(Hopkins, 2019, p. 14) and "we need to be able to imagine positive, feasible, delightful versions of the future before we can create them." Hopkins cites Indian activist Manish Jain, who told him: "Imagination is connected to a sense of abundance."(Hopkins, 2019, p. 60) Other aspects of training the imagination, he writes, are close contact with nature and developing a sense of humour. Hopkins shows examples of how people used imagination and a sense of humour to change the status quo, like Antanas Mockus, university president and then mayor of Bogota, Colombia, who retrained police into mimes in the street and thereby reduced traffic deaths by 50%. Our lack of imagination, Hopkins discovers, is no innocent flaw but an active choice we make. He quotes Madeline Bunting: "Ignoring something . . . requires a form of attention. It costs us attention to ignore something." And that act of ignoring the climate threat, Hopkins adds, "takes a toll" (Hopkins, 2019, p. 60). In other words: When climate disruptions occur elsewhere but are not yet life threatening to the readers themselves, people's energies are used for distracting themselves from that reality rather than imagining – and preparing – a better future instead. Imagination, according to Hopkins, is what is needed to stir us into action. He quotes Gillian Judson, director of the *Imaginative Education Research Group*:

> When you engage imagination, you wake up emotion . . . any educator knows that you need to emotionally engage your students in order for them to have meaningful learning experiences. . . . The things you most remember and understand are those that have affected you.
>
> *(Hopkins, 2019, p. 97)*

Rob Hopkins and his 100 interlocutors from all over the globe envision activism and engagement as part of the solution. For Hopkins, imagining a different future begins with asking questions:

> we need to master the art, it seems to me, of asking questions which address the gravity of our situation yet which also create longing, which evoke a deep and rich sense of the wonders we can still create, rather than shutting it down or putting it into a deep sleep of complacency.
>
> *(Hopkins, 2019, p. 134)*

And he quotes novelist James McKay:

> It's very easy to think of the dystopian ideas. . . . It's almost lazy. Thinking of the good future is actually really hard because you have to envision something that is qualitatively different. Everyone knows what dystopia looks like. It's also exciting, in a dramatic way.
>
> *(Hopkins, 2019, p. 120)*

Hopkins' ideas dovetail with Joshua Rothman's reflections on Kim Stanley Robinson's novel *Ministry for the Future* – a novel the author Diane Cook described to Rothman as

> "activism as much as art"; in a less fragmented society, she said, *The Ministry for the Future* could have played a role like that of Upton Sinclair's *The Jungle* or Rachel Carson's *Silent Spring*. The book tries to do what a news report can't. It wants to offer us the experience of crossing the pass before we cross it – to give us a feeling for the routes we might take.
>
> *(Rothman, 2022)*

Contrary to the "lazy dystopian," Robinson's politics, Rothman suggests, "were perhaps more radical, since he imagined the possibility of an improved world." And further: "His uncool, utopian interests – ecology, equality, democracy, postcapitalism – were prescient." Robinson himself describes his work as "anti-anti-utopian." (Robinson, 2020) What Robinson sought to practise, Rothman writes, "in a phrase popularised by the Marxist philosopher Antonio Gramsci, was "pessimism of the intellect, optimism of the will." Rothman assigns Robinson's novel a status comparable to the novels of Charles Dickens and Elizabeth Gaskell during the Victorian era:

> In *The Ministry for the Future*, societies start to make good choices, in part because citizens revolt against the monied interests that preserve the status quo. But people also thrash about. They grow frustrated, angry, and violent. Some survivors of the Indian heat wave become ecoterrorists and use swarms of drones to crash passenger planes; no one can figure out how to stop the drones, and everyone gets scared. People fly less. They teleconference, or take long-distance trains, or even sail. They work remotely on transatlantic crossings. It's not how we want change to happen. But, in the end, the jet age turns out to have been just that – an age.
>
> *(Rothman, 2022)*

Whether texts that model desirable behaviour actually influence readers' behaviour is one of the many questions investigated by the scholarship of climate writing.

The Scholarship: Ecocriticism and the Environmental Humanities

While climatic changes, storms, floods, and fires have always been a part of human storytelling in cultures across the globe, the unprecedented nature of human-caused climate disruption that began to be felt in the late twentieth century has not only mobilised writers of fiction and nonfiction but also the scholarly engagement with them. The *Association for the Study of Literature and Environment* (ASLE) was founded in 1992 – the year of the Rio summit – with the aim to "inspire and promote intellectual work in environmental literature" by the members of ASLE who were motivated by the threats posed by environmental degradation (Long, n.d.). ASLE incorporated climate action into its work from the very beginning. Unlike other academic associations, ASLE reduced airline travel, for example, by organising its international conferences only bi-annually instead of annually and offering regional networking in between.[16] Although it is unknown how many scholars involved in literary and cultural studies engage in grassroots climate activism, those who do in the transatlantic realm are likely to be found among ASLE members (see for example Ammons, 2010). From around 1990 more and more scholars in the humanities began to incorporate climate-related themes into their work. Many literary scholars began to identify their work as a form of "ecocriticism" – a term that had first been introduced by William Rueckert in 1978 (Rueckert, 1978; see also Glotfelty/Fromm, 1996; Quick Hall/Kirk, 2022). The first international conference of "Environmental Humanities" – a term coined around 2010 (Emmett and Nye, 2017) – was held at the University of Alcalá in Spain in 2018 with the subtitle "Stories, Myths, and Arts to Envision a Change." (Villanueva-Romero et al., 2022; see also Johns-Putra, 2019; Johns-Putra and Sultzbach, 2022). In 2013 Timothy Morton published an influential book suggesting climate change should be studied as a "hyperobject," i.e. an object humans can think about but not see: "Hyperobjects are like that –

like the Dust Bowl, for instance, or the colossal drought in California. We are obliged to do something about them, because we can think them" (Morton, 2015). Indian historian Dipesh Chakrabarty takes yet another approach in *The Climate of History in Planetary Times* (2021), arguing that humans must see themselves from two perspectives at once: the planetary and the global, whereas the globe is a human-centric construction, while a planetary perspective intentionally decentres the human.

All that engagement is not, however, necessarily coupled with concrete calls for climate action. In *The Activist Humanist* Caroline Levine charges that humanists have been avoiding the embrace of practical political action, distracted by what she calls "anti-instrumentalism." Although Levine agrees that literature and humanistic writing has indeed had the ability to "crack the world open to alternative ways of thinking and being – to get in the way of business as usual," (Levine, 2023) it has been taboo to take the next step and suggest how to affect change. Instead, she finds humanists are stuck in a self-indulgent loop of celebrating ambivalence and complexity. Levine criticises this practice as feeding into the neoliberal status quo: "The practice of concluding with calls to ever more complexity and possibility instead of sketching out plans of action is feeding the logic of climate denialism and neoliberal atomization. It is supporting collective inaction" (Levine, 2023, p. 10). Although she does acknowledge that "the particular strengths of the aesthetic humanities – pausing in the face of business as usual, envisioning alternative worlds, and telling many specific stories – remain indispensable"(Ibid.), all that thinking must eventually result in action: "I keep my eye on the practical work that I and others can do, even in the painful conditions of the present, to create more just and sustainable conditions for intergenerational justice" (Levine, 2023, p. 11). Levine borrows Native American scholar Kyle Whyte's concept of "collective continuance" – which she understands as "the establishment of political, cultural, environmental, and economic conditions that allow collective life-worlds to flourish over time" (Levine, 2023, p. 13) to tackle this problem. "Instead of gesturing to unrepresentable futures, I ask: what materials, what agency, what strategies can build conditions for collective continuance here and now?" (Levine, 2023, p. 15). Yes, Levine agrees, "art is valuable because it teaches us to break free from the known world into an unrepresentable otherness to come" (Levine, 2023, p. 19). But what is needed even more, she says, is for us to actually do especially the care-work: "to work for collective continuance will therefore mean revaluing the mundane work that keeps life going over time" (Ibid.). At this late point in the climate crisis, awareness-raising is simply not enough. Indeed, in her final chapter Levine comes back to this:

> One of the mysteries psychologists and communications scholars have tried to unravel is why more people who are fully convinced of the dangers of global warming do not translate those convictions into action. Awareness does not in itself activate the work of change.
>
> *(Levine, 2023, p. 149)*

She ends her volume with a "workbook" for how to get involved in political action.

Using Levine's yardstick as a measure of success, you could argue that the texts that solely aim to describe the danger of a climate changed future, as important as they may be, aren't what is needed to actually tackle the problem. That said, as we have seen in this chapter (and in this section overall), "activating the work of change" can take many different forms, from raising awareness to providing opportunities to mourn and reflect to modelling behaviour modification in the here and now to imagining concrete examples of actions to address specific obstacles to the just transition all the way to describing skills required for survival in a climate-changed world.

Conclusion

Writers of English-language fiction and nonfiction have played a significant role in raising the alarm and spreading awareness about the climate crisis. For a long time, the majority focused on the threat of climate change, and so has the scholarship. However, recent years have seen the arrival of many different genres, formats, and digital forms of climate writing and also different modes of appealing to readers in texts from children's literature to Afrofuturism and beyond. While the majority of cli-fi continues to be concerned with dystopian futures, nonfiction books increasingly include concrete instructions of what to do, and some, though still too few, are devoted to encouraging collective action. As Harvard's Doris Sommer writes:

> It won't do to indulge in romantic dreams about art remaking the world, but neither does it make sense to stop dreaming altogether and stay stuck in cynicism. Between frustrated fantasies and paralysing despair, agency is a modest but relentless call to creative action, one small step at a time.
>
> *(Sommer, 2014, p. 4)*

Ayana Elizabeth Johnson and Katherine K. Wilkinson invite their readers to become part of the "we" in *All We Can Save. Truth, Courage, and Solutions for the Climate Crisis* (2020). Luisa Neubauer and Alexander Repenning dedicate their book to "All Possibilists out there, and those who want to join them" (2023, p. v). Like these authors, given the prevalent binary of hope and despair, Jennifer Atkinson suggests to reframe hope as purpose:

> A good many who work for climate justice do so not because they are convinced they will "win," or even because they feel a particular action will produce results; they do so because fighting for a livable future is the only sane and moral way to live in these times.
>
> *(Atkinson, 2024, p. 311)*

"Telling collective stories that redirect our despair," she concludes, "is an essential first step in building the vision and courage to take that transformative action" (Atkinson, 2024, p. 318). In order to prepare, encourage, and accompany the needed transformation of societies around the globe, all climate writing will have to become more interested and engaged in the collective action that is required to realise a better future.

Acknowledgments

I'd like to thank my son Jonathan von Mering, my niece Laura von Mering, as well as my friends Christiane Zehl-Romero, Debbie Wolozin, and Jennifer Riley for their helpful feedback on drafts of this chapter.

Notes

1 See also Caren Irr's new edited volume *Environmental Futures. An International Literary Anthology* (2024). Climate writing is not restricted to narrative, of course. In addition to a vast array of dramatic writing (see Chapter 14) and film scripts (see Chapter 15) there are also the growing number of anthologies of climate poetry, for example Luisa A. Igloria and Aileen Cassinetto's *Dear Human. At the Edge of Time. Poems on Climate Change in the United States* (2023) and Ben Okri's *Tiger Work: Stories, Essays, and Poems about Climate Change* (2023).
2 https://grist.org/fix/climate-fiction/afrofuturism-to-ecotopia-climate-fiction-glossary/.
3 See *Glass and Gardens: Solarpunk Summers* and *Glass and Gardens: Solarpunk Winters* – two anthologies edited by Ulibarri (2018, 2020).
4 https://grist.org/fix/october-2021-climate-fiction-issue/.
5 https://grist.org/fix/climate-fiction/definitive-climate-fiction-reading-list-cli-fi-books/.
6 https://grist.org/fix/climate-fiction/definitive-climate-fiction-reading-list-cli-fi-books/.
7 https://grist.org/fix/climate-fiction/afrofuturism-to-ecotopia-climate-fiction-glossary/.
8 Simpson et al. (2016). NYC Stands with Standing Rock Collective. https://afsc.org/sites/default/files/documents/Standing%20Rock%20syllabus.pdf
9 https://www.youtube.com/watch?v=zUnTcNzLIVg.
10 Interestingly, nonfiction has become more popular than fiction with American readers for some time now, and of course there is also a growing number of people who do not read books at all (Jones, 2022). Many young people consume climate information solely on social media – a topic that deserves its own volume.
11 https://www.allwecansave.earth/.
12 See https://www.cli-fi.net/. Bloom is often credited for coining the term cli-fi, but in a 2020 blog post Bloom wrote that the term was actually first used by climate deniers poking fun at climate science and Al Gore: https://blogs.timesofisrael.com/the-history-of-the-cli-fi-term-goes-back-to-2009-when-opposing-voices-used-it/.
13 "https://www.tor.com/2020/12/09/lauren-james-launches-climate-fiction-writers-league/.
14 Perhaps a parallel development on the part of readers could be the active proliferation of climate-action focused book clubs.
15 https://transitionnetwork.org/.
16 Recent initiatives by the organisation include, for example, a symposium on *Green Fire: Energy Stories Beyond Extraction* (Florida, May 2024).

References

Primary Sources

Climate Fiction

Atwood, M. (2003). *Oryx and Crake*. Toronto: McClelland & Stewart.
Atwood, M. (2004). *Maddaddam*. Toronto: McClelland & Stewart.
Atwood, M. (2009). *The Year of the Flood*. Toronto: McClelland & Stewart.
Bacigalupi, P. (2009). *The Windup Girl*. San Francisco: Night Shade Books.
Bacigalupi, P. (2015). *The Water Knife*. New York: Knopf.
Ballard, J. G. (1962). *The Drowned World*. New York: Berkley Books.
Ballard, J. G. (1964). *The Burning World*. New York: Berkley Books.
Butler, O. E. (1993). *The Parable of the Sower*. New York: Four Walls Eight Windows.
Butler, O. E. (1998). *Parable of the Talents*. New York: Seven Stories Press.
Gee, M. (2012). *The Ice People*. London: Metro Books.
Itäranta, E. (2014). *Memory of Water*. New York: Harper Voyager.
Kingsolver, B. (2012). *Flight Behavior*. New York: Harper.
Kôláwọlé, S. (2024). *The Road to the Salt Sea*. New York: Amistad.
Le Guin, U. (1975). *The New Atlantis and Other Novellas of Science Fiction*. Portland, OR: Hawthorn Books.
McCarthy, C. (2006). *The Road*. New York: Knopf.

Ngemntu, V. (2022). *The Serpent's Handmaiden.* https://kalaharireview.com/the-serpents-handmaiden-7a4c8b6af7f8

Powers, R. (2018). *The Overstory.* New York: W.W. Norton & Company.

Robinson, K. S. (2018). *New York 2140.* London: Orbit.

Robinson, K. S. (2021). *The Ministry for the Future.* London: Orbit.

Silko, L. M. (1999). *Almanac of the Dead.* New York: Simon & Schuster.

Vandermeer, J. (2014). *Annihilation.* New York: Farrar, Straus, and Girou.

Wyndham, J. (1953/2022). *The Kraken Wakes.* New York: Modern Library.

Anthologies

Cli-Fi Report Global. https://www.cli-fi.net/.

Climate Fiction Writers League. https://climate-fiction.org/.

The Definitive Climate Fiction Reading List. Grist.org/fix.https://grist.org/fix/climate-fiction/definitive-climate-fiction-reading-list-cli-fi-books/.

Eschrich, J. and Miller, C. A. (2018). *The Weight of Light. A Collection of Solar Futures.* Center for Science and the Imagination. Arizona State University. https://csi.asu.edu/books/weight/.

Igloria, L. A. and Cassinetto, A. (2023). *Dear Human. At the Edge of Time. Poems on Climate Change in the United States.* Paloma Press. https://www.asle.org/stay-informed/member-bookshelf/09112023-cassinetto/.

Imarisha, W. and Brown, A. M. (2015). *Octavia's Brood: Science Fiction Stories from Social Justice Movements.* AK Press.

Irr, C., et al. (2024). *Environmental Futures. An International Literary Anthology.* Waltham: Brandeis University Press.

Okri, B. (2023). *Tiger Work: Stories, Essays, and Poems about Climate Change.* Other Press.

Stahl, J. (2021). *The Climate Fiction Issue.* https://grist.org/fix/october-2021-climate-fiction-issue/.

Ulibarri, S. (2018). *Glass and Gardens: Solarpunk Summers.* Albuquerque, New Mexico: World Weaver Press.

Ulibarri, S. (2020). *Glass and Gardens: Solarpunk Winters.* Albuquerque, New Mexico: World Weaver Press.

Nonfiction

Aronoff, K., et al. (2019). *A Planet to Win. Why We Need a Green New Deal.* London and Brooklyn: Verso.

Beavan, C. (2010). *No Impact Man. The Adventures of a Guilty Liberal Who Attempts to Save the Planet, and the Discoveries He Makes about Himself and Our Way of Life in the Process.* London: Picador.

Berners-Lee, M. (2021). *There Is No Planet B.* Cambridge: Cambridge University Press.

De Oliveira, V. M. (2021). *Hospicing Modernity: Facing Humanity's Wrongs and the Implications for Social Activism.* Berkeley, CA: North Atlantic Books.

Fiekowsky, P. and Douglis, C. (2022). *Climate Restoration: The Only Future that Will Sustain the Human Race.* New York: Irvington.

Garcia, E. (2022). *Things You Can Do. How to Fight Climate Change and Reduce Waste.* Berkeley, CA: Ten Speed Press.

Gates, B. (2021). *How to Avoid a Climate Disaster. The Solutions We Have and the Breakthroughs We Need.* New York: Knopf.

Gerhardt, C. (2023). *Sea Change. An Atlas of Islands in a Rising Ocean.* Berkeley, CA: University of California Press.

Gunter, M. M. (2023). *Climate Travels. How Ecotourism Changes Mindsets and Motivates Action.* New York: Columbia University Press.

Hawken, P. (2017). *Project Drawdown.* https://drawdown.org/.

Hayhoe, K. (2021). *Saving Us. A Climate Scientist's Case for Hope and Healing in a Divided World.* New York: One Signal Publisher.

Hopkins, R. (2013). *The Power of Just Doing Stuff. How Local Action Can Change the World.* Cambridge: Green Books.

Hopkins, R. (2019). *From What Is to What If. Unleashing the Power of Imagination to Create the Future We Want*. White River Junction, VT: Chelsea Green Publishing.

Johnson, A. E. (2024). *What If We Get It Right. Visions of Climate Futures*. London: OneWorld.

Johnson, A. E. and Wilkinson, K. K. (2020). *All We Can Save*. London: OneWorld.

Kalmus, P. (2017). *Being the Change. Live Well and Spark a Climate Revolution*. Cabriola Island, Canada: NSP.

Kimmerer, R. W. (2013). *Braiding Sweetgrass. Indigenous Wisdom, Scientific Knowledge, and the Teachings of Plants*. Minneapolis, MN: Milkweed Editions.

Klein, N. (2014). *This Changes Everything*. New York: Simon & Schuster.

Kormos, R. (2024). *Intertwined. Women, Nature, and Climate Justice*. New York: The New Press.

LaDuke, W. (2020). *To Be a Water Protector: The Rise of the Wiindigoo Slayers*. Halifax, Nova Scotia, Canada: Fernwood Publishing.

Macy, J. (2022). *Active Hope: How to Face the Mess We're in with Unexpected Resilience and Creative Power*. Novato, CA: New World Library.

McKibben, B. (1989, 2006). *The End of Nature*. New York: Random House.

McKibben, B. (2012). Global Warming's terrifying new math. In *Rolling Stone Magazine*, July 19. https://www.rollingstone.com/politics/politics-news/global-warmings-terrifying-new-math-188550/

McKibben, B. (2020). *Falter: Has the Human Game Begun to Play Itself Out?* New York, NY: Holt Paperbacks.

McKibben, B. (n.d.) *The Crucial Years*. Substack Newsletter. https://billmckibben.substack.com/

Neubauer, L. and Repenning, A. (2023). *Beginning to End the Climate Crisis. A History of Our Future*. Waltham: Brandeis University Press.

Robinson, K. S. (2020). Dystopias now. The end of the world is over. Now the real work begins. In *Commune Issue 5 Winter*. https://communemag.com/dystopias-now/

Robinson, M. (2018). *Climate Justice. Hope, Resilience, and the Fight for a Sustainable Future*. New York: Bloomsbury.

Roop, H. A. (2023). *The Climate Action Handbook. A Visual Guide to 100 Climate Solutions for Everyone*. New York: Penguin Random House.

Solnit, R. (2020). *Hope in the Dark. Untold Histories. Wild Possibilities*. London: Haymarket.

Stephenson, W. (2016). *What We Are Fighting for Now Is Each Other. Dispatches from the Frontlines of Climate Justice*. Boston: Beacon Press.

Thunberg, G. (2022). *The Climate Book. The Facts and the Solutions*. London: Penguin.

Wallace-Wells, D. (2019). *The Uninhabitable Earth. Life After Warming*. Tim Duggan Books.

White, H. (2024). *Eco-Anxiety. Saving Our Sanity, Our Kids, and Our Future*. New York: Harper.

Williams, J. and Mos-Shogbamimu, S. (2021). *Climate Change Is Racist. Race, Privilege, and the Struggle for Climate Justice*. London: Icon Books.

Wilson-Powell, G. (2023). *365 Ways to Save the Planet*. London: DK.

Children's books

Joyner, A. (2020). *Stand Up! Speak Up! A Story Inspired by the Climate Change Revolution*. Schwartz and Wade. Toronto: Penguin Random House.

Lindstrom, C. and Goade, M. (2020). *We Are Water Protectors*. New York: Roaring Brook Press.

Snyder, G. and Walsh, S. (2022). *Count on Us. Climate Activists from One to a Billion*. Concord, MA: Barefoot Books.

Soontornvat, C. and Jomepour Bell, R. (2022). *To Change a Planet*. New York: Scholastic.

Tucker, Z. and Persico, Z. (2019). *Greta and the Giants, Inspired by Greta Thunberg's Stand to Change the World*. London: Frances Lincoln Children's Books.

Secondary Sources

Ammons, E. (2010). *Brave New Words: How Literature Will Save the Planet*. Iowa City: University of Iowa Press.

Anamata Future News. https://www.flava.co.nz/the-latest/anamata-future-news-web-series/

Andersen, Gregers (2020). Climate Fiction and Cultural Analysis. *A New Perspective on Life in the Anthropocene*. Abingdon: Routledge.

Ashton, E. (2022). *Anthropocene Childhoods. Speculative Fiction, Racialization, and Climate Crisis.* London: Bloomsbury Academic. http://doi.org/10.5040/9781350262416.ch-7

Atkinson, J. (2024). Stories from our future: Beyond the binary of climate hope and grief. In D. J. Rosenthal (ed.) *Teaching the Literature of Climate Change* (pp. 309–319). New York: The Modern Language Association of America.

Bloom, D. (2009). The history of the cli-fi term goes back to 2009 when opposing voices used it. *Times of Israel.* https://blogs.timesofisrael.com/the-history-of-the-cli-fi-term-goes-back-to-2009-when-opposing-voices-used-it/

Brady, A. and Isen, T. (2022). *The World as We Knew It: Dispatches from a Changing Climate.* New York: Catapult.

Chakrabarty, D. (2021). *The Climate of History in Planetary Times.* Chicago, IL: University of Chicago Press.

Emmett, R. S. and Nye, D. E. (2017). *The Environmental Humanities. A Critical Introduction.* Cambridge, MA: The MIT Press. https://doi.org/10.7551/mitpress/10629.001.0001.

Ghosh, A. (2017). *The Great Derangement: Climate Change and the Unthinkable.* London and Chicago: University of Chicago Press.

Glotfelty, C. and Fromm, H. (1996). *The Ecocriticism Reader. Landmarks in Literary Ecology.* Athens, Georgia: University of Georgia Press.

Göpel, M. (2023). *Rethinking Our World: An Invitation to Rescue Our Future.* Lancaster, CA: Scribe US.

Imarisha, W. and Brown, A. M. (2015). *Octavia's Brood: Science Fiction Stories from Social Justice Movements.* Chico, CA: AK Press.

Irr, C. (2024). *Environmental Futures. An International Literary Anthology.* Waltham: Brandeis University Press.

Johnson, A. El. (2024). *What If We Get It Right? Visions of Climate Futures.* New York: Penguin Random House.

Johns-Putra, A. (2019). *Climate and Literature.* Cambridge: Cambridge University Press.

Johns-Putra, A. and Sultzbach, K. (2022). *The Cambridge Companion to Literature and Climate.* Cambride: Cambridge University Press.

Jones, J. M. (2022). Americans reading fewer books than in past in: Gallup. *Economy,* 10 January. https://news.gallup.com/poll/388541/americans-reading-fewer-books-past.aspx.

Kayawe, B. (2023). Dystopia, drought, and hope? African cli-fi takes on climate crisis. *Context,* 9 February. https://www.context.news/climate-risks/dystopia-drought-and-hope-african-cli-fi-takes-on-climate-crisis

Levine, C. (2023). *The Activist Humanist. Form and Method in the Climate Crisis.* Princeton: Princeton University Press.

Liptak, A. (2022). Lauren James launches climate fiction writers league. *Reactormag,* 9 December. https://reactormag.com/lauren-james-launches-climate-fiction-writers-league/

Long, M. C. (n.d.). *The Far Field.* Blog. https://thefarfield.org/department/

Martinez, T. (2024). Changing student perceptions through climate literature. In D. J. Rosenthal (ed.) *Teaching the Literature of Climate Change* (pp. 38–43). New York: The Modern Language Association of America.

Morton, T. (2015). Introducing the idea of hyperobjects. *High Country News,* 19 January. https://www.hcn.org/issues/47-1/introducing-the-idea-of-hyperobjects/

Nielsen, M. (2019). *Interview with Lindsay Ellis and Amy Brady, @HotMessPBS.* https://www.youtube.com/watch?v=zUnTcNzLIVg

Oramus, D. (2023). *Eco-Anxiety in Nuclear Holocaust Fiction and Climate Fiction: Doomsday Clock Narratives.* London: Routledge. https://kalaharireview.com/the-serpents-handmaiden-7a4c8b6af7f8

Quick Hall, K. M. and Kirk, G. (2021). *Mapping Gendered Ecologies. Engaging with and Beyond Ecowomanism and Ecofeminism.* Lanham, MD: Rowman and Littlefield.

Rosenthal, D. J. and De La Mara Molesky, J. (2023). *Cli-Fi and Class: Socioeconomic Justice in Contemporary American Climate Fiction.* Charlottesville: University of Virginia Press.

Rothman, J. (2022). Can science fiction wake us up to our climate reality? *The New Yorker,* 24 January.

Rueckert, W. (1978). Literature and ecology: An experiment in ecocriticism. *Iowa Review,* 9(1), (Winter), 71–86. Reprinted in Glotfelty, C. and Fromm, H. *The Ecocriticism Reader. Landmarks in Literary Ecology* (Athens: Georgia, University of Georgia Press, 1996).

Seidel, M. J. (2022). The first climate fiction masterpiece: On John Wyndham's 1953 novel *The Kraken Wakes*. *LA Review of Books*, 15 October. https://lareviewofbooks.org/article/the-first-climate-fiction-masterpiece-on-john-wyndhams-1953-novel-the-kraken-wakes/.

Simpson, A., et al. (2016). *NYC Stands with Standing Rock*. https://afsc.org/sites/default/files/documents/Standing%20Rock%20syllabus.pdf.

Sommer, D. (2014). *The Work of Art in the World: Civic Agency and Public Humanities*. Durham, NC: Duke University Press.

Streeby, S. (2018). *Imagining the Future of Climate Change: World Making through Science Fiction and Activism*. Berkeley, CA: University of California Press.

Trexler, A. (2015). *Anthropocene Fictions: The Novel in a Time of Climate Change*. Charlottesville: University of Virginia Press.

Trexler, A. (2023). Epilogue: What Has Changed Since *Anthropocene Fictions*? In Debra J. Rosenthal and Jason De La Mara Molesky (eds.) *Cli-Fi and Class: Socioeconomic Justice in Contemporary American Climate Fiction* (pp. 235–243). Charlottesville: University of Virginia Press.

Villanueva-Romero, D., Kerslake, L. and Flys-Junquera, C. (2022). *Imaginative Ecologies. Inspiring Change Through the Humanities*. Leiden and Boston: Brill.

Womack, Y. (2013). *Afrofuturism: The World of Black Sci-Fi and Fantasy Culture*. Chicago: Lawrence Hill Books/Chicago Review Press.

13
WILD HOPE AND CURATORIAL ACTIVISM

Fleur Watson, Wendy Steele,
Naomi Stead, and Katrina Simon

Introduction

In an age of interconnected challenges there is an urgent need to find progressive and creative ways to better understand and take urgent action to address the climate crisis. This chapter explores and celebrates the role of public festivals, exhibitions, and curatorial practices as a creative practice-based and activist response to the climate emergency. To do this it draws insights from *Wild Hope: Conversations for a Planetary Commons,* a large exhibition and public programme held in Naarm/Melbourne as part of the 'Now or Never' Festival in 2023. This case-study serves as both provocation and illustration for how curation and exhibition making contribute to diverse practices of civic engagement and design-led futures. As members of the curatorium – a collective of curators – we see this endeavour as embedded within the shared climate emergency we are seeking to reimagine, renegotiate and resist. As Watson highlights,

> Curating is not an activity that can take place in isolation, either from the wider landscape of design or from the specifics of institutional politics. The audience is the key part of the equation, and the techniques most critical for the curator are those that help them engage their audience.
>
> *(Watson, 2021, p. 12)*

The *Wild Hope* exhibition marked the launch of a novel collaboration between RMIT University and public thinktank Dark Matter Labs (DmL) focused on re-conceiving and protecting our planetary commons: oceans, rivers, soil, forests. As a form of grassroots activism, curation and exhibition-making offer an experimental space for exploring and testing speculative design ideas at the interface of creative practice, performing 'research-in-action' and civic engagement. This is where art, design, and practice research become catalysts for creative civic action and societal change – often overlooked and yet critical in a climate crisis.

The chapter begins with a discussion of progressive forms of creative practice-based curatorship that work across and between the needs of diverse publics. Emergent curatorial roles such as Translator, Activist, and Dramaturge are drawn from earlier work by one of

DOI: 10.4324/9781003396567-15

the authors Fleur Watson – who is both a member of the curatorium and a theorist of the role and possibilities of 'the new curator' more broadly (see Watson, 2021). Drawing from and developing Watson's earlier theoretical framework, we explore critical insights from the *Wild Hope* exhibition: underscoring the necessity for ethical experimentation and creative change in dark times. We conclude the chapter by reinforcing the role of the curatorium as a shared, iterative journey that bridges climate, creative practice, and activism.

Ways of Curating in Climate Change

The term 'curatorial euphoria' (see Mary Anne Staniszewski, 2015) captures the way in which curating has become a verb synonymous with an urbanised and mediated contemporary way of life. What were once seen as everyday activities are now described as curatorial: menus, wardrobes, playlists, and social media feeds are all 'curated' along with – critically, in the context of a 'fake news' era – news, data, and information. As Staniszewski writes,

> What had previously been an activity principally associated with the rarefied domains of aesthetics and museums has now infiltrated a diverse range of territories from the mundane to the esoteric, morphing and multiplying at an exponential pace, serving as a strategy for 'coping' with our networked and globalised world.
>
> *(Staniszewski, 2015, p. 247)*

Galleries, museums, libraries, exhibitions, and festivals remain critical public spaces of engagement but are largely under-represented in the climate activism literature as spaces of radical resistance and inventiveness. Furthermore, in the twenty-first century, the role of the public cultural institution and, in turn, curatorial practice, is rapidly shifting to ask pertinent and vital questions about what it means to curate the contemporary: What is the agency of the curatorium? And how can new decolonised, intercultural, and climate-activist approaches to curating and cultural production grapple with our precarious global condition?

Paola Antonelli – Senior Curator of the Department of Architecture and Design at New York's Museum of Modern Art – proposes that in the twenty-first century public institutions are finding new ways to engage and activate audiences with urgent contemporary issues. In conversation with Watson (2021, p. 275), she suggests:

> It's not relevant anymore to tell people that 'this is the way it is.' Instead, the [curatorial] position may be: 'I have an idea and I think it's interesting now and I'd like to share it and talk about what it might mean.'

Architectural historian and curator Philip Ursprung furthers this notion, stating that it is possible: 'To use an exhibition – not as a means to represent what we already know, but as an opportunity to learn more about what we don't know – open up new terrain' (Ursprung, 2014, in Watson 2021, p. 280).

Expanding on this and despite the ubiquity and appropriation of 'curating' in mainstream culture, a new kind of curator is emerging – leading a movement towards an increasingly expansive, porous, and responsive mode of curatorship and cultural production (see Watson, 2021). Here, the curator-as-activist is 'tuned into' the relationship with the outside world – to environmental, cultural, political, and social contexts, engaged through critical conversations, experiments, and provocations.

Others have addressed similar ideas. In their book *Curating the Future*, Jennifer Newell, Libby Robin, and Kirsten Wehner (2017) emphasise that 'climate change demands urgent transformations in the way we think about ourselves and our world' and curation has a critical role to play in this reframing. They highlight the need for new forms of creative engagement that help bridge deep-seated separations that define the trajectory of modernity: nature vs. culture, coloniser vs. colonised, local vs. global, and authority vs. uncertainty. We would add to this list a fifth entrenched binary: science vs. arts.

Curatorial practice provides a radical, experimental set of methods for 'engaging with and responding to climate change' (Newell et al., 2017, p. 2) and building new types of transdisciplinary or 'un-disciplinary' (see introduction of this handbook) communities of practice. Part of this involves reshaping the conceptual, material, and organisational structures and modes of living that have underpinned institutions such as galleries and museums that help to *produce* and *sustain* the climate crisis. Exhibitions offer curators the opportunity to co-create with artists, designers, and audiences, bringing 'new modes of thinking and understanding, offering associational and synthetic approaches that build abilities to consider how our choice, actions and lives are entangled with the other species and forces of the planet' (Newell et al., 2017, p. 4).

It is within this context that the case for 'The New Curator' in relation to socio-political advocacy and activism has emerged. Following Watson (2021), this curating role includes but is not limited to:

- greater attentiveness to the agency of curators operating both inside and 'outside the museum' and other cultural establishments
- curatorship that investigates and exposes design's value to public life and societal grand challenges, making explicit the need for urgent action on climate change
- deepening the focus on accessibility and inclusiveness in curatorial practice and education
- centring the vital leadership of First Nations-led curators in revealing and recognising the ongoing impact of unceded sovereignty, the frontier wars, and the importance of remembrance and memorial in addressing settler-colonialism
- driving an ethical commitment to new modes of knowledge production through curating creative ideas and instigating opportunities for civic engagement and action

This new wave of curation is a shared process between alternative forms of creative practice and the agency of diverse publics. Watson (2021) describes a series of six emergent curatorial roles or 'moves' and, in this chapter, three of these roles are explored through the lens of the exhibition *Wild Hope*, the 'new curator.' The roles of the *Translator* (Mediator of Research), *Activist* (Curator as Agent) and *Dramaturge* (Event as Performance) frame the following sections with illustrated in-situ examples.

Curating Creative Research for Critical Conversations

Wild Hope: Conversations for a Planetary Commons invited audiences to embrace a radical shift towards 'planetary thinking,' a move vital to the survival of human and non-human life on Earth. The exhibition responded to a provocation from the City of Melbourne in Australia for innovative programming for a new public festival created for a post-Covid city by fusing two festivals that had previously been run as separate events – Knowledge Week and Music Week. The bid for funding and inclusion in this new festival was based on the ideas of planetary commons and civics and the necessity to understand aspects of our shared yet currently disastrously fragmented world in inclusive and deeply connected ways to address the climate crisis. Drawing on the knowledge of a wide range of creative practices addressing climate activism and planetary thinking across Naarm/Melbourne, the successful bid was then expanded to create a curatorium to resource and manage as a collective the exhibition and public program held in performance venues across the cityscape.

The expanded curatorium brought a range of transdisciplinary and creative practice perspectives together in a way that enabled the critical expansion of the initial concept. These perspectives included social science, critical humanities, architectural theory, critical writing, architecture, interior design, landscape architecture and, crucially, curation. To open the process to a wider pool of people engaging with planetary thinking and creative practice, an expression of interest was circulated within the institution that invited submission of ideas, works, and events that engaged with the idea of planetary commons and, with awareness of the understandable prevalence of works focusing on dystopia and crisis, with the ideas of hope. The ambition of the curatorium was to foster these ideas not just by gathering these works on display but by purposefully putting them into relationships with one another and the wider institutions of university and city that supported the exhibition in ways that encourage and stimulate active reflection and engagement.

Through the review of the works proposed, the curatorium worked collaboratively to question the way that different projects and groupings created new insights into the provocation of the planetary commons. Three themes were used to focus attention on sets of interrelated systems and their specific challenges. In the context of a climate emergency, activism around planetary thinking has urgent work to do: to understand as collective our air and atmosphere, land and biodiversity, forests and oceans, river systems and fisheries, ice sheets and 'ecological poles,' the flows of carbon and phosphorous at the planetary scale – all of which must move beyond being managed, owned, and exploited to being understood and governed as a planetary commons. The *Wild Hope* exhibition and integrated program of events explored this vital movement via three curated planetary common's themes exacerbated in climate change,

- waters, oceans, rivers
- carbon, trees, soil
- species, diversity, extinction

The speculative open invitational process raised an initial collection of 57 works, and another group of projects of allied practices were invited to participate to extend the forms of creative expression and experimentation. The iterative process of reviewing and selecting the works that would be included in the exhibition revolved around a collective engagement with the possibilities suggested by the conceptual framing and material presence of the work. This was considered in tandem with challenges and opportunities of organising the works in a distinctive and dramatic exhibition space. This included articulating the exhibition through its interpretive material and activating the exhibition across five weeks with interactive public programming that further extended the opportunities to engage with complex ideas of the linkages between the climate emergency and a more regenerative future at the planetary scale.

In its response to the need for change at the level of the microbe, the city, and the world, the *Wild Hope* exhibition – and the works included in it – champion the planetary and

Figure 13.1 Wild Hope – Conversations for a Planetary Commons, RMIT Design Hub Gallery. 2023.
Source: Photo by Tobias Titz.

regenerative thinking to achieve a more sustainable and equitable future. Comprising over 20 exhibits, including newly conceived works, the ambition was diverse in terms of the materiality, orientation, format, and scale of the exhibits shown, spanning the disciplines of visual art; architecture; landscape; interior and graphic design; fashion, jewellery and textiles; sound; VR and digital design; and material science.

Wild Hope provided a provocation for a series of creative works, research endeavours, conversations, workshops, and participatory programs that brought together researchers, policymakers, creative practitioners, and the public in active, critical, and hopeful exchange (see Figure 13.1).

The exhibition was fittingly and deliberately opened with a First Nations perspective, given the significance of Aboriginal Australia as the most long-standing living culture on earth and hence, by definition, the most 'sustainable' people ever to engage in environmental management. *Meerreeng Karweeyn* – Earth Dance – a purpose-made installation by Indigenous artist and RMIT research fellow Vicki Couzens – brought the visitor's attention to Aboriginal belonging and the interconnectedness of sky, sea, and earth. Couzens's installation was an appeal to humanity and a powerful reminder that listening and learning from Aboriginal ancestral, cultural, and ecological knowledges is imperative in this time of climate and ecological crisis. It was an introduction to potentiality, a possible global manifesto, an ontological journey of discovery towards a way of conceiving our planetary commons.

Moving into the gallery beyond *Meerreeng Karweeyn,* the visitor encountered Jessie French's *To sow the wind and reap the whirlwind* – a series of large-scale 'paintings' made from uncut sheets of the same algae material as the experimental lettering, hung high and moving gently in the space. French created an installation of six such abstract 'paintings' for *Wild Hope,* with each large panel suspended from an aluminium 'robot' winch system with fluctuating movements to mirror the cadence and rhythm of the natural world. This mechanisation is balanced with their organic and textural quality – different organic additives mean their colours and textures are wildly varied, evocative of natural patterns and systems at the micro and macro scale. This work advocates for a revision of our material choices and interactions with the natural world. French's practice reveals the complex relationship between humans and the environment and the potential for choices that reduce our negative impact (see Figure 13.2).

Figure 13.2   Jessie French, 'To sow the wind and reap the whirlwind,' 2023, *Wild Hope* exhibition. RMIT Design Hub Gallery.

Source: Photo by Tobias Titz.

Wild Hope: Conversations for a Planetary Commons pursued a light footprint, a resource-sensitive mode of exhibition making – in response to more common exhibition practices characterised by – amongst other things – very high-waste production. A key example of this was in the exhibition labelling and didactic panels. Instead of the fossil-fuel-based and un-recyclable vinyl lettering typically used in gallery signage, designer Stuart Geddes worked with artist Jessie French to conceive the treatment of introductory texts, replacing plastic-based vinyls with an algae-based sheet material French has invented as part of her creative practice 'Other Matters.' This re-useable and entirely biodegradable material has the potential to dramatically reduce the environmental footprint of exhibition making – as well as being a beautiful, textural, character-filled material that, in the exhibition space, visitors were encouraged to engage with at close range. In addition to the algae lettering, other labels for the work were directly printed onto waste materials recycled from previous gallery exhibitions and upcycling depots. In this way the exhibition was able to produce texts with zero waste – whilst also assisting in the testing and prototyping of a new, sustainable exhibition material with potentially much larger application in the sector.

Curatorially, *Wild Hope* aimed to translate, activate, and 'perform' a diverse range of cross-disciplinary creative works that responded to climate emergency and regenerative futures and explored paradigm-shifting ideas for a movement towards a planetary commons. Significantly, many works were what can be described as nascent and developing creative research works 'in process' or action. Critically, the opportunity was to explore, expose, and advocate for the vital contribution of creative practices in active exchange with diverse publics by maximising the opportunity and reach of the large-scale public festival.

To achieve these objectives, the curatorium organised and framed the space, exhibits, and programmes through the lens of three 'new curator' modes (Watson, 2021): curator as Translator, as Activist, and as Dramaturge. Each mode or 'move' – an important distinction in term that seeks to capture a porous and responsive curatorial process – enabled new relationships to be drawn from and between the collection of works and programs and for the exhibition to become a living demonstration of its ambitions. Each will be outlined in the following sections.

Curator as Translator (Mediator of Research)

New global challenges require new creative research methods and collaborative frameworks that, in turn, demand an expanded way of discussing, debating, and disseminating process and research. In meeting such challenges, the role of the curator as a 'trusted guide' (Antonelli in 'The New Curator,' Watson, 2021, p. 69) and an 'adaptive insider/outsider' (p. 107) enables dialogue and exchange among researchers, practitioners, and audience. Here, the curatorial intent is to mediate and expose the creative research process and to

> create a situation where audiences and experts 'tune in' to one another for moments of encounter, exchange and constructive critique. Curatorially, the imperative is to reveal the incomplete – it is about prototyping, experimentation, failures and successes; and the audience can take an active part in this experience-as-testing-site.
>
> *(Rhodes, in Watson, 2014/2021)*

For *Wild Hope*, the 'curator as translator/mediator of research' takes on an explicit activist agenda: to bring urgent public awareness to the importance of creative research in addressing the challenges of ecological crisis and advocating for a movement towards shared regenerative futures.

It is in this context of curatorial translation that the following works are situated. *Aurum* – a film essay by Georgia Nowak and Eugene Perepletchikov and a major work for *Wild Hope* – was displayed on a large-scale projection screen with the artist's narration available on headphones and with closed captions. Traversing layers of history and mythology, the film examines the complex historical, cultural, and technological entanglements between humans and gold. The extraction of this precious metal is only accomplished with great effort and risk – and the resultant material is both a symbol of power and a unit of value. Whole civilisations have been shaped around the search for, trade in, and valuation of gold. Still an object of desire, gold now takes on a whole new role and functionality – used in tiny trace amounts, it is essential to the electronic circuits and components of today's electronic devices. Its preciousness has taken on an entirely new, previously unthinkable form. *Aurum* tells visual stories of this extraction and valuation of gold – in far-flung locations and in multiple ways. Mapping human culture's magnetic attraction to gold, it also frames the uniquely intrinsic value of this material as a holder of stable value in an increasingly volatile world. As old goldfields are reopened and new open pits spread through the landscape, the devastating ecological cost of extraction must be calculated on a planetary scale.

In contrast, the work of fashion researchers and practitioners D&K (Ricarda Bigolin, Chantal Kirby) with Žiga Testen moves from the scope of the macro to the micro. *Wear Out Mode: Instructions for Commoning Fashion*, emphasises the local and domestic scale of recycled clothing, interacting with a selection of old and 'worn' clothes from different periods, along with other forms of worked textiles – canvas drop-sheets; recycled cotton and polyester fabric; and other found, collected, inherited, and scavenged textiles and trims. It is widely known that the growth of the global fashion industry has had catastrophic social and environmental impacts and that it is untenable in its present form. One simple solution might be for people to simply wear fewer garments more and longer – and yet there are many reasons why contemporary individuals rarely *wear out* fashion. D&K argue that one way to intervene in this destructive cycle is to prolong the use of what we already own and learn how to wear and manage clothes in a way that prevents waste. In *Wear Out Mode*, working with graphic designer Žiga Testen, the researchers capture the catastrophic effects of the global growth of fashion in an inventory of animated poems and AI-mined headlines. Their work proposes modes of wearing with slowness, care, and community and uses 'instructional prototypes' for methods – both material and making – that extend practical use and emotional durability of clothing. As D&K argue: *wearing out* is a metaphor – to create a commoning of fashion.

Specially commissioned for *Wild Hope*, the exhibition journey culminated with a newly commissioned work in Project Room 2 – a multi-channel sound installation by Machine Listening (Sean Dockray, James Parker, and Joel Stern). *Environments 12* is a series of speculative soundscapes that capture 'field recordings' from an imagined future world. The work was conceived as a speculative addition to the once-popular *Environments* series: a sequence of 11 records released between 1969 and 1979 featuring field recordings from natural environments, whose waves and thunderstorms both anticipated and provisioned a mass market for mood-altering nature sounds. The work presented a world in which the environment itself has been technologically updated with loudspeakers and microphones placed everywhere: where the reproduction, synthesis, and management of soundscapes is ubiquitous. Set across a series of imagined historical and contemporary scenes,

Environments 12 was narrated by an ensemble of vocal performers and their generative voice clones. Together, this more-than-human chorus told the stories of 'psychologically ultimate seashores,' reef lullabies, natural symphonies for zoo enclosures, and large language models for whales and crows – and presented, as the artists described, a collection of songs and fables from the ruins of a future history.

Here, the role of the curatorium was to situate, mediate, and translate these diverse research works – all of which were either made or responded directly to the curatorial themes for the exhibition. Complemented with floor talks, listening sessions, and performance lectures, these works created a space for shared exchange where pressing issues could be revealed; explored; and opened for interaction, debate, and discussion. Via the agency and work of a diverse group of creative practitioners, the exhibition used scenarios and narratives to subvert preconceptions of what our cities and communities can be and, vitally, how we might envisage them differently. In this context, there is the potential to reveal something new in our relationship to each other and the earth that remains after the temporal footprint of the exhibit is gone.

The Activist (Curator as Agent)

The role of the 'curatorium as activism' that we argue for here, building upon the earlier work by Watson (2021), is deeply aware of the ecological, social, and political challenges that underpin our current age of disruption and climate crisis. Such a curatorium actively seeks new ways to bring the climate crisis and interconnected challenges to urgent public attention – to resist and transform the extractive, settler-colonial nature of the status quo. This is a deliberate shift from earlier conceptions of the curator as custodian – or even as advocate – to taking up an explicitly activist agenda. In response, acknowledging unceded sovereignty is critical, as is the need to listen and learn from Indigenous leadership, knowledges, and practices and work together to compassionately shape our shared future. Among the complexity of the challenges faced in the contemporary world, the activist agenda of the curator-curatorium is how they can creatively contribute to enable a better future.

An intent for curatorial activism is to catalyse the opportunity of the exhibition environment for active engagement, exchange, and debate. Watson (2021, pp. 71, 234) describes the 'curator as activist' move as one in contrast to the role of custodian or historian in the museological tradition. Instead, here, the curator is 'tuned' and responsive to relationships with the outside world and the social context in which an exhibition is responding as part of a broader agenda of intersectional climate justice and equity. For instance, in *Curatorial Activism: Towards an Ethics of Curating*, Maura Reilly (2018, p. 1)[1] uses the term 'activism' to

> designate the practice of organising art exhibitions with the principal aim of ensuring that certain constituencies of artists are no longer ghettoised or excluded from the master narratives of art . . . and, as such, focuses almost exclusively on work produced by women, artists of colour, non-Euro-Americans, and/or queer artists.

For *Wild Hope*, the curatorial agenda focused on the urgent questions needed to: generate new knowledge; make new and unexpected connections; unravel the unknown; test the speculative; generate debate; and interrogate creative practice's agency in addressing ecological crisis. To this end, the curatorium used the three key planetary commons themes –

waters/oceans/rivers, carbon/trees/soil, species/diversity/extinction – as an overarching curatorial framework to organise, mediate, and communicate the exhibits and programmes while, simultaneously, acknowledging the entangled nature of the themes in a climate crisis.

The first theme – **Water/Oceans/Rivers** – emphasised the symbiotic relationship of humanity with aquatic life and the complexity of human–water entanglements in our constructed world. It encompassed works that explored the movement of wild, untameable hydrological systems (Maj Plemenitas); built upon coded representations of natural systems such as coral reefs (Marc Gibson); presented an immersive VR experience of a flooded metropolis (Patrick Macasaet, Vei Tan, Shuming Ivy Zhou, Zechen Huang); showed glaciers and ice as custodians of deep time and planetary change (Kirsten Haydon); and proposed innovative designs for ocean repair in the context of climate change.

Calcifiers of Change by Pirjo Haikola brought together a suite of new research-driven films created by Pirjo Haikola and Tom Park with an earlier work titled *Urchin Corals* (2020) – an installation and film originally produced for the National Gallery of Victoria's International Triennial. The collective works documented an underwater world in two locations showing significant environmental impacts from human activities on their complex and precariously balanced ecological systems. The two films documented the current state of Port Philip Bay in Melbourne and the Great Barrier Reef, showing devastation and imbalance but also communicating the interactions and agency of Haikola as researcher/designer, whose encounters with the Reef and Bay have led to the design of innovative systems and products for assisting regeneration of damaged marine ecosystems. An adjacent video documented Pirjo Haikola's presentation at *The World Around* design summit event at the Guggenheim Museum (2023) where the design logic of these systems and devices were further exposed to audiences.

The second theme – **Carbon/Trees/Soil** – was curatorially grounded in the vital role of dirt thinking and governance, grassroots practices, and the necessity of deep green politics. Life on Earth depends on soil – from microorganisms to plants and animals. Works exhibited here demonstrated soil as a substance that can awaken our sensorium via sound, smell, touch, and sight (Clare McCracken, Rebecca Najdowksi, Polly Stanton); contemplate the role of ancient forests as elders and ancestors (Alex Le Guillou); use clay, water, and ochre as materials of Aboriginal healing and connection with the natural world (Dean Cross); work with native trees as infrastructure, habitat, refuge, and home (Mark Jacques, Openwork and Sarah Lynn Rees, RMIT ICON Science and the City of Melbourne); and express eco-anxiety and collective grief for lost trees.

Expanding on an earlier body of work, researchers Marnie Badham and Tammy Wong Hulbert, with Ai Yamamoto and George Akl, presented *To the fallen trees*. Installed as a photographic, sound, and written archive, this work was drawn from a performative event that responded to violent windstorms that created devastation across the state of Victoria in 2021, destroying hundreds of mature trees. Large numbers of fallen mountain ash (*Eucalyptus regnans*) were initially left to decay where they fell until the site was cleared by burning the remnants. The initial site-specific performance took place in the company of the fallen trees as local people read poems to the fallen trees, expressing grief and a sense of loss. The public event and the recordings of the shared ritual have created a communal memory and evidence of care and connection to place and to the unique long-lived members of the non-human world that link us to place and time. The work is both poetic and an explicit call to action in response to the increased occurrence of extreme weather events due to climate change.

The third theme – **Species/Diversity/Extinction** – made visible the impact of ecocide and genocide on a planetary commons. Creative works that harnessed data on endangered butterflies and flowers to re-imagine new models of relating to complex living systems (Kate Geck); planning for more-than-human cities and questioning perceptions of public risk (Mark Jacques, RMIT ICON Science); upscaling the potential for reuse and collaborative care for materials through modes of assembly and disassembly (Caitlyn Parry, Helen Duong); providing ecological care for a sandstone quarry on traditional Aboriginal lands (Millie Caitlin, Joseph Norster); and revealing how climate change is deeply entangled in the daily lives of children (David Roussell and Amy Cutter-Mackenzie-Knowles).

~~*50,000 Bees*~~ *Plan B* by Mark Jacques with RMIT ICON Science proposed – and prototyped at 1:1 scale – a research that advocated for more-than-human planning approaches for urban contexts. As the exhibition installation process unfolded, the work became enmeshed in the very social and policy contestation that it addressed – a real-time test of curatorial action. Leveraging the transformation of a barren, uncompleted art installation armature, the harshness of the Design Hub Gallery's predominantly concrete courtyard was enlivened with flowering plants in opportunistically attached armatures. The project initially also included a hive of 50,000 bee pollinators, drawing attention to new planning guidelines that aim to support living infrastructures on walls and roofs, creating opportunities for non-human lifeforms being impacted by development and, in doing so, reducing heat island impacts – all vital ways of rethinking the relationships between human and non-human life in the city. Yet the application to locate the 50,000 bees in the already installed hive was ultimately rejected by the institution's property department due to the perceived risk to humans. Curatorially, the rejection proposed a curatorial and design conundrum – to proceed or not? Inventively – and with appropriate activist sentiment – Jacques and team chose to proceed, calling attention to the University's risk profile which was in direct contradiction with the progressive research it also supported – a conflict reflected for public attention in Jacques' wry title *50,000 Bees Plan B*.

Finding new ways to reveal and translate the complexity of contemporary challenges in climate change is critical, as is creating the space for reimagining the status quo. As Watson (2021) emphasises, in this context, the curatorial intention is purposefully porous, collaborative, and responsive while experimenting with new modes of curation and display that render progressive 'in-process' research 'visible'. . . in order to advance the community's understanding of the currency of creative practice in addressing the entangled challenges of our shared future (p. 234).

Event as Performance (Curator as Dramaturg)

Critical to the curatorium's intent for *Wild Hope* was a foundational commitment to an extensive range of programmes that formed a cornerstone of the early 'expression of interest' process. Researchers and creative practitioners were asked to consider and articulate how their potential contribution to the exhibition should be manifested and mediated within the exhibition environment through an exhibitable work or a performative lens. Rather than being 'complementary' to the works on display, the diverse range of programmes were integral to the ambition for public exchange and conceived by their creators in response to the exhibition's themes and provocation. The spatial design of the exhibition also responded to the value of these temporal events with a dedicated 'programmatic' zone designed into the main gallery (Project Space 1) with a flexible system to allow the rapid adoption of the gallery space with temporal addition of 'light' moveable screens, speakers, and seats, as required.

The notion of the programme or the event 'as performance' is

to actively connect audiences with *encountering* [creative research] ideas, to bring the audience into active participation with the process of making a work and – or – to respond to and activate audiences with the contemporary condition or rapidly changing socio-political issues.

(Watson, 2021, p.248)

The event as performance curatorial mode is generally characterised by its temporal nature – usually a short 'live' duration – and by the creation of a dramaturgical framework that, in turn, creates the curatorial conditions for a series of interactions rather than for direction or choreography.

For *Wild Hope*, the 'event as performance' condition took many varied forms including film screenings, readings, workshops, meal sharing, talks, listening sessions, and performance lectures including an opening keynote lecture that opened the exhibition by Indy Johar of Dark Matter Labs investigating the concept of planetary civics; a screening and panel discussion led by Vicki Couzens of 'To follow the Old Ways'; and a screening of *THE GIANTS* followed by a conversation with the filmmakers Rachael Antony, Laurence Billiet, and eco-activist Bob Brown. PlaceLab, RMIT's urban laboratory, hosted a series of workshops convened by Michael Dunbar and Regen Melbourne to 'measure what matters' in a new 'city portrait' of Melbourne. *Planetary Auditions,* a listening session curated by Joel Stern, with performances by Catherine Ryan, Santiago Rentiera, Sarah Barns, and Machine Listening, explored the intersection of sound, listening, ecology, and AI.

For *Climate Child Imaginaries* David Rousell and Amy Cutter-Mackenzie-Knowles drew on a decade-long research project that inquired directly into how children imagine, perceive, and engage with climate change as a presence in their lives. The installation included photographs taken by children aged 9 to 14 who participated in the research. Some of the original participants, now university age, also joined a public program where they reflected on the experience of being in the research program and how it influenced them and gave them a way to articulate their ideas and concerns. The images capture a mix of beauty and anxiety and a sense of connectedness with the more-than-human world that has a particular poignancy and urgency as it is expressed by young humans who have emerged into a world that is characterised by this abiding concern, enriching the social imaginary.

Between the Cracks was conceived as a spoken word and image-based performance in which Patrick Kelly, Stayci Taylor, Kim Munro, and Angie Black asked visitors, 'Can we take tools from queer thought and practice to aid our collective survival?' Denise Sprinsky and Angela Finn staged a runway performance of fashion items crafted from ocean-waste plastic; Linda Williams led a panel discussion on 'Multispecies Futures & Extinction Imaginaries'; Heike Rahmann, Maud Cassaignau, and Salad Dressing (landscape architects) explored 'rewilding' practices while Jen Lynch and Emily Sia Lian Wong staged an in-situ shared meal centred on care, food, and landscape systems as part of *The Long Table: Garden as Commons.*

In the interactive work *Future Naarm: First Light* – displayed in a gallery in a purpose-designed gaming 'kiosk' – Patrick Macasaet, Vei Tan, Shuming Ivy Zhou, and Zechen Huang (known collectively as the RMIT Architecture Immersive Futures Lab and

Superscale) invited audiences to explore a flooded metropolis transformed by rising sea levels. This immersive, speculative environment harnesses the power of didactic and open-world wandering games to unravel hidden connections and reveal new constellations of ideas about what our future cities could look like. Through the exploration of the interplay among climate, city, culture, and country, *Future Naarm: First Light* prompts reflection on the impact of climate change. It also inspires imagination – we are encouraged to engage in dialogue and envision hopeful narratives for future cultural and built environments. This interactive game is a powerful and provocative catalyst for a collective reimagining of a world where the need for action on climate change is met with creativity and hope in dark times.

Conclusion

This chapter has drawn critical attention to the role of the curator and curatorium as a critical activist response in the shaping of regenerative futures in a climate crisis. As a form of grassroots climate activism, progressive curatorial practices offer an experimental space for exploring, testing, and mediating speculative design ideas at the radical interface of creative practices, performing 'research-in-action' and civic engagement. Activist ideas and creative practices can be explored and opened for immersion, debate, and discussion through research-led exhibitions and large festival programs which can create opportunities for public encounters.

As the case-study of the public exhibition and programme for *Wild Hope: Conversations for a Planetary Commons* in Naarm/Melbourne illustrates, the expanding role of the curator as activist is emerging to question and challenge top-down, extractive practices that have contributed to the global ecological crisis. Curatorship in this context, as Watson (2021, p. 281) writes, must be reimagined in paradigm-shifting ways to respond to questions such as:

- what does it mean to curate the contemporary polycrisis?
- what are the opportunities in the nexus among the ecological, cultural, and virtual environments?
- how can curatorial practice identify, articulate, and reflect upon new ideas and practices that respond to and transform the world in a climate emergency?

In his essay 'The Incomplete Curator: aka Fighting the Delineated Field' Liam Gillick writes (2016, p. 148):

The incomplete curator is aware of shifting curatorial scope. They do not see their work as the production of encyclopaedic knowledge . . . The incomplete curator is part of a curatorial mass . . . pace and discourse are messed around with – to be slowed down or sped up, or kept at the same speed, is the incomplete curator's way to remain in permanent conflict with contradictory flow. They fight hard for a resuscitation of the public domain.

Wild Hope: Conversations for a Planetary Commons was curated to engage, translate, activate, and perform emergent creative research in relation to socio-political responsiveness and climate activism in public performance spaces. The vital leadership of First Nations-led knowledges of Country and the ongoing impact of unceded sovereignty in finding shared pathways centred the paradigm shift needed to address climate change in progress. This is paramount to the curatorial as activist intent.

This notion of 'incompleteness' in relation to curatorial activism and creative practices resonates deeply with the *Wild Hope* project and supports the emergence of a curatorial position that is less 'exposition' and more co-created 'exploration' – to ask challenging and provocative questions and prefigure alternative practices as part of a radical civic and creative process.

Acknowledgments

The authors would like to recognise and sincerely thank all of the artists and researchers who contributed to the *Wild Hope Exhibition and Public Program* and the critical role of the RMIT Design Hub Gallery staff who coordinated and implemented the production and management of the installations and spaces and the keynote events.

Note

1 http://www.artnews.com/2017/11/07/what-is-curatorial-activism/.

References

Gillick, L. (2016). The incomplete curator (aka fighting the delineated field). In P. O'Neill, M. Wilson and L. Steeds (eds.) *The Curatorial Conundrum: What to Study? What to Research? What to Practice?* (p. 148) Cambridge, MA: MIT Press.

Newell, J., Robin, L. and Wehner, K. (2017). *Curating the Future: Museums, Communities and Climate Change*. London: Routledge.

Rhodes, K. and Watson, F. (2021). Curating ideas in action. *Object*, 31 October 2014, as referenced in Watson, F. *The New Curator: Exhibiting Architecture and Design*. London: Routledge.

Staniszewski, M. A. (2015). Some notes on curation, translation, institutionalisation, politicalisation, and transformation. In B. Žerovc (ed.) *When Attitude Becomes the Norm: The Contemporary Curator and Institutional Art* (p. 247). Berlin: Archive Books.

Ursprung, P. (2014). Presence: The light touch of architecture. In *Introductory Essay in Sensing Spaces, Exhibition Catalogue* (pp. 16–46). London: Royal Academy of Arts.

Watson, F. (2021). *The New Curator: Exhibiting Architecture and Design*. London: Routledge.

PART III

Diverse Modes of Organising Different Constituencies

**Introduction to Part III: Diverse Modes
of Organising Different Constituencies**

Grassroots climate activism challenges traditional notions of political engagement and societal change, and has the potential to transcend geographic, cultural, and generational boundaries. The chapters presented in this section, "Diverse Modes of Organising Different Constituencies," highlight how the nature of grassroots climate activism is changing, and how people organise themselves according to different identities, age being the most important category here, in addition to religious affiliation. Contrary to NGO- or corporate-led activism (Allen, 2021), grassroots activism does not follow prescribed patterns. It often has an unpredictable trajectory: often dependent on the willingness of volunteers to spend time and energy on unpaid work. While many join activist groups in order to address the climate crisis, climate activism is also embraced by existing organisations, like churches, temples, or synagogues. This section explores the ways in which climate activists are organising themselves, and the ways in which this complicates our notions of what climate activism looks like on the ground.

While the youth climate movement is the main topic of this section and has been receiving particular attention since 2019 (Dozsa, 2024), we begin with a discussion of movements of elders, in Chapter 14. In "Nanna Mobilities: Elder-led Protest Movements in Australia, the UK and the USA," Prashanti Mayfield focuses on the "Nannas" or elder-led groups engaged in non-violent direct action and protests against fossil fuel extraction in Australia, the UK, and the USA. These groups, such as the Knitting Nannas Against Gas and Greed (KNAG), articulate the intergenerational challenges and responsibilities of the climate crisis. Their community-based activism is enacted through tactics like conversations in public space, awareness and trust building, and bearing witness to the impacts of the carbon industry. The Nannas engage in their local spheres, using crafting to weave connections in their communities and with other activist groups and politicians that can promote change. In this way, the activists respond to immediate impacts on their communities and territories, while remaining aware of struggles in distant but related geographies. Mayfield's ethnographic research demonstrates how these elders find inspiration from the recent wave of youth climate activism, explored in the following chapters, and leverage

DOI: 10.4324/9781003396567-16

their experience and wisdom to mobilise their communities to leave a better world behind and make a difference.

In Chapter 15, "Young People's Climate Activism," Sally Neas, Ann Ward, and Benjamin Bowman explore methodological trends and key themes across contemporary academic literature on young people's climate activism. Following an initial wave of survey-based research of young people and textual analysis of secondary data, this field is experiencing a second wave of qualitative research and a resurgence of emphasis on youth voice. The authors identify the strengths of the existing literature in its exploration of key themes, including the composition, practices, and outcomes of young people's climate activism and the ways young people understand and act on climate change. This chapter also points to several gaps in the literature that arise from a disproportionate focus on certain topics like mass mobilisations, an intensive interest in the individual activist Greta Thunberg, and especially a disproportionate focus on activism in the Global North and in wealthy and white communities.

In Chapter 16, "Pandemic Possibilities: The Corona Crisis as Perceived Opportunity and Threat for Climate Activists," Mattias Wahlström and Lotte Schack examine the implications of the COVID-19 pandemic for the relationship between climate activism and crises. As mentioned in Chapter 1 of this handbook, the COVID-19 pandemic greatly impacted the youth climate movement, and the coming years will show how it finds ways to regroup and reconstitute in the wake of the lockdowns and online organising that greatly diminished its ability to mobilise. Viewing the pandemic as a disruptive event that also created a potential for societal change, this chapter analyses its impact on climate activism and the increased costs of collective action, drawing on empirical data from Finnish and Swedish climate activists and reviewing existing literature. Focusing on individual activists' experiences, the authors explore the diverse impacts of a major crisis within the movement and the personal meanings attached to it. The authors make two arguments. First, that the pandemic imposed constraints on the "affective infrastructure" of the movement while presenting opportunities for reimagining visions of a fossil-free future. Second, that there is no determinate outcome of these kinds of disruptive events: their consequences develop amidst tensions relating to societal context, activists' interpretations of events, and movements' affective infrastructures.

In Chapter 17, "Youth Climate Activists' Practices on Social Media in Belgium and France," Yuliya Samofalova, Andrea Catellani, and Louise-Amélie Cougnon analyse the communication strategies of youth climate activists on social media in Europe, focusing on Belgium and France. The authors explore how digital platforms like TikTok have become a powerful tool for climate activism, given how they enable young activists to build communities, inform the public, and provoke social and political change. The chapter calls for a more multi-platform analysis of online activism, emphasising the need to understand the diverse methods used by youth activists to engage with climate issues. They also stress that studies with a focus on the youth climate movement online still have not sufficiently analysed non-Western contexts. Overall, the authors reveal the growing influence of online activism and its potential to reshape the climate movement's trajectory.

In Chapter 18, "Challenging Politics to Do (and Be) Better: Young People's Climate Activism Role in the Italian Political Landscape," Gabriella Sesti Osséo highlights the gap between youth climate activism and political institutions in Italy. Young climate activists, a group that utilises diverse strategies and practices, are highly important actors within the democratic arena; since late 2018, mobilisations across the globe revitalised global climate activism and gained public attention, and youth-led climate movements played a decisive role in enlivening the new "green wave" in the Global North (see Chapter 1). However, despite their efforts at politicising environmentalism and radicalising its aims, young climate activists' demands are rarely translated into comprehensive policy change in most Western democracies, Sesti Ossèo posits. Through the argument that young climate activists are "norm entrepreneurs" who present themselves as moral authorities to pressure policymakers to face the climate crisis, this chapter explores the strain between climate activism and institutional party politics in Italy. Specifically, it examines youth activists' perceptions of impacts accomplished on specific scales and dimensions, their conceptualisations of politics and diagnosis of climate inaction, and their potential roles in shaping climate politics and policies.

The section closes with Chapter 19, "Christian Communities as Climate Champions," by Reba Elliott. Although the large majority of Christians in the United States lean Republican (Pew Research Center, 2024) and many have shown support for political candidates who deny the scientific consensus on climate change (Zaleha et al, 2015), Elliot discusses practical ways that Christian, in particular Catholic, communities have woven efforts to solve the climate crisis into their everyday practices. She begins the chapter with a brief summary of the main threads of thought in Christian theology that support action on climate change, followed by a discussion of the faith-based non-profit organisations addressing this issue in the United States, including significant interfaith cooperation. Finally, the chapter covers the ways in which these non-profit organisations equip grassroots actors to take up ecological spirituality, sustainability, and advocacy.

The chapters in this part confirm that grassroots climate activists are not homogeneous, but they also highlight the many parallels in the motivations and practices of various activist groups. As such, these chapters demonstrate that scholarship on grassroots climate activism can help broaden our understanding of how people process, experience, and navigate the climate crisis and the various ways in which they strive to affect change.

References

Allen, J. I. (2021). *The New Climate Activism: NGO Authority and Participation in Climate Change Governance*. Toronto: University of Toronto Press.

Dozsa, K. (2024). *Children as Climate Citizens. A Socio-Legal Approach to Public Participation*. London: Routledge.

Pew Research Center (2024). *Changing Partisan Coalitions in a Politically Divided Nation. Party Identification Among Registered Voters, 1994–2023*. https://www.pewresearch.org/politics/2024/04/09/changing-partisan-coalitions-in-a-politically-divided-nation/ [Accessed 14 September 2024].

Zaleha, B. D. (2015). Why conservative Christians don't believe in climate change. In *Bulletin of the Atomic Scientist*, 71(5), 19–30. https://doi.org/10.1177/0096340215599789

14

NANNA MOBILITIES

Elder-led Protest Movements in Australia, the UK and the USA

Prashanti Mayfield

Introduction: Intergenerational Climate Impacts

This chapter focuses on the role of elders in local climate movements, drawing on examples from Australia, the UK and the USA and the antecedents, motivations and practices that underpin their involvement in environmental activism. I explore how the role of the 'elder' is seen from a broader societal perspective and how their involvement in climate activism subverts dominant media and political narratives used to dismiss protestors advocating for environmental protection. In the first section of the chapter, I discuss two 'anti-fracking' and environmental protection movements – the Australian Knitting Nannas Against Gas and Greed (aka KNAG) and the UK-based *Nanashire* – both examples of protest groups that centre elders as a core part of their activist identities and philosophies. Drawing on a combination of ethnographic fieldwork and semi-structured interviews, I show how these groups formed, the actions that they have undertaken, who they include in their work and the aims that they are working towards in their activism.

In the second section of the chapter, I analyse the ways in which elder-led protest movements come together and operate, situating the work of KNAG and Nanashire alongside examples from other elder-led movements, including Th!rd Act (USA), Extinction Rebellion Grandparents and Elders (Australia and the UK) and 1000 Grandmas (USA). This chapter shows how elders form a critical part of the calls for action on climate change. Through discussion of these groups, I highlight key features of elder-led protest movements and the work that they do in the space of climate activism. The chapter concludes with reflections on how these groups work to support climate-change activism both within and for their local communities and the ways in which their work has expanded to both the national and global scales – influencing and inspiring other groups to form and take action in their local communities.

Grandmas Against Gas in Australia and the UK

The extraction of gas as a mechanism for economic prosperity and revitalisation of rural communities is a development strategy that has been pursued by governments in both Australia and the UK – in particular, the mining of 'unconventional gas', also known as shale

DOI: 10.4324/9781003396567-17

gas or coal seam gas, which requires the injection of a combination of water and chemicals into the ground to fracture the rock beds that hold gas to release it, resulting in an extraction method colloquially referred to as 'fracking' (UK Department of Energy and Climate Change, 2016; NSW Department of Resources and Energy, 2016). In both countries the development of this resource has raised concerns from environmental groups, scientists and local communities where developments are proposed due to the dangerous impacts of these forms of extraction that can result in damage to water tables; salinity in soils affecting both forested areas and agriculture; methane gas leakage into the air and local water supplies; earthquakes and tremors impacting buildings and infrastructures; and exposure of local populations to chemicals above safe levels (Hays and Shonkoff, 2016; Deziel et al., 2020; see also Gheorghiu and von Mering in Section IV in this volume).

Reactions to developing unconventional gas resources in areas where gas wells are proposed are typically very polarised. Those in favour argue that the economic and social benefits of unconventional gas extraction outweigh any potential risks that may be there and that these forms of development are needed to ensure the continuance of local communities (Klasic et al., 2022). Those opposed argue that the development of unconventional gas will result in catastrophic outcomes for the local environment and communities surrounding extraction sites, as well as being a significant contributor to greenhouse gas emissions and climate change (Luke and Evensen, 2021; McCrea et al., 2020). Because of the strong push from government and the private sector to pursue development of these resources, often people living in areas affected become politically active in contesting the development and form groups to campaign against them.

In the following sections I discuss the emergence of two such groups, one in Australia and one in the UK. Although the groups started independently of each other, they share strong similarities and solidarity with each other. They are groups that brand themselves as Nannas/Nanas[1] against gas and use the image of the grandmother as a central motif in their protest actions. The materials discussed here are a synthesis of over 55 interviews undertaken with members of both groups throughout 2017.

Knitting Nannas Against Gas and Greed (KNAG): A Disorganisation of Loops

The Knitting Nannas Against Gas and Greed (KNAG) are a horizontal, decentralised environmental protest movement that started in Lismore, Australia in 2012. Initially focused in the Northern Rivers area of New South Wales (NSW) the Nannas were a central part of a campaign to stop the development of CSG in that area, working in collaboration with other groups to hold blockades and meet with government officials, as one Nanna describes:

> At Bentley, we set up our Nanna-Dome and had a project called Nanna-Care and what we would do is every morning go and check the lock-ons and see if they had everything they needed. We would write the numbers of lawyers on their arms in permanent pen, make sure they had toiletry items, hats, sunscreen, water or warm clothes. That was our role. We would also walk around the Blockade, there were a lot of people there, and we just made sure that things went sensibly. De-escalating situations where people were getting a little too passionate. That really made a big difference in Bentley. We are very much loved by the community and that is really something.
>
> *(Nanna A, May 2017)*

As the Nannas' profile grew due to their campaigning, people in other areas asked if they could also become Nannas and over time the movement expanded with several groups, known as 'Nanna Loops', forming across the country. Each loop is formed in response to issues in their local area and responds to issues or developments happening there, while simultaneously contributing to broader environmental campaigns and social issues.

Initially, the Nannas were formed in response to coal seam gas developments, but over the years they have campaigned on a range of climate-change-related issues, including deforestation, water protection, coal mining, Aboriginal and Torres Strait Islander sovereignties and refugee rights. To be a Nanna, you do not have to know how to knit, be a Nanna or identify as a woman, all that is asked is that you subscribe to the *Nannafesto*:

> We peacefully and productively protest against the destruction of our land, air, and water by corporations and/or individuals who seek profit and personal gain from the short-sighted and greedy plunder of our natural resources. We support energy generation from renewable sources, and sustainable use of our other natural resources. We sit, knit, plot, have a yarn and a cuppa, and bear witness to the war against those who try to rape our land and divide our communities.
>
> We want to leave this land no worse than we found it, for our children and grandchildren. They deserve to have a future with a clean and healthy environment, natural beauty and biodiversity, (and don't we have our work cut out for us!)
>
> *(excerpt from the Nannafesto, KNAG, 2016a)*

The Nannafesto works as a unifying statement and rallying point guiding the actions of Nanna Loops and creating an identity that brings them together. Typically, Nanna Loops meet weekly and will often stage 'knit-ins' where they occupy public space and knit as a protest tactic. The Nannas campaign outside of political offices, mining companies and government departments to rally public support and raise awareness of the impacts of mining and industries that 'destroy the air, land and water for the kiddies'. They are also active in political campaigning and advocacy, coordinating petitions, writing submissions and meeting with politicians to push for changes to environmentally destructive actions and policies. The Nannas, over time, have been successful in contributing to changes in government policy on the extraction of coal seam gas in NSW, as well as working with other groups to halt or delay the commencement of new coal and gas mines in locations across the state. However, with the advent of the COVID-19 pandemic economic downturn and the commencement of what the then-federal government of Australia termed a 'gas-led recovery', coal seam gas extraction continues to be an issue they actively campaign on.

Nanashire: A Field, a Tank and 1000 Days of Protest

Nanashire is an anti-fracking group, formed in Lancashire in 2014, contesting the development of shale gas drilling in the county and the UK more broadly (Nana X., November 2017). Nanashire was formed in 2014 when the group undertook the action 'Operation Mothers and Grandmothers' (aka OMG) and occupied a field beside a site in Lancashire where planning applications for shale gas drilling had been filed. The Nanas set up a camp in the field and drank tea and ate cake as the sun rose and the local police were made aware that they were occupying it. The occupation lasted a week and drew attention to the development and brought awareness to their community. The reaction to their protest ranged from bemused and highly supportive to being portrayed as a civil nuisance engaged in anti-social behaviour by the landowner and the mining company. The Nanas were charged with trespass *after* they had evacuated the field, with one of the Nanas being charged as sole defendant, a decision that was not taken lightly:

> We felt if we didn't engage we ran the risk of them re-writing the story of what happened that wasn't true. I decided that if I went and became the named defendant, I owed nothing, I had no possessions as such and was suitably poor. I had no carer responsibilities. And, of course, I was a key instigator, there is no denying that. I had encouraged these women who trusted me to come into this. Later people told us that we probably could have got away with it, but the threat of it was too great to risk.
>
> *(Nana X., November 2017)*

This raised the profile of the Nanas in both their local community and beyond, as the case was reported nationally and this brought broader public support to their cause. The Nanas started to travel across the UK, holding public talks and advising other communities concerned about the impacts of shale gas extraction occurring in their areas. The profile of the Nanas grew to the point that iconic punk fashion designer Vivian Westwood became aware of their actions and offered to help by giving the Nanas a tank which they drove to the door of then-Prime Minister David Cameron's residence with Westwood riding on top of it. As one Nana explained:

> So we went and declared war on fracking outside David Cameron's house. So we rock up in this tank and there is an armed copper outside the house, and he says, 'Nanas, you can't park your tank here, you are going to have to go and park it down the road'. And we said, 'I think you will find we are the ones with the tank and we will take it where ever we want'.
>
> *(Nana Y., November 2017)*

Following this action, membership of Nanashire continued to grow and the group became a constant presence at rallies, drilling site protests and anti-fracking events within Lancashire and beyond (Nana Y., November 2017). In early 2017, the gas mining company commenced development at the PNR site and Nanashire, in collaboration with other community protest groups and local residents, held a continuous protest lasting more than 1,000 days at the roadside opposite the proposed drill site (Nana Z., November 2017). Nanashire used multiple forms of protest tactics during this time including lock-ons, dance, blockading, observation and record keeping, liaising with local and central government authorities, educational events and a weekly silent vigil held on Wednesdays called 'Women in White'. Over time, through their campaigning the Nanas have been successful in contributing to shifts in UK government policy on shale gas extraction and the protection of environmental resources. However, with shifts in government priorities and the economic impacts of the COVID-19 pandemic and Brexit, debates on pursuing shale gas extraction have begun to take more prominence.

Elders in Action: Methods, Tactics and Outcomes

Actions that elders-led activist groups engage in take many forms: engagement in political processes from the local government level through to national and international platforms (Third Act, 2024a); participation in blockades and encampments (KNAG, 2016b); hunger strikes and parliamentary glue-ons (Grandparents for Climate Action, 2024); and public protests and demonstrations in political centres and cities (Extinction Rebellion UK, 2020). In the following sections, I discuss the methods and tactics that elders groups involved in climate-change activism engage in, drawing examples from both KNAG and Nanashire, as well as other groups engaged in environmental protection campaigns from Australia, the UK and the USA.

'Accidental Activists' Subverting Stereotypes

Many people who become involved in protests movements later in life have not previously engaged in political activism or considered themselves to be the 'type' of person who engages in environmental protection campaigns (Ey, 2019). Gelderloos argues that often 'those fighting for clean water, a healthy relationship with the land, or a liveable planet for future generations are deemed extremists' (2022, p. 283) by both governments and the private sector. Further, narratives such as this are deployed in political and media discourse to dismiss, belittle and derail the actions taken by protestors involved in climate activism. Yet older people who become involved in climate protection movements typically defy these labels and the criticisms levelled at younger activists, subverting the stereotype of *who* climate-change protestors are and *why* they have decided to become active. Quite often elders engaged in protest take this image and subvert it. As KNAG (AUS) explain they:

> use the common stereotype of the sweet little old lady to lull the bad guys into a false sense of security . . . the idea of the Knitting Nannas is that of the iron fist in the soft fluffy yellow glove – there aren't many scarier things than a forthright woman in her prime.
>
> *(KNAG, 2016b)*

The symbol of elderly, benign and wise grandparents becomes a tool that can be used in environmental protection and activism to engage people that would not normally listen to younger climate activists. A member of Nanashire (UK) explains the choice to embody the image of the 'Nana' in their activism:

> Most of us are *accidental activists,* we sort of just accidentally opened this Pandora's box and a whole lot of shit came out . . . So, we came up with the tabards . . . that is what we went for, proper Lancashire matriarch, that is what, if you went back years, your nan would be wearing: a head scarf and a tabard. And when shit gets serious, they roll up their sleeves and they get on with it you know. It's our battle armour, that tabard, and the feather dusters are our weapons. I mean, there is nothing less offensive than your nana.
>
> *(Nanna B., October 2017)*

The term 'accidental activists' is a common one amongst the Nanna/Nanas interviewed and is used to describe how, prior to becoming Nanna/Nanas, they had never before attended a protest or identified themselves as politically active. Gearey and Ravenscroft (2019) observe that often a turning point for involvement in environmental activism in older people is an event or experience that triggers a sense of injustice or a need to challenge a status quo, and through doing so they reclaim their political agency and connection with community.

A central concern and motivation present in elders who become engaged in environmental protest movements is concern for the state of the environment that future generations will inhabit and whether it will be liveable. By announcing a development that would impact their local community and environment these elders were motivated to take action and engage in protests. As one member of the UK Extinction Rebellion Grandparents and Elders explains:

> When you're in an emergency you drop what you are doing and grapple with it . . . and this is an emergency, you just have to act very quickly and very radically, otherwise the future for my grandchildren and everybody else's grandchildren and all the lovely kids you see around and the young people is horrible . . . we've been responsible for this mess.
>
> *(Arnold Pease, 93 years old, Extinction Rebellion, September 2020)*

The urgency present in this statement and the care for younger generations expressed is a motivation repeated across elders engaged in environmental activism. Following the theme of intergenerational responsibility, John Lymes, a 91-year-old member of Extinction Rebellion in the UK, when asked why he was protesting, explained:

> It's great that the younger ones are protesting, but it is my generation that have caused all this trouble, so, here I am.
>
> *(John Lymes, 91 years old, Extinction Rebellion, September 2019)*

Similarly, Th!rd Act, a climate-change action network based in the USA founded in 2021, state that:

> as a generation we have unprecedented skills and resources that we can bring to bear. Washington and Wall Street have to listen when we speak, because we vote and because we have a large – maybe an overlarge – share of the country's assets. And many of us have kids and grandkids and great grandkids: we have, in other words, very real reasons to worry and to work.
>
> *(Third Act, 2024)*

Through creating networks and structures that make participation in climate activism accessible and open, Th!rd Act works to mobilise and empower elders to make change in their communities and beyond. Elder-led climate activist groups typically form at a community level and are composed of people that have had similar moments of questioning in regard to the lack of action taken by the government on climate change, or there has been a significant event in their area that has compelled them to take action. The mechanisms and structures through which older people become involved in activism vary. Still, structures and networks such as KNAG and Nanashire, Extinction Rebellion and Th!rd Act provide clear and accessible ways in which they can become involved in actions taking part in their areas that are also connected to collective actions at national and global scales.

Allyship and Standing in Solidarity

A cornerstone of the activism undertaken by many elder led groups is the building of allyship and solidarity – with each other, with younger generations, with other protest groups and with Indigenous communities. In the cases of both KNAG (AUS) and Nanashire (UK), being part of a coalition of other climate action groups, while at the same time retaining full autonomy and independence in their actions and decision-making processes, enabled collective aims to be realised. Because climate-change activism risks exposure to police violence and legal penalties, having trust between members of protest groups is important to maintain solidarity and effectively undertake collective actions. This is where intergenerational capacities can be maximised, as the experiences of older generations meet the energy and abilities of younger generations.

Environmental protection and climate activism movements driven by Indigenous communities are often intergenerational with the actions of younger participants being led by the guidance of elders in their communities (Goodyear-Kaʻōpua, 2017; Whyte, 2017). An example of this is the Dakota Access Pipeline (DAPL) protest that spanned 2016–2017, in North Dakota (USA), where a protest encampment was established to protect the lands and waters of the Standing Rock Sioux from the drilling of an oil pipeline running under the Missouri River (See also Chapter 2 in this handbook). A defining feature of this protest was the coming together of multiple generations in solidarity and the guiding of younger generations by their grandparents and ancestors to protect the water and communities along the river dependent on it (Lane, 2018). The protests gained global attention through social media and news reporting, catalysing many people to come in support of the protest to the encampment and even more to donate resources (Hinzo and Clark, 2019).

The DAPL protests inspired actions in other places as images of what was happening on the frontlines of the blockade spread – with state-based violence being unleashed on protestors practicing non-violent direct action to resist the development. This was the starting point for the USA based movement 1000 Grandmothers, an organisation of grandmothers and elder women engaged in climate-change activism in solidarity with Native American communities, who on their website explain that:

> Moved by the courage and wisdom of the Grandmothers at Standing Rock, a half dozen elder women from Oakland and Berkeley, CA started meeting in October 2016. Our first work together was supporting the Grandmothers at Standing Rock, – who, as water protectors, stood up against the fossil fuel industry's Dakota Access Pipeline. Subsequently, guided by our core group's belief that grandmothers have a deep heart connection to future generations and could be a powerful voice in the Climate Movement, we began organizing elder women locally about the Climate crisis.
>
> *(1000 Grandmothers, 2024a)*

Similar to 1000 Grandmothers, in interviews with members of Lancashire's Anti-Fracking Nanas, the Grandmothers of Standing Rock protesting against the Dakota Access Pipeline were cited as a source of motivation and inspiration to engage in land and water protection actions. Solidarity between the Lancashire Nana movement and the Grandmothers of Standing Rock was further deepened in 2017 through the exchange of items each group had crafted in recognition of the other. These actions show the importance of seeing others taking action in becoming politically active and the interconnection between elder-led climate movements and Indigenous land protection movements. It is through these connections that a literacy, grammar and model for action can be seen in the narratives that shape non-Indigenous climate action groups. It is important for non-Indigenous groups operating on Indigenous lands to recognise and centre the voices and perspectives of the

Indigenous communities whose land they are on and to work in solidarity and ally-ship with them.

Empowering and Educating

Once elders become mobilised in protest movements the sharing of information, connection with other groups and working collaboratively become central features in how groups operate. Learning occurs across multiple scales with people engaged in self education, learning from others in their networks and sharing information with members of the public to increase awareness and drive change on issues they are campaigning on. Larri and Whitehorse (2020) develop the concept of 'Nannagogy' to describe the journey of learning, knowledge sharing and perspective shifting that members of KNAG engage in as they become involved in activism. They argue that through being involved in social movements that older women empower themselves and reject the societal expectations placed on older women, highlighting that 'Activism offers ageing women the opportunity to be active and engaged informal learners' (Larri and Whitehouse, 2020, p. 32).

The forms of learning that the Nannas engage in hinge on conversation, relationships and sharing expertise and skills built over their lives – and through doing so building a collective understanding and vision that underpins their work and activism. In turn the Nannas also work to share their knowledge of mining, fracking and other forms of environmental destruction with the wider community, and through doing so they empower others to act on these issues. These forms of sharing and applying knowledge also apply to other groups involved in climate activism, such as 1000 Grandmothers, who explain that:

> 1000 Grandmothers allows us to augment our individual voices and impact by providing opportunities to act collectively. We find ways to use the particular influence of our being generational grandmothers to strengthen the Climate Justice Movement.
> *(1000 Grandmothers, 2024b)*

The empowerment of knowledge gained through life experience is a resource to elder-led protest movements and in turn is a resource that is used to support action for change. Another example of this is Th!rd Act, who operate as a network of small, community-led groups across America that encourage their members to bring their life experiences to campaigns lobbying businesses and government to take action on climate change, working towards the aim of:

> building a community of experienced Americans over the age of sixty determined to change the world for the better. Together, we use our life experience, skills and resources to build a better tomorrow.
> *(Third Act, 2024b)*

Campaigns that the groups are engaged in include petitioning the banking sector to divest from climate damaging investments, supporting the expansion of voters rights and 'safeguarding' democracy and creating pathways to enable transitions away from fossil fuel dependency. The power of Th!rd Act is in the collective knowledge, experience and influence that its members have as 'experienced Americans' accrued across a lifetime of work and community participation. Th!rd Act creates a space in which people over 60 are encouraged and supported to work together and learn from each other 'to make the changes that must be made to protect our planet and society' (Third Act, 2024a). Learning practices and empowerment through sharing knowledge form the backbone of elder movements, enabling the development of campaigns and actions built from the collective skill sets of group members distilled through a lifetime of practice and experience.

Non-violent Direct Action and Bearing Witness

Non-violent direct action (NVDA) is a key part of climate-change activism and comprises forms of action, intervention and disruption undertaken by protestors to bring attention to an issue or a cause. KNAG (AUS) and Nanashire's (UK) actions are both guided by principles of NVDA in their activisms, ranging from the use of humour in their depictions of Nanna/Nanas characters, knit-ins, camping in fields, use of dance and performance, participating in rallies, blockades, sit-ins and lock-ons, among many other forms of protest. By putting their bodies on the line for their beliefs, the urgency of the issues they campaign on are starkly on display.

An important aspect of NVDA is passive resistance, in that the people who engage in it as a method of protest do not become violent and resist provocations to violence from police and others who may be present at protests. In the case of Nanashire's participation in over 1,000 days of continuous protest, there were moments when tensions would boil over and antagonism between police and protestors would arise. In response to this the Nanas organised the Women in White vigils, as one Nana explains:

> What I like about the Women in White is you only sign up for the 15 minutes that you are there and I would love it if it was something that women all over the place started doing, wearing white and gathering in a public space for 15 minutes silence . . . just to draw attention to where there is conflict and things being ignored: a call for calm could be a thing you do.
>
> *(Nana Z., November 2017)*

By creating space to breathe through silence, the Nanas were able to reset from whatever had happened during the week and start anew. After the vigil had ended, celebratory dancing would be held in the middle of the highway in front of the drill rig to revive the energy of those present and revitalise the focus of the protest.

Visibility is important in climate-change activism and the presence of older people at street protests creates a sense of shock in the media and political reactions to their actions. This can be seen in reactions to the Nanna/Nanas in both Australia and the UK where reporting on their activities in the media is a regular occurrence. Both groups have also featured in parliamentary debates about both unconventional gas extraction policies and proposals to restrict protest rights as a result of their interventions, demonstrating both the popular and political power of NVDA when conducted by elders, as Popovic and Miller (2015) observe:

> The elderly have always been hugely important in successful nonviolent campaigns. They have much time on their hands, and they care about their grandkids more than anyone else in the world.
>
> *(Popovic and Miller, 2015, p. 58)*

An example of older people involved in climate change focussed NVDA are those that have formed groups as part of Extinction Rebellion (XR). There are iterations of XR Grandparents and Elders in Australia and the UK. Methods of protest that they have used to create action on their demands include: blockades in central city places, gluing themselves to public infrastructures, hunger strikes, sit-ins at business and government offices and theatrical/carnivalesque performances to communicate the oncoming impacts of climate change (Grandparents for Climate Action, 2024). Sometimes these actions have resulted in arrests, fines and mandated court appearances, but as one member of the UK branch explains:

> I break the law in order to push the government to keep the law. The government is not keeping its commitments to the Paris agreement and various other protocols to try to keep global warming within 1.5 degrees.
> *(Sue Parker, 77 years old, Extinction Rebellion, September, 2020)*

This statement highlights another central role of the work of the Nanna/Nanas and other elder-led groups: the concept of 'bearing witness'. Through watching, observing, recording and being, these groups work as observers to actions that are undertaken by the state and the private sector that continue to bring environmental harm and contribute to climate change. Through bearing witness to destructive actions, elder-led groups work to shine a light on actions that would otherwise perhaps go unseen, and in the process of doing so they disrupt, curtail and shift the trajectories of these actions and contribute to creating safer climate futures.

Conclusion

Older peoples' involvement in climate activism has expanded rapidly over the past decade and examples of older people becoming increasingly active in environmental and social justice movements can be seen growing across many countries. Elders-movements are characterised and motivated by a deep concern for the intergenerational impacts of climate change and an awareness that the quality of life experienced by future generations may be lower than the one that they experienced through their lifetime. In many of these movements, what is defined as an older person is not a specific age, having a grandchild or being a retiree but more a state of mind and a drive to protect the environment for future generations. The examples of KNAG, Nanashire, 1000 Grandmothers, Th!rd Act and Extinction Rebellion Grandparents and Elders show different ways communities of older people come together and work to enact change.

This chapter highlights commonalities present between the groups – and other elder activist movements more broadly – which include: (1) motivation to act; (2) becoming active; and (3) the types of actions undertaken. Features of these commonalities can be summarised as follows:

1 the *motivations* towards participation in protest actions include:

- wanting to leave the world a better/safer place for their children and grandchildren: 'we are all grandparents, we don't want our children to witness their own extinction'
- wanting to be 'useful' and to 'make a difference' in their communities

- wanting to connect with other people
- having time and resources to participate in actions due to being retired
- having skills and knowledge accumulated over their lifetimes that they are able to use to support their cause

2 examples of where elder activists first encounter the groups and *become active* are:

- in the street, in instances where the group holds vigils or protests in public spaces such as the pavement outside MPs offices or in town/central city squares
- online through social media where a friend or family member may share a post or information about the group
- peer-to-peer where someone they know has also become active and invites them to participate
- through local/regional media reporting

3 forms of protest and *types of action* undertaken include:

- being on the street in differing capacities to engage the public and raise awareness of an issue
- social media and online spaces as a tool to report on actions and communication of issues, using modalities available in this space – such as memes, gifs, informational posts and creating events/actions
- crafting and arts practices – including knitting, dance, painting and poetry
- participation in blockades at mining and forestry sites, including lock-ons and direct actions
- letter writing and petitioning
- meeting with politicians and building alliances with other activists or groups

At a broad level, the point of mobilisation for many older peoples is the growing realisation of the impacts of climate change and its irrevocable impacts on planetary systems. The catalyst for involvement at a more local scale is an event that occurs within their immediate sphere and its impacts are felt at a personal level. The spread of these movements is often done at the one-to-one scale, that is new members join through meeting current members and learning about the experience of other people – and then being able to see themselves undertaking similar actions. Often the idea of being a *protester* or an *activist* is not a role that participants in elders movements have previously identified themselves with. By beginning to participate in community activism they start to see themselves and their roles in their community in a different light and willingly start to lean in to these forms of actions and embody spaces and places that previously they would have considered radical. These elder activist movements, although differing in their composition, formation and focus, are unified by the work of aiming for a common goal of creating a stable and healthy environment for future generations and seeing it as their responsibility to take action in the present to ensure this.

Note

1 There is regional variation in the spelling of Nanna/Nana between Australia and the UK groups; in this chapter the Australian group will be spelled with a double 'n' and the UK group with a singular 'n'.

References

1000 Grandmothers (2024a). 1000 grandmothers. *1000 Grandmothers.* https://www.1000grandmothers.com/ [Accessed 23 January 2024].

1000 Grandmothers (2024b). Our history. *1000 Grandmothers.* https://www.1000grandmothers.com/our-history.html [Accessed 23 January 2024].

Deziel, N. C., Brokovich, E., Grotto, I., Clark, C. J., Barnett-Itzhaki, Z., Broday, D. and Agay-Shay, K. (2020). Unconventional oil and gas development and health outcomes: A scoping review of the epidemiological research. *Environmental Research*, 182, 109124.

Extinction Rebellion UK (2020). *XR Elders – 92 Year Old and 77 Year Old Vicar Arrested.* https://www.youtube.com/watch?v=VxHNc_pDBgA [Accessed 23 January 2024].

Ey, M. (2019). Purling politics: Crafting resistance with the Knitting Nannas against gas. *ACME: An International Journal for Critical Geographies*, 18(4), 956–976.

Gearey, M. and Ravenscroft, N. (2019). The nowtopia of the riverbank: Elder environmental activism. *Environment and Planning E: Nature and Space*, 2(3), 451–464.

Gelderloos, P. (2022). Ecological terror and pacification: Counterinsurgency for the climate crisis. In A. Dunlap and A. Brock (eds.) *Enforcing Ecocide: Power, Policing & Planetary Militarization* (pp. 269–303). Cham: Springer International Publishing.

Goodyear-Ka'ōpua, N. (2017). Protectors of the future, not protestors of the past: Indigenous Pacific activism and Mauna a Wākea. *South Atlantic Quarterly*, 116(1), 184–194.

Grandparents for Climate Action (WA) (2024). *Grandparents for Climate Action – What We Do.* https://sites.google.com/view/grandparents-climate-activists/home/what-we-do [Accessed 23 January 2024].

Hays, J. and Shonkoff, S. B. C. (2016). Toward an understanding of the environmental and public health impacts of unconventional natural gas development: A categorical assessment of the peer-reviewed scientific literature, 2009–2015 Carpenter, D. O. (ed.). *PLoS ONE*, 11(4), e0154164.

Hinzo, A. M. and Clark, L. S. (2019). Digital survivance and Trickster humor: Exploring visual and digital Indigenous epistemologies in the #NoDAPL movement. *Information, Communication & Society*, 22(6), 791–807.

Klasic, M., Schomburg, M., Arnold, G., York, A., Baum, M., Cherin, M., Cliff, S., Kavousi, P., Miller, A. T., Shajari, D., Wang, Y. and Zialcita, L. (2022). A review of community impacts of boom-bust cycles in unconventional oil and gas development. *Energy Research & Social Science*, 93, 102843.

KNAG (2016a). *The Nannafesto.* https://www.knitting-nannas.com/philosphy.php [Accessed 15 April 2018].

KNAG (2016b). What the Knitting Nannas do. *Knitting Nannas Against Gas.* http://www.knitting-nannas.com/what.php [Accessed 20 March 2016].

Lane, T. M. (2018). The frontline of refusal: Indigenous women warriors of Standing Rock. *International Journal of Qualitative Studies in Education*, 31(3), 197–214.

Larri, L. and Whitehouse, H. (2020). Nannagogy: Social movement learning for older women's activism in the gas fields of Australia. *Australian Journal of Adult Learning*, 59(1), 27–52.

Luke, H. and Evensen, D. (2021). After the dust settles: Community resilience legacies of unconventional gas development. *The Extractive Industries and Society*, 8(3), 100856.

McCrea, D. R., Walton, D. A. and Jeanneret, M. T. (2020). An opportunity to say no: Comparing local community attitudes toward onshore unconventional gas development in pre-approval and operational phases. *Resources Policy*, 69, 101824.

Nana X (2017). *Nanashire,* interviewed by Prashanti Mayfield, Blackpool, Lancashire, UK, October 2017.

Nana Y (2017). *Nanashire,* interviewed by Prashanti Mayfield, Blackpool, Lancashire, UK, October 2017.

Nana Z (2017). *Nanashire,* interviewed by Prashanti Mayfield, Blackpool, Lancashire, UK, October 2017.

Nanna A (2017). *Knitting Nannas Against Gas,* interviewed by Prashanti Mayfield, Mullumbimby, NSW, Australia, 2017.

Nanna B (2017). *Knitting Nannas Against Gas,* interviewed by Prashanti Mayfield, Mullumbimby, NSW, Australia, 2017.

NSW Department of Resources and Energy (2016). *How Is Coal Seam Gas Extracted?* http://www.resourcesandenergy.nsw.gov.au/landholders-and-community/coal-seam-gas/the-facts/how-is-coal-seam-gas-extracted [Accessed 3 July 2016].

Popovic, S. and Miller, M. (2015). *Blueprint for Revolution: How to Use Rice Pudding, Lego Men, and Other Non-Violent Techniques to Galvanise Communities, Overthrow Dictators, or Simply Change the World.* Scribe Publications; Penguin Group Australia [distributor]. Scoresby: Brunswick.

Third Act (2024a). Our work. *Third Act.* https://thirdact.org/our-work/ [Accessed 22 January 2024].

Third Act (2024b). About us. *Third Act.* https://thirdact.org/about/ [Accessed 22 January 2024].

UK Department of Energy and Climate Change (2016). *Guidance on Fracking: Developing Shale Oil and Gas in the UK.* https://www.gov.uk/government/publications/about-shale-gas-and-hydraulic-fracturing-fracking/developing-shale-oil-and-gas-in-the-uk [Accessed 3 July 2016].

Whyte, K. (2017). Indigenous climate change studies: Indigenizing futures, decolonizing the anthropocene. *English Language Notes*, 55(1–2), 153–162.

15

YOUNG PEOPLE'S CLIMATE ACTIVISM

Sally Neas, Ann Ward, and Benjamin Bowman

Introduction[1]

Youth-led and youth-centered climate activism has gained momentum and visibility over the past few years. This activism has played a central role in bringing about a "watershed moment" in the politics of climate change (Pickard et al., 2020). The "historical roots, tensions and complexities" in young climate activism remain vitally important, and "the thickest of these roots is environmental justice, a theoretical approach with a long history", especially in the activism of people of color and Indigenous communities (Walker and Bowman, 2022).

This has yielded an emerging body of literature pertaining to young people's climate activism. The vast majority of this literature focuses on activism starting in 2018 and since then. This seems to be connected to the actions of young climate activist Greta Thunberg, who sparked the global #FridaysforFuture movement. This movement encouraged young people to engage in organized school strikes and walkouts (henceforth referred to as the climate strikes) for climate action. In turn, much of this literature is focused on these mobilizations. However, participation in such events constitutes only one activity in a wider web of youth climate activism.

While we agree that the year 2018 reflects a watershed moment in academic literature and global climate politics and that the school strike movement is important, this is problematic. For instance, one could argue that the pivotal year might be 2016 and that the watershed moment was opposition to the Dakota Access Pipeline at Standing Rock. This movement was started by young people, and "spread like wildfire across Turtle Island and the world, moving millions to rise up, speak out and take action" (Estes and Dhillon, 2019, p. 1). The focus on 2018 in the literature also leads to a focus on Greta Thunberg's strike action and work with international organizations like the UN and not, for example, Kathy Jetñil-Kijiner's comparable activism in earlier years (Jetñil-Kijiner, 2014). We are reviewing academic literature, and we agree that academic literature tends to focus on 2018 and beyond. Nevertheless, we recognize that young activism itself is much more heterogeneous in its nature, more expansive in its vision, and has deeper roots than 2018. We discuss the gaps in existing literature later in this chapter.

This review focuses on recent work to make an important contribution to a burgeoning field of international scholarship on young people's climate activism. The current surge of

DOI: 10.4324/9781003396567-18

academic interest in young people's climate activism is inspired by a youth-led, youth-centered approach among activists. We outline recent methodological trends in studying young people's climate activism, detail some of the challenges that arise when studying young people, and review the existing literature and its major themes. Using our review, we indicate significant gaps in this growing literature, which are: a focus on mostly activism by mostly White youth in the Global North, which is separated from other literature on activism in the Global South and in marginalized communities in the Global North; an overemphasis on mass mobilizations; and, stemming from her public visibility in the years around 2018, a hyperfocus on Greta Thunberg. We discuss these themes and challenges and conclude by identifying future opportunities for this vital, international field of research.

Methods for Conducting Literature Review

To account for the various ways scholars have been studying young climate activists, we conducted a literature review. We were interested in gaining a broad understanding of the way people studied young climate activists *and* what they were saying about them, so we used the keyword terms "youth climate strikes", "Fridays for Future", and "youth climate activism". We first decided to exclude any paper that was not published in a peer reviewed journal. There were no geographical limitations on our search, but we only included papers written or translated in English. We recognize the limitations of this approach, but in our search, we came across fewer than five articles that had to be excluded for this reason. We then focused on articles that detailed young climate activism from 2018 and on. Despite the previously acknowledged complexities of considering 2018 a watershed moment for youth climate activism, we decided on 2018 because it was the year that the climate strikes began. Additionally, after doing a precursory literature review, we recognized that 2018 was the year that publications on young climate activism began to increase. We excluded papers that lacked empirical data or that used empirical data but did not pertain to young people engaged in climate activism. We defined climate activism broadly, largely relying on what the researcher considered to be activism, but we excluded activism that was not specific to climate change. Using these exclusion categories, we did a close read of 51 articles that discussed young climate activism after 2018.

We then coded these 51 articles, first by the methods that the authors employed and then by the articles main themes. In coding for methodological approaches, we used the codes "qualitative", "quantitative", "case study", "mixed methods", "text analysis", and "other" which included articles that were primarily literature reviews or more theoretical in nature. This coding informed how we organized the following section describing the methods used for studying young climate activists. In coding for themes, we used the codes "composition of the movement", "reasons for mobilizing", "outcomes of activism", "how young people define and act on climate change", and "perceptions of youth activism". If an article addressed one or more of these components, it was coded accordingly. We used this information to organize the key themes of the literature we reviewed on young climate activists.

Methods for Studying Young People's Climate Activism

The following section explores the various methodologies being used to study young people's climate activism. Of the 51 papers we identified, nearly all of them are research conducted by adults *on* young people (with notable exceptions such as Luna and Mearman, 2020; Mucha et al., 2020; Navne and Skovdal, 2021, discussed further later). Instead, we advocate for an epistemological shift that recognizes young people's claims to knowledge, expertise, and research itself: in other words, more research *in partnership with* young people. We agree with Cutter-Mackenzie and Rousell who call for the recognition of young people's "political agency, creativity and theoretical acumen as legitimate researchers" in the study of climate change (2018, p. 90). Of course, there are significant challenges in partnering with youth on knowledge co-production, including ethical issues, adultism, and the pressure for "fast research" in the neoliberal academy. We detail these challenges further later. Nevertheless, this shift in methodological approach would have important effects on how adult researchers analyze youth climate strikes. Before clarifying *why* this shift matters, it is first important to unpack the ways scholars have been studying the youth climate strikes. While scholars have taken a variety of methodological approaches, very few build in opportunities for collaboration with the young people whom they are studying. This is likely connected to both adultism, by which adults hold power over young people, as well as the practical challenges of conducting participatory work with young people. We highlight these issues further later.

Protest and Online Surveys

One of the more common approaches to documenting the climate strikes, especially on a large scale, is through surveying young climate activists at the climate strikes – or afterward, online. This data provides us with an important baseline understanding of who participated in these strikes. For example, Gaborit (2020) surveyed 1,800 young climate strikers in France, using an approach developed by Quantité Critique. With this methodology, a large number of trained researchers administer questionnaires randomly at these protest events. Wahlström et al. (2019) and de Moor et al. (2020) used a similar methodology in cities across Europe. Called "Caught in the act of protest: contextualizing contestation" or CCC, these scholars use a "pointer" who systematically directs researchers to select participants for surveys and interviews (Walgrave et al., 2016). This data serves as a starting point for scholars who wish to take a more fine-grained approach to studying youth climate strikes.

In addition to quantifying the demographics of these strikes, other scholars have collected quantitative data to parse out motivations and environmental knowledge of the strikers. In an attempt to unpack what drives "pro-environmental activism" of young people, Wallis and Loy (2021) distributed an online survey to 125 young people at a FFF protest in a German city and an additional 418 young people aged 13–26 through an online portal. Similarly, to develop a theoretically driven understanding of youths' motivations to participate in youth climate strikes, Brügger et al. (2020) conducted a cross-sectional online survey that included both strikers and nonstrikers. With 4,057 survey responses, they argue that motivation is tied to activists' feeling as if they are engaging with other like-minded individuals.

One can find an interesting example of collaborative quantitative data collective in Mucha et al. (2020) on resilience and digital protests during COVID-19. Working *with* members of the #FridaysForFuture movement, these scholars co-designed and implemented a survey that sought to unpack the effects of the COVID-19 pandemic on the movement's communication, organization, and mobilization. This kind of collaborative work, where scholars work *with* those who they study to co-create knowledge, should be encouraged across methodological boundaries. It also allows for a more accurate understanding of the activism itself, as it takes these young people's voices, perspectives, and expertise seriously.

Text Analysis

One of the most common ways that scholars have attempted to make sense of the youth climate strikes is through text analysis. As an emergent body of research, text analysis proved useful in producing research relatively quickly, as the data was readily available. Also, text analysis provided available data sources when COVID-19 restrictions limited other data collection methods. Nevertheless, there are epistemological limitations of text analysis, namely that it tends to exclude youth by using adult-centered concepts to analyze mostly adult-oriented texts, producing research outputs that are mostly for the perusal of adults. The text analysis in this literature focuses on one of three sources: speeches given by Greta Thunberg, news coverage of the strikes, or social media posts from climate activists.

Several studies have examined the youth climate movement by analyzing the speeches of prominent youth climate activist Greta Thunberg. For example, using thematic analysis, Holmberg and Alvinius (2020) analyzed five of Greta's speeches and two important climate-related international documents to examine the foci of children's resistance. While it is a response piece, Evensen (2019) also draws on Thunberg's speeches to critique the rhetorical approaches of the #FridaysforFuture movement. Using Thunberg's speeches as a proxy for the messaging of the #FridaysforFuture movement, the author suggests that youth climate activists as a whole need to move beyond the rhetoric of "trust the science" towards more normative, ethical reasonings for tackling the climate crisis.

Studies drawing from newspaper articles are often focused on the way media *frames* the youth climate strikers. For example, Von Zabern and Tulloch (2021) employed rhetorical analysis of media coverage of #FridaysforFuture from three German news outlets. They found seven different kinds of frames used by these news outlets and that the coverage often reproduces existing power structures through marginalizing and depoliticizing the political agenda of protests. Similarly, Bergmann and Ossewaard (2020) analyze German newspapers using critical discourse analysis of one left-wing and one right-wing newspaper in Germany. The authors found that both newspapers primarily use ageist language to delegitimize the youth climate activists.

The final approach to textual analysis uses more diverse materials, including newspaper articles, tweets, social media postings, interviews, and other online materials. For example, Huttunen and Albrecht (2021) analyzed news and social media about Fridays For Future in Finland. Similarly, Kenis (2021) analyzes the way the school strikes in Belgium are framed on websites, social media, television, newspapers, and magazines. Framing their work as a digital ethnography, Wielk and Standlee (2021) used the Twitter feeds of seven prominent climate activists to parse out how young climate activists use social media to create and direct movement communities. A clear limitation of the study, mentioned by the authors in the conclusion, is that this study only represents the perspectives, experiences, and tactics of seven (mostly White, mostly American) youth climate activists. Dealing with a similar challenge, Han and Ahn (2020) analyzed various speeches, interviews, declarations, and online communications from the youth climate movement broadly. The challenge with this kind of approach to studying youth climate activists is that it makes sweeping assumptions about a transnational movement using textual evidence from a relatively narrow set of sources, which the authors themselves acknowledge in conclusion.

There are clear advantages and benefits to textual analysis. In addition to speed and accessibility of data amidst a pandemic, textual analysis can be a way of engaging with youth on their own terms by examining youth-produced media, documenting how youth circumvent mainstream media to shape their own narratives (ex. Huttunen and Albrecht, 2021; Wielk and Standlee, 2021). Textual analysis also offers the opportunity to critique adultist or otherwise problematic portrayals of young activists (ex. Bergmann and Ossewaard, 2020; Huttunen and Albrecht, 2021) At the same time, textual analysis can reinforce incorrect assumptions about the homogeneity of young people (Pickard, 2019) and "view youth as isolated individuals, neglecting the role of adults and communities" (Wood, 2020, p. 217) It can also use adult perceptions to explain the youth climate strikers, rather than young people's experiences and ideas. We ask that scholars who employ this method take care to ensure these data do not lead to overly simplistic claims about young climate activists and consider ways to include young people in this kind of data collection and analysis.

Qualitative Methods

In recent years, there has been an uptick in qualitative scholarship that explores the youth climate strikes. While qualitative data provides its own set of challenges, particularly regarding sample size and representation, these studies give new dimensions to what can otherwise be a flat or reduced description of youth climate activists. This work provides important "thick description" of the experiences and perspectives of these activists (Geertz, 2008).

Focusing on young activists in the UK who were involved in either the school strikes and/ or a group called the Extinction Rebellion, Pickard et al. (2020) draw from a large sample of sixty semi-structured interviews carried out before, during, and after protest actions to try to examine how youth themselves understand their activism. Elsen and Ord (2021) draw from seven in-depth interviews with youth climate activists affiliated with the Fridays For Future strikes. This work explores young people's attitudes toward adult involvement within "youth-led" climate groups. Cattell (2021) used ten in-depth interviews with young climate activists to unpack the ways regular involvement in climate activism shapes their imagined futures. While brief in length, Martiskainen et al. (2020) conducted 10–20-minute interviews with 64 youth activists participating in school strikes across six cities. The interviews– conducted after the strikes had already occurred– add some additional depth to the point-in-time surveys done at the strikes.

Not all scholars who engaged with qualitative methodology drew exclusively from interview data. Haugestad et al. (2021) employed mixed methods in their study of #Fridaysfor-Future protests. Using ethnographic fieldwork, 93 interviews, and survey data from 362 Norwegian high school students, the authors explore the predictive possibility of collective guilt, environmental threat, past protest participation, organized environmentalism, political orientation, and social capital on future protest intentions. In addition to 20 semi-structured interviews with students in Quebec, Dupuis-Deri (2021) draws from letters from school principals addressed to parents and pieces of media to document the way schools can be a place of political conflicts and struggles between students and adults and between students on different sides of the movement. Kettunen (2020) draws upon 4 months of participant observation and 47 interviews with young people between the ages of 15 and 16 to unpack the different ways young people practice environmental politics and construct their environmental citizenship. While not all 47 of her respondents claimed to be active in climate activism, the data set makes an interesting comparison between young people in Finland who did and did not strike. The rich qualitative data collected by these authors allows the youth voice to be an essential part of the overall analysis.

It is important to note two qualitative papers that offer examples for what deeply collaborative knowledge co-production with youth can look like. The articles Luna and Mearman (2020) and Navne and Skovdal (2021) are co-written with individual young activists who serve as lead author in both instances. Both articles include first-person narrative written by the activist and largely unedited, mixed with commentary by both authors. While there are important ethical considerations in such approaches, addressed further later, these papers elucidate methods that legitimately challenge the norms of adult-centered knowledge production.

Earlier, we detailed the different kinds of methodologies implemented by scholars who study young climate activists. These methodologies range from event surveys to textual analysis, to interviews, and other ethnographic approaches, as well as mixed methods. While all of these studies offer valuable information on various components of the climate strikes, we argue that there needs to be deeper consideration of *how* we go about studying young people. First, however, we detail the core themes that this literature has to offer, to set the stage for a discussion of the intimate relationship between what we know about young climate activists and how we know it.

Key Themes in the Literature

While the previous section examines *how* we know about young people's climate activism– the methodologies used to study it– we will now address *what* we know about it. This section explores the various findings available within the research. It is grouped by the various themes we identified within the literature. These themes are: composition of young people engaged in climate activism; pathways into climate activism; outcomes of young people's climate activism; and how young people are understanding and acting on climate change.

Composition of Young People Engaged in Climate Activism

Many scholars have examined the demographics of young climate activists. In Europe, two large quantitative studies gathered data on youth participating in Fridays for the Future protests in March and September of 2019. Findings were similar across the two studies: the biggest age cohort was between 14 and19 years old, and the median age was 21 years old (Wahlstrom et al., 2019; de Moor et al., 2020). Participants are more heavily female, and over 70% come from well-educated families, defined as having at least one parent with a college degree (Ibid).

A similar study found that participants in the climate strikes in the United States are predominately White, female, and from well-educated families (Fisher and Nasrin, 2021a). Interestingly, this data shows a steady increase in the age of participants over time; in the spring of 2019, the median age was 18, then 25 by fall 2019, and 32 by the spring of 2020. In this short time span, adult-led organizations (the Sierra Club and 350) increasingly took leadership roles, squeezing out the leadership of youth-led organizations (Fisher and Nasrin, 2021b). Lorenzoni et al. (2021) also found an increase in older participants over time. This trend of increased adult involvement could lead to co-opting and "watering down" or de-radicalizing young people's climate activism, as research shows that older activists are less radical than their younger counterparts (Bertuzzi, 2019; Elsen and Ord, 2021). One future direction for research is to examine this trend across time and in other geographic locations.

How Young People Come into Climate Action

Researchers have worked to identify the underlying motivations young people have for becoming climate activists. Governmental inaction and the sense that politicians are failing them is a common source of motivation; this is true among youth climate strikers in the UK (Elsen and Ord, 2021; Feldman, 2021; Pickard et al., 2020) and is a common narrative in the Fridays for the Future movement as a whole (Han and Ahn, 2020). Haugestad et al. (2021) and Martiskainen et al. (2020) found a variety of motivations for participation among climate strikers in Canada, the USA, the UK (Martiskainen et al., 2020), and Norway (Haugestad et al., 2021). Both found that concern about the environment and environmental threats and engagement with and desire to affect politics were key motivators. Additional motivators for protest participation included concern for family and future generations; being part of a protest movement; solidarity; anti-capitalism; security; collective guilt; past protest participation; the presence of organized environmentalism; and social capital (Haugestad et al., 2021; Martiskainen et al., 2020). While the aforementioned research examines why youth mobilized, Fisher (2015) discusses the process of coming into activism. Fisher found that many young climate activists have some sort of transformative moment that compels them to commit to activism. For many, this was a moment of seeing environmental destruction. Many youth also held concerns for social justice and nature that proceeded and gave enhanced meaning to their transformative moments. Although sparked by particular moments, youth describe climate activism as a dynamic and ongoing process of continuously recommitting to activism.

Outcomes of Young People's Climate Activism

Research also examines the outcomes of youth activism, both for the youth themselves– how activism impacted their lives– as well as its bigger political ramifications. Regarding the outcomes for youth, many of the findings are not particularly surprising: youth who participated in Fridays for the Future held less exploitative and more preservationist views of nature than students who did not participate (Barbosa et al., 2021). Additionally, students who engaged in activism developed self-efficacy, empowerment, and optimism (Deisenrieder et al., 2020; Cattell, 2021; Elsen and Ord, 2021). Climate activism also enhanced their understanding of climate change, both how it impacts daily life as well as the politics of it (Elsen and Ord, 2021; MacKay et al., 2020). Last, research found that climate activism opened up new opportunities for youth, such as social connections (Elsen and Ord, 2021; MacKay et al., 2020).

While it is important to know that climate activism has positive outcomes for youth, research on the impacts of activism for youth falls into a paradigm that views youth political actors as "citizens-in-training" (Gordon, 2007; Taft, 2015) by viewing activism as a learning experience instead of legitimate political action. Doing so requires that researchers pay more attention to the political impacts of youth climate activism.

And indeed, Fisher and Nasrin (2021a) have made the call to focus on political outcomes. Despite the importance of this call, we could locate only two papers of the 51 identified that address this. Han and Anh (2020) conclude that the youth climate movement has successfully changed the public discourse around climate change and brought attention to global inertia. They have also successfully united disparate groups like students, teachers, unions, and environmentalists. Despite this, the youth climate movement has been unable to garner enough power to enact the systemic change necessary. Kenis (2021) came to similar conclusions, finding that the youth climate movement has succeeded in re-politicizing the issue, thus putting it on the public agenda. However, through "empty demands" (p. 135), Kenis claims that the movement has made itself vulnerable to neutralization and co-option by hegemonic forces.

How Youth Understand Climate Change

A significant portion of the literature on youth climate activism addresses how youth are understanding climate change. Much of this literature suggests that youth view climate change in more radical terms. Pickard et al. (2020) document how youth view climate change as an urgent crisis with potential for catastrophic impacts, thus requiring a radical rebuilding of society. Piispa and Kiilakoski (2021) and Terren and Soler-i-Marti (2021) found that youth view climate change as an issue of justice. Additionally, youth are linking climate change to capitalism (Holmberg and Alvinius, 2020; Pickard et al., 2020) and are willing to put the needs of the environment above those of the economy (Emilsson et al., 2020).

At the same time, young people are not a monolith and not all espouse radical perspectives on climate change– Dupuis-Deri (2021) found a split between more moderate versus radical-left youth within one activist organization. Similarly, Gaborit (2020) also found variance in terms of radicalism, and youth who were further left were more willing to critique capitalism. Piispa and Kiilakoski (2021) found that youth coalesced around the notion of "moderation" and "slowing down", while Huttunen (2021) found that youth favored more responsive politicians over a deeper political revolution– not particularly radical ideas.

Han and Ahn (2020) explored the narrative structure of youth activists' understanding of climate change. In this narrative, the villains (past generations, state leaders, and the media) have passed the burden onto the victims (future generations and the earth) by failing to act. Faced with the gross inaction and negligence of the villains, the victims will not remain mere victims. Instead, they become heroes by seeking justice through collective action. Young people will hold the villains accountable pushing for systemic transformation.

Taken together, this literature paints a picture of youth with more radical and justice-oriented understandings of climate change. Because of this, Bowman (2020) argues that youth climate strikes are typically analyzed through the frames of mainstream environmentalism and engagement approaches but that these frames are inappropriate. Instead, he suggests using Pulido's framing of subaltern environmentalism, distinguished from mainstream environmentalism by economic and social marginalization, by lives structured by domination, and by the centrality of the issue of positionality (Pulido, 1996). Bowman points out Pulido's claim that subaltern environmentalism is characterized by the struggle for environmental justice (Pulido, 1996) and that climate justice is, indeed, central to the demands of young movements like Fridays for Future (2019).

How Youth Are Acting on Climate Change

Literature on youth climate activism also discusses how youth are acting on climate change. Given their more radical understandings of the issue, it is not surprising that much of the literature shows that youth are also taking more radical approaches to action. In particular, three papers found that youth tend to favor collective action and civil disobedience when it comes to climate action (Gaborit, 2020; Kenis, 2021; Pickard et al., 2020). This action tends to be disruptive, but is rooted in non-violence, peace and joy (Pickard et al., 2020). Kenis (2021) comments on how youth activists are using politicized tactics, which have successfully brought climate change to the forefront of public conversations. Overall, the manifestations of activism are largely shaped by availability– youth tend to adopt the politics and tactics of what is already happening in their geographic location (Prendergast et al., 2021; Rainsford and Saunders, 2021).

Just as understandings of climate change differ among youth activists, so do their ideas for action. O'Brien et al. (2018) created a typology that describes the various types of dissent represented by youth climate activists. They categorize youth dissent as dutiful, disruptive, or dangerous. Dutiful dissent represents more normative and non-disruptive approaches, whereas disruptive and dangerous dissent are more radical, disruptive, and non-normative. Similarly, research with French youth climate strikers also found strong support for civil disobedience but differing willingness to engage in it and differing views on the use of violence (Gaborit, 2020).

In regards to youth activism, several scholars (Pickard, 2019; Kettunen, 2020; Dupuis-Déri, 2021) note that youth activists often operate outside of traditional structures, in what Pickard (2019) terms "Do-It-Ourselves" activism. In this, youth "act politically without relying on traditional collective structures, such as political parties and trade unions to inform, organise and mobilise in a top down way" (Pickard, 2019, p. 5). Do-It-Ourselves activism occurs through individual lifestyle politics but also when youth come together to act collectively, such as in the climate strikes.

Another theme regarding how youth are acting on climate change examines the constraints put on activism. Kettunen (2020) found that both inertia among peers and friction between youth and adults created barriers to participation in activism. Dupuis-Déri (2021) also found that, when operating in a school context, adult authority figures often had the final say in decision-making. Nakabuye et al. (2020) talk about the unique challenges faced by African youth activists; these include limited formal climate change and lack of support, especially when demands oppose community economic needs from industries like logging and mining.

In regards to tactics, both Curnow et al. (2021) and Foran et al. (2017) examine how youth use humor, sarcasm, and irony in their activism. Curnow et al. (2021) examine a divestment campaign at the University of Toronto, where activists used what they termed snark– a type of humor that relies on sarcasm, irony, anger, and self-deprecation– to develop identities, express rage, and build agency and solidarity. Foran et al. (2017) studied youth activists who attended COP-19. They found that these youth aimed their activism at pointing out hypocrisy and failures of politicians, often through humorous actions like a lemonade stand to fund climate adaptation. It is likely, given ongoing studies of young people's participation in formal organizations (such as Thew et al., 2020), that more studies of this type are likely to emerge over the next few years.

Finally, two papers examine the use of social media as an organizing tool. Wielk and Standlee (2021) found that youth engaged in coalition building over social media by projecting activist identities, creating narratives that attract and engage other activists and engaging in political conversation using evocative emotional narratives to build connections that resonate. Another paper found that social media was used to create group cohesion and emotional attachment (Brünker et al., 2019).

Discussion

Gaps in the Literature and Opportunities for Expansion

While the past 3 years have yielded important work on young people's climate activism, it is a new body of work, and there are significant gaps and limitations. In the following section, we identify opportunities for expansion.

Moving Beyond White Activism

The first opportunity for expanding on existing work is the gap in the current literature on climate activism in what can be called the Global South and in marginalized communities across the world. More precisely, we not only identify this gap but also a preponderance of data and theory that focuses on the Global North and largely on wealthy and White communities. More deeply, we find a general literature on young activism tends to focus on wealthy, White activism and that research on and with young people from other backgrounds tends to form another separate body of literature. A focus on wealthy, White activism is no surprise: Curnow and Helferty (2018) argue that mainstream environmentalism is a "default white space" and call for an "ethic of accountability" in environmentalism. In current research, most available data focuses exclusively on young people in Europe, the United States, Australia, and New Zealand. This is particularly notable when it comes to examining the composition of young people engaging in climate activism– there have been several large quantitative studies capturing protest participation across Europe (Wahlström et al., 2019; de Moor et al., 2020) and some in the United States (Fisher and Nasrin, 2021b). While these studies are important in establishing the demographics of these movements, the literature neglects participation in most of the world. This paints youth climate activism as centered in wealthy, mostly White countries, when in fact it is a global phenomenon. We argue that researchers must be accountable not only to the unjust impacts of environmental damage but also the unjust distribution of research and knowledge. Thus, one key area for further research is examining the youth-led climate activism in the global South. And indeed, scholars and news media alike have documented youth climate activism in places like the Pacific Islands (Cocco-Klein and Mauger, 2018; Hayward et al., 2020), India (Singh, 2015), and African nations (Nakabuye et al., 2020). However, in centering activism in the Global North, researchers perpetuate misguided notions about who cares about and acts on the environment (Taylor, 1997). The centering of literature on the Global North also impedes full exploration of young people's activism in global, regional, and national climate regimes where "Black lives matter less" (Sealey-Huggins, 2018, p. 102) and in contexts where researchers' direct experiences of climate change may lead them to write with more reflective and autobiographical approaches to research (see Sultana, 2022).

This gap is also notable when it comes to more nuanced topics, like motivations for activism and how youth are understanding and acting on climate change. This is problematic, as there are significant disparities in how climate change is experienced in the Global North and South. While the Global North is responsible for the vast majority of carbon emissions, the Global South will experience the worst impacts of climate change (Shi et al., 2016). Additionally, countries in the Global South are disproportionately exploited for fossil fuel resources and experience a higher level of marginality when it comes to global politics, and young people in the Global South have historically been identified as the "worst affected victims of resource degradation and environmental pollution" (Bajracharya, 1994, p. 41). Thus, there is great opportunity to expand research to include youth in the Global South.

The work done in the Global North can also be expanded. Within the Global North, there are also populations with different relationships to climate change. Groups who already experience marginality in the Global North– communities of color, Indigenous communities, low-income communities, disabled populations, immigrant and migrant communities, and LGBTQ+ communities– are also disproportionately impacted by climate change and fossil fuel extraction, despite typically contributing fewer greenhouse gas emissions (Kaijser and Kronsell, 2013; Routledge et al., 2018). Thus, researchers in the Global North should examine how climate activism is understood and performed differently across lines of race, class, gender identity, sexual orientation, Indigeneity, immigration status, and ability.

More Than Mass Mobilizations

Another key limitation of the current literature is the overemphasis of mass mobilizations. Much of the literature only examines young people engaged in the Fridays for the Future walkouts. Indeed, these protests are an important aspect of youth climate activism, as they are perhaps the most visible elements of this movement and are key in catalyzing action. However, these single-day events are only one element of the climate activism youth are involved in today. In focusing so heavily on them in the literature, research is neglecting the varied activities that youth are engaging in to demonstrate dissent and act for change. This presents a fairly flat and limited image of young people's climate activism. Researchers have documented these varied activities to a degree, such as work capturing everyday dissent (O'Brien et al., 2018), the campus movement for fossil fuel divestment (Curnow et al., 2021), after-school clubs (Dupuis-Déri, 2021) and participation and/or dissent through formalized politics (Foran et al., 2017; MacKay et al., 2020). This work should be expanded. Especially important is capturing localized struggles, such as campaigns against fossil fuel extraction, as this work is often done in low-income communities of color. Thus, in overemphasizing mass mobilizations, research is overemphasizing the work of White and middle-class youth. One important direction is long-term ethnographic work with youth movements, which will be able to present a more nuanced image of youth climate activism and the activities that go along with it.

Moving Beyond Greta Thunberg

Finally, while reviewing this literature, we couldn't help but notice a common trend: a hyperfocus on Swedish climate activist Greta Thunberg. Indeed, invoking Greta Thunberg and her activities that sparked the Fridays for the Future movement seemed a near-ubiquitous way to introduce papers on youth climate activism. However, this framing seriously limits the literature by constraining the true scope and scale of youth climate activism. It obfuscates activities that were happening before Greta's action and outside of her sway. It incorrectly assumes that Greta was the inspiration for much of these activities, when in fact, many youth were already gravely concerned about and taking action on climate change (Cocco-Klein and Mauger, 2018; Fisher, 2015; Foran et al., 2017; Ison, 2009; Lee et al., 2020; O'Brien et al., 2018). Perhaps more problematic than just the "Greta framing" is the "Greta methodology"– using speeches or actions by Greta Thunberg to make broad generalizations about the young people's climate activism. Drawing on the perspectives and actions of a single individual to generalize a global phenomenon constrains an understanding of this movement. To be clear, this critique is not aimed at Thunberg herself– indeed, we celebrate her bold actions, her ability to catalyze a movement, and the important work she is doing to hold those in power accountable– but rather the ways researchers have constrained this literature by hyperfocusing on an individual. Thus, as researchers continue to examine youth climate activism, an important opportunity is to look beyond a singular hero to the varied ways groups and youth are acting on climate change.

Challenges of Studying the Youth-led Movement

In addition to the gaps in the literature named earlier, we also draw attention to the unique challenges that scholars face when studying young people. These challenges exist not only for those studying young climate activists but for scholars studying young people across a variety of contexts. The contested meaning of the category of "youth", practical barriers to studying young people, and adultism in the research process must be considered if scholars of the climate strikes wish to engage in reflexive research ethics (Cordner et al., 2012). We first name and acknowledge these challenges and conclude this chapter with some tentative solutions and possible ways forward.

The Social Construction of "Youth"

One primary challenge of studying young people's activism lies in the way we define the concept of "youth". The terms used to describe young people vary greatly across and within disciplines. The concept of youth is socially constructed (Bourdieu, 1984) and therefore has different meanings in different contexts. This means youth is not a universal category, and neither does it mean the same thing across all social contexts.

While scholars have long acknowledged the social construction of categories like race (Omi and Winant, 1986) and gender (Butler, 1990; Lorber and Farrell, 1991) to name a few, there has been less scholarship that draws attention to the social construction of age (Gordon, 2007). Instead, young people are often described using universal, essentialist categories that reduce "adolescence", "young adulthood", and "childhood" into one monolithic experience. Additionally, tracing the history of the terms adolescent, teenager, and youth, Pickard (2019) highlights that these terms are laden with normative assumptions and are often linked to deviance and delinquency by both scholars and the mainstream media.

These challenges are beyond simple definitional squabbles. Instead, they prose significant methodological problems that need to be made clear. The definition used by the researcher can impact the outcome of the study and the methods chosen. If we continue to rely on essentialist ideas of the concept of "youth", we risk replicating these ideas and, as a result, contributing to an adultist perspective – explained further later – in our own work. In her work on "do-it-yourself democracy", Pickard (2019) provides us with a way forward for dealing with all this complexity. She claims,

> There is however no clear-cut or absolute definition of "young people" and this might go some way to explaining why saying who actually counts as young people is so often omitted in studies on their relationships with politics. But one thing is clear: it is important not to generalize and not to lump all young people together as if they

constituted in some objective way a homogenous demographic collective. . . . Young people have different backgrounds, attitudes and interests; they have diverse political orientations, as well as varying types and levels of participation in politics. At the same time, young people are neither trainee citizens nor just numbers in the dataset I prefer to consider young people as full citizens capable of reasoned thinking with agency.

(27)

In line with this, we argue that scholars studying young climate activists need to be more explicit about the variety that exists across members of this broader movement. To study young people in a way that avoids this essentialism, researchers need to acknowledge the heterogeneity of young people around the world. This means making space for complexity while being clear about the scope of research. This also means *talking to young people* about the definitions researchers use and the boundaries they set around who is and who isn't "youth" or what is or isn't "youth activism".

Practical Challenges of Studying Young People

In addition to the challenges around defining youth, there are also practical challenges to studying young people. For important reasons, young people are often given protected status in the context of research. For example, in the United States, all people under the age of 18 are considered minors, and researchers are usually required to receive parental consent before speaking with these young people. To add to the complexity, these rules change globally.

While these protections are important, parental consent can create problems, like the possibility of forced disclosure between a young person and their parent. Think for example, about a study on young people who identify as LGBTQ+. If a researcher must gain parental consent to speak with any person under the age of 18, that means a researcher could accidentally "out" a young person to their parents by identifying them as a potential participant in this kind of research. While the stakes for young climate activists are different than the example provided above, it *is* possible that young people may not want their parents to know that they skipped school to participate in a climate strike. The challenge of gaining consent, then, becomes one we need to think deeply about when conducting research on young people (Woodgate et al., 2017; Moore et al., 2018). We share these complexities because researchers across the globe who wish to study young people deal with these questions: how do we gain consent and include the voices and perspectives of young people within the confines of important but challenging regulations around researching this population?

Adultism and the Research Process

While other researchers have documented the way adultism plays out in youth-led climate activism (Biswas and Mattheis, 2021; Ritchie, 2021), we argue that one must also be aware of the ways adultism can operate within the research process. Adultism is a bias that regards adults as superior to young people and enables adults to hold power over young people (Checkoway, 1996). Like any kind of systemic inequity, adultism is not relegated to one arena, like between students and teachers. Instead, adultism is structured into the fabric of our world, and as a result, it can appear in the research process in a variety of ways. It can be reinforced in the way researchers define the category of "youth", plan methodologies, or write up and share findings.

In order to challenge this, it is important researchers examine the way scholarship can intentionally or unintentionally reinforce adultism. This often happens through analytical frames. For example, looking at the activism of young people broadly, Earl et al. (2017) draw attention to the often-used "youth deficit model" where scholars make the adultist assumption that young people are not engaged with politics and need adults to do so. This also manifests as the assumption that young people's activism is preparatory work and not legitimate political action (Gordon, 2007; Taft, 2015). We argue that, instead, scholars should consider young people to be "active learners and decision-makers, who have distinct interests and have intersectional identities that include being young [and] need to be studied with agency in mind" (Earl et al., 2017, p. 8). In a similar vein, Biswas and Mattheis (2021) ask observers of young climate strikes to employ a "childist perspective" that takes them seriously as a political action and a form of civil disobedience, rather than labeling the activism as "truancy". While Elsen and Ord (2021) claim that adult activists need to be more aware of the structural power they hold in youth-centered spaces, we argue, along with the scholars cited earlier, that this power imbalance extends to the researchers who are studying young climate activists as well. Being aware of the realities of adultism is an important first step in uprooting it.

A key way to start challenging adultism in research lies in the call we made towards the beginning of this chapter: for the participatory co-production of knowledge alongside young people. As discussed earlier, Luna and Mearman (2020) and Navne and Skovdal (2021) both pose models of deeply collaborative co-research with youth. This, too, comes with ethical challenges, such as the complexities of adults obtaining both consent and assent from youth themselves, given the prominence of adultism. Luna and Mearman (2020) detailed the process of doing so, which included regularly checking in with the young person that they wished to continue the partnership. Furthermore, participatory research is time-consuming for young people and academics alike. On the one hand, this asks more of young activists, who are already juggling multiple demands on their time, such as school, work, and their activism –often without remuneration or direct benefit to them. This is also a challenge for researchers, who operate within neoliberal institutions that place demands for "fast publication". This makes it difficult to devote the time to participatory research.

And indeed, scholars have found that even when adults are aware of and actively trying to uproot adultism, it does not erase all possible challenges. For example, drawing from her research on the Peruvian movement of working children, Jessica Taft (2015) claims that even for well-meaning adults who recognize the agency and authority of young people, the structural power imbalance between adults and young people is difficult to avoid. This reminds us that even with awareness, adultism can creep into spaces where people are actively trying to develop youth-centered methods. Both Luna and Mearman (2020) and Navne and Skovdal (2021) note this possibility even while doing participatory research. In this review, we argue that youth-led and "childist" approaches would not only be beneficial in responding to the opportunities provided by existing literature, but also represent a likely future turn in the academic literature as researchers seek to better understand the youth-centered, youth-led climate action movements that continue to act around the world.

Conclusion

In this review, we explored the development of a growing field of academic literature on young people's climate activism since 2018. We identify that although 2018 does not necessarily mark a watershed moment in activism itself – and that the roots of young people's environmentalist activism run much deeper than this – 2018 can be accurately described as a watershed moment in the study of climate activism and, we suggest, in the ways that adults, including academic researchers, consider young activism. More specifically, we identify a surge of academic work is inspired by a youth-led, youth-centered approach among activists.

The recent methodological trends in studying young people's climate activism offer many opportunities for future research. Among these, we identified the need for research to move beyond a continuing tendency to focus on mass movements among young people and particularly movements of young people in the Global North and in movements where young people are comparatively rich and disproportionately White. We also identify a tendency to focus on the activism of the influential – and now iconic – young activist Greta Thunberg.

We conclude our review with the hope that our work will support future scholars. We have provided detail on the body of work that already exists on young people's climate activism, as well as gaps in the literature where future studies may make a positive contribution. We have also identified several methodological and ethical challenges that arise from the existing literature. Among these, we celebrate a continuing shift towards youth-led and youth-centered methods, including participatory research, as well as youth-centered theory. The responsibility for leading this shift sits not just with researchers, who themselves operate within constraints such as limitations of time, funding, and academic norms. Instead, we hope to contribute to a re-imagining of the systems that produce knowledge. This includes more funding for youth-centered research and increased accountability to youth in making decisions about funding – and the same when it comes to editorial boards.

The young climate activist Brianna Fruean teaches us that "real education sometimes happens outside of the classroom. I think the school climate strikes have really proven that" (Healy, 2020, p. 179). Speaking as adult researchers, we hope that young people's action on climate change can continue to educate us about the concepts, forms, and practices of activism so that we can better uphold young people's voices and more effectively understand and contribute to their action on climate change.

Note

1 A previous version of this article was published in Frontiers in Political Science.

References

Bajracharya, D. (1994). Primary environmental care for sustainable livelihood: A UNICEF perspective. *Childhood*, 2(1–2), 41–55.

Barbosa, R. A., Randler, C. and Robaina, V. (2021). Values and environmental knowledge of student participants of climate strikes: A comparative perspective between Brazil and Germany. *Sustainability*, 13(14), 8010.

Bergmann, Z. and Ossewaarde, R. (2020). Youth climate activists meet environmental governance: Ageist depictions of the FFF movement and Greta Thunberg in German newspaper coverage. *Journal of Multicultural Discourses*, 15(3), 267–290.

Bertuzzi, N. (2019). Political generations and the Italian environmental movement(s): Innovative youth activism and the permanence of collective actors. *American Behavioral Scientist*, 63(11), 1556–1577.

Biswas, T. and Mattheis, N. (2021). Strikingly educational: A child's perspective on children's civil disobedience for climate justice. *Educational Philosophy and Theory*, 54(2), 145–157.

Bourdieu, P. (1984). La jeuness n'est qu'un mot. In P. Bourdieu (ed.) *Questions de Sociologie* (pp. 143–54). Paris France: Les Editions de Minuit.

Bowman, B. (2020). 'They don't quite understand the importance of what we're doing today': The young people's climate strikes as subaltern activism. *Sustain Earth*, 3, 16.

Brügger, A., Gubler, M., Steentjes, K. and Capstick, S. B. (2020). Social identity and risk perception explain participation in the swiss youth climate strikes. *Sustainability*, 12(24), 10605.

Brünker, F., Deitelhoff, F. and Mirbabaie, M. (2019). Collective identity formation on Instagram – Investigating the social movement Fridays for Future. In *Proceedings of the 30th Australasian Conference on Information Systems*. Perth, Australia.

Butler, J. (1990). *Gender Trouble: Feminism and the Subversion of Identity*. New York, NY: Routledge.

Cattell, J. (2021). "Change is coming": Imagined futures, optimism and pessimism among youth climate protesters. *Canadian Journal of Family and Youth/Le Journal Canadien de Famille et de la Jeunesse*, 13(1), 1–17.

Checkoway, B. (1996). Adults as allies. *Partnerships/Community*, 38. https://digitalcommons.unomaha.edu/slcepartnerships/38

Cocco-Klein, S. and Mauger, B. (2018). Children's leadership on climate change: What can we learn from child-led initiatives in the U.S. and the Pacific islands? *Children Youth Environ*, 28, 90–103.

Cordner, A., Ciplet, D., Brown, P. and Morello-Frosch, R. (2012). Reflexive research ethics for environmental health and justice: Academics and movement building. *Social Movement Studies*, 11(2), 161–176.

Curnow, J. and Helferty, A. (2018). Contradictions of solidarity: Whiteness, settler coloniality, and the mainstream environmental movement. *Environment and Society*, 9, 145–163.

Curnow, J., Fernandes, T., Dunphy, S. and Asher, L. (2021). Pedagogies of Snark: Learning through righteous, riotous rage in the youth climate movement. *Gender and Education*, 33, 8.

Cutter-Mackenzie, A. and Rousell, D. (2018). Education for what? Shaping the field of climate change education with children and young people as co-researchers. *Children's Geographies*, 17(1), 90–104.

Deisenrieder, V., Kubisch, S., Keller, L. and Stötter, J. (2020). Bridging the action gap by democratizing climate change education – The case of kidZ 21 in the context of Fridays for future. *Sustainability*, 12(5), 1748.

de Moor, J., Uba, K., Wahlström, M., Wennerhag, M. and De Vydt, M (Eds.). (2020). *Protest for a Future II: Composition, Mobilization and Motives of the Participants in Fridays for Future Climate Protests on 20–27 September, 2019, in 19 Cities Around the World*. https://osf.io/yg9k2

Dupuis-Déri, F. (2021). Youth strike for climate: Resistance of school administrations, conflicts among students, and legitimacy of autonomous civil disobedience – the case of Québec. *A previous version of this article was published in Frontiers in Political Science*, 27.

Earl, J., Maher, T. and Elliott, T. (2017). Youth, activism, and social movements. *Sociology Compass*, 11(4), 1–14.

Elsen, F. and Ord, J. (2021). Enabling empowerment: The role of adults in 'youth led' climate groups. *A previous version of this article was published in Frontiers in Political Science*, 3.

Emilsson, K., Johansson, H. and Wennerhag, M. (2020). Frame disputes or frame consensus? "environment" or "welfare" first amongst climate strike protesters. *Sustainability*, 12(3), 882.

Estes, N. and Dhillon, J. (2019). The Black Snake, #NoDAPL and the rise of a people's movement. In N. Estes and J. Dhillon (eds.) *Standing with Standing Rock: Voices from the #NoDAPL Movement* (pp. 1–12). Minneapolis: Minnesota University Press.

Evensen, D. (2019). The rhetorical limitations of the #FridaysForFuture movement. *Nature Climate Change*, 9, 428–430.

Feldman, H. (2021). Motivators of participation and non-participation in youth environmental protests. *A previous version of this article was published in Frontiers in Political Science*, 3, 1–16.

Fisher, D. and Nasrin, S. (2021a). Climate activism and its effects. *Wiley Interdisciplinary Reviews. Climate Change*, 12(1). https://doi.org/10.1002/wcc.683

Fisher, D. and Nasrin, S. (2021b). Shifting coalitions within the youth climate movement in the US. *Politics and Governance*, 9(2), 112–123.

Fisher, S. R. (2015). Life trajectories of youth committing to climate activism. *Environmental Education Research*, 22(2), 229–247.

Foran, J., Gray, S. and Grosse, C. (2017). "Not yet the end of the world": Political cultures of opposition and creation in the global youth climate justice movement. *Interface: A Journal for and about Social Movements*, 9(2), 353–379.

Fridays for Future. (2019). *Lausanne Climate Declaration: Official Version*. https://drive.google.com/file/d/1Nu8i3BoX7jrdZVeKPQShRycI8j6hvwC0/view

Gaborit, M. (2020). Disobeying in time of disaster: Radicalism in the French climate mobilizations. *Youth and Globalization*, 2(2), 232–250.

Geertz, C. (2008). Thick description: Toward an interpretive theory of culture. In T. Oakes and P. L. Price (eds.) *The Cultural Geography Reader* (pp. 41–51). London: Routledge.

Gordon, H. R. (2007). Allies within and without: How adolescent activists conceptualize ageism and navigate adult power in youth social movements. *Journal of Contemporary Ethnography*, 36(6), 631–668.

Han, H. and Ahn, S. W. (2020). Youth mobilization to stop global climate change: Narratives and impact. *Sustainability*, 12(10), 4127.

Haugestad, C. A., Skauge, A. D., Kunst, J. R. and Power, S. A. (2021). Why do youth participate in climate activism? A mixed-methods investigation of the# FridaysForFuture climate protests. *Journal of Environmental Psychology*, 76, 101647.

Hayward, B., Salili, D. H., Tupuana'i, L. L. and Tualamali'i', J. (2020). It's not "too late": Learning from Pacific Small Island Developing States in a warming world. *WIREs Climate Change*, 11, e612.

Healy, H. (2020). 'Real education happens outside the classroom'?: Pacific climate warrior Brianna Fruean and Anna Taylor of the UK school strikes movement talk about what inspires them and how to avoid activist burnout. *FORUM*, 62(2), 179.

Holmberg, A. and Alvinius, A. (2020). Children's protest in relation to the climate emergency: A qualitative study on a new form of resistance promoting political and social change. *Childhood*, 27, 78–92.

Huttunen, J. (2021). Young rebels who do not want a revolution: The non-participatory preferences of Fridays for Future activists in Finland. *A previous version of this article was published in Frontiers in Political Science*, 3, 1–11.

Huttunen, J. and Albrecht, E. (2021). The framing of environmental citizenship and youth participation in the Fridays for future movement in Finland. *Fennia*, 199, 1.

Ison, N. (2009). At the coal face in Australia: The youth climate movement. In G. Wilson, P. Furniss and R. Kimbowa (eds.) *Environment, Development and Sustainability: Perspectives and Cases from Around the World* (pp. 65–75). Oxford: Oxford University Press.

Jetñil-Kijiner, K. (2014). *Dear Matafele Peinem*. https://www.kathyjetnilkijiner.com/united-nations-climate-summit-opening-ceremony-my-poem-to-my-daughter/

Kaijser, A. and Kronsell, A. (2013). Climate change through the lens of intersectionality. *Environmental Politics*, 23(3), 417–433.

Kenis, A. (2021). Clashing tactics, clashing generations: The politics of the school strikes for climate in Belgium. *Politics and Governance*, 9(2), 135–145.

Kettunen, M. (2020). "We need to make our voices heard": Claiming space for young people's everyday environmental politics in northern Finland. *Nordia Geographical Publications*, 49(5), 32–48.

Lee, K., Gjersoe, N., O'Neil, S. and Barnett, J. (2020). Youth perception of climate change: A narrative synthesis. *WIREs Climate Change*, 11, 1–24.

Lorber, J. and Farrell, S. A. (1991). *The Social Construction of Gender*. Newbury Park, CA: Sage.

Lorenzoni, J., Monsch, G. and Rosset, J. (2021). Challenging climate strikers' youthfulness: The evolution of the generational gap in environmental attitudes since 1999. *A previous version of this article was published in Frontiers in Political Science*, 3, 1–13.

Luna, E. and Mearman, A. (2020). Learning to rebel. *Sustain Earth*, 3, 4.

MacKay, M., Parlee, B. and Karsgaard, C. (2020). Youth engagement in climate change action: Case study on Indigenous youth at COP24. *Sustainability*, 12(16), 6299.

Martiskainen, M., Axon, S., Sovacool, B., Sareen, S., Furszyfer Del Rio, D. and Axon, K. (2020). Contextualizing climate justice activism: Knowledge, emotions, motivations, and actions among climate strikers in six cities. *Global Environmental Change*, 65, 1–18.

Moore, T. P., McArthur, M. and Noble-Carr, D. (2018). More a marathon than a hurdle: Towards children's informed consent in a study on safety. *Qualitative Research*, 18(1), 88–107.

Mucha, W., Soßdorf, A., Ferschinger, L. and Burgi, V. (2020). Fridays For future meets citizen science. Resilience and digital protests in times of Covid-19. *Voluntaris*, 8(2), 261–277.

Nakabuye, H., Nirere, S. and Oladosu, A. T. (2020). Chapter 24: The Fridays for Future movement in Uganda and Nigeria. In C. Henry, J. Rockström and N. Stern (eds.) *Standing Up for a Sustainable World: Voices of Change* (pp. 212–219). Cheltenham: Edward Elgar Publishing.

Navne, D. E. and Skovdal, M. (2021). 'Small steps and small wins' in young people's everyday climate crisis activism. *Children's Geographies*, 19(3), 309–316.

O'Brien, K., Selboe, E. and Browyn, M. H. (2018). Exploring youth activism on climate change: Dutiful, disruptive, and dangerous dissent. *Ecology and Society*, 23(3).

Omi, M. and Winant, H. (1986). *Racial Formations in the United States: From the 1960s to the 1980s*. New York, NY: Routledge and Kegan Paul.

Pickard, S. (2019). Young environmental activists are doing it themselves. *Political Insight*, 10(4), 4–7.

Pickard, S., Bowman, B. and Arya, D. (2020). "We are radical in our kindness": The political socialisation, motivations, demands and protest actions of young environmental activists in Britain. *Youth and Globalization*, 2(2), 251–280.

Piispa, M. and Kiilakoski, T. (2021). Towards climate justice? Young climate activists in Finland on fairness and moderation. *Journal of Youth Studies*, 1–16.

Prendergast, K., Hayward, B., Aoyagi, M., Burningham, K., Hasan, M., Jackson, T., Jha, V., Kuroki, L., Loukianov, A., Mattar, H., Schudel, I., Venn, S. and Yoshida, A. (2021). Youth attitudes and participation in climate protest: An international cities comparison. *A previous version of this article was published in Frontiers in Political Science*, 3, 1–18.

Pulido, L. (1996). *Environmentalism and Economic Justice: Two Chicano Struggles in the Southwest*. University of Arizona Press.

Rainsford, E. and Saunders, C. (2021). Young climate protesters' mobilization availability: Climate marches and school strikes compared. *A previous version of this article was published in Frontiers in Political Science*, 3, 1–12.

Ritchie, J. (2021). Movement from the margins to global recognition: Climate change activism by young people and in particular Indigenous youth. *International Studies in Sociology of Education*, 30, 53–72.

Routledge, P., Cumbers, A. and Driscoll Derickson, K. (2018). States of just transition: Realizing climate justice through and against the state. *Geoforum*, 88, 78–86.

Sealey-Huggins, L. (2018). The climate crisis is a racist crisis: Structural racism, inequality and climate change. In A. Johnson, R. Joseph-Salisbury and B. Kamunge (eds.) *The Fire Now: Anti-Racist Scholarship in Times of Explicit Racial Violence* (pp. 99–116). London: Bloomsbury.

Shi, L., Chu, E., Anguelovski, I., Aylett, A., Debats, J., Goh, K., Schenk, T., Seto, K. C., Dodman, D., Roberts, D., Roberts, J. T. and VanDeveer, S. D. (2016). Roadmap towards justice in urban climate adaptation research. *Nature Climate Change*, 6, 131–137.

Singh, K. (2015). Not politics of the usual: Youth environmental movements. In S. Batabyal (ed.) *Environment, Politics and Activism* (pp. 170–197). Routledge India.

Sultana, F. (2022). The unbearable heaviness of climate coloniality. *Political Geography*. https://doi.org/10.1016/j.polgeo.2022.102638

Taft, J. (2015). "Adults talk too much": Intergenerational dialogue and power in the Peruvian movement of working children. *Childhood*, 22(4), 460–473.

Taylor, D. E. (1997). Women of color, environmental justice, and ecofeminism. In K. Warren (ed.) *Ecofeminism: Women, Culture, Nature* (pp. 38–81). Bloomington, IN: Indiana University Press.

Terren, L. and Soler-i-Marti, R. (2021). "Glocal" and transversal engagement in youth social movements: A Twitter-based case study of Fridays for Future-Barcelona. *A previous version of this article was published in Frontiers in Political Science*. 3, 1–15.

Thew, H., Middlemiss, L. and Paavola, J. (2020). "Youth is not a political position": Exploring justice claims-making in the UN climate change negotiations. *Global Environmental Change*, 61, 102036.

von Zabern, L. and Tulloch, C. D. (2021). Rebel with a cause: The framing of climate change and intergenerational justice in the German press treatment of the Fridays for Future protests. *Media, Culture and Society*, 43(1), 23–47.

Wahlström, M., Kocyba, P., De Vydt, M. and de Moor, J. (eds.) (2019). *Protest for a Future: Composition, Mobilization and Motives of the Participants in Fridays for Future Climate Protests on 15 March, 2019 in 13 European Cities*. https://osf.io/xcnzh/

Walgrave, S., Wouters, R. and Ketelaars, P. (2016). Response problems in the protest survey DESIGN: Evidence from fifty-one protest events in seven countries. *Mobilization: An International Quarterly*, 21(1), 83–104.

Walker, C. and Bowman, B. (2022). Climate activism. In H. Montgomery (ed.) *Oxford Bibliographies in Childhood Studies*. New York: Oxford University Press.

Wallis, H. and Loy, L. S. (2021). What drives pro-environmental activism of young people? A survey study on the Fridays for Future movement. *Journal of Environmental Psychology*, 74, 101581.

Wielk, E. and Standlee, A. (2021). Fighting for their future: An exploratory study of online community building in the youth climate change movement. *Qualitative Sociology Review*, 17(2), 22–37.

Wood, B. (2020). Youth-led climate strikes: Fresh opportunities and enduring challenges for youth research – commentary to Bowman. *Fennia: International Journal of Geography*, 198, 1–2.

Woodgate, R. L., Tennent, P. and Zurba, M. (2017). Navigating ethical challenges in qualitative research with children and youth through sustaining mindful presence. *International Journal of Qualitative Methods*, 16, 1–11.

16
PANDEMIC POSSIBILITIES

The Corona Crisis as Perceived Opportunity and Threat for Climate Activists

Lotte Schack and Mattias Wahlström

Introduction

In the beginning of 2020, at the apex of the thus-far largest global wave of climate protests ever, its momentum was disrupted by the disastrous COVID-19 pandemic. While it is clear that the climate movement became strongly affected, the actual character of this impact appears contradictory. The restrictions on gatherings imposed by many national governments to contain the virus severely constrained movement activities. However, among climate movement intellectuals and organisers, the far-reaching and prompt actions by governments across the world have been used as proof that global joint political action against planetary crises is possible. In this chapter, we understand the 'Corona crisis' as a *disruptive event* that broke everyday habits, thereby creating potential for thoroughgoing societal change and at the same time leading to increased costs of collective action. We explore not only the Corona crisis' effect on the climate movement but also what this tells us about the relationship between grassroots climate activism and crises *in general*. The COVID-19 pandemic is unlikely to be the last health crisis the movement will have to face and other catastrophic events will undoubtedly give rise to both threats and opportunities for climate movements. Analysing the effects of this specific crisis can thus be useful for understanding how other crises will affect the movement as well.

To disentangle the effects of the Corona crisis on the climate movement, this chapter provides an overview of existing literature complemented by an analysis of empirical data on activists in Finland and Sweden. The experiences and reflections in relation to the pandemic by climate activists in these two Global North welfare states do not exhaust those of climate activists globally. However, certain experiences of the pandemic, such as being exposed to some form of state lockdown and fear of being infected, were arguably common to a large share of the global population. Sweden and Finland have previously been comparatively spared from epidemic outbreaks (Torres Munguía et al., 2022), which also makes the pandemic stand out even more.

Focusing on experiences of individual activists instead of organisations or protest events allows us to explore the diverse impact of a major crisis within a movement as well as the meanings that such events have to activists. Using a panel survey conducted in late 2020

DOI: 10.4324/9781003396567-19

and late 2021 among participants in a previous survey round of climate strike demonstrators in September 2019, as well as interviews and observations conducted with Swedish climate activists in 2021, we analyse activists' impressions of the pandemic's effects on the climate struggle and the possibilities for bringing about a fossil-free future.

The chapter is structured as follows. We first theorise the capacity of crises to generate opportunities, as well as posing challenges, for political change. Then, we offer an overview of the existing literature on the pandemic's effects on activism in general and the climate movement in particular. We consider these discussions in relation to data collected with Swedish and Finnish climate activists during and immediately after Corona restrictions were in place. Activists expressed some optimism regarding the potential of the pandemic in showing both how governments are able to implement rapid large-scale changes and that individuals are able to make transitions in their everyday lives. However, the pandemic placed serious constraints on climate activism, thus weakening the 'affective infrastructure' of the movement.

Crisis, Rupture, Event

There is a long-standing acknowledgement in the social movement literature about the sometimes-crucial role played by major events, especially in triggering mobilisation and/or opening opportunities for social change. In historical sociology, it is common to speak of *transformative events*, which are typically defined in post-hoc fashion as those events that can afterwards be shown to have led to social change (Sewell, 1996). A problem with such outcome-dependent definitions is that they provide little guidance for studying phenomena as they happen. An alternative concept is *critical junctures* (Capoccia and Kelemen, 2007; Soifer, 2012; della Porta, 2022), which captures moments when existing social structures are destabilised, which may or may not result in substantial change. However, a major event that radically changes conditions for political mobilisation is something more than a juncture; it is also an occurrence which is widely perceived as standing out from the everyday (Basta, 2018). Some major events (wars, natural disasters, economic crises) not only stand out but also disrupt the taken-for-granted and flow of everyday activities – a *disruption of the quotidian* – which Snow et al. (1998) argue may open up novel possibilities. While such ruptures can have very real and direct effects, they must be interpreted and made sense of in order to become events (Wagner-Pacifici, 2010). Events are therefore inherently socially constructed, even though they may be composed of ever-so-tangible occurrences (Basta, 2018).

Whether conceptualised as transformative events or disruptions of the quotidian, the main thrust within social movement research has been to analyse events as game-changers that facilitate mobilisation. However, major events may also lead to demobilisation, as in the case of movement defeats (Voss, 1996). In the case of the climate movement such an event is exemplified by the COP15 climate summit in Copenhagen, which became widely constructed as a failure and led to a temporary demobilisation among activists (de Moor and Wahlström, 2019; Hadden, 2017). Furthermore, defeat is not the only way in which crises may inhibit mobilisation. Mann and Wainwright (2018) argue that the ecological crisis is in itself a crisis of political imagination. Herbert (2021) argues that climate activists have problems articulating 'imaginative politics', as they find themselves in a 'crisis lock-in' in which the social imaginary has been colonised by 'prevailing norms' (p. 378). This leads activists to have a dualistic view of the present and the future, reserving the former for oppositional struggles and the latter for formulating solutions and alternatives.

In contrast, Ergen and Suckert (2021) argue that crises have the potential to reshape imagined future trajectories. Similar to Snow and colleagues (1998), they define crises as 'exogenous interruptions of routine' which while creating uncertainty also opens up potential for doing things differently. However, this potential is dependent on discursive struggles over the meaning of the crisis. Capturing both the open-ended character of events, as well as their potential for (de)mobilisation, we argue that *disruptive event* is the most neutral yet precise term to conceptualise the Corona pandemic.

The Impact of the Pandemic on Climate Politics

For several progressive intellectuals, the breakout of the pandemic presented a potential to rethink our societies and our way of living. In a commentary from April 2020, Roy (2020) articulated it as seeing the pandemic as a portal. The pandemic, she argued, has along with the death and chaos it has caused also managed to bring the world to a halt, thereby creating a rupture. We should seize this as an opportunity to rethink the organisation of society and instead imagine a different world. Fraser (2021) pointed out how the pandemic has highlighted how we currently find ourselves in 'an epochal crisis', both ecologically but also of 'economy, society and public health' (p. 95). Bhattacharya (2020) similarly argued that the lack of nurses during the pandemic has shown how capitalism constantly jeopardises lives in the pursuit of profit, leading her to state that 'we cannot go back to 'business as usual".

In an interview, Greta Thunberg pointed out that young people had taken responsibility to protect the elderly and that the elderly consequently ought to take responsibility for protecting the younger generations in the context of the climate crisis (Schück, 2021). In contrast, Malm (2020) argues that the responses to the pandemic should not become a blueprint for handling the climate crisis. Despite the harsh measures states introduced, Malm argues that these were exclusively mitigating the effects of the pandemic rather than addressing its roots, failing to prevent more virus outbreaks. Similarly, George Monbiot argued that we should consider the climate-related causes of the pandemic, saying that in contrast to 2008, we should bail out the planet, not the banks (openDemocracy, 2021).

Somehow consistent with the early disagreement on how to interpret the Corona crisis, its consequences for social movements were not uniform (della Porta, 2022). Restrictions or bans on public gatherings created very immediate challenges for movements. In practice, this effect appears to have varied widely both depending on governmental policies, national laws, as well as the character of different political mobilisations. For example, Pressman and Choi-Fitzpatrick (2020) found only limited (early) effects of the pandemic on protest levels in the US, but this can be explained by both the mixed messages from the federal level (read: President Trump) as well as the fact that some mobilisations were specifically about resistance to COVID-19 lockdowns (Kowalewski, 2020). The large number of Black Lives Matter protests in the US and in the rest of the world during 2020 is also a counterexample to the effect of Corona restrictions. Kriesi and Oana (2022), in a study of protest levels in Europe, conclude there was a reduction in the size of protests during the pandemic, whereas the number of protest events was less affected.

Activists also shifted their activism towards mutual aid for citizens affected by the pandemic. While several forms of care initiatives appear to have been taken in civil society in most European countries, the shift in focus by existing civil society groups might have been less common in what Esping Andersen (1990) would term the Nordic welfare regime, compared to the liberal and continental (family-centred) welfare regimes, in which more care responsibility is already from the start located with individuals and their families, for example in Italy (Bosi and Lavizzari, 2023; Esu and Dessì, 2022), Spain (Diz et al., 2023), and the UK (Chevée, 2022), or in contexts of generalised limited social and health security such as the US (Krinsky and Caldwell, 2022). Arampatzi and colleagues (2022) show how, in Athens, the solidarity movement took the form of an *infrastructure*, serving as an alternative to the state's crisis management. Similarly, della Porta and Lavizzari (2022) argue that, during the pandemic, *care* appeared as a bridging meta-frame between different movement concerns, connecting issues such as health rights, the climate, and feminism.

In the case of the climate movement, protests were in many cases cancelled or scaled down, and the protests staged by Fridays For Future (FFF) activists generally complied with pandemic restrictions. Indirectly, the various restrictions and dramatic increases in state expenditure for dealing with the pandemic and its consequences could be interpreted as demonstrating state output capacity under exceptional circumstances. In theoretical terms, this was for some a perceived opening of *political opportunity structures* (POS) for the climate movement (McAdam, 1999; de Moor and Wahlström, 2022). This also concerns the international POS, since international governing bodies and powerful NGOs in many respects demonstrated their capability to introduce radical measures. However, in some respects the pandemic also created challenges for such organisations, including the World Health Organization and the EU, when nation states prioritised their own interests before the international community.

An indirect impact of the pandemic concerns the competition for attention in public arenas between different social problems (Hilgartner and Bosk, 1988). Based on the assumption that there is a limit to how much attention can be paid to social problems at any point in time within an arena (news media, parliamentary politics, etc.), the introduction of a new acute problem is bound to reduce the space for attention to other problems within the arena. The pandemic as a social problem can thereby be regarded as a competitor to the climate issue for public deliberation and concern, in theoretical terms creating a more unfavourable *discursive opportunity structure* (Koopmans and Statham, 1999). While media focus on the pandemic may indeed have had a detrimental effect on attention to climate problems, there has also been *frame bridging* activities in public discourse, linking the climate issue to the pandemic in various ways (Stoddart et al., 2023).

Balancing the possibly reduced discursive opportunities caused by the pandemic, successful connections to the climate issue may have opened up new opportunities for action. Broad changes in lifestyle – most notably limited air travelling – were perceived as indicating that the population indeed can change its habits. Such lifestyle changes – and the importance of not reverting to the ways things were before the pandemic – appear to have influenced future-oriented narratives of climate activists – their *prognostic framing* (Benford and Snow, 2000). Christou and colleagues (2022) show how young Cypriot climate activists attempted to connect the pandemic with the climate crisis by reframing the latter as a 'slow pandemic' (ibid. 11). On one hand, the activists reframed adult ideas of linear temporality through seeing the pandemic not as a step back but as a pathway to a different future. On the other hand, the lack of access to public space forced them to find alternative ways of being present in the public sphere.

No Going Back to Normal: Crisis as Opportunity?

We use the two Nordic countries Finland and Sweden as empirical examples to explore the impact of the pandemic on climate activists. While the restrictions in Finland were similar to those in other European countries, Sweden represents a somewhat different case. Some restrictions on gatherings were imposed, such as limiting the number of attendants at certain events and the amount of people allowed to gather in both public and private spaces. However, Swedish restrictions were relatively relaxed compared with other countries and were to a large part recommendations. Most of the Swedish climate movement, however, strictly followed the regulations, limiting their activities even beyond the state's requirements. Consequently, the climate movement in both countries experienced a dramatic decrease in movement activities.

The analysis builds on data collected 2020–2021. One source is a panel survey disseminated in two waves (September–October 2020 and 2021) to former participants in FFF protests in Sweden and Finland.[1] To further explore activist experiences of the pandemic, we build on interviews with Swedish climate activists conducted during and after Corona restrictions were in place. In addition, we participated in online events of the movement in 2021. The activists of the study come from a range of climate activist groups with the majority active in FFF and the Swedish Society for Nature Conservation (SSNC).

Like the ambiguity described in the previous section, the climate activists in this study were conflicted about the outcomes of the pandemic for the climate struggle. The respondents in the panel survey were cautiously positive as regards the overall effect. In response to the question 'In the long run, do you expect that the Corona crisis will have made it easier or harder for society to sufficiently reduce carbon emissions?' In the 2020 wave, 59% of the respondents answered 'slightly easier' or 'much easier', while only 9% responded 'slightly harder' or 'much harder'. As can be seen in Figure 16.1, the pattern was nearly identical between the countries (the minor differences were not statistically significant).

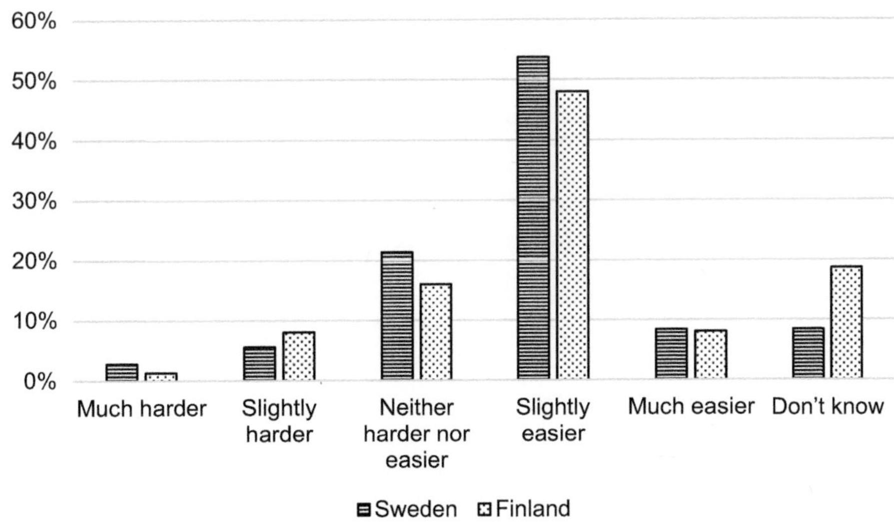

Figure 16.1   Response by country to the question 'In the long run, do you expect that the Corona crisis will have made it easier or harder for society to sufficiently reduce carbon emissions?'

Source: Original Creation

Many respondents took the opportunity to elaborate their response in an open follow-up question. While new, more sustainable habits and developing green technology were regarded as facilitating necessary societal shifts, one of the mentioned countervailing factors was a displacement of the climate issue since all focus had been directed at dealing with the pandemic.

> I believe that society has seen that change can be implemented. . . . People got to see what the world could look like if factories and the like were shut down/got renewable production methods (see China and how the air became healthier when everything was shut down). People were also forced to change their habits, which I hope led to more sustainable changes. At the same time, a lot of focus was taken from other issues when Corona came. The climate crisis had finally started to be addressed for real, but everything was slowed down and forgotten with Corona.
>
> *(Emma, Swedish survey respondent)*

While some of the more hopeful comments emphasised the new and more sustainable habits that people had become accustomed to during the pandemic, others were wary that the 'new normal' would give way to just the 'normal' when the pandemic subsided.

> I feel like people are unwilling or unable to see the connection between the Covid-19 outbreak and global warming. They are waiting for things to return back to 'normal' as stubbornly as a child waiting for Santa. How can you hammer into a whole huge mass's head that things aren't going to be 'normal' because that [sic] 'normal' is unsustainable and unrealistic?
>
> *(Hanna, Finnish survey respondent)*

While the previous question was not posed again in 2021, it is possible to trace the general trend among survey respondents with regards to their hope in the state capacity to address climate change. The respondents to the protest surveys in 2019, as well as the follow-up panels in 2020 and 2021, were all asked to rate their degree of agreement with the statement: 'I feel hopeful about policies being able to address climate change.' For the sake of simplicity, all Swedish and Finnish respondents are counted together in the following diagram (we only include the 2019 respondents who subsequently participated in the panel). We interpret the higher proportion of respondents agreeing 'somewhat' in 2020 as an increased level of doubt caused by the pandemic. However, in 2021, respondents appear to have acquired a generally more pessimistic outlook regarding hope for policy changes (see Figure 16.2).

Perceptions of the ambiguities of the pandemic situation were also reflected in the survey respondents' answers to an open question posed in 2020: 'What lessons do you think the climate movement should draw from the Corona crisis?' Overall, the responses can be categorised into on the one hand observations regarding *opportunities and obstacles* for the climate struggle and on the other hand *tactics and strategies* for the climate movement. Turning first to opportunities and obstacles, these can in turn be divided into those of a general kind (revealed by the pandemic) and those specifically connected to the pandemic itself (see Table 16.1).

The most frequently mentioned general opportunity perceived by the respondents is the potential for change shown by both states and populations in the face of the pandemic.

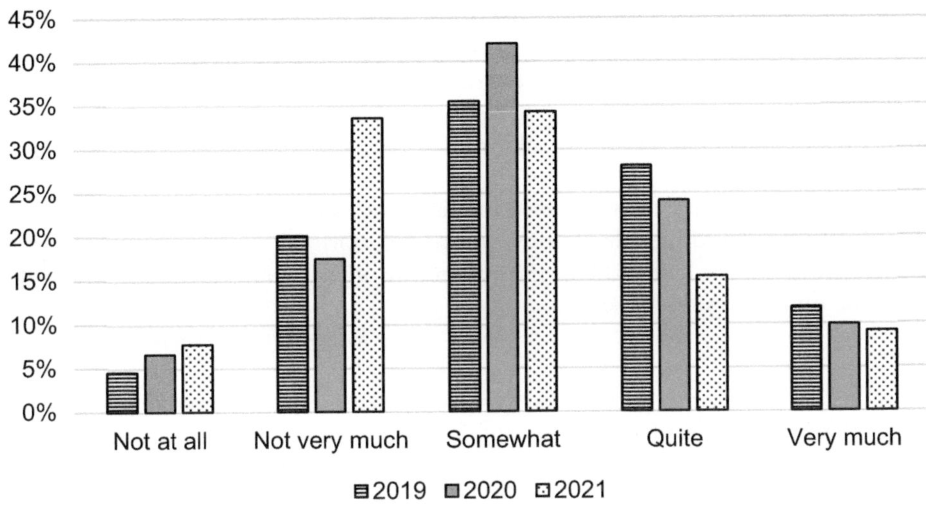

Figure 16.2 Agreement (by year) with the statement 'I feel hopeful about policies being able to address climate change'.

Source: Original creation

Table 16.1 Types of lessons drawn from the Corona crisis about opportunities and obstacles.

	Opportunities	*Obstacles*
General	Political opportunity structures – there is money and state capacity for change People demonstrate that they can change	People are slow to react and easily return to habits Individual lifestyle decisions have a limited impact Economic interests dominate
Specific	Window of opportunity for change – people out of the 'normal'	Climate issue pushed off the agenda

Source: Original creation.

As regards states, this concerns both input and output opportunities, that is, politicians' preparedness to take radical action and states' legal and economic capacity to implement drastic measures (de Moor and Wahlström, 2022). Similarly, many respondents perceived that the climate movement should learn from the fact that the general population has been able to swiftly adapt to new conditions.

> Both the climate movement, the public and politicians have been told in black and white that in less than 1 year, you can change resource allocation, economic priorities and introduce new rules and routines in society based on the Corona pandemic. You can of course also do it to save the planet and the ability of humans and animals to continue living on our planet.
>
> *(Cilla, Swedish survey respondent)*

While many respondents appeared to regard this as a general potential of society revealed by the crisis, others regarded the Corona crisis more as a specific window of opportunity during which various actors in society are more than usually open to change. Implicitly, the opportunities for change may not always be present but are connected to the period of the pandemic, what Snow et al. (1998) would term a 'disruption of the quotidian'.

There is an opportunity for more effective climate communication with the corona crisis: on the one hand, people are now out of the 'normal' anyway and now may be a good time to embrace other changes, and on the other, people have seen what life is like in the midst of major crises.

(Kaisa, Finnish survey respondent)

The interviewed Swedish activists likewise emphasised how the pandemic had shown that radical restructuring of society is possible, contrary to the usual responses to the climate movement's demands. The lifestyle changes brought about by the pandemic were also mentioned by interviewed Swedish FFF activist Edith, who pointed to how people had found joy in climate-friendly activities such as gardening and how some people might have experienced solidarity with their neighbours for the first time in their lives. However, she underlined that it was hard to communicate good things about the pandemic, stressing the need to communicate that although similar large-scale restructuring is needed to address the climate crisis, this would allow for more joyful lives than the Corona restrictions had.

Similarly, where some respondents in the panel study primarily saw opportunities, others identified obstacles that they thought the climate movement should be aware of. This both included politicians' fixation with economic growth as well as ordinary people's unwillingness to change their lifestyles.

The majority of people can't react to a threat unless it's right in front of their face. Even a deadly illness is too far unless someone in your immediate family dies because of it. Scientists can tell people how to act all they want, but most will hang on to the last remnants of their 'normal' a.k.a. the lifestyle they're used to until you have to pull it from their cold, dead hands.

(Hanna, Finnish survey respondent)

The survey respondents also perceived obstacles specific to the pandemic, primarily that the movement needed to struggle to push the climate issue back onto the agenda, after having been sidelined by Corona-related problems.

The last-mentioned topic also relates to the more *tactically and strategically* oriented takeaways by respondents from the Corona crisis. These concerned both the form and content of activism. In terms of *form*, several respondents offered reflections about how to deal with the lockdown and the disadvantage of being prevented from (large) physical assemblies. One proposed solution was more small-scale protest forms:

The climate movement has been dependent on being able to meet physically in public places, it is a big part of the strategy; it creates social pressure and shows that many people think this is important, creates momentum. The corona crisis has made much of this invisible and many have been left alone again. XR in the UK has worked with affinity groups, which has not been done in Sweden. I think this is a big mistake when it comes to making climate movement sustainable in the long term.

(Britta, Swedish survey respondent)

Some respondents urged for further elaboration of online protests to make an impact in the absence of physical protests. Others saw the developments in these respects as already reflecting a broadening of protest forms, which could be advantageous for the future. As regards *content* of upcoming activism, several respondents pointed to class inequalities as well as those resulting from racism or the victimisation of the pandemic. Drawing parallels with the climate struggle, these respondents saw a need to integrate these aspects in the continued climate struggle.

I experience that the climate movement misses how we all are affected differently in crises, something that was noticed in Sweden with, for example, that Somali Swedes and people who are worse off socio-economically were hit harder by Corona due to, for instance, cramped housing, work that cannot be done from home, etc. I think an example from the climate movement is Extinction Rebellion, which has begun to link its activism with Black Lives Matter to show that the climate affects people in other parts of the world which are not majority white.

(Anna, Swedish survey respondent)

A number of respondents also brought up the issue of main targets for the climate movement. While there was a widespread perception (with some disagreement, as we noted earlier) that the Corona crisis had indeed demonstrated the adaptability of the population, several respondents emphasised the central role of the state. First, because the state was seen as having had a crucial role in changing popular behaviours; second, because the Corona crisis was perceived as proving that changes in lifestyles would be far from sufficient to halt or mitigate global warming.

That the reductions in, among other things, carbon dioxide emissions that took place (at most about 17%) required overall political decisions – no individual measures will suffice – we must influence entire structures!

(Erik, Swedish survey respondent)

In summary, whereas respondents acknowledged that the Corona crisis had posed challenges for the climate struggle in various ways, many also saw opportunities. However, such opportunities did not equal unqualified optimism. In our interviews with Swedish climate activists, the constraints placed on movement activities by the Corona restrictions overshadowed the possible opportunities the crisis presented. In the next section, we discuss how the constraints affected the movement, arguing that the pandemic weakened its 'affective infrastructure'.

Affective Infrastructures

One significant challenge presented by the pandemic was the constraints it placed on physical protest activities. Among the participants in the panel survey 2020, 19% had remained active in offline protests, while 45% had shifted to online protest and 36% had stopped participating in climate protest altogether. All the activists that we interviewed in-depth had moved their activities online. While some of them, as described in the previous section, emphasised how the pandemic had shown possibilities for both structural and individual lifestyle changes, the negative impact of the pandemic on group activities took a central place, especially for the younger interviewed activists. Karin from FFF explained that she initially joined the group because she felt alone with her climate anxiety:

> Standing in front of great, dangerous problems is much easier when you are together with people. And then I feel, no matter what people are saying about future catastrophes, then I have friends who are with me. And then, it becomes more like you can use those feelings [of anxiety] to something positive.

While her engagement with other activists did not make her anxieties about the climate crisis completely disappear, it helped her to transform them into political action. Similarly, other activists emphasised the relationships that had come out of their respective activist groups. Lea, active in both her local chapter of FFF and the international coordination group, described this in the following way:

> A lot of friendships and a lot of relationships have spawned out of FFF. I know several people who, without FFF would have never met or probably not fallen in love because like they would have not spent any time with each other . . . . That's, I think, a positive thing, that you have relationships with each other because also in FFF, there's a lot of people who in general are discriminated against in some way by society. . . . [Here] people are appreciated for who they are.

To Lea, a significant part of being active in FFF is the relationships between activists. We understand these relationships as the *affective infrastructure* of the movement (Dean, 2016; Bosworth, 2023). This term highlights that an important part of movements is not only the actors' political or intellectual affinities but also their affective ones: that is, the personal relationships which develop between activists and the emotional energy (Collins, 1993) activists experience through movement activities. According to Dean, just as much as political visions of a different future sustain the daily work of movements, this daily work also sustains the vision of the future, installing a *practical optimism* in actors, as they feel they are actively partaking in the collective struggle to bring about this future. Similarly, Nunes (2023, p. 2) argues that 'an affective infrastructure locks participants in certain affective patterns, sustains the intensity with which these are felt and thus reproduces a certain relation of the group to itself and to its various out groups'. For the young activists in FFF, group activities involving physical co-presence were central to maintaining this infrastructure and its relationships. Karin described her first experience with FFF at one of the organisation's global strikes in 2019:

> It was the first strike I attended, so I wasn't part of organising it, but there were 4000 people in the streets . . . there were so many people and speeches in the town square, and all traffic in the inner city was stopped for us to march, that was amazing . . . I mean, when you're in the streets, it makes people more attentive. . . . It's fun to be there and people see that you're having fun, that we're singing, yelling slogans and so on. We have large banners and when you see people that haven't even thought of

the climate being confronted with it, then it feels more like it is for real. And you feel, this is reality.

What Karin describes here is not just the community with other activists, but also how the demonstrations create an atmosphere that attracts attention and has the potential to draw more people in. According to her, this was impossible to recreate digitally:

We were going to do it [organise a demonstration] right when the pandemic started. . . . We changed it so no people came except for us, but we asked people to come and leave a pair of shoes or multiple pairs of shoes that were going to stand in the square instead of them, and then we held a few symbolic speeches which we livestreamed. It felt quite symbolic but everything digital feels quite meaningless in this way because you are not going to reach new people, it's more the people who are already following, those who think the same.

Maja also highlighted the way digital activities, such as activists posting pictures of themselves with slogans, were unable to reach outsiders to the movement. For both Maja and Karin however, the pandemic's effects on the activists themselves seemed to be the more pressing issue. Karin described how she felt about her climate anxiety during the pandemic:

During the pandemic, I find it harder to think about climate problems. I mean, I am alone. And even when you meet online, it's not the same. It feels a lot like you're talking to yourself and then someone else talks to themselves.

While the community with like-minded activists had alleviated her climate anxiety, this was not transferable to the online sphere. Maja also expressed concern with the way the move to the digital had impacted activists:

It also affects the health of activists, I think. That we're on Zoom for hours and hours and hours every day, that we have so many Zoom-meetings and nothing is allowed to happen IRL, everything is online. We can't meet each other. . . . Because I mean that's a large part of it, that on Friday strikes we meet other activists and talk with other people who care about the same cause. And that's something that's really helping a lot of people with their climate anxiety. It helps us stay motivated, and that disappears. It's really tough. So there are a lot of people who don't have the energy to engage right now.

In this way, the pandemic was an ambivalent event for the movement. While activists cited in the previous section highlighted the way the pandemic had demonstrated the state's ability to impose large-scale structural changes, as well as people's ability to make individual lifestyle changes, the pandemic had devastating effects on the everyday life of movements. We suggest that the relative absence of movement activities during the pandemic had a negative effect on activists' ability to imagine alternative futures. Returning to Herbert's (2021) argument, activists found themselves in a 'crisis lock-in', but here, it was not caused by the knowledge of the climate crisis but by the physical constraints which the pandemic created that in turn weakened the movement's affective infrastructure.

However, not all activists expressed negative feelings about the move to the digital. Older activists were less concerned about the pandemic's impediment of movement activities, which was evident in our interviews with members from the largest non-governmental environmental organisation in Sweden, the Swedish Society for Nature Conservation (SSNC). They expressed enthusiasm about the digital opportunities they had been introduced to because of the pandemic. In an interview, the activist Bengt emphasised how online meetings had made it more accessible and convenient for people to join and had made them more effective. Similarly, being able to have lectures online made it possible to reach more people. Although he mentioned the lack of social activities during the pandemic, this was not as central to his reflections as it was for the young activists in FFF.

A possible reason for this, we suggest, is that the affective infrastructure of SSNC is organised differently from that of FFF. FFF's infrastructure relies on young activists meeting in the town square every Friday where they can share their thoughts and anxieties about the climate (see also Christou et al., 2022). SSNC's main activities are providing information about the environment and the climate and participating as advisors in municipal environmental decisions. Although the way these activities are done changed, they were not made impossible. As Bengt pointed out, they became largely easier.

While the temporary restrictions on mass demonstrations impeded the activities of FFF, civil disobedience started to proliferate amongst Swedish climate activists towards the end of the period of pandemic restrictions. In addition to XR continuing their activities, the group Återställ Våtmarker (Eng. 'Restore Wetlands') rose to prominence. A part of the international A22 network, the group used civil disobedience, mainly road blockades, to pressure the Swedish government to introduce specific climate measures. While the increasing use of civil disobedience arguably reflects a heightened sense of urgency amongst climate activists, we suggest that it could be explained partly as a side-effect of the pandemic restrictions. General bans on public gatherings, such as those imposed during the pandemic, are likely to deter most people that would otherwise have participated in larger protests while at the same time leading activists to shift to smaller transgressive protests. Wahlström and de Moor (2017) demonstrate this pattern in connection with the December 2015 COP21 in Paris. The terrorist attacks in the city just two weeks prior led the state to impose emergency laws banning all public protests. Those who nevertheless protested were those who were prepared to run the risk of repression and to get involved in civil disobedience or direct action. Similarly, the pandemic lockdown led to a shift towards civil disobedience actions, which in turn arguably facilitated maintaining the affective infrastructure among the groups focusing on transgressive protest.

Conclusion

Despite its deadly consequences, the outbreak of the COVID-19 pandemic was initially seen by many as also presenting political opportunities for climate activism. We have shown how many Swedish and Finnish climate activists initially took a carefully optimistic stance and interpreted the pandemic as shifting society out of 'the normal', thereby opening up possibilities for different ways of life and the large-scale societal restructuring necessary for mitigating catastrophic climate change. However, our panel data demonstrated a somewhat muted optimism in this regard further into the pandemic.

The restrictions the pandemic placed on movement activities weakened the movement's 'affective infrastructure' (Dean, 2016), crucial in sustaining movements. The relationships between activists and the affect-laden everyday experiences of political work are what makes it possible for activists to continue struggling for an alternative future. For the interviewed activists, this was what made it possible for them to mentally engage with the climate crisis. The shift to online activism largely undermined this aspect of political engagement for several young climate activists. They appear to have felt alone with their anxieties about the climate, thus finding it harder to engage with climate activism at all. Still, the pandemic's effect on the affective infrastructure varied between groups. For older activists involved in environmental NGOs the pandemic had less wide-ranging consequences. We suggest that this has to do with the organisation of each group's affective infrastructure.

Towards the end of pandemic restrictions, Sweden saw increased civil disobedience from climate activists. We propose that the restrictions could be a contributing factor. Bans on public gatherings deter broad groups of protesters and might stir a minority of activists in the direction of smaller transgressive protests.

In a heating world, pandemics such as COVID-19 are likely to proliferate. Thus, pandemics are in multiple ways bound up with the trajectory of the climate. Furthermore, the occurrence of other kinds of disruptive events will impact the course of climate politics and activism, as demonstrated by the Russian invasion of Ukraine and the war in Gaza. This makes it necessary to understand how activists navigate these disruptive events. The outcomes of these are rarely uniform but formed in the tensions between the societal context, activists' interpretations of the events, and the movements' affective infrastructures.

Note

1 The participants in the panel survey were all respondents to protest surveys of FFF protests in September 2019 in Helsinki (Finland) and Malmö, Gothenburg, and Stockholm (Sweden). Of the original survey respondents, 58% (183 persons) participated in the first wave of the survey and 49% (154 persons) participated in the second wave. For more details on the original protest survey, see de Moor et al. (2020). In this chapter we focus on the survey responses to open questions about the implications of the pandemic for the climate movement.

References

Arampatzi, A., Hara, K. and Dimitris, P. (2022). Re-thinking solidarity movements as infrastructure during the COVID-19 pandemic crisis: Insights from athens. *Social Movement Studies*, 1–17.

Basta, K. (2018). The social construction of transformative political events. *Comparative Political Studies*, 51(10), 1243–1278.

Benford, R. and Snow, D. (2000). Framing processes and social movements: An overview and assessment. *Annual Review of Sociology*, 26(1), 611–639.

Bhattacharya, T. (2020). Social reproduction theory and why we need it to make sense of the COVID-19 crisis. *URPE* (blog), 2 April. https://urpe.org/2020/04/02/social-reproduction-theory-and-why-we-need-it-to-make-sense-of-the-corona-virus-crisis/

Bosi, L. and Lavizzari, A. (2023). How young activists responded to the first wave of the Covid-19 crisis in Italy: Variations across trajectories of participation. *Mobilization: An International Quarterly*, 28(1), 89–107.

Bosworth, K. (2023). What is 'Affective Infrastructure'? *Dialogues in Human Geography*, 13(1), 54–72.

Capoccia, G. and Kelemen, D. (2007). The study of critical junctures: Theory, narrative, and counterfactuals in historical institutionalism. *World Politics*, 59(3), 341–369.

Chevée, A. (2022). Mutual aid in North London during the Covid-19 pandemic. *Social Movement Studies*, 21(4), 413–419. https://doi.org/10.1080/14742837.2021.1890574

Christou, G., Theodorou, E. and Spyrou, S. (2022). 'The slow pandemic': Youth's climate activism and the stakes for youth movements under Covid-19. *Children's Geographies*, 1–14.

Collins, R. (1993). Emotional energy as the common denominator of rational action. *Rationality and Society*, 5(2), 203–230.

Dean, J. (2016). *Crowds and Party*. Brooklyn, NY: Verso.

della Porta, D. (2022). *Contentious Politics in Emergency Critical Junctures. Progressive Social Movements during the Pandemic*. Cambridge: Cambridge University Press.

della Porta, D. and Lavizzari, A. (2022). "Framing health and care: Legacies and innovation during the pandemic. *Social Movement Studies*, 1–18. https://doi.org/10.1080/14742837.2022.2134109

de Moor, J., Uba, K., Wahlström, M., Wennerhag, M. and De Vydt, M. (eds.) (2020). *Protest for a Future II: Composition, Mobilization and Motives of the Participants in Fridays for Future Climate Protests on 20–27 September, 2019, in 19 Cities Around the World*. https://osf.io/asruw/

de Moor, J. and Wahlström, M. (2019). Narrating political opportunities: Explaining strategic adaptation in the climate movement. *Theory & Society*, 48(3), 419–451.

de Moor, J. and Wahlström, M. (2022). Environmental movements and their political context. In M. Grasso and M. Giugni (eds.) *The Routledge Handbook of Environmental Movements* (pp. 263–277). Oxon: Routledge.

Diz, C., Estévez, B. and Martínez-Buján, R. (2023). Caring democracy now: Neighborhood support networks in the wake of the 15-M. *Social Movement Studies*, 22(3), 361–380.

Ergen, T. and Suckert, L. (2021). 'Crises' as catalysts for more sustainable futures? *Economic Sociology*, 22(2), 10.

Esping-Andersen, G. (1990). *The Three Worlds of Welfare Capitalism*. Princeton, NJ: Princeton University Press.

Esu, A. and Dessì, V. (2022). Recasting solidarity during the COVID-19 pandemic: A case study. *Social Movement Studies*, 1–17. https://doi.org/10.1080/14742837.2022.2134105

Fraser, N. (2021). Climates of capital: For a trans-environmental eco-socialism. *New Left Review*, 34.

Hadden, J. (2017). Learning from defeat: The strategic reorientation of the US climate movement. In C. Cassegård, L. Soneryd, H. Thörn and Å. Wettergren (eds.) *Climate Action in a Globalizing World* (pp. 143–164). London: Routledge.

Herbert, J. (2021). The socio-ecological imagination: Young environmental activists constructing transformation in an era of crisis. *Area*, 53(2), 373–380. https://doi.org/10.1111/area.12704

Hilgartner, S. and Bosk, C. L. (1988). The rise and fall of social problems: A public arenas model. *American Journal of Sociology*, 94(1), 53–78. http://www.jstor.org/stable/2781022

Koopmans, R. and Statham, P. (1999). Ethnic and civic conceptions of nationhood and the differential success of the extreme right in Germany and Italy. In M. Giugni, D. McAdam and C. Tilly (eds.) *How Social Movements Matter* (pp. 225–251). Minneapolis: University of Minnesota Press.

Kowalewski, M. (2020). Street protests in times of COVID-19: Adjusting tactics and marching "as usual". *Social Movement Studies*. http://doi.org/10.1080/14742837.2020.1843014.

Kriesi, H. and Oana, I. (2022). Protest in unlikely times: Dynamics of collective mobilization in Europe during the COVID-19 crisis. *Journal of European Public Policy*, 30(4), 740–765.

Krinsky, J. and Caldwell, H. (2022). Resilience, reworking and resistance in New York City. In B. Bringel and G. Pleyers (eds.) *Social Movements and Politics During COVID-19: Crisis, Solidarity and Change in a Global Pandemic* (pp. 177–86). Bristol: Bristol University Press.

Malm, A. (2020). *Corona, Climate, Chronic Emergency: War Communism in the Twenty-First Century*. London and New York: Verso.

Mann, G. and Wainwright, J. (2018). *Climate Leviathan: A Political Theory of Our Planetary Future*. London: Verso Books.

McAdam, D. (1999). *Political Process and the Development of Black Insurgency, 1930–1970*. (2nd ed.). Chicago, IL; Chichester: University of Chicago Press; Wiley.

Nunes, R. (2023). Affective infrastructures and political organisation. *Dialogues in Human Geography*, February, 20438206221144824. https://doi.org/10.1177/20438206221144823

openDemocracy, dir. (2021). *George Monbiot: 'It's Time to Bail Out the Planet'*. https://www.youtube.com/watch?v=-0guuB34MKM.

Pressman, J. and Choi-Fitzpatrick, A. (2020). Covid19 and protest repertoires in the United States: An initial description of limited change. *Social Movement Studies*. http://doi.org/10.1080/147428 37.2020.1860743

Roy, A. (2020). Arundhati Roy: 'The pandemic is a portal. *Financial Times*, 3 April. https://www.ft.com/content/10d8f5e8-74eb-11ea-95fe-fcd274e920ca

Schück, H. (2021). 'Thunberg: Tungt ansvar på äldre efter pandemin' [Thunberg: Heavy responsibility weighs on those older after the pandemic]. *Tidningen Syre*, 30 June. https://tidningensyre.se/2021/30-juni-2021/thunberg-tungt-ansvar-pa-aldre-efter-pandemin/

Sewell, W. (1996). Historical events as transformations of structures: Inventing revolution at the Bastille. *Theory and Society*, 25, 841–881.

Snow, D., Cress, D., Downey, L. and Jones, A. (1998). Disrupting the 'Quotidian:' Reconceptualizing the relationship between breakdown and the emergence of collective action. *Mobilization: An International Quarterly*, 3(1), 1–22.

Soifer, H. D. (2012). The causal logic of critical junctures. *Comparative Political Studies*, 45(12), 1572–1597.

Stoddart, M., Ramos, H., Foster, K. and Ylä-Anttila, T. (2023). Competing crises? Media coverage and framing of climate change during the COVID-19 pandemic. *Environmental Communication*, 17(3), 276–292. https://doi.org/10.1080/17524032.2021.1969978.

Torres Munguía, J., Badarau, F., Pavez, L., Martínez-Zarzoso, I. and Wacker, K. M. (2022). A global dataset of pandemic- and epidemic-prone disease outbreaks. *Scientific Data*, 9(1), 683. http://doi.org/10.1038/s41597-022-01797-2

Voss, K. (1996). The collapse of a social movement: The interplay of mobilizing structures, framing, and political opportunities in the Knights of Labor. In D. McAdam, J. D. McCarthy and M. N. Zald (eds.) *Comparative Perspectives on Social Movements*. Cambridge: Cambridge University Press.

Wagner-Pacifici, R. (2010). Theorizing the restlessness of events. *American Journal of Sociology*, 115(5), 1351–1386.

Wahlström, M. and de Moor, J. (2017). Governing dissent in a state of emergency: Police and protester interactions in the global space of the COP. In C. Cassegård, L. Soneryd, H. Thörn and Å. Wettergren (eds.) *Climate Action in a Globalizing World* (pp. 57–80). New York: Routledge.

17

YOUTH CLIMATE ACTIVISTS' PRACTICES ON SOCIAL MEDIA IN BELGIUM AND FRANCE

Yuliya Samofalova, Andrea Catellani, and Louise Amelie Cougnon

Online Climate Activism

From Online Social Movements to Online Climate Activism

Over the last decades, digital media have influenced the organisation of social movements, which have passed from digitally facilitated to digitally organised movements. In comparison to digitally facilitated movements, whose actions primarily happen offline but are supported by the use of digital technologies, digitally organised movements are predominantly organised online (Firer-Blaess, 2016). In this context, the internet and social media (hereafter, SM) platforms – which is, online platforms that are specifically designed to "facilitate socialisation between . . . users" (Aichner et al., 2021, p. 2015) – play an important role for social movements (Jeppesen, 2021) and contribute to the development of online social movements. In earlier works, online social movements were defined as "the adoption and use by social movements and community activists of new information and communication technologies, such as the Internet and the World Wide Web" (Hara and Huang, 2011, p. 2). However, with the proliferation of digital technologies, online social movements are not limited to the use of websites and email anymore and have adapted to various digital applications. Since the beginning of the 2000s, their activism has been referred to as "hashtag activism" (Boulianne et al., 2020), "digital activism" (George and Leidner, 2019), "online social movements" (Hara and Huang, 2011), "cyberactivism" (McCaughey and Ayers, 2003), or "e-movements" (Earl and Schussman, 2002). A more recent definition of online activism is "to take action over the internet in a digital environment in the form of the advocacy of a goal, organising around this goal in order to achieve it, transmitting relevant messages to the masses, lobbying, boycotting, and site blackout" (Gezmen, 2022, p. 149). Although online activism of climate movements is closely related to offline activism (Greijdanus et al., 2020), in this chapter, we only focus on online activism, understood as organised online activities of groups that can be active both online and offline. We will retain the following definition of online activism: a digital dimension of social movements intended to build communities and inform and engage more people around particular societal problems.

DOI: 10.4324/9781003396567-20

A substantial body of literature has focused on online activism via SM from different perspectives. Lewis et al. (2014) investigated how SM facilitated online protests from a sociological perspective, in particular looking at fundraising and recruitment. More recent works have focused on hashtag-based activist movements, such as *#MeToo* or *#BlackLivesMatter* (Hush, 2020). Although numerous studies have investigated online activism on SM, the platforms' potential for mobilisation is "a contested issue" (Kavada, 2012, p. 31). For Askanius and Uldam (2011), Roth-Cohen (2019), and Kowasch et al. (2021), such communication can create a sense of community. However, it may not be sufficient to motivate individuals "to take to the streets" (Kavada, 2012, p. 31). Indeed, Jeppesen (2021) argues, such digital technologies cannot be neutral and are embedded with the values and interests of their content creators and users: SM can therefore both initiate and limit social movements.

As a social movement, climate activists aim "to change the way people think, and so behave, on the one hand", and try to impact "social institutions and collective decisions, on the other" (Dryzek, 2022, p. 188). In general, climate activism encompasses a variety of approaches, from grassroots campaigns and protests to lobbying and policy advocacy (Sabherwal et al., 2021). Online climate activities, including the use of SM, have become a critical means to combat global climate change (Chen et al., 2020) and to mobilise activists to take up the climate-change agenda (Askanius and Uldam, 2011). In this context, online climate activism refers to the use of digital technologies to create and direct social movement communities toward greater collective action for climate-change mitigation (Wielk and Standlee, 2021). As any kind of online activism, online climate activism takes various forms – SM campaigns, online petitions, digital art, and virtual protests, described as the hierarchy of online activism by George and Leidner (2019). Online climate activism provides an opportunity for people who may not be able to participate in physical forms of activism: it allows them to get involved and make their voices heard.

Online climate activism shares the goals of offline climate activism: to inspire people to act on climate change. SM campaigns have become a significant occasion for people to talk about news, debates, actions, and social inputs related to climate change (Askanius and Uldam, 2011). Particularly, SM is a means for sharing knowledge about climate change, engaging activist communities, and providing space for discussing the issue with others (Anderson, 2017). Although climate change is a known phenomenon, online communication campaigns provide information on problems or impacts associated with climate change (Vu et al., 2021). Thus, the participants of such online activism can also be considered "as agents of change and transformative learning processes" (Kowasch et al., 2021, p. 2).

Climate activists use SM to build communities and mobilise followers around their movement (Wielk and Standlee, 2021). They use narrative framing and linguistic conventions to create a collective identity within a diverse generation (Wielk and Standlee, 2021; Belotti et al., 2022). In addition, online climate activism can reach international audiences and organise protests by varying the use of different platforms (Hopke and Paris, 2022), which play a role of networking agents within the environmental protest space: they constitute organising mechanisms within specific protests (Segerberg and Bennett, 2011).

Open to larger publics, SM provide an opportunity for "new voices" and actors to engage in climate-change debates. For instance, on Twitter (now X), these actors endorse political actions and policy changes by addressing issues around climate change, such as social justice (Terren and Soler-i-Martí, 2021; Chen et al., 2022). On Instagram, eco-influencers contribute to producing a change in society through their own example (Cornelio et al., 2021). Climate activist movements can use SM platforms differently regarding their goals and target audiences. For instance, FFF activists prefer Twitter to address institutions and mass media. Newer SM platforms, such as TikTok, are more frequently used to reach younger individuals (Belotti et al., 2022). Pearce and colleagues (2020) highlighted the affordances of different SM platforms and the differences in their use for climate-change communication. In this chapter, we provide examples of how different online climate activist groups used SM platforms; however, we do not investigate the specificities of each platform used by climate activists.

Climate strikes constitute a significant part of climate movements and political activism (de Moor et al., 2021) for which SM can help in the organisation or call-to-action process. In this case, online climate activism is correlated with offline actions (Boulianne and Theocharis, 2020). In addition, climate activism can bring together (cf. parts 2.2.1 and 2.2.2) activist groups (Belotti et al., 2022), considering that social justice movements, such as BlackLivesMatter, Free Uygurs, LGBTQ+, and climate justice movements tend to congregate online as well (Kowasch et al., 2021). The civil disobedience actions of recent climate movements – also called "radical" (Dryzek, 2022) – demand that political leaders "take radical socio-ecological transformation measures" (Kowasch et al., 2021, p. 1). Thus, online climate activism can potentially create significant social and political change, both on SM and beyond (Belotti et al., 2022).

On the individual level, the impacts of online climate activism lie within the development of personal traits and the emotional condition of the participants. Climate movements on SM can enhance self-confidence and self-identification with the movements by providing "the feeling of being part of a 'family'", which can be a motivating factor for youth to participate (Kowasch et al., 2021, p. 3); unfortunately, they can also emphasise feelings of anxiety, fears, and other disorders (Tyson et al., 2021).

As research has demonstrated, online activism can be considered as a separate form of activism, which is focused on communication about societal problems and on mobilisation of larger audiences. In particular, climate activism online has multiple impacts on individuals and societies: it raises awareness about climate-related issues among different audiences, mobilises people for climate actions and strikes, and contributes to the change within social and political systems. Additionally, online climate activism improves the understanding of the problem, self-identification within the movement, and emotional connection with it.

Evolution of Global Climate Activism Online: NGOs' and Youth Climate Movements

Online climate activism has evolved due to a combination of changes in the broader environmental movement and its online activities. In the early days of the internet, environmental activists used emails and online forums to share information, organise protests, and advocate for policy change (Merry, 2012). As "mediators between scientific expertise and the public" (Vu et al., 2021, p. 92), environmental organisations such as Greenpeace provided information on environmental issues and campaigns on their websites, as well as tools for online activism, for example, email petitions and online fundraising. The SM platforms changed that by providing a powerful new tool for online climate activism that allows activists to reach a broader audience quickly, easily, and almost at no cost (Segerberg and Bennett, 2011).

Previous studies reflect on the evolution of climate activism on Twitter (now X), YouTube, Facebook, Instagram, and TikTok. Segerberg and Bennett (2011, p. 198) analysed Twitter posts with climate-related hashtags – #cop15 and #thewave – posted in 2009, and they named the period "2009 Twitter Revolutions", which evoked the potential effect of SM on politics and political decisions. Numerous other studies were based on Twitter messages at the same moment and in the following years (Boulianne et al., 2020; Chen et al., 2022; Padilla-Castillo and Rodríguez-Hernández, 2023). A first systematic study of climate activism around the COP-15 meetings in Copenhagen analysed YouTube videos, including videos by the "Never Trust a Cop" (NTAC) network. These videos served as a strategy to access mass media and as a call to action: the activists used the platform as a free and easy channel to reach a large audience, although without a carefully crafted media strategy (Askanius and Uldam, 2011). However, this lack of a strategy has caused problems in the current constantly evolving media landscape: today most NGOs and activist movements are forced to deal with platform algorithms to become more visible and to reach a broader public with a real SM strategy, which consists of "a goal-directed planning process for creating user-generated content" (Effing and Spil, 2016, p. 2) and integrating SM resources to achieve the NGO's goals.

The year 2019 was extraordinary "in terms of the unprecedented scale and coordination of mobilizations on the climate crisis" (de Moor et al., 2021, p. 619). The starting point of this mobilisation was the protests of Greta Thunberg, who stood in front of the Swedish parliament every Friday in August 2018. This resulted in a series of protests all over the world, the creation of the official hashtag "#FridaysForFuture" for these protests, and, subsequently, the establishment of the subgroups Fridays for Future (FFF) and Youth for Climate (YfC). The movement initiated by Thunberg encouraged individuals and environmental NGOs to launch various online campaigns in 2019, such as #ForClimate. This campaign on TikTok and other short videos related to climate-change issues were at the centre of the analysis by Hautea et al. (2021), who investigated the frames used by creators on TikTok to communicate to their audiences.

The street protests were interrupted due to the COVID-19 pandemic, with much activism moving online (Fisher and Nasrin, 2021). Hopke and Paris (2022) compared the use of SM by the #FridaysForFuture movement to the activists of #noDAPL, a US climate campaign against the Dakota Access Pipeline: they both included "powerful visuals" in their online campaigns "to draw attention to their demands and generate mainstream media coverage" and reach transnational audiences (Hopke and Paris, 2022, p. 367).

Alongside the 2019 movement related to Thunberg, NGOs' strategies on SM have been studied (O'Neill and Nicholson-Cole, 2009; Vu et al., 2021), in particular Greenpeace strategies on Facebook from a perspective of communication studies (Katz-Kimchi and Manosevitch, 2015), on Twitter from a perspective of political science (Gupta et al., 2018), on Instagram from a linguistic perspective (Lippert, 2022), and on Twitter from the perspective of economics (Gezmen, 2022). Human personalities, besides Thunberg (Andersson, 2021; Molder et al., 2021; Park et al., 2021; Arce-García et al., 2023), also emerged in SM publications. This is the case with the African continent, which has fuelled Communication and Psychology studies, notably about Vanessa Nakate – a climate activist from Uganda (Barnes, 2021).

In conclusion, global climate activism online is a phenomenon that appeared with the development of online communication and escalated with the rise of SM. We have seen that hashtag-based data allowed researchers to investigate the structure of the global or local branches of the movement, identify the accounts of opinion leaders, and investigate communication strategies and the evolution of the movement. The data obtained from particular accounts provided insights on organisational or individual communication. This data collected for previous studies was used to investigate the strategies for effective communication and mobilisation of different audiences online and it provided significant insights on the role of particular individuals for the development of the movement.

Belgian and French Youth Climate Activist Movements Online

Youth Climate Activism in Belgium and France: From Offline to Online Practices

The emergence of online climate activism has transformed the landscape of the global climate movement, particularly among young activists (Wielk and Standlee, 2021). As stated in the previous part, the protests of Greta Thunberg inspired a large number of young people, in particular teenagers (Belotti et al., 2022), to join the action to make politicians and those who are responsible for social change take an active position on climate-change mitigation. Since then, many subgroups of the movement have emerged worldwide under various names. In the European Union, these include FFF, YfC, Youth Strike for Climate, and Youth Climate Strike (Figure 17.1: different colours distinguish the major groups of climate activists in each country, as of May 2023).

Figure 17.1 provides important information about the diversity of climate activists' movements' denominations in Europe, but it also hides the diversity of designations within one country and their various names online. The initial name of the movement, Fridays For

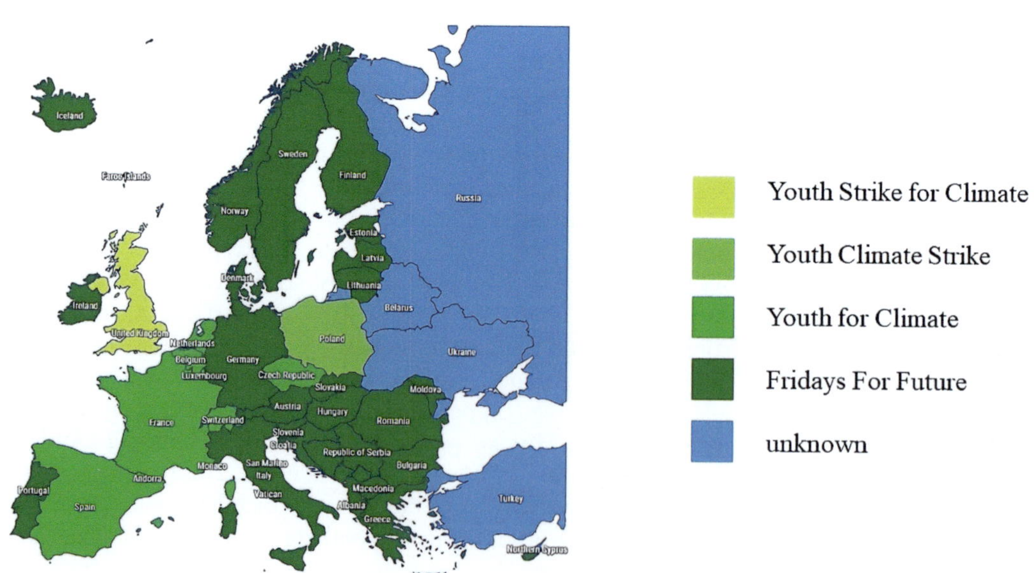

Figure 17.1   Denominations of youth climate activism in Europe.

Source: Created with Visme by the authors.

Future, resulted from a hashtag #FridaysForFuture in many countries, such as Germany, Norway, or Italy, where the movement is still represented under this name (fourth colour line, Figure 17.1). However, this is not the case in other countries, such as Belgium, Czech Republic, France, Poland, Spain, or the UK. In Belgium, Youth for Climate (third colour line, Figure 17.1) is the main denomination for youth climate activism, although it represents the same FFF international movement. In France, the situation is more complex: there are two names of youth activist groups, YfC and FFF, where YfC is a representative body of all youth climate activist groups in the country, including FFF. According to the information on the YfC official website in France (*Youth for Climate FR*, 2023), YfC in France represents the international FFF movement where young people gather "to act for the environment and our future". Youth climate activist movements in France maintain their activities under different names, organise some of their environmental campaigns independently, and communicate via separate websites and SM pages, which could confuse their possible audiences or people interested in youth climate activism. Despite their diversity, all of the listed youth climate activist movements have roughly the same roots and goals.

Scholars have outlined the online mobilisation of the participants in all European countries; however, research has been mostly focused on comparative observations of numerous climate activist groups to provide a global view on their communication. This is the case with Wahlström et al. (2019) who compared Austria, Belgium, Germany, Italy, the Netherlands, Poland, Sweden, Switzerland, and the UK to reveal the potential of SM as an information channel and a space to raise awareness about political issues. Another comparative study was conducted by Sorce and Dumitrica (2021) who collected hashtag-based data from Facebook in different European countries, including Belgium and France, and examined it through the lens of a thematic framing and network analysis in a crisis situation.

According to Orsini et al. (2021), the history of the modern youth climate movement in Belgium started on 2 December 2018 with a strike for climate in Brussels' streets. January 2019 in Belgium is considered a starting point in the organisation of weekly strikes and the movement itself, whose aim was "to limit the consequences of the climate and biodiversity crisis by uniting as a movement and exert political pressure while raising awareness in society" (*Youth for Climate BE*, 2023). Anuna De Wever and Adélaïde Charlier are the representative figures of the bilingual Belgian movement, which has its official website and pages on Facebook, Instagram, TikTok, Twitter, and YouTube.

Orsini et al. (2021) underlined important aspects of Belgian mobilisations for climate, such as the perception of insufficient national and political ambition to climate-change mitigation, and the complexity of its internal heterogeneity:

> While Youth for Climate was born in Flanders, the movement developed mostly in francophone Belgium, where it is supported by several institutionalised organisations. . . . Beyond this institutionalised dimension, the national movement comprises a wide variety of activist groups.
>
> *(Orsini et al., 2021, p. 5)*

The movement leaders themselves emphasised the lack of structural organisation and systematic action to achieve the Paris Agreement, and they identified another challenge for the movement: online threats and hate speech toward young activists, including gender-based harassment (Heyden et al., 2020).

As suggested by Tyson et al. (2021, p. 21) in general, younger adults "report seeing climate-related content on SM platforms to a greater extent than older generations, which means that online climate activists may be younger than climate activists". Yet online practices do not seem to have an impact on the importance of social class as a mobilising factor of the climate activists in Belgium: a recent study among young people has concluded that young Belgians from higher classes seem to be more conscious about climate issues (Ducol et al., 2022).

YfC in Belgium has been analysed from four dimensions: the record of its actions, organisational choices, form of commitment, and its relationship to politics (Dumas, 2020). Vossen (2021) assessed the movement's communication about "green issues" on Facebook and Instagram in 2019 with a critical discourse approach. The author noted the radicality of the individual level of politicisation of the participants, who are positioning themselves as subjects of environmental struggle and focusing on collective and structural change (Vossen, 2021, p. 8).

The YfC movement appeared in France in January 2019, a month after Belgium. It is described as a French branch of the FFF movement, including 130 local groups by now, which are located mostly in big cities and have their profiles on SM in addition to the general pages of the French YfC. Youth activists themselves consider the French YfC website as a digital platform that lists locally organised actions by school pupils and students and allows them to gather beyond global actions (Derolez et al., 2019). According to the information provided on their website, YfC is affiliated with no other associations, organisations, or political structures. The aims of the movement include mobilisation for climate and social justice and protection of the environment and biosphere (*Youth for Climate FR*, 2023). The main initiatives of the movement are protests, marches, and other activities to support climate-change mitigation, such as discussions, debates, workshops, or specific campaigns to demand polluters and governments act. The communication about these actions is spread on the website and SM of global and local branches of the movement. The movement is present on Facebook, Instagram, LinkedIn, TikTok, and Twitter.

The last few years have witnessed a surge of interest in investigating the youth climate movements in France. Through survey data from the 2019 climate strikes in Paris, Gaborit (2020) has shown that radical individuals in France have significant cultural capital and left-wing familial political socialisation, influencing their awareness of the climate emergency. Pachocińska (2020), who concentrated on slogans from youth climate marches in France from a discursive perspective, presented the collective ethos that emphasises the role of SM on the construction of the community in France. Finally, Wagener (2022) investigated hashtag-based data from Twitter, which allowed the author to identify specific Twitter accounts of several French activists like Cyril Dion and Claire Nouvian. In France, activists recognise SM as the simplest way to mobilise people, provided that this type of online communication is in good synergy with other forms of engagement, such as manifestations (Derolez et al., 2019). One specificity of the French movement is that their messages are centralised: it directly addresses the President of the French Republic in the publications instead of addressing local institutions (Chaillou and Monti-Lalaubie, 2020).

To summarise, the youth climate movements that appeared in Belgium and France after Thunberg's protests in 2018 have been much discussed in recent literature. In Belgium, scholars focused on the structure and the problems associated with the movement's organisation, as well as their use of SM. Few of these works prioritised the communication of the YfC online and highlighted the movement's politicisation and connections to global social justice movements. The existent research on the online YfC movement in France emphasised the movement's politicisation, its identification of opinion leaders, and the role of SM for mobilisation. Most of the research investigated the offline YfC movement: there is a lack of studies emphasising the online communication of youth climate activist groups in France. This could be explained by the report of ELABE (2022) showing that more than 70% of French citizens (from a sample of 1,005 participants) do not understand the climate activist movements in general, and 80% of the respondents believe that the actions organised by the movements are not useful for ecological transition. The people aged 35 or younger represent 41% of the respondents supporting climate activist movements.

To complement the previous research and to better understand online youth climate activism in Belgium and in France, we suggest a comparative case study of YfC communication on two more recent SM, Instagram and TikTok. In the next part, we provide the context for the research and the most significant results.

Youth for Climate Online Practices in Belgium and France

The YfC movement in Belgium and France is widely present on SM. On Instagram, they have their official central profiles as well as the accounts of local groups. For example, in Belgium they include at least seven local Instagram accounts of the movement: in Antwerp, Brussels, Bruges, Leuven, Louvain-La-Neuve, Mechelen, and Ostend. In France, we observed 20 local accounts of the movement on the same SM, including @youthforclimate_colmar, @yfc_valenciennes, @yfcrouen, @yfc_nevers, @yfc.rennes, and other accounts. On TikTok, however, the YfC movement in both countries communicates through one official account. Although the two SM can be used for different purposes (Masciantonio et al., 2021), the platform differences are not an object of this study. In this part, we reflect on the communication on Instagram and TikTok by the youth climate movements in two countries without going into the details of SM algorithms and affordances.

We conducted a comparative case study to understand how the youth climate activist groups in Belgium and France communicate on Instagram and TikTok. We analysed the communication of the official accounts by the YfC on Instagram and TikTok in the two countries. Our data included 215 Instagram posts and 119 TikTok videos (Table 17.1). The corpus data covers the period from January 2019 to January 2022 – the peak of the COVID-19 crisis. The data were collected with the help of Vurku services.

Our aim was to observe and compare how the main accounts of youth activist groups communicate on the selected SM platforms. More precisely, we looked at the main actors represented as responsible for climate-change mitigation and the obstacles to pro-environmental action on different modes of Instagram posts. By visual mode here we understand the images and the videos posted by the account; the verbal mode covers the captions

Table 17.1 Collected and analysed SM data.

Country	Instagram account name	TikTok account name	Analysed Instagram posts	Analysed TikTok posts
Belgium	youth.for.climate	youthforclimatebe	114	7
France	youthforclimatefr	youthforclimatefr	101	112

Source: Original creation.

accompanying the posted visuals. To obtain a broader sense of the data, we conducted a multimodal content analysis, including a codebook of five questions:

1 is the post related to climate change and environmental topics?
2 which climate action is proposed in the post?
3 who is responsible to act for the climate and environment?
4 who is criticised for their anti-environmental behaviour?
5 which obstacles to pro-environmental behaviour are mentioned in the post?

All posts were encoded by a project researcher. The results of this exploration and the most significant examples are presented in the following parts.

Youth for Climate Communication on Instagram and TikTok in Belgium

The results of multimodal content analysis of the posts by the Belgian YfC movement on Instagram and TikTok showed that the main action in favour of climate-change mitigation was joining the climate movement: 58.77% of analysed Instagram posts and 42.56% of TikTok videos represented this action. The dramatic image illustrating the climate and COVID-19 crises from Figure 17.2 exemplifies this call to join the movement, with the text "COME JOIN US" in the image. Other forms of pro-environmental actions are related to food, energy, and the transportation sector, and most of them are closely connected to political decisions. For instance, making public transport more affordable and providing support to innovative and sustainable projects in agriculture were mentioned in relation to transport and food topics respectively. Rather large and vague actions such as changing the system or the economy are proposed in 8% of Instagram posts. The post from Figure 17.2 represents an example of "rebuilding our economy" action as stated in the image.

Figure 17.2   Youth for climate Belgium post, Instagram, 24 August 2020.

The Belgian team of the movement proposed fewer environmental actions on TikTok: 31% of the posts do not include a call for action or proposition of any solution to the climate crisis.

According to the Belgian team of the YfC both on TikTok and Instagram, the activists and "you", the reader of the post, are the ones who should act for the climate: they are most frequently represented in the calls to action in the posts' texts and in the images illustrating climate marches. Although in 29% of Instagram posts and 14% of TikTok posts politicians and governments were also represented as the actors to mitigate the effects of climate change, their political (in)-action is mentioned as the main obstacle to climate-change mitigation in 43% of TikTok videos and 15% of Instagram posts. This difference in the percentage is related to a highly different number of analysed posts of each account. The Instagram posts by the Belgian YfC tended to criticise the government and the politicians for not respecting the Paris Agreement.

As put forward by Heyden et al. (2020), the youth climate activists in Belgium are faced with online threats and gender-based harassment. However, this violence is not specific to climate activism and is shared with other social movements, such as LGBTQ+ and #BlackLivesMatter. On their Instagram posts, YfC explains the role of equality for climate action in the post's text "because the fight against racism and climate change are the same" as mentioned in the post on 21 March 2020 (Figure 17.3). The image from Figure 17.3 illustrates the demand for women of the world to unite.

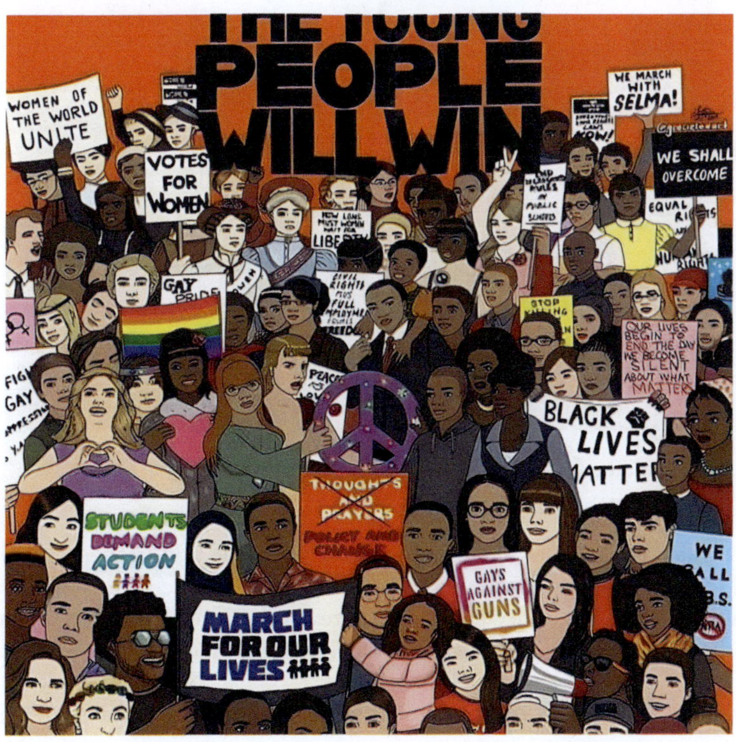

Figure 17.3   Youth for climate Belgium post, Instagram, 21 March 2020.

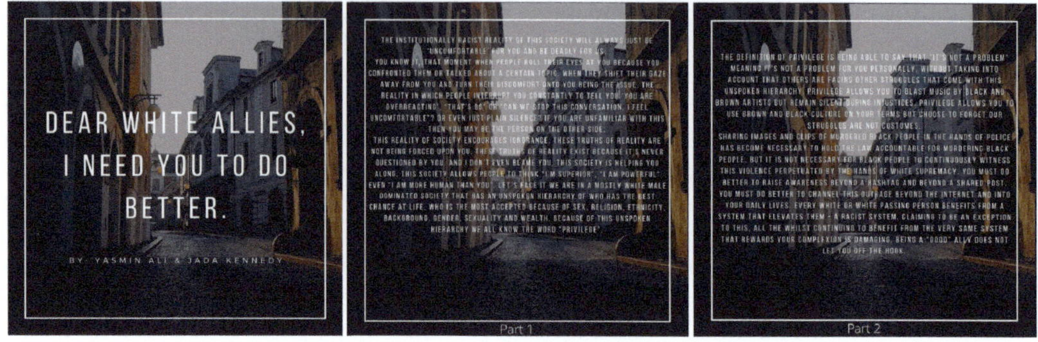

Figure 17.4　Youth for climate Belgium post, Instagram, 31 May 2020.

Figure 17.5　Youth for climate Belgium post, Instagram, 7 June 2020.

The visual and verbal modes in Figure 17.3 highlight the importance of support and collaboration of the Belgian youth climate movement with other social movements. Other Instagram posts of the movement reflected on privileges in the post's visuals (cf. Figure 17.4) and documented the movement's participation in the #BlackLivesMatter protests (cf. Figure 17.5).

Youth for Climate Communication on Instagram and TikTok in France

In this part, we review the communication of the YfC France on Instagram ($n =101$) and TikTok ($n = 112$) in 2019–2022 and provide a brief synthesis and most significant examples. As in Belgium, the French YfC movement emphasises joining climate action as an important solution to the climate crisis: this action is present in 57% of Instagram posts and 25% of their TikTok videos. However, the solutions related to the food sector, which are creating shared gardens or eating less meat, are more frequent in the French corpus on Instagram. We suppose this can be related to the particular relation to food in this country as introduced by DeSoucey (2010). The movement attempts to raise awareness about environmental problems by using storytelling techniques, statistics, and infographics about the causes and impacts of climate change, for example, the impact of the farming industry (cf. Figure 17.6) and the possible pro-environmental actions.

Most of the posts on Instagram and TikTok videos indicate activists, people who read the posts, and everybody in the world as the main actors for the climate. Although the movement in France considers itself politically neutral, many images on Instagram and videos on TikTok YfC pages were highly critical of existing politics, the government, and politicians, such as Emmanuel Macron, as demonstrated in Figure 17.7.

Figure 17.6 Youth for climate France post, Instagram, 11 June 2020.[1]

Figure 17.7 Youth for Climate France, Instagram, 10 April 2022.[2]

Figure 17.8 Youth for Climate France, TikTok video, 13 November 2021.[3]

In their TikTok videos posted around the presidential election in April 2022, YfC warns on the one hand about the inadequate solutions proposed by the candidates and on the other hand of the government's actions, which according to the movement are not sufficient to overcome the climate-change problem. A controversial video (Figure 17.8) picturing President Macron at the beginning shows the negative effects of climate change on nature and biodiversity. It ends with the list of negative measures for climate and environment taken during the last years. In the end, the video criticises everybody who believes in this "destructive system".

On both SM platforms, communication is more spontaneous than strategically organised. On Instagram, the majority of the posts are posted in relation to climate strikes or important events, for example, IPCC reports. For example, eight posts (8% of all analysed posts) covered the climate strike on 25–26 March, 2022. They provided a map of the mobilisations (Figure 17.9), explained with infographics the importance of the mobilisation (Figure 17.10), showed with videos how one could prepare for the protest (Figure 17.11), and documented the mobilisation in different cities (Figure 17.12).

As in Belgium, social justice and climate justice in France are intrinsically connected in the posts of the activist movement on Instagram and TikTok. Several posts follow the same structure as Figure 17.9: a cover image explaining the theme of the post which is followed by eight or nine images capturing the text that explains the important questions around the subject; the sources used to prepare the publication are always mentioned in the caption of the post.

Figure 17.9   Youth for climate France post, Instagram, 2 March 2022.[4]

Figure 17.10   Youth for climate France post, Instagram, 3 March 2022.[5]

Posts on Instagram are more informative than on TikTok, where videos are extremely critical of the government policies and actions, the greenwashing by some industries and companies considered as polluting ones, and people stigmatising climate activists. The accounts use the same hashtags, such as #climatejustice, #socialjustice, and #lgbt, but the communication strategy is different. Some videos on TikTok criticise or denounce people

Figure 17.11   Youth for climate France post, Instagram, 11 March 2022.

Figure 17.12   Youth for climate France post, Instagram, 21 April 2022.

who take racist or anti-feminist positions. An example of such a video is illustrated by a post on 14 May 2021, where an activist denounces people who believe that "feminism does not include gender minorities". The denouncement is shown in the expressed emotions and gestures and is accompanied by a cheerful fast melody. This melody is used in more than ten thousand TikTok videos, which frequently disapprove of other people's opinions or actions.

Conclusion and Future Perspectives

As a social movement, climate activism has a long history, which started before the internet and proliferated with the development of digital platforms. In this chapter, we drew particular attention to youth climate activism online, which was initially organised by young individuals but turned into an intergenerational movement. It is very difficult to identify the portraits of the actors of this movement, besides opinion leaders, due to its continuous transformation. In general, the movement's presence on SM aims to provide information about climate-change issues, motivate people to act, and contribute to social and political changes.

The modern climate activist movement online has increased with the climate strikes in 2018 in particular in Europe with the appearance of various local groups of FFF, which are named differently but have similar objectives. Most studies focus on the movement's organisation and communication on a basis of single-platform studies from hashtags and textual components, mostly on Twitter. Following Padilla-Castillo and Rodríguez-Hernández (2023), we see a high potential for further comparative studies of this movement's communication based on multiple platforms; also visual ones, for instance, YouTube, Instagram, or TikTok. This will provide a better understanding of the online practices regarding different platform affordances and a clearer picture of obstacles or challenges to climate-change mitigation concerning the audiences' opinions on each platform.

In this work, we synthesised the existent research on organisation and communication of the YfC movement in Europe and in particular in Belgium and France. Additionally, we provided a case study focusing on the online practices of the Belgian and French YfC on two SMs, Instagram and TikTok, which connect with different audiences. While the form of the messages can be different – a series of images on Instagram versus a short video with dramatic scenery or music on TikTok – the movement in both countries attempts to raise awareness about the climate-change problem, to be connected to other social problems, for instance, social justice, and to draw attention to important political actions. Political figures and governments are largely criticised on both platforms in the two countries. Future research on online practices of the YfC movement could focus on particular campaigns of the movement by comparing online data to ethnographic observations from offline protests, for example.

Like Hopke (2024), we believe that the recent research on online climate activist movements requires more rapid and real-time observations of communication on SM. Combining qualitative, quantitative, and computational approaches, such as machine learning and artificial intelligence, could do this. However, future studies should consider the key weaknesses of this integration beforehand, such as (re-)introducing empiricist epistemologies, ignoring bias for precision, and the environmental impacts of cloud computing services (Schäfer and Hase, 2022).

Even though the movement is present in all European countries, studies dedicated to online practices of the movement investigated countries in Northern, Western, and Southern

Europe. More research should be done on youth climate movements' online practices of Eastern European countries and in developing regions of the world. Such comparative studies could bring a new perspective on online practices that would provide a more nuanced and culture-related picture of online activism of a multi-faceted worldwide youth climate movement.

Notes

1 Image on the left, translation: Why does meat pollute? Image in the centre, translation: And while nearly a billion human beings suffer from malnutrition, 2/3 of the earth's land surface is devoted to livestock breeding or the production of animal feed. By eating meat rather than cereals, we feed 10 times fewer people for the same cultivated area. Image on the right: These millions of hectares used for livestock consume enormous amounts of water, both to grow the cereals and to feed the animals. For example, to produce one kilo of beef, no less than 13,000 litres of water are needed, compared with just 1,000 for wheat!
2 Translation: Macron's track record. 53% of the 169 ecological measures taken since 2017 are harmful to the planet. Here is a summary of some of them. Source: Reporterre.
3 Translation: Climate change: Total known as early as 1971. To believe in a destructive system is to be complicit in its destruction. Macron, a year of agricultural and environmental "deception" according to Attac.
4 Translation: Already multiple mobilisations. 25 March.
5 Translation: Why protest on 25 March?

References

Aichner, T., et al. (2021). Twenty-five years of social media: A review of social media applications and definitions from 1994 to 2019. *Cyberpsychology, Behavior, and Social Networking*, 24(4), 215–222. http://doi.org/10.1089/cyber.2020.0134

Anderson, A. A. (2017). Effects of social media use on climate change opinion, knowledge, and behavior. In *Oxford Research Encyclopedia of Climate Science* (pp. 1–20). http://doi.org/10.1093/acrefore/9780190228620.013.369

Andersson, M. (2021). The climate of climate change: Impoliteness as a hallmark of homophily in YouTube comment threads on Greta Thunberg's environmental activism. *Journal of Pragmatics*, 178, 93–107. http://doi.org/10.1016/j.pragma.2021.03.003

Arce-García, S., Díaz-Campo, J. and Cambronero-Saiz, B. (2023). Online hate speech and emotions on Twitter: A case study of Greta Thunberg at the UN climate change conference COP25 in 2019. *Social Network Analysis and Mining*, 13(1), 48. http://doi.org/10.1007/s13278-023-01052-5

Askanius, T. and Uldam, J. (2011). Online social media for radical politics: Climate change activism on YouTube. *International Journal of Electronic Governance*, 4(1/2), 69–84. http://doi.org/10.1504/IJEG.2011.041708

Barnes, B. R. (2021). Reimagining African women youth climate activism: The case of Vanessa Nakate. *Sustainability*, 13(23), 13214. http://doi.org/10.3390/su132313214

Belotti, F., et al. (2022). Youth activism for climate on and beyond social media: Insights from FridaysForFuture-Rome. *The International Journal of Press/Politics*, 27(3), 718–737. http://doi.org/10.1177/19401612211072776

Boulianne, S., Lalancette, M. and Ilkiw, D. (2020). "School strike 4 climate": Social media and the international youth protest on climate change. *Media and Communication*, 8(2), 208–218. http://doi.org/10.17645/mac.v8i2.2768

Boulianne, S. and Theocharis, Y. (2020). Young people, digital media, and engagement: A meta-analysis of research. *Social Science Computer Review*, 38(2), 111–127. http://doi.org/10.1177/0894439318814190

Chaillou, A. and Monti-Lalaubie, M. (2020). Jeunes pour le climat: en coulisses, ça continue! *Revue Projet*, 375(2), 44–49. http://doi.org/10.3917/pro.375.0044

Chen, B., et al. (2020). Motivation analysis of online green users: Evidence from Chinese "Ant Forest". *Frontiers in Psychology*, 11, 1–9.

Chen, K., et al. (2022). How climate movement actors and news media frame climate change and strike: Evidence from analyzing Twitter and news media discourse from 2018 to 2021. *The International Journal of Press/Politics*, 28(2), 1–31. http://doi.org/10.1177/19401612221106405

Cornelio, G. S., Ardèvol, E. and Martorell, S. (2021). Environmental influencers on Instagram: Connections and frictions between activism, lifestyles and consumption. *AoIR Selected Papers of Internet Research* [Preprint]. http://doi.org/10.5210/spir.v2021i0.12238.

de Moor, J., et al. (2021). New kids on the block: Taking stock of the recent cycle of climate activism. *Social Movement Studies*, 20(5), 619–625. http://doi.org/10.1080/14742837.2020.1836617

Derolez, M., Lazare, L. and de Mullenheim, A. (2019). Pour un métissage des pratiques de mobilisation. *Revue Projet*, 371(4), 12–19. http://doi.org/10.3917/pro.371.0012

DeSoucey, M. (2010). Gastronationalism: Food traditions and authenticity politics in the European Union. *American Sociological Review*, 75(3), 432–455. http://doi.org/10.1177/0003122410372226

Dryzek, J. S. (2022). *Politics of the Earth* (4th ed.). New York: Oxford University Press.

Ducol, L., et al. (2022). Jeunes, Communication & Climat. Diversité des enjeux climatiques auprès des 15–24 ans en Belgique. *preprint. SocArXiv*. http://doi.org/10.31235/osf.io/87psm

Dumas, A. (2020). *Le Mouvement Youth for Climate, Un Mouvement 'Spontané. In Quelles conséquences sur la mobilisation?* Master Thesis. Université de Liège. http://hdl.handle.net/2268.2/9944 [Accessed 14 April 2023].

Earl, J. and Schussman, A. (2002). The new site of activism: on-line organizations, movement entrepreneurs, and the changing location of social movement decision making. In P. G. Coy (ed.) *Consensus Decision Making, Northern Ireland and Indigenous Movements* (pp. 155–187). Leeds: Emerald Group Publishing Limited (Research in Social Movements, Conflicts and Change). http://doi.org/10.1016/S0163-786X(03)80024-1

Effing, R. and Spil, T. A. M. (2016). The social strategy cone: Towards a framework for evaluating social media strategies. *International Journal of Information Management*, 36(1), 1–8. http://doi.org/10.1016/j.ijinfomgt.2015.07.009

ELABE (2022). *Les Français et l'environnement*. https://elabe.fr/environnement/ [Accessed 14 January 2023].

Firer-Blaess, S. (2016). *The Collective Identity of Anonymous: Web of Meanings in a Digitally Enabled Movement*. PhD dissertation. Acta Universitatis Upsaliensis. https://urn.kb.se/resolve?urn=urn:nbn:se:uu:diva-292734 [Accessed 2 October 2023].

Fisher, D. R. and Nasrin, S. (2021). Climate activism and its effects. *WIREs Climate Change*, 12(1), e683. http://doi.org/10.1002/wcc.683

Gaborit, M. (2020). Disobeying in time of disaster: Radicalism in the French climate mobilizations. *Youth and Globalization*, 2(2), 232–250. http://doi.org/10.1163/25895745-02020006

George, J. J. and Leidner, D. E. (2019). From clicktivism to hacktivism: Understanding digital activism. *Information and Organization*, 29(3), 1–45. http://doi.org/10.1016/j.infoandorg.2019.04.001

Gezmen, B. (2022). Digital activist movements for energy resources: The case of Greenpeace Turkey. In H. Dinçer and S. Yüksel (eds.) *Sustainability in Energy Business and Finance: Approaches and Developments in the Energy Market* (pp. 145–158). Cham: Springer International Publishing (Contributions to Finance and Accounting). http://doi.org/10.1007/978-3-030-94051-5_13

Greijdanus, H., et al. (2020). The psychology of online activism and social movements: Relations between online and offline collective action. *Current Opinion in Psychology*, 35, 49–54. http://doi.org/10.1016/j.copsyc.2020.03.003

Gupta, K., Ripberger, J. and Wehde, W. (2018). Advocacy group messaging on social media: Using the narrative policy framework to study Twitter messages about nuclear energy policy in the United States. *Policy Studies Journal*, 46(1), 119–136. http://doi.org/10.1111/psj.12176

Hara, N. and Huang, B.-Y. (2011). Online social movements. *Annual Review of Information Science and Technology*, 45(1), 489–522. http://doi.org/10.1002/aris.2011.1440450117

Hautea, S., et al. (2021). Showing they care (or don't): Affective publics and ambivalent climate activism on TikTok. *Social Media + Society*, 7(2), 1–14. http://doi.org/10.1177/20563051211012344

Heyden, A. D. W. V. der, Neubauer, L. and Heyden, K. van der (2020). Fridays for Future – FFF Europe and beyond. In C. Henry, J. Rockström and N. Stern (eds.) *Standing Up for a Sustainable World* (pp. 197–211). Cheltenham: Edward Elgar Publishing Limited. http://doi.org/10.4337/9781800371781.00035

Hopke, J. (2024). Tweeting on a rapidly warming planet: Environmental communication social media research. In A. Carvalho and T. R. Peterson (eds.) *Handbook of Environmental Communication*. De Gruyter Mouton. https://www.researchgate.net/publication/369553743_Tweeting_on_a_Rapidly_Warming_Planet_ Environmental_Communication_Social_Media_Research.

Hopke, J. and Paris, L. (2022). Environmental social movements and social media. In B. Takahashi et al. (eds.) *The Handbook of International Trends in Environmental Communication* (pp. 357–372). Oxford: Routledge. http://doi.org/10.4324/9780367275204-26

Hush, A. (2020). What's in a hashtag? Mapping the disjunct between Australian campus sexual assault activism and #MeToo. *Australian Feminist Studies*, 35(105), 293–309. http://doi.org/10.1 080/08164649.2020.1843997

Jeppesen, S. (2021). Intersectional technopolitics in social movement and media activism. *International Journal of Communication*, 15, 1961–1983.

Katz-Kimchi, M. and Manosevitch, I. (2015). Mobilizing Facebook users against Facebook's energy policy: The case of Greenpeace unfriend coal campaign. *Environmental Communication*, 9(2), 248–267. http://doi.org/10.1080/17524032.2014.993413

Kavada, A. (2012). Engagement, bonding, and identity across multiple platforms: Avaaz on Facebook, YouTube, and MySpace. *MedieKultur: Journal of Media and Communication Research*, 28(52), 28–48. http://doi.org/10.7146/mediekultur.v28i52.5486

Kowasch, M., et al. (2021). Climate youth activism initiatives: Motivations and aims, and the potential to integrate climate activism into ESD and transformative learning. *Sustainability*, 13(21), 1–25. http://doi.org/10.3390/su132111581

Lewis, K., Gray, K. and Meierhenrich, J. (2014). The structure of online activism. *Sociological Science*, 1, 1–9. http://doi.org/10.15195/v1.a1

Lippert, E. (2022). Argumentative strategies and neologisms in Greenpeace communication: Ecocide and climaticide on Instagram. *Neologica: Néologie et environnement*, 16, 173–202. http://doi.org/10.48611/isbn.978-2-406-13219-6.p.0173

Masciantonio, A. et al. (2021). Don't put all social network sites in one basket: Facebook, Instagram, Twitter, TikTok, and their relations with well-being during the COVID-19 pandemic. *PLoS ONE*, 16(3), 1–14. http://doi.org/10.1371/journal.pone.0248384

McCaughey, M. and Ayers, M. D. (2003). *Cyberactivism: Online Activism in Theory and Practice*. New York: Routledge.

Merry, M. K. (2012). Environmental groups' communication strategies in multiple media. *Environmental Politics*, 21(1), 49–69. http://doi.org/10.1080/09644016.2011.643368

Molder, A., et al. (2021). Framing the global youth climate movement: A qualitative content analysis of Greta Thunberg's moral, hopeful, and motivational framing on Instagram. *The International Journal of Press/Politics*, 27(3), 668–695. http://doi.org/10.1177/19401612211055691

O'Neill, S. and Nicholson-Cole, S. (2009). "Fear won't do it": Promoting positive engagement with climate change through visual and iconic representations. *Science Communication*, 30(3), 355–379. http://doi.org/10.1177/1075547008329201

Orsini, A., Cobut, L. and Gaborit, M. (2021). Climate change acts non-adoption as potential for renewed expertise and climate activism: The Belgian case. *Climate Policy*, 21(9), 1205–1217. http://doi.org/10.1080/14693062.2021.1978052

Pachocińska, E. (2020). Les slogans des jeunes dans les marches pour le climat en France (2018–2019) et la construction de l'identité collective. *Academic Journal of Modern Philology*, 9, 143–153.

Padilla-Castillo, G. and Rodríguez-Hernández, J. (2023). International youth movements for climate change: The #FridaysForFuture case on Twitter. *Sustainability*, 15(1), 1–12. http://doi.org/10.3390/su15010268

Park, C. S., Liu, Q. and Kaye, B. K. (2021). Analysis of ageism, sexism, and ableism in user comments on YouTube videos about climate activist Greta Thunberg. *Social Media + Society*, 7(3), 1–14. http://doi.org/10.1177/20563051211036059

Pearce, W., et al. (2020). Visual cross-platform analysis: Digital methods to research social media images. *Information, Communication & Society*, 23(2), 161–180. http://doi.org/10.1080/13691 18X.2018.1486871

Roth-Cohen, O. (2019). #MeToo empowerment through media: A new multiple model for predicting attitudes toward media campaigns. *International Journal of Communication*, 13, 5427–5443.

Sabherwal, A., et al. (2021). The Greta Thunberg effect: Familiarity with Greta Thunberg predicts intentions to engage in climate activism in the United States. *Journal of Applied Social Psychology*, 51(4), 321–333. http://doi.org/10.1111/jasp.12737

Schäfer, M. S. and Hase, V. (2022). Computational methods for the analysis of climate change communication: Towards an integrative and reflexive approach. *WIREs Climate Change*, 14(2), 1–10. http://doi.org/10.1002/wcc.806

Segerberg, A. and Bennett, W. L. (2011). Social media and the organization of collective action: Using Twitter to explore the ecologies of two climate change protests. *The Communication Review*, 14(3), 197–215. http://doi.org/10.1080/10714421.2011.597250

Sorce, G. and Dumitrica, D. (2021). #fighteverycrisis: Pandemic shifts in Fridays for Future's protest communication frames. *Environmental Communication*, 17(3), 263–275. http://doi.org/10.1080/17524032.2021.1948435

Terren, L. and Soler-i-Martí, R. (2021). "Glocal" and transversal engagement in youth social movements: A Twitter-based case study of Fridays for future-Barcelona. *Frontiers in Political Science*, 3, 1–15. http://doi.org/10.3389/fpos.2021.635822

Tyson, A., Kennedy, B. and Funk, C. (2021). *Gen Z, Millennials Stand Out for Climate Change Activism, Social Media Engagement with Issue*. Washington, DC: Pew Research Center. https://www.pewresearch.org/science/2021/05/26/gen-z-millennials-stand-out-for-climate-change-activism-social-media-engagement-with-issue/ [Accessed 3 November 2023].

Vossen, K. C. (2021). Youth and the (de-)politisation of discourses on "green" issues: Between eco-civism and eco-activism. In *Virtual General Conference*. European Consortium of Political Research (ECPR), 30 August 2021–3, September 2021. https://dial.uclouvain.be/pr/boreal/object/boreal:253314 [Accessed 15 March 2023].

Vu, H. T., et al. (2021). Social media and environmental activism: Framing climate change on Facebook by global NGOs. *Science Communication*, 43(1), 91–115. http://doi.org/10.1177/1075547020971644

Wagener, A. (2022). *Climat et Environnement en France (2017–2022)*. Impakt Faktor. https://sysdiscours.hypotheses.org/files/2022/03/Rapport-Le-Climat-en-France-2017-2022.pdf [Accessed 30 March 2023].

Wahlström, M., et al. (2019). *Protest for a Future: Composition, Mobilization and Motives of the Participants in Fridays for Future Climate Protests on 15 March, 2019 in 13 European Cities*. https://protestinstitut.eu/wp-content/uploads/2019/07/20190709_Protest-for-a-future_GCS-Descriptive-Report.pdf [Accessed 16 July 2023].

Wielk, E. and Standlee, A. (2021). Fighting for their future: An exploratory study of online community building in the youth climate change movement. *Qualitative Sociology Review*, 17(2), 22–37. http://doi.org/10.18778/1733-8077.17.2.02

Youth for Climate BE (2023). *Youth for Climate Official Website*. https://youthforclimate.be/about [Accessed 2 January 2023].

Youth for Climate FR (2023). *Youth for Climate France Official Website*. https://youthforclimate.fr/qui-sommes-nous/ [Accessed 1 February 2023].

18

CHALLENGING POLITICS TO DO (AND BE) BETTER

Young People's Climate Activism in the Italian Political Landscape

Gabriella Sesti Osséo

Introduction

Since late 2018, several mobilisations across the globe have revolved around the topic of the climate crisis and ecological breakdown, revitalising global climate activism and gaining public attention (Wahlström et al., 2019; De Moor et al., 2020; Sisco et al., 2021; Della Porta and Portos, 2023; Svensson and Wahlström, 2023). Youth-led climate movements played a decisive role in enlivening the new "green wave" in the Global North (Neas et al., 2022), acting as "norm entrepreneurs" (Sunstein, 1996; Spaiser and Stefan, 2020) and presenting themselves as moral authorities to pressure policymakers to face the climate crisis and convince masses of citizens to do the same. By implementing a variegated action repertoire, young climate activists perform as relevant actors within the democratic arena. However, despite the efforts at politicising environmentalism and radicalising its aims, climate activists' demands are rarely translated into comprehensive policy change in Western democracies.

This chapter explores the strain between climate activism and institutional party politics in Italy by looking at (1) activists' perception of impact accomplished on specific scales and dimensions; (2) activists' conceptualisation of politics and diagnosis of climate inaction and (3) activists' potential role in shaping climate politics and policies. The methodological approach of Youth Participatory Action Research (YPAR)[1] facilitated the investigation of such research questions. Data have been generated and collected through qualitative methods, specifically via ethnography, interviews and focus groups, part of which will be reported and analysed later in this chapter.

In the following, the Fridays for Future (FFF), also known as School Strikes for Climate (S4CS) movement, will serve as an example of the post-2018 emerging social movements (1) that revolved around the climate crisis, (2) whose geographical origin and core base are located in the Global North and (3) that are led by youth. Indeed, FFF is an international movement established in 2018, as a spontaneous grassroots initiative after the protest staged by the pupil Greta Thunberg – who then was aged 15 – in August 2018 outside the Swedish parliament, asking the government to meet the commitments of the Paris Climate Agreement. FFF is now a global climate movement whose goal is to "put moral pressure on policymakers, to make them listen to the scientists, and then to take forceful action to

DOI: 10.4324/9781003396567-21

limit global warming".[2] The movement mainly comprises school or university students, children and young people who skip classes, mostly on Fridays, to demand climate action from governments and economic leaders. Thus, most of the academic literature on this subject agrees almost unanimously in defining FFF as a full-fledged youth movement in which Generation Z (1997–2012) plays a leading role, together with a small part of Millennials (1981–96) and Generation Alpha (2012–present) (Fisher and Nasrin, 2021; Biswas, 2021). FFF displays a denoting example of contemporary youth climate activism both on a symbolic and a material level of analysis. Although every social movement rises on the shoulders of others, FFF shows an authentic and lively identity that captured the attention of the media and public opinion all around the world, particularly in the Global North, where its core base is located.

In this chapter, I will concentrate on the Italian branch of FFF, a unique and peculiar case when compared to other national counterparts active around Europe and to other contemporary climate movements or previous forms of environmental mobilisations active around Italy. Specifically, since its formation, on the one hand, in terms of estimated population size,[3] FFF Italy represents the largest group of FFF after Germany within the European continent.[4] FFF is also considered the leading youth-led actor in the context of Italian contemporary climate activism. By means of the climate strike, FFF contributed to providing a connective action (van Stekelenburg and Klandermans, 2017) for the variegated constellation of locally unwanted land use (LULU) movements that, according to Loris Caruso, constitutes "one of the forms of collective action which in recent years have grown more in terms of diffusion and mobilization ability" (Caruso, 2015). In addition to that, due to its position in the Mediterranean, Italy is considered a "climate hotspot". On the one hand, Italy has counted an increasing number of extreme weather events and unprecedented drought and has recorded the hottest temperature in 2022 since 1800 (Brunetti, 2022), and on the other hand, it is still one of the world's main financiers of fossil fuels and hosts of one of the weakest green parties among the EU countries in terms of electoral consensus (Carter, 2013; Muller-Rommel and Poguntke, 2013; Rhodes, 1992, 1995; van Haute, 2016). Italy offers a vantage point to observe the unfolding of the climate crisis and its discontent.

Challenging Social Norms Through Moral Authority: Young Prophets in Times of Denial

In the intricate Italian political context, FFF stands out as a powerful agent of change, marking a significant turning point in the perception of and approach to climate issues. In the following section, I examine the positioning of FFF within this political landscape by exploring the role that young activists have carved out and exploring their significance as not just climate advocates but also as catalysts for young people's political engagement, "prophets of truth" within the broader society and crucial bridge-builders connecting varied experiences of past and contemporary forms of Italian environmentalism.

Indeed, FFF provided a tailored space of expression for politically aware young people, who framed their youth as a political position. Youth cannot be described as monolithic (Sloam, 2014). We cannot assume that they share the same background. With "political position", we highlight the invisible barriers to political participation and some common conditions that characterise younger cohorts, such as the status (studentship or low-paid workforce), the dwelling in mostly adult-controlled contexts and the lack of opportunity to articulate their preferred claims and to gain political experience in institutional politics.

The differences in class structure and stratification, as well as those of race, ethnicity, gender, sexual identity and ability, which often are twisted into real inequalities for those who belong to minorities, are a matter of fact. However, as members of their generation, within the movement, young climate activists share a similar diagnostic framing and tone of voice (Benford and Snow, 2000).

Young climate activists in FFF Italy introduce themselves to the general public as recognisable "prophets" of truth and "oracles" of the climate. This aspect is visible both at the discursive and symbolic level. Concerning the first, often in their public appearances, young activists engage in the polarisation of the discourse around the climate crisis by constructing moral dilemmas. Regarding the second, they replaced the traditional symbolic public image of white-bearded men in formal clothes speaking the truth with young – and often female – subjects, and through that they gained media attention (especially in 2019). The most evident, personalised, media-covered counter-symbol was Greta Thunberg: an underage neurodivergent girl. However, counter-symbols are not necessarily personified. The climate strike represents a practice historically characteristic of workers, which is now replaced by students. If, on the one hand, this allowed the rising youth climate movement to gain attention, on the other hand, it has been the target of dismissing discourses. Climate activists' function as prophets has been ridiculed systematically by some MPs and Heads of State or government and climate deniers. Overall, adults who are in a position of power often deny young activists political subjectivity through practices of depoliticisation and delegitimation and insist that the transition towards adulthood constitutes the *conditio sine qua* for individuals' political recognition (Thew et al., 2020). As a result, they often trivialise, tokenise and patronise young activists, enforcing an age-based power dynamic and considering them merely "recipients" of policies, even though they are also stakeholders and active participants in political life (Such and Walker, 2005).

According to the national context where they operate, in relation to the old generation of environmental movements and NGOs, youth climate activists act also as bridge-builders between different forms of environmentalism. Namely, traditional environmentalism[5] was absorbed within a fresher, more captivating and radical twenty-first-century science-based version of it: the "environmentalism of the malcontent"[6] (Arsel, 2022). The overpowering emphasis on the climate crisis made them realise that coalition building is necessary for more radical justice-oriented demands and ambitious time targets, as well as for less-green capitalist and individual solutions. By employing such a role of bridge-builders, young activists seem appealing to everyone who gravitates to the world of ecology, seeing in them the chance for a "rebranding" of the environmental cause.

Being the oracles of the climate truth or bridge-builders between social worlds, young climate activists have gradually been recognised as experts in the field. This is clear for climate education in schools – where teachers ask activists to organise classes – or for the information sphere, as several activists engaged in the realisation of self-produced podcasts, didactic books and articles in the newspaper or digital magazines. When meeting with institutional figures personified by adults in formal clothes, young activists demonstrate a higher preparation when it comes to understanding the political and social aspects of the climate crisis and knowledge of constructive solutions. Debunking the fragility of many of their arguments and unveiling "institutional greenwashing", they hold the political elites accountable and demand they take responsibility. This unveils the paradox of having experts lacking practical experience and experienced individuals with the proficiency of beginners.

Ultimately, young climate activists seemed to succeed in gaining recognition as moral authorities, enabling them to drive changes in social norms and their societal acceptability. However, the subsequent challenge that arises is understanding to what extent the activists have affected policy change and how they relate to institutional party politics. To address this question, it is crucial to delve into key concepts that derive from the literature on social movements' consequences and apply them to the Italian background.

Young Italian Activists Navigating Limits and Opportunities in Shaping Climate Policies

By outlining the contemporary Italian political opportunity structure (POS), a distinct divergence emerges between climate activism and institutional party politics with long range strategies which oscillate among hostility, negotiation and co-optation (Della Porta and Diani, 2006, p. 214). The notion of POS was first defined by Peter Eisinger (1973) as "the degree to which groups are likely to be able to gain access to power and to manipulate the political system". In other terms, POS reveals the "capacity of the political system to convert demands into public policy" and, thus, "shapes the points of access and inclusion in the policymaking process" (Kitschelt, 1986). The interactions between the actors who constitute the country's polity might create a "window of opportunity" (Kingdon and Stano, 1984, pp. 173–204) for "policy change in general and the intervention of social movement actors in particular" (Kriesi, 2004). According to Marco Giugni (2008), "social movements can have a substantial impact on public policy to the extent that the structure of political opportunities (e.g.: the political alignments within the institutional arenas, and the role of public opinion) are favourable to them".

Although FFF Italy defined itself as "political but non-partisan", the movement's demands and approaches ensure its political reference across the leftist spectrum. What stands out is a contradictory scenario where social mobilisation revolving around the climate crisis is high, but its capacity to successfully produce policy change is low. The potential causes of this gap are many and hard to identify, especially because it is an ongoing process. However, the youth climate movements' perspective can help shine a light on this particular short circuit that occurs in the space between the inputs and the outputs.

FFF has a strategy-oriented nature, rather than an identity-oriented one (Pizzorno, 1978). Having policy change as its primary goal, climate activists direct their actions mainly toward powerholders and policymakers. Hence, the election period provides a special opportunity to do so. The 2022 Parliamentary elections held historical significance for the Italian political system. They signified not only the first post-pandemic electoral process but also an exceptional convergence of events, with a summer electoral campaign unfolding amidst record-breaking heatwaves, ultimately resulting in the formation of the most far-right government in the history of the Italian Republic. Those elections constituted a critical juncture for developing a more dialogical and less confrontational relationship between the young climate activists and institutional party politics. This specific time window represented a crucial moment for FFF activists for several reasons: (1) many FFF activists were first-time voters in 2022 or first-time voters in general elections; (2) the election day also represented a moment for evaluation of the movement's work in terms of consciousness-raising around the climate issue on the general public, thus indirectly testing their agency to influence voting behaviour; (3) the emerging legislature would be in charge of the first half of the decade that is left to limit average global temperature to under +1.5 degrees; (4) it was an opportunity to influence public debate and insert the theme of climate into the electoral process, stressing substantial incoherences in party programs and giving visibility to social and climate justice proposals; (5) it was a juncture in which the movement reformulated its position towards institutional politics and conventional forms of political participation.

Contact between activists and institutional party politics remains complicated, especially considering the fast-evolving POS under which it occurs. After the election, FFF activists faced the challenge of: (1) lack of gatekeepers[7] within the institutional system; (2) the formation of the most far-right government of republican history with a large parliamentary majority; (3) an institutional discourse that delegitimises and depoliticises activists' work, often showing a patronising and devaluing attitude; (4) a rising criminalisation of climate activism through the adoption of punitive laws.

Such conditions have boosted young activists to engage in a more variegated action repertoire. In fact, most FFF activists engage in forms of unconventional political participation (Pitti, 2018), without excluding or refusing the conventional ones. On the one hand, they keep constructively criticising the authority of political institutions by demystifying arguments, denouncing contradictions, or verifying the implementation of their promises related to climate politics; on the other hand, they eventually acknowledge party politics as a valuable tool for climate action and decide to run for office.

Indeed, FFF tried not only to raise awareness among the general public but also to influence media, political debates, electoral campaigns and results and advocated for climate justice through policy proposals. As FFF behaves like a strategy-oriented (externally directed) rather than identity-oriented (internally directed) type of movement, it is even more relevant to explore to what extent FFF Italy achieved its declared goal of influencing the policymaking community in taking climate action. This is particularly true considering that the movement's claims have been "parliamentarised" and that over its four years of life it has shown how the timing of its claims-making comes to "depend more closely on the rhythms of parliamentary discussion and governmental action" (Kriesi, 2004, p. 67).

To evaluate FFF's long-term impact will require longitudinal studies. Social scientists have been debating about potential indicators to measure success or variables to consider when analysing the outcomes of grassroots mobilizations. Some scholars prefer to concentrate on the biographical and life-course consequences rather than the cultural and political ones (Giugni, 2008; Vestergren et al., 2017). As political and social outcomes trajectories could be multidimensional and dependent on mediating factors, most of the time, the operationalisation of the hypothesis is complex and might lead to "tentative answers" (Rucht, 1999) or contradictory findings. Indeed, the effects of social movements could often be indirect, unintended or unrelated to their explicit claims (Giugni, 1998). However, it is possible to assess young activists' perceptions of effective actions and achieved changes, which ultimately establish the preliminary analysis for an evaluation of the repertoire of action engaged in as well as of the attitude adopted towards institutional and party politics. The following results of a participatory action study with the use of focus groups can help provide further insights.

Activists' Perception of Change. A Self-evaluation on the Efficacy of Climate Movement Actions

The focus groups' work was articulated in nine sessions organised with local branches of FFF Italy. Three of these, respectively, belonged to the Northern, Central and Southern part of Italy,[8] although they were chronologically conducted in a different spatial order. The sampling of cases included the FFF local group of Torino, Brescia, Pavia, Pistoia, Roma, Civitavecchia, Bari, Aversa and Catania. The fieldwork included this number of focus groups intending: (1) to build a more representative sample of the Italian FFF movement; (2) to incorporate diverse clusters in terms of latitudes, city dimensions and size of the group; (3) to compare and identify potential patterns or variability of responses. Altogether, focus groups involved 42 FFF young activists (21 female, 21 male), with an average age of 20 years old, having participants between 16 to 27 and a diverse level of seniority as activists.

The activity consisted of an exercise aimed to explore young climate activists' perceptions of their agency to affect change across several impact scales. Participants had ten minutes to individually write down all the impactful actions according to their personal criteria, including those in which they participated and those implemented by FFF nationwide and worldwide. After a small break, I unveiled a grid inspired by and adapted from the one drawn by Amelia Clarke and Ilona Dougherty (2010). The grid was composed of two axes: on one side, the level of impact (namely: individual, local/internal, national or European/international) and on the other, the strategy of change implemented (awareness-raising, influencing or direct action). Participants were asked to place their own post-its in a specific table's square.

Table 18.1 represents an overview of the data resulting from the nine focus groups. It includes the most recurring features which have been grouped in accordance with participants' placement. As summarised in the grid, young climate activists perceived that their individual and collective actions as a movement were particularly successful in terms of:

1 raising awareness by turning the climate crisis into a popular issue, making science more accessible through media presence, local events and climate education projects
2 influence by ensuring the presence of the climate crisis issue in political agendas, debates, public discourse, electoral campaigns, parties' electoral programs and reinvigorating green parties across Europe
3 direct impact by engaging in conventional politics by running for office, mobilising people to vote for the climate, producing policy proposals and framing the climate as a problem of social justice

When focusing on the national level, young activists do not identify relevant policy change, apart from the modification of the Constitutional Charter via L. Cost. 1/2022.[9] During the exercise, participants explored this problem and diagnosed it in some cases in the party politics' decay or the moral degradation among politicians, in others in the functioning of representative democracy and its institutions. In both cases, activists denounced the ineffectiveness of the political system in receiving their demands and acting towards their fulfilment.

In the following section, I will collect FFF activists' critical conceptualisation of politics and its normative meaning, together with their diagnostic analysis of climate inaction. Through the analysis of this data, it is possible to better grasp the strain between climate activism and institutional party politics in Italy as well as identify the potential future development of such complex relations.

Table 18.1 Overview of data resulting from nine focus groups, including the most recurring features which have been grouped in accordance with participants' placement.

Strategy of change/level of impact	Individual	Local	National	EU/international
Raising awareness	Political consumerism Self-education Personal development	Schools and other places of socialisation Public spaces Increase of hope and a sense of possibility	Media presence Recognition as experts with political agency Infiltration with other local groups or movements	Culture change Accessible science Cool sustainability Climate change as massive issue Voice of a generation
Influence	Being a multiplier of change among family members and friends Places of socialisation adopt sustainable practices Other people's voting behaviour Community inclusion Validation of one's own feelings	Political agenda setting Local policies and local councillors Connection with other environmental movements Nationalisation of local issues Localisation of national issues	Climate crisis in political agendas, debates, electoral campaigns, parties' electoral programs Reconciliation between youth and institutional politics Interest of politicians in activists Generate public opinion "Convergence" between the movement and labour struggle	First-time demonstration for many young people Accountable politicians Stronger EU Greens EU Green Deal Greta as a symbol of counter-values
Direct impact	Diet change (vegetarian or vegan) Public mobility over private Second-hand purchases Boycott Activism Role model Change of bank Peer support	Candidateship Discontinuation of climate-damaging projects or locally undesirable land uses Mental health improvements Consistent presence within the territory	Termination of climate-damaging projects Pressure policymakers Constitution reform (L. Cost. 1/2022) Policy proposals Formation of Left/Green Alliance party Just transition framework	Outcomes of COP26 and COP27 Politicisation of the climate crisis Climate litigation Creation of other environmental and climate movements EU Commission sensitive to the topic of climate crisis More ambitious climate targets

Source: Original creation.

"This Is All Wrong". Activists' Conceptualisations and Normative Meanings of the Political System

As discussed earlier, young climate activists are challenging the normative core of the climate crisis. They question its fundamental assumptions, revealing the webs of dysfunction and domination within the socioeconomic and political order. This section explores how the activists inspire change by naming systemic flaws and advocating for the profound remaking of history. To do so, we will explore their views by analysing their discourse, namely their critique of the political system, political élites, as well as their definition of generational tensions, political representation and attitude towards institutional and party politics.

As young climate activists' provocation applies to social norms, the normative action engaged questions also about the actual meaning of the political in its broader sense. FFF activists have a resolute opinion of what "true" politics should be:

> I do not want to associate politics with power advantages and talk shows. It is a horrible thing. That is not politics. . . . FFF is politics in its true and pure sense. It is closely intertwined with sociability and humanity. . . . FFF insists on creating a sense of community. . . . It is so beautiful. What unites people is sacred to me.
>
> *(Arianna, 24 y/o)*

Young climate activists also defined the role and the purpose of political institutions:

> If you look at the Latin etymology of the term "minister," it is the one who serves, the one who puts others above themselves. . . . Ministers should be of service.
>
> *(Lorenza, 20 y/o)*

FFF activists denounced politics' essential betrayal of its own raison d'être:

> Politics is supposed to be the tool aiming to solve society's problems. Instead, it seems like it is doing the opposite: creating problems. For years.
>
> *(Thomas, 21 y/o)*

> Right now, there is no politics of the *polis*, but only the politics of those seated in power.
>
> *(Pietro, 17 y/o)*

Building on their opinions about "true" politics, young climate activists extend their critique of the current democratic system and its functioning.

Critique of the Political System and Functioning

The analysis now delves into the activists' perspectives on political institutions, revealing a collective sentiment that characterises the current democratic system as distorted and broken. Namely, they identify a disruption in the cycle between inputs and outputs: civic demands used to remain stuck in a "blackbox" (Easton, 1965) without being converted into policies (Wellstead et al., 2013). Indeed, many climate activists describe politics as a realm "dominated by immobility" (Massimiliano, 25 y/o) or inefficiency.[10] Nicolò (22 y/o) proposes, as a representative example, the presence of a significant delay in the reception of social and cultural change already underway:

> Social change is much faster than political change. . . . Politics is not able to keep up with the extremely fast-changing needs of citizens *and* global developments because of its cumbersome and time-consuming inner workings. . . . For example, in my opinion, when it comes to climate change and its related problems, there is already an ongoing huge social change, as people are now much more aware of the topic. However, we cannot see political change in progress yet.

Such a disruption might come from what Samuele (22 y/o) has identified as "a misunderstanding and wrong attitude based on suspicion" between activists and power holders as soon as they confront each other. The origin of this distrust is traced by the young environmentalists to class-based division. On several occasions, when discussing political élites, activists used the juxtaposition between the two subjects, "we" and "they",[11] identified as separated and opposed not only for the difference of their role within the political system but also for their collective class belonging. Young activists charge political élites of "having and promoting the same interests of big corporations" (Mathias, 21 y/o), which makes them "lose contact with reality" (Arianna, 24 y/o). Indeed, when talking about the world-system analysis, young activists refer to these élites with the term "ruling class".

> Politics has lost much of its public dimension and has become entangled in one of capitalism and corporate greed. Our politicians are often members of such companies, which as a result influences the parties' leadership and the choices of policymakers at all levels.
>
> *(Maria Domenica, 23 y/o).*

> Political elites do not want to break these immoral imbalances based on profit and power. They do not want to act with responsibility and admit that they are acting in their own interests. Political élites are detached from people, as though they are restricted to their specific caste. Even during elections, we see all these seemingly different parties . . . act in the same way in the end. . . . There is no radical change as they share the same mentality, which is the same as the upper echelons of society.
>
> *(Giulia M., 19 y/o)*

Critique of Élites

The image that FFF activists draw of political elites is severe. When discussing the quality of candidates and elected politicians, young activists diagnose what they have defined as a significant lack of competence and knowledge about the complexity of contemporary political issues, including the climate crisis.

> I am unaware of the number of politicians that know of something other than plastic bottles and cups. . . . This might lead to "unaware greenwashing" – by that I mean, the government might commit to solving specific superficial problems but in reality, their effects are just drops in the ocean. An example is banning things like disposable

cutlery. Similar to what certain companies do, intentionally or not . . . they try to make it look like something is being done, but actually, it is almost useless.

(Massimiliano, 25 y/o).

These people are not prepared for this type of job. . . . Unfortunately, they do not have an example of a past generation that has had to deal with increasingly massive pollution and a brutal climate crisis that has taken place over 50 years. As a result, they too do not know how to act.

(Maria Domenica, 23 y/o).

Climate activists show concern about politicians' limited awareness of environmental issues and inadequate preparedness for dealing with challenges posed by the climate crisis. According to them, this generally induces the adoption of superficial measures that create an illusion of progress with minimum impact or political inaction. According to research participants, such incompetence can be intentional or not, as not having the example of past generations who have successfully navigated similar cases leads to a potential lack of knowledge on effective action.

Generational Tension

According to climate activists, the poor quality of politicians is due to a combination of factors. The most recurrent one is generational, even if it intersects with other factors on a more systemic level. Because of their age, activists perceive that politicians do not experience the same level of anxiety, fear, frustration and anger as the youth concerning the future and the climate. Activists affirm having an embodied and rational understanding of how the course of their lifetime will be impacted by the effects of politicians' inaction regarding the climate crisis. According to the research participants, adult power holders do not share the same feeling of urgency; indeed they lack what Bjork (20 y/o) has called an "emotive connection with the problem and, most importantly, the people who suffer or will suffer for it". As affect and emotions represent significant drivers of climate-change perceptions and actions, they facilitate a significant commitment to climate action (Brosch, 2021; Stanley et al., 2021). Therefore, when these emotions are missing, it is harder to understand how pervasive the climate crisis is and why it should be considered a priority. Thus, the numbness and lack of empathy reinforce the incapacity to recognise the problem and, subsequently, to address it adequately. Young climate activists perceive such a difference in sensitivity and pragmatism as a generational gap. Being young in itself is understood to be a condition that eases a peculiar way of feeling political issues and "doing politics", different from the already existing and established modalities. In this context, being young represents an aspect of identity; thus, FFF activists feel the need to affirm their diversity by distinguishing themselves not only from other climate movements or *disobbedienti* or *antagonisti* militant groups[12] but especially from contemporary institutional politics.

Activists defined institutional politics as a "system of rotation" (Mattia, 19 y/o) where "low-quality candidates take turns" (Marco M., 20 y/o) alternating each other, "just like the seasons" (Elia, 21 y/o). In this sense, as concluded by Mathias (21 y/o), the urgently needed political change appears "impossible if requested by those who are already an active part of the problem". One solution to stop such a cycle of things might be to "have a generational change in politics" (Livia, 18 y/o). Also due to this interpretation of the facts, they present themselves as an actor with a political agenda on the democratic stage. In doing so, on the one hand, they persuade peers that they are capable and eligible to participate in politics (Vromen et al., 2015), reinvigorating youth political participation and reconciling youth with institutional politics.

> FFF helped young people to get back into politics. Such a dialogue was non-existent, apart from the baronage within political parties, universities and all the so-called "high" circles.
>
> *(Maria Domenica, 23 y/o)*

On the other hand, they convince power holders of their condition of rights and expertise without compromising on their role of bridge-builders or, as Elisa (16 y/o) defines it, the role of "repairers":

> Italian politics . . . is not a complete loss; it should not be totally demolished. Italian politics has a good foundation, an excellent theoretical basis and also a good practical implementation. It just needs to be repaired. We need to fix a little here and a little there, but it is not true that everything is falling apart – it is a bit of bullshit, to be honest. . . . Complaining is okay, but also recognising that not everything is going to shit is necessary. It just needs to be fixed.

Political Underrepresentation

Since young activists acknowledge the centrality of conventional politics as one of the most powerful tools to tackle the climate crisis, some of them decided to run for office, mainly for local councils.[13] For those who decided to run, the candidacy represents both a way to overcome the impasse and respond to their need for self-representation in politics:

> At a time when the great ideologies have collapsed, and youth can no longer believe in parties, factions, etcetera . . . we create something to believe in. Thus, we, young people, are the ones who draw what society is, what humanity looks like and what we hope the future will be. It is natural that there are differences among young people engaged in politics. . . . However, the fact that young people are being questioned in ever-growing discourses . . . is nonetheless significant in my opinion. Because if we have to inherit this world, we must also make it within our reach.
>
> *(Maria Domenica, 23 y/o)*

> We are not in government. This is why they are not doing enough for the climate crisis. Because we are not there! There is only male, pale and stale. We are not represented in any way. Of course, they do nothing about the climate crisis. Why should they care?
>
> *(Arianna, 24 y/o)*

This need for self-representation is a direct result of the lack of representativeness between contemporary Italian society and political élites of which young activists are keenly aware. Namely, the most-mentioned subjects considered underrepresented in politics are women and young people. While venting her frustration, Arianna (24 y/o) says: "I am fed up with white straight, cisgender males who do not even bother to put themselves in someone else's shoes". Livia (18 y/o) suggests, "we need a politics made up of young people, not septuagenarians who say everything is going to be okay". The point is made even more evident by Fiore (18 y/o) who ironically asserts: "It is hard to believe that governments will take climate action seriously when the youngest governor is 72 years old, having been born before the steam-engine".

For this reason, stepping into politics has been slowly considered a worthwhile strategic option in the fight for climate justice in Italy. The possibility of engaging in party politics periodically animates movements' internal debates, also at the moment of writing. Indeed, the movement is more open to dialogue and recognises overall that antagonising or dismissing institutional politics is everything but advantageous. Borrowing Carlo's (16 y/o) definition, FFF remains "a grounded, pragmatic and non-rhetorical movement".

Activists with seniority within the movement have initially looked at institutional representatives mainly as knowledgeable but unwilling to take care of the climate crisis; later, they considered them unskilled and unqualified.

> When the movement was born, it seemed absurd to imagine climate activists interviewing political candidates. Then it happened: we went and talked to them and ended up unsatisfied. Because we realised we were the only ones well-prepared.
>
> *(Marzio, 25 y/o)*

> A fundamental problem at the base of everything, and one of the things we care about and work on as FFF, is the youth's lack of political skills. . . . We need to combine political skills and environmental knowledge; otherwise, the government will not make decisions in line with the needs of the younger generations.
>
> *(Salvatore, 18 y/o).*

Likewise, if at the beginning of their activism, young environmentalists perceived politics as a distant reality, in the future they may start to feel like it is an area easy to approach and step into.

> I would never have thought in my life "yes, one day I could become a politician" even at the level of my small town. Now I am thinking about it".
>
> *(Thomas, 21 y/o)*

A Developing Attitude Towards Institutional and Party Politics

The strain to adjust or revisit the movement's repertoire of action and to assume a more-or-less adverse attitude towards institutional politics reflects two main changes that occurred over time, on a collective and individual level, across the FFF movement.

1 The collective change consists of the gradual politicisation of the climate crisis. Specifically, the acknowledgement of the structural causes that fuel climate and biodiversity collapse made the movement more radical (as it aims to change the roots of the problem). Concurrently, in light of the time constraints of the physical world, we are more aware of the necessity to work with the tools available (i.e. institutional politics). The

latter aspect reminds us of the old and open debate about reform or revolution, which was also explored during the focus groups. Alice Q. (17 y/o) expresses this dilemma by wondering "whether we actually need a revolution and have the time to do it or, instead, we should put pressure on existing governments to change things". According to different personal sensitivities, activists gave different answers to whether "the Master's tools will ever dismantle the Master's house" (Lorde, 1984). Though, because all of them consider the climate crisis rightfully as a comprehensive and complex phenomenon, the general feeling is that every tool is valid and that people with a climate education and ecological sensitivity should be everywhere.

2 The individual change that occurred over time was the further development of the dimension of injustice within the climate crisis and the deepening understanding of the political nature of its solutions. Quoting Alice F. (22 y/o): "both on a personal level and a movement level: our movement was born as 'let's get rid of plastic', we got to being 'against capitalism'. It means that there was a process in action". Along the activists' journey, the comprehension of the climate issue was not the only conceptualisation to evolve but also the reflection on their political agency (for better or for worse) both as climate activists and young people, the meaning of radicality or the relative perception of violence. Using Simone's (27 y/o) words:

> When I hear the word "radical", I combine two thoughts: "root of the problem" and "extremism". I started to interpret "radical" as something that related to the "root of the problem" not too long ago – and it has unburdened the word "radical" a bit. . . . Once you see it in this light, you say: "damn, that's the way to take action and solve problems" because if you don't intervene on the roots, the problems will reappear. . . . As a person who has always reflected on these things but has never had peers to discuss these topics, the word "radical" has always meant "extremism," which in a society that does not tolerate extremes, it frequently elicits an opposing reaction. Still, we really like being at the centre, being moderate, and doing things calmly . . . it is a powerful thing . . . a reflection that perhaps is not often done.

Nevertheless, political parties and FFF mutually recognised their role within the democratic arena.[14] In a way, the relationship between the movement and political parties took the form of co-dependency. On the one hand, climate activists need institutional politics to ensure mitigation and adaptation policies for a just ecological transition. On the other hand, political parties need climate activists to reinforce their electoral power, reinvigorate their image and get others on the bandwagon of the green wave, which ironically originated from their political inefficiency in tackling the climate crisis. However, in the specific context of the climate issue, even if the theoretical rules of the role-playing game of the two actors are well-known (Goldstone, 2003; van Stekelenburg et al., 2013) and move among the lines of "contention and convention" (McAdam and Tarrow, 2010), the democratic circuit seems interrupted or broken: movements have sometimes substituted parties as mediators of citizens' needs. As noticed by Lorenza (20 y/o), "the issue of the climate crisis has been brought to attention more by social movements than by political parties".

Some parties have tried to recruit candidates from among the movement's ranks to regain credibility and reinforce the relationship with the community of young activists. In some cases, the invitation to join the party has been met with resentment and other times as an opportunity to bring the fight for climate justice someplace other than the streets. In any case, young activists expressed their discontent with the "missed" evolution of parties in framing and addressing the climate crisis for what it is. This is particularly valid when considering the Italian Greens, a striking example of a "heated kinship" (Berker and Pollex, 2021; Corry and Reiner, 2021; Sloam et al., 2022) in which the more affinity FFF feels, the more conflictual their relationship is.

Indeed, on the one hand, FFF debunks the media's simplification of the equation between the movement itself and the party by clarifying their nature as two different subjects. On the other hand, the Green Party draws on the movement as an energy tank, an opportunity for rebranding the party and a source for recruiting new candidates. The result is that, on the one side, the movement does not consider the Green Party as politically reliable and ethically intact; on the other side, the Green Party jumped on the bandwagon of the "green wave", agitated by the movement. As the latter put at the top of its political priorities the issue of climate protection, activists' frustration of not gaining much influence over it is prominent compared to other political parties' interlocutors. FFF activists recognised that the Green Party did not make the same progress as environmental movements, even blaming them for not being based in science while also being attached to old environmentalism:

> Those who identify themselves in the left and centre-left clash with the absence of radicality – intended as the will to intervene at the root of the problem in an adequate manner, as well as with substantial immobility. Parties still think about ecologism in the "classical way". For example, looking at the record of the Greens in Italy . . . they are somewhat anchored in the most outdated positions . . . without being able to acknowledge the most urgent issues. . . . However, for those who identify more within the right wing, their so-called right-wing ecology is also incapable of structural thinking, in a way that goes beyond their own country. . . . Perhaps they focus a lot on recycling or environmental protection – interpreted in nationalist terms as Motherland – such as forests, and parks . . . but when the consequences of those actions have little direct effect locally, but unbeknownst to them, greater effect globally they give no importance to the effort. We must overcome this inability to not think in a systemic and long-term way and closer to a global perspective.
>
> *(Lorenza, 20 y/o)*

Overall, young climate activists navigate a complex relationship with institutional politics, critiquing its actors and functioning. The tension between remaining outside the party political sphere and actively engaging with it becomes particularly pronounced during critical electoral events, like the parliamentary election that took place in September 2022 in Italy. Despite their critique, young activists maintain an optimistic and constructive spirit. Some are actively stepping into party politics, showcasing a commitment to reshaping the system from within. This ongoing transformative journey underscores the activists' potential to inspire positive change in addressing the pressing challenges of the climate crisis.

Conclusions

This chapter explored the strain between climate activism and institutional party politics in Italy by adopting young climate activists' perspective. Young people played a key and catalysing role in the rise of the most recent wave of climate activism and in putting moral pressure on policymakers. They presented themselves on the global stage by claiming their own victimhood of human-induced climate-related crimes, dramatically introducing the generational aspect as an unavoidable dimension of justice. They did so by experiencing and promoting an intellectual and emotional understanding.

FFF Italy stands out as a relevant example of the broader phenomenon of young people's climate activism and as a specific case to observe climate activists' perspectives on the deeper meaning of politics and their relations with institutional party politics. The data collected sheds light on the activists' perception of the achievements and on the challenges faced by the movement. If, on the one hand, FFF activists believed they had successfully raised awareness of the climate crisis and influenced social norms through their moral power, on the other hand, they were not convinced that they truly affected policy change. In fact, FFF activists succeeded in impacting only the first stage of the policy process, namely: problem emergence, agenda setting and consideration of policy options and not the entire cycle (Adelle et al., 2012; Benson and Jordan, 2015). According to the activists, this happened because of a lack of gatekeepers within the party system as well as a lack of like-minded politicians with relevant electoral power.

As the time dimension has a crucial value in mitigating the devastating consequences of the climate crisis, young activists will likely face internal and external challenges. On the one side, a potential demotivation of the members of the movement and a resultant demobilisation might occur. On the other side, the movement risks losing public attention while being criminalised, dismissed and patronised. This unravelling scenario suggests FFF and the broader climate justice movement should rethink its theory of change as well as its relations with institutional party politics.

The post-2018 climate movements represented polarising agents (Hutter and Kriesi, 2020) who gave a boost that re-politicised the climate crisis (Marquardt and Lederer, 2022). This constitutes an opportunity also for institutional political players to structure a new "cleavage" in party competition (Lipset and Rokkan, 1967). Nevertheless, political parties took advantage of it only on symbolic and discursive levels and not on material and policy-oriented ones.

As foreseen by climate scientists, global heating will intensify its effects on people's lives over time, provoking a variety of democratic challenges and conflicts and a potential new axis of contentious politics (Tilly and Tarrow, 2015). Further research and longitudinal studies will help in understanding the development of climate activists' role in such a vital political battleground.

Acknowledgments

This chapter would not be possible if it was not for all the Fridays for Future activists who agreed with sincere joy and enthusiasm to participate in the research. I felt privileged and grateful for the relationship of trust we built together. In alphabetic order, thanks to: Arianna, Bjork, Carola, Chiara O., Chiara T., Carlo, Elia, Elisa, Fiore, Giacomo P., Giada, Giulia M., Giulia S., Luca, Livia, Ludovica, Massimiliano, Mattia, Marco L., Maria Domenica, Marianna, Mariarita, Martina, Marzio, Michele, Mina, Nicolò, Roberto, Ruggero, Salvatore, Samuele, Sara C., Simone, Thomas and Valentina. Additionally, special thanks to the group coordinators Alice F., Alice Q., Emanuele Akira, Ferdinando, Giovanni, Marco M., Mathias, Pietro and Sara D., who were great helpers and organisers, taking care of group composition and the arrangement of the logistics. Last, thank you also to the Fridays for Future activists who did not join the focus groups but showed interest in my work and found the time to exchange views with me: Agnese, Francesca, Giacomo Z., Giorgio, Giulia G., Lorenzo, Luca S., Sofia. Each one of you has been irreplaceable, just like you are in the fight for climate justice.

Note: According to the activists' preferences, participants' names reported are in some cases real, while in others they are pseudonymised. When possible, I decided to state participants' names as a sign of acknowledgement of their authorship.

Notes

1 Youth Participatory Action Research (YPAR) approach is a collaborative, interactive, maieutic and practice-oriented technique in which young participants are involved as co-producers in some or all stages of the research.
2 Fridays for Future. *Who We Are*. Fridaysforfuture.Org. Accessed 16 June 2021. Retrieved from https://fridaysforfuture.org/what-we-do/who-we-are/
3 Because of the spontaneous and constantly evolving nature of social movements, defining their population size is often hard and might draw only a temporary picture of the situation. The data we are referring to here should be considered approximative and dicey. They have been collected via interviews with national referees and have not been crossed with other sources.
4 At the moment of writing, Italy and Germany count the same number of activists nationwide, which is approximately 250–300 people. However, the number and the type of distribution of local groups are different: FFF Germany is articulated in around 100 active local groups, while FFF Italy hosts about 40–50 local cells. In both cases, the local groups are mainly present in bigger cities, have a considerable degree of autonomy and experienced a regression of participation between 2020 and 2021, resulting from the COVID-19 outbreak.
5 Traditional environmentalism was the hegemonic or dominant sector of the movement, mainly concerned with narrowly defined quality-of-life issues and other elitist matters like endangered species and wilderness preservation. This form of environmentalism generally trusts lifestyle solutions to address ecological and climate crises and embraces a dualistic ontology of humans and nature as different realms. Within traditional environmentalism, the dimension of justice is absent or not prioritised. As it is mainly composed of western, white upper- or middle-class urban citizens, the issues of survival, poverty, housing, global economic inequalities and social justice are not considered part of the environmental and climate struggle.
6 For Arsel et al. (2015, p. 17), "the 'environmentalism of the malcontent' concept refers to the type of environmentally themed activism whose motivations are primarily – though of course not exclusively – animated by a political economic posture that is not informed by direct or immediate personal livelihood concerns". Here, the shared malcontent acts as "a mobilizer itself implies a camaraderie and solidarity against the state" (Akbulut, 2014, p. 20). So, the environmentalism of the malcontent requires by definition cross-class collaborations (Arsel, 2022).
7 According to David Easton, the gatekeepers are "not only those who initiate a demand by first voicing it; the term also designates those whose actions, once a demand is moving through the channels of the system, at some point have the opportunity to determine its destiny". Thus, gatekeepers represent "the key structural elements in determining what the raw materials of the political process will be" (Easton, 1965, p. 88).
8 The reference is to the orthodox division of conventional country's political maps. The three mentioned areas do not necessarily correspond to a rigid political-geographical category. For example, Civitavecchia's socio-economical background is similar to a Southern reality rather than one of Central Italy, despite its position.
9 The constitutional reform establishes the recognition of the fundamental value of the environment, explicitly stating its status as a constitutionally protected common good. The reform primarily affects Article 9 of the Constitution, which has been enriched with a new third paragraph, according to which the Republic protects – in addition to the landscape and the nation's historical and artistic heritage – "the environment, biodiversity, and ecosystems, also in the interest of future generations". It continues with the provision that "the State regulates the methods and forms of

animal protection". Article 41 of the Constitution has been also subject to amendments, which now expressly state that private economic initiative cannot be carried out in a manner that damages health and the environment, as well as safety, freedom and human dignity. Furthermore, the last paragraph of the mentioned article includes "environmental purposes" alongside the social purposes towards which public and private economic activity can be directed and coordinated.

10 Among the causes of inefficiencies, every focus group has mentioned the issue of the short-term horizon and different timing of politics and physics. In the Italian political system, this appears particularly relevant, as in Roberto's (21 y/o) words: "one government's work might be completely messed up by the subsequent government".

11 The we/they moves across the lines of the age difference (young/old), class ("normal-average people"/ reach) and social role (activist/power holder). Some early studies on Greta Thunberg's communication (Martínez García, 2020) underline the same juxtaposition but with the first person singular "I".

12 *Disobbedienti* and *antagonisti* groups emerged within the political tradition of *Centri Sociali Autogestiti* – CSA (Autonomous Social Centers). *Antagonisti* engage in confrontational processes and tend to consider their social centre as a microcosm, separate from the external world and its logic, whereas *Disobbedienti*, whilst sharing with *Antagonisti* conflicting practices, also build "bottom-up welfare" and are less self-referential and more open to exchanges with the existing social reality (Becucci, 2003).

13 So far, some FFF activists decided to run as candidates in Brescia, Monza, Torino, Genova and Massa for the local elections and in Lombardia and Lazio for the regional ones.

14 This has been confirmed by the participation of MP candidates in public meetings with climate activists at the local level. Right-wing parties never showed up on this occasion, except for the centre-right Azione – Italia Viva.

References

Adelle, C., Jordan, A. and Turnpenny, J. (2012). Policy making. In A. Jordan and V. Gravey (eds.) *Environmental Policy in the EU. Actors, Institutions and Processes* (pp. 235–252). Routledge.

Akbulut, B. (2014). Neither poor nor rich but "Malcontent": An anatomy of contemporary environmentalisms. *Marmara Üniversitesi Siyasal Bilimler Dergisi*, 2, 9–24.

Arsel, M. (2022). Climate change and class conflict in the Anthropocene: Sink or swim together? *The Journal of Peasant Studies*, 1–29.

Arsel, M., Akbulut, B., Adaman, F. (2015). Environmentalism of the malcontent: Anatomy of an anti-coal power plant struggle in Turkey. *Journal of Peasant Studies*, 42, 371–395.

Becucci, S. (2003). Pratiche di sovversione sociale: il movimento dei disobbedienti. *Quaderni di Sociologia*, 5–20. https://doi.org/10.4000/qds.1159

Benford, R. D. and Snow, D. A. (2000). Framing processes and social movements: An overview and assessment. *Annual Review of Sociology*, 26(1), 611–639.

Benson, D. and Jordan, A. (2015). Environmental policy: Protection and regulation. In J. D. Wright (eds.) *International Encyclopedia of the Social & Behavioral Sciences* (pp. 778–783). Elsevier.

Berker, L. E. and Pollex, J. (2021). Friend or foe? – Comparing party reactions to Fridays for Future in a party system polarised between AfD and Green Party. *Zeitschrift Für Vergleichende Politikwissenschaft*, 15(2), 1–19.

Biswas, T. (2021). Letting teach: Gen Z as socio-political educators in an overheated world. *Frontiers in Political Science*, 28.

Brosch, T. (2021). Affect and emotions as drivers of climate change perception and action: A review. *Current Opinion in Behavioral Sciences*, 42, 15–21. https://doi.org/10.1016/j.cobeha.2021.02.001

Brunetti, M. (2022). *Climate Monitoring for Italy*. CNR-ISAC.

Carter, N. (2013). Greening the mainstream: Party politics and the environment. *Environmental Politics*, 22, 73–94. https://doi.org/10.1080/09644016.2013.755391

Caruso, L. (2015). Theories of the political process, political opportunities structure and local mobilizations. The case of Italy. *Sociologica*. https://doi.org/10.2383/82471

Clarke, A., Dougherty, I. (2010). Youth-led social entrepreneurship: Enabling social change. *International Review of Entrepreneurship*, 8(2), 1–23.

Corry, O. and Reiner, D. (2021). Protests and policies: How radical social movement activists engage with climate policy dilemmas. *Sociology*, 55(1), 197–217.

Della Porta, D. and Diani, M. (2006). *Social Movements: An Introduction* (2nd ed.). Blackwell Publishing.

Della Porta, D. and Portos, M. (2023). Rich kids of Europe? Social basis and strategic choices in the climate activism of Fridays for Future. *Italian Political Science Review/Rivista Italiana di Scienza Politica*, 53(1), 24–49.

De Moor, J., Uba, K., Wahlström, M., Wennerhag, M. and De Vydt, M. (2020). *Protest for a Future II: Composition, Mobilization and Motives of the Participants in Fridays for Future Climate Protests on 20–27 September, 2019, in 19 Cities Around the World*. https://osf.io/3hcxs/download

Easton, D. (1965). *A Systems Analysis of Political Life*. John Wiley & Sons.

Eisinger, P. K. (1973). The conditions of protest behavior in American cities. *American Political Science Review*, 67(1), 11–28.

Fisher, D. R. and Nasrin, S. (2021). Shifting coalitions within the youth climate movement in the US. *Politics and Governance*, 9(2), 112–123.

Giugni, M. (1998). Was it worth the effort? The outcomes and consequences of social movements. *Annual Review of Sociology*, 24, 371–393. https://doi.org/10.1146/annurev.soc.24.1.371

Giugni, M. (2008). Political, biographical, and cultural consequences of social movements: Consequences of social movements. *Sociology Compass*, 2, 1582–1600. https://doi.org/10.1111/j.1751-9020.2008.00152.x

Goldstone, J. A. (eds.) (2003). *States, Parties, and Social Movements, Cambridge Studies in Contentious Politics*. New York: Cambridge University Press.

Hutter, S. and Kriesi, H. (2020). Politicizing Europe in times of crisis. In J. Zeitlin and F. Nicoli (eds.) *The European Union Beyond the Polycrisis?* (pp. 34–55). London: Routledge.

Kingdon, J. W. and Stano, E. (1984). *Agendas, Alternatives, and Public Policies* (Vol. 45). Boston: Little, Brown.

Kitschelt, H. P. (1986). Political opportunity structures and political protest: Anti-nuclear movements in four democracies. *British Journal of Political Science*, 16(1), 57–85.

Kriesi, H. (2004). Political context and opportunity. In *The Blackwell Companion to Social Movements* (pp. 67–90). Blackwell Publishing Ltd.

Lipset, S. M. and Rokkan, S. (eds.) (1967). *Party Systems and Voter Alignments: Cross-National Perspectives* (Vol. 7). New York: Free Press.

Lorde, A. (1984). The Master's tools will never dismantle the master's house. In A. Lorde (ed.) *Sister Outsider: Essays and Speeches* (pp. 110–114). Crossing Press Berkeley.

Marquardt, J. and Lederer, M. (2022). Politicizing climate change in times of populism: An introduction. *Environmental Politics*, 31(5), 735–754.

Martínez García, A. B. (2020). Constructing an activist self: Greta Thunberg's climate activism as life writing. *Prose Studies*, 41, 349–366.

McAdam, D. and Tarrow, S. (2010). Ballots and Barricades: On the reciprocal relationship between elections and social movements. *Perspectives on Politics*, 8, 529–542. https://doi.org/10.1017/S1537592710001234

Muller-Rommel, F. and Poguntke, T. (2013). *Green Parties in National Governments*. Routledge.

Neas, S., Ward, A. and Bowman, B. (2022). Young people's climate activism: A review of the literature. *Frontiers in Political Science*, 4, 940876.

Pitti, I. (2018). *Youth and Unconventional Political Engagement*. Springer.

Pizzorno, A. (1978). Political exchange and collective identity in industrial conflict. In C. Crouch and A. Pizzorno (eds.) *The Resurgence of Class Conflict in Western Europe since 1968: Volume 2: Comparative Analyses* (pp. 277–298). London: Palgrave Macmillan.

Rhodes, M. (1992). Piazza or palazzo? The Italian Greens and the 1992 elections. *Environmental Politics*, 1, 437–442. https://doi.org/10.1080/09644019208414034

Rhodes, M. (1995). The Italian Greens: Struggling for survival. *Environmental Politics*, 4, 305–312.

Rucht, D. (1999). The impact of environmental movements in western societies. In M. Giugni, D. McAdam and C. Tilly (eds.) *How Social Movements Matter* (pp. 204–224). University of Minnesota Press.

Sisco, M. R., Pianta, S., Weber, E. U. and Bosetti, V. (2021). Global climate marches sharply raise attention to climate change: Analysis of climate search behavior in 46 countries. *Journal of Environmental Psychology*, 75, 101596.

Sloam, J. (2014). New voice, less equal: The civic and political engagement of young people in the United States and Europe. *Comparative Political Studies*, 47, 663–688. https://doi.org/10.1177/0010414012453441

Sloam, J., Pickard, S. and Henn, M. (2022). Young people and environmental activism: The transformation of democratic politics. *Journal of Youth Studies*, 25(6), 683–691.

Spaiser, V. and Stefan, C. (2020). "How dare you?" – The normative challenge posed by Fridays for Future. *SSRN Journal*. https://doi.org/10.2139/ssrn.3581404

Stanley, S. K., Hogg, T. L., Leviston, Z. and Walker, I. (2021). From anger to action: Differential impacts of eco-anxiety, eco-depression, and eco-anger on climate action and wellbeing. *The Journal of Climate Change and Health*, 1, 100003.

Such, E. and Walker, R. (2005). Young citizens or policy objects? Children in the 'rights and responsibilities' debate. *Journal of Social Policy*, 34, 39–57.

Sunstein, C. R. (1996). Social norms and social roles. *Columbia Law Review*, 96, 903. https://doi.org/10.2307/1123430

Svensson, A. and Wahlström, M. (2023). Climate change or what? Prognostic framing by Fridays for Future protesters. *Social Movement Studies*, 22(1), 1–22.

Thew, H., Middlemiss, L. and Paavola, J. (2020). "Youth is not a political position": Exploring justice claims-making in the UN climate change negotiations. *Global Environmental Change*, 61, 102036. https://doi.org/10.1016/j.gloenvcha.2020.102036

Tilly, C. and Tarrow, S. G. (2015). *Contentious Politics*. Oxford: Oxford University Press.

van Haute, E. (ed.) (2016). *Green Parties in Europe*. Routledge.

van Stekelenburg, J. and Klandermans, B. (2017). Protesting youth: Collective and connective action participation compared. *Zeitschrift für Psychologie*, 225, 336–346. https://doi.org/10.1027/2151-2604/a000300.

van Stekelenburg, J., Roggeband, C. and Klandermans, B. (eds.) (2013). *The Future of Social Movement Research: Dynamics, Mechanisms, and Processes*. Minneapolis: University of Minnesota Press.

Vestergren, S., Drury, J. and Chiriac, E. H. (2017). The biographical consequences of protest and activism: A systematic review and a new typology. *Social Movement Studies*, 16, 203–221. https://doi.org/10.1080/14742837.2016.1252665

Vromen, A., Xenos, M. A. and Loader, B. (2015). Young people, social media and connective action: From organisational maintenance to everyday political talk. *Journal of Youth Studies*, 18, 80–100. https://doi.org/10.1080/13676261.2014.933198

Wahlström, M., Sommer, M., Kocyba, P., De Vydt, M., De Moor, J., Davies, S. and Buzogany, A. (2019). *Protest for a Future: Composition, Mobilization and Motives of the Participants in Fridays for Future Climate Protests* on 15 March, 2019 in 13 European *Cities*. https://protestinstitut.eu/wp-content/uploads/2019/07/20190709_Protest-for-a-future_GCS-Descriptive-Report.pdf

Wellstead, A. M., Howlett, M. and Rayner, J. (2013). The neglect of governance in forest sector vulnerability assessments: Structural-functionalism and "black box"; problems in climate change adaptation planning. *Ecology & Society*, 18, art23. https://doi.org/10.5751/ES-05685-180323

19
CHRISTIAN COMMUNITIES AS CLIMATE CHAMPIONS

Reba Elliott

In 2015, Pope Francis sent a pair of shoes to Paris. Negotiators of the Paris Climate Agreement were hammering out the details of their global accord on climate change, and with police cancelling a planned street march due to security concerns, grassroots leaders came up with a last-minute solution: shoes instead of feet. Laid out in an orderly grid on the streets of Paris, thousands of pairs of empty shoes provided a haunting (and media-friendly) reminder of the real people UN negotiators serve. UN chief Ban Ki-moon sent a pair. So did Hollywood actors. More than 10,000 ordinary people from around the world joined them (Rowling and Mackinnon, 2015). Pope Francis' well-worn black shoes appeared in many newspapers that covered the story and were shared on social media. The pope's symbolic presence in Paris, along with his publication of *Laudato Si'*, a landmark document on climate change and ecology published just before the conference, captured the public imagination (Shaw, 2015; McCallum, 2019; Pou-Amérigo, 2018; Van Brempt, 2015). This serves as just one example how people of faith, with their millennia-old traditions and often conservative mindsets, are able to galvanise attention as unexpected allies in the fight for climate solutions.

This sense of widespread public pressure was key to the victory in Paris. In order to achieve the milestone of the Paris Climate Agreement, negotiators had to feel the pressure of public accountability. And not just pressure from climate activists, but from everyone. Along with a global climate march in cities around the world and thousands of media stories covering these events, Pope Francis' shoes helped exert the necessary public pressure on UN negotiators that pushed the Paris Climate Agreement across the finish line.

Leadership from high-profile figures like Pope Francis continues to be relevant today. But it is not only the figureheads who matter. A highly distributed movement of grassroots faith leaders across the world is taking action. Such actions include and go far beyond preaching and prayer. Within Catholicism alone, over 350 institutions (Laudato Si' Movement, 2023a) have divested from fossil fuels. That this is more than symbolism but deeply impactful work in the transition away from fossil fuels can be seen in Shell's acknowledgement of divestment as a "material risk" (Shell, 2017) to its model of continued fossil fuel extraction.

DOI: 10.4324/9781003396567-22

Each of these divestment decisions was led by champions within their institutions. This chapter explores how Christian actors in the United States find their way to the climate movement, what support and resources they have, and how they are taking action.

Climate Change and Christianity

About two-thirds of American adults are Christian (Pew Research Center, 2022). Although the Christian population is declining, scholars generally agree that "Christianity influences America's political climate" and that even those Americans who are not Christian "are likely to have been impacted and encountered it in some way" (Burnidge, 2022). Building on this significant cultural presence, organised communities of Christians have successfully driven major shifts in American culture. The focus, reach, and organising power of Christian communities have been remarkably effective at moving public opinion and policy on issues across the political spectrum, from advocating for civil rights for African Americans (Marsh, 2008) to opposing access to abortion (Lewis, 2017; Peters and Kamitsuka, 2022). In recent decades, Christian communities have increasingly explored whether action on climate change is an issue they should take up.

To engage any issue within a faith that has a two-millennia-long intellectual history, substantial intellectual justification is needed. Many Christians feel comfortable taking action only when they know that the action is aligned with the tenets of their faith. A network of theologians and scholars has provided that intellectual framework. These thinkers work through dedicated networks such as the Yale Forum on Religion and Ecology, religious-founded universities such as Georgetown University and Baylor University, divinity schools such as those at Duke University, and other centres of learning. Thousands of pages have been written on why Christians should engage in environmental protection. Excellent introductions are available in *The Christian Future and the Fate of Earth* (Berry, 2011), *Earthspirit: A Handbook for Nurturing an Ecological Christianity* (Dowd, 1991), *Laudato Si'* (Franciscus, 2017), *Stewards of the Earth: Christianity and Creation Care* (Christianity Today, 2022), and *Stewards of Eden: What Scripture Says About the Environment and Why It Matters* (Richter, 2020).

For ease of reference, we might summarise three common approaches supporting Christian activism on climate change as "stewardship," "honour," and "justice." These are not mutually exclusive ideological territories but rather three strands of thought that are often woven together in the worldviews of Christians who support action on climate change.

The *stewardship* approach promotes an understanding of humanity as the appointed caretakers of the natural world, which ultimately belongs to God. "Being created in God's image endows people with the authority to act in behalf of the true owner of the world," who is God (Reichenbach and Anderson, 2022, pp. 48–49). A concern for the viability and usefulness of natural resources across the generations infuses this perspective. "Stewards are servants who, in being commanded to serve (*abad*), benefit that over which they rule," (Reichenbach and Anderson, 2022, p. 54) and Christians who hold this worldview see humanity as the long-term custodian of the earth.

Relatedly, an *honour* approach promotes caring for nature because nature is created by God and is an expression of God's creativity. It is related to but distinct from a *stewardship* approach because it sees caring for God's creation as being good in itself, beyond any utility that the natural world provides. This perspective is rooted in the many scriptures that communicate God's connection to nature, such as the moment in Genesis when "God looked at all of this creation, and proclaimed that this was good – very good" (The Holy Bible: Genesis 1:31). Christians who embrace this worldview believe that the Earth is "the material-vital aspect of God's creation" (Santmire, 1985, p. 11), and humanity therefore cares for creation as an act of honouring the creator.

Finally, a *justice* approach highlights the many scriptures in which God instructs humanity to care for the most vulnerable members of society. This approach connects those scriptures with science on climate change's disproportionate harms to the poorest. Christians who embrace this worldview hold that "as followers of Jesus, we need to respond to the suffering of those most directly affected by the degradation of God's creation" (Borse et al., 2022, p. 19).

Faith-environment Non-governmental Organisations

Building on the work of theologians and scholars, a network of non-governmental organisations (NGOs) has developed a grassroots Christian climate movement. These faith-environment NGOs have achieved two things that are crucial for making progress in grassroots movements (McAdam, 2017). First, they have developed frameworks for community organising, with standing structures in place to activate networks of leaders (Sauer et al., 2021) who are trusted messengers (Webster et al., 2020) within their communities. Second, they have developed "collective processes of interpretation, attribution, and social construction" (McAdam, 2017, p. 193), the shared stories that help community activists find meaning in their relationship to big challenges like climate change.

There are hundreds of local faith and environment NGOs. These organisations do critical work by building relationships and taking action with relatively small but deeply committed groups that are leading change on the local level.

A few organisations have successfully galvanised nationwide movements, coordinating distributed community organising at scale. Some of these organisations reach Christian communities as part of a wider interfaith effort. These interfaith NGOs include Interfaith Power and Light, GreenFaith, the National Religious Partnership for the Environment, Blessed Tomorrow, the National Religious Coalition on Creation Care, and the Center for Earth Ethics. Other organisations are ecumenical, meaning that they engage multiple Christian denominations under the broad umbrella of Christianity. These include A Rocha, Por la Creación, and, to some extent, Creation Justice Ministries, which is Christian in focus but which counts other faith traditions among its partners. Finally, another group of organisations engages single faith traditions within Christianity. This group includes Laudato Si' Movement and Catholic Climate Covenant (which work with Catholic communities), the Evangelical Environmental Network (which works with Evangelical communities), and Green the Church (which works within Black church traditions).

Across the board, all faith-environment NGOs engage a variety of leaders. This includes grassroots actors, denominational heads, and institutional partners, affiliates, or members, with an important nuance around hierarchy in Black churches, within which leadership can sometimes be more distributed. The NGOs often promote action under the banner of a faith-relevant term, such as "creation care" or "justice and peace." These NGOs sometimes offer training that gives volunteers more credibility, equipping them to become "champions," "animators," "leaders," "ambassadors," or other signifiers of authority.

Taking Action

Through years of trial and error, faith-environment NGOs have identified the methods, networks, and messages that are crucial to engaging communities in climate action. They maintain frequent contact with the people who are on their "lists," typically via email but also via phone calls, texting tools, and social media. Anecdotally, the organisations that engage people of faith often have higher levels of engagement through their communications efforts than the nonprofit industry average. The communications that these NGOs send offer both ongoing encouragement and practical resources for volunteers to address climate change within their congregations, schools, or other communities.

Congregations and other communities typically become aware of volunteers' activities when the volunteers organise one-time events such as a prayer service or film screening or when they undertake ongoing work such as developing an environmental education program for young people. The approval of the church's leading pastor(s) for these activities is an important signal to all members that the norms of the community include caring for the environment (Sparkman et al., 2021).

The volunteers who are active in Christian communities often express significant passion and a sense of personal calling towards their activism, saying things like "I felt compelled to contribute" (Jose, 2023) and "faith has been my compass on this incredible journey towards environmental stewardship and sustainable living" (Henkel, 2023). In addition to a sense of calling, a sense of unity also pervades much of the Christian work on climate change. These leaders speak about "collective responsibility" (George, 2023) and about building community support to "move us all forward to greater stewardship of our natural world" (Mountain Sky Conference, 2023).

With a strong sense of personal commitment and community support, it is not surprising that Christians' engagement on climate change tends to be durable, with many active in local volunteering over the course of many years, according to anecdotal conversations. Grassroots actors typically engage in sustainability, spirituality, or policy. All three of these avenues can have widespread repercussions for the community as a whole.

Spirituality

A common theme across all environmental activities undertaken by Christian communities is an interest in spirituality. Many people speak of having an ecological conversion, one in which the person begins to perceive that acting on climate change is an essential way to demonstrate his or her faith in the world today. For these Christians, an ecological conversion "is less about the turn to nature and more about a responsible and loving attitude towards it, inspired by an appropriate appreciation of the particular role of human beings" (Deane-Drummond, 2012, p. 88). This deep and lasting change in perspective influences many decisions, from daily sustainability habits to policy and voting preferences.

All of the faith and environment NGOs address spirituality. One area of work for these organisations is equipping local community volunteers to organise shared acts of prayer in their communities. Engaging in shared acts of prayer has two outcomes. First, this can be a deeply moving experience, one that transforms the person and inspires or renews a commitment to care for the Earth. Participants often believe that their prayer is efficacious and good in itself and hope that it will lead to real change in climate policies and carbon-intensive lifestyles. Second, prayer bonds actors together. The experience of sharing one's hopes and fears and of committing to something bigger together is a powerful way to create the sense of unity that is important for activist groups to exist (Blee, 2012). From a pragmatic standpoint, this sense of unity is essential to creating strong, durable networks of actors.

Organising prayer services is a relatively low-bar entry point for many volunteers, one that allows them to engage their communities in action on climate change in a way that is familiar to the community and that has no more logistical hurdles than finding a time, place, and group of people to participate. While many local volunteers are willing to undertake this work, they need resources and training. As examples of the type of support provided to volunteers, Laudato Si' Movement provides a monthly prayer guide for volunteers to use in their churches, training for volunteers to facilitate ecological prayer retreats, and a "community contemplation" program that connects volunteers to each other to pray together online (Laudato Si' Movement, 2023b). Like many others, this NGO also provides special materials that equip volunteers to connect ecological themes to annual Christian celebrations such as Lent and Advent (Laudato Si' Movement, 2023b).

A prayer celebration that deserves special mention is the Feast of St. Francis, a figure from the twelfth century who remains beloved across Christian traditions. St. Francis was known for his care for the natural world and is frequently cited by Christians who are active on climate change as a figure of inspiration. As a saint, he has a dedicated day (October 4) to reflect on his work and life story. As an example of how NGOs build on this day to promote wider community reflection, Catholic Climate Covenant provides materials that equip volunteers to engage their communities in St. Francis events (Catholic Climate Covenant, 2023).

Another area of spiritual work for these organisations is equipping pastors to speak about climate change in their sermons. While activism within congregations often begins with community volunteers, pastors do sometimes independently arrive at the conclusion that the climate crisis should be a focal point for their teaching. Given the complexity of climate science, the sensitive nature of this issue in public discourse, and the depth of thinking on how climate change relates to Christianity, many pastors seek guidance from faith-environment NGOs. The resources provided to pastors range from multi-week education courses to written materials such as bulletin inserts and sermon outlines to media materials such as videos and posters.

As examples of how organisations fulfil this need, Creation Justice Ministries runs a regular pastor training course (Creation Justice Ministries, 2023), and Green the Church offers summits that bring pastors and other church leaders together and organises ongoing training seminars on everything from green theology to sustainable practices to building power for political change (Green the Church, 2023).

Many Christian communities take action on spirituality first and then progress to other actions in sustainability and advocacy.

Sustainability

Christian actors' work in sustainability can encompass a wide range of activities, many of which are focused on improving facilities. Facilities improved typically include churches, educational institutions, and communities of religious women and men. The activities range from low-bar, less impactful projects (such as installing recycling bins) to high-bar, more impactful projects (such as installing solar panels). Highly impactful projects such as installing solar panels and increasing energy efficiency can make a real difference in emissions. There are approximately 350,000 church facilities in the US, according to the US Religion

Census (Association of Statisticians of American Religious Bodies, 2020), and decreasing emissions across this massive portfolio would be significant. As an example of how NGOs are encouraging emissions reductions in these facilities, Catholic Climate Covenant offers a "Catholic Energies" program that has installed solar panels on Catholic institutional buildings, realising "40 projects in 14 states" (Heatherington, 2023).

However, the greatest impact of sustainability efforts may lie beyond the reduction of emissions. Activities to make any facilities upgrades typically require the approval of religious authorities such as the church pastor or school principal. To secure this approval, community volunteers often deliver sustained internal activism, bringing together a committee, holding meetings with members, exploring the budgetary implications of facility improvements, and finally successfully making the case for change. Because the decision to upgrade facilities involves so many stakeholders, it has long-term consequences across programming in many institutions. The curriculum of religious education classes, the prayer and song selections that raise community awareness of the ecological crisis, and even the policy focus of the church might change as a result of this internal activism.

Certifying sustainability improvements allows actors to continue making the case for change beyond the initial decision-making period. Certification regimes, which are optional, typically involve both a long-term commitment from the participating institution and a public statement of its accomplishments, often made by posting a banner or plaque on the grounds of the facility or by publishing statements via the institution's communications channels. Certification provides a way for the church or religious school to communicate that climate action is an ongoing priority for the community. This has ripple effects. Many churches and schools are highly aware of what others in their area are doing, and certification programs can inspire greater ambition across a local area.

There are several options for certification programs offered by faith-environment NGOs. Many faith activists are familiar with Interfaith Power and Light's "Cool Congregations" program (Interfaith Power and Light, 2023a), which provides guidance, a carbon calculator, and certification for a 10% or greater reduction in greenhouse gas emissions. The Environmental Protection Agency offers a free program, "ENERGY STAR for Congregations" (United States Environmental Protection Agency, 2023), which is not a certification program per se but does offer many of the benefits of certification. These include tools, training, and support for places of worship, including a benchmarking tool, "Portfolio Manager," that allows places of worship to track improvements over time.

Among the most ambitious of the certification programs may be the Laudato Si' Action Platform (Dicastery for Promoting Integral Human Development, 2023). This platform is a project of the Vatican's environmental and social office – its Dicastery for Promoting Integral Human Development – and has operational support from Laudato Si' Movement. The platform advices about the most impactful ways to take action across seven strategic areas, which is adjusted to the needs of the user, and offers certification for institutions, helping promote greater ambition in the community as a whole.

Advocacy

Christian communities advocate on a wide range of policy issues, from local to national to global questions. Across the board, Christians' ability to turn out voters and to craft public discourse has been seen as bringing substantial power to the issues they embrace, with past examples including civil rights for African Americans (Marsh, 2008) and opposition to access to abortion (Lewis, 2017; Peters et al., 2022), as discussed earlier.

The potential of faith communities to promote pro-climate policies is enormous, and faith NGOs play a crucial role in elevating awareness of climate policies and in equipping actors to address them. In general, faith and environment NGOs are classified as 501(c)3 organisations in the US federal taxation system. This means that they must strictly limit their work on lobbying. Rather than promoting or opposing specific candidates for office, these organisations typically support general policy perspectives or specific policy proposals and ensure that policymakers are aware of faith-based demands for action.

In the area of supporting policy, these NGOs work in a wide variety of ways. Most faith-environment NGOs organise frequent educational efforts, such as webinars and conference calls, to inform members about policy issues. As one example, Interfaith Power and Light manages a Faith Votes campaign that provides a guide to policy choices that protect the environment and human dignity and encourages people to pledge to vote in line with these priorities (Interfaith Power and Light, 2023). As another example, the National Religious Partnership for the Environment regularly promotes statements or webinars on national policy issues from partner groups and other allied faith-based organisations (National Religious Partnership for the Environment, 2023).

In the area of connecting with policymakers, many of the faith-environment NGOs support both online and in-person efforts. As examples, Catholic Climate Covenant coordinates joint statements from Catholic organisations urging policymakers to support key policy decisions (Catholic Climate Covenant, 2023), the state affiliates of Interfaith Power and Light organise visits of their members to the offices of elected officials (Interfaith Power and Light, 2023), and a decades-long program of the National Religious Coalition on Creation Care, the annual National Prayer Breakfast for Creation Care, brings religious leaders together for a day of meetings with members of Congress (National Religious Coalition on Creation Care, 2023).

While national organisations often encourage the US Congress to develop stronger policies, smaller faith organisations are often deeply effective on more local policy issues. Work on climate justice, for example, can be powerfully addressed at the local level. State and local organisations are able to sustain a focus on particular power plants, coal ash ponds, highways, and other infrastructure assets and are able to pressure elected officials through direct relationships with communities that are aware of the issues and likely to vote. The state affiliates of Interfaith Power and Light, including in Alabama, Illinois, and Georgia, are among the leaders of this work (Alabama Interfaith Power and Light, 2023; Faith in Place, 2023; Georgia Interfaith Power and Light, 2023).

A handful of US faith-environment NGOs work on the global level, including Green-Faith and Laudato Si' Movement. GreenFaith organises mobilisations in Europe, Africa, and Asia, collaborates with UN bodies, and promotes the Fossil Fuel Nonproliferation Treaty. Laudato Si' Movement supports policy change in Hispano-America, Brazil, Western Europe, Eastern Europe, Eastern Africa, South Africa, and the Asia-Pacific region. Examples of its national policy work include supporting the creation of a Polish environmental ministry, protecting Indigenous land stewardship in Ecuador, and organising events with political candidates to discuss coal-based electricity in the Philippines (Laudato Si' Movement, 2023b). Laudato Si' Movement also organises high-level summits in Rome and at various climate and biodiversity COPs and coordinates a body of work on the Fossil Fuel Nonproliferation Treaty, which the Vatican has supported (Laudato Si' Movement, 2023b).

This work by Laudato Si' Movement and GreenFaith demonstrates the extent to which action on climate change is supported by people of faith around the world.

Across the board, Christians' awareness of policy issues – and policymakers' awareness of their interest – pressures officials to commit to support a just energy transition. But direct advocacy is not the only way that faith actors encourage policy shifts. Story-telling on social media and in the press can be as effective as or even more effective than direct engagement.

There is a perception that action on climate is a niche interest, one embraced only by youth activists and far-left liberals. When this is the public perception, it can be all too easy for decision makers to dismiss demands for ambitious climate policies. Action by people of faith communicates a very different story. It suggests that a tremendous swath of the American public is following this issue and expecting accountability from its elected officials. This provides a framework for centrist policymakers who would like to act on climate change but who are reluctant to align themselves with the approach of "leftist" activists. Contributions by Christian actors bring unique and urgently needed strengths to the larger movement of climate advocates.

These news stories are not reaching faith actors alone. Instead, they are reaching the broad spectrum of people who are curious about why unexpected leaders like Pope Francis are speaking publicly about the climate crisis. This sense of a massive movement among unexpected allies constitutes a powerful argument for elected officials who are keen to keep their jobs.

Conclusion

Support from people of faith brings something important to the climate movement. Pope Francis' visible leadership ahead of the Paris climate summit is just one example of the role that people of faith have in changing business as usual for the planet. While highly visible leaders such as Pope Francis can have an outsized impact, it is important that ordinary members of faith communities are taking action.

A grassroots movement of such climate actors who both draw inspiration from and build on the power of their faith communities has been seeded across the United States and in other countries. These leaders are praying, changing lifestyles, and advocating for moral climate policies. Drawing on these strong networks, faith-based environmental NGOs equip actors to take actions that contribute to broad cultural change and generate demand for accountability for policymakers.

References

Alabama Interfaith Power & Light (2023). *Alabama Interfaith Power & Light*. https://www.alabamaipl.org/ [Accessed 1 December 2023].

Association of Statisticians of American Religious Bodies (2020). *US Religion Census*. https://www.usreligioncensus.org/ [Accessed 1 December 2023].

Berry, T. (2011). *The Christian Future and the Fate of the Earth*. Maryknoll: Orbis Books.

Blee, K. (2012). *Democracy in the Making: How Activist Groups Form*. New York: Oxford University Press.

Borse, D., et al. (2022). *Loving the Least of These*. Washington, DC: National Association of Evangelicals. https://www.nae.org/loving-the-least-of-these/ [Accessed 1 December 2023].

Burnidge, C. (2022). Religious impressions on American politics. *Northern Iowan*, 7 April, https://www.northerniowan.com/16865/opinion/religious-impressions-on-american-politics/.

Catholic Climate Covenant (2023). *Catholic Climate Covenant*. https://catholicclimatecovenant.org/ [Accessed 1 December 2023].

Christianity Today (ed.) (2022). *Stewards of the Earth: Christianity and Creation Care*. Bellingham: Lexham Press.

Creation Justice Ministries (2023). *Creation Justice Ministries*. https://www.creationjustice.org/ [Accessed 1 December 2023].

Deane-Drummond, C. (2012). Ecological conversion in a changing climate: An ecumenical perspective on ecological solidarity. *International Journal of Orthodox Theology*, 3(1), 78–104.

Dicastery for Promoting Integral Human Development (2023). *Laudato Si' Action Platform*. https://laudatosiactionplatform.org/ [Accessed 1 December 2023].

Dowd, M. (1991). *Earthspirit: A Handbook for Nurturing an Ecological Christianity*. Waterford: Twenty Third Publications.

Faith in Place (2023). *Faith in Place*. https://www.faithinplace.org/ [Accessed 1 December 2023].

Franciscus (2017). *Laudato Si': On Care for Our Common Home*. Rome: Libreria Editrice Vaticana.

George, Z. (2023). Catholic groups join sea of protesters in New York march against fossil fuels. *National Catholic Reporter*, 19 September. https://www.ncronline.org/earthbeat/justice/catholic-groups-join-sea-protesters-new-york-march-against-fossil-fuels.

Georgia Interfaith Power and Light (2023). *Georgia Interfaith Power and Light*. https://gipl.org/ [Accessed 1 December 2023].

Green the Church (2023). *Green the Church*. https://www.greenthechurch.org/ [Accessed 1 December 2023].

Heatherington, K. (2023). US Catholic dioceses are leading the '.green revolution' championed by popes. *The Boston Pilot*, 1 September. https://www.thebostonpilot.com/article.php?ID=195476 [Accessed 1 December 2023].

Henkel, L. (2023). Cycling into season of creation. *Interviewed by Laudato Si' Movement, Laudato Si' Movement*, 12 September. https://laudatosimovement.org/news/cycling-into-season-of-creation/.

The Holy Bible (2007). *Genesis. 1:31. The Inclusive Bible: The First Egalitarian Translation*. Lanham: Rowman and Littlefield Publishers, Inc.

Interfaith Power and Light (2023a). *Cool Congregations*. https://interfaithpowerandlight.org/coolcongregations/get-certified/ [Accessed 1 December 2023].

Interfaith Power and Light (2023b). *Faith Votes*. https://interfaithpowerandlight.org/faithclimatejustice voter/ [Accessed 1 December 2023].

Interfaith Power and Light (2023c). *Interfaith Power and Light*. https://interfaithpowerandlight.org/ [Accessed 1 December 2023].

Jose, J. (2023). Community Resilience and Empowerment: Lessons from a Divine Encounter in the Philippines. *Laudato Si' Action Platform*, 13 June. https://laudatosiactionplatform.org/community-resilience-and-empowerment-lessons-from-a-divine-encounter-in-the-philippines/.

Laudato Si' Movement (2023a). *Divestment*. https://laudatosimovement.org/divestment/ [Accessed 1 December 2023].

Laudato Si' Movement (2023b). *Laudato Si' Movement*. https://laudatosimovement.org/ [Accessed 1 December 2023].

Lewis, A. (2017). *The Rights Turn in Conservative Christian Politics: How Abortion Transformed the Culture Wars*. New York: Cambridge University Press.

Marsh, C. (2008). *The Beloved Community: How Faith Shapes Social Justice from the Civil Rights Movement to Today*. New York: Basic Books.

McAdam, D. (2017). Social movement theory and the prospects for climate change activism in the United States. *Annual Review of Political Science*, 20, 189–207. https://www.annualreviews.org/doi/full/10.1146/annurev-polisci-052615-025801 [Accessed 1 December 2023].

McCallum, M. (2019). Perspective: Global country-by-country response of public interest in the environment to the papal encyclical, *Laudato Si'*. *Biological Conservation*, 235, 209–225.

Mountain Sky Conference (2023). *Creation Justice*. https://www.mtnskyumc.org/creation-justice [Accessed 1 December 2023].

National Religious Coalition on Creation Care (2023). *National Religious Coalition on Creation Care*. https://nrccc.org/ [Accessed 1 December 2023].

National Religious Partnership for the Environment (2023). *National Religious Partnership for the Environment*. http://www.nrpe.org/ [Accessed 1 December 2023].

Peters, R. and Kamitsuka, M. (ed.) (2022). *T&T Clark Reader in Abortion and Religion: Jewish, Christian, and Muslim Perspectives*. New York: Bloomsbury.

Pew Research Center (2022). Modeling the Future of Religion in America. *Pew Research*, 13 September. https://www.pewresearch.org/religion/2022/09/13/modeling-the-future-of-religion-in-america/ [Accessed 1 December 2023].

Pou-Amérigo, M.-J. (2018). Framing 'Green Pope' Francis: Newspaper coverage of encyclical *Laudato Si'* in the United States and the United Kingdom. *Church, Communication, and Culture*, 3(2), 136–151.

Reichenbach, B. and Anderson, V. (2002). *On Behalf of God: A Christian Ethic for Biology*. Eugene: Wipf & Stock Publishers.

Richter, S. (2020). *Stewards of Eden: What Scripture Says about the Environment and Why It Matters*. Westmont: InterVarsity Press.

Rowling, M. and Mackinnon, M. (2015). Pope Francis puts his shoes in the centre of Paris to protest against climate change. *Independent*, 30 November. https://www.independent.co.uk/news/people/pope-francis-puts-his-shoes-in-the-centre-of-paris-to-protest-against-climate-change-a6754151.html [Accessed 1 December 2023].

Santmire, H. (1985). *The Travail of Nature: The Ambiguous Ecological Promise of Christian Theology*. Philadelphia: Fortress Press.

Sauer, K., et al. (2021). Six minutes to promote change: People, not facts, alter students' perceptions on climate change. *Academic Practice in Ecology and Evolution*, 11, 5790–5802.

Shaw, M. (2015). *Paris Earth Summit: Visual Activism Rises to the Climate*. https://www.readingthepictures.org/2015/12/the-paris-earth-summit-visual-activism-rises-to-the-climate/ [Accessed 1 December 2023].

Shell (2017). *Strategic Report: Shell Annual Report and Form 20-F 2017*. https://reports.shell.com/annual-report/2017/servicepages/downloads/files/strategic_report_shell_ar17.pdf [Accessed 1 December 2023].

Sparkman, G., Howe, L. and Walton, G. (2021). How social norms are often a barrier to addressing climate change but can be part of the solution. *Behavioural Public Policy*, 5(4), 528–555. https://www.cambridge.org/core/journals/behavioural-public-policy/article/abs/how-social-norms-are-often-a-barrier-to-addressing-climate-change-but-can-be-part-of-the-solution/90BCDC030D23B9CE5078FB025EEBECCB [Accessed 1 December 2023].

United States Environmental Protection Agency (2023). *ENERGY STAR for Congregations*. https://www.energystar.gov/buildings/resources_audience/congregations [Accessed 1 December 2023].

Van Brempt, K. (2015). *What We Can't Do, Francis Can*. https://www.euractiv.com/section/climate-environment/opinion/what-we-can-t-do-francis-can/ [Accessed 1 December 2023].

Webster, R., et al. (2020). *Communicating Climate Change During the COVID-19 Crisis: What the Evidence Says*. Oxford: Climate Outreach. https://orca.cardiff.ac.uk/id/eprint/135308/1/Climate-Outreach-Communicating-climate-change-during-Covid-19-crisis%20(3).pdf [Accessed 1 December 2023].

PART IV

Civil Disobedience, Litigation, and the Suppression of Climate Activism

Introduction to Part IV: Civil Disobedience, Litigation, and the Suppression of Climate Activism

Thanks to a long history of activism and organising in support of causes from women's suffrage to labour and civil rights, grassroots climate activists have a large and multifaceted toolbox of tactics and strategies at their disposal (350.org, 2022; Hunter, 2019; Sharp, 1973; TFC, n.d.). In addition to marches and petitions, that toolbox includes legislative action and litigation, direct action, and civil disobedience, among others (Burkett, 2016). While it is spectacular civil disobedience that typically garners public and media attention (Engler and Engler, 2016) – and some climate activist groups such as The Last Generation, Climate Defiance, or Just Stop Oil explicitly embrace civil disobedience as a primary tactic (Blazevic et al., 2020; Extinction Rebellion, 2019; Prakash, 2020) – most activists employ a wide range of different approaches in their activism and often coordinate across different groups.

This section takes a closer look at the strategies of direct action and civil disobedience that grassroots climate activists engage in, highlighting that climate activists are not only breaking the law to draw attention to governments' climate inaction; they are also holding authorities accountable to follow and improve existing climate laws. Litigation is a fast-growing form of climate activism. In fact, in countries like China, where open protest is not tolerated, litigation may be one of the few existing forms of climate activism (see Chapter 22). Although criminalisation of climate activists has always been more pronounced in Global South countries, activists in Global North countries, too, are increasingly facing more draconian police brutality, harsh prison sentences (Climate Rights International, 2024) and legal challenges (see Chapter 23), spurring the necessity for organisations like the Climate Disobedience Center to support activists through legal challenges (climatedisobedience.org/). At the same time, plaintiffs have been able to achieve major wins for climate litigation, from the Dutch Urgenda case that forced the Dutch government to meet its climate targets (Urgenda, 2019), to the German constitutional court issuing a ruling in support of future generations (Neubauer *v.* Germany, 2020), to the Swiss climate elders winning their case at the European Court of Human Rights (Klimaseniorinnen Schweiz/Greenpeace, 2024), giving hope for the just transitions that activists are working hard to achieve.

DOI: 10.4324/9781003396567-23

While grassroots climate activists engaging in civil disobedience and litigation aim to pressure governments to act, climate scientists and those who engage in research related to climate change have often themselves been the target of government pressure (Oreskes and Conway, 2010). Additionally, governments who deny the scientific consensus on climate change can use their power to exert political influence on scholars and research institutes, for example, by directing or withholding resources (see Chapter 24). This part of the handbook takes a closer look at these issues to highlight intersecting challenges and opportunities for grassroots climate activism.

In Chapter 20, "Extinction Rebellion and Non-violent Civil Disobedience", Linda Williams and Damien Rudd delve into the history and strategies of Extinction Rebellion (XR), tracing its roots from Britain to its emergence as a global climate movement. XR's approach to mass grassroots non-violent civil disobedience has been both innovative and controversial. The authors examine the moments of discreditation the movement faced in public discourse, as they were labelled as "extremists" and "eco-terrorists" or criticised for having a privileged and elitist composition. The chapter discusses how XR methods and tactics resonated with – and led to strong engagement among – youth activists. It also examines concerns that emerge out of the heavy-handed responses by authorities to their actions. This chapter also explores XR's engagement with non-hierarchical structures, effective media communication, and theatrical performances, highlighting the challenges and successes of a global movement rooted in civil disobedience.

Daniel Hofinger and Maximilian Becker, in Chapter 21, "System Change, Not Climate Change: The Climate Justice Movement in Germany 2008–2021", trace the emergence of climate activist networks in Germany. Based on their own participation in these movements, they examine their experiences of civil disobedience led by the anti-coal initiative Ende Gelände and the broader youth-led movement, Fridays for Future. The authors describe how disruptive happenings like climate camps in lignite mining areas involve practices of networking, education, action, and "living utopias". The chapter explores the challenges and successes of grassroots democracy and consensus-building within these movements, highlighting the role of (queer) feminist orientations in addressing internal power imbalances. The slogan "System Change – Not Climate Change" serves as a rallying cry, encapsulating the grassroots ethos that drives climate justice efforts.

Chapter 22, "Unburnable Coal and Unlabelled Climate Activism in China", by Bowen Gu and Juan Liu, focuses on China's coal production and its impact on climate activism. Despite China's heavy reliance on coal, activists and NGOs are challenging the industry's environmental, social, and health impacts. The authors frame these engagements as "unlabelled" climate activism, given the grassroots nature of protagonists' tactics, their seemingly distant relations to the international climate justice movement, and an overall lack of institutional support. The chapter examines the diverse tactics employed by activists, from direct confrontation and environmental public litigation to creative and performative interventions. This chapter illustrates the unique challenges climate activists face in China, a context where activism is often decentralised and lacks formal recognition.

In Chapter 23, "Beyond Protest: How Legal Actions Drive Climate Justice", Lea Maine-Klingst, Maria Antonia Tigre and Hermann E. Ott from *Client Earth* draw on experiences from Europe and Latin America to analyse the use of climate litigation as a strategy for climate activism. The authors explore rights-based litigation by looking at different cases that highlight the possibilities of using legal pathways for more environmental and climate protection. These cases include youth activists and the elderly in Europe, and strategic litigation and land rights litigation in Latin America. Additionally, the chapter addresses some of the perils climate activists face in Europe and Latin America, and how the law has been used to mitigate or exacerbate these perils.

In Chapter 24, "'Academic Freedom' v. Climate Change Denial: How the Politics of Research Funding Shapes the Possibilities for Researching Grassroots Activism", Rob Watts, Judith Bessant, Stewart Jackson, Michelle Catanzaro, Faith Gordon and Philippa Collin focus on the politicisation of research on young people and climate action. Specifically, they examine efforts by the Australian government in 2021 to prevent such research by vetoing a project that was approved by the national research council for funding. They highlight how the political influence of climate-sceptic think tanks and carbon capital has been overt and covert, shaping the conditions in which grassroots climate activism operates – with effects on how it can be researched and understood. While research on grassroots environmental and climate action, often with participants themselves, has made a valuable contribution to explaining, informing, and amplifying efforts by various movements and organisations, the authors explore the ways in which the scale of grassroots activism has far outweighed the research funded and conducted on climate politics. In discussing broader issues about academic freedom and climate denial, the authors argue that the politics of accessing research funding directly intersects with the politics of climate activism and climate scepticism in ways that have significant implications for praxis in grassroots climate activism – and democracy.

The chapters in this part shed light on the varying conditions under which grassroots climate activists operate across countries, including the legal and, at times, existential perils. They also aid in understanding how activist practices have at times breached the gap and become institutionalised in legal systems through litigation.

References

350.org (2022). *Trainings.* https://trainings.350.org [Accessed 6 May 2024].

Blazevic, S., et al. (2020). We shine bright: Organizing in hope and song. In V. Prakash and G. Girgenti (eds.) *Winning the Green New Deal: Why We Must, How We Can.* New York: Simon & Schuster.

Burkett, M. (2016). Climate disobedience. *Duke Environmental Law & Policy Forum,* 27(1), 1–50.

Climate Rights International (2024). *On Thin Ice. Disproportionate Responses to Climate Change Protestors in Democratic Countries.* https://cri.org/reports/on-thin-ice/

Engler, M. and Engler, P. (2016). *This Is an Uprising: How Nonviolent Revolt Is Shaping the Twenty-First Century.* New York: Nation Books.

Extinction Rebellion (ed.) (2019). *This Is Not a Drill: An Extinction Rebellion Handbook.* London: Penguin.

Hunter, D. (2019). *Climate Resistance Handbook. Or, I Was Part of a Climate Action, Now What?* Boston: 350.org

Klimaseniorinnen (2024). *Climate Action. We Won.* https://en.klimaseniorinnen.ch/

Neubauer v. Germany. (2020). https://climatecasechart.com/non-us-case/neubauer-et-al-v-germany/

Oreskes, N. and Conway, E. (2010). *Merchants of Doubt: How a Handful of Scientists Obscured the Truth on Issues from Tobacco Smoke to Global Warming.* London: Bloomsbury.

Prakash, V. (2020). People power and political power. In V. Prakash and G. Girgenti (eds.) *Winning the Green New Deal: Why We Must, How We Can.* New York: Simon & Schuster.

Sharp, G. (1973). *The Politics of Nonviolent Action. Part Two: The Methods of Nonviolent Action.* Boston: Porter Sargent.

TFC (n.d.). *Tools. Training for Change.* https://www.trainingforchange.org/tools/ [Accessed 6 May 2024].

Urgenda. (2019). *The Urgenda Climate Case Against the Dutch Government.* https://www.urgenda.nl/en/themas/climate-case/

20

EXTINCTION REBELLION AND NON-VIOLENT CIVIL DISOBEDIENCE

Linda Williams[1] and Damien Rudd

Introduction

The IPCC has warned 'There is a rapidly closing window of opportunity to secure a liveable and sustainable future for all. . . . The choices and actions implemented in this decade will have impacts now and for thousands of years'.[2] As the environment becomes increasingly unstable through extreme heat, fire, flood and turbulent storms, the climate emergency impacts people everywhere: not least the poor of the developing world who are not only more vulnerable to the escalating effects of climate change but are also least responsible for it (Yusoff, 2017; Nixon, 2013).

Meanwhile, wealthy, industrially developed countries have consistently resisted too much disruption to economies reliant on a 'business as usual' approach to energy and consumption. In this chapter we explore examples of civil disobedience in grassroots climate activism that have sought to draw attention to this disjuncture, in particular the role of high-profile groups such as Extinction Rebellion (XR) who have gained significant international recognition.

The first sections of this chapter will focus on the XR movement and its development in Britain and other wealthy nations in Europe, Australasia and North America. It has also recently emerged in Africa and in the carbon intensive, politically conservative regimes of the middle east where there are now small groups of activists in Lebanon, Qatar and Israel.[3] XR is now recognised as a decentralised, global movement with some 1,029 autonomous groups acting across 88 countries, from Antarctica to Zimbabwe. Yet, critiques of the movement often depict XR as a predominantly white, middle-class movement, largely composed of privileged people from wealthy countries engaged in civil disruption.

We then discuss the movement's tactics and successes around urban disruption, before turning to how XR's savvy media campaigns and urban theatre have prompted a global momentum in climate activism. An important strategic turn occurred when the movement realigned itself to gain momentum in shifting public opinion: not, in our view, as a regressive response to legal crackdowns but as a necessary step towards social transformation. The chapter concludes with a discussion of the ways activist groups navigate civil disobedience.

DOI: 10.4324/9781003396567-24

The Rise of Extinction Rebellion (XR)

One of XR's founders, Roger Hallam, has explained how the ideas for XR emerged from a conversation between himself and Gail Bradbrook in 2018. At the time, they were both members of the environmental group *Rising Up!* – but were frustrated by the lack of real social change it had achieved. Hallam's initial conversation with Bradbrook was focused on the 'unfolding environmental crisis, and the inability of the current system to even recognise the oncoming chaos, and theories about how to transform politics' and it was this conversation that laid the conceptual foundations for the Extinction Rebellion movement (Taylor, 2020). Bradbrook took up this discussion with fellow climate activists, and during a crucial meeting that year in Stroud, UK, a group of about 15 seasoned campaigners, activists and academics reached a consensus. After years of addressing various smaller-scale issues, they decided it was time, as Bradbrook put it, to 'go for the *big one*': the climate crisis. As another co-founder Simon Bramwell recalled: 'we decided to throw all of our energy and intelligence at something that could change the planet' (Taylor, 2020). And from that moment, XR was formed.

The founders agreed the new movement should be more than just an 'online phenomenon' and that its success would require members with the capacity to self-organise, engage in and recruit community support to rebel against prevailing government inaction on climate change. According to Bradbrook, the group also agreed that to resonate with a broader public across a wide political spectrum, XR would need a much broader social appeal beyond a 'leftist echo chamber' (Taylor, 2020). Rather than affiliating with any political party, XR should remain fiercely independent.

The founding members of XR defined the scale of the climate-change crisis and the consequences of inaction (Taylor, 2020), which led to three demands of the UK government[4]:

1 TELL THE TRUTH. Governments must tell the truth by declaring a climate and ecological emergency, working with other institutions to communicate the urgency for change
2 ACT NOW. Governments must act now to halt biodiversity loss and reduce greenhouse gas emissions to net zero by 2025
3 GO BEYOND POLITICS. Governments must create and be led by the decisions of a Citizens' Assembly on climate and ecological justice

The group also outlined ten principles for any group intending to organise autonomously in the name of XR.[5] In this way, the founders argued, 'power is decentralised, meaning that there is no need to ask for permission from a central group or authority' (XR, 2023).

XR's remarkable ascendancy from its modest inception to international activist phenomena in less than 2 years can be traced through an escalating series of public actions, with each action increasing the movement's membership and scale through media attention. The first of XR's official actions was the occupation of Greenpeace's London offices on 17 October 2018 'to encourage their members to participate in mass civil disobedience as the only remaining alternative to avert the worst of the catastrophe' (Taylor, 2018). The following day, more than 1,000 XR members rallied outside the Houses of Parliament in London to hear the Declaration of Rebellion announced in response to the UK government's inaction on the climate crisis with the goal of 'peaceful civil disobedience, traffic disruption, and symbolic criminal actions' (Extinction Rebellion UK, 2018). Speakers at the rally included prominent activists Greta Thunberg and George Monbiot as well as several members of the UK Green Party, and 15 people were arrested.

Just a month later in November 2018 XR staged Rebellion Day, when some 6,000 protesters occupied five bridges on the River Thames, resulting in 70 arrests (Taylor and Gayle, 2018). By April 2019, XR's most iconic action lasted 11 days and resulted in a total of 1,130 arrests. By May that year, the public discussions following these cumulative events resulted in a consequential, albeit symbolic, act by the British Parliament in the formal declaration of a climate emergency. These and later actions had the effect of disseminating XR's ideas, methods and membership from local climate action groups into a global movement.

In 2019, the English journalist George Monbiot publicised his arrest at an XR rally by explaining that his protest on climate change and 'the gathering collapse of our life support systems' was based on a long, respectable tradition of nonviolent civil disobedience including the suffragettes, the American civil rights movement and the Polish and East German Democracy movements.[6] XR has been particularly effective in drawing on the public imagination around figures committed to principles of pacifist resistance, particularly the grassroots activism of thousands of people in Gandhi's long *Salt March* in 1930 against the repressive salt taxes enforced by the British.

Civil disobedience has a long history, and XR's publicity campaigns have frequently called attention to other civil disobedience movements, such as the activism for the rights of African Americans inspired by the religious leadership of Martin Luther King Jr, who himself referred to the Athenian civic activism of Socrates in the fifth century BCE.[7] In particular, XR representatives often cite the courageous example of the Freedom Riders of 1961 who endured violent attacks by their opponents to ride on the buses from Washington DC to Jackson, Mississippi to claim the desegregation of public transport in the American South. The Freedom Riders were committed to the humanist ethics of egalitarianism and civil rights, combined with Christian virtues of self-sacrifice for the common good. It was a wide, grassroots movement engaged in the kind of deliberate and persistent resistance that led to so many arrests that it overwhelmed the capacity of police management, the court system and prisons.

While civil disobedience can be considered as contempt for the law, King regarded the rejection of morally unjust legislation as a form of respect for the law:

> Any man who breaks a law that conscience tells him is unjust and willingly accepts the penalty by staying in jail to arouse the conscience of the community on the injustice of the law is at that moment expressing the very highest respect for the law.[8]

In moral philosophy, this is seen as a way of distinguishing between legality and legitimacy (Rawls, 1971), where disobedience is a last resort deployed only after all other options have been exhausted. From this perspective, the aim is to rectify rather than dismantle the existing legal framework, thereby maintaining a form of fidelity to the law: a view endorsed in *Extinction Rebellion: Insights from the inside* where in 'a democratic society, one is justified in disobeying the law only when other parliament alternatives have been exhausted, and the injustice being protested against is grave' (Read and Alexander, 2020, p. 43).

Extinction Rebellion (XR) – Tactics and Successes

By highlighting the common values with long traditions of pacifist resistance, environmental stewardship, compassion and personal sacrifice, XR has broadened its base by collaborating with various religious communities. Members of XR's *Faith Bridge*, for example, have participated in public vigils for the environment, conceived as a way of re-enacting liturgical practices such as those in Muslim prayers, Christian liturgies and Jewish Seders. By engaging with a broad spectrum of beliefs unified by a sense of environmental stewardship, this kind of activism also presents a way XR could expand the ethnic representation and class base of its membership, which has been criticised as too narrow.[9] As one parish priest of the Christian Climate Action group saw it: the ecological crisis is also a 'spiritual crisis', necessitating personal responses based on a coalition of religion, ritual and rebellion.[10]

At the same time, XR has also been particularly successful in how it has mobilised less-traditional, non-hierarchical organisational structures to produce effective media and communication strategies where relatively autonomous 'circles' are able to target diverse media platforms. In England, national media coordinators produce YouTube tutorials on how to 'split workload between team members as the team scales, and how to decentralise power within the team'. There are two media leaders in each media team targeting four major media platforms: the press and mainstream media; spokespeople and public figures[11]; content providers such as writers or photographers and not least, social media campaigns on platforms such as Facebook, Instagram, YouTube and Twitter.[12]

Such social media strategies appear to confirm the findings of a recent study that climate-change activists engaged in direct action outside the established political system often have policy knowledge and agendas comparable to or surpassing those active within the mainstream system (Cory and Reiner, 2021). Moreover, since visual media prevail in contemporary communication systems, XR's pithy use of visual symbols (an hourglass within a circle) combined with the performative tactic of urban street theatre are well suited to its media campaigns. The well-publicised performances by *Red Brigade* groups are now recognised internationally as a powerful visual sign of Extinction Rebellion events.

The first troupe of Red Brigade performers that emerged from the *Invisible Circus* street theatre collective in Bristol has since become synonymous with the public identity of XR. Members in local activist cells follow XR YouTube tutorials on how to make the costumes that are designed in deliberate contrast to everyday urban dress codes, along with how to enact XR's silent choreography.[13] XR's use of colour as urban theatre is a compelling way of drawing public attention: whether as figures in black protesting oil extraction companies, as people dressed in brilliant green as 'spirits' of dying forests or in the crimson costumes and slow choreography of the Red Brigade as a way of conveying the symbolic references to 'the common blood we share with all species, that unifies us and makes us'.[14]

In contemporary urban culture, red is also the colour of alert, warning or emergency, and hence in recent ecocritical thought it has been associated with the apocalyptic imaginary of the Anthropocene (Menely and Ronda, 2013). As such, XR's choice of white-faced actors dressed in vibrant red calls on deep cultural associations that work effectively against the typically grey backdrop of the city. Apart from the personal risk of performing in tightly policed public space, they are designed to convey deep mourning for an endangered planet: an emotional state that only the most cynical would accuse the protestors of manufacturing. Whether politically effective or not, the collective public enactment of loss, mourning or grief requires the kind of commitment Michael Hardt once described as a form of 'affective labour' (Hardt, 1999). For all the now-familiar imagery that renders XR actors anonymous, this is street theatre aimed to reveal the terrifying instability of our world, rather than a predictable script, and it requires significant emotional commitment from activists.

Figure 20.1 The Red Brigade, Extinction Rebellion rally Kurfürstendamm, Berlin, October 2019.
Source: Leonhard Lenz, image in public domain.

At first, XR rapidly gained public support, because its means of non-violent civil disobedience were enacted through peaceful demonstrations and creative street theatre. Yet, as XR activism took more strident steps to undermine the offices of multinational companies invested in fossil fuels, and particularly as more frequent demonstrations began to turn to civic disruption by 'blocking roads, using fake blood, recreating funeral marches, and surprise nakedness' (Hensby, 2019, p. 1), popular support for XR began to shift.[15] In response, on the first of January 2023, XR released a fourth aim on Twitter which was to:

> halt our tactics of public disruption. Instead, we call on everyone to help us disrupt our corrupt government.

This was supplemented by a call for a mass public rally dubbed 'The Big One' on 21 April 2023 outside the Houses of Parliament in London. For some commentators, XR had become a victim of its own success in escalating its effective media-smart tactics. But rather than limiting its horizons to urban events designed to dramatise climate change and raise environmental awareness, XR's agenda was always more ambitious, since its aim ultimately is to produce the kind of social change that will transform the world through nonviolent struggle against climate change. One of the key tenets of XR's theory of change is that mobilising just 3.5% of the population can lead to political transformation and system change:

> Our media messaging is based on research by Erica Chenoweth and Maria J. Stephan, which demonstrates that to achieve social change the active and sustained participation of just 3.5 per cent of the population is needed. It's that 3.5 per cent that we want to engage.[16]

Commentators such as journalist Nafeez Ahmed have questioned this as a means for nonviolent change, noting that all 323 cases in Chenoweth's study were formed in resistance to repressive regimes, occupations, or movements advocating for secession (Ahmed, 2019). In Ahmed's view, the so-called 3.5% rule can only be considered meaningful when applied to 'forms of political resistance against regimes involved in considerable, highly visible, domestic repression, by mobilising communities directly affected by that repression' (Ahmed, 2019). The study, which was not peer-reviewed, did not include many examples where democracy was overthrown, and perhaps most notably, there were no instances of successful nonviolent movements aimed at overthrowing a Western liberal democracy. Berglund and Schmidt also argue that the 3.5% claim fails to stand up to scrutiny when applied to climate action since this specific percentage does not actually feature in Chenoweth and Stephan's 2011 work (Berglund and Schmidt, 2020, p. 84) or in a later associated dataset (Chenoweth and Lewis, 2013).[17] Yet this claim continues to have a prominent role in XR's theory of social transformation through nonviolent civil disobedience in the context of climate activism.

Urban Disruption: Limitations/Critiques/Crackdowns

Social responses to XR have resulted in a coeval development of tactics and strategies not only within the movement itself but also in the legal frameworks it encounters. XR activist strategies have involved a dialectical process of action and counter-reactions in their aims to provoke institutions, legal systems and the public to reconsider and reconfigure their own positions on the issue of civil disobedience. The following critiques, limitations and responses to XR highlight the tension between the urgency of its climate activist agenda and the societal and economic structures it seeks to change.

A well-publicised criticism of XR is raised against its tactics of obstructing roads, bridges and public transport, which generally inconveniences the public and thus significantly hinders widespread public support. Several studies have offered support for disruptive civil disobedience as an effective tactic, a result that nonetheless contradicts public sentiment, at least in the UK. A YouGov survey conducted in February 2023 showed that 78% of British respondents believed disruptive protests negatively impact the key objectives of activists (though some academic studies have disputed this).[18] Some even suggest that, because such tactics often receive damning media coverage and backlash from people on the street, XR could be actively limiting progressive action on climate change. As such, XR's emphasis on the kind of direct action that could lead to arrest overlooks the significance of establishing wider social and political coalitions to effect meaningful change.

Many have argued that deeper public support is needed along with a more multifaceted approach encompassing electoral politics and policy advocacy along with more local, grassroots action. A case in point was an event in October 2019, when XR activists targeted

London's bustling public transport network during rush hour. Several XR protesters scaled a train disrupting the daily commute for thousands of people, including many from disadvantaged areas. This provoked public resentment and a media backlash that led to XR's public acknowledgement that the action had been misguided.[19] As we mentioned earlier, this sense of the need to shift to deeper public support prompted XR to publicly announce on its website that it would 'shift away from public disruption as a primary tactic' and instead 'prioritizes attendance over arrest and relationships over roadblocks'.[20]

Perhaps one of the most enduring critiques of XR, however, focuses on the charge of diversity and inclusivity within the movement itself. One recent study of XR membership, for example, revealed that it comprised mainly highly educated, middle-class members (Saunders et al., 2020). There are many less 'media savvy' people who, as geographer Karen Bell (2019) has pointed out, have not gained the same kind of media attention as XR's sophisticated media tactics,[21] and this can be a source of resentment. XR's civil disobedience has led to the charge that it disproportionately favours disaffected yet privileged white people who are less likely than people of colour to face violent encounters with law enforcement.[22]

In one sense, it could be reasonably argued that activism protesting climate change should be led by those who represent the rich countries most responsible for how it impacts everyone else. Yet since London is the most ethnically diverse region in the UK, with vast inequalities of wealth, many have argued that XR has failed to represent the ethnic diversity of the city.[23] While many XR activists claim they are inspired by the black civil rights activists of the 1960s who were prepared to face arrest, this does not prevent the perception of XR as a movement largely orchestrated by – and for – white, middle-class individuals.[24] It should be acknowledged that the question of the need for greater inclusivity in activism is not aimed exclusively at XR but instead at how many environmental movements in the Global North do not reflect the diversity that exists within and across communities.

Besides the issue of diversity, XR has also faced criticism for its claim to 'go beyond politics' by refusing to align itself with existing democratically elected political parties in preference for what it sees as more inclusive 'citizens' assemblies composed of randomly selected individuals chosen to represent the democratic process and thus combining diverse perspectives typically excluded from traditional political processes. XR contends this will result in better-informed policies catering to a wider range of the global populace, yet many have seen this as a significant weakness in the movement's ability to effect radical systems change. XR's refusal to engage with the current political institutions and parties is criticised as limiting its ability to achieve meaningful outcomes.

XR has faced criticism for not adopting a more pronounced anti-capitalist position in its challenges to the prevailing economic and political systems that prioritise ever-greater growth and consumption (Dhillon, 2019). XR has countered this by pointing out that, though it does not explicitly demand the dismantling of capitalism, it argues the current system is ill-equipped to deliver the urgent structural changes required to address the climate crisis. If XR's advocacy for citizens' assemblies and net-zero emissions by 2025 implies the need for systemic change, the movement favours mass mobilisation rather than overthrowing the current socio-economic system as the key to environmental transformation. One of XR's three key demands for the UK to go 'Beyond Politics' by creating a Citizens' Assembly has been perceived by some as unfeasible within the given time frame.[25] Yet XR defends this target, while acknowledging it is highly ambitious and would require a mass mobilisation on the 'scale of a World War'.[26]

XR and its members have been characterised in the press and by the UK government as an 'anarchist'[27] and 'extremist'[28] movement, with its members seen as 'criminals',[29] 'green fanatics',[30] and even as responsible for 'eco-terrorism'.[31] This practice of discrediting XR can be understood as a form of 'othering' aimed at portraying the activists as anarchists who resist all democratic process and who reject all social values. This is the representation of XR activists as the 'ultimate "other" in the realm of political movements, representing in the mind of the collective imagination a violent, left-wing extremism akin to hooliganism' (Berglund and Schmidt, 2020, p. 11).

Other claims that XR constitutes a 'climate cult' originate largely from right-wing media outlets,[32] a charge often repeated by British conservative politicians. In this context, XR's call for swift, comprehensive action on climate change is characterised as a neo-Puritan ethos that strives to impose stringent moral imperatives that are not widely shared. The charge of puritanical 'virtue signalling' is also brought to bear on XR's advocacy of personal sacrifices such as advocating tactical mass arrests or encouraging personal sacrifices in relinquishing carbon-intensive practices such as consuming red meat, travelling by air, or other forms of conspicuous consumption, which are rejected as a fanatical religious asceticism rather than practices designed to avert a climate emergency. This extends to how XR's emphasis on the inevitability of global crisis appears to take an apocalyptic tone, which some have referred to as secular climate eschatology and the belief that the world will meet a catastrophic end that only the most radical actions have the capacity to avert. In response, XR supporters counter that such accusations are meant to undermine the movement, asserting that an apocalyptic tone is justified given the severity of the impending crisis.

The UK government, as well as other governments in countries including Australia, have begun to crack down on disruptive protests, particularly on XR and other groups engaged in urban disruption, such as Insulate Britain or Just Stop Oil, by introducing stricter regulations on protests and granting police additional powers. After numerous defeats in the House of Lords, in May 2022 the Police, Crime, Sentencing and Courts Act[33] was passed in the UK, intensifying the severity of punishments for public nuisance and highway obstruction offences and increasing the likelihood of protesters being imprisoned. The act also aims to restrict marches by allowing police to impose strict conditions on public events if they anticipate serious public disorder, property damage or community disruption, including the disruptive use of noise, thus effectively diminishing the range of demonstrations, especially larger protests with more attendees.

The British government's Policing Act (Police, Crime, Sentencing and Courts Act) came into effect in April 2022. Part three of this act made legislative changes to laws on public order that effectively lower the threshold on what constitutes public disruption (including noise). It gave police much wider powers in arresting protestors for public nuisance,[34] while giving the courts the authority for significantly more serious punishment.[35] The serious consequences of this new Police Act recently made news in April 2023, when two Just Stop Oil protesters, Morgan Trowland and Marcus Decker, were each sentenced to over two and a half years imprisonment for the public nuisance of blocking all traffic for 40 hours after they scaled Queen Elizabeth II Bridge in Greater London. The sentence handed down by Judge Shane Collery was clearly aimed as an exemplary deterrent: 'You have to be punished for the chaos you caused and to deter others from copying you'.

Some months before this, however, XR itself appeared to have reached a roadblock in how to recruit wider support by re-evaluating its strategies for public disruption. As a result, in January 2023, XR posted a New Year's Resolution on its website[36] and Twitter campaign proclaiming its aim to 'halt our tactics of public disruption. Instead, we call on everyone to help us disrupt our corrupt government'. Hence the old tactic of DISRUPTION TO ~~PEOPLE~~ was to be replaced by DISRUPTION TO NARRATIVES OF POWER with a public invitation to 'Change Your Future' by assembling at the British Houses of Parliament on 21 April – a call for 100,000 people to assemble in protest in 100 days. This was noted in the press the following day when the *Guardian* reported that XR would change its tactics of smashing windows and people gluing themselves to artworks.[37]

The Future of Extinction Rebellion

XR consistently emphasises the importance of nonviolent civil disobedience, a practice contrasting markedly with the heavy-handed response by British Counter Terrorist Police in 2019 in their widely circulated document comparing XR to violent Jihadist movements and Neo-Nazi activists.[38] Though that document was redacted after exposure in the press, its categories of concern are revealing because they appeared less focused on the disruption XR causes to urban traffic or business than with the vulnerability of young people to XR described as:

> An anti-establishment philosophy that seeks system change underlies its activism; the group attracts to its events school-age children and adults unlikely to be aware of this.[39]

The fear that the XR movement could incite a kind of emotional contagion attractive to young people – especially in response to a willingness to face arrest – seems to be a primary political concern. When coupled with the wide appeal of the School Strike for Climate movement, XR's calls to action appear to offer a 'next step' to children distressed by rapid environmental degradation. This is of concern to parents, since, as trained legal observers noted in 2019, in Australian cities (as in England) there has been a disproportionately heavy-handed response by police to XR demonstrations.[40]

Apart from the clear infringement of civil liberties suppressing protests, more utilitarian arguments have been made that protest functions as a social safety valve, allowing individuals to voice their concerns and grievances in the public domain.[41] On this view, punitive laws that restrict peaceful protest could inadvertently push activists toward adopting more radical and violent forms of action. The Swedish writer Andreas Malm is one of the most prominent voices criticising non-violent direct action by advocating a radical flank thesis in which pushing boundaries thereby makes moderate factions seem more socially acceptable (Malm, 2021, p. 48). This potential radicalisation could manifest in various ways, from targeted sabotage and property destruction to potentially harmful activities that may put both activists and others at risk.

Taking a longer view, disruptions to work and everyday routines may eventually seem minor when compared to the catastrophic possibilities of global climate change. As we noted earlier, however, while frightening future scenarios may galvanise some to engage in disruptions to civic order, in some ways XR has been a victim of its own success since its more radical tactics also make people fearful, particularly of the radicalisation of children. XR's identification of mass grassroots non-violent civil disobedience as the essential factor of social transformation is insightful and, in our view, the key to its continued success.

Notes

1 This chapter was supported by the Australian Research Council [LP220100191]
2 Headline Statements – CI {2.1, 2.2, 3.1, 3.2, 3.3, 3.4, 4.1, 4.2, 4.3, 4.4, 4.5, 4.6, 4.7, 4.8} https://www.ipcc.ch/report/ar6/syr/resources/spm-headline-statements.
3 https://www.facebook.com/ExtinctionRebellionSouthAfrica/ Ashraf Hendricks Ground Up News/ South Africa https://www.groundup.org.za/article/protesters-cover-themselves-in-fake-blood-oil/ *Inside Arabia* Voice of the Arab People (December, 2019) /Adele Suliman *Reuters News Now* (November, 2019) https://insidearabia.com/extinction-rebellion-movement-mushrooms-in-the-middle-east / Freya Pratty 'The Teenage activists bringing the climate crisis to the middle east' (September 11, 2019) *Al-Monitor https://www.al-monitor.com/originals/2019/09/lebanon-extinction-rebellion-protests-climate-change.html*
4 With the title: 'Heading for Extinction and What to Do About It'
5 1. WE HAVE A SHARED VISION OF CHANGE. Creating a world that is fit for the next 7 generations to live in.
   2. WE SET OUR MISSION ON WHAT IS NECESSARY. Mobilising 3.5% of the population to achieve system change – such as 'momentum-driven organising'.
   3. WE NEED A REGENERATIVE CULTURE. Creating a culture which is healthy, resilient and adaptable.
   4. WE OPENLY CHALLENGE OURSELVES AND THIS TOXIC SYSTEM. Leaving our comfort zones to take action for change.
   5. WE VALUE REFLECTING AND LEARNING. Following a cycle of action, reflection, learning, and planning for more action. Learning from other movements and contexts as well as our own experiences.
   6. WE WELCOME EVERYONE AND EVERY PART OF EVERYONE. Working actively to create safer and more accessible spaces.
   7. WE ACTIVELY MITIGATE FOR POWER. Breaking down hierarchies of power for more equitable participation.
   8. WE AVOID BLAMING AND SHAMING. We live in a toxic system, but no one individual is to blame.
   9. WE ARE A NONVIOLENT NETWORK. Using nonviolent strategy and tactics as the most effective way to bring about change.
   10. WE ARE BASED ON AUTONOMY AND DECENTRALISATION. We collectively create the structures we need to challenge power.
6 Monbiot, George 2019, Wed 16th October https://www.theguardian.com/commentisfree/2019/oct/16/i-aim-to-get-arrested-climate-protesters
7 King referred to Socrates in his and again in his famous speech *I've Been to the Mountaintop* in 1968 the day before he was assassinated.
8 (Excerpt from Letter from Birmingham City Jail, April 16, 1963)
9 https://extinctionrebellion.uk/event/faith-bridge-at-the-xr-international-rebellion-uk/http://www.earthvigil.co.uk/xr-faith-bridge/
10 https://christianclimateaction.org/
11 In the UK this has included the ex-Archbishop of Canterbury, Rowan Williams, the actor Emma Thompson, musician David Byrne and well-known artist Grayson Perry.
12 https://www.youtube.com/watch?v=lCm_c93HrJM <Accessed April 20, 2021>
13 These costumes and mask-like stage makeup recall the masks and robes of classical Greek theatre, while in contrast the hats and gloves recall medieval dress.
14 XR – Invisible Theatre – Red Rebel Brigades as described by the founder of Invisible Circus, Doug Francisco. https://www.facebook.com/groups/1518920178239149/permalink/1548662935264873/ < Accessed May 30, 2021> An XR strategy of street-theatre as a sombre and mournful response to environmental crisis XR compares with 'Greek chorus, Buto dancers, Victorian funeral mourners'. In western cultures, the colour red has long signified blood and sacrifice, while in recent political history, it has continued to signify courage, particularly in revolutionary struggle: in the red caps of the French revolutionaries, the red vanguard of the Soviet Bolsheviks or the Chinese red guards.

15 https://yougov.co.uk/topics/politics/explore/not-for-profit/Extinction_Rebellion. The poll established that though 75% of the British public interviewed had heard of XR, 41% of people interviewed actively disliked it, 18% liked it and 15% were neutral. This compares unfavourably with WWF for example, of which 95% were aware, 75% liked and 5% disliked or with the more activist inclined Greenpeace of which 91% had heard, 50% liked, 24% were neutral and 17% disliked.

16 This Is Not A Drill, 2019

17 As Chenowth's estimates of numbers involved in civil resistance campaigns are not presented as a percentage of general populations, the 3.5% claim cited by XR appears to be an extrapolation of Chenoweth's public talks and opinion pieces, such as her 2013 Tedx talk and a 2017 article in the *Guardian*.

18 *The majority of the public believe protests rarely, if ever, make a difference | YouGov* n.d., yougov.co.uk. https://yougov.co.uk/topics/politics/articles-reports/2023/02/14/majority-public-believe-protests-rarely-if-ever-ma

   Social Change and Protests, a study undertaken by Apollo Academic Surveys, showed that 69% of experts considered 'the strategic use of nonviolent disruptive tactics' to be effective for issues like climate change, while another study, *Does Climate Protest Work? Partisanship, Protest, and Sentiment Pools*, corroborates the effectiveness of disruptive action but within a US context. The research showed that disruptive civil disobedience had a positive impact, specifically among Democrats, while interestingly the study also found no 'backfire' effects for any type of protest among any political group.

19 https://www.theguardian.com/environment/2019/oct/20/extinction-rebellion-tube-protest-was-a-mistake

20 https://extinctionrebellion.uk/2022/12/31/we-quit/

21 https://www.theguardian.com/environment/2019/oct/11/a-working-class-green-movement-is-out-there-but-not-getting-the-credit-it-deserves

22 https://www.independent.co.uk/voices/extinction-rebellion-arrests-london-protests-climate-change-people-of-colour-global-south-a8879846.html

23 GOV.UK Ethnicity Facts and Figures (2021): 46.2% of residents identified with Asian, black, mixed or 'other' ethnic groups and a further 17.0% with white ethnic minorities <https://www.ethnicity-facts-figures.service.gov.uk/uk-population-by-ethnicity/national-and-regional-populations/regional-ethnic-diversity/latest#:~:text=2021%20Census%20data%20for%20England,17.0%25%20with%20white%20ethnic%20minorities>

24 https://ecohustler.com/culture/extinction-rebellion-and-the-inclusivity-problem

25 https://www.bbc.com/news/science-environment-47947775

26 XR. (2020). *The Emergency*. rebellion.global. https://rebellion.global/the-emergency/

27 https://www.telegraph.co.uk/news/2023/04/22/surrendering-to-eco-anarchists-threatens-democracy-itself/

28 https://www.bbc.com/news/uk-51071959

29 https://www.theguardian.com/environment/2020/sep/08/extinction-rebellion-criminals-threaten-uks-way-of-life-says-priti-patel

30 https://www.dailymail.co.uk/debate/article-7541965/DOUGLAS-MURRAY-listen-bunch-anarchists-work-fire-hose.html

31 https://www.mailplus.co.uk/edition/comment/98462/extinction-rebellion-activists-are-terrorists

32 Such as British papers the *Telegraph* or the *Spectator*.

33 https://bills.parliament.uk/bills/2839

34 https://www.legislation.gov.uk/ukpga/2022/32/part/3/enacted

35 The offence of serious harm to property; serious annoyance or inconvenience to people in the Magistrates Court, could carry a sentence of up 12 months in prison, an unlimited fine or both and sentence by in the Crown Court can be up to 10 years in prison, an unlimited fine or both.

36 https://extinctionrebellion.uk/the-big-one/#why-100-days

37 https://www.theguardian.com/world/2023/jan/01/extinction-rebellion-announces-move-away-from-disruptive-tactics#:~:text=Extinction%20Rebellion%20announces%20move%20away%20from%20disruptive%20tactics,-This%20article%20is&text=The%20climate%20protest%20group%20Extinction,in%202023%2C%20it%20has%20announced.

38 https://www.theguardian.com/uk-news/2020/jan/10/xr-extinction-rebellion-listed-extremist-ideology-police-prevent-scheme-guidance

39 On 28 January 2020, Jamie Grierson and Vikram Dodd reported in an article in the *Guardian* 'Terror police list that included Extinction Rebellion was shared across government' that the document was sent to the Home Office, the Department for Education, NHS England, the Ministry of Defence, HM Prison Service, Probation Service and Ofsted, as well as 20 local authorities, five police forces and Counter Terrorism Policing headquarters (CTPHQ) in London. https://www.theguardian.com/environment/2020/jan/27/terror-police-list-extinction-rebellion-shared-across-government [Accessed February 2021]

40 In Melbourne, many wanting to join the protest were prevented from joining by a street cordon which (despite the legal right to peaceful public assembly) was constructed and defended by mounted police 'in the interests of public safety'. On Saturday 14 September 2019 Melbourne Activist Legal Support (MALS) fielded a team of eight trained Legal Observers at the 'Princes Bridge Block and Dance' protest event that took place on St Kilda Road between the Flinders Street intersection and the Victorian Arts Centre precinct. Their Statement of Concern – and a response by Victoria Police – can be accessed at:

   https://melbactivistlegal.org.au/2019/09/16/statement-of-concern-the-policing-of-extinction-rebellion/

   [Accessed May 2021]. In 2019 similar over-reactions by police were observed in Brisbane for example: https://www.theguardian.com/australia-news/2019/oct/10/extinction-rebellion-labor-members-say-chilling-mass-arrests-have-echoes-of-bjelke-petersen-era

   And in Sydney: https://www.theguardian.com/environment/2019/oct/10/extinction-rebellion-scott-ludlam-has-absurd-bail-conditions-dismissed-by-judge

41 https://www.lag.org.uk/article/211406/environmental-crisis-and-civil-rights.

References

Ahmed, N. (2019). The flawed social science behind Extinction Rebellion's change strategy. *Medium*, 29 October. https://medium.com/insurge-intelligence/the-flawed-science-behind-extinction-rebellions-change-strategy-af077b9abb4d

Bell, K. (2019). *Working Class Environmentalism. An Agenda for a Just and Fair Transition to Sustainability*. London: Palgrave Macmillan.

Berglund, O. and Schmidt, D. (2020). *Extinction Rebellion and Climate Change Activism*. Cham, Switzerland: Palgrave Macmillan.

Chenoweth, E. (2013). *The Success of Nonviolent Civil Resistance: Erica Chenoweth at TEDxBoulder*. United States: YouTube. https://www.youtube.com/watch?v=YJSehRlU34w

Chenoweth, E. (2017). It may only take 3.5% of the population to topple a dictator – With civil resistance. *The Guardian*, 1 February. https://www.theguardian.com/commentisfree/2017/feb/01/worried-american-democracy-study-activist-techniques

Chenoweth, E. and Lewis, O. A. (2013). Nonviolent and violent campaigns and outcomes dataset, v.2.0 Denver. *Journal of Peace Research*, 50(3), 415–423.

Chenoweth, E. and Stephan, M. J. (2011). *Why Civil Resistance Works*. Chichester: Columbia University Press.

Cory, O. and Reiner, D. (2021). Protests and policies: How radical social movement activists engage with climate policy dilemmas. *Sociology*, 55(1), 197–217.

Dhillon, A. (2019). Extinction Rebellion must decide if it is anti-capitalist- and this greenwashing mining company shows us why. *Independent*, Wednesday, 23 October. https://www.independent.co.uk/voices/extinction-rebellion-climate-crisis-bhp-mining-coal-colombia-a9167601.html [Accessed 11 May 2021].

Extinction Rebellion UK (2018). *Rebellion Against UK Government's Criminal Inaction on Climate Breakdown*, 26 October. https://extinctionrebellion.uk/2018/10/26/rebellion-against-uk-governments-criminal-inaction-on-climate-breakdown/

Francisco, D. (2019). *Red Rebel Brigade. Head Dress*. https://www.youtube.com/watch?v=_qIfSNP2nuo Costume and Movement https://www.youtube.com/watch?v=hGPv0wALfWw Red Rebel Warm Up https://www.youtube.com/watch?v=Dp_3xgwDZNo

Hardt, M. (1999). Affective labour. *boundary* 2, 26(2) (Summer), 89–100.

Hensby, A. (2019). Extinction rebellion: Disruption and arrests can bring social change. *The Conversation*. https://theconversation.com/extinction-rebellion-disruption-and-arrests-can-bring-social-change-115741

Malm, A. (2021). The moral case for destroying fossil fuel infrastructure. *The Guardian*, 18 November.

Menely, T. and Ronda, M. (2013). Red. In Jeffrey Jerome Cohen (ed.) *Prismatic Ecology: Ecotheory beyond Green* (pp. 22–41). Minneapolis: University of Minnesota Press.

Nixon, R. (2013). *Slow Violence and the Environmentalism of the Poor*. Cambridge, MA: Harvard University Press.

Rawls, J. (1971). *A Theory of Justice*. Harvard, MA: Belknap Press.

Read, R. and Alexander, S. (2020). *Extinction Rebellion: Insights from the Inside*. Melbourne: Simplicity Collective.

Saunders, C., Doherty, B. and Hayes, G. (2020). A new climate movement? Extinction Rebellion's activists in profile. *Technical Report*, July.

Sharp, G. (1973). *The Politics of Nonviolent Action*. Boston: Porter Sargent.

Taylor, M. (2018). "We have a duty to act": Hundreds ready to go to jail over climate crisis. *The Guardian*, 27 October. https://www.theguardian.com/environment/2018/oct/26/we-have-a-duty-to-act-hundreds-ready-to-go-to-jail-over-climate-crisis

Taylor, M. (2020). The evolution of extinction Rebellion. *The Guardian*, 4 August. https://www.theguardian.com/environment/2020/aug/04/evolution-of-extinction-rebellion-climate-emergency-protest-coronavirus-pandemic

Taylor, M. and Gayle, D. (2018). Dozens arrested after climate protest blocks five London bridges. *The Guardian*. https://www.theguardian.com/environment/2018/nov/17/thousands-gather-to-block-london-bridges-in-climate-rebellion

XR. (2020). *The Emergency*. rebellion.global. https://rebellion.global/the-emergency/

XR. (2023). *About Us*. rebellion.global. https://rebellion.global/about-us/

Yusoff, K. (2017). *A Billion Black Anthropocenes or None*. Minneapolis: Minnesota University Press.

21
SYSTEM CHANGE, NOT CLIMATE CHANGE

The Climate Justice Movement in Germany 2008–2021

Daniel Hofinger and Maximilian Becker[1]

Germany – land of the *Energiewende* (energy transition), sustainability, and waste separation. Or the land of lignite, highways, and global fossil fuel imports? For decades, German climate and energy policy has been caught between the interests of powerful corporations and the pressure of recalcitrant social movements. This article traces the development and politics of the German climate justice movement from the climate summit in Copenhagen in 2009 to the climate ruling of the Federal Constitutional Court in 2021.

During this period, numerous new actors emerged, the mobilisation capacity and societal relevance of the climate movement has increased overall, and the movement's thematic horizons have multiplied. Still, Germany's increasingly professional climate justice movement was only able to achieve a handful of significant political successes between 2009 and 2021, and Germany is still not acting in line with the 1.5 degrees Celsius temperature rise limit of the Paris Climate Agreement. Nevertheless, the movement had a broad mobilisation potential, it had the power of images and narrative on its side, and it made the need for climate protection the new consensus for the population as a whole. Fundamental to the climate justice movement in Germany (and anywhere) is the realisation that those societies and groups that contribute the least to climate change are also those that are the first and most affected by the crisis. In this sense, it reminds German society of its international obligation and responsibility in climate protection and of the historical and ongoing (neo-) colonial causes of Germany's political and economic power and its climate-destroying consumption of resources. At demonstrations, in solidarity actions and in its priorities, the movement emphasises its own interconnectedness with climate struggles elsewhere in the world, wherefrom activists learn and take inspiration.

We, the authors of this text, have been active in the German climate justice movement for many years and know each other through our work with Ende Gelände and the protests in the Rhenish lignite mining area. We both live in Germany, are white, and have completed a university degree. We do not experience discrimination. This text is based on experiences made alone or together in the movement and on numerous conversations with other activists. That said, within the scope of this article, we could not mention all of the many currents and influences within the broader German movement for climate justice. We apologise for any omissions.

DOI: 10.4324/9781003396567-25

"What Do We Want? Climate Justice"

"Mr Obama – what will you tell your grandchildren?" Large, illuminated billboards in the wintry streets of Copenhagen attracted the attention of the more than 10,000 state delegates who had gathered in the Bella Centre for the World Climate Conference. They were supposed to finally agree on a binding roadmap to save the world's climate, 14 years after the first global climate negotiations.

Tens of thousands of climate activists prepared for large demonstrations and actions of civil disobedience throughout the city. Following the tightening of laws, the Danish police took over 2,000 of them into preventive custody, stopped numerous activities, and deliberately injured activists. Nevertheless, the city experienced fierce protests with up to 100,000 participants. The conference itself did not come to rest either. As the conflicts between the interests of rich industrialised countries of the Global North and the countries already affected or threatened by climate change were too great, the conference ended in a diplomatic fiasco. Without an adopted final text and without an agreed path to stabilise the atmosphere, "Flopenhagen" represented a low point in global efforts to achieve climate justice. State actors and global NGOs failed in their strategy to negotiate a consensus between those responsible for the climate crisis and those directly affected by it. At the same time, the summit protestors and their networks of climate justice groups coined the slogan "system change – not climate change", which has since become a motto of the climate movement and summarises the core idea of climate justice: if climate change is to be contained, structural, political, economic, and social changes are needed as well as the combination of climate and redistribution policies. Yet, how exactly, by what means, and in which areas "the system" should be changed has been a perplexing headache ever since.

The disappointment over the failed climate conference led activists and European civil society to search for more effective ways to intervene in the political process. Instead of trying to influence international government negotiations, many felt it would be more effective to make the concrete sites of climate injustice and destruction the starting point for actions and protests. In Germany this meant targeting the planned construction and expansions of coal plants and existing coal mines (Sander, 2016).

Climate Camps in the Rhenish Coal Fields

Since 2010, the educational, cultural, and protest events known as *Klimacamps* (climate camps) have been held annually in the Rhenish lignite mining area in the west German federal state of North Rhine-Westphalia. Located between the metropolitan area of Cologne and the Netherlands, the coalfield is operated by the energy giant RWE. It consists of three open-cast mines (*Hambach, Garzweiler* and *Inden*) and, with the power plant units in Neurath, Niederaußem, and Weißweiler, it is home to three of the five largest single emitters of CO_2 in Europe. The open-cast mines are the largest holes in Europe with dimensions of up to 8 × 10 kilometres (5 by 6.2 miles) and 450 metres (500 yards) depth. To keep them dry, the groundwater is pumped out over large areas, with repercussions as far away as the Netherlands. Dozens of villages and small towns have been destroyed and over 40,000 people have been resettled since the 1950s to clear the path for the mines as they devoured the landscape. The coal is transported from the open-cast mines to the power stations on a dedicated railway network of over 300 km (190 miles) length. With emissions of 100 million (!) tonnes of CO_2 at the time, it was Europe's largest source of CO_2 and became a focal point of the German climate justice movement for over a decade.

The idea of climate camps was developed during protests against road construction in the UK (Frenzel, 2011) and is based on four pillars: networking, education, action, and living utopia. The latter signifies the movement's prefigurative politics of anticipating the future implementation of goals for society through individual and collective action in the here and now. The camps and most of the activists at the time were rather anarchistic and "hippiesque" in character. Due to the movement's tender beginnings, disputes about its direction and fields of action were on the agenda. The experiences of civil society institutions in the Rhenish mining area, which had already been fighting against opencast coal mining for decades in citizens' associations and court cases, were joined by those who brought a younger and more radical spectrum from the *Castor* transports against nuclear power, from the *Gendreck-Weg* campaign against genetically modified crops, from the *Kelsterbach Forest* protests against the expansion of the Frankfurt airport, and from the anti-*G8* protests of Heiligendamm against neoliberal globalisation to the Rhineland. In the first direct blockade actions that originated from the climate camp in 2010, a few dozen people blocked the public access road to a coal-fired power plant and the transport tracks of the coal trains under the slogan "Whoever digs a pit for others . . .".[2] The network *ausgeCO₂hlt*, founded in 2011, was the first networking platform for long-term protests in the Rhineland coalfield and supported the organisation of climate camps and protests around RWE's annual general meeting.

Building on the EarthFirst! network's tradition[3] of direct action, activists occupied the Hambach Forest, which was threatened with deforestation for an open-cast mine, at a "Forest not Coal" festival in spring 2012. This permanent physical presence of people critical of coal and capitalism revitalised and radicalised the fading resistance of local citizens' initiatives. Those involved modelled their actions on the anti-nuclear *Castor*-protests in the North German Wendland region on the one hand and the protests against the expansion of an airport in Notre-Dame-Des-Landes in western France on the other (Häußermann and Wollny, 2016). In the winter of 2012, the Hambach Forest was cleared by police for the first time. A person trapped in a self-built, six-metre-deep tunnel system delayed the eviction for four days. Immediately after the eviction, the forest was reoccupied in several places – a clear sign that the resistance would be permanent. By now, the action form of forest occupation has become firmly anchored in the German climate justice movement and a number of forests throughout Germany have been occupied in recent years (CrimethInk, 2021).

Lusatia in eastern Germany, close to the border with Poland, is home to the second-largest German lignite coalfield, where climate camps have been held regularly since 2011. In 2013, people chained themselves in so-called lock-on actions to coal tracks near the Welzow-Süd open-cast mine. As part of the "Reclaim Power Tour", activists cycled from the Lusatian climate camp to the climate camp in the Rhineland, symbolically and personally linking the climate struggles in the coalfields. This networking was an essential building block in forming the anti-coal movement as a supra-regional and nationwide movement. As the highlight of the 2014 climate camp, thousands of activists took part in a human chain across the border river Neiße, sending a clear signal for international cooperation in the climate justice movement. This networking continues to this day in the joint struggles against Polish opencast mining and the Turów coal-fired power plant in Poland.[4]

Starting from the climate camp in the Rhineland in 2013, climate justice activists occupied the state party headquarters of Bündnis 90/Die Grünen in Düsseldorf, state capital of North Rhine-Westphalia. The party emerged from West German environmental, peace, feminist, and anti-nuclear movements in 1980 and has since been involved in numerous state and federal governments. The office occupation was triggered by the decision by one of the party's regional branches to retroactively legalise the illegally built Datteln IV black coal-fired power plant in a planning procedure instead of opposing its commissioning with full force. The anarchist-influenced movement around Hambach Forest rejected political parties in general, and anti-nuclear veterans who had already felt betrayed when the Greens supposedly took too long to phase out nuclear power were greatly disappointed by the Greens' reactions to the protest action. The party's then state chairwoman brought legal charges against the climate activists, which she did not withdraw despite the judge's insistence at the main hearing. This resulted in a court case that attracted a great deal of attention in the media and within the movement. It made it clear within the climate justice movement for the next few years that the Green Party was not a reliable partner in this fight. This assessment was reinforced at the end of 2020 by disputes over the Dannenröder Forest in the federal state of Hesse, where a forest occupation was evicted by police with Green government participation to make way for the construction of a motorway.[5]

The climate camps have always been places where the movement's core ideals and direction were debated. In their political decision-making processes, the camps aim for grassroots democracy and consensus-building. This is intended to reduce power imbalances among the participants on the one hand and to mobilise them to implement decisions through active involvement on the other. This model was put through its paces, for example, in a multi-year discussion process at the Rhineland midsummer climate camps about whether it was appropriate for cis men to move around the camp shirtless. The conflict symbolised internal movement disputes between a reform-oriented focus solely on coal and CO_2 on the one hand and the demand for broader and fundamental social change on the other. The debate ultimately ended with the decision that everyone had to cover their upper bodies at least partially. The experiences gained in this debate formed part of the foundation for the movement's later more normalised positioning as (queer) feminist, as a reflection of its own sexist structures was placed on the movement's agenda. The thematic work at the camps was always internationally oriented and influenced by discussions in other movements. In 2013, for example, several hundred people with an agricultural focus attended the climate camp in the Rhineland in cooperation with the European constellation for peasant farmers and landless people called "Reclaim the Fields!". Two years later, hundreds of people, this time with an academic background, took part first in the summer school "Degrowth konkret: Klimagerechtigkeit" and then in Ende Gelände's first large-scale action.

The climate camps and actions of the early 2010s resulted in networking and expansion of the climate movement. It established itself as a young movement that campaigned for an intersectional interpretation of climate justice and used new forms of action to draw attention to German coal power as Europe's largest source of CO_2. First experiences with direct action, the positive media response, and the continuous influx of people gave the movement hope for the future.

"Auf Geht's – Ab Geht's – Ende Gelände"

By 2015, the climate movement began to plan larger and more impressive actions. After human chains, demonstrations, and other forms of protest in previous years, it was time to go one step further with a mass civil disobedience action. In August 2015, the Ende Gelände ("Here and no further") action alliance blocked the Rhenish Garzweiler open-cast mine for the first time.

The chosen form of action proved to be a great success. The central building block of the campaign was the motto "We say what we do and we do what we say". In contrast to clandestinely organised blockade actions by hardened activists on the one hand and purely legal large-scale demonstrations by environmental associations on the other, Ende Gelände began to talk about blockading an opencast mine on lecture tours, in established political organisations, and on social networks 6 months before the actual action. This public announcement of the action created an attention and mobilisation window that extended the action period significantly. Months before the blockade, there was a debate in the media about the legitimacy of civil disobedience in the face of the impending climate catastrophe. Activists, NGOs, and philosophers were thus able to provide a protective shield of social legitimation for the action before it occurred.[6] Numerous action training sessions were aimed at enabling people with little experience to participate at Ende Gelände. Through a clearly agreed and communicated "action consensus", the organisers were able to conclusively "say what we are doing" to potential participants and the interested public (media). Likewise, participants were able to trust the organisation and prepare themselves to "do what we say". Ende Gelände adapted the idea of an action consensus from the anti-nuclear movement and it supported the formation of the alliance's identity. An insurrectionist-militant subset of people within Ende Gelände rejected the commitment to (only) one particular form of action to contain organic rage and instead pleaded for a so-called diversity of tactics. They took part in some of the larger choreography of the protest days with their own actions (see later mention) and later founded their own organisation, Zucker im Tank, which specialised in clandestine, small-group actions (see Zucker im Tank, 2022).

Through sophisticated action choreography, various "finger" action groups, each consisting of several hundred people, activists succeeded in accessing the open-cast coal mine and, in some cases, the coal excavators located several kilometres inside the mine. In the choreographies of the action days, Ende Gelände combined various forms of action as an "escalation ladder". Even if the tactical approach of the police and the coal company's private security service was not exactly predictable, the activists could consciously decide where they wanted to participate based on the expected course and approach of the individual fingers. Communication between organisers and activists was always a balancing act between the greatest possible transparency, trust-building, and accountability, as well as the necessary secrecy of tactical plans at certain points. Nevertheless, activists were able to know, for example, which finger was planning a long and calm march from the climate camp to the coal mine and which finger was expecting to have to cross police lines at certain narrow points in the terrain. Experienced small groups and representatives of militant tendencies were also able to deliberately and autonomously block police supply routes or cause a disabling short circuit on the route of the coal trains, thereby enhancing the safety of other activists who would later stage mass sit-in blockades of these tracks.

The organisers of Ende Gelände knew that their political strategy of publicly breaking the law depended on popular legitimacy. They therefore cooperated with various environmental associations and civil society organisations that were often themselves dissatisfied with the impact of purely legal methods. At the same time, for legal and media reasons, these organisations were unable or unwilling to call for civil disobedience directly. Instead, they issued public letters of solidarity, in which their prominent representatives expressed their understanding of the actions of Ende Gelände and participation in the actions. In some cases, executives of large NGOs also took part in the actions "as private individuals". Another essential feature was that the large environmental and nature conservation

Figure 21.1   Ende Gelände action in Lusatia 2016: Activists in white protective suits block a coal excavator.

Source: Photo by Tim Wagner.

organisations held legally registered demonstrations in the area of the open-cast mines, while Ende Gelände blocked them. This enabled participation in the protests without confrontation with the police and without having to walk quickly through open terrain. At the same time, many Ende Gelände activists also took part in these protest marches in order to get closer to the open-cast mines and to find a way from the demonstration across fields and through forests at tactically relevant points.

The novel and spectacular images created during the Ende Gelände actions have characterised the media debate on the coal phase-out ever since. For the climate movement, the experience of the first Ende Gelände action in 2015 was so encouraging that the alliance, which was initially only founded for this one action, decided to organise further actions together and set up local groups.[7]

Following the success of 2015, Ende Gelände organised another spectacular action the following year, this time in the Lusatian lignite mining region in eastern Germany. Almost 4,000 activists took part in the action, far more than in the previous year. The action took place at a time when the operator of the opencast mines and power plants there, the Swedish company Vattenfall, was looking for a buyer for its German coal division. The alliance chose the slogan "We are the investment risk" as a clear message to the potential buyer. The call to action also reflected the radical climate justice movement's now firmly established anti-capitalist self-image (Sivanesan, 2016). The action met with far greater rejection from the local population and politicians in Lusatia than the Ende Gelände protests in the

Rhineland. The often harsh criticism clearly reflected the power of the coal industry and the coal miners' union in the region. Politically, a right-wing hegemony had prevailed in Lusatia since at least 2015. In addition, the reactions showed how disconnected from the local population the (inter-)national radical climate movement and its actions were at the time (Lindenberg and Müller, September 2016). With the later emergence of Fridays for Future and the popularisation of the climate movement, this changed – at least in part. While activists in Lusatia in 2016 were still labelled "terrorists" by the conservative governing party CDU, the media less strongly questioned the legitimacy of civil disobedience during later actions (Kreutzfeld, 2016).

"Hambi? Bleibt!"

Although large parts of the movement were critical of the international climate negotiations following the experiences of Copenhagen in 2009, the meetings have repeatedly provided important impetus for the orientation of movement activities. This was also the case in November 2017, when the climate conference known as COP23 was held in Germany because the Pacific island state of Fiji, which was actually hosting the conference, lacked the material resources for a world climate conference on its own territory. For this reason, the conference was held at the UN site in Bonn – fewer than 50 kilometres (30 miles) away from the Rhenish coalfield. The cynicism that the mines were contributing to rising sea levels and thus threatened the very existence of the Republic of Fiji was a catalyst for the climate justice movement: a broad alliance organised a demonstration with over 25,000 participants in Bonn, and hundreds of events took place at the alternative "People's Climate Summit". A blockade of the supply line inside (!) a coal power plant by activists from the group "Never trust a COP" (a slogan that was also used in Copenhagen) meant that several units of the Weißweiler power plant had to be completely shut down, for which the operator RWE is still demanding two million euros in compensation from protesters today.[8] The Pacific Climate Warriors held a traditional *sevusevu* ceremony in Manheim – a village that was to be demolished for the Hambach coal mine – and handed over large, artistically crafted cloth flowers to representatives of people affected by German mining and climate activists to express their solidarity with the struggles in the communities responsible for the global climate crisis (Jetñil-Kijiner, 2017). At the same time, thousands once again stood in front of the huge coal excavators inside the mine and sent a clear message: while international climate diplomacy is failing, the movement is taking the coal phase-out into its own hands.

Images of blocked coal excavators went viral in the global press and contrasted with Germany's image as a "climate pioneer" and "land of the energy transition". The movement combined this re-framing with an innovative leap in protest strategy: while the residents of the Rhenish mining area were primarily concerned with the local ecological and health damage caused by coal mining, the climate camps linked the mining area with nationwide climate movements. Ende Gelände connected the German movement with European comrades[9] and, ultimately, Germany's reputation in the community of states was at stake during COP23. This spatial escalation and re-scaling of the protests also seemed politically necessary: where the coal company RWE had bought the support and silence of local politicians (Zacher, 2023), the German national government had to be persuaded to phase out coal under international pressure.

At the same time, the situation at Hambach Forest continued to escalate. In the winter of 2017, ongoing protests and a temporary court judgement based on a lawsuit filed by the environmental organisation BUND (Friends of the Earth) for the first time halted the logging that takes place every winter – a motivating example for the movement for future protests. After 6 years of continuous forest occupation, around 200 activists lived permanently in around 50 tree houses in the forest. They tried out various forms of action and attracted the attention of the European eco-anarchist movement. Local residents, onlookers, and activists came together on monthly walks through the forest. The unique aesthetics of the forest occupation and its experimental way of living became the heart of the German climate movement. As open-cast mining removes additional parts of the forest every year, the immediacy of direct action enveloped the occupation: forest or mine – you can't have both. In 2018, the remaining part of the forest had become so small that it could easily have been destroyed in one clearing period. It was all or nothing. In September 2018, one of the largest police operations in the history of the Federal Republic of Germany with thousands of units took place to clear hundreds of activists and their tree houses. Despite massive police violence, the eviction progressed slowly and the daily protests in and around the forest grew. On 19 September 2018, student and documentary film-maker Steffen Meyn died in a tragic fall while documenting the eviction of the tree houses for a film project. After the eviction was completed despite this caesura, a court banned the clearing of the forest on 5 October 2018. Again following a complaint by BUND (Friends of the Earth), the court ruled that the endangered Bechstein's bat, found in Hambach Forest, is to be protected and that the old forest could therefore not be logged. The following day, at least 50,000 people celebrated their success at a demonstration and rebuilt the first of the tree houses that still exist today (Weiermann, 2019). As an anti-coal movement in Germany, the climate movement had become a powerful actor capable of mobilisation that the existing power structure could no longer ignore (Hofinger, 2019).

The preservation of the Hambach Forest is THE success of the German climate movement of the last 10 years. By reducing the size of the open-cast mine, 1.1 billion (!) tons of coal and thus 1,100 million tons of CO_2 remain in the ground and one village was saved from destruction. The "Hambi Bleibt" movement had a global impact and inspired numerous countries' struggles against fossil fuel extraction projects. Building on years of struggles for "discourse" and "awareness", the Hambach Forest movement was finally able to achieve material success.

The events in Hambach Forest coincided with the work of the so-called Coal Commission. The German government tasked it in June 2018 with preparing a law for the coal phase-out. Individual non-governmental organisations and people affected by opencast mining were invited to participate in this commission, which led to major debates within the climate movement about the most effective theory of change: where is cooperation between movement institutions and the state important for translating movement pressure into real political change? The commission's disappointing results in terms of climate policy confirmed the fears of many actors that parts of the climate movement had turned into a fig leaf for a policy of climate injustice (Behrmann, 2019). The envisaged coal phase-out in 2038 and the billions in benefits for the coal industry associated with it could only be interpreted as a declaration of climate bankruptcy. While the climate

Figure 21.2   Tree houses in Hambach Forest. Up to 90 of these constructions, some of which were connected to each other, could be found in the forest, as well as huts accessible from the ground.

Source: Photo by Tim Wagner.

movement perceived the process and results of the coal commission as a slap in the face, it became an export model for the government: subsequently the Czech Republic and other countries implemented a coal commission based on the German model to negotiate a coal phase-out.

"Our Future – Not Your Business"

Based on the protest by schoolgirl Greta Thunberg in Stockholm, who began to stand in front of the Swedish parliament in August 2018, first daily, then weekly, with a poster, Fridays for Future became a global climate movement of young people that also brought hundreds of thousands onto the streets in Germany. With their strikes, the students made clear that they were fed up once and for all with adults burning up their future. The Fridays for Future protests in Germany gained astonishing momentum in their very first year. In the Rhineland in particular, many pupils who initiated the first school strikes at the end of 2018 had already participated in the demonstrations and actions to preserve Hambach Forest in the summer. The possibility of a school strike for the climate was being discussed in the fall as the students were convinced that adults would not adequately solve the generational problem of the climate crisis. The starting signal was given by a group of pupils in Kiel in northern Germany and young people led by Luisa

Neubauer in Berlin, who independently called for the first protests in December 2018. They used WhatsApp groups to organise the first school strike, which took place in seven German cities in the first week, and began setting up a long-term organisation the very next weekend. The conference call with activists from all over Germany that took place two days after the first school strike started a series of weekly organising calls that still continues 5 years after it began.

Within a few weeks, the Fridays for Future protests grew from a few individual students to thousands who took to the streets every Friday instead of going to school. Meanwhile, a fierce debate broke out in the German media and among politicians about the legitimacy of the school strike. In the first few months, simply staying away from school was a disobedient breach of the rules that demanded a great deal of courage. The first school strikers had to explain and defend to their parents and families, teachers, and classmates why they were disrupting the normal course of the school day in order to draw attention to a problem outside of school in the narrower sense. These pioneers of the school strike proved through their actions that the school strike was possible and that the real consequences were less severe than feared. By surviving the initial repression and carrying on, they broke the fear barrier that many social movements experience. As a result, more and more students dared to take part in the weekly protest. In January 2019, the movement mobilised 10,000 students from all over Germany to Berlin to march on the occasion of the final meeting of the Coal Commission. During the first internationally organised school strike on 15 March 2019, 300,000 people took to the streets in Germany, primarily pupils and students. Increasingly, adults also supported the protests and joined the protests as Parents for Future, Psychologists for Future, or Architects for Future in a kind of "Everyone for Future" umbrella. This broadening of support quickly normalised the school strike, which on the one hand made the demand for real climate protection acceptable to the majority but on the other hand, in the perception of core activists in particular, took away the sharpness and oppositional rebellion of the protests.

The protests made the climate crisis the dominant political issue in Germany in 2019 and the demand for genuine climate protection gained majority support. As a result, the German government set up a new cabinet committee on climate in March 2019 to discuss and decide on measures to meet climate targets over several months. During this period, the weekly school strikes organised by Fridays for Future continued and mobilised more and more people.

The Ende Gelände mass action in June 2019 showed just how close ties between the young climate movement around Fridays for Future and the more radical movement around Ende Gelände had become in the wake of the disappointing results of the Coal Commission and the frustration over the lack of further climate protection measures: In addition to a nationally mobilised climate strike on a Friday in Aachen, an hour west of Cologne on the edge of the Rhenish coalfield, Fridays for Future had also registered a demonstration near the Garzweiler coal mine for the following Saturday. When this demonstration reached the edge of the open-cast mine, more than 1,500 activists from Ende Gelände's golden finger suddenly broke out of the demonstration. They streamed into the mine as they were cheered on by tens of thousands of demonstrators. At the same time, numerous other Ende Gelände action groups blocked coal infrastructure in various parts of the Rhenish mining area. The so-called colourful finger, consisting of wheelchair users, children, physically disabled people and people who had signed a cease and desist declaration from RWE or who could not or did not want to walk for miles, blocked a central public access road to an opencast mine for the first time. This moment of collaboration resulted from years of movement building, listening, learning, and experimenting. It was no longer just mass absences from school that were now considered disobedient rule-breaking; entering a coal mine, with its increased potential for police repression and subsequent prosecution, was now also a legitimate part of Generation Climate's repertoire of actions. The question of whether the capitalist economy should be criticised and overcome as the cause of climate destruction continues to be controversially discussed in Fridays for Future. FFF demonstrates great integrative power in keeping these currents together and uniting them productively, instead of giving free rein to their centrifugal forces and allowing them to splinter into different sub-groups.

Fridays for Future reached the peak of its mass mobilisation on 20 September 2019, when almost 1.4 million people took to the streets in over 550 sites across Germany as part of a Global Climate Strike. In addition to the success of the sheer unbelievable size of the demonstrations – it was probably the largest demonstration in the history of the Federal Republic of Germany to that date – which took place in even the smallest villages, this day also marked a major realpolitik defeat for the movement. For, at the same time, the Federal Government's Climate Cabinet Committee presented the key points of its much-heralded "Climate Protection Program 2030". Among other things, by making far-reaching concessions to the automotive industry, it did little to reduce emissions and came across as sheer mockery to all climate activists. A few months later, nine young people brought a lawsuit against the Climate Protection Act before the Federal Constitutional Court and were surprisingly vindicated in April 2021: the judges ruled that the Climate Protection Act was in parts unconstitutionally weak because it disregarded the civil liberties of future generations.

The disappointment with the original climate protection law at the moment of its supposed greatest triumph in autumn 2019 also marked the temporary end of the growth-oriented mobilisation phase of the climate justice movement in Germany and was the beginning of a crisis of meaning that was exacerbated by the COVID-19 pandemic and lasted until the end of the period under review in this text (ausgeCO2hlt, 2023).

"United We Stand – Divided We Fall"

In addition to the ongoing dispute over coal power, the climate movement increasingly focused on a second fight following the protests at Hambach Forest: in September 2019, tens of thousands demonstrated in Frankfurt am Main on the occasion of the International Motor Show (IAA) for a change in transport policy and blocked the car fans' jubilant celebration with the new movement actor Sand im Getriebe (sand in the gears) (Aljets, 2019).

With the occupation of the Dannenröder Forest in Hesse, the "Forest not Asphalt" alliance created important momentum in 2020. Although the protests around the "Danni" did not prevent the clearing of the centuries-old forest in favour of the expansion of the Autobahn 47, the occupation gave rise to an independent mobility movement against the car and its industrial and ideological status in Germany.[10] It soon put the 800 additional kilometres of new freeways to be built in Germany by 2030 on the movement's potential protest map. Further forest occupations quickly sprang up at numerous construction sites and planned routes throughout the country, particularly around the extremely costly expansion of the Autobahn 100 in the centre of Berlin. They represented an important step in the development of the climate movement from an anti-coal movement into a thematically diverse climate justice movement. In 2020, a promising collaboration between Fridays for Future and the trade union ver.di also emerged with regard to the mobility transition, and since then they have been working together for the expansion of public transport and good working conditions for public transport employees. There has been a great deal of solidarity, diverse actions, and joint strikes, making this cooperation a hopeful start for a class-conscious climate and mobility justice movement.

The anti-capitalist climate movement became increasingly differentiated with initiatives against the aviation, car, and gas industries, against coal imports and against fossil agricultural fertilisers in a very dynamic phase between the clearing of Hambach Forest and the start of the coronavirus pandemic (Häußermann, 2018). In addition, Extinction Rebellion emerged as another actor that attracted a lot of attention with its sometimes spectacular inner-city blockade actions but was also criticised for anti-Semitic statements and a lack of clear political positioning within the movement (Häußermann et al., 2024). This differentiation led to high coordination and communication efforts and increasingly complex and confusing internal movement structures.

Within these networks, the hegemonically white climate movement began an in-depth confrontation with racism within its own ranks. An expression of this internal racism is also the fact that anti-racist work was largely initiated by self-organisation of activists affected by racism – such as the Black Earth BIPoC Environmental and Climate Justice Collective – and had to be pushed through against sometimes fierce resistance from white activists.[11] Building on this, there was an anti-racist finger of action at the Ende Gelände protests in the Rhenish lignite mining area in fall 2020 – which took place between Covid waves – and the organisation of a digital BIPoC climate conference in fall 2020.[12]

Fridays for Future was able to unite large sections of society behind the demand for more climate protection. More and more social groups organised as "Parents . . .", "Scientists . . .", "Entrepreneurs . . .", or "Faggots for Future". Although this mobilisation repeatedly brought hundreds of thousands of people onto the streets, after the unifying moment of preserving the Hambach Forest and the legal victory in Germany's highest court, there were no further high-level and clear-cut successes.

Over the past 15 years, a strong and well-connected climate justice movement has emerged in Germany and around the world. The increasing gravity of the climate crisis and the unwillingness of governments to act have made a major contribution to this. However, it was also the thousands of activists who used creative forms of action such as occupations, civil disobedience, and school strikes to turn the issue of climate justice into a central political battlefield. All of these publicly visible events, recorded in texts like this one, would not have happened without countless people who spend days at their desks writing organisational emails, creating healthy group dynamics in political contexts, or had sleepless nights thinking about the climate movement. Our deepest appreciation goes to all these people in the "background" of our social movement.

Notes

1 We want to thank Pauline Brünger for her contributions to this chapter.
2 A proverb based on a biblical quote – "Whoever digs a pit for another, will fall into it" (Proverbs 26:27).
3 For a critical perspective on EarthFirst! see Bierl (2014).
4 See https://deutsch.radio.cz/internationaler-protest-gegen-ausbau-der-kohlefoerderung-turow-8132520

5 See https://taz.de/Die-Gruenen-und-der-Dannenroeder-Forst/!5723948/
6 Regarding strategy, organisation, and practicalities of mass media publicity at Ende Gelände and the German climate movement see: Hedwig A. Lindholm (2020): *Handbuch Pressearbeit – soziale Bewegungen schreiben Geschichte(n)*. Münster (Unrast). The presswork of the German climate justice movement was significantly inspired by New Economy Organizers (2019a, 2019b) from the United Kingdom, as well as Reinsborough and Canning (2017), from the United States.
7 For the history and experiences of Ende Gelände, see *Ende Gelände*, 2022.
8 https://wedontshutup.org/en/
9 As part of "Ende Gelände goes Europe", there were notable collaborations with Code Rood from the Netherlands, Limity Jsme Me from the Czech Republic, Power Beyond Border from the UK, No Grandi Navi from Venice/Italy, and Oboz dla klimatu from Poland.
10 The creation of this movement was also kick-started by the "Road Raging Zine" by Timo Luthmann. Online: https://ia801805.us.archive.org/2/items/road-raging-zine-leseversion/Road_Raging_Zine_Leseversion.pdf and https://roadraging.blackblogs.org/
11 Kanakische Welle (2019).
12 https://bipoclimatejusticenetwork.org/conference-2020/

References

Aljets, J. (2019). *Podcast she drives mobility #14 with Katja Diehl (Sonderfolge) Guest: Janna Aljets*. https://katja-diehl.de/she-drives-mobility-14-sonderfolge-gast-janna-aljets/ [Accessed 28 January 2024].

ausgeCO2hlt (2023). *Jenseits von Hoffnung & Zweifel. Gedanken zum Widerstand in der Klimakrise*. Münster: Unrast.

Behrmann, I. (2019). *Zu unserer Hoffnung stehen. Klimapolitik zwischen radikalem Systemwandel und realpolitischen Prozessen*. https://blog.interventionistische-linke.org/klima/zu-unserer-hoffnung-stehen [Accessed 28 January 2024].

Bierl, P. (2014). *Grüne Braune. Umwelt-, Tier- und Heimatschutz von rechts*. Münster: Unrast.

CrimethInk (2021). *Die Waldbesetzungsbewegung in Deutschland. Taktiken, Strategien und Kultur des Widerstands*. https://de.crimethinc.com/2021/03/10/die-waldbesetzungsbewegung-in-deutschland-taktiken-strategien-und-kultur-des-widerstands [Accessed 28 January 2024].

Ende Gelände (2022). *We Shut Shit Down*. Hamburg: Nautilus.

Frenzel, F. (2011). Entlegene Orte in der Mitte der Gesellschaft. In A. Brunnengräber (Hrsg.) *Zivilisierung des Klimaregimes*. s.l.: VS Verlag für Sozialwissenschaften.

Häußermann, D. (2018). *Rebellion, Aufstände, Klima-Alarm*. www.klimareporter.de/protest/rebellion-aufstaende-klima-alarm [Accessed 28 January 2024].

Häußermann, D., Kapfinger, M., Parekh, P. and Schmelzer, M. (2024). Eine Bewegung, viele Generationen. Entwicklungen in der Klimabewegung im deutschsprachigen Raum. In *Grebenjak, Manuel: Kipppunkte. Strategien im Ökosystem der Klimabewegung*. Münster: Unrast.

Häußermann, D. and Wollny, L. (2016). Anti-Kohle-Bewegung. Gegen Klimawandel, Kapitalismus und Wachstum! In K. N. Ö. and. D. Postwachstumsgesellschaften (Hrsg.) *Degrowth in Bewegung(en)*. München: oekom.

Hofinger, D. (2019). *Aus einer "Illusion" wurde Realität*. https://www.klimareporter.de/protest/aus-einer-illusion-wurde-realitaet [Accessed 28 January 2024].

Jetñil-Kijiner, K. (2017). *After COP23: A Sevusevu in German Rain*. https://www.kathyjetnilkijiner.com/after-cop23-a-sevusevu-in-german-rain/.

Kanakische Welle (2019). Wie weiß ist (deutscher) Klimaaktivismus? *Podcast*, 20 September 2019. https://open.spotify.com/episode/55sEX3CDZIS0nKmmVuYEWi [Accessed 3 September 2024].

Kreutzfeld, M. (2016). *Braunkohle-Fans drehen völlig durch*. https://taz.de/Kommentar-Klimaprotest/!5304723 [Accessed 28 January 2024].

Lindenberg, H. and Müller, T. (2016). Ende Gelände im Gerechtigkeitsdilemma. Warum der Kohleausstieg nicht bis 2040 warten kann. *Zeitschrift Luxemburg*, September. https://zeitschrift-luxemburg.de/artikel/ende-gelaende-gerechtigkeitsdilemma/ [Accessed 28 January 2024].

New Economy Organizers (ed.) (2019a). *Press Officer's Handbook.* https://www.neweconomyorganisers.org/work/support-resources/toolkits [Accessed 3 September 2024].

New Economy Organizers (ed.) (2019b). *Spokesperson Handbook*. https://www.neweconomyorganisers.org/work/support-resources/toolkits [Accessed 3 September 2024].

Reinsborough, P. and Canning, D. (2017). *Re:Imagining Change. How to Use Story-Based Strategy to Win Campaigns, Build Movements and Change the World*. Oakland: PM Press.

Sander, H. (2016). *Die Klimagerechtigkeitsbewegung in Deutschland. Entwicklung und Perspektiven*. https://www.rosalux.de/fileadmin/rls_uploads/pdfs/Studien/Onlinestudie_Klimagerechtigkeit.pdf [Accessed 28 January 2024].

Sivanesan, S. (2016). *Mass action media: Ende Gelände, Break Free 2016*. https://unprojects.org.au/article/mass-action-media-ende-gelande-break-free-2016/ [Accessed 28 January 2024].

Weiermann, S. (2019). *#HAMBIBLEIBT! Eine Reportage vom Kampf um einen Wald, der das Land erschütterte*. https://www.rosalux.de/fileadmin/rls_uploads/pdfs/sonst_publikationen/Broschur_Hambibleibt_web.pdf [Accessed 28 January 2024].

Zacher, T. (2023). *Stadt Kerpen und RWE: "Zukunft verkauft"?* https://www1.wdr.de/nachrichten/landespolitik/rahmenvereinbarung-kerpen-rwe-manheimer-bucht-100.html.

Zucker im Tank (2022). *Glitzer im Kohlestaub. Vom Kampf um Klimagerechtigkeit und Autonomie*. Berlin und Hamburg: Assoziation A.

22

UNBURNABLE COAL AND UNLABELLED CLIMATE ACTIVISM IN CHINA

Bowen Gu and Juan Liu

Introduction

The Global Climate Strike led by Greta Thunberg in September 2019 is said to have involved 7.6 million people across 185 countries (350.org, 2019). Yet, it is difficult to think of a name or face when we discuss climate activism in China. Ou Hongyi, the first and only young Chinese climate striker that attempted to engage in Greta-inspired Fridays for Future climate strikes, was told that she cannot return to school unless she stops her activism (Standaert, 2020). In the absence of climate marches or strikes, the climate justice movement in China takes a different shape compared to other countries and regions. Civil activism with the "climate change" label in China is generally linked to government or market actors to widen their political space (Schröder, 2011). For instance, some NGOs with a focus on climate change actively participate in the policymaking process for climate legislation. Others engage in promoting climate solutions such as agroecology, climate-smart agriculture, and low-carbon commuting. Despite being on the rise over the past years, the few existing studies on climate justice in China have mainly taken a top-down perspective to investigate climate policymaking and legislative climate governance, with little discussion on grassroots climate justice movements. As Wu (2019, p. 197) argued, "climate justice is not just an unfinished, but also a non-started cause in China so far".

Climate justice is a concept that has been extended from "environmental justice" to the climate field, where recognition, procedural, and distribution justice are considered the main pillars (Schlosberg and Collins, 2014). In China's context, the discussion of climate justice goes beyond inter-nation and inter-generation dimensions and has its implications within the nation, where there has been a lack of recognition of the unequal distribution of climate impacts shouldered by vulnerable groups and communities (Cao, 2016; Wu, 2019). In response to such "invisible" inequality and injustices, we attempt to address the question of whether there is grassroots climate activism in China in this chapter. If so, how does grassroots climate activism in China manifest? What role does grassroots climate activism play in China? Are they aligned or connected with global climate activism?

This chapter grounds its inquiry in two case studies in the Global Atlas of Environmental Justice (EJAtlas). The EJAtlas is co-created with scholarly and activist knowledge based on

DOI: 10.4324/9781003396567-26

a set of more than 100 data fields (Temper et al., 2015, 2018; Martinez-Alier, 2023). The case entries in the EJAtlas are based on secondary data collected from various data sources, enabling readers to cross-check and reduce possible bias in the description of the context, process, and outcomes of conflicts (Temper et al., 2015). Selected sources include academic papers, legal proceedings, mainstream media reports, grey literature, published evidence from impacted communities, and government documents. Before a case is published, each entry is cross-checked for consistency and quality control by a permanent team of moderators based at the Autonomous University of Barcelona (for further methodological details, see Temper et al., 2015). We also draw insights from fieldwork that the first author conducted in China in 2021, which included interviews with NGOs that work on environmental and climate-change issues in China and the Belt and Road Initiative.

The chapter is organised as follows: the next section provides a brief introduction on coal and climate justice in the context of China and the role of civil society in the environmental and climate governance system in China. This is followed by two case studies from the EJAtlas to demonstrate the "unlabelled" climate activism in China and an additional example of grassroots activism towards Chinese coal financiers.

Coal, Civil Society, and Climate Justice

China ranks second based on cumulative CO_2 emissions since 1850, accounting for 11.4% of the global total (Evans, 2021). China's reliance on coal is one of the main causes for its "leading position" in cumulative carbon emissions, which also poses significant and enduring challenges for the prospects of global climate goals. As of 2022, coal still makes up more than 50% of China's total primary energy use. Globally, there has been a growing trend of anti-coal movements. These range from the fossil fuel divestment movement organised by 350.org to local movements that protest against individual fossil fuel infrastructure projects like coal plants. In China, the world's largest coal producer and consumer, the anti-coal movement seems to be absent.

China is known as a top-down, centralised state, which is reflected in its environmental and climate governance system. Indeed, environmental activism in China is also referred to as "embedded activism" and "resigned activism" (Ho, 2007; Lora-Wainwright, 2017), since it is viewed as deeply embedded in the country's socio-political context. In previous years, mass NIMBY protests have taken place in response to proposals of polluting infrastructures, such as paraxylene (PX) plants and incinerators. In recent years, with increasing attention to social stability, there has been limited space for mass protest. However, this does not mean that the coal value chain (including coal mining, coal transport, coal-fired power plants, coking, as well as coal chemical plants, such as coal-to-gas and coal-to-oil conversion facilities) is uncontested.

On the other hand, there tends to be a gap between the global concern for increasing GHG emissions and climate change and the domestic concerns for energy security to sustain and drive development in the Global South (Boulle, 2019). This makes it even more challenging for grassroots NGOs to take anti-coal action, which could imply energy and social security risks. As a result, NGOs in China have to be strategic about their approach towards climate justice. Taking climate litigation as an example, in contrast to the emerging climate litigation scene in the Global North, it has been argued that the Global South has taken a so-called stealth approach towards climate litigation (Lin and Peel, 2021; Roy, 2021). This is exemplified in China, where environmental public interest litigation cases have been filed under less-controversial claims such as air and atmospheric pollution (Li, 2019; Zhao, 2019), which to some extent coincide with the Chinese government's strategy to explore the synergy between air pollution and climate governance.

Considering China's high dependence on coal and its climate implications, it could be argued that any grassroots activism related to the coal value chain could be considered as a direct or indirect approach towards climate justice in China's context. Thus, this chapter will continue to explore to what extent and how grassroots climate activism manifests in China through the lens of two coal-related grassroots activism case studies. Through investigating how grassroots activism has been contesting the socio-environmental impacts of coal, we attempt to understand the various forms, tactics, and outcomes of "unlabelled" climate activism in China.

In addition, the role of civil society – or the participation "from below" – has evolved over the past years and serves as an integral part of the seemingly top-down environmental governance system in China. Wu (2009) indicated that a number of civil society organisations in China have played a crucial role in climate governance and promoting climate justice, with examples of NGOs such as the youth organisation China Youth Climate Action Network (CYCAN), Global Environmental Institute (GEI), and the Greenovation Hub (GHub). The NGOs, including grassroots and government-organised NGOs (GONGOs), participate in climate governance through different approaches and with distinct goals. These range from awareness raising among youth and the general public (e.g. the International Youth Summit on Energy and Climate Change, IYSECC, organised by CYCAN), pushing for transparent information disclosure from corporations and government bodies related to pollutant and GHG emissions control, monitoring compliance and restoration outcomes, and public interest litigation to policy advocacy and contributing to legislation. The following case study will provide an example of how different grassroots NGOs navigate the windows of opportunity to reach various objectives towards climate justice from below.

Public Interest Litigation and Climate Justice

In April 2023, the "first climate litigation" in China concluded. The plaintiff, China's first grassroots environmental NGO, Friends of Nature (FON), reached a mediation agreement with the defendant, the Gansu branch of State Grid, after almost 7 years of litigation. FON was founded in 1993 with the mission to create a platform for fostering public participation and action in environmental protection. As of 2023, FON has a network of more than 40,000 volunteers and 18 member groups organised by theme and location across China (FON, 2023). Over the years, FON has worked on environmental protection and conservation through various approaches including public education, research and investigation, public interest litigation (PIL), and policy advocacy. It is also one of those rare NGOs with in-house environmental lawyers, with PIL as an important pillar of its work. In August 2016, FON sued State Grid Gansu Electric Power Co, the provincial grid enterprise in Gansu, China, on the grounds that the company refused to purchase all the electricity produced from wind and solar generators (EJAtlas, 2021). This phenomenon is also known as wind and solar curtailment, which means the reduction of output of a renewable source below what it could have otherwise produced within its geographic boundary (Zhu et al., 2018). Besides, as the amount that was not produced with renewable energy sources was substituted with coal-fired power, FON indicated that the company's behaviour led to the increase of air pollution and GHG emissions, which has direct implications for climate change. This case is thus also considered the first climate-change litigation case in China and included in the 2019 White Paper on China's Environmental Resource Trial published by the Supreme People's Court as an example of cases that relate to climate-change response and mitigation (EJAtlas, 2021).

FON directly referred to Article 14 of the Renewable Energy Law 2005/2009 (related to mandatory connection and full purchase guarantee) to reiterate that grid companies are obliged to sign purchase contracts with renewable electricity generators and purchase their electricity in the full amount. The main claims of the plaintiff FON included: (1) the defendant shall purchase the electricity generated by wind and solar generators within its coverage in the full amount to cease environmental damage caused by electricity generation from coal-fired plants; (2) the defendant shall pay for the environmental damages caused by electricity generation through coal instead of wind and solar between 1 January 2015 and 30 June 2016, which is at least CNY 1.718 billion (~USD 0.24 billion) according to preliminary calculation of environmental loss compensation; (3) the defendant shall issue an apology to the public in national and provincial media; (4) the defendant shall bear all litigation costs and attorney fees (Zhang, 2019).

The litigation went through a long and bumpy process. Almost 2 years after the case had been filed by FON, the Lanzhou Intermediate Court refused the claim of FON in August 2018, on the grounds that "State Grid Gansu Electric Power, as a power grid company that purchases, sells and deploys power supply, is not a power generation company". According to the court, the grid company did not directly contribute to environmental pollution or ecological damage through its own activities. As a result, FON appealed to the Gansu Higher People's Court, which overturned the decision of the Lanzhou Intermediate People's Court and ordered the case to be handled by the Gansu Kuangqu People's Court due to the court's exposure to environmental resource issues. According to the mediation result, in addition to raising the awareness of the authorities, the case led to the commitment of Gansu Electric Power to invest at least CNY 913 million (~USD 129 million) in the construction of power grids that facilitate the renewable energy development and enhance renewable energy power generation and transmission capabilities.

In many other countries, litigation against coal mining or coal-fired power corporations is seen as a common tactic in anti-coal movements and climate related litigation. In this case, instead of directing its litigation towards the coal-fired power producer, FON opted for an indirect path, suing the power grid company, which is part of the long and complex value chain involved in coal-fired power generation, transmission, and distribution. However, this innovative approach also bears risks. As explained by the director of FON Liu Jinmei, who was quoted by Lin (2023) in an article in *China Dialogue*, "In China, environmental public interest lawsuits still start with the concept of a tort – there needs to be harm or a significant risk of harm. That doesn't work for climate cases".

Beyond the litigation, another NGO, China Biodiversity Conservation and Green Development Foundation (CBCGDF), intervened through the information disclosure approach. A CBCGDF representative sent a letter in August 2016 to the parent company of State Grid Gansu Electric Power Co, State Grid Corporation of China (State Grid), to address the issues of their local branches' involvement with wind and solar curtailment. In September 2016, CBCGDF received a response from State Grid, which gave the slowdown of the electricity demand, the lack of capacity to adjust the peak demand in certain regions, and the lack of inter-provincial transmission pipelines as the main reasons for failing to guarantee full purchase of renewable energy power.

From the perspective of the nature of the plaintiff and defendant, this case of public interest litigation corresponds to the pillar of "holding fossil fuel corporations accountable for the central role they play in contributing to global warming" in the definition of climate justice coined by CorpWatch back in 1999 (Bruno et al., 1999, p. 5). However, in contrast to many litigation cases outside of China that focus on stopping fossil fuel projects, one interviewee from the NGO indicated that the primary goal of this public interest litigation is not to target the corporation but to "enhance broader transition and policy advocacy on the issue of renewable curtailment and climate change" (Personal communication,[1] July 2021). The attorney representing FON pointed out that one of the objectives of the litigation is to send a message to the government through the grid companies that there's a need for better guarantees for the full purchase of renewable energy power generation (EJAtlas, 2021). Another NGO representative said,

> This (wind and solar curtailment) is a prominent issue in the industry. However, this is not addressed adequately due to various reasons. It is not a matter of technology, but a game of interests. In the end, it is a matter of breaking the (power) imbalance and revealing the issue to the public.
>
> *(Personal communication, July 2021)*

The long duration of this litigation reveals some challenges in pursuing climate justice in China's context. This includes the imbalance of capacity in terms of legal expertise and human resources between the NGOs and the corporations, the lack of environmental expertise of the court, and the absence of institutional support. These are also representative of bottlenecks for grassroots participation in the broader climate governance system in China. The interviews with civil society representatives revealed that the lack of climate legislation is one of the barriers for civil society organisations to file climate litigation (Personal communications, July 2021). Another factor hindering climate justice activism in China that has been highlighted in the existing literature is the lack of linkage between social grievances and climate change (Harris et al., 2013; Wu, 2019). This also explains the rationale behind FON's strategy of filing public interest litigation to sue the corporation and promote broader policy advocacy.

NGOs' participation in public interest litigation does not have a very long history in China. This is primarily built upon two legislative decisions, namely the amendment of the Civil Procedure Law (CPL) in 2012 and the revised Environmental Protection Law (EPL) in 2015. With the new EPL taking effect as of 1 January 2015, environmental NGOs that have operated for 5 years without violating the law became eligible to file public interest litigation. Building upon almost 10 years' experience, NGOs still have a much longer way to go. This first climate litigation case opened new channels for dialogue with corporations and government bodies and demonstrated civil society's willingness to participate in climate governance. On the other hand, it also echoes what Naomi Novik (2006, p. 35) wrote: "Justice is expensive". Climate justice is not only expensive in material terms but also in terms of time and patience.

Performance Art and Climate Justice

Art plays an important role in environmental movements around the world. As illustrated in the EJAtlas, more than 450 out of the 3,900 environmental justice movement cases documented in the database (as of December 2023) involved some forms of artistic and creative action. In China, where activism is increasingly suppressed, one artist, working under the pseudonym Brother Nut, has adopted artistic narratives to oppose "destruction wreaked by the Greenhouse Gangsters at every step of the production and distribution process" and provide "assistance to communities threatened or impacted by climate change", which are another two pillars in CorpWatch's definition of climate justice (Bruno et al., 1999, p. 5).

The environmental, social, and health impacts of the coal value chain have been well studied, including its impact on groundwater (Greenpeace, 2013; Yang and Qian, 2019). Against such impacts local communities and civil society organisations have mobilised around the world. However, challenges remain, given that it takes a long time to prove causal relationships between the coal mining activities and the health burdens on local communities and their livestock. For example, it takes years for patterns of certain chronic diseases to emerge. In addition, many coal mines and coal-fired power plants are located in remote areas in northern China, which makes it even more difficult to expose such impacts and hold the corporations accountable. Xiaohaotu Township is one of these places.

The township's name, "Xiaohaotu", is derived from Mongolian and means small place located in a windy desert area with rich groundwater resources (EJAtlas, 2020). In addition to abundant groundwater, Xiaohaotu Township is also rich in coal and natural gas resources. Within 30 kilometres of the township, there are three coal mines and one gas field in operation (EJAtlas, 2020). The discovery and exploration of coal and natural gas has dramatically changed the groundwater access situation in Xiaohaotu. Located on the southern edge of the Maowusu Desert on the border between Inner Mongolia and Shaanxi Province and home to approximately 14,500 people, Xiaohaotu Township and its water struggles would hardly be known by people outside the area if not for the art project that Brother Nut initiated in June 2018. Brother Nut used 10,000 bottles of clean drinking water from one of China's most popular bottled spring water brands, Nongfu Spring (which literally means the water that farmers drink), and exchanged the water with five tons of polluted drinking water from the local villagers that had suffered from water pollution caused by the coal mines for many years (EJAtlas, 2020). The artist then transported the bottles to Beijing and Xi'an in preparation for an exhibition in each city. The exhibition in Beijing (see Figure 22.1) was soon suspended by the authorities, ostensibly due to trademark infringement and a complaint from the bottled water company. In addition, Brother Nut invited a heavy metal band to give a concert at Xiaohaotu. The concert took place with just instrumental music, without lyrics or audience (the original idea to get 100 sheep from local villagers that also suffered from the poor water quality as an audience was not implemented). The artist indicated that it was a sarcastic way to use the pun rhetoric of "heavy metal" to raise awareness about the heavy metal pollution in the local drinking water.

The water contamination issue at Xiaohaotu received massive attention on social media thanks to Brother Nut's art project. Shortly after the exhibition of Brother Nut was circulated on social media, the local authorities started an investigation into the quality of drinking water sources, which revealed high levels of iron and manganese (in one case exceeding the permitted maximum 4.2 times). The water contamination issue at Xiaohaotu was also mentioned during the regular press conference on environmental issues held by the Ministry of Ecology and Environment. In response, the local government implemented measures to improve the water quality for villages in Xiaohaotu, including digging deep wells and installing water purification equipment, as well as digging three drainage channels to divert the water discharged from the three coal mines. By September 2018, 3,805

Figure 22.1   The exhibition "Drink water like the farmers", which displayed 9,000 bottles of polluted water from Xiaohaotu. The polluted water was put into the bottles of the mineral water brand "Nongfu Spring", meaning the spring water that farmers drink.

Source: Photo by Brother Nut.

water purification machines had been installed across 12 villages in Xiaohaotu and 12 wells of 120-metre depth had been dug to supply drinking water for local villagers. On the other hand, the coal mines continued their operations in the neighbourhood of Xiaohaotu.

It has been argued that art, especially visual art, can contribute to the climate justice movement in at least three ways, including providing a channel for marginalised communities to share their voice, serving as a rallying point and information sharing tool, and fighting against extractive industries (Adams, 2019). Brother Nut's performance art corresponds to all three aspects. In addition, he raised public awareness through social media, which forced the local and central government to respond and take action. When effective, this alliance of art and media, including social media, could lead to third-party and public monitoring of corporations and government bodies. On the other hand, being exposed to the spotlight also put Brother Nut's further activities under scrutiny by the authorities, which limited follow-up activities in the area. This is also a dilemma that grassroots activists face in China.

More recently in the summer of 2023, Brother Nut started a new initiative amidst major heat waves sweeping across China. He used "Global warming redemption vouchers"

Figure 22.2　An example of a "Global warming redemption voucher".

Source: Designed by Peng Mitai.

(see Figure 22.2) to call for public awareness of the difficulties faced by vulnerable groups and workers and the responsibilities of Chinese and international high-carbon-emitting companies towards climate change. The plan was to let the public "purchase" the redemption vouchers on behalf of these companies, thereby forcing the companies to react and take up climate responsibility through social pressure. Some volunteers supported the project by purchasing drinking water and heat stroke prevention equipment for outdoor workers with the raised funds. While Brother Nut's plan was to raise CNY 100,000 through crowdfunding, less than 10% of the goal was reached.

Pursuing climate justice is a long-term effort. Although Brother Nut's performance art addresses drinking water access and heat stroke prevention issues shouldered by vulnerable groups, it will take continuous monitoring of the corporations' activities and increased public and corporate awareness of climate (in)justice to address the issue. At present, coal consumption continues to rise in China and on the global level (IEA, 2023). Coal mining and fossil fuel companies will continue to impact local communities' livelihoods and the ecological environment in their neighbourhoods as well as global climate prospects until they are shut down for good.

"Global China" and Cross-border Climate Justice

The previous two cases demonstrate domestic efforts to tackle environmental and climate injustice in the coal value chain in China. Scholars have argued that the consideration of climate justice in China should take into account both intra-country and inter-country dimensions, as well as the intergenerational aspect (Cao, 2016; Wu, 2019). Since the "going out" strategy in the 2000s that encouraged Chinese corporations to invest abroad and the launch of the Belt and Road Initiative (BRI) in 2013, China has increasingly extended its coal value chain beyond its borders. While President Xi Jinping announced China's commitment to stop overseas coal-fired power financing at the UN General Assembly in September 2021

(Ministry of Foreign Affairs PRC, 2021), the socio-environmental and climate legacies of coal-fired power projects that have been planned, are being constructed, or are already in operation will persist (Rogelja, 2020; Gu, 2024).

Although not the focus of this chapter, it is important to take note of grassroots efforts beyond China's border that contest the socio-environmental and climate impacts of this cross-border coal value chain. "Go clean ICBC" is an example of such efforts that join forces with the broader fossil fuel divestment movement. A coalition of international NGOs established the initiative to oppose the coal finance of the Industrial and Commercial Bank of China (ICBC), which is the largest bank in the world and one of the largest financiers of the top 30 coal power companies. On the initiative's website, it shows on the one hand the portfolio of coal finance of ICBC and on the other hand examples of ICBC's commitment, including financing renewable energy projects and withdrawal of finance for the Lamu coal project in Kenya (Go clean ICBC, 2023).

As one interviewee from one of the "Go clean ICBC" coalition members indicated, "When it comes to campaigns such as 'Go clean ICBC', we have very little leverage. But we are doing it anyway, in solidarity with international organisations" (Personal communication, June 2021). While facing similar barriers as domestic NGOs in China, in terms of capacity, imbalance of power, and lack of institutional support, the NGOs outside of China strengthened their capacity to counter the power imbalance through building a broader international coalition. The initiative also adapted their narrative to align with the "language" of the Chinese government's commitment and the bank's policy, such as "reaching Net Zero by 2060", to gain leverage in their engagement with ICBC. According to one coalition member, this helps them with gaining more negotiating power (personal communication, June 2021).

Conclusion

In recent years, local communities, activists, and NGOs in China have turned to various tools and approaches to contest projects in the coal value chain. These initiatives are usually not labelled as climate activism but nevertheless are closely connected to the global climate movement. This chapter provides an overview of the "unlabelled" climate activism in China, including legal and artistic initiatives, as well as the visions they advocate for and the dilemmas they face.

In a context where the media is closely scrutinised and mass protest suppressed, NGOs and activists have navigated through the most feasible channels to reach their goals, ranging from raising the voices of local communities to policy advocacy for future legislation. Poderati and Ou (2021) proposed a hybrid approach, combining top-down and bottom-up initiatives, to tackle climate governance in China. This chapter echoes and reinforces this proposal with two concrete examples. Through different tactics and channels, they counter the irresponsible behaviours of corporations in the fossil fuel supply chain, including but not limited to coal mining and grid companies. In addition, the domestic efforts in China have been complemented by NGOs abroad that have formed coalitions to contest Chinese financiers' involvement in coal financing. Whether it is legal or artistic, domestic or abroad, the NGOs and activists have increasingly incorporated the narratives of the authorities, especially the Chinese government's policy and commitment, in their campaign and activism, which provides them with more leverage against the corporations.

Being "unlabelled" also means that grassroots activism in China lacks institutional support and seems disconnected from the broader international climate justice movement. However, the grievances that they attempt to respond to – and their objectives and visions – are aligned with climate justice principles and to some extent share the same challenges as climate activists in other countries in the Global South. This calls for the need for further study to better understand those "peripheral" and "indirect" climate justice movements that may not directly refer to "climate" or "climate justice" in their name but are built upon recognizing the complexity and intersectionality of the climate crisis in their local or regional context and can potentially help move climate action forward even in countries where open confrontation with authorities is extremely difficult if not impossible.

Note

1 This and the following personal communications quoted in the *Public interest litigation and climate justice* section were translated from Chinese to English by the authors.

References

350.org. (2019). 7.6 million people demand action after week of climate strikes [online]. *350.org*. https://350.org/7-million-people-demand-action-after-week-of-climate-strikes/ [Accessed 29 June 2023].

Adams, A. (2019). Photographs, performance, and protest: The fight for climate justice through art. In P. E. Perkins (ed.) *Local Activism for Global Climate Justice: The Great Lakes Watershed* (1st ed.). London: Routledge.

Boulle, M. (2019). The hazy rise of coal in Kenya: The actors, interests, and discursive contradictions shaping Kenya's electricity future. *Energy Research & Social Science*, 56, 101205. https://doi.org/10.1016/j.erss.2019.05.015

Bruno, K., Karliner, J. and Brotsky, C. (1999). Greenhouse gangsters vs. climate justice. *Corp Watch*. https://www.corpwatch.org/sites/default/files/Greenhouse%20Gangsters.pdf [Accessed 30 June 2023].

Cao, M. (2016). The legal standpoint and strategy of China to participate in international climate governance: From the perspective of climate justice. *Chinese Legal Science*, 1, 29–48.

EJAtlas (2020). *Water Pollution Caused by Coal Mines and Gas Field at Xiaohaotu, Shaanxi, China*. https://ejatlas.org/conflict/water-pollution-caused-by-coal-mines-and-gas-field-at-xiaohaotu-shaanxi-china [Accessed 30 June 2023].

EJAtlas (2021). Friends of Nature sued State Grid for refusing to purchase clean energy in Gansu, China. *EJAtlas*. https://ejatlas.org/conflict/friends-of-nature-sued-state-grid-for-refusing-to-purchase-clean-energy-in-gansu-china [Accessed 4 July 2022].

Evans, S. (2021). *Carbon Brief Website*. https://www.carbonbrief.org/analysis-which-countries-are-historically-responsible-for-climate-change/ [Accessed 15 November 2023]

FON (2023). *Friends of Nature Website*. https://www.fon.org.cn/ [Accessed 11 November 2023].

Go Clean ICBC (2023). https://gocleanicbc.org/ [Accessed 30 June 2023].

Greenpeace (2013). *Thirsty Coal 2: Shenhua's Water Grab*. https://issuu.com/greenpeace_eastasia/docs/thirsty_coal_2_shenhua_s_water_grab

Gu, B. (2024). Black gold and green BRI – A grounded analysis of Chinese investment in coal-fired power plants in Indonesia. *The Extractive Industries and Society*, 17, 101411. https://doi.org/10.1016/j.exis.2024.101411

Harris, P. G., Chow, A. S. Y. and Karlsson, R. (2013). China and climate justice: Moving beyond statism. *International Environmental Agreements: Politics, Law and Economics*, 13(3), 291–305. https://doi.org/10.1007/s10784-012-9189-7

Ho, P. (2007). Embedded activism and political change in a semiauthoritarian context, *China Information*, 21(2), 187–209. https://doi.org/10.1177/0920203X07079643

IEA (2023). *Global Coal Demand Set to Remain at Record Levels in 2023*. https://www.iea.org/news/global-coal-demand-set-to-remain-at-record-levels-in-2023 [Accessed 8 January 2024].

Li, J. (2019). Climate change litigation: A promising pathway to climate justice in China? *Virginia Environmental Law Journal*, 37(2), 132–170. https://www.jstor.org/stable/26742667

Lin, J. and Peel, J. (2021). The farmer or the hero litigator?: Modes of climate litigation in the Global South. In C. Rodríguez-Garavito (ed.) *Litigating the Climate Emergency: How Human Rights, Courts, and Legal Mobilization Can Bolster Climate Action*. Cambridge: Cambridge University Press.

Lin, Z. (2023). From smog to carbon: Chinese NGOs in transition. *China Dialogue*. https://chinadialogue.net/en/climate/from-smog-to-carbon-chinese-ngos-in-transition/ [Accessed 29 June 2023].

Lora-Wainwright, A. (2017). *Resigned Activism: Living with Pollution in Rural China*. Cambridge, MA: The MIT Press.

Martínez-Alier, J. (2023). *Land, Water, Air and Freedom: The Making of World Movements for Environmental Justice*. Edward Elgar Publishing. https://doi.org/10.4337/9781035312771

Ministry of Foreign Affairs of the PRC (2021). Xi Jinping attends the general debate of the 76th session of the United Nations general assembly and delivers an important speech. *Ministry of Foreign Affairs of the People's Republic of China*. https://www.fmprc.gov.cn/mfa_eng/wjb_663304/zzjg_663340/gjs_665170/gjsxw_665172/202109/t20210923_9580159.html [Accessed 25 June 2023].

Novik, N. (2006). *Black Powder War* (p. 35). Harper Voyager.

Poderati, G. and Ou, S. (2021). Tackling climate change in China: A hybrid approach. *Chinese Journal of Environmental Law*, 5(2), 141–171. https://doi.org/10.1163/24686042-12340070

Rogelja, I. (2020). Concrete and coal: China's infrastructural assemblages in the Balkans. *Political Geography*, 81, 102220. https://doi.org/10.1016/J.POLGEO.2020.102220

Roy, B. (2021). *Koyla Kahini. The Political Ecology of Coal in India*. https://ddd.uab.cat/pub/tesis/2021/hdl_10803_672611/brro1de1.pdf

Schlosberg, D. and Collins, L. B. (2014). From environmental to climate justice: Climate change and the discourse of environmental justice. *WIREs Climate Change*, 5(3), 359–374. https://doi.org/10.1002/wcc.275

Schröder, P. (2011). Civil climate change activism in China–More than meets the eye. *German Asia Foundation/Asienstiftung*. https://us.boell.org/sites/default/files/climate_change_activism_in_china.pdf [Accessed 30 January 2024]

Standaert, M. (2020). China's first climate striker warned: Give it up or you can't go back to school. *The Guardian*. https://www.theguardian.com/world/2020/jul/20/chinas-first-climate-striker-cant-return-to-school [Accessed 30 June 2023].

Temper, L., del Bene, D. and Martinez-Alier, J. (2015). Mapping the frontiers and front lines of global environmental justice: The EJAtlas. *Journal of Political Ecology*, 22(1). https://doi.org/10.2458/v22i1.21108

Temper, L., et al. (2018). The global environmental justice atlas (EJAtlas): Ecological distribution conflicts as forces for sustainability. *Sustainability Science*, 13(3), 573–584. https://doi.org/10.1007/s11625-018-0563-4

Wu, F. (2009). *Environmental Activism in China: 15 Years in Review, 1994–2008*. https://www.harvard-yenching.org/wp-content/uploads/legacy_files/featurefiles/WU%20Fengshi_Environmental%20Civil%20Society%20in%20China2.pdf [Accessed 29 June 2023].

Wu, F. (2019). China: Climate justice without a social movement? In Bhavnani et al. (eds.) *Climate Future: Re-imagining Global Climate Justice* (pp. 190–199). London: Zed Books. https://ssrn.com/abstract=3017678

Yang, Z. and Qian, X. (2019). *Risky Business: Growth of Water-Intensive Coal Conversion Projects in Western China*. https://www.scribd.com/document/437023177/Risky-Business-Growth-of-Water-Intensive-Coal-Conversion-Projects-in-Western-China [Accessed 25 June 2023].

Zhang, H. (2019). Prioritizing access of renewable energy to the grid in China: Regulatory mechanisms and challenges for implementation. *Chinese Journal of Environmental Law*, 3(2), 167–202. https://doi.org/10.1163/24686042-12340041

Zhao, Y. (2019). Potential pathways for climate change litigation in China: Empirical analysis of 41 public interest litigation cases involving air pollution. *Journal of Shandong University (Philosophy and Social Sciences)*, 6, 26–35.

Zhu, M., Qi, Y., Belis, D., Lu, J. and Kerremans, B. (2018). The China wind paradox: The role of state-owned enterprises in wind power investment versus wind curtailment. *Energy Policy*, 127, 200–212. https://doi.org/10.1016/j.enpol.2018.10.059

23

BEYOND PROTEST

How Legal Actions Drive Climate Justice

Lea Main-Klingst, Hermann E. Ott, and Maria Antonia Tigre

Introduction

In April 2023, the Intergovernmental Panel on Climate Change (IPCC) published its most recent overview of the science on climate change. This so-called Synthesis Report made clear the urgency of near-term climate action, stressing that: "[t]here is a rapidly closing window of opportunity to secure a liveable and sustainable future for all" (IPCC, AR6 Synthesis Report, para. C.1) and that current political measures are wholly insufficient. Finally, the report highlighted the crucial role of civil society in effective global climate action. As part of this vital role, the IPCC's Sixth Assessment Report acknowledged the significant impact of climate litigation on climate governance. It recognised climate litigation as a vital avenue for actors to shape climate policy beyond the formal UNFCCC processes (Dubash et al., 2022).

Activism and civil society engagement are often understood or defined as actions to bring about political or social change (Cambridge Dictionary). As regards activists, this includes challenging powerful systems and actors (both political and corporate), which can entail personal risks (Das, 2022).

This challenge of systems and actors is where climate activism and environmental litigation, including climate litigation, run parallel. Further, they both seek to bring about political and/or societal change. Thus, activism and litigation share a starting point, which is, recognising the insufficiency of existing legislative and administrative frameworks for environmental and climate protection – and an end goal. Their approaches may be very different yet complementary at times (Gerstetter, 2022, p. 116).

Since 2015, global climate litigation has more than doubled in numbers, with over 1,000 cases filed in the past 6 years (Setzer and Higham, 2021; UNEP, 2023). Around half of the 2,431 cases in the Sabin Center's climate-change litigation databases (updated until September 2023) have had direct outcomes favourable to climate action (Setzer and Higham, 2023). This chapter explores litigation as a form of climate activism by examining different societal/interest groups using legal pathways for more environmental and climate protection.

A wide variety of legal resources and arguments have been used to advance climate action through litigation. For example, UNEP's 2023 report highlights six categories

DOI: 10.4324/9781003396567-27

where most cases can be placed: (1) cases relying on human rights, (2) domestic enforcement cases, (3) keeping fossil fuels in the ground and protecting carbon sinks, (4) corporate liability and responsibility, (5) climate disclosures and greenwashing, and (6) failure to adapt and the impacts of adaptation (UNEP, 2023, p. 26). In each of these categories, cases can be filed at the international, regional, or national level (UNEP, 2023, pp. 29–43).

Due to the geographies the authors have worked and lived in, this chapter draws on examples from Europe and Latin America. First, to demonstrate how geographical context and circumstances are evidenced in legal strategies and approaches. Second, in light of the fact that land and environmental defenders and activists have faced and continue to face threats to their lives and liberty, this chapter will also look at litigation as a tool to defend the defenders. The chapter aims to: (1) show the interconnectedness between the different forms of activism and (2) provide some examples of how legal intervention is being used as a tool to curb the climate crisis around the globe. With this goal, the chapter shows how litigation and other forms of legal intervention are being innovatively used to address the climate crisis. The next section, following this introduction, provides background on rights-based climate litigation, specifically focusing on age and activism and land-rights climate litigation as examples of rights-based strategies activists use. The final section discusses some of the challenges climate defenders, litigants, and litigators face and how the law is adapting to address these.

Rights-based Climate Litigation

Analysis of global climate litigation has shown that most cases brought against States are based on their positive or negative obligations to respect, prospect, and fulfil human rights (UNEP, 2023, p. 23). Rights-based cases are the most "visible" of categories of climate cases, since they emphasise the human angle of the climate crisis (UNEP, 2023, p. 23). At the same time, the human rights system offers a well-established judicial infrastructure, allowing individuals to file claims and offering remedies and enforcement of rights. Therefore, the first part of this section discusses prominent examples brought by different actors, relying on national and international human and fundamental rights' frameworks. In each case exemplified here, the plaintiffs relied on their age as a factor, making them particularly vulnerable to the climate crisis. The second part looks at rights-based litigation in Latin America and the role of land and territory defence in Indigenous and traditional groups' litigation.

(1) Age and Activism

The effects of the climate crisis are not felt equally. Evidence shows that health-related impacts of extreme temperatures can be traced along age, pre-existing medical conditions, and socio-economic lines (*KlimaSeniorinnen*, 2020, para 2.4.2). While *"old age exacerbates negative health outcomes"* and older people are more likely to die from the effects of heatwaves, children, especially those with illnesses, are at particular risk of heat stress (*KlimaSeniorinnen*, 2020). The adverse impacts of climate change and environmental pollution also have adverse effects on the development of children. Young people are further disproportionately impacted as they will live longer. Age is thus a defining factor in establishing vulnerability.

Youth Activists

Some of the most prominent cases of current and past climate litigation were brought by or on behalf of children and youth. As mentioned earlier, the climate crisis impacts children's human rights regarding health, especially their future(s) and ability to live in a safe and healthy environment. That is why these developments – and perhaps the choice of legal tools – are unsurprising, considering that children – due to their age and voting age restrictions – often find themselves and their interests excluded from the political processes that lead to climate and environmental policies. These cases often rely on the youth's unique vulnerability to climate harm and the principle of intergenerational equity, including claims of equal treatment (UNEP, 2023, p. 41).

In September 2020, six Portuguese youths filed an application with the European Court of Human Rights, the Court tasked with overseeing the implementation and examination of alleged violations of the European Convention on Human Rights (*Duarte Agostinho and Others v. Portugal and 32 Other States*, [2020]). The convention and court were created under the Council of Europe, an international organisation established after World War II, aiming to uphold and protect European human rights. It currently has 46 Member States, including all European Union Member States.

The youths' application aims to make carbon emissions reductions and mitigation measures legally enforceable through the European Convention on Human Rights. Basing their claims on Article 2 (right to life) and Article 8 (right to private and family life) of the Convention, the applicants argue that the 33 states their application is addressed to – this being the EU's 27 Member States as well as Norway, Russia, Switzerland, Turkey, Ukraine, and the United Kingdom – are collectively responsible for most of Europe's emissions. This assessment is based on the states' respective contributions to climate change through emissions, the export of fossil fuels, and the import of goods whose production is linked to emissions (*Duarte Agostinho and Others v. Portugal and 32 Other States*, [2020], para. 9).

Portugal, the home state of the six applicants, recorded unprecedented absolute temperatures of 44 degrees Celsius and above in both 2018 and 2019 (*Duarte Agostinho and Others v. Portugal and 32 Other States*, [2020], expert report, p. 540). These temperatures resulted in an increase in adverse impacts on human health, including respiratory diseases, heat mortality, and exacerbating the various respiratory conditions of the four applicants (*Duarte Agostinho and Others v. Portugal and 32 Other States*, [2020], expert report, pp. 560–561). At the same time, the heatwaves curtailed their ability to enjoy the outdoors (*Duarte Agostinho and Others v. Portugal and 32 Other States*, [2020], expert report, p. 473). These heat waves could also be linked to an increase in wildfires (*Duarte Agostinho and Others v. Portugal and 32 Other States*, [2020], expert report, p. 543), which in one case prevented one applicant from attending school for over a week, due to the amount of smoke in the air. These developments and the connected restrictions on the enjoyment of their human rights lead to anxiety amongst the applicants. As noted in scientific reports, all these "jeopardise the future for children on this planet" (*Duarte Agostinho and Others v. Portugal and 32 Other States*, [2020], Joint WHO, UNICEF and Lancet report).

In accordance with the jurisprudence of the European Court of Human Rights, Articles 2 and 8 of the convention require states to, inter alia, put in place a legislative and administrative framework that effectively deters the threats to the right to life and right to private and family life. Applicants also made clear that both articles – and the protection guaranteed thereunder – apply to risks and harms that may only materialise in the future, including several decades into the future. This is of particular relevance considering the future impacts of the climate crisis. To emphasise their vulnerable status, the youths relied on age as a status under Article 14 ECHR, the prohibition of discrimination. They also relied on the UN Convention on the Rights of the Child (United Nations, [1990]).

This international convention has previously been invoked by another group of children in the context of the climate crisis. In September 2019, 16 complainants under the age of 18 – including well-known youth climate activist Greta Thunberg – filed five simultaneous complaints under the Optional Protocol to the Convention on the Rights of the Child (*Sacchi et al. v. Argentina, et al.*, [2019]). The convention is the most widely ratified human rights treaty. Its Optional Protocol establishes the individual right to petition. The complainants stem from 12 different countries – Brazil, France, Germany, India, Marshall Islands, Nigeria, Palau, South Africa, Sweden, Tunisia, and the United States of America – of which only five ratified the Optional Protocol. The complaints were thus addressed to five of the largest historic emitters, Argentina, Brazil, France, Germany, and Turkey.

Relying on the right to life (Article 5), the right to the highest attainable standard of health (Article 24), and the right to enjoy culture (Article 30) in conjunction with Article 3 (best interests of the child), the complainants asked the committee to find that climate change amounted to a children's crisis. It was submitted that the respondent states knowingly acted "in disregard of the available scientific evidence regarding the measures needed to prevent and mitigate climate change" (*Sacchi et al. v. Argentina et al*, [2019], para 3.8). The complaint sought to have the respondent states amend their national laws and policies to ensure proper mitigation and adaptation measures in complete alignment with the best interests of the child. To the complainants, this meant that children had to be consulted and heard in all efforts to curb the climate crisis. Their approach also sought to ensure intergenerational justice for children (Tigre and Lichet, 2021).

In October 2021, the committee rendered its five, almost identical decisions. While the complaints were unsuccessful, because they were found not to have exhausted domestic remedies, the committee made some helpful findings. First and foremost, the committee found that, despite the global character of the climate crisis, "states parties still carry individual responsibility for their own acts or omissions in relation to climate change and their contribution to it" (*Sacchi et al. v. Argentina et al*, [2019], Decision, para. 10.8). Further, the committee found that this responsibility did not end with a state's border(s). Rather, it adopted the approach developed by the Inter-American Court of Human Rights (*Advisory Opinion 23/17*, [2016]), namely, where a state has "effective control" over the activities causing harm and impacting the enjoyment of human rights of persons outside its territory, such persons are considered to be under the state's jurisdiction, which may trigger state responsibility (*Sacchi et al. v. Argentina et al*, [2019], decision, para. 10.5).

The committee recognised that children are particularly affected by climate change:

> both in terms of how they experience its effects and the potential of climate change to have an impact on them throughout their lifetimes, particularly if immediate action is not taken. Due to the particular impact on children, and the recognition by States parties to the Convention that children are entitled to special safeguards, including appropriate legal protection, States have heightened obligations to protect children from foreseeable harm.
>
> *(Sacchi et al., v. Argentina et al., [2019], decision, para. 10.13)*

Thus, while the case did not proceed for procedural reasons, the Committee's findings have nonetheless been beneficial in developing the jurisprudence on children's' rights in the climate crisis and extraterritorial jurisdiction of States. It is a great example of how legal interventions do not have to be "won" to have a beneficial impact in moving the needle further. It exemplifies the interplay between climate activism and legal intervention.

The case also led to the drafting of the Committee's General Comment No. 26 ("GC26," OCHCR, 2023), focusing on children's rights and the environment, particularly in the context of climate change. GC26 emphasises the urgent need to address the adverse impacts of environmental degradation on children's rights while providing authoritative guidance to states on mitigating environmental harm, especially concerning climate change. GC26 underscores the importance of child participation in environmental decision-making and highlights children as "agents of change." It recognises states' obligations to protect children from environmental degradation as a form of "structural violence against children." These recognitions enhance children's arguments in climate litigation cases (Tigre and Iliopoulos, 2023).

Similarly, the German Constitutional Court, in its decision of *Neubauer et al. v. Germany*, found that the German government's climate protection measures were insufficient to protect the rights of current but particularly future generations, as it represented a too significant inhibition of their rights and freedoms in the future (*Neubauer et al. v. Germany*, [2021]). The Constitutional Court partially upheld the complaint because it found that the German Climate Protection Law was insufficient insofar as it pushed decisive action into the future, thus disfavouring the younger generations. While generally perceived as a favourable outcome and now frequently invoked as the legal basis for more climate protection by the German government, the constitutional court's decision has triggered a number of additional litigation, as follow-up action by the German government has been insufficient. For example, sectoral emissions reduction targets were unmet and/or scrapped altogether (ClientEarth, 2023). A close link and interplay between civil society – and especially youth – activism and litigation can be observed here.

At the international level, a group of students from the Pacific Islands were the initiators behind the effort of getting an advisory opinion from the International Court of Justice (ICJ) – often referred to as the world's highest court – on the interplay between the climate crisis and State obligations (Horne et al., 2023). In 2019, the Pacific Islands Students Fighting Climate Change (PISFCC) was founded to persuade the leaders of the Pacific Island Forum to bring a United Nations' General Assembly (UNGA) resolution, requesting an advisory opinion on climate change from the ICJ. This is a particularly pertinent example of the interplay of climate activism and legal intervention, as the youth activists specifically sought out legal means and attached large public campaigns to this work. In March 2023, the UNGA adopted a resolution requesting an advisory opinion from the International Court of Justice (ICJ) (Tigre and Carrillo Bañuelos, 2023). Again, youth activism was thus crucial in pressing forward global action on the climate crisis, whilst also devising novel legal strategies to do so. Following the deadlines for written submissions in January and April 2024, the ICJ will likely hold public hearings in the second half of 2024. The final Advisory Opinion has the potential to provide an enormous push for climate and environmental obligations on states.

Elderly

Relying on similar scientific evidence highlighting not only the vulnerability of children but also that of the elderly as regards heat-related deaths, a group of elderly Swiss women has submitted an application to the European Court of Human Rights (*KlimaSeniorinnen*, 2020, para 3). They too invoked art. 2 and 8 of the European Convention and submitted that the respondent (i.e., Switzerland) had failed and continues to fail to put into place the necessary and required measures to protect their human rights. Their claim is aimed at the positive obligations of states – meaning the duty to protect and fulfil – to protect against increased heatwaves' negative effects on elderly women. They ask the court to find that the respondent has to establish a legislative and administrative framework in accordance with international climate law and the best available science. As mentioned earlier, old age, like young age, renders persons particularly vulnerable to the effects of the climate crisis. With old age, negative health outcomes increase, and older people are more likely to die due to the impact of heatwaves. In the case of *KlimaSeniorinnen*, age intersects with biological sex. Elderly women are particularly susceptible to a "real risk not only of severe physical and mental impairment from heat-induced illness, but of death" (*KlimaSeniorinnen*, 2020, Oral statement, para. 11). While their health concerns are different, both age groups rely on similar scientific evidence and arguments to prove their particular vulnerability and the violation of their rights. And both vulnerable groups resort to legal means, including the European Court of Human Rights, to see their rights upheld.

(2) Land Rights and Climate Litigation

The prevalence of climate litigation in Latin America has witnessed significant growth in recent years (Tigre et al., 2023b). Notably, climate cases in the region heavily rely on human rights. This reliance can be attributed to a combination of factors, including the widespread recognition of the constitutional right to a healthy environment (Tigre, 2023), the existence of broad legal standing rules, the direct incorporation of international law into the legal system, and the prominent role played by the Inter-American System of Human Rights within the region (Auz, 2022). The proliferation of multilateral environmental agreements (MEAs) has inspired a similar expansion of domestic environmental norms (Tigre, 2021).

As a result, the region counts on a rich jurisprudence of environmental cases, as well as a tradition of strategic (rights-based) litigation of economic and social rights. These factors positively inform and influence the proliferation of climate litigation (Batros and Khan, 2023). Additionally, shared characteristics in constitutional frameworks foster a regional narrative in climate litigation, where the success of one country's legal proceedings can serve as a precedent for others. The next section delves into this background, showing how climate litigation has been used in the region to advance climate justice and some of the pitfalls of this strategy.

Using Litigation as a Strategy for Climate Action

Against this backdrop, land rights activists, as well as environmental and climate advocates, have *strategically* embraced litigation as a means to advance climate protection goals (Batros and Khan, 2022).[1] This implies that legal cases are conceived as part of a broader strategy to effect change in the realm of climate action. Law, alongside other tactics in activism, is utilised as a tool to facilitate societal transformation. A few cases have, as such, been strategically filed to induce state action on climate change.

This strategy includes, for example, *Future Generations v. Ministry of the Environment and Others*, in which several youth plaintiffs challenged the lack of implementation of policies to curb deforestation and comply with Colombia's commitments to reduce GHG emissions at the international level (*Future Generations v. Ministry of the Environment and Others*, [2018]). The case successfully relied on constitutional human rights norms to advance positive enforcement of a domestic and international framework of climate law. The Supreme Court ruled in favour of the plaintiffs, recognising the Amazon rainforest as a subject of rights and specifically requesting the government to establish a plan to address deforestation in the Amazon. Significantly, the decision has not been enforced yet, and the NGO that brought the case, Dejusticia, has continuously pushed for its implementation.

Another successful example of how strategic litigation has been used to advance climate action is found in Brazil. In *PSB et al. v. Brazil (on the Climate Fund)*, the Brazilian Supreme Court ruled that the executive government – at the time under the Bolsonaro administration, which openly questioned climate change – had a constitutional obligation to mitigate climate change and could not refuse to implement measures mandated by law (Tigre and Setzer, 2023). Significantly, the decision recognised the Paris Agreement as a human rights treaty based on the constitutional right to a healthy environment. This recognition is crucial because it allows activists to use the human rights legal framework to ensure compliance with environmental and climate protection. In addition, human rights treaties have supralegal status, meaning they rank above domestic laws and policies. As such, any norms that are contrary to MEAs can be declared unconstitutional if they do not comply with their mandates. This includes, for example, any commitments under the Paris Agreement, such as the obligation to comply with the rule of progression in nationally determined contributions (Tigre and Setzer, 2023). The case was brought by political parties due to a particularity of the Brazilian procedural rules. However, civil society organisations heavily pushed for the case to be urgently added to the "green agenda" of the court, along with a few other environmental cases.

These examples show how rights-based litigation has been employed as a tactic in a broader advocacy strategy. In theory, courts provide a platform where all parties can seek a fair resolution to their disputes. Some commentators contend that courts can "equalise power imbalances in society and enforce the rule of law" (*Surge in court cases*, 2021). Thus, climate litigation serves as a prominent means of climate activism in Latin America, allowing activists to leverage legal mechanisms to challenge existing norms, advocate for stronger climate policies, and hold governments and corporations accountable for their actions or inactions concerning climate-change mitigation and adaptation.

In addition, experts highlight that the outcome of a trial should not be the sole measure of victory – this was also evidenced above in the section on youth litigation in Europe (Johnson, 2022). The attention garnered through activism accompanying climate lawsuits can be equally vital as a courtroom win or loss (Batros and Khan, 2023). Litigation has the potential to draw media attention to specific issues, mobilise public interest groups, and inspire ordinary citizens to become activists themselves (Posner, 2007). The objectives of litigation may extend beyond prevailing in court; they may involve obtaining information through the discovery process, compelling defendants to adopt a formal stance on the public record, or securing specific factual or legal findings from the court, even if the plaintiffs do not ultimately win the case (Bateman, 2022).

As a general regional trend, cases have often been filed to induce enforcement of domestic environmental and climate legislation and the failures of governments to implement measures upholding a country's climate pledges (Nationally Determined Contributions

(NDCs) submitted pursuant to the Paris Agreement, specifically in relation to land use change and forestry (Peel and Lin, 2019; Setzer and Benjamin, 2019; Setzer and Benjamin, 2020; Bouwer, 2018; Tigre et al., 2023a). However, there are challenges to this approach. First, enforcement of decisions has been problematic, with governments taking years to implement the remedies requested due to governance and resource constraints. Second, strategic cases often fail to account for the local risks communities face in light of climate change. This aspect is addressed next.

Challenges of Climate Litigation to Address Local Climate Gaps

Crucially, climate litigation in Latin America is also intrinsically linked to the protection of land use and land rights. Agriculture, forestry, and land use collectively contribute to approximately 46% of the region's GHG emissions (Vergara et al., 2020). Activities such as overgrazing and deforestation (often illegal) exacerbate GHG emissions, directly impacting communities that depend on the land for their livelihoods. Consequently, these communities, particularly Indigenous groups and traditional communities, experience the repercussions of climate change with greater intensity (WMO, 2022). In this context, climate litigation in Latin America is inherently intertwined with land use and the wellbeing of traditional communities. Additionally, the impacts of climate change are felt more acutely where other human rights and socio-economic concerns have also been lacking. As such, Latin America is already experiencing "the tip of the climate spear" (World Bank, 2021).

Nonetheless, the communities most impacted often lack access to justice for multiple reasons. First, communities may lack the skills, money, and knowledge to bring forward a case without assistance from civil society organisations, public prosecutors, or pro bono lawyers. Litigation has been criticised as an "elitist" strategy, wherein the agency of local communities is supplanted by legal professionals who may lack comprehensive understanding of the intricacies of the land and its significance in the lives of these communities (Duffy, 2018). If cases are brought, these are likely not to be framed as climate cases, because the communities, the lawyers, or the judges lack the proper expertise in climate law. Second, despite significant advances in climate attribution science in the last decade, this has not trickled down to the needs of communities that require evidence to submit to courts to establish causation (Stuart-Smith et al., 2021; Otto et al., 2022). Third, considering the sluggish pace of judicial proceedings, activists may have a better chance of achieving their ultimate goals by combining litigation efforts with other forms of activism. And fourth, environmental and climate activists, including lawyers, face significant personal risks in Latin America, known as the deadliest region for environmental defenders, as detailed in the following section.

Defending the Defenders

Environmental defenders are pivotal to the protection of the environment and to ensuring sustainability (United Nations, 2019). These individuals, including Indigenous people, peasants or fisherfolks, environmental activists, or journalists, actively defend the environment in various ways (Scheidel et al., 2020). The movements pushed forward by environmental defenders are crucial to ensuring a just environmental future. However, their activism can be costly. Defenders face murders; physical violence; severe environmental, health, and cultural impacts; social stigma; and other forms of repression (Navas et al., 2018). To ensure the sustainability of these movements, the law must provide for legal responses to ensure their protection.

In 2022, the UNECE's first Special Rapporteur on environmental defenders under the Aarhus Convention was elected ("World first Special Rapporteur," 2022). The Special Rapporteur's mandate includes a rapid response mechanism, allowing for complaints to be directly submitted to the Special Rapporteur. The establishment of the mandate underlines the increased threats environmental defenders face. A similar mechanism in Latin America (under the Escazú Agreement architecture) would be extremely useful to allow for redress mechanisms beyond the state.

(1) *Latin and South America*

Latin America stands out as an exceedingly perilous region for environmental and climate defenders. As revealed by a survey conducted by Global Witness, a staggering 68% of the murders targeting these defenders over the past decade occurred in Latin America, with Brazil, Colombia, Mexico, and Honduras accounting for the highest number of fatalities (McGrath, 2022). International NGO Global Witness has documented a total of 1,733 murders of climate defenders since 2012 (Hines, 2022). This alarming statistic becomes even more significant when considering that Indigenous communities, comprising just 5% of the global population, bear the brunt of these attacks. In 2021, over 40% of fatal assaults targeted Indigenous people, predominantly in Mexico, Colombia, Nicaragua, and Peru (Hines, 2022). Tragically, Indigenous populations often find themselves embroiled in struggles to safeguard their lands against the encroachment of mining, oil, logging, and hydropower developers (McGrath, 2022). This data has significant consequences for using climate litigation as a strategy.

In 2021, the Escazú Agreement entered into force, bringing several environmental rights and duties to Latin America and the Caribbean (LAC). The Escazú Agreement, currently signed by 24 countries, has 15 parties who have ratified and internalised the agreement. Significantly, Escazú is the first environmental agreement that fully integrates human rights (Tigre, 2024). As one of its pillars, it ensures the protection of environmental and land defenders (Tigre, 2021). The agreement seeks to guarantee a safe environment for these individuals, groups, and organisations, recognising the importance of their vital work in protecting the environment and human rights. To ensure their protection, the Escazú Agreement mandates that member countries take effective and timely measures. These measures are aimed at preventing, investigating, and punishing any attacks, threats, or intimidation suffered by human rights defenders in environmental matters. The agreement acknowledges the risks and dangers faced by those who advocate for environmental justice and ensures that governments in the region proactively address any potential threats to their safety. By requiring prompt and comprehensive action, the Escazú Agreement sends a clear message that the protection of environmental and land defenders is a priority. This commitment aims to create an enabling environment for these defenders to continue their essential work without fear of reprisals, fostering a more sustainable and just future for all in the region. As such, the Escazú Agreement paves the way for new opportunities for activists to leverage litigation as part of their advocacy plan (Peel and Lin, 2019). However, implementation of these provisions still lies ahead.

The Global Network for Human Rights and the Environment (GNHRE) Principles for the Effective Implementation of the Escazú Agreement suggests the following measures to ensure compliance with this mandate.

20. In order to secure the dignity, integrity, safety and life of environmental defenders, pursuant to Article 9 of the Escazú Agreement, State Parties must commit to providing a secure environment in which environmental defenders can exercise their duties. Hence, State Parties should guarantee effective remedies and transparent processes that prevent impunity. State Parties should hold accountable State and non-State actors whether individual or corporate if their actions infringe on the rights of defenders.

21. To this end, State Parties can devise mechanisms to systematise and widely diffuse data concerning the threats and actions taken against environmental defenders, so that society is conscious regarding the subject.

22. State Parties should encourage an active role for national human rights institutions by developing early warning systems or other tools that will allow for a secure environment in which defenders can carry out their work. In this sense, regional and international human rights and environmental bodies must be active in providing a platform for environmental defenders.

(GNHRE)

Implementation is a high priority. In 2022, the First Annual Forum on Human Rights Defenders in Environmental Matters in Latin America and the Caribbean was held. The Forum aimed to facilitate dialogue on the challenges faced by environmental human rights defenders in the region, bringing together parties to the Escazú Agreement, experts, and the public, especially defenders and Indigenous Peoples. Topics included the situation of defenders, good practices in promoting their rights, protection mechanisms under the Escazú Agreement, and discussions on the action plan for defenders. The forum's official report contains a summary of panels, discussions, and proposals to inform the preparation of the action plan on human rights defenders of the Escazú Agreement, which will likely be adopted in the next few years (Report of the First Annual Forum on Human Rights Defenders in Environmental Matters in Latin America and the Caribbean, 2022). As with every new agreement, establishing rules of procedures and other rules related to its implementation takes time. The Escazú's Conference of the Parties (COP) meets every year, and these types of documents are approved yearly.

(2) Europe

While environmental defenders and activists across Europe do not face the same levels of physical violence, their work and activism are increasingly being challenged and undermined. Simultaneously, activists are testing the limits of the law, in some instances consciously violating law, to seek judicial clarification on whether their actions are allowed in the climate emergency.

In March 2022, a German district court addressed the question of whether climate protests were covered by the fundamental rights protection of the German Constitution – the Basic Law. The defendant had been charged with trespassing, which is entering someone's land or property without permission, when protesting on the site of a lignite mine (*District*

Court Mönchengladbach-Rheydt, [2022]). The activist's defence was based on the German Constitution's Articles 8(1) freedom of assembly; 5(1)(1) freedom of expression; and 4(3) freedom of conscience. While the District Court found that the entering of the site fulfilled the elements of the crime under Section 123 (1) of the German Criminal Code, the Court considered this to be neither "unlawful" nor "without authority." Instead, it held that climate change was an issue of "paramount importance." In the words of the District Court, the political decision to cease lignite mining – lignite mining itself an activity recognised as harmful – and the subsequent decision to delay its phase-out process, nonetheless, could place individuals in a state of moral dilemma, wishing to protest. The Court thereby recognised that protests against climate-damaging measures could be covered by the freedom of conscience guaranteed by the German Constitution (*District Court Mönchengladbach-Rheydt*, [2022]).

A district court in Flensburg, Northern Germany acquitted an activist on similar grounds. The activist had previously occupied trees to protest the announced clearance of the forest for building a hotel. Even though the court found that the criminal law allegation of trespass had been met, it held the trespass not to be illegal. The court found that the defendant's goal of protecting the forest outweighed the interests of the hotel investors. Here, the climate crisis was held to be a "justifiable emergency" (Wolf, 2022). The same was held by a district court in Berlin. The judge, asked to rule on protests by the climate group "Letzte Generation" (last generation), found that the climate crisis was an "objectively urgent situation" and "scientifically undeniable," justifying the protests. The judge also considered that the "moderate political progress" on the climate crisis should be considered in the evaluation of the protests (Kühn, 2022).

In stark contrast is the case of Lützerath. The hamlet of Lützerath was to be cleared to allow for the mining of lignite that lies beneath it. Over an extended period of time, climate activists tried to stop the clearing, living in emptied and abandoned buildings for months. These were later cleared on grounds of unauthorised stay of persons ("OVG, 2023"). Activism is also being curtailed by numerous judgments ordering preventive detention of climate activists and protesters (Bauer, 2022).

In France, environmental group Les Soulèvements de la Terre (The Earth Uprisings Collective) was dissolved by authorities in June 2023 on allegations of incitement of violence. The group had previously protested a number of French government measures on climate and the environment (Stroehlein, 2023). The cases demonstrate how both individuals as well as groups can face (criminal) charges for their activism.

Another risk faced by environmental defenders, including organisations, are so called SLAPPs: Strategic Lawsuits against Public Participation (Mańko, 2023). As the name indicates, such lawsuits, often brought by corporate and/or private actors, aim at curtailing public participation in civil society, as they present large financial and time burdens. Especially in the case of civil society organisations, SLAPPs can put such actors at great financial risk. In 2022, NGO Earthrights International published research indicating that globally over 150 environmental activists and defenders were targeted by the fossil fuel industry through the medium of intimidating lawsuits ("As climate crisis intensifies," 2022).

At the same time, activists are fighting back. On 27 June 2023, in the case of *Bryan and Others v Russia*, the European Court of Human Rights held that a violation of rights to liberty and security and freedom of expression had occurred (*Bryan and Others v. Russia*, [2023]). In 2013, Greenpeace activists protested a Russian offshore drilling platform, during which two activists climbed the platform and launched dinghies. Although Russia had ceased to be a Party to the European Convention at the time the judgement was rendered, the Court found that it still had jurisdiction to deal with the case, as the facts giving rise to the alleged violations of the convention had taken place before the date on which Russia ceased to be a party to the European Convention.

Conclusion

The urgency of near-term climate action highlighted by the IPCC's Synthesis Report underscores the crucial role of civil society in global climate governance. Climate litigation has emerged as a prominent strategy employed by activists and environmental defenders to advance climate protection and hold governments and corporations accountable. By strategically utilising legal mechanisms, activists aim to challenge existing norms, advocate for stronger climate policies, further develop the law, and create societal transformation.

However, climate litigation also faces significant limitations. At the level of public international law, there simply is no judicial infrastructure to bring claims against governments or corporations. The only available route, in most cases, is the resort to human rights courts and tribunals.

Access to justice remains a considerable obstacle, with marginalised communities often lacking the resources and expertise to bring forward cases. Additionally, enforcing legal decisions can be challenging, and the slow pace of judicial proceedings may necessitate the combination of litigation with other forms of activism to achieve desired outcomes. Despite these challenges, climate litigation has proven to be a very effective tool in advancing climate justice and shaping climate policy in Europe, Latin America, and other regions. It complements and synergises with other forms of activism, ultimately contributing to the broader goal of addressing the climate crisis and protecting the rights of present and future generations.

Note

Climate litigation is a rapidly evolving field, often outpacing academic scholarship. Since the drafting of this chapter and as the book went to print, significant decisions were handed down in the **Klimaseniorinnen** and **Duarte Agostinho** cases, both of which are referenced here. These rulings not only reflect the dynamic nature of climate litigation but also underscore the importance of ongoing analysis to understand the implications of such landmark decisions in the broader legal and social context.

1 Batros and Khan note that there is no single definition of what "strategic litigation" is. However, it is broadly understood as cases that seek to achieve "broader change beyond the direct interests of the plaintiffs in the case or the remedies sought by them – typically changes to policy, social norms, or corporate behaviour."

References

As climate crisis intensifies, fossil fuel companies seek to silence their critics (2022). *EarthRights International*. https://earthrights.org/media_release/as-climate-crisis-intensifies-fossil-fuel-companies-seek-to-silence-their-critics/ [Accessed 28 September 2023].

Auz, J. (2022). Human rights-based climate litigation: A Latin American cartography. *Journal of Human Rights and the Environment*, 13(1), 114–136. https://doi.org/10.4337/jhre.2022.01.05 [Accessed 28 September 2023].

Bateman, J. (2022). Why climate lawsuits are surging. *BBC Future*. https://www.bbc.com/future/article/20211207-the-legal-battle-against-climate-change [Accessed 28 September 2023].

Batros, B. and Khan, T. (2023). Thinking strategically about climate litigation. In *Litigating the Climate Emergency: How Human Rights, Courts, and Legal Mobilisation Can Bolster Climate Action*. Cambridge: Cambridge University Press.

Bauer, M. (2022). *Klimaprotest: Warum die Präventivhaft in Bayern strittig ist, Tagesschau*. https://www.tagesschau.de/inland/gesellschaft/praeventivhaft-klima-protest-bayern-101.html [Accessed 28 September 2023].

Bouwer, K. (2018). The unsexy future of climate change litigation. *Journal of Environmental Law*, 30, 483.

Bryan and Others v. Russia [2023], 22515/14.

ClientEarth (2023). *Climate Protection Needs Sector Goals and Implementation at Project Level*. https://www.clientearth.de/aktuelles/ressourcen/klimaschutz-braucht-sektorziele-und-umsetzung-auf-vorhabensebene/ [Accessed 1 October 2023].

Das, L. (2022). *I Asked Greenpeace Volunteers to Define Activism. Here's What They Said in Greenpeace UK*. https://www.greenpeace.org.uk/news/what-is-activism-meaning/ [Accessed 28 September 2023].

Duarte Agostinho and Others v. Portugal and 32 Other States [2020], 39371/20.

Dubash, N. K., Mitchell, C., Boasson, E. L., Borbor-Cordova, M. J., Fifita, S., Haites, E., et al. (2022). Chapter 13: National and subnational policies and institutions. In *Climate Change 2022: Mitigation of Climate Change. Working Group III Contribution to the Sixth Assessment Report of the Intergovernmental Panel on Climate Change*. Geneva: Intergovernmental Panel on Climate Change. https://www.ipcc.ch/report/ar6/wg3/downloads/report/IPCC_AR6_WGIII_Chapter_13.pdf [Accessed 1 October 2023].

Duffy, H. (2018). *Strategic Human Rights Litigation: Understanding and Maximising Impact*. Oxford Hart Publishing.

District Court Mönchengladbach-Rheydt, Urt. v. 14.3.2022–21 Cs – 721 Js 44/22–69/22 [2022]. *Future Generations v. Ministry of the Environment and Others* [2018], 11001. Corte Suprema de Justicia [C.S.J.] [Supreme Court], Sala de Casación Civil, abril 5, 2018, M.P.: L.A. Tolosa Villabona, Expediente 11001-22-03-000-2018-00319-01 (Colom.).

Gerstetter, C. (2022). Im Gerichtssaal und auf der Straße – unterschiedliche Wege, ein Ziel? Zum Verhältnis von politischem Aktivismus für Klimagerechtigkeit und Klimaklagen. *juridikum zeitschrift für kritik*, 1, 116–123.

Hines, A. (2022). *Decade of Defiance: Ten Years of Reporting Land and Environmental Activism Worldwide, Global Witness*. https://www.globalwitness.org/en/campaigns/environmental-activists/decade-defiance/ [Accessed 28 September 2023].

Horne, K., Tigre, M. A. and Gerrard, M. (2023). *Status Report on Principles of International and Human Rights Law Relevant to Climate Change*. https://scholarship.law.columbia.edu/faculty_scholarship/3924

IPCC. *Synthesis Report of the IPCC Sixth Assessment Report (AR6)*. https://report.ipcc.ch/ar6syr/pdf/IPCC_AR6_SYR_SPM.pdf [Accessed 1 October 2023].

Johnson, D. (2022). Across the globe, those harmed by climate change are turning to courts. *Eco*. https://www.eco-business.com/news/across-the-globe-those-harmed-by-climate-change-are-turning-to-courts/ [Accessed 28 September 2023].

KlimaSeniorinnen (2020). *Letter to the Registrar, European Court of Human Rights*. https://www.klimaseniorinnen.ch/wp-content/uploads/2020/11/201126_Application_ECtHR_KlimaSeniorinnen_extract_anonymised-2.pdf [Accessed 30 September 2023].

Kühn, T. (2022). *Blockaden der letzten Generation: Richterlicher Widerstand, taz.de*. https://taz.de/Blockaden-der-Letzten-Generation/!5890597/ [Accessed 28 September 2023].

Mańko, R. (2023). *Strategic Lawsuits Against Public Participation (SLAPPs)*. European Parliament.

McGrath, M. (2022). Over 1,700 environment activists killed in decade – report. *BBC News*, 29 September. https://www.bbc.com/news/science-environment-63064471 [Accessed 28 September 2023].

Navas, G., Mingorria, S. and Aguilar-González, B. (2018). Violence in environmental conflicts: The need for a multidimensional approach. *Sustainability Science*, 13(3), 649–660. http://doi.org/10.1007/s11625-018-0551-8

OCHCR (2023). *General Comment No. 26 (2023) on Children's Rights and the Environment with a Special Focus on Climate Change.* https://www.ohchr.org/en/documents/general-comments-and-recommendations/general-comment-no-24-2019-childrens-rights-child [Accessed 1 October 2023].

Otto, F. E. L., Minnerop, P., Raju, E., Harrington, L. J., Stuart-Smith, R. F. and Boyd, E., et al. (2022). Causality and the fate of climate litigation: The role of the social superstructure narrative. *Global Policy*, 13, 736– 750. https://doi.org/10.1111/1758-5899.13113

OVG (2023). Lützerath darf geräumt werden (zum Betretungs- und Aufenthaltsverbot). *Legal Tribune Online.* https://www.lto.de/recht/nachrichten/n/ovg-nrw-luetzerath-5b1423-raeumung-allgemeinverfgung-bestaetigt-antrag-einstweiliger-rechtsschutz-erfolglos-klima-aktivismus/ [Accessed 28 September 2023].

Peel, J. and Lin, J. (2019). Transnational climate litigation: The contribution of the Global South. *American Journal of International Law*, 113(4), 679–726. http://doi.org/10.1017/ajil.2019.48

Posner, E. A. (2007). Climate change and international human rights litigation: A critical appraisal. *University of Pennsylvania Law Review*, 155, 1925, 1932.

Sacchi et al. v. Argentina et al. (2019). Communication No. 104/2019 (Argentina), Communication No. 105/2019 (Brazil), Communication No. 106/2019 (France), Communication No. 107/2019 (Germany), Communication No. 108/2019 (Turkey).

Scheidel, A., et al. (2020). Environmental conflicts and defenders: A global overview. *Global Environmental Change*, 63, 102104. http://doi.org/10.1016/j.gloenvcha.2020.102104.

Setzer, J. and Benjamin, L. (2019). Climate litigation in the Global South: Constraints and innovations. *Transnational Environmental Law*, 8, 1.

Setzer, J. and Benjamin, L. (2020). Climate change litigation in the Global South: Filling in gaps. *AJIL Unbound*, 114, 56.

Setzer, J. and Higham, C. (2021). *Global Trends in Climate Litigation: 2021 Snapshot, Grantham Research Institute on Climate Change and the Environment.* https://www.lse.ac.uk/granthaminstitute/publication/global-trends-in-climate-litigation-2021-snapshot/ [Accessed 28 September 2023].

Setzer, J. and Higham, C. (2023). *Global Trends in Climate Litigation: 2023 Snapshot, Grantham Research Institute on Climate Change and the Environment.* https://www.lse.ac.uk/granthaminstitute/publication/global-trends-in-climate-change-litigation-2023-snapshot/ [Accessed 28 September 2023].

Stroehlein, A. (2023). *Climate Crisis: France Shoots the Messenger, Human Rights Watch.* https://www.hrw.org/the-day-in-human-rights/2023/06/23 [Accessed 28 September 2023].

Stuart-Smith, R. F., Otto, F. E. L., Saad, A. I., Lisi, G., Minnerop, P., Lauta, K. C., et al. (2021). Filling the evidentiary gap in climate litigation. *Nature Climate Change*, 11(8), 651– 655.

Surge in Court Cases Over Climate Change Shows Increasing Role of Litigation in Addressing the Climate Crisis. (2021). United Nations Environment Programme. https://www.unep.org/news-and-stories/press-release/surge-court-cases-over-climate-change-shows-increasing-role [Accessed 28 September 2023].

Tigre, M. A. (2021). *Six Pillars of the Escazú Agreement.* https://gnhre.org/community/the-six-pillars-of-the-escazu-agreement-part-1/ [Accessed 1 October 2023].

Tigre, M. A. (2024). Developing a right to a healthy environment in LAC: Compliance through the inter-American system and the Escazú Agreement. In C. Voigt and C. Foster (eds.) *Non-compliance Mechanism Versus International Courts and Tribunals.* Cambridge: Cambridge University Press.

Tigre, M. A. and Carrillo Bañuelos, J. A. (2023). The ICJ's advisory opinion on climate change: What happens now? In *Climate Law: A Sabin Center Blog.* Sabin Center for Climate Change Law, 29 March. https://blogs.law.columbia.edu/climatechange/2023/03/29/the-icjs-advisory-opinion-on-climate-change-what-happens-now/ [Accessed 28 September 2023].

Tigre, M. A. and Iliopoulos, A. (2023). Climate litigation and children's rights: Unpacking the CRC's new general comment. *Climate Law: A Sabin Center Blog.* Sabin Center for Climate Change Law, 7 September. https://blogs.law.columbia.edu/climatechange/2023/09/07/climate-litigation-and-childrens-rights-unpacking-the-crcs-new-general-comment/ [Accessed 28 September 2023].

Tigre, M. A. and Lichet, V. (2021). The CRC decision in Sacchi v. Argentina. *ASIL Insights*, 25(26). https://www.asil.org/insights/volume/25/issue/26 [Accessed 28 September 2023].

Tigre, M. A. and Setzer, J. (2023). Human rights and climate change for climate litigation in Brazil and beyond: An analysis of the climate fund decision. *The Georgetown Journal of International*

Law, 54(4), 1–53. https://scholarship.law.columbia.edu/sabin_climate_change/219/ [Accessed 8 September 2024].

Tigre, M. A., Urzola, N. and Goodman, A. (2023a). Climate litigation in Latin America: Is the region quietly leading a revolution? *Journal of Human Rights and the Environment*, 14(1), 67–93. https://doi.org/10.4337/jhre.2023.01.04

Tigre, M. A. et al. (2023b). *Just Transition Litigation in Latin America: An Initial Categorization of Climate Litigation Cases Amid the Energy Transition*. Sabin Center for Climage Change Law, January 2023. https://scholarship.law.columbia.edu/sabin_climate_change/197/ [Accessed 8 September 2024].

UN (2019). Recognizing the contribution of environmental human rights defenders to the enjoyment of human rights, environmental protection and sustainable development. *United Nations. Human Rights Council Resolution 40/L.22*. un.org

United Nations Environment Programme (2023). *Global Climate Litigation Report: 2023 Status Review*. Nairobi.

Vergara, W., et al. (2020). *Forests and Agriculture: The Key to Latin America's Climate Commitments*, INITIATIVE 20X20. https://initiative20x20.org/news/ndcs-latin-america-nature-based-solutions#:~:text=Agriculture%2C%20forestry%20and%20land%20use,of%20its%20employment%20in%202019 [Accessed 28 September 2023].

Wolf, J. (2022). *Klimaschutz als rechtfertigender Notstand, Verfassungsblog*. https://verfassungsblog.de/klimaschutz-als-rechtfertigender-notstand/ [Accessed 28 September 2023].

World Bank (2021). *Promoting Climate Change Action in Latin America and the Caribbean*. https://www.worldbank.org/en/results/2021/04/14/promoting-climate-change-action-in-latin-america-and-the-caribbean

World Meteorological Organization (WMO) (2022). *State of the Climate in Latin America and the Caribbean 2021*. https://library.wmo.int/idurl/4/58014.

World's First Special Rapporteur on Environmental Defenders Elected Under the Aarhus Convention (2022). UNECE. https://unece.org/environment/press/worlds-first-special-rapporteur-environmental-defenders-elected-under-aarhus [Accessed 28 September 2023].

24

'ACADEMIC FREEDOM' V. CLIMATE CHANGE DENIAL

How the Politics of Research Funding Shapes the Possibilities for Researching Grassroots Activism

Rob Watts, Judith Bessant, Stewart Jackson, Michelle Catanzaro, Faith Gordon and Philippa Collin

Climate change has occupied the minds of many citizens, political parties and politicians at least since the 1992 Rio Earth Summit that produced the UN Framework Convention on Climate Change. Despite the weight of evidence supporting President Obama's claim in 2014 that 'Climate change is a fact', climate-sceptic think tanks and other interests have wielded inordinate political power, influencing relevant government national and international policy in advanced democracies, such as Australia (Obama, 2014; Jacques et al., 2008: Plehwe, 2021; Lucas, 2021).

As the state and the major parties in countries such as Australia have failed to make the substantial policy changes required to militate the worst effects of climate change, citizens have increasingly taken action. Grassroots activism, perhaps most usefully characterised by Burgmann and Baer (2012) as 'politics from below', has generally been driven by environmental NGOs and other actors (Goodman and Morton, 2023; Collin et al., 2020). Since the early 2000s a network of grassroots and larger climate and environment groups and NGOs formed coalitions and shared campaigns and actions (Goodman and Morton, 2023). Among them are large-scale youth-led organisations such as the Australian Youth Climate Coalition, Seed and, most recently, the School Strike 4 Climate network (Hilder and Collin, 2022). Amongst their actions were calls for specific policy changes and opposition to local developments, such as new mining activity. While not always decisively positive in their outcomes, this kind of pressure led to action on climate change becoming a key policy issue in the 2022 Australian federal election. We also saw the emergence of a pro-climate action candidate support platform – Climate 200 – that supported climate-aware independent candidates running for election (Turner, 2021; Hendriks and Reid, 2022).

Over the past 20 years, research on grassroots environmental and climate action, often with participants themselves, has made a valuable contribution to explaining, informing and amplifying efforts by various movements and organisations. Even so in Australia, there has been more grassroots activism than there has been research on climate politics. While recognition, interest and support for youth climate activism in civil society have significantly

DOI: 10.4324/9781003396567-28

increased in the last 5 years (Judd, 2022), political and funding support for research into this growing phenomenon has not.

While there are a multitude of challenges when conducting research on social movements and grassroot climate action (Gillan and Pickerill, 2012; Chesters, 2014), public funding of research to better understand what is taking place has not been given priority. Moreover, as we discovered, the politics of accessing research funding directly intersects with the politics of climate activism and climate scepticism in ways that have significant implications for praxis in grassroots climate activism. Our reflexive approach to research in this field as onto-ethical practice considers the political economy of research funding to be entirely entangled with our politics as activist academics and the role we believe research plays in explaining, advancing and amplifying grassroots climate activism *as* world making (Pearse et al., 2010, p. 78; Collin and Matthews, 2021, p. 126).

In this chapter we focus on the politicisation of research on young people and climate action. Specifically, we focus on efforts by the Australian government to prevent such research by vetoing a project approved by the national research council for funding on this topic. We note how this occurred in the context of increasing public frustration especially with coalition governments' refusal after 2013 to take decisive action on climate change. We use an Australian case study involving a national government minister vetoing funding for our research into student climate politics as a springboard for discussing broader academic freedom and climate denial issues. While there are local particularities, we suggest there are sufficient similarities between the Australian context and other liberal democracies that make the outcomes from this case relevant to those interested in climate politics and the state of contemporary democratic government.

We begin with a brief survey of Australian research of climate politics before outlining the event that sparked the production of this chapter.

Grassroots Activism and Research in Australia

Amidst a surge of grassroots climate action over the past 20 years, as has been the case in many other countries, social movements and their participants have been studied in Australia (Hutton and Connors, 1999; Crowley, 2020). Studies have traced the climate movement in Australia and experiences of participants (e.g. Rosewarne et al., 2013; Verlie, 2021), specific campaigns (e.g. Hintjens, 2000), key organisations (e.g. Hilder and Collin, 2022) and the various strategies that have been used (Hintjens, 2000; Pearse et al., 2010).

Until relatively recently, there was little research focussed on young and female climate activists. However, in the last decade, particularly since the Fridays for Future movement and the mass mobilisation of young people, a body of research has examined how and why young people and especially girls are participating in climate action (Collin, 2015, pp. 104–108; Collin and Matthews, 2021; Feldman, 2021; Arnot et al., 2023; Bessant et al., 2023). There has also been research into the media representations and other discourses of participants (O'Neill, 2013; Collin and Matthews, 2021; Mayes and Hartup, 2022), the role of humour in protest actions (Mayes and Centre, 2023; Hee et al., 2022) and the visual language of the movement (Catanzaro and Collin, 2021). Finally, literature has begun examining responses of the state and others (e.g. mining corporations themselves) that include moves to criminalise protest dissent (e.g. Lee, 2021; Gulliver et al., 2023).

While we are scholars with a long-standing interest in youth research we also see ourselves as active participants in climate politics. Amongst other things this has involved acknowledging and reflecting on our roles as activist-researchers. As Pearse et al. (2010, p. 78) observe, there is value in being mindful of what it means to be both involved and detached. It is a practice that requires 'activist practice, as much as research method'. While Pearse et al. (2010) point toward Bourdieu and Wacquant (1992) and Riach (2009), we are also guided by Touraine (1981, p. 27) on being politically present researchers and political actors.

Our Story

Between 2019 and early 2021, we (the authors) were part of a group of seven academics working in five Australian universities who applied for an Australian Research Council (ARC) 'Discovery Project' grant to study the student climate movement. The ARC administers the National Competitive Grants Program, the major source of research funds for academics in the social sciences, humanities and sciences. With a success rate of 19% of applications, these awards are hard to get and extremely competitive. Around this scheme swirls the vague but comforting idea that the program is above politics, because it is based on rigorous anonymous peer review, characterised by objectivity, expertise and good process (Kelly et al., 2014, p. 227). Determining which applications are funded is based on an assessment process led by the ARC College of Experts, consisting of 213 scholars who are experts and international leaders in their fields.

This concept of peer review, redolent of notions of academic freedom and university autonomy, is central to the idea that all successful ARC applications need to meet 'the highest standards': an assessment based on scholarly criteria including originality, significance, feasibility and benefit. More contentious criteria, determined by experts and 'others', specify that ARC grant applications must also 'represent value for money' and meet the prescribed 'National Interest Test', a vague term frequently invoked by politicians in Australia, seeking support for domestic policy and political objectives.

Our grant application sought funding for an interdisciplinary study of Australian climate action by young people, mostly school students, associated with the School Strike 4 Climate. The 3-year project was designed to understand who these young people were, why they were mobilising and how they were expressing their political commitments. The project also aimed to generate vital new knowledge about how youth politics and political participation was evolving in a time of crises – and what the implications of this might be for contemporary democracies.

In the Weberian tradition of *wertfrei* (value free) social science, we took our political sympathies for student climate activists seriously (Hammersley, 2016).[1] Our political sympathies reflected our concern that Australia – a country highly dependent on fossil fuels – has one of the highest levels of per capita greenhouse emission in the world. Moreover, since the mid-1990s, every successive Australian government has behaved recklessly by failing to implement effective climate-change mitigation policies. That policy of stonewalling reflected decades of increasingly intense political conflict over climate-change policy. Our research proposal was also political since it reflected evidence that increasing numbers of young Australians played a major role in reviving contentious climate politics. Unbeknown to us, our project was connected to a much broader political context.

On 24 December 2021 we learned, via Twitter, of both our 'success' and failure. On the one hand, we were delighted to learn the ARC recommended our project for funding.[2] On the other hand, we were stunned to also read that the acting Federal Minister for Education, Stuart Robert, vetoed the ARC's recommendation that our project be supported (Collin et al., 2022). In addition to our own project, Minister Robert vetoed five others recommended for funding in that grant round. The only rationale offered by the minister for doing this was that they did 'not demonstrate value for taxpayers' money' or 'contribute to the national interest'. While contentious, this move was entirely legal according to the enabling legislation under which the ARC functioned, the *Australian Research Council Act* 2001, which provides for a ministerial veto. Further, Minister Robert's veto of projects recommended by the ARC for funding in 2021 was not the first time this happened: two earlier Liberal National Coalition Education Ministers had vetoed a total of 16 projects.

The Problem Outlined

Unlike previous ministerial interventions, Minister Robert's veto triggered widespread critical response. Senior university leaders and academics voiced their outrage. Tens of thousands of academics around the world signed two separate petitions condemning political censorship of humanities and climate research. The ARC College of Experts members declared they were 'angry and heartsore' and resigned from the College of Experts 'in protest'. Francis and Sims (2022) pointed out that the way three coalition ministers had vetoed ARC recommendations was not possible in other 'liberal-democracies like Canada, Britain, New Zealand and the United States'. Greens Senator Mehreen Faruqi initiated a Senate Inquiry.

Critics of Robert's veto emphasised how such ministerial interventions affected the capacity of universities to pursue significant, independent research. Many also drew attention to how these Ministerial vetoes undermined the ARC's review processes designed to enhance the capacity of universities to discharge their obligation to society as critical sites and 'autonomous institutions' above and beyond the reach of political authority. This notion is a central feature of the *Magna Charta Universitatum* (Magna Charta Universitatum Observatory, 1988) signed by hundreds of universities world-wide.[3] That charter declared that 'to meet the needs of the world around it, university research and teaching must be *morally and intellectually independent of all political authority and intellectually independent of all political authority and economic power*' (emphasis added). Ostensibly, even Australian universities operating under the Higher Education Support Act 2003 (pp. 19–115) are required to have a policy upholding 'free intellectual inquiry in relation to learning, teaching and research' as a condition of being registered as a university. So, how should we understand the force of this appeal to notions of university autonomy and academic freedom in a context where climate politics are contentious *and* polarised and where appeals to the rule of law principle run off in troubling ways?

Minister Robert's veto was a lawful exercise of his ministerial authority under the enabling legislation that had set up the ARC. Moreover, attempts to remove the ministerial veto power collapsed quickly in 2022. This was a reminder that the academic right to engage in autonomous scholarly practices was not favoured by lawmakers or *even by the ARC itself*. Of the 80 submissions received regarding a private member's bill to abolish the veto power, only three organisations opposed the bill: the ARC, the Federal Department of Education and the neo-liberal thinktank the Institute of Public Affairs. The bill was rejected by the senate committee with government senators arguing that 'removing ministerial discretion would affect the minister's ability to fulfil their obligations under the *Public Governance, Performance and Accountability Act* 2013'. Opposition ALP Senator Kim Carr concurred, arguing 'The ministerial veto contained in the ARC Act is a mechanism to facilitate the accountability of executive government' (Lu, 2022). Acknowledging her defeat, Senator Feruqi, who sponsored the bill, stated that 'Politics has trumped good policymaking as the government and Labor have refused to concede their political power to interfere with individual research grants' (Lu, 2022).

Politics was indeed at play at every stage of this saga, culminating in the ministerial veto. But how should we understand this idea that 'politics' was in play? How does this bode, especially for research into grassroots climate activism? What are the implications for activist-scholars and indeed for academic research that is concerned with advancing the capacity of humanity to respond ethically to a planetary crisis?

We start from the premise that a political crisis is playing out in the 'carbon capitalist economic system'. In this crisis, liberal-democracy is pitted *against* sustainability (Frankel, 2021). Liberal-democracy is a system in which every few years the citizens choose between various factions of a neo-liberal political elite which, since 1992, has enabled Australian governments to ignore global warming while blocking effective policies that would mitigate greenhouse gas emissions. What this means is that, centre-right or centre-left, Greens parties and various extra-parliamentary social movements which emerged from mid-2018 like the School Strike 4 Climate or Fridays for Future have failed to force the introduction of anything like the effective policies that would mitigate a climate catastrophe.

Suppose this is the central political crisis of our time. In that case, it foregrounds the need to address three questions that relate to the practice and research of grassroots political action: first, what does the ministerial veto of 2021 imply about the 'politics' of climate in Australia? Second, what does that veto and responses to it by Australian universities and others say about relations between the theory and practice of academic freedom, ideas of university autonomy and climate change, given Australia's liberal political tradition? Third: does the ministerial veto say anything about the contradictions operating in the liberal tradition which sees politics as the exercise of government informed and constrained by the rule of law principle?

The Veto . . .

Minister Robert never provided a public rationale for his decision to veto the ARC grant applications recommended for funding by the ARC. What we know about Robert, however, enables us to make some reasonable inferences.

Like a small number of other centre-right Australian politicians, Robert is a Pentecostalist. Robert appears to support the Seven Mountains mandate, a little-known Christian pentecostalist movement that aims to influence seven key areas of society. Those seven key sectors are: education, religion, family, business, government/military, arts/entertainment and media. Those who follow the Seven Mountains mandate believe that 'before Christ can return' the church must take control of these 'seven spheres for the glory of Christ'. Since 2017, supporters have organised an annual 'Church and State Summit', which brings together prominent Christian figures from politics, business, the media and academia. The annual Australian Church and State Summit is 'a gathering of Bible-believing Christians to discover and declare that the Kingdom of God is where the greatest opportunity for blessing and flourishing is found' (Church and State, 2023). The avowed aim of the Summit is to turn a 'culture of injustice and foolishness into a nation governed with justice and wisdom'. While the summit's website paints the event as neither politically right nor left, speakers at these Annual Summits have generally come from the far-right and include the 'A-listers'.[4]

The topics are predictable and include the need to respond to attacks on western civilisation, defend family values and traditional authority and call out the 'scientific conspiracy to exaggerate the threat of climate warming'. At the 2023 Summit speakers:

> railed against 'the transgender issue', abortion and climate action. They also referenced the Great Reset conspiracy alongside perceived attacks on Christian values and a decline of Western civilization.
>
> *(Denien, 2023)*

Under the like-minded leadership of Scott Morrison (Australian prime minister until mid-2022), Robert proved a staunch opponent of climate-change policies and a defender of the coalition government's 'climate policies'.[5] In 2022 he voted six times against climate legislation introduced by the Labor Albanese government (They Vote for you, 2022). It seems that with Morrison, Robert believes there is nothing humans can do to avert global warming: this is best left to God to address (Martin, 2021). Again, like Morrison, it is likely that Robert regards social media, heavily relied on by young climate activists, as the work of 'the evil one', in reference to the Devil (Martin, 2021). There is no reason to doubt that, given the opportunity to veto an ARC project on climate change, young people and their social media use, Robert did not want to let the opportunity to do God's work slip from his hands. Robert's intervention makes further sense considering a broader political culture in Australia that has been deeply polarised for more than four decades.[6]

The Contentious Character of Climate Politics

In 1992 it seemed to bode well that, following the first UN Earth Summit in Rio de Janeiro in 1992, the centre-left Paul Keating Labor government (1991–1997) expressed its support for strong, binding greenhouse gas emissions reduction targets with bi-partisan support. However, that political agreement eroded quickly and, since the mid-1990s, Australian governments of all persuasions have proved unable or unwilling to adopt policies likely to avert or even mitigate the increasingly serious threat posed by global warming.

The election of the conservative John Howard (Liberal/National Party coalition) government (1997–2007) saw climate politics become more polarised along ideological and party-political lines (Fielding et al., 2012; Manne, 2011), with the government first threatening to derail the Kyoto Protocol in 1997, before simply refusing to ratify it. Though the Kevin Rudd (Labour) government (2007–2011) ratified the Kyoto Protocol in 2007, Australia's climate policy collapsed to such a degree that it cost Rudd his leadership, and global warming became one of the most divisive issues in Australian politics (Copland, 2020). Having passed the *Clean Energy Act* 2011 to introduce a carbon pricing scheme, the Julia Gillard Labour Government lost the 2013 federal election, and Commonwealth government action on climate change thereafter ceased entirely.

Prime Minister Tony Abbott in the new conservative government (2013–2022) declared that 'climate science was crap'. The government denied the impact of climate change. It quickly passed the *Carbon Tax Repeal Act* in July 2014 to abolish the major policy mechanism introduced by the former labour government to address climate change. The Abbott government and successive conservative governments blocked any serious policy measures designed to reduce Australia's reliance on fossil fuels, including the export of coal and gas or the upscaling of Australia's energy distribution networks to accommodate distributed renewable power systems. It is difficult not to conclude that a lack of climate leadership from the major political parties has meant that for decades policies designed to reduce greenhouse gas emissions fell radically short of what was needed to avert catastrophe (UNEP, 2021).

This record of policy inertia reflects the success of a campaign by a 'climate-change counter-movement' involving carbon-based industries, together with peak bodies, think tanks, publishers, media outlets and politicians to capture the state (Lucas, 2021). Lucas' empirical account demonstrates how covert networks of political influence have been carefully constructed over many years by Australia's fossil fuel and resource extraction industries to further their own political and financial interests and to block effective climate-change policies.[7] The result is a case of state capture by the fossil fuel industry (Lucas, 2021). This is evidenced by continuous expansion of the carbon industries and ever-increasing greenhouse gas emissions. State capture is also signified by the way the fossil fuel and resource extraction industries have managed to successfully repeal a national carbon price, prevented the introduction of policies promoting electric vehicles, demonised efforts to phase out coal exports, and overcome a state-wide moratorium on coal seam gas extraction. They have also been instrumental in weakening environmental regulations, strengthening anti-protest laws to directly curtail social movements and ensuring the removal of two insufficiently compliant prime ministers (Kevin Rudd (ALP) and Malcolm Turnbull (Coalition) (Wilkinson, 2020). In addition to this, the ability of corporations to shape the policy of governments of all stripes, and, in the case of labour, the power of workers and unions employed in the carbon industries is not accidental. Rather, it results from a large, successful strategy of a covert network of lobbyists intent on thwarting the development of a general consensus supporting urgent policy action to combat climate warming (Kelly, 2019). While a clear majority of Australians (60%) want urgent action (Lowy Institute, 2018) the polarised political and public debate on how to address climate change has been further exacerbated by powerful media players encouraging doubts or scepticism about the science of climate change to avert effective climate-change mitigation policies (e.g., Manne, 2011; Beeson and McDonald, 2013) and demonising grassroots activists. As a result, the broader public debate about the policy questions posed by climate science has also been equally polarised.

In short, the past decades have seen a successful and well-coordinated political campaign designed to block effective climate-change mitigation policies and spread enough doubt about climate science to create a polarised climate politics. This raises questions about the politics of academics upset by the 2021 ministerial veto of climate activism research.

The Fugitive Status of Academic Freedom

Academics in Australian universities have long promoted a conception of 'academic freedom' as central to any defensible idea of the university (Evans and Stone, 2021; Swannie, 2022). Many understood this as requiring that universities be free from legal regulation and intrusion by the state, even as such a notion became unrealistic. There are two points to be made about the idea that Australian universities are autonomous entities and that

Australian academics either value or enjoy 'academic freedom' understood as freedom from government regulation and/or as 'freedom of speech'.

In November 2018, the Morrison coalition government commissioned Robert French, a former Chief Justice of the High Court and the Chancellor of University of Western Australia, to report on the state of 'academic freedom' in Australian universities. The review was in response to provocations from right-wing commentators that a network of 'woke' Marxists, feminists and post-modernists were cancelling people who were not 'woke' thereby creating a crisis of 'free speech' (e.g., Lesh, 2018; Albrechtsen, 2020) – of which French (2019) found no evidence. French pointed out that academics may enjoy 'academic freedom' *up to a point*, but there is no absolute 'right to free speech' in Australia or in its universities. This distinction has been made by others, for example, in America (e.g. Scott, 2017; Post, 2017). French remains wedded to a positivist liberal legal tradition and did not recommend introducing a bill guaranteeing a right to 'academic freedom' and 'free speech'. French reminds us that the Commonwealth government regulates the higher education sector through state and territory regulatory statutes. The best that French recommended be done is that universities voluntarily adopt a Model Code embedded in higher education providers' institutional regulations or policies, a draft version of which he provided.

More generally, according to Gelber (2002, 2019) Australia is an odd liberal-democracy where freedom of speech leads a fugitive existence. Australia has no Charter of Human Rights Act or constitutional Bill of Rights. Even so it signed off on the UN International Covenant on Civil and Political Rights (ICCPR) which requires that Australia introduce the ICCPR as domestic legislation. Doing this would have built in some protection for the right to freedom of speech. The High Court found an 'imputed right to free political expression' while others appeal to the notion that Australian common law protects free speech (High Court of Australia 1992; Cass, 1993, p. 230).[8] However, as climate activists know, governments repeatedly introduce legislation (including everything from the *Public Service Act* 1999 to a host of anti-discrimination laws (as will be discussed later) abrogating rights to freedom of political speech or expression by public servants or citizens. The High Court routinely finds this kind of legislation to be a lawful exercise of government authority. However, a more crucial question remains: to what extent can academics in the modern university practice academic freedom? All this brings us to our second point.

Australian universities have been subjected to a neoliberal policy assault since 1988–1989 (Bessant, 1995; Kayrooz et al., 2001). There is now a sizable literature on the 'neoliberal effect' in Australian higher education – some considering it benign (Marginson and Considine, 2000); most are more critical (Hil, 2015; Watts, 2016; Connell, 2019). What Connell (2019) calls a 'neoliberal policy cascade' heralded deep cuts in government-funding as a prelude to the expectation that universities would increase their student intakes and fund that increase by reintroducing tuition fees backed by a student loan scheme. But has that policy assault affected the idea that academics are committed to 'nurturing critical thought' and 'advancing knowledge' or that form of 'academic freedom' best expressed as 'public scholarship'?

On one reading, 'public scholarship' involves scholarly and creative work which produces 'public goods' like accessible and valuable research and transformative teaching (Cohen and Yapa, 2005, p. 1) How well do the actual practices of Australian academics conform with this notion? A case could be made for an ever-widening gap between the theory and practice of 'public scholarship' (Andrews, 2007). However, we need to bypass the nostalgic narrative that once upon a time we had 'real universities' and now we don't

(Coady, 2000; Gaita, 2012). This idea of academia is characterised by elitism and social irrelevance threatened by its nemesis, a mass, instrumental institution working in servitude to the market. Academic teaching or research are treated as private matters, best conducted 'outside of the public gaze and at a distance from public affairs' (Andrews, 2007, p. 61).

Instead, 'public scholarship' is a practice which discharges an obligation framed as positive freedom to contribute to various public goods – and which is prepared to accept that this is a deeply political exercise defined in Sluga's (2014) terms as contested conversations about the good life and the just society. On this account the work of teachers and students is framed 'not as the isolated, self-indulgent actions of a campus segregated from society, but as the contributions of scholar-citizens with membership in a larger community' (Andrews, 2007, p. 1) We need to imagine a practice of academic freedom conceived less as being *free from* interference and much more in terms of being 'free to' pursue public goods by engaging in those practices constitutive of critical inquiry which involve, in Arendt (2000) and Foucault's (2001) terms, speaking 'truth to power'. Is this, however, something our liberal-democracy wants or supports in its researchers, including with regard to sustainability, climate change or activism on any other politically contentious issue?

The Crisis of Liberalism and Its Legitimacy

The crisis which pits liberal democracy against sustainability highlights a crisis within liberal-democracies. The idea that liberal-democracies value tolerance, freedom of speech and difference has long been commonplace in political theory. Western liberalism, which begins with Hobbes and Locke, relied on the idea that citizens have a right to express diverse religious and other opinions. Courtesy of Constant and J.S. Mill in the nineteenth century and modern liberals like Berlin (1969) and Rawls, this evolved into the proposition that human progress was best served by protecting and encouraging a plurality of attitudes and ways of living (Nederman and Larsen, 1996, pp. 2–3). However, as we discovered, Australia's liberal state has little taste for publicly funded inquiry that may illuminate dissent.

The polarisation of Australian climate politics has been characterised by persistent efforts to denigrate and even criminalise dissent by climate activists and those who seek to understand or support them. The result has been a growing reliance on the security-state to restrain climate activism. Increasing powers for police and national security forces have enabled a greater range of surveillance and control measures to be applied to protesters. While such measures are said to be justified due to potential terror threats, the reality is that most climate protests have been peaceful and generally non-disruptive. This is true of Extinction Rebellion (XR) and Fridays for Future, new grassroots movements offering new kinds of protest.

Four states of Australia (NSW, Victoria, Queensland and Tasmania) have all recently criminalised protest or significantly increased penalties for existing offences. Queensland was the first to pass such laws in 2019, outlawing activities that disrupted traffic and the use of devices that might otherwise hinder the clearing of roads. In New South Wales (NSW) similar legislation (NSW Government, 2022) established a special Strike Force Guard and allows police to arrest and fine anyone proceeding on a transport thoroughfare with the intent to disrupt traffic unless authorised by the police. The NSW legislation has been used against protesters blocking rail links and in 2022–2023 to conduct active surveillance of gathering protesters. Victoria's Andrews Labor government introduced the Sustainable Forests Timber Amendment (Timber Harvesting Safety Zones) Bill 2022 targeting forest activists by sanctioning anyone attempting to prevent native forest logging. The Tasmanian Police Offences Amendment (Workplace Protection) Bill provides for large fines and prison time for protestors who disrupt workplaces. This legislation breaches the UN International Covenant on Civil and Political Rights (ICCPR) to which Australia is a signatory. At the time of writing, Australia has neither a parliamentary Charter of Human Rights nor a constitutional Bill of Rights. This anomaly leaves the status of Australians' right to freedom of political speech in limbo.

Further, why wouldn't such states encourage research that aims to understand engagement and dissent of all kinds? One answer is that the liberal tradition has drawn excessively on theorists like Rawls and Habermas, treating consensus as more important than dissent (Anderson, 2005, p. xv). Another answer is that, because liberal-democratic political systems are said to provide enough politically expressive channels like open media, freedom of expression, or free and open elections, there is no need for dissent or protest. However, this ignores the evidence adduced earlier of the possibility that states have been captured by special interests like fossil fuel industries intent on shutting down genuinely contested politics (Lucas, 2021).[9] That notwithstanding and given that liberal-democratic governments are supposed to uphold the rule of law and protect everyone's rights to free speech and other civic or human rights, anyone who openly challenges these 'rules of the game' is up to no good. Worse, people who ignore or bypass the available channels must be doing so because they are 'irrational', 'deviant', criminally inclined, or worse, treasonous. At stake here is the implied notion that citizens should accept the legitimacy of the liberal-democratic state and always obey the state and its laws.

Yet, calling for compliance with the principle of legitimacy cannot make these kinds of contradictions operating in liberal-democracies disappear (Watts, 2020). The evidence shows that in Australia successive governments have been protecting fossil fuel interests, not people's interests – even though fossil fuel interests pose the greater threat to the public. The politics of climate reveal that academic freedom or university autonomy are meaningless ideas and that the political always trumps the principle said to define the liberal ethos: the rule of law.

In our own time, some theorists noticed how the liberal rule of law actually works. Modern liberal-democracies use the law to do what they want to do: they can enact legislation enabling them to do what they want even if it egregiously breaches fundamental human rights. What we might call 'the political' trumps the law by using the law. Government policymaking and legislation provides plenty of legal 'black holes' and legal 'grey holes' (Dyzenhaus, 2006; Vermeule, 2009). Legal black holes 'arise when statutes or legal rules either explicitly exempt the executive from the rule of law requirements or explicitly exclude judicial review of executive action' (Dyzenhaus, 2006, p. 3). Grey holes 'arise when there are some legal constraints on executive action . . . but the[y] are so insubstantial that they pretty well permit government to do as it pleases' (Dyzenhaus, 2006, p. 42). Both kinds of holes are instances of the political will of sovereign power: allowing the executive arm of government to do as it pleases – to activists and scholars alike.

This is how we need to understand the legislative basis for the ministerial veto: the Australian state persistently uses the rule of law to protect the political will of the executive from being required to uphold the rule of rights and freedoms.

Conclusion

Research on grassroots climate activism always takes place in broader contexts in which the politics of research funding, of the researchers themselves, industry, community and the government of the day are inextricably bound. Contrary to the myth of liberal-democracy, western governments generally prefer consensus to dissensus and promote obedience to the law by stigmatising or criminalising those who dissent. The veto by the Australian government and the defence of its use by all the major political parties in 2022 reminds us that liberal-democracy is antagonistic to sustainability. This is an insight Australian academics interested in exercising their freedom of speech, their academic freedom on questions like climate change, or other contentious issues need to reflect on. While Australia claims to be a liberal-democracy, its citizens have never been guaranteed any serious right to freedom of political expression, while the enjoyment of academic freedom ended many decades ago. If anything, the ministerial exercise of a right to veto an ARC-funded research project addressing young people's climate-change activism was a reminder that governments regularly rely on legal 'black' and 'grey' holes to negate the rights of citizens – or in this case those of academics (Mann, 2018). The exercise of the Ministerial veto that cancelled the ARC's recommendation that this project be funded, like the recent passage of anti-protest laws, is a reminder that these laws are actually 'spaces where law and liberal proceduralism speak and operate *in excess*' (Johns, 2005, p. 631 (emphasis added).[10] Research can – and should – play a key role in explaining, advancing and amplifying grassroots activism for action on climate change. This inevitably involves a reflexive practice that reckons directly with the politics of denial, deferral and delay.

Acknowledgments

After the veto we revised and resubmitted the research application 'New Possibilities: Young People and Democratic Renewal' which was funded by the Australian Research Council Discovery Program Grants (DP230101704) in the 2022 round, following a change of government.

We would like to thank the student collaborators who have contributed to this research.

Notes

1 As Martyn Hammersley notes, Weber emphasised that his advocacy for value-free social science does not imply that research should be neutral in relation to – or free from influence by – all values. He clearly argued that science is necessarily committed to epistemic values while social inquiry must depend upon the adoption of particular value-relevance frameworks (Hammersley, 2016).
2 The success rate in 2021 for this form of research grant was 17%.
3 In Europe, where the *Magna Charta Universitatum* (1988) originated in countries like France, Italy and Germany, it operates on the basis that academics are civil servants. Life tenure, for example, ensures that as civil servants academics promote the neutrality of public administration and the legal and economic independence of civil servants (Anzivino et al., 2020).
4 Speakers include Murdoch columnist Miranda Devine, Queensland Senator Amanda Stoker, Senator Eric Abetz, Queensland LNP state MP Fiona Simpson, Cardinal George Pell, Australian Christian Lobby head Martyn Iles and Senator Matt Canavan, who proudly declared that he had taken 'a public role in arguing for the biblical definition of marriage during the public debate on the topic in 2017'. Former Sydney media star Alan Jones provides infrastructure support to the summit.
5 Of Robert's 748 speeches recorded in Hansard between 2007–2023, the only Parliamentary speech Robert made on the climate issue involved him defending the Morrison government's record on climate policy in early 2023.

6 Robert's parliamentary register of interests records that his family and his self-managed superfund held shares in 18 ASX-listed companies, nearly half of which were in mining companies. Other irregularities were discovered later. He gave a parliamentary speech largely written by a Gold Coast developer in 2012 and failed to register receiving two Rolex watches worth $50,000 each in 2013 from a Chinese billionaire. In 2017 it was revealed he had put his parent's names on company documents as company directors without their knowledge.

7 Lucas' work is based on careful tracking of political donations, lobbying and revolving-door employment by 160 current or former senior politicians, political staffers and bureaucrats who currently or formerly held key positions in the energy and resource portfolios and who had also previously or subsequently worked for a fossil energy, electricity, or resource company or industry peak body.

8 In 1992 the Australian High Court (in *Australian Capital Television and Others v. The Commonwealth* (1992) 177 CLR 106), held that because the Australian Constitution enshrined a system of representative government, this implied the right to freedom of communication on 'political' matters.

9 Lucas notes that the energy and resource industries are reluctant to publicly admit they are 86% foreign owned. It is noteworthy that the 'Big Four' accountancy firms, (KPMG, PwC, Deloitte and Ernst and Young) were paid more than $3.1 billion in consultancy fees by the Federal Coalition Government between 2012 and 2018 replacing the public service workforce courtesy of successive waves of austerity and 'rationalization' (Lucas, 2021, p. 4).

10 Let one shocking instance stand as evidence of this generalisation. In March 2023, the *Guardian* revealed that a 13-year-old Indigenous child held on remand in Townsville's Cleveland youth detention centre had been held in solitary confinement for 45 days, with one stretch lasting 22 days. At the time, Queensland's 'lawmakers' were debating government youth justice laws, including plans to bypass Queensland's Human Rights Charter and criminalise bail breaches by children.

References

Albrechtsen, J. (2020). Intellectual freedom? Only if your values are 'aligned'. *The Australian*, 20 February. https://www.theaustralian.com.au/inquirer/intellectual-freedom-only-if-your-values-are-aligned/news-story/6665eedc434bc01622e80b2b8859ef9a

Anderson, P. (2005). Arms and Rights, *New Left Review*, 31 (Jan-Feb)

Andrews, S. (2007). *'Compromising Positions': Relations of Power and Freedom in the Australian University*. Unpublished PhD Thesis. Melbourne: RMIT University.

Anzivino, M., Ceravolo, F. and Rostvan, M. (2020). The two dimensions of Italian academics' public engagement. *Higher Education*, 81, 107–125.

Arendt, H. (2000). Truth and politics. In P. Baier (ed.) *The Portable Hannah Arendt* (pp. 296–313). New York: Vintage.

Arnot, G., Thomas, S., Pitt, H. and Warner, E. (2023). "It shows we are serious": Young people in Australia discuss climate justice protests as a mechanism for climate change advocacy and action. *Australian and New Zealand Journal of Public Health*, 47(3), 1–7.

Beeson, M. and McDonald, M. (2013). The politics of climate change in Australia. *Australian Journal of Politics and History*, 59(3), 331–348.

Berlin, I. (1969). Two concepts of liberty. In I. Berlin (ed.) *Four Essays on Liberty* (pp. 118–172). London: Oxford University Press.

Bessant, J. (1995). Corporate management and its penetration of university administration and government. *Australian Universities Review*, 1, 59–62.

Bessant, J., Collin, J. and Watts, R. (2023). 'Blah, Blah, Blah. . .[not] business as usual': Politics through the lens of young female climate leaders. *Australian Journal of Political Science*. http://doi.org/10.1080/10361146.2023.2224239

Bourdieu, P. and Wacquant, L. J. D. (1992). *An Invitation to Reflexive Sociology*. Chicago: University of Chicago Press.

Burgmann, V. and Baer, H. (2012). *Climate Politics and the Climate Movement in Australia*. North Melbourne: Melbourne University Press.

Cass, D. (1993). Through the looking glass: The high court and the right to free speech. *Public Law Review*, 4, 229–247.

Catanzaro, M. and Collin, P. (2021). Kids communicating climate change: Learning from the visual language of the SchoolStrike4Climate protests. *Educational Review*, 75(1), 9–32.

Chesters, G. (2014). Social movements and the ethics of knowledge production. In K. Gillan and J. Pickerill (eds.) *Research Ethics and Social Movements: Scholarship, Activism and Knowledge Production*. Oxford: Routledge.

Church and State (2023). https://www.churchandstate.com.au

Coady, A. (2000). Universities and the ideals of inquiry. In A. Coady (ed.) *Why Universities Matter* (pp. 73–84). Sydney: Allen & Unwin.

Cohen, J. and Yapa, Y. (2005). Introduction: What is public scholarship? *Blueprint for Public Scholarship at Penn State.* www.publicscholarship.psu.edu/PDFs/blueprint.pdf/

Collin, P. (2015). *Young Citizens and Political Participation in a Digital Society: Addressing the Democratic Disconnect.* Basingstoke: Palgrave Macmillan.

Collin, P., Churchill, B., Gordon, F., Bessant, J., Cantanzaro, M., Watts, R. and Jackson, S. (2022). 'Disappointment and disbelief' after Morrison government vetoes research into student climate activism. *The Conversation*, 13 January. https://theconversation.com/disappointment-and-disbelief-after-morrison-government-vetoes-research-into-student-climate-activism-174699

Collin, P. and Matthews, I. (2021). School strike for climate: Australian students renegotiating citizenship. In J. Bessant, A. Mejia Mesinas and S. Pickard (eds.) (2020) *When Students Protest, Vol. 1: Secondary School Students* (pp. 125–144). Lanham: Rowman and Littlefield.

Collin, P., Matthews, I., Churchill, B. and Jackson, S. (2020). Australia. In Joost de Moor, Katrin Uba, Mattias Wahlstrom, Magnus Wennerhag and Michiel De Vydt (eds.) *Protest for a Future II: Composition, Mobilization and Motives of the Participants in Fridays for Future Climate Protests on 20–27 September, 2019, in 19 Cities Around the World* (pp. 35–51). Australia: Centre for Open Science.

Commonwealth of Australia (2003). *Higher Education Support Act 2003 No. 149, 2003 Compilation No. 72.* https://www.legislation.gov.au/Details/C2020C00197.

Connell, R. (2019). *The Good University: What Universities Actually Do and Why It's Time for Radical Change.* London: Zed Books.

Copland, S. (2020). Anti-politics and global climate inaction: The case of the Australian carbon tax. *Critical Sociology*, 46(4–5), 623–641.

Crowley, C. (2020). The politics of the environment in Australia. In J. Lewis and A. Tiernan (eds.) *The Oxford Handbook of Australian Politics.* Hobart: University of Tasmania.

Denien, M. (2023). 'Long March': The right wing Christian plan to Infiltrate politics. *Sydney Morning Herald*, 6 March. https://www.smh.com.au/politics/federal/long-march-the-right-wing-christian-plan-to-infiltrate-politics-20230306-p5cpod.html

Dyzenhaus, D. (2006). *The Constitution of Law: Legality in a Time of Emergency.* Cambridge: Cambridge University Press.

Evans, C. and Stone, A. (2021). *Open Minds: Academic Freedom and Freedom of Speech in Australia.* Bundoora: La Trobe University Press.

Feldman, H. (2021). Motivators of participation and non-participation in youth environmental protests. *Frontiers in Political Science*, 3, 662687.

Fielding, K., Head, B., Laffan, W., Western, M. and Hoegh-Guldberg, O. (2012). Australian politicians beliefs about climate change: Political partisanship and political ideology. *Environmental Politics*, 21(5), 712–733.

Foucault, M. (2001). *Fearless Speech.* Edited by J. Pearson. Los Angeles: Semiotext(e).

Francis, A. and Sims, A. (2022). We resigned from the ARC college of experts – political meddling gave us no choice. *The Guardian*, 2 February. https://www.theguardian.com/commentisfree/2022/feb/02/we-resigned-from-the-arc-college-of-experts-political-meddling-gave-us-no-choice

Frankel, B. (2021). *Democracy Versus Sustainability: Inequality, Material Footprints and Post Carbon Futures.* Melbourne: Greenmeadows.

French, R. (2019). *Independent Review of Freedom of Speech in Australian Higher Education Providers.* Canberra: DESE.

Gaita, R. (2012). To civilize the city? *Meanjin*, 71(1), 29–49.

Gelber, K. (2002). *Speaking Back: The Free Speech Versus Hate Speech Debate.* Amsterdam: John Benjamins.

Gelber, K. (2019). The precarious protection of free speech in Australia: The *Banerji* case. *Australian Journal of Human Rights*, 25(3), 511–519.

Gillan, K. and Pickerill, J. (2012). The difficult and hopeful ethics of research on, and with, social movements. *Social Movement Studies*, 11(2), 133–143.

Goodman, J. and Morton, T. (2023). Climate movements in Germany, India, and Australia: Dynamics of transition, transformation, and emergency. *Globalizations*. http://doi.org/10.1080/147477 31.2023.2215101

Gulliver, R., Banks, R., Fielding, K. and Louis, W. (2023). The criminalization of climate change protest. *Contention*, 11(1), 24–54.

Hammersley, M. (2016). On the role of values in social research: Weber vindicated? *Sociological Research Online*, 22(1), 7–21.

Hee, M., Jürgens, A. S., Fiadotava, A., Judd, K. and Feldman, H. R. (2022). Communicating urgency through humor: School Strike 4 climate protest placards. *Journal of Science Communication*, 21(5), A02.

Hendriks, C. and Reid, R. (2022). The rise and impact of Australia's movement for community independents. In A. Gauja, M. Sawer and J. Shepherd (eds.) *Watershed: The 2022 Australian Federal Election*. Canberra: ANU Press.

High Court of Australia. (1992, September 30). *Australian Capital Television Pty Ltd & New South Wales v Commonwealth [1992] HCA 45*. CLR 106, 177.

Hil, R. (2015). *Selling Students Short: Why You Won't Get the University Education You Deserve*. Sydney: Allen and Unwin.

Hilder, C. and Collin, P. (2022). The role of youth-led activist organisations for contemporary climate activism: The case of the Australian Youth Climate Coalition. *Journal of Youth Studies*, 25(6), 793–811.

Hintjens, H. (2000). Environmental direct action in Australia: The case of Jabiluka Mine. *Community Development Journal*, 35(4), 377–390.

Hutton, D. and Connors, L. (1999). *A History of the Australian Environment Movement*. New York: Cambridge University Press.

Jacques, P., Dunlap, R. and Freeman, M. (2008). The organisation of denial: Conservative think tanks and environmental scepticism. *Environmental Politics*, 17(3), 349–385.

Johns, F. (2005). Guantánamo Bay and the Annihilation of the exception. *European Journal of International Law*, 16(4), 613–635.

Judd, B. (2022). Young people are on front line of climate change, but they're not accepting it as an inevitability. *ABC News Online*, 26 April. https://www.abc.net.au/news/2022-04-26/young-people-climate-change-what-does-future-hold/100950770

Kayrooz, C., Kinnear, P. and Preston, P. (2001). *Academic Freedom and Commercialisation of Australian Universities: Perceptions and Experiences of Social Scientists*. Discussion Paper Number 37 (March). Canberra: Australia Institute.

Kelly, D. (2019). *Political Troglodytes and Economic Lunatics: The Hard Right in Australia*. Bundoora: LaTrobe University Press.

Kelly, J., Sadeghieh, T. and Adeli, K. (2014). Peer review in scientific publications: Benefits, critiques, & a survival guide. *EJIFCC*, 25(3), 227–243.

Lee, M. (2021). Policing the pedal rebels: A case study of environmental activism under COVID-19. *International Journal for Crime, Justice and Social Democracy*, 10(2), 156–168.

Lesh, M. (2018). *Free Speech on Campus Audit 2018*. Melbourne: Institute of Public Affairs.

Lowy Institute (2018). @018 Lowy Institute Poll. https://www.lowyinstitute.org/publications/2018-lowy-institute-poll

Lu, D. (2022). Federal bill to prevent ministers vetoing research grants rejected by Senate committee. *The Guardian*, 22 March. https://www.theguardian.com/australia-news/2022/mar/22/federal-bill-to-prevent-ministers-vetoing-research-grants-rejected-by-senate-committee

Lucas, A. (2021). Investigating networks of corporate influence on government decision-making: The case of Australia's climate change and energy policies. *Energy Research and Social Science*, 81.

Magna Charta Universitatum Observatory (1988). https://www.magna-charta.org.

Mann, I. (2018). Maritime legal Black Holes: Migration and rightlessness in international law. *European Journal of International Law*, 29(2), 347–372.

Manne, R. (2011). Bad news: Murdoch's *Australian* and the shaping of the nation. *Quarterly Essay*, 43(September).

Marginson, S. and Considine, M. (2000). *The Enterprise University: Power, Governance and Reinvention in Australia*. Melbourne: Cambridge University Press.

Martin, S. (2021). Scott Morrison tells Christian conference he was called to do God's work as prime minister. *The Guardian*, 26 April. https://www.theguardian.com/australia-news/2021/apr/26/scott-morrison-tells-christian-conference-he-was-called-to-do-gods-work-as-prime-minister.

Mayes, E. and Center, E. (2023). Learning with student climate strikers' humour. *Environmental Education Research*, 29(4), 520–538.

Mayes, E. and Hartrup, M. (2022). News coverage of the school strike for climate movement in Australia: The politics of representing young strikers' emotion. *Journal of Youth Studies*, 25(7), 994–1016.

Nederman, C. and Larsen, J. (1996). *Difference and Dissnet: Theories of Toleration in Medieval and EARLY MODERN EUROPE*. Lanham, MD: Rowman and Littlefield.

NSW Government (2022). *Roads and Crimes Legislation Amendment Act, 2022 No 7*. https://legislation.nsw.gov.au/view/pdf/asmade/act-2022-7

Obama, B. (2014). *2014 state of the union address*. Address to the Joint Houses, 28 January. https://obamawhitehouse.archives.gov/the-press-office/2014/01/28/president-barack-obamas-state-union-address.

O'Neill, S. J. (2013). Image matters: Climate change imagery in US, UK and Australian newspapers. *Geoforum*, 49, 10–19.

Pearse, R., Goodman, J. and Rosewarne, S. (2010). Researching direct action against carbon emissions: A digital ethnography of climate agency. *Cosmopolitan Civil Societies: An Interdisciplinary Journal*, 2(3), 76–103.

Plehwe, D. (2021). Think tanks and the politics of climate change. In D. Abelson and C. Rastrick (eds.) *Handbook on Think Tanks in Public Policy* (pp. 150–165). Cheltenham: Edward Elgar Publishing.

Post, R. (2017). The classic first amendment traditions under stress: Freedom of speech and the university. *Yale Law School Public Law Research Paper no. 619*. https://papers.ssrn.com/sol3/papers.cfm?abstract_id=3044434

Riach, K. (2009). Exploring participant-centred reflexivity in the research interview. *Sociology*, 43(2), 356–370.

Rosewarne, S., Goodman, J. and Pearse, R. (2013). *Climate Action Upsurge: The Ethnography of Climate Movement Politics*. Abingdon: Routledge.

Scott, J. (2017). On free speech and academic freedom. *American Academy of Arts and Sciences*. Summer. https://www.amacad.org/news/free-speech-and-academic-freedom

Sluga, H. (2014). *Politics and the Search for the Common Good*. Cambridge: Cambridge University Press.

Swannie, B. (2022). Protection from institutional censorship: An essential aspect of academic freedom. *UNSW Law Journal*, 45(4), 1489–1512.

They vote for You. (2022). *Stuart Robert Voted Consistently against the Paris Agreement*. https://theyvoteforyou.org.au/people/representatives/fadden/stuart_robert/policies/183

Touraine, A. (1981). *The Voice and the Eye: An Analysis of Social Movements*. Translated by A. Duff. New York: Cambridge University Press.

Turner, B. (2021). Simon Holmes à Court: 'If it works, the payoff will be enormous'. *Sydney Morning Herald Good Weekend*, 30 October. https://www.smh.com.au/national/simon-holmes-court-if-it-works-the-payoff-will-be-enormous-20210916-p58sbn.html

UNEP (United Nations Environment Programme). (2021). Emissions gap report 2021: The heat is on – a world of climate promises not yet delivered. *Nairobi*. https://www.unep.org/resources/emissions-gap-report-2021

Verlie, B. (2021). *Learning to Live with Climate Change: From Anxiety to Transformation*. Abingdon: Routledge.

Vermeule, A. (2009). Our Schmittian administrative law. *Harvard Law Review*, 122, 1098–1152.

Watts, R. (2016). *Public Universities, Managerialism and the Value of Higher Education*. London: Palgrave-Macmillan.

Watts, R. (2020). *Criminalising Dissent: Liberal States, Dissent and the Problem of Legitimacy*. Abingdon: Routledge.

Wilkinson, M. (2020). *The Carbon Club*. Melbourne: Allen & Unwin.

PART V

Grassroots Critical Perspectives on Climate Actions

Introduction to Part V: Grassroots Critical Perspectives on Climate Actions

The climate movement is not essentialist or monolithic, nor will it stop evolving: it is characterised by continual exploration, critical reflexivity and engagement. A crucial feature of climate activism is the ongoing efforts of activists to grapple with the climate crisis in all of its complexity. Moreover, it is often at the grassroots where critiques are articulated and new directions are initiated. In this part of the handbook, we zoom in on some of these critical perspectives: critiques of large-scale renewable energy projects in Global South countries that often disregard local community input and consent and seem mainly to serve to prolong excessive resource use by Global North countries (Zeglam; Groupp); the fact that, in some contexts, environmental activism cannot be assumed to be "progressive," given how it is imbricated with right-wing ideologies (Ellis and Ichinkhorloo) or when it has to find ways to form coalitions outside of government influence (Karakadzai; Chirisa); and the climate movement's relative failure to embrace grassroots initiatives against animal agriculture and its problematic embrace of regenerative agriculture (Voss-Andreae).

In Chapter 25, "Grassroots Climate Activism in North Africa: Local, Personal and Radical," Haneen Ali Zeglam begins with an examination of the climate crisis and the diverse forms of grassroots activism that have emerged in response in North Africa. In regions with limited civic engagement due to political repression, grassroots systems of organisation are crucial sites of resistance, and they often operate outside traditional state and civil society structures. These activists connect local struggles with global challenges, confronting issues of systemic failure, privatisation, corruption, and diminished accountability. Zeglam's work contributes an important critique to the global climate agenda, as it exposes the impact of large-scale renewable energy projects on local communities and environments. The chapter also explores the complexities of advocating for improved local environmental conditions while promoting alternative visions for a more equal and decentralised future.

In Chapter 26, "Transnational Grassroots Networks and Renewable Energy Governance," Emilia Groupp presents a synthesis of her ethnographic research on the renewable energy industry and 31 case studies of grassroots activism focused on large-scale wind development. In so doing, Groupp argues that the specific political and economic landscapes of climate change activism and renewable energy development have engendered

DOI: 10.4324/9781003396567-29

opportunities for historically marginalised actors to leverage new forms of political power, based around shared concerns with the environmental and social impacts of large-scale renewable energy sites. She describes these networks as having predominantly been formed among environmental organisations, labour unions, and rural stakeholders whose environments, livelihoods, and social worlds stand to be significantly altered by new sites of energy extraction. This chapter explores the kinds of transnational grassroots networks that have emerged, how they impose different forms of governance over an increasingly transnational renewable energy industry, what power dynamics emerge, and in what ways these networks challenge the seemingly dominant influence of international financial institutions and multinational energy companies.

In Chapter 27, "Community-led Responses to Climate-induced Disasters in Zimbabwe: Towards a Politico-community-based Model" Thomas Karakadzai and Innocent Chirisa study urban, rural, and informal settlements in Zimbabwe to describe the struggle of communities which are already experiencing great hardship as a result of climate change. From 1900 to 2020, Zimbabwe saw a range of natural disasters – including droughts, epidemic diseases, floods, and storms – and thus great need for action. Yet, while state and non-profit organisations were managing the crises with limited success, the government, which has its own ideas about how to solve the problem, views grassroots climate activists as antagonists. The authors explain how, as a consequence, civil society organisations have had to form nongovernmental coalitions with other stakeholders to raise awareness of the need for climate change activism and more effective climate action.

In Chapter 28, "Climate Activism and Environmental Politics in Mongolia", Joe Ellis and Byambabaatar Ichinkhorloo provide a case study exploring the complex interplay between extractive industries and Mongolian modes of living, which are situated within diverse landscapes, from mountains to steppes and deserts. The chapter reveals how environmental activism emerges from acts seen as transgressions against sacred land, often triggered by mining activities that disrupt "traditional" landscapes, and it sheds light on the intersections between environmental movements and right-wing political networks in Mongolia. Highlighting ethnography's role in challenging prevalent assumptions, the authors problematise the relationship between climate activism and progressive politics. They also outline the need for decolonial frameworks that can help us to understand the struggles and agendas of Mongolian climate activists, and more broadly, the plural contexts of Global South and Indigenous geographies. As such, Ellis and Ichinkhorloo speak to the importance of thinking critically about the various kinds of politics that are being engendered as the climate crisis unfolds.

In Chapter 29, "The Climate Movement and Animal Agriculture; Grassroots vs Greenwashing" Adriana Voss-Andreae critically considers the role of grassroots climate activist organisations, such as Fridays for Future Germany, in shifting social norms and building political momentum toward reducing meat consumption and production. Her analysis of 20 US- and Europe-based climate or environmental organisations' online materials finds that, while the majority recognise the need for food system transformation, their approaches vary widely, and while most acknowledge industrial animal agriculture as a major contributor to environmental degradation, only a third advocate for a transition to a more plant-based food system, despite the scientific consensus on it being imperative. The chapter concludes that transforming the food system requires challenging entrenched corporate power, redefining social norms, and aligning principles of justice, equality, and compassion. Throughout the chapter, Voss-Andreae highlights the role that grassroots climate and environmental organisations can play in addressing the significant impact of animal agriculture on the climate and ecological crises.

Each chapter in this part brings a distinct perspective, illuminating the growing constellation of trends and challenges in grassroots climate activism. Overall, they offer a rich tapestry of grassroots activism across different geographies, revealing its diversity and dynamism. They showcase the innovative tactics, unique challenges, and varying outcomes of grassroots activism, underscoring the importance of a holistic, polysemic, and inclusive approach. As such, they speak to how this handbook showcases how grassroots climate activism emerges, evolves, and interacts with unique social and environmental landscapes (Apostolopoulou et al., 2022).

Reference

Apostolopoulou, E., Bormpoudakis, D., Chatzipavlidis, A., Cortes-Vazquez, J. A., Florea, I., Gearey, M., Levy, J., Loginova, J., Ordner, J., Partridge, T., Pizarro, A., Rhoades, H., Symons, K., Veríssimo, C. and Wahby, N. (2022). Radical social innovations and the spatialities of grassroots activism: Navigating pathways for tackling inequality and reinventing the commons. *Journal of Political Ecology*, 29, 143–188.

25

GRASSROOTS CLIMATE ACTIVISM IN NORTH AFRICA

Local, Personal and Radical

Haneen Ali Zeglam

Introduction

Across North Africa, communities are being increasingly confronted with the realities of the growing climate crisis. The region is warming at a rate of twice the global average, hugely impacting ecosystems, inhabitants and livelihoods. Given the current trajectory of rising temperatures, desertification, soil degradation and ecological depletion, it is estimated that North Africa will become uninhabitable by the end of the century (Pal and Eltahir, 2015; Broom, 2019; Wehrey and Fawal, 2022).

North Africa's climate crisis has been – and continues to be – further exacerbated by decades of poor governance (Moneer, 2020) and socioeconomic trends such as rapid population growth, rising urbanisation and worsening rates of poverty. Against this backdrop of growing environmental threat and social tension, the Arab Spring of 2011 lit a spark of activism that went on to engulf the region. Even after significant democratic back-sliding, the discourse of rights articulated so strongly in 2011 continues to live on, shaping all forms of activism in the region to this day.

Given this context, it is no surprise that grassroots climate activism is prominent across North Africa. The 2011 revolutions and subsequent periods of social unrest (Morocco during 2015; Algeria during 2019; etc.) have provided frames, means and motivations for climate activists working since. Throughout the region, people are protesting the growing climate crisis – and its impact on their lives and livelihoods – through an inherently political discourse of their rights as citizens – to safety, to health and to environmental protection. Despite this, climate activism remains largely neglected within the study of activism in North Africa, with academics and commentators alike focusing on more explicitly 'political' forms of protest. By understanding environmental demands as peripheral within the region to the demands of democracy, social justice and economic change, we fail to see how different discourses of activism rely on each other and speak to the same visions of the future. Climate activism in North Africa is distinctly political and belongs at the centre of the study of evolving forms, modes and themes of activism in the region.

In order to define the scope of this chapter early on, two definitions are needed: that of "North Africa" and that of "grassroots climate activism". Both of these terms can be

DOI: 10.4324/9781003396567-30

defined varyingly, but within this work at least, a focus on North Africa means a focus on Morocco, Algeria, Tunisia and Egypt. All four of these states have different contexts, norms, and trajectories but some broad trends remain comparable. Libya, unfortunately, is largely omitted. Given the last decade of upheaval and civil war, little to no substantial literature exists on environmental engagement in Libya (Orrnert, 2020). However, the overwhelming response of the Libyan people to the catastrophic floods in Derna of September 2023 suggest that instances of climate-based mobilisation are likely to increase. As for this chapter's working definition of "grassroots climate activism", the first section will be dedicated to unpacking that term within the context of North Africa.

This chapter is structured by questions that it hopes to, if not answer fully, then at least explore, contributing to the work on both contemporary climate activism and North African activism more broadly. In the course of my research, I spoke to a number of environmental journalists and activists within the region between April and December 2023. I am grateful to all of them for their time and their insight. An important note here is that it might be more in line with my participants' perceptions of themselves to refer to them as "people who engage/engaged in forms of activism" as opposed to "activists". One key finding of this chapter is that grassroots climate activism in the region is largely spearheaded by local residents and community members mobilising around local issues concerning their environment, health, and livelihoods, many of whom don't identify explicitly as "activists".

As Hamouchene and Sandwell (2023) rightly argue, much of the work on the ecological crisis and the energy transition in North Africa is dominated by top-down, market-based prescriptions and solutions, overlooking questions of local agency, resistance and mobilisation. In sharp contrast to this, this chapter hopes to centre commonly overlooked "voices from below" as both an epistemological choice and a methodological one. This chapter also draws on a body of media and social media coverage of various episodes of climate-based mobilisation.

However, the limitations of both of these approaches are evident. Many movements were never institutionalised and remained largely leaderless, making it challenging at times to find relevant actors to speak to, especially sometimes years after the principal movement dispersed. Moreover, although much has been made of the so-called social media revolutions of the Arab Spring and the enormous power of social media in enabling new modes of activism, it also can't be denied that much grassroots activism still occurred offline throughout the last decade, leaving significant gaps in this chapter's purview.

Finally, although this chapter hopes to provide a comprehensive overview of the state of grassroots climate activism in the region, it also seeks to advance a two-fold argument: the importance of looking at the grassroots level in order to understand trends in activism more broadly; and the increasing importance of understanding climate activism as indivisible from other forms of activism in the Middle East and North Africa (MENA) that typically receive more scholarly and public attention.

What Does Grassroots Climate Activism Look Like in North Africa?

The history of repressive regimes in the region and the limited nature of civic space means that, for a long time, much of the most meaningful activism in North Africa operated on a very grassroots level, with little institutional backing in the form of civil society organisations, lobby groups or formal campaigns. This history shapes environmental activism to this day, even post-Arab Spring, with most outbreaks of activism often remaining very local and largely leaderless, relying on strong social ties and the mutual trust of a community (Loschi, 2019).

In her work on Tunisian activism, Pepicelli (2021) defines forms of "eco-resistance" as ranging from the work of associations and NGOs to organised social movements and also social "non-movements", using a phrase coined by Asef Bayat to help us understand various means of activism in politically closed settings. In this chapter, the latter two categories will be used as a working definition of grassroots climate activism, limiting the scope to non-movements and social movements that have arisen from the ground up, not ones initiated by pre-existing environmental groups and institutions.

Bayat's concept of "non-movements" provides us with an analytical framework for understanding the "collective action" of "non-collective actors" (2013, p. 20). Although lacking in formal leadership and organisational structures, the marginalised and disenfranchised can still bring about social and political change. Bayat argues that, by focusing almost exclusively on formal social movements and civil society work, scholars miss other hugely significant forms of social activism, especially within the Middle East and North Africa's exclusionary political systems (Sowers, 2018b).

In contrast to the formally organised and ideologically grounded forms of activism more common in the West, "non-movements" tend to be action-oriented and more limited in ambition and scope (Pourmokhtari, 2015). Engaging in incremental, grassroots forms of activism by, for example, occupying the land around a harmful mine or tapping electricity from municipal power lines can still be a valid and vital form of environmental activism and protesting socio-economic conditions. Many grassroots activists in the region don't set out to be activists and work without many of the trappings of activism we've come to recognise. Rather than working for a movement or a mission, many communities are brought together to work for their lives and livelihoods, in increasingly hostile conditions.

While Bayat focuses primarily on the urban subaltern, arguably his ideas can still be applied to grassroots climate activism in the region, despite its predominantly rural nature. Rural communities across Morocco, Tunisia, Algeria and Egypt often bear the brunt of the unfolding climate crisis and the degradation of their local environment to a greater extent than their urban counterparts. Moreover, when discussing disenfranchisement and the lack of institutional mechanisms for activism and protest, rural communities are often further marginalised from networks of power and patronage that are centred in the capital cities.

The activism these communities most often engage in is what Sowers terms "popular resistance campaigns" (Sowers, 2018a). These actions are often spontaneous and small-scale disruptions, aimed at forcing those responsible to negotiate and accommodate demands. Forms of protest regularly involve ordinary citizens blocking roads and public places, striking, reaching out to local media, petitioning local officials and generally pressuring authorities (*ibid*). Moneer (2019, p. 42) studies how activists are introducing "alternative modes of action to express their dissent, without resorting to direct or open confrontation with power". Examples include occupying public spaces, street art and even performances, all of which can't be contained as easily as more organised forms of political protest. However, despite the limited scope of both the reach of these movements and their demands, this chapter argues that these moments of protest are fundamental to understanding the present and future of climate activism in the Global South.

How Does Grassroots Climate Activism Fit Into the Landscape of State-led and Civil Society-led Climate Initiatives in North Africa?

Given the understanding of grassroots activism defined earlier, it is clearly distinguishable from state- and civil-society-led work on the environment. However, it remains important to sketch out the boundaries, priorities and efficacy of each type of climate work within North Africa to help us better understand the broader context within which grassroots climate activists are working.

The priority of the North African states has been preparing the region for a future as a primary exporter of renewable energy and green hydrogen. The growing desire to wean off Russian gas has seen Europe increasingly turn to North Africa to supply its current need for fossil fuel-based energy and its future desire for green alternatives. However, as North African regimes promote investment, infrastructure and climate initiatives that advance this view of the region as an international leader in sustainable energy, they are also domestically side-lining grassroots climate activists and their causes.

Like elsewhere, North Africa sees a gulf between the priorities of state and corporation-led climate initiatives and the work and demands of activists on the ground. Many of the mega-projects that are celebrated as helping build a green and sustainable future are guided by trade prospects and geopolitical considerations and in many cases can be seen to have actively harmed the environment of local communities. Morocco, for example, is often hailed as a regional leader in combating climate change, having signed numerous international climate conventions and embarking on a 10-year Solar Plan in 2009 to become a leading solar power producer. The state announced the construction of solar power plants across the country, mainly in rural and marginalised regions, and by 2019, three power plants (Noor I, II and III) in south-central Morocco had become operational (Haddad et al., 2022). However, despite the understood importance of renewable energy production, the Solar Plan predicated a groundswell of grassroots activism against the plants and the damage they were causing to the local environment and communities. A particular concern was the potential water consumption of the Noor Plants, in an area where drought is common (Hamouchene, 2016b; Aoui et al., 2020). The local community also felt aggrieved by many of the practicalities of the plan, which saw local land being bought by the government for a fraction of market value, with no apparent input from or advantage to the local community (Rignall, 2016; Aoui et al., 2020). More abstractly, many in the community questioned the principle of their local environment being turned upside down in order to allow the government to sell solar energy to Europe, while those within a few kilometres of the site continue to either have to pay extortionate costs for their own electricity or go without (Salime, 2021).

State-led mega-projects such as this often support the maintenance of exclusionary political regimes and are often a key focus around which grassroots activism revolves. While the transition away from fossil fuels is a necessary component of the fight against climate change, the centralised ownership and control of energy by the state and corporate actors often significantly disadvantages and disenfranchises local communities, leading to a contradictory phenomenon of grassroots climate activists sometimes opposing climate-saving projects of the state, despite their ostensibly aligned goal of environmental protection.

The strained relationship between governments and grassroots climate activists becomes most visible when the regimes engage with international climate initiatives. The hosting of COP22 in Morocco in 2016, and COP27 in Egypt in 2023, largely excluded local climate activists from voicing any real and necessary calls for domestic change. Climate activists at the time protested that the state was using climate talks as an opportunity to "greenwash the environmental crimes of the state" (Hamouchene, 2016a) against its own citizens. Activists organised an anti-COP22 conference called "System Change, not Climate Change", articulating their belief that climate change and environmental damage in North Africa would be most effectively aided by systemic socio-political changes: more inclusive democracy, greater agency for rural citizens and prioritising the health of local communities over economic gain.

Activists who found themselves barred from attending COP27 in Egypt also organised a simultaneous "climate camp" in Tunisia. These alternative spaces proved necessary in allowing alternative discourses on climate action in the region to be heard. While world leaders met, the repression of climate activists across North Africa reached new peaks. Hardly any were able to attend the conference or even protest outside. Reportedly, not a single activist from Egypt, Mali, or Morocco received accreditation to attend the conference (Lakhani, 2022).

The exclusion of many civil society organisations (CSOs) from COP27, and more broadly from having an effective role in society, also demonstrates the increasing importance of grassroots climate activism. The authoritarian pushbacks in Egypt and Tunisia (in 2013 and 2019, respectively), as well as the cautious approach of the Moroccan and Algerian regimes to civic activism, leave limited room on the whole for CSOs to flourish. The threat to civil society is especially serious in Egypt, with the state recently revising laws to increase control over organisations – especially those in receipt of foreign funding (Halawa, 2017) – and regularly attempting to delegitimise CSOs and NGOs as foreign agents (Orrnert, 2020). Many organisations have had their funds frozen and travel bans imposed on employees.

As a result, activists or would-be activists are discouraged from aligning themselves with formal organisations, fearing arrest and harassment. While environmental activism might once have been more tolerated by the regimes than explicitly political activists, the increasing use of a political language of rights and agency to frame environmental demands means that climate activists are now a perceived threat. As of June 2023, 38 environmental activists in Tunisia are facing prosecution and prison sentences (Crispino, 2023) and describe themselves as being up against "a mafia" (*ibid*) of hostile private companies and state agencies.

The most sensitive environmental issues are those that target the government's failure to protect citizen's rights and the security of the environment, including issues relating to water scarcity, industrial pollution, and environmental harm stemming from corporate and state business (Human Rights Watch, 2022). Not only do these causes trace back to the regime, but they also have the power to unite a lot of anger. Rather than fighting for abstract causes like "democracy" and "equality", clear examples of environmental neglect and the damage caused to the health and livelihoods of citizens provide a clear way of articulating the need for more inclusive and responsive forms of government.

During Morocco's hosting of COP22, Shearlaw (2016) reported various forms of state intimidation of activists. Around COP27, Human Rights Watch (2022) released a similarly damning report on the harassment of climate CSOs and activists in Egypt, concluding that the government had "severely curtailed groups' ability to carry out independent policy, advocacy, and field work essential to protecting the country's natural environment".

Harassment, bureaucratic hurdles, funding challenges, and imprisonment are only some of the difficulties cited.

The few climate-based CSOs and NGOs that continue to operate despite the atmosphere of repression have smaller budgets, fewer staff and fewer projects than any other form of civil society in the region, according to a survey by the Mediterranean Dialogue for Rights and Equality (Schwartzstein, 2022). The few that seem to operate openly and successfully are, more often than not, associations and movements that "support the status quo, advocate conservative reforms or are simply apolitical" (Hawthorne, 2005, p. 83).

Given the limited potential of civil society work and the exclusionary climate policy of the regimes themselves, grassroots movements increasingly operate within individual communities, with campaigners focusing on conspicuous, local grievances such as poor waste management or failing water infrastructure. As repression tightens, so do the forms of activism and the scope of causes championed. As Halawa (2017, p. 17) puts it, "gone are the days of fighting for nationwide or regional causes". Local causes and networks with fluid, non-institutional forms of mobilisation can still be significantly effective while having the advantage of being harder to stifle.

With all of this in mind, this chapter argues that it is more important to study grassroots activism in the region than ever before. Against a backdrop of state greenwashing and civil society repression, grassroots climate activism is perceived among activists and citizens as the only remaining 'authentic' form of activism, free from corporate influences, regime restrictions, or institutional manipulation. Grassroots activism remains the singular way for ordinary citizens of North Africa to protest the increasing pressures of the climate crisis on their everyday lives.

How Has Grassroots Climate Activism Changed Since 2011?

Although grassroots activism has a long and powerful history in North Africa, in order to understand the forms it takes today, it is necessary to reflect on the changes the last decade has wrought. In 2011, three countries underwent complete revolutions, and although the Moroccan and Algerian regimes remained standing, they could not fail to feel the effects of such a seismic period. 2011 changed the identity of the region as a whole – the prominent discourses, frames and ideas that arose then remain influential to this day. Even after subsequent democratic back-sliding, the way people "think about, and engage in, the everyday conduct of politics" (Sowers, 2012, p. 230) has undoubtedly changed.

Practically, the social and political upheaval of 2011 allowed for the creation of new political openings, with climate activists learning to exploit the new infrastructure provided by the revolutions to advance their protests (Orrnert, 2020). The uprisings of 2011 were the first in the region to heavily make use of social media, allowing for a more unfiltered form of expression. The impact that this social media revolution had on activism in the region to this day is demonstrated by a recent poll by the European Institute of the Mediterranean showing that North African citizens perceive social media as the second most meaningful type of activism, only after public demonstrations (Moneer, 2020). Interestingly, the polling suggests that the least effective mode of activism, in the minds of citizens, is reaching out to international audiences, further supporting the notion that grassroots activism in the region is increasingly inward looking and locally based. In contrast, climate activism in the Global North, which often benefits from greater access to resources, technology, and more prominent media platforms, tends to have a more globalised scope, with a focus on international climate agreements, policy advocacy and influencing global climate action.

In a more abstract sense, the political changes of 2011 also tied activism of all kinds more firmly to a broad discourse of political, social, and economic rights for citizens. The vocabulary of environmental protests and activism continues to be constructed through that of the 2011 revolutions (Pepicelli, 2021), having become both broader and more politicised (Sowers, 2018b). Even local environmental campaigns now build on widespread concerns with privatisation, state and corporate corruption and a lack of accountability. In advocating for their local environment, citizens are now also advocating for transformative visions of the future, in which the current socio-political norms are overturned.

In bringing the politicised aspects of climate activism to the forefront, 2011 also saw climate movements shift from understanding environmental concerns as a matter of personal behaviour to systemic state and corporate failures. Prior to 2011, North African regimes took up 'the environment' as a soft policy issue, believing it to be relatively uncontroversial domestically and well received internationally. They worked to portray environmental issues as almost wholly a matter of personal, moral responsibility, with a focus on campaigns around single-use plastics and littering.

The fall of these regimes and the reworking of political consciousness in 2011 saw this idea collapse; local environmental issues were suddenly being articulated as yet another sign of institutional failure and systemic corruption (Salman, 2021). Movements born out of local environmental concerns such as water pollution (in Gabes), toxic landfills (in Djerba) or coal mining (in cities across Egypt) began shifting the understanding of sustainability beyond cleanliness and aesthetics to much more radical notions of government failure, systemic injustice and the "disposability of certain bodies at the expense of others" (Salman, 2021, p. 2). As the democratic gains of 2011 are increasingly reversed, this new articulation of climate activism is increasingly perceived as a political threat. Mobilisation that now targets state projects or sources of state revenue is decidedly less welcome than prior campaigns that pushed for more recycling bins in public spaces (Sowers, 2018a).

Countless examples of grassroots climate campaigns help demonstrate the impact of 2011's revolutionary discourse on contemporary activism. In Egypt the anti-coal movement not only made use of the "Tahrir networks" of activist alliances forged in the revolution but also framed the renewal of coal imports as a continuation of the Mubarak-era priorities, prioritising big business interests over public health and the local environment (Zayed and Sowers, 2014).

In Tunisia, post-2011 climate campaigns have also been a way for people to express their frustrations with bad governance, low quality of life, corruption and marginalisation (Moneer, 2020). Pepicelli (2021) writes how during protests against the phosphate industry in Gabes and Sfax, protestors adopted slogans clearly modelled on famous Arabic chants of the Arab Spring; "The people want a clean environment" instead of "The people want the fall of the regime". Or "Pollution, leave!" based on the famous "Ben Ali, dégage! [Ben Ali, leave]".

In Algeria, 2011 did not see a successful revolution but nonetheless outbreaks of social unrest have occurred periodically over the last decade, knitting together discourses of broader political change and smaller, localised environmental protests. Protests against the government's fracking plans took on discourses of democracy and national sovereignty. Protestors alternated chants of "There will be no gas, there will be no oil, tell France to exploit [shale gas] in Paris" and "There won't be shale gas, [President] Tebboune is not legitimate", uniting ideas of environmental damage and misuse with the government's lack of authority in the eyes of citizens (Belakhdar, 2020).

Sowers (2018a) argues that because the type and distribution of environmental problems reflects the political and economic policies of the states, the demands are inherently political. Looking briefly at these examples and the ways in which grassroots climate activists express themselves and their concerns in the language of social, political and economic justice, it becomes clear that we cannot talk about environmental activism in North Africa without addressing questions of popular sovereignty and state-society relations (Hamouchene, 2020).

In all of these instances – and many others – climate activists in North Africa strongly articulated their grievances and concern for their local environment using a language of political rights and the desire for more democratic decision-making. This rhetoric ties into the growing recognition of the importance of procedural justice in the climate transition, which focuses on ensuring that the groups most affected by climate change have meaningful opportunities to design and implement climate responses that work best for them. A recurring feature within green movements around the world is an emphasis on radical democracy as a political imaginary that demands a "decentred and egalitarian society that is in harmony with nature" and "involves grassroots participation" (Machin, 2022, p. 552). This argument is intensely relevant to North Africa, where climate activists take a seemingly small-scale local environmental concern and use it to articulate much more expansive, radical visions of change.

These shifting and evolving discourses, imaginaries, and frames that have taken off with new momentum since 2011 demand the close attention of climate scholars. Rather than transposing imported notions of "climate justice" to North Africa uncritically, we must instead situate such ideas more precisely in the context of the region, focusing on the specific issues that directly affect the lives and livelihoods of residents and citizens, and on how those issues interact with the local political, economic and social context. Climate causes in North Africa – and the grassroots activists who champion them – cannot be divorced from the situations within which they work, respond and mobilise.

How Do Grassroots Climate Activists in North Africa Understand Their Activism?

Within the newly politicised space of climate activism, contemporary North African climate activists further understand their activism as profoundly anti-extractivist, inextricably linked to notions of personal rights as citizens, and heavily centred on local and directly pressing issues.

Locally Rooted

Most grassroots climate activism in North Africa is based around locally rooted, specific issues often tied to concerns of environmental damage, depletion or misuse at the hands of state, or corporate projects. Hamouchene (2020) argues that while climate activism in the West is more focused on ideology, activists in North Africa are more likely to be fighting to win material benefits or changes to their everyday life and local environment. This emphasis on the local and tangible, as opposed to the global and abstract, also means that many grassroots climate activists in North Africa are less likely to define themselves as "activists" by identity; instead, many of the local movements are led by people who perceive themselves as citizens and residents first and foremost and victims at the hands of an environmental crime, as opposed to activists for the sake of the cause. There is evidence that environmental protests are more concentrated in areas most affected by pollution and the climate crisis (Pepicelli, 2021) and often more inclusive of all social classes than their counterparts in the West (Orrnert, 2020, p. 10), which largely remain the realm of the university-educated and relatively privileged middle classes (della Porta and Portos, 2021).

In Salman's work (2021) on Tunisia, she remarks how the leaders of the Manish Msab ("I am not a dumpsite") protests in Agareb, Tunisia, a historically marginalised area, defined themselves as local residents, not activists. She notes that they "insisted they were no environmental experts" (2021, p. 10); to them, technical details of toxicity levels, soil composition and standards of industrial filters were of little concern. Toxicity, to the local community, was more about the negative consequences they experienced every day. Moreover, the activist leaders expressed what Salman calls a "deep scepticism" (*ibid*) about the broader climate activism landscape; they refused to institutionalise the movement and worried about being accused of capitalising on their neighbourhood's suffering.

Referring back to the notion of the political imaginaries of the green movements in North Africa, it ought to be noted here that another key dimension within this vision is often the notion of decentralisation (Fischer, 2017; Machin, 2022). This idea of a return to the local is echoed in climate activism around the world with the well-known "think globally, act locally" (Giugni and Grasso, 2015), but in North Africa it is less of a spoken parameter and more of a natural state of being. Given the limited civic and political space in North Africa, grassroots activism such as this is increasingly the only viable form of expressing environmental concerns. As civil society is increasingly circumvented, activism in the region is shifting more inward, both out of necessity but also due to lack of regard for the role of the international community (Halawa, 2017).

Emerging out of local environmental hazards and concerns, movements then often become formulated heavily around the rights of local residents, more so perhaps than abstract notions of rights of the environment. The environmental concerns that see the most local mobilisation are those around effects on public health and livelihoods (Sowers, 2018a). When environmental hazards are protested, more often than not the ecological element is presented as secondary to more pressing socio-economic issues such as jobs, the distribution of wealth and resources, and a desire for more popular inclusion in decision-making processes.

In Morocco, some of the longest-running environmental protests against the silver mines at Imider are similarly centred on a highly localised issue and based on the rights of local residents. The area's water resources have been irreparably damaged, with huge effects on local livestock and agriculture. Soil around Imider has also been found to contain high levels of pollutants, including arsenic, cadmium and lead (Elia, 2020). Protests have erupted sporadically since the opening of the mine in 1978 but they began to take off in earnest in 2011, when local households found themselves without water for weeks while the pipeline to the mines was drawing 12 times the daily consumption of all of Imider's residents (Salime, 2019). The residents mobilised and successfully shut down the pipe, protesting the damage to their health and livelihoods. This direct action turned into the longest sit-in protest encampment in Moroccan history, and it still remains, despite the national and international coverage, profoundly local. The protest was not only sparked by the concerns of local residents but was also sustained and led by the local community for years.

In Egypt, protests against the hazards posed by a large fertiliser plant in Damietta also focused almost exclusively on the harmful effects on the local community, once again including water shortages, damage to the fishing industry, agricultural harm and pollution. The call of local activists was clear; the state was unacceptably catering to foreign industry at the expense of local health and rights. Damietta in this way was, like many grassroots climate movements in North Africa, "a local protest with national overtones" (Sowers, 2012, p. 147).

Anti-extractivist

Most of the issues that spark climate-based protest in North Africa are the result of some form of either state or corporate exploitation, neglect or monopolisation of natural resources and local environments. Therefore, not only do activists often rally against the detrimental *effects* of these interests and industries on their own health, livelihoods and environments but they also increasingly protest and question the *rights* of these entities to even embark on such projects. When targeting the practices of their own state, climate activists take on a generically anti-extractivist stance and, when opposing foreign corporations or state interests, the rhetoric taps into a historically strong anti-colonial discourse in the region. Either way, climate activism often targets a well-defined "enemy" accused of environmental injustice against the local community. These campaigns are born from specific issues, with specific opponents and specific demands.

In Algeria, the government's shale gas exploration agreements with a number of multinational firms sparked strong grassroots opposition. The protests at Ain Salah quickly spread across the south of the country, drawing not only on the anger caused by the perceived "selling out" of their region by the government but also the feeling of historic economic neglect and political marginalisation. The involvement of international firms allowed the campaign to frame its opposition to fracking in a number of ways: in environmental terms, in terms of personal rights and hazards to health, but perhaps most importantly in terms of a nationalist anti-neocolonialism. The understanding that France had banned fracking domestically – but was seemingly happy to do it in Algeria – sparked fury (Petitjean and Chapelle, 2016). Climate activism in this case was not just a matter of protesting environmental damage but of protesting environmental injustice.

The work of energy corporations and the petrochemical industry in similarly rural and marginalised areas of Tunisia (Tozeur, Gabes) sparked similar contestation (Orrnert, 2020). Phosphate extraction especially, in the cities of Gabes and Gafsa, has led to long-standing confrontation between local residents and foreign corporations. Protests often take the form of strikes and demonstrations, rail/road blockades and sit-ins, and physical occupations of sites and buildings (Rousselin, 2018).

In Morocco, the "We Are Not Trash" campaign, in response to the government's decision to import trash from Italy to burn for energy, accused the state of being complicit in more than an act of climate injustice but in a form of neocolonialism. The complaints of the protestors were many; not only was there concern for the broader climate impact of burning litter but the local impact on pollution and the general lives of the community. The central discourse, however, soon arose to be a question of profound injustice: why had the Moroccan government been so quick to discard their standard of living in the face of a cheap energy source? Why should Moroccans be forced to deal with the detrimental impact of burning Italian rubbish? On top of environmental concerns, the symbolism of importing Europe's garbage provided much of the focus of the activism.

Similar questions of injustice are often raised in Morocco by climate activists protesting their own government, with the anti-colonialism directed at foreign corporations or states evolving into a more domestic incarnation of anti-capitalism, this time targeting the central government's own exploitation of local communities and environments in the pursuit of international trade. The government's Noor solar energy project has seen this discourse of state "land-grabbing" and colonial-adjacent tactics become central to the activism that has emerged in response. There has also been a selection of journalistic and academic work on the construction of the power plants that addresses how the government's initial acquisition of land drew on colonial strategies of dispossession (Rignall, 2016, Hamouchene, 2016a). Local residents did not receive a fair price for the land, did not see any development projects financed and, most importantly, the clean energy produced is bound straight for European consumers. Rignall (2016) argues that the activism that emerged around the power plants was centred on this idea of a "subordinate position" in a global "economy of repair" (2016, p. 542).

Despite the advantages of more solar energy production to both Morocco and the wider, global climate, the means to attain more solar power can – and should – still be held up for scrutiny. Somewhat ironically, campaigns of grassroots climate activists in North Africa sometimes find themselves in opposition to state climate initiatives. In this example, the Moroccan government's solar power plants are a clear example of how progressive narratives of renewable energy can still be entrenched in patterns of global inequality, injustice and marginalisation.

The case of North Africa is a stark example that demonstrates the importance of applying the principles and ideas of climate justice to climate mitigation and adaptation measures, as well as to our understanding of climate culpability. Not only are North African communities bearing the brunt of the climate crisis, but they are also bearing the most significant burden of various decarbonisation strategies. Large-scale renewable energy projects or carbon offset initiatives may lead to land grabbing, the displacement of Indigenous communities or the exploitation of local resources without proper consultation or compensation, as has been the case with Morocco's solar power initiatives. Across North Africa we find cases of climate activists protesting against climate mitigation and decarbonisation projects that fail to ensure a just transition towards a sustainable future. Not all climate adaption measures are necessarily in line with the principles of climate justice, as grassroots activists are increasingly highlighting.

Conclusion

Given the general atmosphere of activism across North Africa and the growing climate crisis, grassroots climate activism is becoming ever more important across the region. Grassroots activists hold a unique position, working often completely outside of the spheres of state and civil society climate activism. When addressing their mobilisation, it is important to bear in mind this context of state repression of the civic and political space that pushes civil society organisations to the margins and leaves activists with little to no space other than grassroots mobilisation and sporadic protest. In this way, to understand activism in North Africa is to look at the grassroots.

Despite the argument throughout this chapter that climate activism in North Africa emerges from and centres on local issues, it is also global in discursive scope, taking on some of the biggest behemoths of our time: the neoliberalism of North African states and the neo-colonialist attitudes and climate policies of the West. Despite being primarily born out of local communities uniting around specific issues, the activism addressed in this chapter is also radical in the egalitarian, decentralised visions of the future it articulates.

Grassroots climate activism forces external observers to rethink a lot of our implicit assumptions, for example, the mistaken belief that any move towards renewable energy is to be welcomed. Instead, grassroots campaigners target both state initiatives that sacrifice local communities and environments to claim a stake in the renewable energy market, and the capitalist ethos and practices of corporations that are to blame for the misuse of fossil fuels (Hamouchene, 2016b). By engaging in simple, local protests, grassroots climate activists are radically challenging our ways of producing and distributing things, our consumption patterns and fundamental issues of equity and justice (*ibid*).

References

Aoui, A., Amrani, M. and Rignall, K. (2020). Global aspirations and local realities of solar energy in Morocco. *Middle East Report*, 296. https://merip.org/2020/10/global-aspirations-and-local-realities-of-solar-energy-in-morocco-296/ [Accessed 15 April 2023].

Bayat, A. (2013). *Life as Politics: How Ordinary People Change the Middle East* (2nd ed.). Stanford: Stanford University Press.

Belakhdar, N. (2020). "Algeria is not for sale!" Mobilizing against fracking in the Sahara. *Middle East Report*, 296. https://merip.org/2020/10/algeria-is-not-for-sale-mobilizing-against-fracking-in-the-sahara/ [Accessed 30 April 2023].

Broom, D. (2019). How the Middle East Is Suffering on the Front Lines of Climate Change. *World Economic Forum*. https://www.weforum.org/agenda/2019/04/middle-east-front-lines-climate-change-mena/ [Accessed 30 April 2023].

Crispino, I. (2023). In Tunisia, Environmental Activists on Trial. *Inkyfada*. https://inkyfada.com/en/2023/08/18/environmental-activists-trial-tunisia/ [Accessed 30 April 2023].

della Porta, D. and Portos, M. (2021). Rich Kids of Europe? Social basis and strategic choices in the climate activism of Fridays for Future. *Italian Political Science Review*, 53(1), 24–49. http://doi.org/10.1017/ipo.2021.54.

Elia, C. (2020). Water defender Moha Tawja fights for the right to water in Morocco. *Water Grabbing Conservatory*. https://www.lifegate.com/water-defender-moha-tawja [Accessed 18 April 2023].

Fischer, F. (2017). *Climate Crisis and the Democratic Prospect: Participatory Governance in Sustainable Communities*. Oxford: Oxford University Press.

Giugni, M. and Grasso, M. (2015). Environmental movements: Heterogeneity, transformation, and institutionalization. *Annual Review of Environment and Resources*, 40, 337–361. http://doi.org/10.1146/annurev-environ-102014-021327.

Haddad, C., et al. (2022). Imagined inclusions into a 'green modernisation': Local politics and global visions of Morocco's renewable energy transition. *Third World Quarterly*, 43(2), 393–413. http://doi.org/10.1080/01436597.2021.2014315.

Halawa, H. (2017). Egypt and the middle east: Adapting to tragedy. In R. Youngs (ed.) *Global Civic Activism in Flux* (pp. 13–18) [Online]. Carnegie Endowment for International Peace. https://carnegieendowment.org/research/2017/03/global-civic-activism-in-flux?lang=en¢er=europe [Accessed 6 September 2024].

Hamouchene, H. (2016a). We can't tackle environmental and social justice in Morocco if we don't talk about Western Sahara. *Middle East Eye*. https://www.middleeasteye.net/opinion/we-cant-tackle-environmental-and-social-justice-morocco-if-we-dont-talk-about-western [Accessed 4 May 2023].

Hamouchene, H. (2016b). The Ouarzazate solar plant in Morocco: Triumphal 'green'. *Al Jadaliyya*. https://www.jadaliyya.com/Details/33115/The-Ouarzazate-Solar-Plant-in-Morocco-Triumphal-%60Green%60-Capitalism-and-the-Privatization-of-Nature [Accessed 6 September 2023].

Hamouchene, H. (2020). Environmental and climate justice in North Africa. *Review of African Political Economy*. https://roape.net/2020/03/20/environmental-and-climate-justice-in-north-africa/ [Accessed 30 April 2023].

Hamouchene, H. and Sandwell, K. (2023). The challenges of climate change and just energy transition in North Africa. *Al Jadaliyya*. https://www.jadaliyya.com/Details/44770 [Accessed 30 April 2023].

Hawthorne, A. (2005). Is civil society the answer? In T. Carothers and M. Ottaway (eds). *Uncharted Journey: Promoting Democracy in the Middle East* (pp. 81–114). Washington: Carnegie Endowment for International Peace.

Human Rights Watch. (2022). Egypt: Government Undermining Environmental Groups. https://www.hrw.org/news/2022/09/12/egypt-government-undermining-environmental-groups [Accessed 30 April 2023].

Lakhani, N. (2022). African activists struggle to attend UN climate talks in Egypt. *The Guardian*. https://www.theguardian.com/environment/2022/oct/06/cop27-african-activists-climate-crisis [Accessed 17 April 2023].

Loschi, C. (2019). Local mobilisations and the formation of environmental networks in a democratizing Tunisia. *Social Movement Studies*, 18(1), 93–112. http://doi.org/10.1080/14742837.2018.1540974

Machin, A. (2022). Green democracy: Political imaginaries of environmental movements. In M. Grasso and M. Giugni (eds.) *The Routledge Handbook of Environmental Movements* (pp. 552–564). New York and London: Routledge.

Moneer, Z. (2019). Waging a war not only on coal but much more: Types of youth activism among Egyptians against the coal movement. In S. Yom et al (eds.) *POMEPS Studies 36: Youth Politics in the Middle East and North Africa* pp 41–49 [Online]. POMEPS. https://pomeps.org/pomeps-studies-36-youth-politics-in-the-middle-east-and-north-africa [Accessed 6 September 2024].

Moneer, Z. (2020). Environmental activism in the post-arab spring: It is not about a mere clean environment. *IEMed*. https://www.iemed.org/publication/environmental-activism-in-the-post-arab-spring-it-is-not-about-a-mere-clean-environment/ [Accessed 30 April 2023].

Orrnert, A. (2020). Drivers and barriers to environmental engagement in the MENA region. *K4D Helpdesk Report, 832*. Brighton, UK: Institute of Development Studies. https://www.gov.uk/research-for-development-outputs/drivers-and-barriers-to-environmental-engagement-in-the-mena-region [Accessed 17 April 2023].

Pal, J. and Eltahir, E. (2015). Future temperature in Southwest Asia projected to exceed a threshold for human adaptability. *Nature News*. https://www.nature.com/articles/nclimate2833 [Accessed 17 April 2023].

Petitjean, O. and Chapelle, S. (2016). Shale gas: How Algerians rallied against the regime and foreign oil companies. *Multinationals Observatory*. https://multinationales.org/en/investigations/shale-gas-how-algerians-rallied-against-the-regime-and-foreign-oil-companies-181/ [Accessed 4 September 2024].

Pepicelli, R. (2021). "People want a clean environment": Historical roots of the environmental crisis and the emergence of eco-resistances in Tunisia. *Studi Magrebini*, 19(1), 37–62. http://doi.org/10.1163/2590034X-12340039

Pourmokhtari, N. (2015). Non-movements as social activism. *Against the Current, 176*. https://againstthecurrent.org/atc176/p4428/ [Accessed 15 April 2023].

Rignall, K. (2016). Solar Power, state power, and the politics of energy transition in pre-saharan Morocco. *Environment and Planning A: Economy and Space*, 48(3), 540–557. http://doi.org/10.1177/0308518x15619176

Rousselin, M. (2018). A study in dispossession: The political ecology of phosphate in Tunisia. *Journal of Political Ecology*, 25(1), 20–39. http://doi.org/10.2458/v25i1.22006

Salime, Z. (2019). Protest camp as counter-archive at a Moroccan silver mine. *Middle East Report, 291*. https://merip.org/2019/09/protest-camp-as-counter-archive-at-a-moroccan-silver-mine/ [Accessed 15 April 2023].

Salime, Z. (2021). Life in the Vicinity of Morocco's Noor Solar Energy Project. *Middle East Report, 298*. https://merip.org/2021/04/life-in-the-vicinity-of-moroccos-noor-solar-energy-project-2/ [Accessed 15 April 2023].

Salman, L. (2021). Environmentalism after decentralization: The local politics of solid waste management in Tunisia. *Arab Reform Initiative*. https://www.arab-reform.net/publication/environmentalism-after-decentralization-the-local-politics-of-solid-waste-management-in-tunisia/ [Accessed 17 April 2023].

Schwartzstein, P. (2022). COP27 in sharm: Few opportunities and more challenges for MENA environmentalists. *New Security Beat*. https://www.newsecuritybeat.org/2022/03/cop-27-sharm-opportunities-challenges-mena-environmentalist [Accessed 15 April 2023].

Shearlaw, M. (2016). Moroccan activists plan protests to coincide with UN climate summit. *The Guardian*. https://www.theguardian.com/world/2016/nov/05/moroccan-activistsplan-protests-un-climate-summit-marrakech [Accessed 30 April 2023].

Sowers, J. (2012). *Environmental Politics in Egypt: Activists, Experts and the State*. New York and London: Routledge.

Sowers, J. (2018a). Environmental activism in the Middle East: Prospects and challenges. *Humanities Future*. https://humanitiesfutures.org/papers/environmental-activism-in-the-middle-east-prospects-and-challenges/ [Accessed 30 April 2023].

Sowers, J. (2018b). Environmental activism in the Middle East and North Africa. In H. Verhoeven (ed.) *Environmental Politics in the Middle East* (pp. 27–52). Oxford: Oxford University Press.

Wehrey, F. and Fawal, N. (2022). Cascading climate effects in the Middle East and North Africa. *Carnegie Endowment*. https://carnegieendowment.org/2022/02/24/cascading-climate-effects-in-middle-east-and-north-africa-adapting-through-inclusive-governance-pub-86510 [Accessed 15 April 2023].

Zayed, D. and Sowers, J. (2014). The campaign against coal in Egypt. *Middle East Report, 271*. https://merip.org/2014/07/the-campaign-against-coal-in-egypt/ [Accessed 30 April 2023].

26

TRANSNATIONAL GRASSROOTS NETWORKS AND RENEWABLE ENERGY GOVERNANCE

Emilia Groupp

Introduction

Renewable energy development has become a key element in the worldwide effort to address climate change. At the same time, there has been increasing attention to the social and environmental impacts of the large-scale expansion of new energy projects, from solar and wind to wave energy and biofuels. These concerns have predominantly come from environmental organisations, labour unions, and local stakeholders whose environments, livelihoods, and social worlds stand to be significantly altered by large-scale energy projects (e.g., Allison et al., 2014; Abramsky, 2010).[1] While disputes between stakeholders and renewable energy companies are often dismissed as a conflict between pragmatism and idealism, local versus global outlooks, or the short-term focus of conservationism versus the long-term necessity of the energy transition,[2] few actors involved in these movements would deny the importance of renewable energy. Rather, their goal is to shape where – and under what conditions – new forms of energy are produced to ensure that the benefits of transitions to new energy sources are equitably distributed across and between societies. Such movements are not just a reaction to renewable energy development but critical engagements that highlight histories of environmental and energy injustice and the necessity of ensuring the fair distribution of benefits in future energy systems.[3]

With the large-scale expansion of renewable energy development over the last few decades, the transnational implications of these projects have taken on altogether new scales. Renewable energy sites, once relatively small scale and community oriented, are now growing ever larger as companies try to maximise profit and economies of scale. The mass production of materials for renewable energy technologies has also brought attention to labour and environmental inequities along global supply chains. Transnational financial institutions, including hedge funds and development banks, are increasingly funding large-scale renewable energy projects, raising questions regarding accountability and the dispute processes available to transnational actors. Some renewable energy projects are export oriented,[4] inviting scepticism as to who will benefit from these developments. Moreover, the construction and ownership of renewable energy facilities around the world has come to be dominated by large multinational energy companies working from headquarters in

DOI: 10.4324/9781003396567-31

the United States, Northern Europe, and the United Arab Emirates, for instance. This has added to concerns of social justice and equity, particularly with regards to the ways in which many of these projects emerge from much longer relations of colonial extraction and exploitation. Thus, renewable energy development is firmly rooted in global circulations and (unevenly) distributed impacts.

These global implications have given rise to different kinds of transnational coalitions seeking to shape the development of renewable energies for more equitable outcomes. Yet, while studies of transnational networks in relation to renewable energy development have largely focused on the role of formal institutions in governance, there has been less attention to transnational networks formed by grassroots actors, despite the growing number of such initiatives around the world. *Grassroots transnational networks* refer to informal and decentralised associations or movements that originate from local communities, organisations, or individuals in different countries and which collaborate on shared causes or issues of transnational significance. These networks tend to lack centralised structures and hierarchical decision-making, favouring flexibility (Batliwala, 2002). Although more difficult to trace than the well-documented work of institutions, transnational grassroots actors are increasingly active in shaping new energy geographies.

As such, this chapter aims to answer the following questions: what kinds of transnational grassroots networks have emerged in response to renewable energy development? How do these networks impose different forms of governance over an increasingly transnational renewable energy industry? What power dynamics emerge, and in what ways are these networks challenging the influence of international financial institutions and multinational energy companies?

The analysis here is threefold. Following a summary of research methods, I provide an overview of the role of transnational activist networks in climate activism more generally and the several important shifts that have led to the pronounced role of grassroots actors in climate change governance. I then trace the major themes around which transnational grassroots actors have coalesced relating to renewable energy development and the movement for a just energy transition, demonstrating the ways in which large-scale renewable energy projects are increasingly negotiated and governed through transnational networks. Finally, I provide an analysis based on ethnographic research on the renewable energy industry, in addition to an examination of 31 case studies of conflicts over wind development in the Global South, to assess if and how transnational networks are challenging the power of international financial institutions and multinational energy companies. Drawing insights from this synthesis of case studies, ethnographic data, and literature, I advance the argument that climate activism has shaped the specific political and economic landscapes of renewable energy development in ways that have engendered opportunities for previously marginalised actors to leverage new forms of political power.

Methods

This chapter draws on over two years of ethnographic research[5] with renewable energy companies, environmentalists, and rural stakeholders in the United Kingdom and Tunisia, as well as a review of research on transnational networks that have emerged in response to large-scale renewable energy development. From 2021 to 2023 I conducted research on the work of renewable energy companies to develop large-scale solar and wind energy facilities in North Africa and their socio-spatial implications.[6] I spent the first year of this project (2021–2022) conducting ethnographic research at a multinational renewable energy company in the UK, working as an unpaid intern primarily on site development. I spent the second year (2022–2023) in Tunisia, jointly focusing on the work of European renewable energy companies to develop renewable energy sites as well as their socio-spatial implications.

To confirm my initial findings regarding the role of transnational networks in governing renewable energy sites, I sampled interviews from 11 of the largest renewable energy development companies operating transnationally, headquartered in Europe, the Emirates, Canada, and the United States. Those with whom I spoke had extensive experience working one-on-one with transnational organisations and grassroots actors on specific renewable energy projects. Additionally, I interviewed 14 individuals working for transnational conservationist organisations and 17 individuals working at small grassroots environmental organisations in Tunisia and Brazil. I also interviewed nine actors working at state-level agencies who dealt with renewable energy projects and were familiar with the role of environmentalists working within transnational networks in Tunisia and the UK. Finally, I attended three renewable energy industry conferences in London, Lisbon, and Tunis in 2021 and 2022, and I draw on further documentary sources including Environmental Impact Assessments (EIAs), news reports, and internal communications relating to the broader renewable energy industry.[7]

In order to ascertain whether my findings drawn from qualitative research are consistent outside of my specific research sites, I have included an analysis of 31 case studies of grassroots organising around wind projects in the Global South. Although engagements between grassroots actors and energy companies have emerged over many different kinds of sites, I concentrated exclusively on wind energy sites to maintain a manageable and focused analysis. Twenty-eight case studies were drawn from the Environmental Justice Atlas (Ej-Atlas), and three additional cases were included from field research, for a total of 31 case studies (the full list of cases is provided in Table 26.2. Case Studies of Wind Conflicts). The Ej-Atlas is a shared platform and database to document and study cases of grassroot activism emerging from the uneven global distribution of environmental burdens (Temper et al., 2015). The Ej-Atlas database documents both qualitative and quantitative data on each conflict including details of the issues at stake, the actors involved, impacted stakeholders, those involved in mobilisation, outcomes (if known), and any additional information necessary to assess cases. Each case study is reviewed by an internal board, which assures accuracy before addition to the database.

For this analysis, I selected cases based on three criteria: (1) the wind projects are utility-scale (at least 10MW), (2) the projects are located in regions that are politically and economically less powerful (most of which fall under the category of the 'Global South'),[8] and (3) the cases have reached some form of outcome in order to ascertain the impact of grassroots organising. I categorised outcomes into one of the following results categories: (1) cancelled, (2) suspended or delayed, (3) in operation, and (4) in operation but delayed, altered, or suspended after construction. As a result of the third criterion, most of the conflicts selected began over a decade ago. Each case was recorded along with the conflict start date, the company involved, key issues at stake, funding organisations, the outcome of the project, whether a new Environmental Impact Assessment was required, and whether or not funding was withdrawn as a result of the conflict. The cases were also cross-referenced with the Global Energy Monitor,[9] an open-access data repository that tracks energy projects around the world, to ensure that the most up-to-date status of the sites was included in the assessment. Finally, I reviewed and verified the cases drawn from the EJ-Atlas and in some instances updated their status depending on available information. It is important to note that the cases used in my analysis do not represent a conclusive list of all conflicts or negotiations over large-scale renewable energy projects, as there are a considerable number at stages too early to determine outcomes, in addition to those which are not documented.

Transnational Activist Networks and Grassroots Climate Change Governance

Scholars and activists have long critiqued the failure of top-down approaches to the challenges posed by climate change, particularly international regulatory efforts that are seen as largely ineffective in initiating meaningful change.[10] On the other hand, local actions are often dismissed as hopelessly futile in the face of the global scale of climate change.[11] Transnational networks[12] are challenging this dilemma, however, working to stitch together local actions that are resulting in global forms of intervention, governance,[13] and transformation.

Transnational advocacy networks, which function by using the media and direct communication to bring attention to issues in order to pressure governments or companies to alter their behaviour, are perhaps one of the best-known organisational forms to have shaped the climate movement (Hadden, 2015; Ciplet et al., 2015) and have a central place in the history of environmental activism (Keck and Sikkink, 1998). What is distinct – and powerful – about transnational advocacy networks is the ways in which they mobilise different publics to take interest in an issue that may not directly affect them. The impact of transnational activist networks on both domestic and global climate change politics have been as significant as they have been geographically expansive. They have been successful in bringing attention to environmental impacts that cross political boundaries in a way that nationally organised groups could not, such as air pollution and water security. Some of the earliest and best-known transnational advocacy networks arose in opposition to fossil fuel production, including coal mining and oil drilling (Keck and Sikkink, 1998, p. 147).

In the 1990s and 2000s, transnational activist networks largely worked through the institutional channels of UN bodies and large NGOs to bring attention to climate change and other environmental issues, leading this period to be referred to as a 'mobilisation from above' (Brecher, 2015), with networks predominantly initiated in the Global North (Bulkeley et al., 2014). However, beginning in the mid-2000s, there was a marked change in the climate movement, which was characterised by a growing number of grassroots movements in both the Global North and South (Almeida, 2019). The 2000s also saw the emergence of transnational grassroots movements organised under the burgeoning concept of climate justice, which has since seen a rise in voices from historically marginalised groups and regions (e.g., Tokar, 2019). This trend points towards changing strategies amongst some climate activists who are distancing themselves from international institutions and policy development, and focusing instead on specific interventions at localised sites of concern (cf. Dietz and Garrelts, 2014; Osofsky and Levit, 2008). Finally, the 2010s saw another significant shift in the expansion of partnerships between conventional international non-governmental organisations (INGOs), transnational grassroots movements, and civil society organisations (Dupuits, 2021) which sought to bypass the state altogether.

There are many reasons for this emerging reorientation, including a general frustration with the lack of progress by states along with a belief that some policies are incompatible with actual practices. Others have argued that policies and international agreements concerning emissions have merely provided a way for states and the private sector to avoid taking serious action on climate change (Okereke, 2010). This includes the critique that some climate change solutions pursued in capitalist frameworks merely elide the underlying causes of climate change itself and perpetuate environmental degradation and social inequities. There are also those, such as Indigenous communities and other marginalised groups, who have been altogether excluded from decision-making processes and have thus opted to pursue more transformative and inclusive approaches to climate change action (Harlan et al., 2015). Finally, there are movements that could be said to have focused on the local from the start, such as community energy projects and sustainable food systems that have gained salience alongside the climate movement (e.g., Brinker and Satchwell, 2020; Klein and Coffey, 2016).

In addition to a growing orientation towards local initiatives, there has been an increased focus on what has been termed grassroots climate change governance.[14] Initially, transnational efforts that were focused on climate change in the 1990s concentrated their efforts on influencing state responses to climate change (Betsill and Corell, 2008; Newell and Bulkeley, 2017). Beginning in the 2000s, there was a marked shift towards engaging directly in climate change governance, with the rise of initiatives and coalitions that differed in many crucial ways from state and multilateral climate governance (e.g., A. Jordan et al., 2018; Rootes et al., 2012). Since the 2021 United Nations Climate Change Conference in Glasgow (COP26), the Global Climate Action Portal,[15] which has been tracking the progress of non-state actors and initiatives in climate change governance, currently lists 3,443 organisations, 282 regions, and 11,354 cities, as well as 149 transnational coalitions. An emerging body of literature stemming from political science, anthropology, and sociology has also documented a growing number of transnational networks engaged in climate change governance involving grassroots actors (e.g., A. Jordan et al., 2018; Bulkeley et al., 2014; Salazar, 2023).[16]

While there has been an expanding focus on grassroots networks in climate change governance, we have less of an understanding of their role specifically in energy governance. Additionally, amongst transnational initiatives, there are striking imbalances in the geographies of networks, which tend to be concentrated in the Global North (Kaiser, 2022). Some have voiced concerns that these networks may merely recreate power inequities between the Global South and industrialised North (Andonova, 2021), and the role of actors in what can be characterised as the Global South remains a key area of research (Chan et al., 2018). This chapter thus helps to carve out some preliminary answers to these gaps and concerns.

Taking into account these broader shifts in transnational climate change governance, the next section focuses on the major themes around which transnational grassroots networks have increasingly coalesced and their expanding role in governing various sites of renewable energy development.

Transnational Grassroots Networks and Renewable Energy Development

While advocacy for renewable energy has a long history, there has been a surge in political organising, international agreements, and technological advancements aimed at reducing greenhouse gas emissions through transitioning to low-carbon energy sources over the last 20 years. The introduction of targets for renewable energy in the transport, heating, and electricity sectors in highly industrialised countries has also contributed to the expansion of what is now a multi-billion-dollar global renewable energy industry.[17] As a result, there has been a proliferation of renewable energy installations worldwide, from solar farms and wind turbines to hydropower and geothermal projects.

There are currently over 25,000 utility-scale wind projects and 16,000 utility-scale solar sites planned, under construction or in operation globally (Global Energy Monitor, 2023).[18] These numbers are expected to grow considerably over the next 5 years, accelerated as a result of emissions reduction goals as well as conflicts such as the Russian invasion of Ukraine and disruptions of oil and gas supplies (International Energy Association, 2022). Moreover, while most renewable energy development has been concentrated in the United States, China, and the European Union, there has been a significant increase in both the number of sites planned and under construction in Latin America, Southwest Asia, and Africa. Between 2016 and 2021, renewable energy production doubled in emerging and developing economies[19] and is expected to triple before 2027 (International Energy Association, 2022). Of the total number of utility-scale solar and wind sites, just over 4,000 large-scale solar sites and another 4,000 large-scale wind projects are located in emerging and developing economies. The majority of these sites are being developed by multinational companies headquartered in the Global North (Global Energy Monitor, 2023).[20]

The rapidly expanding renewable energy industry has raised concerns with regard to the impacts of large-scale projects on adjacent communities and environments (e.g., Allison et al., 2014); conflicts around land use, land grabbing, and displacement (Scheidel and Sorman, 2012; Ott et al., 2023); labour exploitation along global supply chains (e.g., Stop Uyghur Genocide, The Helena Kennedy Center for International Justice, and Unison, 2022); and the handling of toxic waste and other materials during the production process and after sites have been retired (e.g., Mulvaney, 2016). Additionally, a large number of projects have in recent years been sited in areas with long histories of capitalist and colonial exploitation, exacerbating existing social and environmental inequities and raising critical questions about ongoing resource extraction in these regions. These issues have prompted a variety of grassroots coalitions across civil society actors, academic researchers, stakeholders, and NGOs who are campaigning not only for a transition away from fossil fuels but a socially and environmentally *just transition*: a response to climate change that centres environmental justice and does not recreate familiar patterns of violence, dispossession, and environmental harm.

Some of the earliest organising around the just transition concept was focused on labour rights, beginning first as national movements and later taken up by global organisations (Stevis and Felli, 2020). Labour exploitation has become a major focal point of transnational networks and campaigns, particularly over the loss of jobs with the retirement of previous energy sites (such as coal mines or nuclear facilities) as well as labour exploitation along the global supply chain. Broad-based concerns over labour rights and the opportunity to shape new kinds of labour dynamics alongside the energy transition have given rise to international networks of unions working to advance democratic control and social ownership of energy. For instance, Trade Unions for Energy Democracy[21] includes members from 37 countries in the Global North and South and facilitates information sharing and collective organising.[22] South-South labour networks have also emerged with a focus on challenging labour exploitation and demanding a focus on sustainable employment alongside the energy transition.[23]

Specific incidents have given rise to ad-hoc networks focused on events of environmental or social harm. For instance, evidence that China, the world's largest exporter of solar panels, has used (and possibly is still relying on) forced Uyghur labour to produce solar energy components led to transnational coalitions between Uyghur activists, researchers, anti-slavery organisations, and labour unions to push for a ban on solar technology imports from China. This movement persuaded the United States, Canada, the UK, Australia, and the EU to impose various sanctions on Chinese solar panels and components produced from Uyghur labour. Governments and public agencies of the countries who are leading renewable energy development around the world now increasingly require a review of company sourcing records to ensure that they are not engaging in forced labour. NGOs like Solar Scorecard[24] publish yearly rankings of solar manufactures in an effort to increase transparency along the supply chain and expose the companies that are purchasing components which have not passed environmental and social standards.

Networks of grassroots actors have organised to push for public ownership of energy systems. These networks contend that democratic control over energy systems is pivotal to creating a more equitable and sustainable energy landscape. They challenge the dominance of profit-driven corporations and advocate for community ownership, cooperative models, and public control over renewable energy infrastructure. In essence, they envision an energy sector where decision-making power is decentralised, where local communities have a say in how their energy is produced and distributed, and where the benefits are shared collectively. Municipalities and city networks have been especially prominent in not only climate change governance more broadly (e.g., Betsill and Bulkeley, 2004) but also in radically re-claiming energy production and distribution under public ownership. These networks share resources, research, and guidelines for developing community energy programs (e.g., Diezmartínez and Zhang, 2023; Brinker and Satchwell, 2020; Klein and Coffey, 2016; Trumbull et al., 2020).

Transnational coalitions have emerged from struggles focused on sovereignty over land and natural resources, as well as renewable energy development in the context of ongoing colonial occupation. Large-scale wind and solar developments in the Western Sahara, which is currently occupied by Morocco, being built for export, has given rise to transnational networks challenging 'green extractivism' (Western Sahara Resource Watch, 2022). Across Latin America, internal displacements and violations of Indigenous rights in the course of renewable energy development has similarly produced transnational coalitions both within and beyond the Americas to challenge human rights abuses and Indigenous dispossession. The vast majority of transnational action, however, has focused on – or been framed by – the environmental implications of renewable energy development.[25] Over the last decade, activists have coalesced with regard to manufacturing practices and handling of waste following the retirement of renewable energy sites.

For instance, several movements grew out of the discovery that silicon tetrachloride – a highly toxic material used to fabricate solar panels – waste was being dumped in rural communities by manufacturers in China. International coalitions formed, including one between Greenpeace and the Chinese Renewable Energy Industry Association, which worked to distribute information and demand closer scrutiny to practices involved in renewable energy development. Another incident in 2011 involving the dumping of hydrofluoric acid in rural areas in China led to villagers storming and occupying Jinko Solar's factory in Zhejiang Province, which incited global backlash against the solar industry. While headlines of environmental harm from fossil fuel extraction and burning appear daily in the world's presses, with seemingly mild reactions, allegations of toxic waste dumping along the solar energy supply chain had major repercussions.

Unlike allegations of forced labour or Indigenous rights violations, evidence that the solar energy industry was aggravating environmental harm impacted the industry severely. This is primarily because the renewable energy industry is largely funded by investors in the socially responsible investment community and the green attributes of renewable energy development draw substantial private and public funding (Mulvaney, 2019). As a result, these incidents led to the widespread withdrawal of funding, increased opposition to solar, and a drop in the price of solar companies' stock (*Ibid.*,19, 33). This is now a well-known pattern of critique across many different kinds of 'green' energy projects and has significantly shaped the larger political landscape of renewable energy development.

For example, allegations of environmental harm as a result of the expansion of the biofuel (or agrofuels) sector has given rise to transnational campaigns and coalitions, particularly between Europe and Southeast Asia. Opposition movements to deforestation and land dispossession by rural peasants and Indigenous peoples across Southeast Asia merged with environmental justice activists in Europe to form powerful transnational movements against destructive biofuel practices. Over 250 groups were actively campaigning to prevent the inclusion of mandatory targets for biofuels in the European Union in 2010, including 1,200 European municipalities (Pye, 2010). These ongoing campaigns have largely prevented the inclusion of biofuels into mandatory EU energy targets and led to a major shift in how biofuels are perceived by publics around the world (*Ibid.*). Like critics of solar energy supply chains, the campaign led to new forms of governance over biofuel production, shaping standards regarding labour, land use, and environmental impacts.

The increased attention to the social and environmental impacts of renewable energy development has given rise to another form of transnational action that merits attention for its role in shaping the political and economic landscape of renewable energy development. Shareholder activists, long assumed to act primarily in the interest of increasing the value of a company, have more recently adopted radical strategies aimed at holding companies accountable for their contributions to climate change (Sandhu, 2022; Tillotson et al., 2023). In particular, shareholder activism has increasingly pressured companies to transition away from fossil fuels, fund clean-up activities, and invest in alternative energies.[26] Shareholders of both renewable energy companies and the institutions that fund renewable energy development have reshaped corporate practices and increased attention to the environmental and social implications of projects (Barko et al., 2022). Energy companies remain critically aware that their actions are under close scrutiny by their shareholders, whose decisions can radically shape internal power dynamics and even determine the future survival of corporations.

Finally, it is important to note that there has been a marked rise in transnational countermovements, some of which have taken the form of global advocacy movements to both deny climate change and advocate *against* renewable energy development (Dunlap and McCright, 2015, p. 318). These networks add to the dynamics of the political landscape of renewable energy development in that they often invoke evidence of environmental harm (as a result of renewable energies) to bolster arguments against the energy transition.[27] They have often sought to mimic the strategies of transnational climate activist networks, and it is sometimes difficult to identify their politics without close attention to parent organisations or funding sources (which can often be linked back to the oil and gas industries).[28] This dynamic has similarly shaped the context in which grassroots actors and renewable energy companies struggle to shape future energy systems.

It is thus clear that the political context of renewable energy development is influenced by the dominant identity of energy companies as 'sustainable,' 'green,' or 'environmentally friendly.' Allegations of environmental harm – and to some extent social damages – threaten renewable energy companies with significant reputational risk and can result in the withdrawal of project support from shareholders, funding institutions, and public approval (cf. Wright and Nyberg, 2015). While reputational concerns have also produced informal forms of governance over the oil and gas industry (Gillies, 2010), the renewable energy industry faces a much higher bar. Detrimental impacts to the environment can fuel anti-renewables campaigns by oil and gas lobbies, impacting the work of all actors in favour of renewable energies, regardless of their particular political orientation. In response, energy companies have increasingly attempted to manage accusations of greenwashing or environmental harm by forming relationships with environmental NGOs in an effort to pre-empt criticism (cf. Wright and Nyberg, 2015, p. 60).[29] This strategy has not always benefited the industry outright, however, and environmental organisations have as a result become more involved in directly shaping the development of energy sites.[30]

This larger context of transnational networks that are both shaping and operating within the political and economic landscape of renewable energy development and climate activism provides a foundation for exploring specific case studies of engagements among grassroots activists, environmental NGOs, and multinational renewable energy corporations.

Ethnographic Research in the Renewable Energy Industry

Between 2021 and 2022 I conducted ethnographic research on the renewable energy industry in London, while working as an unpaid intern for a multinational renewable energy company. On a cold, foggy morning in March, I overheard a loud conversation between a renewable energy developer[31] and an environmentalist associated with a transnational NGO arguing about a wind farm that was to be built in a rural region of Brazil. The stress in their voices caught my attention, and their argument had to do with the fact that the financial backing for a wind farm had recently been halted by the bank funding the project, according to the energy developer, 'because of a few birds and an angry farmer.' Curious, I pursued the environmentalist once they had finished their conversation to find out more.

It turned out that the argument had been over the fact that a group of people who were living near a proposed wind site in Brazil – including farmers, local politicians, residents, and a small environmental organisation – had determined that the wind project being developed by the energy company would impact a sensitive ecological area. The ecological area was not just valued for its biodiversity and the presence of endangered species, but it provided important resources for residents and farmers who collectively used the land seasonally for both farming and foraging. The group had attempted to contact the energy company directly and had been organising against state authorities with little success. Eventually, they had reached out to the larger NGO stationed in London. The environmentalist at the NGO (whom I overheard arguing with the developer) then called the bank funding the project to raise the residents' concerns over its environmental – and subsequently social – impacts. The bank promptly put a hold on the funding until the energy company agreed to consult with the Brazilian environmentalists and farmers.

The grassroots actors were not against the energy project in principle; rather, they had forwarded specific requests, including a proposal to move the wind farm to a new location to avoid harm to protected bird species and land uses, in addition to increased investment in the local community. The energy company had countered, offering to buy up new land to replace what would no longer be accessible and to fund a breeding program for the protected birds that would be 'impacted' (a type of program which Brazilian environmentalists pointed out had a historical record of failure). The company also offered to install new technologies to monitor the environment and specie populations, including AI programs that (they claimed) would be able to identify specific species of birds and trigger an automatic wind turbine shutdown if they came within a certain distance of the wind farm (the effectiveness of which the NGO in London similarly challenged). The grassroots organisers had refused and reformulated their demands, while the bank maintained that it would not fund the project until the energy company had reached an agreement with the Brazilian stakeholders.

This case challenged my assumptions about the power dynamics of the energy industry and also raised questions as to why, how, and in what contexts grassroots actors were able to leverage such influence over transnational corporations and financial institutions. Over the next several months, I increasingly found that these transnational networks among grassroots actors, NGOs, and energy companies were widespread and well known throughout the renewable energy industry. However, many of the engagements and negotiations I documented were not publicised or pursued through traditional legal venues, leaving no records with which to trace these networks. Even environmental impact assessments, which often list stakeholders (including international NGOs involved), seldom mention the negotiations that are often a part of how the development of these sites proceed. Because renewable energy sites take on average five years to develop, these negotiations often continue for years, sometimes even delaying site development for a decade. From interviews, I heard of similar networks initiated in North and South America, East and North Africa, India, and several countries in the Mashreq[32] and southwest Asia regions. Despite their seemingly ad-hoc nature, these transnational networks were surprisingly common around the world where renewable energy companies were operating and shared similar characteristics despite differences in local political landscapes. At the same time, they were distinct from the well-known forms of transnational advocacy (Keck and Sikkink, 1998) in specific ways.

Most networks were initiated by rural grassroots actors living near proposed energy sites who had made attempts to reach out directly to energy companies and were either ignored or failed to come to an agreement. They then connected with larger NGOs operating transnationally, typically one in the (Global North) country that hosted the headquarters of the energy company, which would pass along messages to financial institutions. The concerns of the grassroots actors varied, but most were framed with regards to the environmental impacts of new energy infrastructures, in addition to their contributions (or lack thereof) to local socio-economic wellbeing. Grassroots actors rarely opposed the idea of renewable energy sites altogether, and in fact, many were robust supporters of both renewable energy and rural infrastructure projects. However, their concerns had to do with details regarding the location, design, and operation of infrastructures, along with the flows of capital resulting from the production of new energy resources. In this vein, their goals focused on extending forms of governance over the design and operation of renewable energy sites, rather than any straightforward rejection of the projects.

Grassroots activists not only initiated these networks but played a lead role throughout the entire negotiating process. Apart from the fact that their knowledge of local environments, political contexts, and community needs informed the specific demands they forwarded, they would observe and document the practices of companies to ensure compliance once agreements were met. Some grassroots environmentalists would even use drones to conduct surveillance on sites to monitor companies and share real-time video footage with larger NGOs and financial institutions funding the projects. Those environmentalists associated with NGOs who facilitated contact with energy companies and financial institutions admitted that they could not possibly know about many of these projects or their potential impact to specific environments, and neither could they delineate the specific practices that would govern energy infrastructures effectively in each case.

Energy company employees with whom I spoke about these networks would often complain about the heterogeneity of both the processes and actors they would need to engage at any given site. Every engagement was slightly different, with distinct norms, requests, demands, politics, and cultural specificities. One developer commented:

> The process is difficult and hard to navigate. As an organisation we want to engage with them and do the right thing but the terms of engagement change depending on who it is you are talking with, within the same org[anisation]. There is no clear process, belief, or structure to follow. It's more based on the individual's belief system.[33]

Most developers I spoke with expressed similar frustrations about having to adapt to local demands that in no way could be streamlined or anticipated, which often worked, in their perspective, to slow down projects by months or even years.

Yet, despite the tension between energy companies and rural stakeholders, one of the defining features of these networks was that negotiations and conflicts were rarely publicised. This is not to say they were secret, as environmentalists, grassroots actors, and energy companies were happy to discuss them with me. However, actors engaged in these networks would often deliberately avoid engaging the media or legal venues and instead sought to negotiate directly with renewable energy companies and banks, engaging in forms of contentious politics that are far less visible and much harder to trace. While transnational advocacy networks traditionally focused on changing the behaviour and practices of states and corporations through policy or public pressure (Brunnengräber, 2014, p. 270), the aim of these coalitions was instead to quietly govern.

This choice was not one imposed by energy companies, for instance, in the way corporations make use of non-disclosure agreements or alternative dispute resolution processes to silence dissent (Nader, 2018). In fact, renewable energy companies quietly (and not so quietly) protest the power of transnational networks in different ways, attempting to reduce their influence without publicly denouncing them. For instance, a CEO of a large renewable energy company publicly argued that the industry should propose legislation that would limit the ability of 'non-local actors' from intervening in renewable energy projects.[34] Framed initially (and perhaps misleadingly) as a concern over the ability of the oil industry to prevent the development of renewable energy projects, it became obvious that he was referring to environmentalists and grassroots activists leveraging power through transnational networks. That the head of a global renewable energy company would propose actions to legally block the work of these networks hints at the extent of their influence.

This dynamic – with a focus on direct governance and a tendency to avoid the media – stems from the broader landscape of evolving climate activism and renewable energy development, as well as actors' perceptions of what these politics are. Amongst both the NGOs and local environmentalists whom I interviewed, there was a certain fatigue with transnational media campaigns. Even while some of those with whom I spoke would pursue traditional forms of transnational advocacy, they admitted that the power of public campaigns had probably passed their peak era of effectiveness. While powerful tools at the height of transnational advocacy campaigns in the 1990s and early 2000s, the general view was that people are bombarded with so much information on a daily basis that it is difficult to cultivate sustained engagement. One British conservationist at a large NGO involved in these networks lamented,

> there is always something more dramatic in the news to distract people, it is rare that we can get the same kind of reactions that we got twenty, ten or even five years ago. Sometimes we still try a campaign if all else fails, but it is worth questioning whether that is the best strategy anymore.[35]

Additionally, there was a general disinterest in any kind of campaign to design and implement more formal policy that would govern renewable energy sites in transnational contexts. Both grassroots actors and those who worked at larger NGOs expressed little faith that energy companies would follow guidelines or standards without direct, local intervention. One Tunisian environmentalist I spoke with suggested that this concern with direct oversight had emerged from the legacy of the oil and gas industry, which has historically committed serious harms against inhabitants and environments at sites of production with little accountability. Moreover, once the damage had been done, it was extremely difficult to restore environments and even more difficult to penalise foreign companies for their actions after the fact. Intervening in the early development of sites was seen as a key strategy to avoid irreparable harms in the future. Thus, amongst NGOs and grassroots activists, there is a sense of urgency in shaping where and under what conditions renewable energy sites are developed.

Another explanation for why these networks and details of conflicts are rarely made public is because of actors' views on the larger political context. Both environmentalists and renewable energy industry personnel shared a remarkably consistent perception of global climate change politics as being divided into a strict dichotomy between those who believed in climate change as a human-induced (and solvable) problem and those who did not; between renewable energy and fossil fuels; between the left and the right; between those who want to protect the environment and those who want to destroy it. In other words, there was a categorical separation between two, unequivocal positions. This shared belief provided a foundation for ongoing relationships between environmentalists and renewable energy companies and mutual interest in avoiding publicity.

In this vein, both NGOs and local activists were wary of condemning the renewable energy industry outright, as they felt that this could provide fuel for anti-renewables groups to counter the work to develop renewable energies *and* protect environments. For rural actors and environmentalists, publicising concerns relating to renewable energy development could result in a backlash from multiple sources that would discredit their concerns.

This view was certainly not unfounded and was in part shaped by previous media coverage of environmentalists or even individuals who opposed specific renewable energy projects. Reporters would often characterise those questioning the development of climate mitigation infrastructures as working against the climate movement or simply oblivious to the urgency of the energy transition.[36] Yet, most actors working in these spaces were fierce supporters of renewables but felt that the media would not be able to capture the subtlety of their concerns. The renewable energy industry maintained a similar position, and they work diligently to avoid outright clashes with environmentalists, such as those that might end up in the news or in litigation, as this could result in not only a cancelled project but also serious harm to their public image and reputation. These specific perceptions of climate change politics, situated within a changing political-economic landscape of climate activism, have come together in such a way as to increase the political power of grassroots actors.

While a clear picture of grassroots governance working through transnational networks emerges from ethnographic research, to what extent do these patterns exist outside these specific field sites? Next, I turn to an analysis of case studies to understand if these specific dynamics are more broadly present at sites of renewable energy development around the world.

Case Studies

Drawing on 28 case studies of wind energy conflicts documented by the Environmental Justice Atlas (EjAtlas), in addition to three cases I documented during fieldwork, I analysed a total of 31 case studies of grassroots organising around wind projects in the Global South (See Table 26.2 Case Studies of Wind Conflicts for complete list). Countries include Kenya, Tunisia, Western Sahara, Chile, Albania, Mexico, South Africa, India, Colombia, Honduras, Brazil, and Turkey, and each of these sites was located in rural areas. This broad geographic representation minimises the potential for random variation and increases the likelihood of capturing trends or patterns.

The results show a clear pattern (Table 26.1). Grassroots movements that did not include an international coalition or network resulted in very few changes in the outcome of the final project. Of the 12 cases that did not involve a transnational coalition, only one project was suspended, and the rest of the projects proceeded as planned (Figure 26.1). However, for those grassroots movements which did involve an international coalition, these engagements resulted in significantly more diverse outcomes. Of the 19 conflicts which involved a transnational network, nine projects were cancelled, three were suspended or delayed, four proceeded, and an

Table 26.1 Case studies analysis.

Movements with international networks		Movements without international networks	
Cancelled	9	Cancelled	0
Suspended, delayed	3	Suspended, delayed	1
In operation	4	In operation	11
In operation, but delayed, altered, or suspended after construction	3	In operation, but delayed, altered, or suspended after construction	0
Total cases	19	Total cases	12

Source: Original creation.

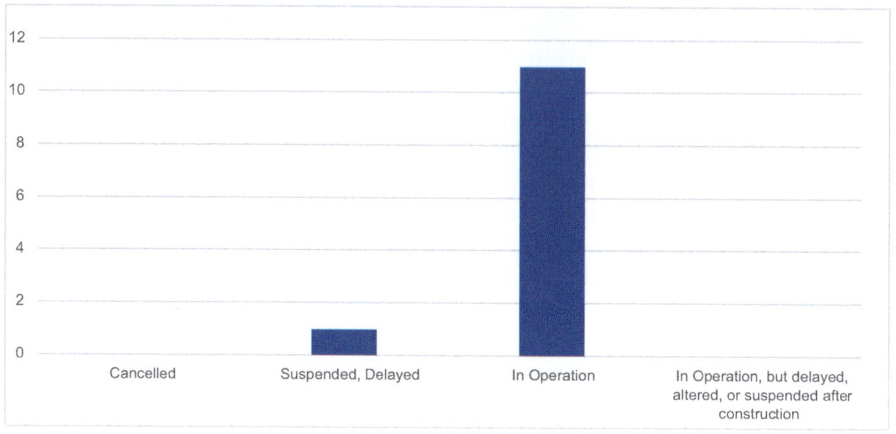

Figure 26.1 Grassroots organising without international networks.

Source: Original creation.

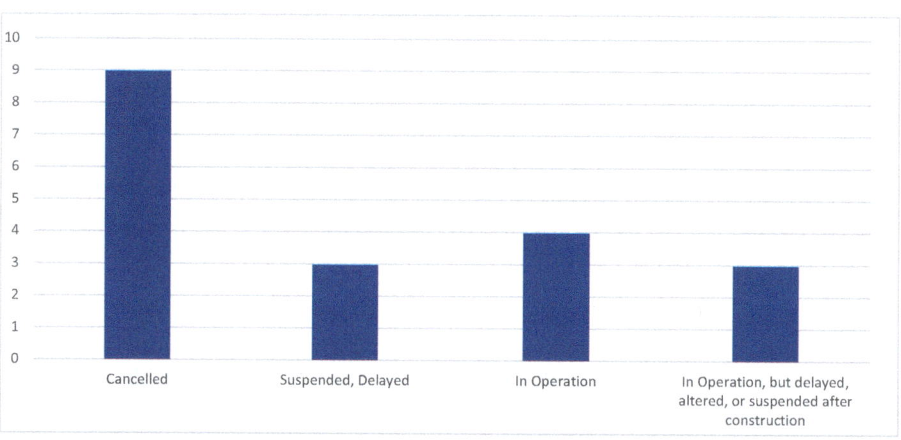

Figure 26.2 Grassroots organising involving international networks.

Source: Original creation.

additional three proceeded but were altered or suspended after construction (Figure 26.2). Seventeen of the nineteen grassroots initiatives that involved a transnational network also included a Global North renewable energy company. Additionally, banks or funding institutions withdrew funding for seven of the 31 projects, six of which involved a transnational network.[37]

While this is a modest sample size, these case studies represent a diversity of sociopolitical contexts, providing a sufficiently representative subset of data to enable meaningful conclusions. It is clear that those grassroots movements which mobilised transnational networks achieved a much more dynamic field of political engagement. Projects were cancelled, delayed, suspended, or even dismantled after construction as a result of grassroots organising and transnational networks, suggesting that these movements were able to challenge the power of multinational energy companies and international financial institutions to a greater extent than those movements which lacked a transnational component.

Table 26.2 Case studies of wind conflicts.

Conflict start date	Name of project	Site location	Companies involved	Company headquarters	Key issues	International networks involved?	Funding institutions	Outcome	New environmental impact assessment required?	Funding withdrawn?
2015	Parque eólico Pililín	Chile	Acciona Energía	Spain	Predominantly environmental impacts	Yes	Unconfirmed but likely the Inter-American Development Bank	Cancelled	Yes	Unknown
2014	Mersinli Wind Farm	Turkey	3E Enerji and Nordex	Turkey and Germany	Predominantly environmental impacts	No	European Bank for Reconstruction and Development	In operation	Yes	Yes
2011	Caetité region wind farms	Brazil	Ibedrola, Renova	Brazil and Spain	Predominantly environmental impacts	No	Unknown	In operation and in construction	No	Unknown
2010	Alegria I and II	Brazil	New Energy, BioEnergy	Brazil	Predominantly environmental impacts	No	The Brazilian Development Bank (BNDES)	In operation	No	No
2010	Chiloé wind power project	Brazil	EcoPower	Chile and Sweden	Predominantly environmental impacts	Yes	Swedish funds	Under negotiation, likely cancelled	Yes	Yes
2019	Canudos wind complex	Brazil	Voltalia	France	Predominantly environmental impacts	Yes	Banco do Nordeste The Brazilian Development Bank (BNDES) European Bank for Reconstruction and Development	Suspended	Yes	Yes

(*Continued*)

Table 26.2 (Continued)

Conflict start date	Name of project	Site location	Companies involved	Company headquarters	Key issues	International networks involved?	Funding institutions	Outcome	New environmental impact assessment required?	Funding withdrawn?
2006	Lake Turkana Wind Power Project	Kenya	Norfund, Vestas, IDC	Norway, Denmark, Finland, Netherlands	Predominantly social impacts	Yes	European Investment Bank	In operation (delayed)	Unknown	Yes, World Bank backed out. Others reassessed project. Google backed out.
2011	Cerro de Hula wind project	Honduras	Ibedrola, Gamesa	Spain and Costa Rica	Both environmental and social impacts	No	Partly funded by World Bank (for extension of project)	In operation	Unknown	No
2013	Tarfaya Windfarm Complex	Western Sahara	GDF, Nareva, Siemens, Suez Energy	Morocco, Germany, France	Predominantly social impacts	Yes	IFC	In operation	No	No
2016	Foum El Oued Wind Farm	Western Sahara	Siemens, Enel, Nareva	Morocco, Italy, Germany	Predominantly social impacts	Yes	Moroccan and European sources	In operation	Unknown	No
1985	Wind farms in and near Bhimashankar Wildlife Sanctuary, including the Andhra Lake Wind Farm	India	CLP, Enercon	Hong Kong, Germany, India	Both environmental and social impacts	No	Private funders	In operation	Unknown	Unknown

(Continued)

Table 26.2 (Continued)

Conflict start date	Name of project	Site location	Companies involved	Company headquarters	Key issues	International networks involved?	Funding institutions	Outcome	New environmental impact assessment required?	Funding withdrawn?
2016	Parque eólico en Kimbilá	Yucatán, Mexico	Elecnor	Spain	Both environmental and social impacts	Yes	Private funders	Suspended, negotiations	No	No
2009	Karaburun Wind	Turkey	Lodos Electricity, Enercon	Turkey, Germany	Predominantly environmental impacts	No	Landesbank Baden-Württemberg	In operation	No	No
2013	Lamu Cordisons Wind Power Project (also called Baharini Wind Power Project)	Kenya	Cordison, then Kenwinds	US, Belgium	Both environmental and social impacts	Yes	Partly funded by International Finance Corporation	Cancelled	Yes	Yes
2014	Kinangop Wind Park	Kenya	EcoGen, KenGen	UK and Kenya	Predominantly environmental impacts	Yes	Norfund, Standard Bank South Africa, African Investment Fund	Cancelled	No	Yes
2012	Parque Eólico PIER de Iberdrola en Puebla	Mexico	Ibedrola	Spain	Predominantly environmental impacts	No	Nacional Financiera S.N.C. (NAFIN)	In operation	No	No
2016	Parque Eólico Tizimín	Mexico	Avant Energy	United States	Predominantly environmental impacts	No	Spanish Official Credit Institute (ICO), Banco Sabadell, and the Mexican national development bank, Nacional Financiera (NAFIN)	In operation	No	No

(*Continued*)

Table 26.2 (Continued)

Conflict start date	Name of project	Site location	Companies involved	Company headquarters	Key issues	International networks involved?	Funding institutions	Outcome	New environmental impact assessment required?	Funding withdrawn?
2018	Parque Eólico Chicxulub en ejido de Ixil (aka Progresso Wind Farm, Peninsula Wind Farm)	Mexico	Grupo, Elawan Energy	Spain	Predominantly environmental impacts	Yes	Climate Bonds	In operation	No	No
2012	Baleia Wind Power Complex (aka Acaraú wind farm)	Brazil	Furnas	Brazil	Predominantly environmental impacts	No	Unknown	Delayed	Yes.	Unknown
2001	Wind farms in the Koyna Sanctuary	India	Suzlon; Enercon, others	India, Germany	Predominantly environmental impacts	Yes	Unknown	Cancelled	Unknown	Unknown
2007	Mareña Renovables in San Dionisio del Mar	Mexico	Mareña Renovables	Spain	Predominantly environmental impacts	Yes	Inter-American Development Bank (IADB)	Cancelled	No	Yes
2007	Suzlon wind farm in Dhule	India	BP India (British Petroleum)	UK	Both environmental and social impacts	Yes	Rabobank (Netherlands)	In operation, delayed, project size reduced by 90% of original plan	Yes	No/partly

(*Continued*)

Table 26.2 (Continued)

Conflict start date	Name of project	Site location	Companies involved	Company headquarters	Key issues	International networks involved?	Funding institutions	Outcome	New environmental impact assessment required?	Funding withdrawn?
2012	Parque Eólico Dzilam Bravo (aka Eolica Golfo 1)	Mexico	Eólica del Golfo 1 S.A.P.I. de C.V. Vive Energía Envision Energy International Ltd.	Mexico and China	Predominantly environmental impacts	Yes	Unknown	In operation	No	Unknown
2010	Wind Farms in the sustainable development reserve of Ponta do Tubarão	Brazil	BioEnergy, New Energy Options	Brazil	Predominantly environmental impacts	No	Banco Nacional de Desenvolvimento Econômico e Social (BNDES)	In operation	No	No
2005	Suzlon Energy wind farms in Kutch District	India	Suzlon Energy	India	Both environmental and social impacts	No (EJ atlas says yes, but cannot find any evidence)	Unknown	In operation	No	No
2001	Parque eólico Jepirachi	Colombia	Empresas Públicas de Medellín Industrias Ectricol Deutsche WindGuard	Germany, Colombia	Predominantly environmental impacts	Yes	World Bank	Operated, then dismantled for non-compliance	Yes	No

(Continued)

Table 26.2 (Continued)

Conflict start date	Name of project	Site location	Companies involved	Company headquarters	Key issues	International networks involved?	Funding institutions	Outcome	New environmental impact assessment required?	Funding withdrawn?
2008	Wind Farm installation in the protected area of Karaburuni Peninsula	Albania	Moncada Energy Group	Italy	Predominantly environmental impacts	Yes	Unknown	Cancelled	Unknown	Unknown
2010	Chiloé Wind Power Project	Chile	Ecopower	Belgium	Predominantly environmental impacts	Yes	Possibly Swedish financer	Cancelled	Unknown	Unknown
2020	Faro del Sur wind farm	Chile	Enel	Italy	Predominantly environmental impacts	Yes	Unknown	Cancelled	No	Unknown
2020	Inyanda-Roodeplaat wind farm	South Africa	Inyanda Energy Projects (Pty) Ltd	South Africa	Predominantly environmental impacts	Yes	Unknown	Cancelled	No	Unknown
2011	Sidi Daoud Wind Farm	Tunisia	STEG	Tunisia	Predominantly social impacts	No	Partly supported by World Bank	In operation	No	No

Source: Original creation.

Conclusion

This chapter began by tracing the evolution of transnational networks engaged in climate activism, from transnational advocacy networks to the increasing prevalence of grassroots actors involved in climate governance. These movements have shaped the larger political landscapes of renewable energy development and have given rise to novel governance spaces (Bulkeley et al., 2014b) that have allowed grassroots actors to leverage new forms of political power in their engagements with renewable energy companies and international financial institutions. While transnational networks are important conduits of power inequities, the cases presented here indicate that they can also be leveraged to challenge historical disparities. In this vein, grassroots transnational networks exemplify the rising potential for collective action across borders that highlights the interconnectedness of what are simultaneously local and global concerns. As such, rural grassroots actors are not just passive observers but active architects of a more just energy future.

In this vein, international corporations and financial institutions, often perceived to be out of reach of any kind of local intervention, are increasingly pulled into the orbit of locally based governance and intervention. At the same time, the success of these transnational networks may be limited in terms of establishing broader standards or international policies that can be used to regulate the energy industry, given their highly localised focus on specific contexts and somewhat ad-hoc nature. While I met several environmentalists who hoped to establish formal standards for the industry that could be legally imposed, there was an acknowledgment that it would be difficult to locate a legal venue or transnational authority to hold companies accountable. This is in addition to the increasing scepticism of formal governance structures that have often failed marginalised communities.

This is, of course, one of the greatest challenges of climate change policy in general when it comes to actions and actors that cross national boundaries and in efforts to govern transnational flows. At the same time, there are some indications that hint towards the creation of broader norms.[38] Large NGOs have come to play an additional role beyond just functioning as a conduit through which local actors can reach energy companies and financial institutions. These organisations, which work with activists around the world, also retain information on each negotiation and its outcome; details which they could pass on to other grassroots environmental organisations or other locally based actors who could learn and adapt their strategies based on previous cases. Thus, even while the specific projects may change, NGOs maintained an informal archive of standards and policies that grassroots actors can mobilise in their negotiations with the renewable energy industry.

These grassroots efforts, operating within a complex web of transnational networks, offer a unique perspective on the evolving landscape of climate activism. As the world continues to struggle to transition away from fossil fuels, these initiatives serve as a reminder that the path to a more environmentally and socially just future requires more than just the adoption of alternative energies; it demands a radical reorientation away from the past practices and politics that have shaped our current energy dilemmas.

Ethics Statement

This study was approved by the Stanford University Institutional Review Board under Expedited Review (eProtocol #62525). Informed consent was obtained from all participants. The author does not have any financial interests that are directly or indirectly related to the projects under study.

Acknowledgments

This research was supported by grants from the Social Science Research Council International Dissertation Research Fellowship, the Stanford University Europe Center, and the Stanford Institute for Research in the Social Sciences.

Notes

1 There are a large number of studies that have addressed what is sometimes called 'public perceptions of renewables,' or conflicts stemming from renewable energy development. Many of these are focused on highly industrialised contexts such as Western Europe and the United States, although there is a growing focus on the Global South as projects in these regions increase.
2 See for instance, Berezow (2021); Chait (2022); Demsas (2022); Gibbens (2022); Meigs (2002); Osaka (2022); Pino (2022); and Spector (2015).
3 Although diverse in form and stemming from existing (and complex) political and economic contexts, movements focusing on the development of energy sites are sometimes organised under the concept of a just energy transition (e.g., Heffron and McCauley, 2018; Baker and Phillips, 2019; Baker, 2021), which goes beyond simply shifting from fossil fuels to renewable energy sources. Rather, this approach entails a deliberate focus to ensure that the benefits of transitions to new energy sources are equitably distributed across and between societies. Many of these efforts can also be grouped under the larger umbrella of *energy justice*, an established and growing focus of activism and academic research around the world that focuses on the multi-layered impacts of energy systems. The concept of energy justice focuses on how to design energy systems that centre justice and equity. Such projects have also been framed as an *environmental justice* approach by some scholars and activists (e.g., Levenda et al., 2021). See also High and Smith (2019).
4 For instance, most biofuel production is export-oriented, and several large-scale renewables projects are being developed for export across national borders. Examples include plans to export wind and solar generated electricity from North Africa to European consumers; Australia's plan to export solar generated electricity to Singapore; and many existing hydropower plants that export energy to neighbouring countries, for instance, from Mozambique to South Africa.
5 Ethnographic research methods include different forms of interview, participant observation, and long-term immersive fieldwork.
6 Anthropological attention to energy has been central in explaining how energy regimes unevenly distribute forms of social, political, and economic power. While the majority of anthropological attention to energy production, distribution, and consumption has focused on fossil fuel production, distribution, and consumption, there is an increasing focus on renewables and alternative energy experiments. These include: Franquesa, 2018, Alonso Serna, 2022, Friede and Lehmann, 2016, Nadaï and Labussière, 2010, Dracklé and Krauss, 2011, and Hughes, 2021 on wind; Whitington, 2018 and Lord, 2016 on hydropower; Rignall, 2016, Jensen, 2019, Cross and Murray, 2018, Lennon, 2021 on solar; and Watts, 2018 on wave power (and other energy experiments).
7 Fieldnotes, interview transcripts, and documentary sources were coded qualitatively using MAXQDA (maxqda.com) to identify common patterns and themes, and follow-up interviews were conducted to confirm key findings. Names, places, and specific events have been de-identified to protect the confidentiality of research participants and specific sites under discussion.
8 There are not always straightforward distinctions between the Global North/South or West/East, and there are different arguments about which countries fall into these categories. For simplicity, this study includes those countries with a GDP per capita of < US$30,000 as 'Global South.' China was excluded because of the inability to verify information regarding grassroots organising in the country.
9 globalenergymonitor.org
10 For instance, the 1997 Kyoto Protocol, the 2009 Copenhagen Climate Conference, and the 2015 Paris Agreement have been criticised for the general lack of legally binding commitments, enforcement mechanisms, and processes for inclusivity, and the failure to address the evolving dynamics of emissions.
11 For a review of these two conflicting approaches, see Adams (2015).

12 Transnational here refers to connections between actors in two or more countries. Transnational networks can take on many forms and involve different kinds of coalitions, including among subnational governments, NGOs, businesses, and a range of non-state actors.

13 Governance here refers to the processes, mechanisms, relations, and organisations through which actors influence decision-making in specific arenas. Anthropological approaches to governance generally differ from a policy or International Relations conception of governance by including a focus on informal and non-state actors in the governance of different arenas, for instance, human rights, global health, or the environment. For an introduction to anthropological approaches to governance, see Eckert et al. (2012). For an overview of approaches to environmental governance, see Lemos and Agrawal (2006).

14 Grassroots climate change governance refers to the bottom-up, community-driven, and decentralised processes through which individuals, local groups, and communities engage in climate action and decision-making.

15 https://climateaction.unfccc.int/

16 These networks include villages, universities, small-scale NGOs, municipalities, and city governments, such as the Cities for Climate Protection program (Betsill and Bulkeley, 2004) and C40 (Lin, 2018).

17 By some estimates, it is now a multi-trillion-dollar industry. See International Energy Agency (2023).

18 Utility-scale is defined by the Global Energy Monitor as 10MW and above. Sites that were listed as cancelled, retired, mothballed, and shelved were removed from these totals.

19 The International Energy Agency determines 'emerging and developing economies' as the countries which cannot be classified as advanced economies. China and Turkey are not included in the emerging and developing economies category because of their distinctive market and investment dynamics.

20 The Global Energy Monitor provides detailed, up-to-date data on renewable energy projects, including size, location, and the companies involved.

21 www.tuedglobal.org

22 Networks of labour unions are not necessarily new in relation to the energy transition. Earlier networks of workers' unions also helped shape the policy landscape in the United States and the EU for the expansion of renewable energy (e.g., Vasi, 2009).

23 Many of these grassroots networks came together at the Our Future is Public conference in Chile in 2023 to shape, discuss, develop, and share strategies for reclaiming public services. Energy was a major focus and resulted in The Energy Democracy Declaration.

24 https://solarscorecard.org/

25 Although it is difficult to track all conflicts and negotiations surrounding renewable energy development, particularly given that some conflicts and negotiations are deliberately kept out of the media, the Global Atlas of Environmental Justice tracks local disputes emerging from various energy-related developments and can give a good estimation of the major issues cited by grassroots actors and civil society organisations involved in disputes.

26 For instance, in 2021, a tiny activist hedge fund managed to gain seats on ExxonMobile's Board of Directors during a campaign in which it criticised Exxon's failure to address its contribution to climate change and its lack of investments in renewable energies. By garnering broad support for their stance, climate activists won seats on the governing body of one of the world's largest and dirtiest fossil fuel corporations (Hiller and Herbst-Bayliss, 2021). See also Cronin (2019).

27 These include both local community organisers all the way up to the oil and gas industry. For instance, the toxicity of solar is often cited as a leading reason by opponents of the energy transition and climate deniers not to develop alternative energy. For example, the Heartland Institute, an oil-funded climate denial organisation, publishes regularly on the toxicity of solar, using the case of silicon tetrachloride dumping as a key example. Other grassroots organisers, like the Long Island-based community organisation Stop the Solar Farms in Calverton, have organised based on the environmental impacts of solar to 'stop international energy companies and greedy politicians' (www.stopthesolarfarms.com). It's important to note that while these reports and concerns over solar's toxicity are not wrong and are often used to draw attention to the fact that there are inequities in the way energy sites are being built, climate change deniers typically argue against *any* transition to alternative energies. Opponents to wind development similarly span from grassroots organisations to transnational coalitions to think tanks backed by powerful oil lobbies, and their

arguments predominantly focus on the negative environmental implications of large-scale wind. Some transnational networks against wind include the European Platform Against Windfarms and the North American Platform Against Wind Power (NA-PAW).

28 Some transnational networks that have played key roles in diffusing denial include the Atlas Economic Research Foundation, the Economic Freedom Network, and the International Policy Network. The latter has also founded the Civil Society Coalition on Climate Change, which consists of grassroots organisations and individuals in over 40 nations who share the goal of actively denying climate change.

29 This strategy is increasingly used by a wide variety of companies in the energy transition space, including electric vehicle manufacturing, the airline industry, and electric utilities.

30 Transnational environmental organisations are not only shaping current practices and extending new forms of governance over renewable energy development; they also played significant roles in the formation of the renewable energy industry, advocating for renewable energy, opposing oil and gas monopolies, and pushing for policies and political support that would allow the widespread development of alternative energies (e.g., Vasi, 2009; Sine and Lee, 2009).

31 The employees at renewable energy companies who carry out the bulk of the work to develop renewable energy sites – including negotiating access to land, acquiring permits, and conferring with environmentalists – are typically referred to as business or project developers in the industry. I refer to them as energy developers here to make it easier for readers to follow along.

32 The region includes the states of Bahrain, Egypt, Iraq, Jordan, Kuwait, Lebanon, Oman, Qatar, Saudi Arabia, Sudan, the State of Palestine, Syria, the United Arab Emirates, and Yemen.

33 Interview #48, 16 January 2022.

34 Field note. Large Scale Solar Europe Conference, Lisbon, March 2022.

35 Interview #62, 2 February 2022. I conducted this interview just as Russia was invading Ukraine, which had captured the attention of the media and general public at the time. This may have impacted this particular individual's view, but I have heard similar sentiments across a time period of several years.

36 For some examples, see Berezow, 2021; Chait, 2022; Gibbens, 2022; Osaka, 2022; Pino, 2022; Simon, 2022; Spector, 2015.

37 There could easily have been more banks and funding institutions that withdrew funding, but it was not possible to verify this information for each case.

38 Norms have been thought of by some as informal governance, and they have also served as the beginnings of international law. See for instance, Levit, 2005.

References

Abramsky, K. (ed.) (2010). *Sparking a Worldwide Energy Revolution: Social Struggles in the Transition to a Post-Petrol World*. Oakland, CA and Edinburgh: AK Press.

Adams, D. (2015). Why we cannot wait: Transnational networks as a viable solution to climate change policy. *Santa Clara Journal of International Law,* 13(2), 307.

Allison, T. D., Root, T. L. and Frumhoff, P. C. (2014). Thinking globally and siting locally – renewable energy and biodiversity in a rapidly warming world. *Climatic Change*, 126(1), 1–6. https://doi.org/10.1007/s10584-014-1127-y

Almeida, P. (2019). Climate justice and sustained transnational mobilization. *Globalizations*, 16(7), 973–979. https://doi.org/10.1080/14747731.2019.1651518.

Andonova, L. B. (2021). Clean energy and the hybridization of global governance. In J. C. W. Pevehouse, K. Raustiala and M. N. Barnett (eds.) *Global Governance in a World of Change* (pp. 288–310). Cambridge: Cambridge University Press. https://doi.org/10.1017/9781108915199.011.

Baker, L. and Phillips, J. (2019). Tensions in the transition: The politics of electricity distribution in South Africa. *Environment and Planning C: Politics and Space*, 37(1), 177–196. https://doi.org/10.1177/2399654418778590

Baker, S. H. (2021). *Revolutionary Power: An Activist's Guide to the Energy Transition*. Washington, DC: Island Press.

Barko, T., Cremers, M. and Renneboog, L. (2022). Shareholder engagement on environmental, social, and governance performance. *Journal of Business Ethics*, 180(2), 777–812. https://doi.org/10.1007/s10551-021-04850-z

Batliwala, S. (2002). Grassroots movements as transnational actors: Implications for global civil society. *Voluntas: International Journal of Voluntary and Nonprofit Organizations*, 13(4), 393–409.

Berezow, A. (2021). Environmentalists want renewable energy, but not power lines. *American Council on Science and Health*, 12 February. https://www.acsh.org/news/2021/02/12/environmentalists-want-renewable-energy-not-power-lines-15341

Betsill, M. M. and Bulkeley, H. (2004). Transnational networks and global environmental governance: The cities for climate protection program. *International Studies Quarterly*, 48(2), 471–493. https://doi.org/10.1111/j.0020-8833.2004.00310.x

Betsill, M. M. and Corell, E. (eds.) (2008). *NGO Diplomacy: The Influence of Nongovernmental Organizations in International Environmental Negotiations*. Cambridge, MA and London: MIT Press.

Brecher, J. (2015). *Climate Insurgency: A Strategy for Survival*. Boulder: Paradigm Publishers.

Brinker, L. and Satchwell, A. J. (2020). A comparative review of municipal energy business models in Germany, California, and Great Britain: Institutional context and forms of energy decentralization. *Renewable and Sustainable Energy Reviews*, 119(March), 109521. https://doi.org/10.1016/j.rser.2019.109521

Brunnengräber, A. (2014). Between pragmatism and radicalization: NGOs and Social movements in international climate politics. In M. Dietz and H. Garrelts (eds.) *Routledge Handbook of the Climate Change Movement*. Abingdon and New York: Routledge.

Bulkeley, H., Andonova, L. B., Betsill, M. M., Compagnon, D., Hale, T., Hoffmann, M. J., Newell, P., Paterson, M., Roger, C. and VanDeveer, S. D. (2014). *Transnational Climate Change Governance*. Cambridge: Cambridge University Press, July. https://doi.org/10.1017/CBO9781107706033

Chait, J. (2022). Why are environmental activists trying to block green energy? *Intelligencer*, 9 September. https://nymag.com/intelligencer/2022/09/why-are-environmental-activists-trying-to-stop-green-energy.html.

Chan, S., Falkner, R., Goldberg, M. and van Asselt, H. (2018). Effective and geographically balanced? An output-based assessment of non-state climate actions. *Climate Policy*, 18(1), 24–35. https://doi.org/10.1080/14693062.2016.1248343

Ciplet, D., Roberts, J. T. and Khan, M. R. (2015). *Power in a Warming World: The New Global Politics of Climate Change and the Remaking of Environmental Inequality*. Cambridge, MA and London: The MIT Press.

Cronin, D. (2019). Activist investors force change in the oil industry. *Carbon Tracker Initiative*, 12 June. https://carbontracker.org/activist-investors-force-change-in-the-oil-industry/

Cross, J. and Murray, D. (2018). The afterlives of solar power: Waste and repair off the grid in Kenya. *Energy Research & Social Science*, 44(October), 100–109. https://doi.org/10.1016/j.erss.2018.04.034

Demsas, J. (2022). Not everyone should have a say. *The Atlantic*, 19 October, sec. Ideas. https://www.theatlantic.com/ideas/archive/2022/10/environmentalists-nimby-permitting-reform-nepa/671775/

Dietz, M. and Garrelts, H. (eds.) (2014). *Routledge Handbook of the Climate Change Movement*. Routledge International Handbooks. Abingdon and New York: Routledge.

Diezmartínez, C. V. and Zhang, A. (2023). Powering just energy transitions: A review of the justice implications of community choice aggregation. *Energy Research & Social Science*, 103 (September), 103221. https://doi.org/10.1016/j.erss.2023.103221

Dracklé, D. and Krauss, W. (2011). Ethnographies of wind and power. *Anthropology News*, 52(5), 9. https://doi.org/10.1111/j.1556-3502.2011.52509.x

Dunlap, R. E. and McCright, A. (2015). Challenging climate change: The denial countermovement. In R. E. Dunlap and R. J. Brulle (eds.) *Climate Change and Society: Sociological Perspectives* (pp. 300–332). New York: Oxford University Press.

Dupuits, E. (2021). Reversing climatisation: Transnational grassroots networks and territorial security discourse in a fragmented global climate governance. *International Politics*, 58(4), 563–581. https://doi.org/10.1057/s41311-020-00256-2

Eckert, J., Behrends, A. and Dafinger, A. (2012). Governance – and the state: An anthropological approach. *EthnoScripts*, 14(1), 14–34.

Franquesa, J. (2018). *Power Struggles: Dignity, Value, and the Renewable Energy Frontier in Spain*. New Anthropologies of Europe. Bloomington, IN: Indiana University Press.

Friede, S. and Lehmann, R. (2016). Consultas, corporations, and governance in Tehuantepec, Mexico. *Peace Review*, 28(1), 84–92. https://doi.org/10.1080/10402659.2016.1130388

Gibbens, S. (2022). Activists fear a new threat to biodiversity – Renewable energy. *National Geographic*, 26 May, sec. Environment. https://www.nationalgeographic.com/environment/article/activists-fear-biodiversity-threat-from-renewable-energy.

Gillies, A. (2010). Reputational concerns and the emergence of oil sector transparency as an international norm. *International Studies Quarterly*, 54(1), 103–126. https://doi.org/10.1111/j.1468-2478.2009.00579.x

Global Energy Monitor (2023). https://globalenergymonitor.org/.

Hadden, J. (2015). *Networks in Contention: The Divisive Politics of Climate Change*. Cambridge Studies in Contentious Politics. Cambridge: Cambridge University Press. https://doi.org/10.1017/CBO9781316105542

Harlan, S., Pellow, D., Roberts, J. T., Bell, S. E., Holt, W. and Nagel, J. (2015). Climate justice and inequality. In R. E. Dunlap and R. J. Brulle (eds.) *Climate Change and Society: Sociological Perspectives* (pp. 127–165). Oxford: Oxford University Press.

Heffron, R. J. and McCauley, D. (2018). What is the 'just transition'? *Geoforum*, 88(January), 74–77. https://doi.org/10.1016/j.geoforum.2017.11.016

High, M. M. and Smith, J. M. (2019). Introduction: The ethical constitution of energy dilemmas. *Journal of the Royal Anthropological Institute*, 25(S1), 9–28. https://doi.org/10.1111/1467-9655.13012

Hiller, J. and Herbst-Bayliss, S. (2021). Exxon loses board seats to activist hedge fund in landmark climate vote. *Reuters*, May 27, 2021, sec. Sustainable Business. https://www.reuters.com/business/sustainable-business/shareholder-activism-reaches-milestone-exxon-board-vote-nears-end-2021-05-26/

Hughes, D. M. (2021). *Who Owns the Wind? Climate Crisis and the Hope of Renewable Energy?* London and New York: Verso.

International Energy Agency. (2023). *Overview and Key Findings – World Energy Investment 2023 – Analysis*. https://www.iea.org/reports/world-energy-investment-2023/overview-and-key-findings

International Energy Association. (2022). *Executive Summary – Renewables 2022*. Paris. https://www.iea.org/reports/renewables-2022

Jensen, C. (2019). Here comes the sun?: Experimenting with Cambodian energy infrastructures. In K. Hetherington (ed.) *Infrastructure, Environment, and Life in the Anthropocene* (pp. 216–235). Durham, NC: Duke University Press.

Jordan, A., Huitema, D., van Asselt, H. and Forster, J. (eds.) (2018). *Governing Climate Change: Polycentricity in Action?* Cambridge: Cambridge University Press. https://doi.org/10.1017/9781108284646

Kaiser, C. (2022). Rethinking polycentricity: On the North–South imbalances in transnational climate change governance. *International Environmental Agreements: Politics, Law and Economics*, 22(4), 693–713. https://doi.org/10.1007/s10784-022-09579-2

Keck, M. E. and Sikkink, K. (1998). *Activists beyond Borders: Advocacy Networks in International Politics*. Ithaca, NY: Cornell University Press.

Klein, S J. W. and Coffey, S. (2016). Building a sustainable energy future, one community at a time. *Renewable and Sustainable Energy Reviews*, 60(July), 867–880. https://doi.org/10.1016/j.rser.2016.01.129

Lemos, M. C. and Agrawal, A. (2006). Environmental governance. *Annual Review of Environment and Resources*, 31(1), 297–325. https://doi.org/10.1146/annurev.energy.31.042605.135621

Lennon, M. (2021). Energy transitions in a time of intersecting precarities: From reductive environmentalism to antiracist praxis. *Energy Research & Social Science*, 73(March), 101930. https://doi.org/10.1016/j.erss.2021.101930

Levenda, A. M., Behrsin, I. and Disano, F. (2021). Renewable energy for whom? A global systematic review of the environmental justice implications of renewable energy technologies. *Energy Research & Social Science*, 71(January), 101837. https://doi.org/10.1016/j.erss.2020.101837

Levit, J. K. (2005). A bottom-up approach to international lawmaking: The tale of three trade finance instruments. *Yale Journal of International Law*, 30, 125–209.

Lin, J. (2018). *Governing Climate Change: Global Cities and Transnational Lawmaking*. Cambridge Studies on Environment, Energy and Natural Resources Governance. New York: Cambridge University Press.

Lord, A. (2016). Making a 'Hydropower Nation': Subjectivity, mobility, and work in the nepalese hydroscape. *HIMALAYA*, 13.

Meigs, J. B. (2002). The green war on clean energy. *City Journal* (blog). https://www.city-journal.org/article/the-green-war-on-clean-energy/

Mulvaney, D. (2016). Energy and global production networks. In T. Van de Graaf, B. K. Sovacool, A. Ghosh, F. Kern and M. T. Klare (eds.) *The Palgrave Handbook of the International Political Economy of Energy* (pp. 621–640). London: Palgrave Macmillan.

Mulvaney, D. (2019). *Solar Power: Innovation, Sustainability, and Environmental Justice*. Oakland, CA: University of California Press.

Nadaï, A. and Labussière, O. (2010). Birds, wind and the making of wind power landscapes in Aude, Southern France. *Landscape Research*, 35(2), 209–233. https://doi.org/10.1080/01426390903557964

Nader, L. (2018). The ADR explosion – The implications of rhetoric in legal reform. In *Contrarian Anthropology: The Unwritten Rules of Academia*. New York: Berghahn.

Newell, P. and Bulkeley, H. (2017). Landscape for change? International climate policy and energy transitions: Evidence from sub-Saharan Africa. *Climate Policy*, 17(5), 650–663. https://doi.org/10.1080/14693062.2016.1173003

Okereke, C. (2010). Climate justice and the international regime. *Wiley Interdisciplinary Reviews: Climate Change*, 1(3), 462–474. https://doi.org/10.1002/wcc.52

Osaka, S. (2022). To fight climate change, environmentalists may have to give up a core belief. *Washington Post*, 4 September. https://www.washingtonpost.com/climate-environment/2022/09/02/fight-climate-greens-have-embrace-big-energy-projects-fast/

Osofsky, H. M. and Levit, J. K. (2008). The scale of networks? Local climate change coalitions. *Chicago Journal of International Law*, 8(2), 409-436 https://papers.ssrn.com/abstract=1018310

Ott, A. S., Sorman, A. H., Avila, S., Del Bene, D., Ott, J., Sorman, A., Avila, S. and Del Bene, D. (2023). Renewables grabbing: Land and resource appropriations in the global energy transition. In A. Neef, C. Ngin, T. Moreda and S. Mollett (eds.) *Routledge Handbook of Global Land and Resource Grabbing*. Abingdon and New York: Routledge.

Pino, D. (2022). Environmentalists against green energy. *National Review*, 21 February. https://www.nationalreview.com/corner/environmentalists-against-green-energy/

Pye, O. (2010). The biofuel connection – transnational activism and the palm oil boom. *Journal of Peasant Studies*, 37(4), 851–874. https://doi.org/10.1080/03066150.2010.512461

Rignall, K. E. (2016). Solar power, state power, and the politics of energy transition in pre-Saharan Morocco. *Environment and Planning A: Economy and Space*, 48(3), 540–557. https://doi.org/10.1177/0308518X15619176

Rootes, C., Zito, A. and Barry, J. (2012). Climate change, national politics and grassroots action: An introduction. *Environmental Politics*, 21(5), 677–690. https://doi.org/10.1080/09644016.2012.720098

Salazar, V. (2023). From practices to praxis: ASEAN's transnational climate governance networks as communities of practice. *Journal of Current Southeast Asian Affairs*, 42(2), 190-215. https://doi.org/10.1177/18681034231167443

Sandhu, B. (2022). The return of Robin Hood: A new kind of shareholder activism. In C. Liao (ed.) *Corporate Law and Sustainability from the Next Generation of Lawyers* (pp. 197–220). Montreal: McGill-Queen's University Press.

Scheidel, A. and Sorman, A. H. (2012). Energy transitions and the global land rush: Ultimate drivers and persistent consequences. *Global Environmental Change, Global Transformations, Social Metabolism and the Dynamics of Socio-Environmental Conflicts*, 22(3), 588–595. https://doi.org/10.1016/j.gloenvcha.2011.12.005

Serna, L. A. (2022). Land grabbing or value grabbing? Land rent and wind energy in the Isthmus of Tehuantepec, Oaxaca. *Competition & Change*, 26(3–4), 487–503. https://doi.org/10.1177/10245294211018966

Simon, J. (2022). Misinformation is derailing renewable energy projects across the United States. *NPR*, 28 March 2022, sec. Climate. https://www.npr.org/2022/03/28/1086790531/renewable-energy-projects-wind-energy-solar-energy-climate-change-misinformation

Sine, W. D. and Lee B. H. (2009). Tilting at windmills? The environmental movement and the emergence of the U.S. wind energy sector. *Administrative Science Quarterly*, 54(1), 123–155. https://doi.org/10.2189/asqu.2009.54.1.123

Spector, J. (2015). Even environmentalists are cautious about 100% renewable energy plans. *Bloomberg. Com*, 20 July. https://www.bloomberg.com/news/articles/2015-07-20/the-environmentalist-case-against-100-renewable-energy-plans

Stevis, D. and Felli, R. (2020). Planetary just transition? How inclusive and how just? *Earth System Governance*, 6, 100065, 1–11. https://doi.org/10.1016/j.esg.2020.100065

Stop Uyghur Genocide, The Helena Kennedy Center for International Justice, and Unison. (2022). *Dirty Energy: Sourcing Solar Energy without Uyghur Forced Labour*. Sheffield Halam University.

Temper, L., del Bene, D. and Martinez-Alier, J. (2015). Mapping the frontiers and front lines of global environmental justice: The EJAtlas. *Journal of Political Ecology*, 22(1), 255–278. https://doi.org/10.2458/v22i1.21108

Tillotson, P., Slade, R., Staffell, I. and Halttunen, K. (2023). Deactivating climate activism? The seven strategies oil and gas majors use to counter rising shareholder action. *Energy Research & Social Science*, 103(September), 103190. https://doi.org/10.1016/j.erss.2023.103190

Tokar, B. (2019). On the evolution and continuing development of the climate justice movement. In T. Jafry (eds) *Routledge Handbook of Climate Justice*. Abingdon and New York: Routledge.

Trumbull, K., Gattaciecca, J. and DeShazo, J. R. (2020). *The Role of Community Choice Aggregators in Advancing Clean Energy Transitions*. Los Angeles, CA: UCLA Luskin Center for Innovation.

Vasi, I. (2009). Social movements and industry development: The environmental movement's impact on the wind energy industry. *Mobilization: An International Quarterly*, 14(3), 315–336. https://doi.org/10.17813/maiq.14.3.j534128155107051

Watts, L. (2018). *Energy at the End of the World: An Orkney Islands Saga*. Infrastructures. Cambridge, MA: The MIT Press.

Western Sahara Resource Watch. (2022). Western sahara resource watch. *Dirty Green Energy on Occupied Land*, 14 April. https://wsrw.org/en/news/renewable-energy.

Whitington, J. (2018). *Anthropogenic Rivers: The Production of Uncertainty in Lao Hydropower*. Expertise: Cultures and Technologies of Knowledge. Ithaca: Cornell University Press.

Wright, C. and Nyberg, D. (2015). *Climate Change, Capitalism, and Corporations: Processes of Creative Self-Destruction*. Cambridge: Cambridge University Press.

27

COMMUNITY-LED RESPONSES TO CLIMATE-INDUCED DISASTERS IN ZIMBABWE

Towards a Politico-community-based Model

Thomas Karakadzai and Innocent Chirisa

Introduction

Grassroot communities in urban, peri-urban, rural, and informal settlements in Zimbabwe are among the most susceptible to the effects of climate change. Even if the disasters brought on by climate change keep happening, grassroots communities in Zimbabwe's urban, rural, and informal settlements are still unable to confront the issue (Chirisa and Karakadzai, 2021). Since then, emergency treatments – which are just transient and neither strong nor long-lasting – have started to be mainstreamed by humanitarian organisations (World Food Programme, 2022). This study sheds light on the tactics and role of grassroots communities in combating climate change, as well as the struggles Zimbabwean communities have faced to maintain their space and rights in the face of this phenomenon. This chapter is predominantly qualitative in nature. In order to draw conclusions about the experiences of community-led activism towards climate change, the study uses document review, key informant interviews, and focus groups conducted with grassroots community organisations like the Zimbabwe Homeless Peoples' Federation and the Zimbabwe Young Peoples Federation, non-state actors like Accountability Lab and Climate Action Network Zimbabwe (CAN), and state actors such as the City of Harare and the Civil Protection Unit. From January to March 2023, three focus groups with the Zimbabwe Homeless People's Federation and the Zimbabwe Young People's Federation were held in Harare. Between January and March of 2023, four key informant interviews were carried out with representatives from the Accountability Lab, the City of Harare, the Civil Protection Unit, and Climate Action Network Zimbabwe (CAN).

This chapter is organised as follows: The section that follows the introduction provides a description of the climate-change trends and disaster context in Zimbabwe. The next issues discussed include Zimbabwe's political situation and how it affects the way grassroots communities are responding to climate change. A conclusion follows an overview of Zimbabwe's grassroots climate advocacy projects. The chapter ends by arguing that community-led grassroots coalitions involving state and nonstate entities are essential to advancing climate-change policies and practices that improve climate justice.

DOI: 10.4324/9781003396567-32

Characterisation of Zimbabwe's Context

Sub-Saharan Africa, where Zimbabwe is located, is home to several climate-change hotspots, where strong physical and ecological effects of climate change intersect with large populations of poor and vulnerable communities (Godfrey and Tunhuma, 2020). Currently, more than 11 million people are experiencing crisis or emergency levels of food insecurity due to deepening drought and climate crisis impacts in Sub-Saharan Africa (ibid). Developing nations and their cities, like Zimbabwe, are among those most in danger from the effects of climate change, as argued in projections by the Intergovernmental Panel on Climate Change (IPCC, 2023). Over the past century, Zimbabwe has seen a variety of natural disasters, including floods, storms, pandemic diseases, and droughts. Over 20 million people were affected by the country's 7 drought disasters, 22 disease episodes, 12 floods, and 5 storms between 1900 and 2017, which caused $950 million in damage and killed an estimated 7,000 people (CARE, 2019). Droughts have been seen to increasingly impact more individuals overall, and they are also responsible for more economic loss. The nation is one of six where impoverished individuals are overexposed to natural disasters, such as droughts, storms, and flooding (50% more probable than non-poor people over the past century in Zimbabwe) (World Bank Group, 2021a). Cyclone Idai, which struck Zimbabwe in 2019 and devastated the country on 14–15 March, left 270,000 people in severe need of humanitarian aid (Oxfam International, 2023). The storm generated high winds and heavy precipitation that caused rivers to overflow and destroyed harvests, leaving millions of people without houses and means of support (World Bank Group, 2021a). In February of 2017, Cyclone Dineo struck the nation, producing flooding and landslides that resulted in at least 246 fatalities and the displacement of homes. At least 136 people were killed in floods and landslides brought on by Cyclone Eline in February of 2000. The Mavhura Cyclone of 2020 caused thousands of people to be displaced and at least two deaths.

Despite the losses experienced from the climate related disasters, Krasovskaia (2005) and Terpstra and Gutteling (2008) note that community people were thought to contribute very little to the effort of protecting against the effects of disasters. Zimbabwe's climate change approaches have historically been state-led and exclusive and have frequently neglected the efforts of other groups (Mitlin, 2022). The same criticism is raised by Watson (2009), who claims that the current state-led, centralised methods of urban administration contribute to social and geographic exclusion, anti-poor policies, and a lack of action to achieve urban sustainability and climate resilient cities. In retrospect, this implies that the community is a victim that can only respond to crises in a reactive manner (Santiago et al., 2014). De Goyet notes that

> The myth that the affected population is too shocked and helpless to take responsibility for their own survival is superseded by the reality that on the contrary, many find new strength during an emergency.
>
> *(2000, p. 763)*

There have been more community-led responses to climate-change-related disasters over the past 10 years. The "historical roots, tensions and complexity" of climate activism still have a significant amount of weight, and "the thickest of these roots is environmental justice, a theoretical approach with a long history," particularly in the activism of people of colour and Indigenous communities (Neas et al., 2022, p. 1). Indigenous knowledge is a crucial component of Priority 3 of the Hyogo Framework for Action (2005–2015), which is focused on education and knowledge, according to the United Nations (UNISDR, 2013). One of the important tasks listed under Priority 3 Action is the preservation of pertinent traditional and Indigenous knowledge and cultural treasures. Since the 1970s, there has been an increase in research showing that local knowledge and practices can enhance disaster preparedness (Dekens, 2007; Alcántara-Ayala, 2004; Battista and Baas, 2004; Campbell, 2009; Chan and Parker, 1996).

Grassroots climate action has become an essential part of dealing with, or managing the effects and repercussions of, climate change in communities and settlements. Because of pre-existing elements including their geographic locations, social stratification, and building systems, communities living in urban, rural, and informal settlements have been particularly vulnerable to climate-related calamities. However, many public sector organisations continue to offer services that are either insufficient or fragmented (Ouellette et al., 1999). The Middle East, North Africa, and Southern Africa's civil society and grassroots activism have been tightly controlled, which has contributed to the region's ranking as one of the least democratic regions in the world (Adhikari et al., 2017). Even though the institutional framework for governing climate issues has made some progress, there are still significant issues with horizontal and vertical coordination, including institutional overlap with other intersectoral platforms like disaster risk reduction. Additionally, in fragile states, institutions for reducing climate risk and promoting adaptation may be very weak or non-existent (Hartmann and Sugulle, 2009; Sietz et al., 2011). The activist groups trying to develop alternative models for social change and climate change are hampered by institutional constraints.

Numerous grassroots organisations have sprouted up all over the world since the early 2000s to address the issue of unresponsive governments (Flores and Samuel, 2019). By applying pressure to their government and forcing it to respond, these grassroots organisations aim to address the needs of their underserved groups, such as the urban poor (Maru, 2010). Seyfang and Smith (2007, p. 585) characterise grassroots innovations as "innovative networks of activists and organisations that lead bottom-up solutions for sustainable development; solutions that respond to the local situation and the interests and values of the communities involved." Through the theoretical lens of self-organisation, an expanding area in the study of community-based grassroots innovations, one can examine the intricate organisational processes of grassroots activism (Hasanov and Beaumont, 2016). Power dynamics, the degree of citizen organisation, participatory abilities, political will, and a lack of financial resources are only a few of the things Gaventa and Valderrama (1999) name as obstacles to participation for citizens in general and members of disadvantaged sections in particular.

The Political Characterisation in Zimbabwe

This section looks at Zimbabwe's political environment as a whole and how community-led, grassroots climate-change advocacy is affected by it. There is a long history of entanglements among the official government, activists, and social movements for justice (Biti, 2014).

The national government of Zimbabwe views grassroots climate action in Zimbabwe as a collection of regime change agendas, and it has encountered some difficulties in trying to obtain authorisation for demonstrations. According to Ureke (2016), the word "activism" has a connection to political movements and is tied to unrest in the political system. If Zimbabweans heeded the invitation to connect with people's real-world experiences of climate change, it would lead to a surprising redefinition of the terms politics and activism (Nolas et al., 2017). Zimbabwe's current national political structure greatly affects and controls how decisions or strategies to deal with climate action are implemented at the local level. Zimbabwe's political shift from colonial minority rule to post-colonial majority rule occurred on 18 April 1980, as a result of various, interconnected, but not necessarily mutually supporting factors. At a period when one-party states were popular in post-colonial Africa, independence also unquestionably established multi-party politics as the country's political framework (Lewanika, 2017).

Intense contestations with ambiguous outcomes permeated many dualities at the national and local level, politically and developmentally, and these were characteristics of the post-colonial state in formation/transition (Lewanika, 2022). The underpinnings of Zimbabwe's national political settlement go further back than the country's political independence. "Stakeholders" and "stockholders" are the people who run the nation, and their actions have an impact on how the nation's politics are conducted. Due to their sacrifices made during the liberation fight, the country's stockholders are considered to be its "owners," whereas stakeholders are characterised as having an interest but not ownership (Lewanika, 2022).

In 2016, then commander of Zimbabwe's Defence Forces, Constantino Chiwenga, warned that: "We [the military] are stockholders of the country. Some are stakeholders. Stakeholders come and go, but stockholders have nowhere to go, so we come with it [the country]" (Lewanika, 2022). This aspect of the political climate has caused a different transitional era in Zimbabwe, one that is marked by a shifting ruling party and the formation of a new political order that includes the president's allies among the political, military, and business elites. Instead, we learn that political participation may be found in everyday actions as well as in everyday settings through connecting with the pasts, present, and anticipated futures of activism (Nolas et al., 2017). A more complex knowledge of individual biographies and the cultural and historical circumstances in which political engagement occurs results from the methodological diversity in the scientific investigation of political activism (Andrews, 2017; Pickard et al., 2020). Punitive regulations that control protests and activity in Zimbabwe were created as a result of the country's post-2000s political and economic crises, and the line between political activism and climate activism blurred more and more (Chibuwe, 2021). The political mobilisation, organising, and practise taking place at the national level directly affect grassroots organisations' plans to engage in grassroots climate activism in Zimbabwe.

Given the political structure described earlier, the current state of climate activism in Zimbabwe is complex, involving a variety of organisations including those from civil society, non-governmental organisations, the private sector, organisations led by young people and women, and community-based groups. These groups are active in a variety of settings, including urban areas, rural areas, and informal settlements. These players have only been able to make piecemeal efforts to address climate change advocacy because they lack a clear framework or strategy. Resources and programmes for climate action are distributed differently across rural, urban, and informal settlements, depending on the state of the nation's politics.

The ruling party – Zimbabwe African National Union–Patriotic Front (ZANU PF) – and the opposition Citizens for Coalition Change (CCC), which represent different locales and regions, have current members of parliament, councillors, and shadow members of parliament and council who frequently stifle development efforts from social movements and activist organisations. That their work is seen as threatening regime change poses challenges to grassroots organisations aiming to conduct climate activism; for example, they are required to undergo a review process through the District Development Coordinator (DDC) and the Provincial Development Coordinator (PDC). The political system in Zimbabwe and the power structures both had an impact on the various types of settlements. The national institutions, benefits, and rent space are influenced by the interests of the various members of the national leader's block (Lewanika, 2022). This means that individuals with political connections and/or clout profit disproportionately from opportunities that institutions and authority restrict and close off to others who don't support their political agenda.

Furthermore, both formal and unofficial actors are heavily involved in the grassroots climate advocacy space. The formal actors are members of the national government, especially the ruling party, who think that matters relating to combating climate change should only be handled formally through COPs and other international forums. High-ranking delegates (government officials) frequently attend these forums year after year, but Zimbabwe's response to climate challenges has been slow to develop.

Current Efforts of Grassroot Climate Activism in Zimbabwe

Solutions to the current climate problem must take into consideration the intricate and varied realm of international politics, as it has grown into a humanitarian and political emergency (Steinke, 2022; Martiskainen et al., 2020). This section examines the ways in which coalitions could address climate injustice in Zimbabwe's informal settlements, which are already disadvantaged and excluded from planning and policymaking processes.

This study is significant for three key reasons. First, this section helps practitioners working for different local authorities, as well as communities in informal settlements devastated by climate-related disasters, by providing a platform for coalition between diverse state and non-state actors. Second, it sheds light on how grassroots climate coalitions in Zimbabwe might be mainstreamed and included in the nation's current fragmented policies. Third, putting the study's recommendations into practice will increase the effectiveness and efficiency of reacting to climate-related disasters, which are now hampered by delays and uncertainty. To aid communities, local government representatives, and other relevant parties in coping with current and upcoming disasters, the grassroots climate coalition would offer a practical resource package of policy recommendations that will help not only cities in Zimbabwe but many other African cities that are trying to develop resilient informal settlements that would be developed through research on this topic.

The efficacy and efficiency of responding to climate-related disasters, which are currently hindered by delays and uncertainty, would be improved by implementing the study's suggestions. Even with the implementation of the Sustainable Development Goals (SDGs) (13, 11, and 10) – which commit to reducing inequality (Goal 10), acting swiftly to halt climate change and its harmful effects (Goal 13), and creating inclusive, resilient, sustainable cities (Goal 11) – climate-change approaches have historically been exclusive and state-led and have ignored the efforts of grassroots communities. This section outlines the grassroots organisations' current activities in Zimbabwe to combat climate change. The overview of informal settlements across Zimbabwe is also described.

Overview of the Informal Settlements in Zimbabwe

Different typologies of informal settlements with a heterogeneous population define Zimbabwe. The manner of acquisition, location, jurisdiction, and tenure status of the informal settlements are taken into consideration while classifying them. In Zimbabwe, there are several informal settlements which are located in the peri-urban areas, where the boundaries of the settlement are a bone of contention and there are always evictions (Manyowa et al., 2023). Of the informal settlements across Zimbabwe, 80% are located in the peri-urban areas while those located under municipal or council land comprise 20% (Dialogue on Shelter and Zimbabwe Homeless People's Federation, 2014). Figure 27.1 depicts the informal settlement distribution across Zimbabwe.

The informal settlements in Zimbabwe are situated in proximity to the country's major cities. The people who live in these informal settlements work in the big cities and return to their homes at night. Large tracts of land, especially those in the peri-urban areas, have ownership disputes and are often vulnerable to climate change in general, placing a pressing demand on policymakers, planners, and managers to comprehend the environmental dimensions of development (Karakadzai and Mpahlo, 2022). This has sparked the creation of grassroot coalitions and strategies amongst the informal settlement dwellers to address the vulnerabilities of climate change.

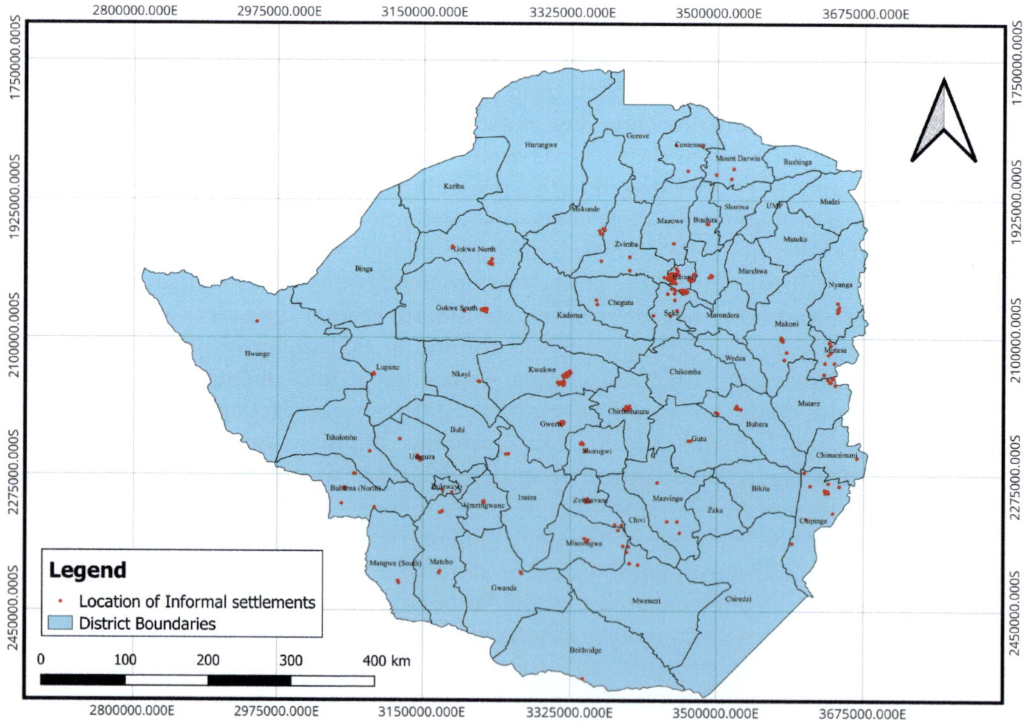

Figure 27.1 Map showing the distribution of informal settlements in Zimbabwe.

Source: Authors' own creation (2024).

Climate Initiatives in Informal Settlements

Informal settlements spreading across Zimbabwe have been devastated by climate change (Gumbodete and Magondo, 2023). In order to combat climate change, this has led to the formation of reform coalitions with the government, local governments, other civil actors, and the private sector. The grassroots communities, who frequently visited government offices with the intention of working together to address climate-related disasters, came up with the idea for the coalition. However, the process was stifled along the way because government ministries and councils refused to associate themselves with informal settlements or other grassroots organisations. Nevertheless, the grassroots communities persisted until agreements were signed by state actors, including ministry departments and councils, outlining a framework for the parties to work together to address climate-related disasters. Memorandums of Understandings between stakeholders continued to be established, giving the grassroots organisation an operating mandate to carry out and lead climate-change programs in Zimbabwe. For instance, the City of Harare, the Zimbabwe Homeless People's Federation, and the Civil Protection Unit have established climate response projects. These programs are designed to guarantee an immediate response to any humanitarian difficulties that occasionally develop in the informal settlements. In collaboration with the Civil Protection Unit, the Environmental Management Agency launched its educational programs on climate change in informal settlements in 2020 and 2021 (Government of Zimbabwe, 2015). Many co-production projects and programs, including those in Dzivarasekwa, Epworth, Mabvuku-Tafara, Victoria Ranch, Mucheke, Cherry-Bank, Gweru, Gimboki, Hopley, and Churu Farm, to name a few, were carried out as a result of these collaborations. A number of housing and hazard-proofing actions have been concluded through this grassroots organisation (ZIHOPFE) as a result of collective community efforts. As a result of the grassroots organisation's successful work, numerous interventions have been carried out throughout the cities of Zimbabwe. One of the best ways to help communities adapt to climate change is to build or modify homes in the majority of informal settlements throughout Zimbabwe, so that the shacks can endure extreme weather occurrences.

Urban Area Climate Initiatives

Multiple actors are heavily involved in the urban cluster and in driving grassroots climate activism. To join the campaign in the Global Fight to End Fossil Fuels activities, organisations like the Climate Action Network Zimbabwe (CAN) organised the September Global Days of Action in 2023. CAN coordinated protests that took a variety of forms, including public speaking engagements, art installations, marches, sit-ins, occupations, discussions, meetings, civil disobedience, and online mobilisations. Events and actions can be centred on any of the threats the fossil fuel business poses to our ecosystems and communities, as well as those that support the industry, such as governments, financial institutions, and other industries.

Additionally, Youth Lead (in collaboration with Youth Excel, Accountability Lab, Atlas Corp, Grassroots Soccer, Dialogue on Shelter for the Homeless Trust, and YEO 2030) launched on International Youth Day calls on young changemakers worldwide to unite to dismantle barriers, foster diversity, and reshape power dynamics within societies. The youth in urban areas across Zimbabwe participated in this action. Several activists who are residents from different artistic fields provided the support for online action, from strategy to execution. This art space was designed more as a production area for groups to organise and carry out acts rather than as a social area.

Because different organisations have distinct mandates that drive their main objectives for how they are trying to reduce the consequences and vulnerabilities of climate change in their respective locations, the main efforts toward climate activism are also fragmented. For instance, in Zimbabwe, players in urban, rural, and informal settlements do not agree on the main concerns across the different geographies. This is due to the fact that each of these actors has unique demands that must be satisfied while also allowing them to survive day to day in the environment. In the case of grassroots activity in rural regions, banning the use of fossil fuels is ineffective since the rural people in Zimbabwe have no other options, therefore they depend on forestry, which contributes to climate change. However, there is no single voice guiding the movement or the grassroots organisation in metropolitan areas.

Despite these attempts by urban private organisations and civil society, obtaining permission for grassroots activism is a difficult process. The Zimbabwe Republic Police is reluctant to offer permits to various parties to stage protests about climate-change issues in metropolitan areas. Grassroots activism is frequently seen as having a political bent and as being supported by multilateral organisations with a goal of bringing about regime change.

Coalition Reforms Amongst Grassroot Organisations

Given the background and experiences of negative perceptions towards grassroot climate activism, various actors from the community-based organisations (CBOs), civil society organisations, private sector, individuals, academia, and businesspeople are forming coalitions which are meant to amplify their voices on climate-change activism in Zimbabwe. The coalitions are operationalised through a number of Memorandums of Understandings after different actors have joined hands with organisations who have interest in pushing the climate activism agenda. The perspective of grassroots climate activism coalitions offers a normative orientation that highlights the need for radical and systemic change to address enduring social, environmental, and economic issues and to move deliberately towards resilient, sustainable cities over the long term (Hölscher et al., 2020; Ruiz-Mallén et al., 2022). "Grassroots climate activism coalitions" means reform that is initiated primarily by communities based on, and oriented to, closing the gaps that come with top-down approaches from central or local government or foreign donors (Adhikari et al., 2017). The core agenda for the formation of the grassroots climate activism coalitions is to counter the external responses from either local or central government or foreign donors that have been exclusionary, insufficient, and not effective in addressing climate change vulnerabilities in communities (Skaidrė, 2019). Reform coalitions have the potential to handle the complex climate issues in the towns and cities of the Global South, as argued by authors like Win (2004), Collard et al. (2021), and Mitlin (2022). However, in Zimbabwe the government is punitive towards the formation and operation of grassroots climate activism coalitions (Bureau of Democracy, Human Rights, and Labor, 2020). They are often scrutinised and viewed as a machinery for political regime change rather than climate change agents. This perception has resulted in stifling progress in climate change issues by the grassroots communities, whose only hope lies with the central, local, and humanitarian organisations that respond to climate change in times of crisis (Chowdhury et al., 2018; Zakariah and Kismartini, 2018).

Conclusion

This chapter described the experiences of climate-related disasters that occurred in Zimbabwe between 1900 and 2020 and the catastrophic effects they had on people's lives and livelihoods. However, the Civil Protection Unit, a government body, has played a major role in mitigating disasters brought on by climate change. The Civil Protection Unit's efforts have fallen short and are excluding the impacted communities. Due to this, there is an urgent need for grassroots climate action organisations from the corporate sector, academia, civil society organisations, non-governmental organisations, and the private sector. The Zimbabwean government, which saw the activists as supporters of a political uprising or regime change, did not take well to this campaign.

As a result, Zimbabwe's efforts to achieve climate justice have been thwarted by the ruling government's introduction of punitive measures that require scrutiny of each and every civil society organisation that advocates for addressing climate change. Additionally, Zimbabwe's current national political arrangements have an impact on how local communities establish their own boundaries. The boundaries of the grassroots organisations are important in determining whether grassroots activism is actually autonomously self-organised because they separate the inside from the outside and make it possible to determine where and how the limits of a self-organising entity's autonomy are drawn (Serugento et al., 2005). However, the democratic values, civic involvement, and governance of grassroots climate action, which is said to improve social and environmental performance within their networks, are still up for debate (Dupuis and Gillon, 2009). The meaning connected to climate change is significantly reliant on varying and multifaceted interests, worries, and values (Feldman, 2021; Rossi, 2017; Goodman et al., 2010; Morris and Kirwan, 2011; Abass et al., 2018). In order to overcome the national government of Zimbabwe's reluctance, civil society organisations are forming coalitions with other stakeholders who embrace the same core values and are motivated to raise awareness of climate-change activism. Through Memorandums of Understanding with like-minded entities around the nation, these organisations have put the coalition reforms into practise.

References

Abass, R., Mensah, A. and Fosu-Mensah, B. (2018). The role of formal and informal institutions in smallholder agricultural adaptation: The case of Lawra and Nandom Districts, Ghana. *West African Journal of Applied Ecology*, 26(SI), 56–72.

Adhikari, A., Bradlow, B., Heller, P., Li, K., Pheiffer, C., Schrank, A., and Walton, M. (2017). *Grassroots Reform in the Global South*. Research and Innovation Grants Working Papers Series, 21 September. Brown University and Institute of International Education. https://pdf.usaid.gov/pdf_docs/PA00N1RF.pdf

Alcántara-Ayala, I. (2004). Flowing mountains in Mexico. *Mountain Research and Development*, 24(1), 10–13. https://doi.org/10.1659/0276-4741(2004)024[0010:FMIM]2.0.CO;2

Andrews, M. (2017). Enduring ideals: Revisiting lifetimes of commitment twenty-five years later. *Contemporary Social Science*, 12(1–2), 153–163. https://doi.org/10.1080/21582041.2017.1325923

Battista, F. and Baas, S. (2004). *The Role of Local Institutions in Reducing Vulnerability to Recurrent Natural Disasters and in Sustainable Livelihoods Development*. Consolidated Report on Case Studies and Workshop Findings and Recommendations. Rome: Food and Agriculture Organization of the United Nations (FAO). https://www.fao.org/4/ae190e/ae190e00.htm

Biti, T. (2014). *Rebuilding Zimbabwe: Lessons from a Coalition Government*. Center for Global Development. https://www.cgdev.org/sites/default/files/Tendai-Biti-Zimbabwe-Sept-2015.pdf

Bureau of Democracy, Human Rights, and Labor. (2020). *2020 Country Reports on Human Rights Practices: Zimbabwe*. U.S. Department of State. https://www.state.gov/reports/2020-country-reports-on-human-rights-practices/zimbabwe/

Campbell, J. (2009). Islandness: Vulnerability and resilience in Oceania. *The International Journal of Research into Island Cultures*, 3(1), 85–97. https://hdl.handle.net/10289/2898

CARE. (2019, 18 March). *Cyclone Idai and Floods Cause Massive Destruction, Deaths in Mozambique, Zimbabwe, and Malawi*. CARE International. https://reliefweb.int/report/mozambique/cyclone-idai-and-floods-cause-massive-destruction-deaths-mozambique-zimbabwe-and

Chan, N. W. and Parker, D. J. (1996). Response to dynamic flood hazard factors in Peninsular Malaysia. *The Geographical Journal*, 162(3), 313–325. https://doi.org/10.2307/3059653

Chibuwe, A. (2021). The nexus of journalism and political activism in post-2000 Zimbabwe: A field theory critique. *Journal of Applied Journalism & Media Studies*, 10(3), 275–296. https://doi.org/10.1386/ajms_00033_1

Chirisa, I. and Karakadzai, T. (2021). Urban development management in light of the risks and disasters caused by climate change. In A. R. Matamanda, V. Nel and I. Chirisa (eds.). *Urban Geography in Postcolonial Zimbabwe*. The Urban Book Series. Cham: Springer.

Chowdhury, R., Kourula, A. and Siltaoja, M. (2018). Power of paradox: Grassroots' organizations' legitimacy strategies over time. *Business and Society*, 60(2), 420–453. https://doi.org/10.1177/0007650318816954

Collard, M., Goodfellow, T. and Adebi Asante, L. (2021). *Uneven Development, Politics and Governance in Urban Africa: An Analytical Literature Review*. ACRC Working Paper 2021–02. Manchester: African Cities Research Consortium, The University of Manchester. https://www.african-cities.org/wp-content/uploads/2021/12/ACRC_Working-Paper-2_November-2021.pdf

de Goyet, C. de V. (2000). Stop propagating disaster myths. *The Lancet*, 356(9231), 762–764. https://doi.org/10.1016/S0140-6736(00)02642-8

Dekens, J. (2007). *Local Knowledge for Disaster Preparedness: A Literature Review*. Kathmandu: International Centre for Integrated Mountain Development (ICIMOD). https://doi.org/10.53055/ICIMOD.474

Dialogue on Shelter and Zimbabwe Homeless People's Federation. (2014). *Harare Slum Profiles Report, Edition 2*. https://www.iied.org/g03861

Dupuis, E. M. and Gillon, S. (2009). Alternative modes of governance: Organic as civic engagement. *Agriculture and Human Values*, 26, 43–56. https://doi.org/10.1007/s10460-008-9180-7

Feldman, H. (2021). Motivators of participation and non-participation in youth environmental protests. *Frontiers in Political Science*, 3, 662687. https://doi.org/10.3389/fpos.2021.662687

Flores, W. and Samuel, J. (2019). Grassroots organisations and the sustainable development goals: No one left behind? *BMJ: British Medical Journal*, 365, 2269. https://doi.org/10.1136/bmj.l2269

Gaventa, J. and Valderrama, C. (1999, 21–24 June). Participation, citizenship and local governance. Background note prepared for workshop on 'strengthening participation in local governance'. *Institute of Development Studies*. https://www.participatorymethods.org/resource/participation-citizenship-and-local-governance

Godfrey, S. and Tunhuma, F. A. (2020). *The Climate Crisis: Climate Change Impacts, Trends and Vulnerabilities of Children in Sub Sahara Africa*. Nairobi: United Nations Children's Fund Eastern and Southern Africa Regional Office. https://www.unicef.org/esa/reports/climate-crisis

Goodman, M. K., Maye, D. and Holloway, L. (2010). Ethical foodscapes? Premises, promises, and possibilities. *Environment and Planning A: Economy and Space*, 42(8), 1782–1796. https://doi.org/10.1068/a43290

Government of Zimbabwe. (2015). *Zimbabwe's National Climate Change Response Strategy*. Ministry of Environment, Water and Climate. https://www.fao.org/faolex/results/details/en/c/LEX-FAOC169511/

Government of Zimbabwe. (2021, 11 March). *Update on Identification and Quantification of All Irregular and Dysfunctional Settlements Including Flood-Prone Areas*. Term of Reference No. 3.

Gumbodete, P. and Magondo, I. (2023, 10 November). Zim battles climate change shocks – report. *Newsday*. https://www.newsday.co.zw/local-news/article/200019299/zim-battles-climate-change-shocks-report

Hartmann, I. and Sugulle, A. J. (2009). *The Impact of Climate Change on Pastoral Societies of Somaliland*. Report of Candlelight for Health, Education & Environment. https://www.unisdr.org/files/13863_FinaldraftEffectsofclimatechangeonp.pdf

Hasanov, M. and Beaumont, J. (2016). The value of collective intentionality for understanding urban self-organization. *Urban Research & Practice*, 9(3), 231–249. https://doi.org/10.1080/17535069.2016.1149978

Hölscher, K., Frantzeskaki, N., Pedde, S. and Holman, I. (2020). Agency capacities to implement transition pathways under high-end scenarios. In K. Hölscher and N. Frantzeskaki (eds.) *Transformative Climate Governance: A Capacities Perspective to Systematise, Evaluate and Guide Climate Action* (pp. 381–416). Cham: Palgrave Macmillan.

IPCC (2023). Summary for policymakers. In Core Writing Team, H. Lee and J. Romero (eds.) *Climate Change 2023: Synthesis Report. Contribution of Working Groups I, II and III to the Sixth Assessment Report of the Intergovernmental Panel on Climate Change* (pp. 1–34). Geneva: IPCC. https://doi.org/10.59327/IPCC/AR6-9789291691647.001

Karakadzai, T. and Mpahlo, R. (2022). Interrogating emerging land access and tenure documentation in Zimbabwe's informal and semi-formal settlements. *Journal of Urban Systems and Innovations for Resilience in Zimbabwe*, 2022(Special Issue 1), 67–85.

Krasovskaia, I. (2005). *Perception of Flood Hazard in Countries of the North Sea Region of Europe.* FLOWS WP2A-1 Report. Oslo: Norwegian Water Resources and Energy Directorate. https://publikasjoner.nve.no/flows/2005/reportwp2a_1.pdf

Lewanika, M. (2017, 09 January). How Zimbabwe can embrace the future of work. *Africa at LSE Blog.* blogs.lse.ac.uk/africaatlse/2017/01/09/how-zimbabwe-can-embrace-the-future-of-work/

Lewanika, M. (2022). *Political Settlements Mapping Note.* Unpublished ACRC Harare Report. Manchester: African Cities Research Consortium, The University of Manchester.

Manyowa, T., Muganyi, S. and Nyamangara, T. (2023). *Informal Settlement Communities as first Responders to Disasters: Covid-19 Pandemic Experiences in Harare.* Policy Brief September 2023. Manchester: African Cities Research Consortium, The University of Manchester. https://www.african-cities.org/wp-content/uploads/2023/09/ACRC_Covid_Collective_Covid-19-pandemic-experiences-in-Harare.pdf

Martiskainen, M., Axon, S., Sovacool, B., Sareen, S., Furszyfer Del Rio, D. and Axon, K. (2020). Contextualizing climate justice activism: Knowledge, emotions, motivations, and actions among climate strikers in six cities. *Global Environmental Change*, 65, 102180, 1–18. https://doi.org/10.1016/j.gloenvcha.2020.102180

Maru, V. A. (2010). Allies unknown: Social accountability and legal empowerment. *Health and Human Rights*, 12(1), 83–93.

Mitlin, D. (2022). The contribution of reform coalitions to inclusion and equity: Lessons from urban social movements. *Area Development and Policy*, 8(1), 1–26. https://doi.org/10.1080/23792949.2022.2148548

Morris, C. and Kirwan, J. (2011). Ecological embeddedness: An interrogation and refinement of the concept within the context of alternative food networks in the UK. *Journal of Rural Studies*, 27(3), 322–330. https://doi.org/10.1016/j.jrurstud.2011.03.004

Neas, S., Ward, A. and Bowman, B. (2022). Young people's climate activism: A review of the literature. *Frontiers in Political Science*, 4, 940876. https://doi.org/10.3389/fpos.2022.940876

Nolas, S. M., Varvantakis, C. and Aruldoss, V. (2017). Political activism across the life course. *Contemporary Social Science*, 12(1–2), 1–12. https://doi.org/10.1080/21582041.2017.1336566

Ouellette, P. M., Lazear, K. and Chambers, K. (1999). Action leadership: The development of an approach to leadership enhancement for grassroots community leaders in children's mental health. *The Journal of Behavioral Health Services & Research*, 26(2), 171–184. https://doi.org/10.1007/BF02287489

Oxfam International. (2023). *After the Storm: One Year on from Cyclone Idai.* https://www.oxfam.org/en/after-storm-one-year-cyclone-idai

Pickard, S., Bowman, B. and Arya, D. (2020). "We are radical in our kindness": The political socialisation, motivations, demands and protest actions of young environmental activists in Britain. *Youth and Globalization*, 2(2), 251–280. https://doi.org/10.1163/25895745-02020007

Rossi, A. (2017). Beyond food provisioning: The transformative potential of grassroots innovation around food. *Agriculture*, 7(1), 6. https://doi.org/10.3390/agriculture7010006

Ruiz-Mallén, I., March, H. and Satorras, M. (eds.). (2022). *Urban Resilience to the Climate Emergency: Unravelling the Transformative Potential of Institutional and Grassroots Initiatives.* The Urban Book Series. Cham: Springer.

Santiago, A. M., Soska, T. and Gutierrez, L. (2014). Responding to community crises: The emerging role of community practice. *Journal of Community Practice*, 22(3), 275–280. https://doi.org/10.1080/10705422.2014.949508

Serugento, D. G., Gleizes, M. P. and Karageorgos, A. (2005). Self-organization in multi-agent systems. *The Knowledge Engineering Review*, 20(2), 165–189. https://doi.org/10.1017/S0269888905000494

Seyfang, G. and Smith, A. (2007). Grassroots innovations for sustainable development: Towards a new research and policy agenda. *Environmental Politics*, 16(4), 584–603. https://doi.org/10.1080/09644010701419121

Sietz, D., Boschütz, M. and Klein, R. J. T. (2011). Mainstreaming climate adaptation into development assistance: Rationale, institutional barriers and opportunities in Mozambique. *Environmental Science and Policy*, 14(4), 493–502. https://doi.org/10.1016/j.envsci.2011.01.001

Skaidrė, Ž. (2019). Grassroots activism and sustainable development. In W. L. Filho (ed.). *Encyclopedia of Sustainability in Higher Education*. Cham: Springer.

Steinke, A. (2022, 16 October). At a turning point – climate change and the responsibilities of humanitarian actors. *Public Anthropologist*. https://publicanthropologist.cmi.no/2022/10/16/climate-change-is-the-game-change-to-the-humanitarian-world/

Terpstra, T. and Gutteling, J. M. (2008). Households' perceived responsibility in flood risk management in the Netherlands. *International Journal of Water Resources Development*, 24(4), 555–565. https://doi.org/10.1080/07900620801923385

UNISDR (2013). *Implementation of the Hyogo Framework for Action: Summary of Reports 2007-2013*. Geneva: United Nations Office for Disaster Risk Reduction. https://www.undrr.org/publication/implementation-hyogo-framework-action-summary-reports-2007-2013

Ureke, O. (2016). State interference, para-politics and editorial control: The political economy of 'Mirrorgate' in Zimbabwe. *Journal of African Media Studies*, 8(1), 17–34. https://doi.org/10.1386/jams.8.1.17_1

Watson, V. (2009). Seeing from the South: Refocusing urban planning on the globe's central urban issues. *Urban Studies*, 46(11), 2259–2275. https://doi.org/10.1177/0042098009342598

Win, E. J. (2004). When sharing female identity is not enough: Coalition building in the midst of political polarisation in Zimbabwe. *Gender and Development*, 12(1), 19–27. https://doi.org/10.1080/13552070410001726486

World Bank Group (2021a). *Climate Change Knowledge Portal for Development Practitioners and Policy Makers*. https://climateknowledgeportal.worldbank.org/

World Bank Group (2021b). *Climate Risk Profile: Zimbabwe*. Washington, DC: World Bank Group. https://climateknowledgeportal.worldbank.org/sites/default/files/2021-05/14956-WB_Zimbabwe%20Country%20Profile-WEB%20%281%29.pdf

World Food Programme (2022). *World Food Programme Zimbabwe Highlights – 2021 Annual Country Report Highlights*. https://www.wfp.org/publications/world-food-programme-zimbabwe-2021-annual-country-report-highlights

Zakariah, H. W. and Kismartini, K. (2018). Community participation in flood disaster management in Sumbawa regency (case study in Songkar Village). *E3S Web of Conferences*, 73, 08004. https://doi.org/10.1051/e3sconf/20187308004

ZimStat (2023). *Zimbabwe 2022 Population and Housing, Census Report Volume 1*. Zimbabwe National Statistics Agency. https://www.zimstat.co.zw/wp-content/uploads/Demography/Census/2022_PHC_Report_27012023_Final.pdf

28

CLIMATE ACTIVISM AND ENVIRONMENTAL POLITICS IN MONGOLIA

Joe Ellis and Byambabaatar Ichinkhorloo

Introduction

This chapter presents ethnographic material on Mongolian environmental movements to provide an account of climate activism in the country. It shows how the protection of the environment has extreme political salience. More than this, the manner in which this plays out within both environmental activism and the wider population forces us to question the logic we assume to be present in global climate politics. This overview will unsettle an often-implicit assumption in contemporary accounts: that social movements seeking to protect the environment or reduce the effects of climate change are *automatically* progressive, or somehow left-wing. The chapter shows how particular moral conceptions of the landscape and environment can in turn generate forms of climate activism rather different from those observed in the Global North. The material from Mongolia is used to hint at a more general comparative point: that the coming political struggle will not so much be between those who seek to prevent runaway climate change and those who ignore it, but rather what kind of politics (and restructuring of societies) emerges as a response to climate change as an unavoidable fact.

The chapter is born out of a collaboration between us. We are both anthropologists whose primary research is carried out in Mongolia. Joe is based at the University of Cambridge and has conducted around 3 years of ethnographic fieldwork in the country, including research on 'ethnic' conflicts around land use in the context of climate change. Byambabaatar is based at the National University of Mongolia and has conducted 12 years of extensive fieldwork focusing on environmental conflict, pastoralism, and climate change. In 2012–2015, he interviewed 92 participants for his doctoral fieldwork[1] on the impact of the global environmental agenda on Mongolia and local people in three different places – Uvs, Arkhnagai, and Dundgovi provinces – representing Mongolia's western mountainous, central steppe, and southern desert regions. This research was the basis for his further detailed analysis[2] of how local communities transform and respond to right- and left-wing ideologies and politics, including postsocialism, neoliberalism, environmentalism, and developmentalism. He continued to track environmental issues and movements by interviewing 78 participants in Uvs, Central, and Arkhangai provinces in 2014–2016; over 140 urban and rural people focusing on the social

DOI: 10.4324/9781003396567-33

and political impact of extractive industries in Ulaanbaatar, South Gobi, and Sukhbaatar provinces in 2019–2021; and 18 household-level-deep interviews in Arkhangai in 2022. He has interviewed environmental movements in all his research and observed Enkhbat of Radical Movement, all heads of Bosoo Khukh Mongol, board members of Orlogo River Movements, and many other activists by phone and via social media.

This chapter aims to combine Byambabataar's extensive participant observation and interview-based work on environmental movements in the country with Joe's attempt to situate this empirical material in wider disciplinary debates on global climate activism. The chapter proceeds by introducing readers to the ways in which climate change and activism have been thought about in anthropology, as well as some concerns the authors have. It then takes the reader through a range of ethnographic and empirical cases from Mongolia in order to demonstrate the specificities of climate activism in the country. It concludes with a consideration of what Mongolia might tell us about how to think about this topic outside of the Global North and its implications for climate activism more generally.

Climate Change in Anthropology and the Over-familiar Political

The fact that discussions of climate change have become increasingly central within the social sciences in general – and anthropology in particular – is seemingly so obvious that it (almost) goes without saying. The social sciences have both sought to lend insight into the causes and effects of climatic shifts, while also rethinking our theoretical assumptions, as necessitated by the recognition of the mutual impact of humanity and the climate on one another. In anthropology (as in many academic disciplines), this is most famously manifested in the conceptual dominance that the 'Anthropocene' has come to hold. The use and impact of this concept in anthropology is multifaceted but includes the creation of an explanatory context for global political and social change (Moore, 2015) and an acceleration of long-held interests of the entanglements between humans and non-humans within the social world (Chua and Fair, 2019). Yet a lingering question has been what exactly an *ethnographic* approach can provide to understandings of global climate change.

A significant amount of work that engages with that question involves examining the varied responses to the growing impacts of environmental change in particular cultural settings. This kind of scholarship seeks to describe and reconcile the infinitely varied effects of – and responses to – a single global phenomenon (Hastrup, 2009; Crate and Nuttall, 2009; Ulturgasheva and Bodenhorn, 2022). The characteristic anthropological insight here is to highlight the lived and everyday realities of the impact of climate change and to show how they are far from uniform. Unsurprisingly, this includes a growing corpus of work on the political responses and forms of climate activism in the global south.

Perhaps most famously Marisol de la Cadena (2015) has provided an account of the political strategies of Quechua speaking people in Peru. She brilliantly shows how political responses to ecological crises from Amerindian peoples draw upon a cosmological and ontological foundation that goes beyond liberal or leftist conceptions of climate activism. Her work thus represents a double form of emancipation. The description and empowerment of Indigenous activism in responding and combating the consequences of climatic change is accompanied by the recognition of political concepts as they are understood by Indigenous activists themselves. Cosmological 'earth beings' are as much in play in this kind of climate activism as are scientific measures of ecological destruction. This exemplifies a wider corpus of work that seeks both the political empowerment of Indigenous climate activists in the Global South *and* the conceptual self-determination of people whose politics may draw upon sources thought well 'outside' politics in the Global North (Escobar, 2018; De la Cadena and Blaser, 2018; see O'Reilly et al., 2018 for a comprehensive review).

In essence, due to its focus on comparisons across global societies, ethnographic attention to climate activism has revealed the importance of taking seriously a far wider range of political and ontological logics that motivate the strategies of environmentalists than is found within Euro-American liberal politics. Undoubtedly, this has the potential to lay the foundation for building the kinds of transnational solidarity required for addressing a global crisis. Yet, for an anthropologist, there is perhaps a nagging doubt that this is all a little too familiar. The ethnographic record of climate activism in the Global South described earlier largely seems to consist of the strategies of people seeking to reconcile and mitigate ecological destruction. These are further described as political projects that seek to empower the marginalised, combat economic and political inequalities, and fight the human suffering that climate change emerges from and generates. However, given its commitment to empirical realities and complexities of lived politics, an obvious anthropological question arises. It can most crudely be summarised as 'why would climate activism automatically be about the emancipation of humans?' Put another way, is there any logical reason climate activism should produce political outcomes that a leftist-green consensus would consider 'good'?

As a provocation, one might wonder if the kinds of climate activism that are being recognised in the Global South are merely the forms with outcomes that are politically acceptable to those engaged with the subject in the Global North. Surely if there are forms of environmental politics that draw upon differing philosophical and civilisational resources that result in outcomes that fit with our rose-tinted glasses of climate activism, there should be others with very different political ramifications. This is important because to understand how to enable climate activism requires us to fully grasp its consequences. Yet more than this, to assume all such political forms necessarily overlap with our values would be to commit an act of intellectual and material colonialism of the very kind that has been said to partly lie behind the climate crisis we see today (Davis and Todd, 2017). The decolonisation of the climate struggle must mean ideas from the Global South are taken seriously regardless of whether we find them desirable or not.

In what follows, this chapter presents an overview of certain elements of climate activism and environmental politics in Mongolia. It attempts to capture it in its true complexity, embracing contradictions. This will provide an impetus to consider ways to incorporate the diversity of real existing climate politics in Mongolia and elsewhere into our understanding of the local and global strategies for weathering the coming storms.

Resource Extraction in Mongolia

Mongolia is a landlocked country bordering Russia in the North and China in the South. Following a period of over 200 years of Qing Dynasty rule, it declared its independence in 1911. A decade of political turbulence and armed struggle followed, ultimately resulting in a second declaration of independence in 1921 with the backing of the Soviet Union and the Red Army. For the next 70 years Mongolia was ruled by a socialist government heavily dependent upon the USSR for economic aid. The period saw a rapid program of agricultural rationalisation and collectivisation, industrialisation, and sedentarisation. Access to education and public health was greatly expanded during this time. However, the socialist period also saw bursts of political repression and violence, particularly in the 1930s; the political and religious elite were especially targeted. Following the end of socialism in the USSR and Mongolia at the start of the 1990s, Mongolia transitioned to multiparty democracy and underwent a period of mass privatisation and neoliberal 'shock therapy' initiated by the world bank, IMF, and Mongolian political elites. This created significant social and economic turbulence that lasted well into the 2000s.

In the mid 2000s the discovery of significant fossil fuel and mineral deposits (coal, copper, and gold) led to increased attention from multinational mining companies (Ichinkhorloo, 2021). Most famously, a combined copper and gold mine in the far south of the country known as Oyu Tolgoi was projected to have an annual output of *one third* of the entire GDP of Mongolia and is described as the largest single mining project in the world. This project was 'shared' between the government of Mongolia and Rio Tinto, a multinational mining conglomerate. Its cost has already passed $10 billion.

This mine, along with numerous others to exploit significant coal, copper, and gold reserves for export to the resource hungry neighbouring economy of China, has formed the basis of governmental policy in Mongolia for the last 15 years. Through extensive government borrowing and agreements with foreign mining companies, successive governments have sold a promise of rapid development. This has included improved infrastructure and poverty alleviation across the country, as well as housing construction for the over a million people living in the '*ger* districts' of the capital.

Progress in this regard has been mixed. While there has been success in reducing poverty rates and improving cross-country road infrastructure, as well as a huge construction boom in the capital, the story is not straightforward. Inequality has increased in Mongolia, and though the middle class has gained access to newly built apartments, they remain out of reach for many of the country's poor, who often feel they have benefited little from the mining boom. Constant corruption scandals reinforce the view that resource extraction has largely benefited a small group of politicians, business owners, and their families, who are accused of 'eating money' (embezzlement) and stashing their money overseas while sending their children to be educated in Europe and America. When combined with a series of economic crises and bailouts by the IMF, discontent regarding how much non-Mongolian companies are profiting from its resource wealth and doubts as to how mining impacts Mongolia's significant population of animal herders make it clear that mining is contentious in Mongolia even *before* one considers the environmental impact of resource extraction and burning coal. Indeed, we can see how Mongolia is far from a 'peripheral' player in processes of climate change. The certainty of extractive logics of capitalism, embraced by the national government with the justification of the country's economic prosperity, feed into climatic shifts in varied ways. Mongolia is now the primary supplier of coking coal for blast furnaces used in steel in China. Steel production accounts for 7% of global energy sector CO_2 emissions, and China produces over half of the world's steel. It is also deeply involved in proposed shifts to more copper-intensive forms of green energy, designed to reduce global emissions yet still resting upon and requiring the extraction of ever-expanding amounts of resources in the country. Perhaps unsurprisingly then, a range of environmental activist movements have emerged in the last decade.

From Environmental to Climate Activism

It is important to recognise that, while movements in Mongolia often draw attention to processes of climate change and thus place their actions in a climate activist register, they emerge from localised environmental conflicts. Accordingly, an immediate anthropological insight is that an understanding of climate activism, particularly in the Global South, coun-terintuitively needs to be wary of framing such movements as 'global' at the outset. This is a concrete example of a concern we raised in the opening of the chapter: the danger of describing all forms of environmental and climate activism across the world in terms of a particular concept of climate struggle that dominates in the Global North. Instead, we need to consider what concepts of planetary climate shifts environmental movements are actually engaging in rather than assuming that we already know. Put another way, the study of climate activism in the Global South has to leave open the possibility that the 'climate' may mean radically different things across the world. As such, this chapter now moves on to a description of 'environmental activism' that has emerged in particular places and times in Mongolia. It then considers the moment of generalisation and translation when these environmental movements engage with ideas of a more global register and thus become 'climate activism'. By beginning with the localised environmental activism and examining this moment of scaling up, we also examine what is brought to the local from the global. As will be shown, part of what is brought can be complex and sometimes have troubling, political implications.

Mette High has provided one important anthropological account of mining in Mongolia. Focusing on informal and small-scale 'ninja' gold miners, she gives an account of the complex moral landscape that gold miners, traders, and impacted animal herders operate within (High, 2017). In particular, she examines how gold miners were understood, through their excavations of the ground, to have 'angered spirits and transgressed taboos of the land' (High, 2021, p. 5). She thus developed a concept of a Mongolian 'cosmoeconomy' in which the economic motivations, logics, and outcomes of gold mining had to be understood as part of a wider moral framework which included insulted spirits. This has important implications for our understanding of political and activist responses to mining. To this end, Rebecca Empson (2018) argued that environmental protest movements that emerged in the 2010s were in part responses to such moral violations. Empson draws upon ethnographic accounts of how groups of herders mounted legal and protest challenges to mining activities that would result in the degradation of pastureland. Crucially, these protests justified their opposition to mining activities on the basis that they were the rightful occupants of the land due to their historical worship of land master spirits that held the sole authority to determine land use. Here, then, environmental protest emerges from a particular notion of pasture rights as mediated by spirits that abhor mining activity, putting them into conflict with the forms of private land ownership that mining requires (*Ibid*). What these accounts amount to is a description of the way in which climate activism in Mongolia, particularly concerning the environmental and climatic impact of mining, emerges from, and is structured by, a moral and cosmological logic that determines appropriate use of land (see also Humphrey, 1995). In order to understand climate activism, then, a full account of this foundation is needed. In what follows we provide an ethnographic sketch of the kinds of more recent environmental and climate activism in Mongolia as they arise from this context.

Environmental movements in rural Mongolia often emerge from people's desire to preserve or improve local ways of living that connect people and communities with nature. This desire is produced when local communities meet urban people and external institutions that affect local lifestyles and renew people's relationship with nature. The nature of this relationship is specific to the country. For example, Mongolian pastoralists tend to interpret it as the cohabitation of people and nature; it is materialised by local rituals and customs that regulate access to and use of natural resources, as outlined earlier. When communities encounter these new participants – such as the state, mining companies, and new resource users – in their localities, conflict can develop and cause tension in the community. These divisions lead to the establishment of separate groups that either defend their former connection with nature or support newly emerged interests in collaboration with new participants. Local groups quickly form environmental movements to resist or support changes in their social life and customs when confronted with other competitors.

Internal and external competition over access to the environment and natural resources lead to conflicts; local people seek to reach a consensus in order to change how they deal with the environment and create rules in the form of rituals, taboos, and customary activities. Those environmental movements with integrated local beliefs and values are often from local communities who are encouraged and supported by national and international players. In contrast to the rural movements, urban movements become a part of global environmental networks: they join climate programmes and agendas, designed by multilateral and intergovernmental organisations. These movements advocate for environmental justice: they argue for equal access to – and equal distribution of – environmental resources, and they protest against air pollution, toxic chemical leaks, and the localised impacts of climate change. As will be seen, they additionally defend and support nationalist identities, which rest on the sacredness of nature.

Rural Mongolia

Many environmental movements in rural Mongolia emerged at the grassroots community level and may have aims that go beyond that of just protecting the environment. For example, the environmental movement in Umnugovi district of Uvs province in Western Mongolia has changed its goals and tactics many times depending on the issues raised by the public. In the beginning, the movement protested against perceived outsiders – a mining company and artisanal miners – who were extracting gold from the banks of the Orlogo river in 2009. Nine people, mainly local schoolteachers and subdistrict officers, organised local herding families and female doctors to protect the river. While they were successful in stopping the resource extraction, 30 men were arrested according to local and movement people. The second wave of the protest began in 2012, when new companies came to extract gold in exchange for promised rehabilitation of the damaged land. This time, the movement encountered not only the external mining companies but also national politicians who backed the companies; the local community became divided between pro-artisanal mining and pro-pastoralists. In addition, the Orlogo river protection movement began to produce NGOs and other political movements; leadership represented different interests and started to be composed of local politicians and businesspeople. In the 2016 and 2020 local elections, environmental issues became the key agenda to win local elections. As a result of the local election in 2020, former members of the environmental movement took power in the district, representing different political parties. In summary, a reaction against resource extraction emerged from both economic interests and customary norms and transformed into a more formal kind of democratic politics, ultimately resulting in a power shift in local elections. This transformation has continued and has expanded to the national scale but in a way that also brings in international actors.

Built upon the foundation described earlier, a new generation of movements is emerging in this region of Mongolia. It centres around a new hydropower station called

'Erdeneburen' that is planned in Khovd, a neighbouring province. This is a Chinese-funded initiative within the 'Belt and Road' international development project. While its intention is to provide renewable energy and lessen Mongolia's dependence upon coal, a whole range of environmental concerns have been raised about its impact on the ecosystem of the Khovd river basin – a dynamic that is common to these kinds of projects. Mr Tsedenbal, a former member of the Orlogo river movement and current governor of the Uliast subdistrict, is leading a protest against this hydropower station. In an official letter of complaint to the Chinese government and the Chinese Export-Import Bank, sent via the Chinese Ambassador to Mongolia in December 2021, he demanded that funding for the hydropower station be withdrawn due to its potential impact upon herders' livelihoods. But Mongolia's Minister of Justice accused him of treason and the public debate became fractious.

The online debate about the 'Erdeneburen' hydropower station brought up questions about whether it was appropriate to take money from China to fund renewable energy development, given the Chinese economy's own ongoing contribution to carbon emissions. Yet, more than this, the debate drew upon and reinforced negative views that many Mongolians have of China. Some of this was simply criticism of the Chinese government, but it also related to sinophobic ideas about Chinese people as a whole. As Franck Billé (2016) has most systematically described, sinophobic sentiments remain widespread in Mongolia, and this issue resulted in the circulation of further claims that only 'true' Mongols should have the right to pursue development projects in the country. Here, then, a local environmental movement that drew on economic concerns to do with herders *and* cosmological norms surrounding rivers quickly transformed into one that had international and global claims about China's contribution to climate change. Yet it also articulated sinophobic and exclusionary political claims that one would not associate with left-progressive climate activism. This point will become clearer with an account of climate movements in urban Mongolia.

Urban Movements

Perhaps the most (in)famous environmental group that operated in Mongolia's capital Ulaanbaatar was the 'Fire Nation' Coalition. As has been extensively described by Dulam Bumochir (2020), on 16 September 2013 a group of activists approached the entrance of the parliament building. This was a small group that included herders who had travelled from the countryside and united with urban actors. Armed with hunting rifles and hand grenades, they appeared to attempt to gain access to the state parliament building. While no one was hurt, a shot was fired and all participants in the action were arrested. Later, the police reported that explosives were found in the area, including near the Ministry of the Environment and Green Development (*Ibid*). The men were found guilty of attempted terrorism and crimes against the state.

These events were contentious and there was disagreement over whether their actions could rightly be described as 'green terrorism'. This is partly because their leader, Munkhbayar Tsetsegee, was an influential figure who marshalled significant popular support from sections of the rural and urban population of the country. Indeed, Munkhbayar gained global recognition when he was awarded the Goldman Environmental Prize in 2007. Whilst in prison for the actions described earlier, Munkhbayar began to articulate a set of political ideologies. Initially committed to the defence of sacred rivers in rural Mongolia, he developed a platform that called for the replacement of the current 'illegitimate' state in Mongolia with a return to a kind of traditionalist and imperial governance structure modelled on the Mongol empire. In this platform the sacredness of nature and power of the Mongol heavens is used to justify an environmental movement that transforms into a set of ideas about the need to form a new kind of authoritarian state; Mongols are recognised as the only rightful inheritors of its resources.

The fire nation movement was an exemple of a more widespread process whereby environmental movements that attract attention from global climate activists nonetheless intersect with existing right-wing networks and draw upon notions of the need to protect sacred nature alongside notions of Mongol supremacy. This can be seen in another environmental movement that has received significant attention: the 'Bosoo Khukh Mongol' group, which translates to 'Stand up Blue Mongols'. The name in this case refers to the colour of the eternal blue heavens that are seen to have ordained the Mongol empire, and which are often entangled with ideas of the divine supremacy of the Mongol people. Officially established on 27 May 2013, Bosoo Khukh Mongol was formed to resist perceived environmental crimes and pollution in Mongolia, especially in relation to illegal mines. Many news outlets, including the *Times*, the *Guardian*, and *Reuters* reported on this movement's connection with Mongolia's far-right nationalist movements such as Dayar Mongol, White Swastika, Fire Nation, and Blue Mongolia (AFP, 2013; Branigan, 2010; Moxley, 2009). In the late 1990s, these vigilante groups took action against the many Chinese restaurants and businesses that opened and used Chinese characters on signs and documents, which was technically unlawful at that time. This vigilante action was not without public support, and the vigilante groups were a transformation of former street hooliganism collectives found in 1980s socialist Mongolia. They refocused their attention from Chinese restaurants to Chinese construction companies and mines in the late 2000s, as highlighted by the significant number of Bosoo Khukh Mongol members connected to them. Bosoo Khukh Mongol announced that it will conduct vigilante actions against any unlawful environmental activities, especially environmental crimes and pollution conducted by foreign-financed companies.

After its founding in May 2013, Bosoo Khukh Mongol started chapters in 16 provinces and 10 districts. In its first 4 years (2013–2017), Bosoo Khukh Mongol was involved in violent activities under the leadership of the first leader. But the next leader changed tactics and conducted legal activities, making Bosoo Khukh Mongol very popular among the public. In our interviews, Gankhuyag, the head of the movement and other board members of this movement informed us that local people across Mongolia have provided Bosoo Khukh Mongol with thousands of reports about illegal mining activity and environmental crimes (also see video interviews in Mongolian by Ulaan Bal[3] and Battushig on youtube.com and tug.mn).

As of 2018, Bosoo Khukh Mongol members had visited hundreds of coal and gold mining sites and filed lawsuits against – and successfully cancelled – 70 mining licences. The former head, Gankhuyag, claimed in an interview that 20–70 members of the movement visit different areas of the country. They meet with mining companies, local people, and officials, and they are often followed by Mongolian police. The movement checks if the companies are breaking laws and regulations. If they find any illegal activity, they sue the company in court. Its most successful campaign was to create, alongside government working groups and other environmental movements, the national discussion forum 'Mother Nature-Responsible Mines', which attracted participants from all provinces in 2019. Another success was their lawsuit against Badrakh-Energy, a Uranium mining company, with 66% of its investment by Areva, a majority French-owned multinational and 33% investment by the Mongolian government since 27 January 2018. This uranium was intended to be exported to the French nuclear energy industry. By March 2023, Bosoo Khukh Mongol had participated in 67 court sessions; the court cancelled 9 out of 15 licences owned by Barakh-Energy.

Over the years, the reputation of Bosoo Khukh Mongol has increased significantly among the public; many of its members announced that they would run for the parliamentary election in 2020. However, according to the heads of the movement, their successes brought counter-attacks from mine owners as well as politicians who accused movement leaders of fabricating criminal cases. Due to this, many lost their right to run for election in early 2019 and the movement split into Bosoo Khukh Mongol and Bosoo Mongol in 2020. These accusations also pointed out the involvement of its members in previous vigilante right wing groups. In this case, while these violent practices morphed into legal methods, they draw upon the same ideological ecosphere as the fire nation discussed earlier. This situation becomes even more complicated as these kinds of environmental groups engage with more global organisations seeking to combat climate change, as will be outlined later.

Another key group operating in the movement ecosystem of urban Mongolia is The Radical Reform Movement, which demanded the reform of mineral laws in 2006. Their protest started with a demonstration protesting bus fare increases in response to the resignation of the government in 2005. After repeated protests in the streets in January and February 2006, they stormed the opposition party building. However, in an interview with the authors, D. Enkhbat, a general coordinator of the movement, said this movement and its members did not have any environmental goals until they met with senior professors, lawyers, and former government officials, whom they call elders. The elders approached key activists and invited them to discussions about society, ethics, justice, equality, environmental issues, and the extractive industry. After reading materials given to them and a month-long period of meetings and training, Radical Reform, led by S. Ganbaatar and D. Enkhbat, focused their protest and demonstration on mineral law revision, licences, and environmental issues, and they targeted the Oyu Tolgoi copper mine run by Rio Tinto described earlier in this chapter. Demonstrations, hunger strikes, police arrests, and violence followed. Finally, a joint working group was established by newly appointed Prime Minister M. Enkhbold to bring the protesters into formal consultation with the government in 2006.

When the protestors were invited to join the joint working group, many movements left due to suspicions regarding the government's motives, according to Ganbaatar and Enkhbat. The 'Radical Reform' and 'My Mongolian land and soil' stayed. Backed by the Elders and legal advice from professors at the National University of Mongolia and the University of Science and Technology, the Radical movement negotiated and worked with government officials to draft a new mineral law, which included many environmental protection and compliance issues. The Elders also connected Radical Reform with German legal and civil society experts who worked in the Bundestag; they taught them how to conduct public consultation and about the importance of information disclosure, according to D. Enkhbat, the former coordinator of this movement. The Radical Reform cooperates closely with Mongolia's 'neoliberal' democratic party and occasionally with global environmental organisations such as the Asia Foundation, Snow Leopard Trust, and WWF. Ganbaatar Sainkhuu, the head of this movement, was twice elected to parliament and co-authored an article on resource nationalism that criticised Mongolian government policy (Blunt and

Sainkhuu, 2015), before he ran for president in 2017. D. Enkhbat also highlighted in an interview with one of the authors how government and agency officials were obedient to the working group and supplied all necessary information such as company-related information, registration, taxes, licences, agreements, and deposit-related details that were undisclosed to the public.

As a result of this monitoring, the government revised operating agreements with the biggest mines in Mongolia. Even today, the Elders are respected by the public and many environmental movements. According to the coordinator of Radical Reform, their protests awakened society and stopped the issuing of all type of mining licences in 2010–2012, cancelled 106 exploration licences for environment protection in 2013, and renewed licensing regulations and procedures six times in 2006–2018 to improve transparency, in response to public and extractive industry demands. Importantly, Radical Reform collaborated with Bosoo Khukh Mongol, Snow Leopard Trust, and political parties in 2019 and organised a huge protest against a mining company in the Gobi region. Here then, internationally focused actors from Germany back in 2006–2007 and global environmental organisations in 2012–2019 were seemingly unwittingly pulled into offering advice to an environmental movement that had complex connections with right-wing politics. Accordingly, engagements from global climate movements may unwittingly reinforce national political processes far beyond their intentions or wishes. This case is hardly unique in Mongolia.

The environmental projects and programmes of international organisations, including multilateral and non-governmental organisations, have directly intervened in and shaped Mongolia's environmental politics (Ichinkhorloo, 2021). The ~60 organisations have implemented over 1,000 projects in Mongolia. The Mongolian government adopted over 100 policy papers followed by new laws and regulations drafted and lobbied by these organisations including UNDP, World Bank, UNEP, ADB, WWF, WCS, environmental programs of the universities, and international non-governmental organisations and individual climate activists.

The main indicator of the project's successes was the policy impacts these projects brought to Mongolia and the number of people or species that benefited and were affected. Unfortunately, many of these policies and laws were contradictory due to the different goals of these organisations. The failure to reach agreement was evident in Mongolia and it led to clashes in the environmental politics of these global players. Part of this failure stemmed from a misrecognition by international organisations of what it meant for Mongolian environmental movements to think globally about the climate.

Conclusion

This chapter has provided a sketch of environmental movements in Mongolia, the specific context they emerge from, and the processes by which they engage with international organisations that attempt to recruit them as contributors to the global climate struggle. It was suggested that many of these movements arise in response to violations of proper conduct in the Mongolian landscape. Mining in particular is thought to transgress the interests of a sacred conception of nature, in which land and water spirits are angered by the environmental impact of such modes of extraction.

Critically, this cosmological ideology also contains very particular ideas about the 'correct' and 'traditional' modes of subsistence of the Mongols, their right to land use, and imperial ideas of Mongol supremacy that reference the Mongol empire. When environmental movements emerge from such contexts, they articulate a pre-existing right-wing ideology that tends towards a struggle for the formation of a 'pure Mongol' authoritarian state. As such, environmental activism emerges within a network of far-right actors that have a longer history in Mongolia. In turn, as international actors seek to advise and encourage these movements to think globally and, in effect, transform environmental activists into climate movements, there are consequences both for that new global thinking and internal politics within Mongolia. Climate activism then risks becoming a kind of manifestly sinophobic process which also seeks a new form of state with environmentally justified absolute rule and exclusionary politics of 'pure Mongol' supremacy. In a country of numerous Mongol and non-Mongol minorities and people of mixed heritage, this is undoubtedly a dangerous situation.

This chapter has argued that in order to mount a truly global climate struggle it is necessary to seriously consider the kind of environmental and climate activism in the Global South. This kind of decolonisation is both an absolute necessity and reveals further challenges. In Mongolia, there are climate movements that might appear as logical allies in such a struggle. Yet, due to the particular cosmological and political histories of these movements, one has to think again about whether the ultimate social aims of such actors are in line with the value system of the global climate movement. Protecting the environment does not automatically produce desirable outcomes. What an anthropological and comparative approach ultimately reveals is that our descriptions of climate activism around the world must leave room for *critical* engagement as well as support. We must engage with movements and take their ideas seriously. Yet, it also makes clear that those involved in the climate struggle cannot limit their efforts just to enabling all those who appear to be engaged in the same project to prevent the worst outcomes of climate change. Instead, a vision must be provided of what political and social outcomes ought to look like as we restructure our global societies to mitigate – and adapt to – climate change. Preventing climate change cannot be a cause in itself but rather an impetus to re-imagine what a just kind of politics can be as it becomes our reality.

Notes

1 His dissertation 'Development Intervention in Environmental and Agricultural Sectors in Mongolia' focused on how rural and urban people were affected by and responded to the development agenda and environmental programmes.
2 His interviews were carried out as a part of: the 'Green Gold' project, funded by SDC in 2014–2016; 'Mediation Model for Sustainable Infrastructure Development – Scaling up Praxis from Mongolia to Central Asia' by University of Oxford, funded by GCRF Grant ES/S000798/1 in 2019–2021; and 'Using Sound to Advance Conceptual Frameworks of Resilience of Integrated Grassland-Pastoralist Systems' by Purdue University, funded by NSF Grant 2009669 in 2022.
3 Interview with Bosoo Khukh Mongol by Ulaan Bal journalist on 15 January 2019 at https://youtu.be/JS_2RPwf1YM?si=e9p207uIM_N4w98i and interview with Gankhuyag by B.Battushig on June 14, 2020 at https://fb.watch/q-ZwMSZ-_f/.

References

AFP. (2013). Mongolia's "eco-Nazis" target foreign miners. *Bangkok Post*, 10 October. https://www.bangkokpost.com/world/374004/mongolia-eco-nazis-target-foreign-miners.
Billé, F. (2016). *Sinophobia: Anxiety, Violence and the Making of Mongolian Identity*. Honolulu: University of Hawaii Press.

Blunt, P. and Sainkhuu, G. (2015). Serendipity and stealth, resistance and retribution: Policy development in the mongolian mining sector. *Progress in Development Studies*, 15(4), 383–399.

Branigan, T. (2010). Mongolian Neo-Nazis: Anti-Chinese Sentiment Fuels Rise of Ultra-Nationalism. *The Guardian*, 2 August. https://www.theguardian.com/world/2010/aug/02/mongolia-far-right [Accessed 24 March 2024].

Bumochir, D. (2020). *The State, Popular Mobilisation and Gold Mining in Mongolia*. London: UCL Press.

Chua, L and Fair, H. (2019). Anthropocene. *The Open Encyclopaedia of Anthropology*. http://doi.org/10.29164/19anthro.

Crate, S. and Nuttall, A. M. (2009). *Anthropology and Climate Change: From Encounters to Actions*. Walnut Creek: Left Coast Press.

Davis, H. and Todd, Z. (2017). On the importance of a date, or, decolonizing the anthropocene. *ACME: An International Journal for Critical Geographies*, 16(4), 761–780.

De la Cadena, M. (2015). *Earth Beings: Ecologies of Practice Across Andean Worlds*. Durham: Duke University Press.

De la Cadena, M. and Blaser, M. (2018). *A World of Many Worlds*. Durham: Duke University Press.

Empson, R. (2018). Claiming resources, honouring debts: The cosmoeconomics of Mongolia's mineral economy. *Ethnos*, 83(6), 263–282.

Escobar, A. (2018). *Designs for the Pluriverse: Radical Interdependence, Autonomy and the Making of Worlds*. Durham: Duke University Press.

Hastrup, K. (2009). The nomadic landscape: People in a changing environment. *Danish Journal of Geography*, 109(2), 181–189.

High, M. (2017). *Fear and Fortune: Spirit Worlds and Emerging Economies in the Mongolian Gold Rush*. Ithaca: Cornell University Press.

High, M. (2021). Changing the conversation on energy at the centre for energy ethics. *Energy Humanities*. https://www.energyhumanities.ca/news/changing-the-conversation-on-energy-at-the-centre-for-energy-ethics.

Humphrey, C. (1995). Chiefly and shamanist landscapes in Mongolia. In E. Hirsch and M. O'Hanlon (eds.) *The Anthropology of Landscape: Perspectives on Place and Space* (pp. 135–162). Oxford: Oxford University Press.

Ichinkhorloo, B. (2021). Contestations over mining policies and mineral ownership in Mongolia from the socialist period to the present. In T. Sternberg, K. Toktomushev and B. Ichinkhorloo (eds.) *The Impact of Mining Lifecycles in Mongolia and Kyrgyzstan: Political, Social, Environmental and Cultural Contexts* (pp. 33–46). London: Routledge.

Moore, A. (2015). Anthropocene anthropology: Reconceptualizing global contemporary change. *Journal of the Royal Anthropological Institute*, 22(1), 27–46.

Moxley, M. (2009). The neo-Nazis of Mongolia: Swastikas against China. *Time Magazine*, 27 July. https://content.time.com/time/subscriber/article/0,33009,1910893,00.html [Accessed 24 March 2024].

O'Reilly, J., Isenhour, C., McElwee, P. and Orlove, B. (2018). Climate change: Expanding anthropological possibilities. *Annual Review of Anthropology*, 49(13), 13–29.

Ulturgasheva, O. and Bodenhorn, B. (2022). *Risky Futures: Climate, Geopolitics and Local Realities in the Uncertain Circumpolar North*. Oxford: Berghahn.

29

CLIMATE ACTIVISM AND ANIMAL AGRICULTURE

Grassroots vs. Greenwashing

Adriana Voss-Andreae

Introduction: Animal Agriculture and the Climate and Ecological Crises

Our food system is a key driver of the climate crisis, as well as the leading cause of the larger environmental crisis. It is estimated that, if left unchecked, global warming from agriculture alone, most (~70%) of which comes from the production and consumption of animals and animal products, will add ~1 degree Celsius of global heating by the end of the century, thereby heating the planet well past the key target of 1.5 degrees Celcius (Ivanovich et al., 2023; IPBES, 2019). Animal agriculture also uses up ~ 80% of all global farmland, while providing just 18% of our food calories (or 37% of protein) (Poore and Nemecek, 2018; Ritchie, 2019) and is responsible for a combined estimate of 14–20% of all global anthropogenic greenhouse gas (GHG) emissions (Goodland and Anhang, 2009; Gerber et al., 2018; Twine, 2021; Xu et al., 2021). Animal agriculture is also the major driver of natural ecosystems loss, precipitous biodiversity loss, dangerous levels of pollution, depletion of fresh water sources and the decreasing resilience and stability of the global food supply itself (Shukla et al., 2019).

Worldwide consumption of meat has more than doubled in the past 20 years (Ritchie and Roser, 2019; Searchinger et al., 2019). Research indicates that even if we immediately stopped burning fossil fuels, animal agriculture alone will eat up 80% of the entire global allowable GHG budget (set in the Paris Climate Accords) in just the next 20–30 years, risking mass starvation and environmental catastrophe in much of the world (GRAIN and IATP, 2018; Clark, 2020; Ivanovich et al., 2023). Alarmingly, by most predictions, global demand for beef is expected to increase by about 90% and demand for all meat by 70% by 2050 (Ivanovich et al., 2023).

This chapter highlights the role of grassroots climate activist organisations, such as Fridays for Future Germany, in shifting social norms and building political momentum toward reducing meat consumption and production. The websites (and other online materials) of 20 US- and/or Europe-based climate/environmental NGOs and organisations – a cross-section of the much larger and more diverse climate movement, many of which also have numerous international offices, chapters, activists and/or affiliate organisations across the world – were examined for educational materials, campaigns and/or demands related to animal agriculture. Although virtually all the organisations describe themselves as eliciting

DOI: 10.4324/9781003396567-34

grassroots support or building a grassroots movement, they use differing strategies, tactics and approaches. The organisations examined are:

- 350.org
- Audubon Society
- Center for Biological Diversity (CBD)
- Citizens' Climate Lobby (CCL)
- Climate Justice Alliance (CJA)
- The Climate Reality Project (CRP)
- Earthjustice
- Environmental Defense Fund (EDF)
- Extinction Rebellion (XR)/Animal Rebellion (AR)
- Fridays for Future (FFF)
- Friends of the Earth (FoE)
- Food & Water Watch
- Greenpeace
- National Wildlife Federation (NWF)
- Natural Resources Defense Council (NRDC)
- Rainforest Action Network (RAN)
- Sierra Club
- Sunrise Movement
- The Nature Conservancy (TNC)
- World Wildlife Fund (WWF)

In doing so this chapter raises the question: how are US- and Europe-based climate and environmental organisations prioritising this issue, if at all? And, are their proposed solutions based in science and the underlying values and principles of climate justice and equity?

The Need to Shift to a Plant-based Food System

Leading scientists have stated in more ways than one that eating a plant-based diet is the most important contribution every individual can make to reversing the climate crisis and probably the single biggest way to reduce your environmental impact, "far bigger than cutting down on your flights or buying an electric car" (Carrington, 2018; Foer, 2020). The science is very clear that a substantial shift toward a largely plant-based food system is urgently needed as a key component to addressing the climate crisis.

Indeed, if we all shifted to a plant-based diet, it would "free up 75% of agricultural land for restoring and rewilding, while still plentifully feeding the world" (Poore and Nemecek, 2018). One study estimated that a rapid global phaseout of animal agriculture has the potential to stabilise GHG levels for 30 years and offset nearly 70% of CO_2 emissions this century (Eisen and Brown, 2022). Such a phaseout would also have numerous critical environmental co-benefits, including cleaner waterways, vital aquifer replenishment, critical habitat restoration and improved air quality.

In the most rigorous study of its kind, the EAT-Lancet Commission developed global scientific targets for healthy diets and sustainable food production to keep our food system aligned with UN Sustainable Development Goals and the Paris Agreement (Willett et al, 2019). Based on extensive epidemiological evidence, the commission proposed boundaries that global food production should stay within to decrease the risk of irreversible and potentially catastrophic shifts in the earth's systems. The commission described our current global food production as threatening climate stability and ecosystem resilience, calling it the "single largest driver of environmental degradation and transgression of planetary boundaries. Taken together the outcome is dire. A radical transformation of the global food system is urgently needed" (Willett et al., 2019). It also recognised food to be "the single strongest lever to optimise human health and environmental sustainability on Earth" and "a defining issue of the 21st century" that will require "a new agricultural revolution that is . . . driven by sustainability and system innovation."

The key component is the need to cut meat almost entirely out of our diets and to double the global consumption of fruits, vegetables, nuts and legumes (Willett et al., 2019). The commission's planetary health diet guidelines calls for North Americans "to eat 84% less red meat but six times more beans and lentils" and Europeans to eat "77% less red meat and 15 times more nuts and seeds" (Carrington, 2019). The planetary health diet is largely plant-based (with most of the dietary protein coming from pulses and nuts), allowing for one weekly serving of meat and a couple servings of fish and/or a small amount of dairy and eggs. "We are talking about a way of eating that can be healthy, flavorful and enjoyable" explained Willet (Carrington, 2019). This dietary shift would be a "win-win", according to the scientists, preventing over 11 million deaths (1 out of 5 adult deaths) a year from unhealthy diets, the leading cause of illness worldwide, while preventing the collapse of the natural world that humanity depends upon (Willett et al., 2019).

A recent report published by Harvard Law School's Animal Law and Policy Program points out how "much of the political focus has been on the energy transition, however a food transition is also needed – especially . . . from livestock production" and surveyed over 200 climate scientists and sustainable agriculture experts across 48 countries (Harwatt et al., 2024). They found that a consensus (~80%) of experts agreed it is important that absolute livestock numbers peak globally by 2025 followed by a rapid decline and that diets shift from livestock-derived foods to plant-based alternatives, in both high- and middle-income countries. Most also agreed that financial assistance should be provided to farmers to convert their practices away from livestock production and to rewild the land. They conclude that

> the experts surveyed suggest a clear pathway ahead – one which departs significantly from current trends. To align with the Paris Agreement, global emissions from livestock production must decline by 50% during the next 6 years, and this must be accomplished without negatively impacting farmed animal welfare or increasing the number of farmed animals. High producing and consuming nations must lead the efforts, and policies to support a deep and rapid transition away from livestock production and consumption will be needed.
>
> *(Harwatt et al., 2024)*

Red to Green: Is the Climate Movement Advocating for a Plant-based Shift?

The research here suggests that most climate organisations recognise the need for a food system change as a necessary part of addressing the global climate and ecological crises. There are few, however, that specifically make reference to the harms of industrial animal agriculture and/or support policies that promote plant-based diets. Of these organisations, the Sunrise Movement, a US grassroots youth-led climate justice organisation, prominently features the Green New Deal's (GND) demands on its website, including the "right to clean air and water" and "access to healthy food" for all and, under "honouring Indigenous rights and sovereignty," also includes "growing productive food without depleting the nutrients of the soil with monocultures" (Sunrise Movement, 2024). However, they make no demands related to the harms caused by our current food system or any mention of animal agriculture.

CJA, a US-based organisational alliance of frontline communities and organisations centred on a "Just Transition" away from "extractive systems and political oppression" and toward "regenerative and equitable economies," also advocates for food sovereignty as the "right of peoples to healthy and culturally appropriate food produced through ecologically sound and sustainable methods, and their right to define their own food and agriculture systems" (CJA, 2024). Like the Sunrise Movement, CJA makes no specific mention of animal agriculture or its disproportionate impact on frontline communities in the US and globally. However, they do link to a definition of food sovereignty with a rejection of "methods that harm beneficial ecosystem functions, that depend on energy intensive monocultures and livestock factories, destructive fishing practices and other industrialised production methods, which damage the environment and contribute to global warming" (Neyeleni, 2007).

The majority of the climate organisations examined do call out industrial animal agriculture as a leading contributor to the climate and environmental crisis, with two-thirds placing significant emphasis on the issue in their work or campaigns. However, a closer look elucidates that the organisations differ widely in their approaches and strategies, sometimes with completely opposing views on whether a particular climate solution is effective or, instead, helps a polluting industry "greenwash" its image. Two strategies highlight this divide: "regenerative" animal agriculture (in particular grazing cattle) vs. a shift to a plant-based food system. Whereas 85% of the organisations mention regenerative animal agriculture as a climate solution and 70% are even strongly in support (with only 5% significantly in opposition), shifting to a largely plant-based food system is mentioned only by 55% of the organisations, with 30% expressing strong support. In other words: there is far more support for retaining animal agriculture in some form than for supporting a shift to a plant-based food system among climate organisations despite the science clearly indicating that animal agriculture, including "regenerative grazing", have grave impacts on biodiversity, the environment and the climate (Garnett et al., 2017; Hayek and Garrett, 2018; Filazzola et al., 2020; Zhang et al., 2023; Monbiot, 2022; Dutkiewicz and Rosenberg, 2021).

Several climate activist organisations that make scant or no mention of he need to shift toward a plant-based system do, however, have grassroots affiliates that do make such mention include it. A notable example is Animal Rebellion (AR), a former grassroots affiliate of Extinction Rebellion (XR). AR has worked to draw the public's attention to the connections between meat and dairy consumption and the climate crises. In just a few years, AR successfully managed to rapidly gain a great deal of publicity and notoriety in UK and international mainstream news outlets for their form of visually dramatic and controversy-stirring direct-action protests. A few examples of their actions include blocking dairy trucks that temporarily stopped milk supplies in parts of England (Zeldin-O'Neill, 2022), blockading the UK's largest meat market (Taylor, 2019), occupying tables at high-end meat and seafood restaurants (BBC, 2022) and blockading McDonalds distribution sites (McSweeney, 2021). Notably, for varied reasons, people in the UK consumed less meat in 2022 than at any point in the prior 50+ years (Goodier and Sunnemark, 2023).

Other organisations, such as CBD, FoE, Greenpeace and RAN, which have long been promoting a plant-based shift as "exponentially more healthy for our bodies and the planet" and encouraging the "consumption of fewer animal products which have a huge climate footprint and drive gross human and animal rights abuses" (RAN, 2014), immediately embraced and widely shared the key recommendations of the EAT-Lancet report. These and other organisations also instituted their own plant-forward procurement policies (NRDC, 2019; CRP, 2019) and/or went a step further by encouraging local grassroots campaigns and providing them with toolkits, case studies and/or other guidance to enact procurement policies at schools, in government institutions, in Universities and in workplaces (FoE, 2017; Hamerschlag and Kraus-Polk, 2017; NWF, 2012; Greenpeace, 2022).

Recognising that "what we decide to eat, as individuals and as a global society [is] one of the most powerful tools we have in the fight against climate change and environmental destruction," Greenpeace launched a global campaign calling for a 50% reduction of meat and dairy and a significant increase of plant-based foods in both production and consumption by 2050 (Greenpeace, 2018). They also call on governments to remove subsidies and other policies that support industrial meat and dairy production and instead support farmers in making the shift. A study CBD conducted with scientists at the Universities of Michigan and Tulane found that cutting 90% of beef consumption and replacing 50% of other animal-based foods with plant-based foods in the US alone would be the rough emissions equivalent of taking nearly half the entire world's cars off the roads (CBD, 2020). They also crafted a policy guide to support these steps that can be undertaken at all levels of government, with key recommendations for decision-makers.

WWF's extensive report "Bending the Curve: The Restorative Power of Planet-Based Diets" acknowledges that "the scientific evidence to support nature's decline is unequivocal – it is being destroyed by us at a rate unprecedented in human history" and that

> far from being alarmist, the dire warnings . . . this report and many others show that there is now overwhelming scientific evidence that our past actions are catching up with us. . . . A shift toward more plant-based foods could reduce global biodiversity loss by nearly half (vegan diet).
>
> *(WWF, 2020b)*

WWF's work on moving the world toward a plant-based shift includes developing a framework convention on food systems to spur international commitments and hold countries legally accountable to them. However, in WWF's Business Case report, for example, released within a year of their Bending the Curve report, they support a US dairy initiative (whose collaborators also include big global meat companies Syngenta and TNC (NZI, 2024)), boasting that "it is clear that many people in the world could benefit from more animal protein, not less. With so many cultures and diets globally, WWF is reluctant to tell people what to eat" (WWF, 2020). This not only runs entirely counter to the aims of their own report(s) but also supports the use of factory farm biogas digesters and other schemes that ultimately help entrench the industrial animal agriculture model (F&WW, 2024).

FoE Europe's "Meat Atlas: facts and figures about the animals we eat," seeks to "support all those who seek climate justice and food sovereignty, and who want to protect nature" by examining a large cross-section of interconnected adverse impacts of the animal agriculture industry in the EU (FoE, 2021). Several Atlas chapters also discuss vegan and vegetarian meat alternatives that "are gaining popularity fast" and that "young people in

Germany – the 'Fridays for Future generation' – eat less meat than their elders. Their attitudes and habits are likely to steer food consumption and policy in the coming decades."

Among their actions, the European FFF movement sent a joint open letter to leaders calling for the transformation of one of the biggest subsidy programs in the world (€58 billion/year), the EU Common Agricultural Policy (CAP), stating that

> over 3600 scientists are calling for action on CAP. . . . Halving production and consumption of animal products could reduce emissions by up to 40% . . . And above all, a regional, fair and livelihood securing agriculture will tackle injustice and deliver a future perspective for farming.
>
> *(FFF, 2020)*

FFF and other youth activists from 33 countries also wrote an open letter to world leaders at COP26, stating that "the youth of today are not only inheriting a world that is burning, flooding and melting, but we are being born into it. . . . [We are] calling for a Plant Based Treaty that phases out animal agriculture and ensures a fair and just transition towards a plant-based food system" while "actively restoring key ecosystems and reforesting the Earth" (Plant Based Treaty, 2021).

It is notable that, while FFF in Germany and other parts of Europe have made reducing animal agriculture and meat consumption one of their core climate justice demands, which has had a profound influence on shifting cultural food norms, FFF US' website makes no such demands (FFF US, 2023), which is "in line with many US environmental organisations, most of which say we need to move away from a meat-heavy food system but mention it sparingly" (Torrella, 2022). It is also in line with findings here, with only ~30% of the climate organisations working on or making significant mentions of the need to reduce animal agriculture or meat consumption – particularly in the past few years – despite the overwhelming scientific consensus that supports it.

Germany and the FFF Movement: A Model for Plant-based Grassroots-driven Change?

The youth-led FFF movement in Europe, which in some respects is comparable to the Sunrise Movement in the US (Nilsen, 2019), has been remarkably effective at placing the climate crisis on the political agenda (Haunss and Sommer, 2021). Its 2019 Climate Action Day mobilised more than six million people worldwide (Taylor et al., 2019). The FFF's German website lists the movement's demands, which includes: the internalisation of costs that the agricultural industry has externalised; the transition of at least 30% of agricultural lands to more sustainable agricultural systems by 2030; a sustainable diet that is accessible to all; improved working conditions and worker pay in the food processing industry; establishing a framework in which "livestock numbers and meat consumption are halved by 2035"; mandating land use policy that allows a maximum of one livestock unit/acre; creating "economic incentives to increase consumption of plant-based foods" and removing economic incentives that currently favour unsustainable products (FFF Germany, n.d.).

In one of their most impressive feats, FFF has been credited with success in shifting consumer behaviour in some European countries, including reducing meat consumption, a trend that runs counter to virtually everywhere else on the planet, where meat consumption is quickly rising (Fritz et al., 2023; Torrella, 2022). In Germany, for example, where the FFF movement is popular and where over 16% of young people surveyed said they take part in it to some degree, FFF has become an "important [driver] for the rise in plant-based eating habits," (FoE, 2021) in a country that has become "one of the few places in the world where meat consumption is decreasing – and fast" (Torrella, 2022). A 2021 survey found that

> it is clear that many (especially young) people no longer want to accept the profit-driven damage caused by the meat industry and are increasingly interested in and committed to climate, sustainability, animal welfare and food sovereignty causes . . . forgoing meat is the trend among adolescents and young adults

~13% of whom do not eat meat, a rate twice that of Germany's population as a whole (FoE, 2021).

> Around one-third of the vegetarians and vegans in the survey had switched to meat-free diets only in the previous year. . . Of those young people who do eat meat, 44 percent say they want to reduce their intake in the future

including 25% who eat meat but have already reduced their meat consumption (FoE, 2021). The survey also found that

> people who eat little meat tend to be more concerned about the environment, and especially about nutrition and animal welfare. . . . Among vegans, 75% say they are part of the climate movement; almost 50% of vegetarians say the same, while only 15% of omnivores do so.

Notably, 70–80% of young Germans also "see the state as having a shared responsibility for sustainable diets." This has led to a virtuous spiral where increased youth-driven demand for plant-based foods has been met with an increase in plant-based options in supermarkets and restaurants across the country, which, in turn, has made it easier for even more people to access and choose plant-based options.

For young adults "it's more like a political statement to eat less meat or to eat no meat at all. The 'eat less meat' sentiment already appears to be taken seriously in some corners of Germany's federal government" (Torrella, 2022). Two German ministers have stepped up to call for a reduction in meat consumption and the country's minister of food and agriculture "recently listed shifting diets to be more plant-based as the first of four priorities in the agency's forthcoming nutrition strategy plan." FFF's grassroots success of beginning to significantly influence public opinion, shift cultural norms and build political power around a plant-based shift in Germany offers hope that transformative change towards a more just, sustainable food system is possible.

Conclusion: Climate Activism and Plant-based Politics

The science is clear: animal agriculture, no matter how it is done, is a major contributor to the climate crisis and the single largest driver of ecological breakdown, biodiversity loss and mass species extinction worldwide. These compounding crises are disproportionately impacting the world's most vulnerable populations and are leading to growing humanitarian catastrophes of unprecedented magnitude.

Of the climate organisations examined, those who promote technology- and market-based solutions and center partnerships with large agribusiness are helping big polluting corporations greenwash regenerative animal agriculture and other animal agriculture schemes that only entrench and serve to accelerate the current disastrous industrial animal agriculture model. Only 30% of climate activist organisations examined actively promote a rapid and just transition to a more plant-based food system, despite overwhelming scientific consensus that such a transition is imperative, particularly in wealthier nations with meat-heavy diets such as the US and Europe. This is a real problem.

The organisations fighting the industrial model and promoting a just transition tend to be more grassroots oriented and centre frontline communities. The fierce backlash by big agribusiness, especially within the US culture war narrative, may partly explain why some grassroots organisations, for example FFF in Europe, promote a plant-based transition, while comparable organisations in the US, for example FFF US and Sunrise Movement, have not. Overall, however, it is the most grassroots groups, such as FFF Germany, that are providing the most powerful movement models for shifting the narrative and social norms.

Ultimately, transforming our food system to one that sustains life, like addressing the climate crisis itself, will take more than switching forms of animal agriculture or passing policies that rely on and/or entrench business as usual. Our current economic paradigm, with its stark inequalities, disregards environmental (and social) externalities, which are already surpassing planetary boundaries. Politically, there is a need to challenge concentrated power and corporate influence that erode democratic values and prioritise profit over the ecological systems that support life. Culturally, we must re-examine and redefine our societal norms to promote values of justice, compassion and interconnectedness. Our daily habits, including dietary choices, reflect broader societal attitudes and must align with values and principles that honour basic human, animal and nature's rights.

As articulated by scholar and civil rights activist Angela Davis,

> The fact that we can sit down and eat a piece of chicken without thinking about the horrendous conditions under which chickens are industrially bred in this country is a sign of the dangers of capitalism, how capitalism has colonised our minds . . . part of a revolutionary perspective [is how] we not only discover more compassionate relations with human beings but how can we develop compassionate relations with the other creatures with whom we share this planet.
>
> *(27th Empowering Women of Color Conference, 2012)*

Installed at the Washington Gladden Social Justice Park in Columbus (Ohio), this 10' (3 m) tall public sculpture, with its

> quantum wave-like structure serves as a dual metaphor: how a small group of diverse -yet unified- people can build movements for social justice, creating ripples that emanate and can grow into waves of transformational change; it is also a metaphor that despite differences in race, religion, gender, sex, ability, age, and species, we are all part of a single living ecosystem and inter-related structure of reality. When we cause harm and suffering to others, we are simultaneously inflicting suffering upon ourselves. If we can transcend the ultimately artificial notion of being separate from each other, and instead act as part of the same oneness, a single garment of destiny, we would be capable of healing ourselves and the world from the existential crises we are facing.

Figure 29.1   Our Single Garment of Destiny, 2021.
Source: Julian Voss-Andreae (www.JulianVossAndreae.com).

References

27th Empowering Women of Color Conference (2012). *On Revolution: A Conversation Between Grace Lee Boggs and Angela Davis.* https://www.radioproject.org/2012/02/grace-lee-boggs-berkeley/

350.org (n.d.). *We Fight for a World Beyond Fossil Fuels.* https://350.org/?r=US&c=NA [Accessed 1 March 2024].

BBC (2022). *Knightsbridge Steakhouse: Animal Rebellion Activists Removed from Site.* https://www.bbc.com/news/uk-england-london-63850883

Carrington, D. (2018). *Avoiding Meat and Dairy Is 'Single Biggest Way' to Reduce Your Impact on Earth.* https://www.theguardian.com/environment/2018/may/31/avoiding-meat-and-dairy-is-single-biggest-way-to-reduce-your-impact-on-earth

Carrington, D. (2019). *New Plant-Focused Diet Would 'Transform' Planet's Future, Say Scientists.* https://www.theguardian.com/environment/2019/jan/16/new-plant-focused-diet-would-transform-planets-future-say-scientists [Accessed 2023].

CBD (2020). *Appetite for Change: A Policy Guide to Reducing Greenhouse Gas Emissions of U.S. Diets By 2030.* https://www.biologicaldiversity.org/programs/population_and_sustainability/climate/Climate-Diet-report.html

CJA (2024). *Food Sovereignty: Climate Justice Alliance.* https://climatejusticealliance.org/workgroup/food-sovereignty/ [Accessed January 2024].

Clark, M. A. (2020). Global food system emissions could preclude achieving the 1.5° and 2°C climate change targets. *Science*, 370(6517), 705–708.

CRP (2019). *Eat Your Veggies: Ditch the Meat.* https://www.climaterealityproject.org/blog/dishing-plate-impact

Dutkiewicz, J. and Rosenberg, G. N. (2021). *The Myth of Regenerative Ranching.* https://newrepublic.com/article/163735/myth-regenerative-ranching [Accessed 2023].

Eisen, M. B. and Brown, P. O. (2022). Rapid global phaseout of animal agriculture has the potential to stabilize greenhouse gas levels for 30 years and offset 68 percent of CO2 emissions this century. *PLoS Climate*, 1(2), e0000010.

FFF (2020). *Change The Cap! – Open Letter to the EU.* https://fridaysforfuture.org/change-the-cap/

FFF Germany (n.d.). *Forderungen Im Bereich Landwirtschaft.* https://fridaysforfuture.de/forderungen/landwirtschaft/ [Accessed 1 December 2023].

FFF US (2023). *National Demands.* https://fridaysforfutureusa.org/demands/

Filazzola, A., Brown, C., Dettlaff, M. A. and Batbaatar, A. (2020). The effects of livestock grazing on biodiversity are multi-trophic: A meta-analysis. *Ecology Letters*, 23(8), 1177–1311.

FoE (2017). *Meat of the Matter: A Municipal Guide to Climate-Friendly Food Purchasing.* https://foe.org/wp-content/uploads/2017/12/MunicipalReport_ko_120117_v2-1.pdf

FoE (2021). *Meat Atlas: Facts and Figures about the Animals We Eat.* https://eu.boell.org/sites/default/files/2021-09/MeatAtlas2021_final_web.pdf

Foer, J. S. (2020). *The End of Meat Is Here.* https://www.nytimes.com/2020/05/21/opinion/coronavirus-meat-vegetarianism.html

Fritz, L., Hansmann, R., Dalimier, B. and Binder, C. R. (2023). Perceived impacts of the Fridays for Future climate movement on environmental concern and behaviour in Switzerland. *Sustainability Science*, 18, 2219–2244.

F&WW (2024). *The Big Oil and Big Ag Ponzi Scheme: Factory Farm Gas.* https://www.foodandwaterwatch.org/wp-content/uploads/2024/01/RPT2_2401_GreenwashingBiogas-WEB3.pdf [Accessed January 2024].

Garcia, D., Galaz, V. and Daume, S. (2019). EATLancet vs yes2meat: The digital backlash to the planetary health diet. *Lancet*, 349(10215), 2153–2154.

Garnett, T., et al. (2017). *Grazed and Confused? Ruminating on Cattle, Grazing Systems, Methane, Nitrous Oxide, the Soil Carbon Sequestration Question – and What It All Means for Greenhouse Gas Emissions.* s.l.: FCRN, University of Oxford.

Gerber, P., et al. (2018). *The Global Livestock Environmental Assessment Model, Version 2.0, Revision 5.* s.l.: Food and Agriculture Organization of the United Nations (FAO).

Goodier, M. and Sunnemark, V. (2023). *UK Meat Consumption at Lowest Level Since Records Began, Data Reveals.* https://www.theguardian.com/environment/2023/oct/24/uk-meat-consumption-lowest-level-since-record-began-data-reveal

Goodland, R. and Anhang, J. (2009). *Livestock and Climate Change: What If the Key Actors in Climate Change Are Cows, Pigs, and Chickens?* s.l.: World Watch.

GRAIN & IATP (2018). *Emissions Impossible: How Big Meat and Dairy Are Heating Up the Planet.* https://www.iatp.org/emissions-impossible

Greenpeace (2018). *Greenpeace Calls for Decrease in Meat and Dairy Production and Consumption for a Healthier Planet.* https://www.greenpeace.org/international/press-release/15111/greenpeace-calls-for-decrease-in-meat-and-dairy-production-and-consumption-for-a-healthier-planet/

Greenpeace (2022). *The Spanish Factory Farming Industry Is Trying to Silence Us.* https://www.greenpeace.org/international/story/54303/the-spanish-factory-farming-industry-is-trying-to-silence-us/

Hamerschlag, K. and Kraus-Polk, J. (2017). *Shrinking the Carbon and Water Footprint of School Food: A Recipe for Combating Climate Change.* https://foe.org/wp-content/uploads/2017/11/FOE_FoodPrintReport_7F.pdf

Harwatt, H., Hayek, M. N., Behrens, P. and Ripple, W. J. (2024). *Options for a Paris-Compliant Livestock Sector: Timeframes, Targets and Trajectories for Livestock Sector Emissions from a Survey of Climate Scientists.* https://animal.law.harvard.edu/wp-content/uploads/Paris-compliant-livestock-report.pdf

Haunss, S. and Sommer, M. (2021). Fridays for Future – Die Jugend gegen den Klimawandel. Konturen der weltweiten Protestbewegung. *Johns Hopkins University Press*, 44(3), 634–636.

Hayek, M. N. and Garrett, R. D. (2018). Nationwide shift to grass-fed beef requires larger cattle population. *Environmental Research Letters*, 13(8).

IPBES (2019). *UN Report: Nature's Dangerous Decline 'Unprecedented'; Species Extinction Rates 'Accelerating'.* https://www.un.org/sustainabledevelopment/blog/2019/05/nature-decline-unprecedented-report/ [Accessed 2023].

Ivanovich, C. C., Sun, T., Gordon, D. R. and Ocko, I. (2023). Future warming from global food consumption. *Nature Climate Change*, 13, 297–302.

McSweeney, E. (2021). *Animal and Climate Activists Blockade McDonald's Distribution Centers Across England.* https://www.cnn.com/2021/05/22/uk/animal-rebellion-mcdonalds-gbr-intl-scli/index.html [Accessed 2023].

Monbiot, G. (2022). *Regenesis: Feeding the World Without Devouring the Planet* (p. 79). s.l.: Penguin Books.

Neyeleni (2007). *Nyeleni 2007 – Forum for Food Sovereignty: Synthesis Report.* https://nyeleni.org/IMG/pdf/31Mar2007NyeleniSynthesisReport-en.pdf [Accessed January 2024].

Nilsen, E. (2019). *The New Face of Climate Activism Is Young, Angry – and Effective.* https://www.vox.com/the-highlight/2019/9/10/20847401/sunrise-movement-climate-change-activist-millennials-global-warming

NRDC (2019). *Plant-Based Diet: Healthier for Us and Our Planet.* https://www.nrdc.org/bio/maria-mccain/plant-based-diet-healthier-us-and-our-planet#:~:text=One%20critical%20step%20is%20adopting,70%25%20from%20now%20until%202050

NWF (2012). *Climate-Savvy Menus: Eating Lower on the Food Chain.* https://blog.nwf.org/2009/06/climate-savvy-menus-eating-lower-on-the-food-chain/

NZI (2024). *Full List of Partners.* https://www.usdairy.com/getmedia/96832ca1-947b-49b4-90e9-a3aa6bbe803f/U-S-Dairy-Net-Zero-Initiative-Partners.pdf

Plant Based Treaty (2021). *Youth Open Letter: We Need a Plant Based Treaty Now!.* https://plantbasedtreaty.org/youth-letter/

Poore, J. and Nemecek, T. (2018). Reducing food's environmental impacts through producers and consumers. *Science*, 360(6392), 987–992.

RAN (2014). *An Ode to Real Food.* https://www.ran.org/the-understory/an_ode_to_real_food/.

Ritchie, H. (2019). *Half of the World's Habitable Land Is Used for Agriculture.* https://ourworldindata.org/global-land-for-agriculture [Accessed September 2023].

Ritchie, H. and Roser, M. (2019). *Our World in Data: Meat and Dairy Production.* https://ourworldindata.org/meat-production#meat-production-by-animal [Accessed May 2023].

Searchinger, T., Waite, R., Hanson, C. and Ranganathan, J. (2019). *Creating a Sustainable Food Future: A Menu of Solutions to Feed Nearly 10 Billion People by 2050.* Washington, DC: World Resources Institute.

Shukla, J., Skea, E. and Calvo, V. B. (2019). *IPCC, 2019: Climate Change and Land: An IPCC Special Report on Climate Change, Desertification, Land Degradation, Sustainable Land Management,*

Food Security, and Greenhouse Gas Fluxes in Terrestrial Ecosystems. s.l.: The Intergovernmental Panel on Climate Change, United Nations.

Sunrise Movement (2024). *We Demand a Green New Deal for All to Stop the Climate Crisis.* https://www.sunrisemovement.org/our-demands/#stop [Accessed January 2024].

Taylor, M. (2019). *This Article Is More Than 4 Years Old Animal Rebellion Activists to Blockade UK's Biggest Meat Market.* https://www.theguardian.com/environment/2019/aug/16/animal-rebellion-extinction-activists-to-blockade-smithfield-meat-market-environment

Taylor, M., Watts, J. and Bartlett, J. (2019). *Climate Crisis: 6 Million People Join Latest Wave of Global Protests.* https://www.theguardian.com/environment/2019/sep/27/climate-crisis-6-million-people-j oin-latest-wave-of-worldwide-protests

TNC (2022). *Regenerative Grazing Lands.* https://www.nature.org/en-us/what-we-do/our-priorities/provide-food-and-water-sustainably/food-and-water-stories/sustainable-grazing-lands/

TNC (2023). *Regenerative Food Systems.* https://www.nature.org/en-us/what-we-do/our-priorities/provide-food-and-water-sustainably/food-and-water-stories/regenerative-food-systems/

Torrella, K. (2022). *How Germany Is Kicking Its Meat Habit.* https://www.vox.com/future-perfect/23273338/germany-less-meat-plant-based-vegan-vegetarian-flexitarian.

Twine, R. (2021). Emissions from animal agriculture – 16.5% is the new minimum figure. *Sustainability*, 13(11), 6276.

Vallone, S. and Lambin, E. F. (2023). Public policies and vested interests preserve the animal farming status quo at the expense of animal product analogs. *One Earth*, 6, 1213–1226.

Willett, W., et al. (2019). Food in the Anthropocene: The EAT–lancet commission on healthy diets from sustainable food systems. *The Lancet*, 393(10170), 447–492.

WWF (2020a). *An Environmental and Economic Path Toward Net Zero Dairy Farm Emissions.* https://files.worldwildlife.org/wwfcmsprod/files/Publication/file/570hr7ghl5_Net_Zero_Dairy_Business_Case_v10.pdf

WWF (2020b). *Bending the Curve: The Restorative Power of Planet-Based Diets.* https://files.worldwildlife.org/wwfcmsprod/files/Publication/file/7b5iok5vqz_Bending_the_Curve__The_Restorative_Power_of_Planet_Based_Diets_FULL_REPORT_FINAL.pdf.pdf

WWF (2020c). *Is Biogas a "Green" Energy Source?* https://www.worldwildlife.org/blogs/sustainability-works/posts/is-biogas-a-green-energy-source [Accessed 2023].

Xu, X., et al. (2021). Global Greenhouse Gas Emissions from Animal-Based Foods Are Twice Those of Plant-Based Foods. *Nature Food*, 2, 724–732.

Zeldin-O'Neill, S. (2022). *Animal Rebellion Activists Stop Milk Supply in Parts of England.* https://www.theguardian.com/environment/2022/sep/04/animal-rebellion-activists-stop-milk-supply-parts-england

Zhang, M., et al. (2023). Experimental impacts of grazing on grassland biodiversity and function are explained by aridity. *Nature Communications*, 14(5040).

PART VI

Fighting Against False Solutions and Prefiguring Alternatives

Introduction to Part VI: Fighting Against False Solutions and Prefiguring Alternatives

Climate activists may not always agree on the best path forward, as we saw in Part V, but a growing number of activists are organising, even across differences and long distances, to address climate injustices and fight against what they view as false solutions. People are mobilising at the grassroots level to provide humane and equitable answers to critical questions such as – What should be done? Who should be doing it?

Talk of climate solutions is typically dominated by discussions of technology: renewable energy (Bush, 2020; De Lucia, 2014; Morris and Jungjohann, 2016 as well as chapters 25 and 26 in this handbook), retrofitting buildings (Felius et al., 2023), even questionable experiments in geo-engineering (Sovacool et al., 2023), as well as investing in public transportation (Abdallah, 2017; 2023), lifestyle changes involving the adoption of plant-based diets (see Chapter 29 in this volume), or reducing air travel (Ahmad, 2017). The chapters in this part explore the various ways that grassroots climate activists try to move things forward, often against all odds. How do local activists connect across long distances to make their voices heard in halls of power that are often a continent away? What does pushing for equitable solutions look like under difficult conditions, in different parts of the world? How do climate activists do their work in the face of extreme hardship in sub-Saharan Africa? What are the motivations that grassroots climate activists share as they endeavour to actually live a climate-friendly lifestyle? How can documentary and feature films portray solutions to the climate crisis in a way that encourages climate activism? And what about climate activists who try to engage in "insider activism" within their institutions – in the private or in the public sector? As they describe the challenges grassroots climate activists face in these various circumstances, chapters in this section prefigure "a world of solutions."

In Chapter 30, "No Greenwashing of Fossil Gas – Creating a Grassroots Transatlantic Climate Bridge Against False Solutions," Andy Gheorghiu and Sabine von Mering describe fossil gas as a major blind spot. Despite over a decade of scientific research suggesting the contrary, gas continues to be touted as a clean alternative to oil and coal. Growing commitments to addressing the climate crisis and increasingly vocal opposition

DOI: 10.4324/9781003396567-35

notwithstanding, fossil gas infrastructure is seeing record build-out around the globe, including on both sides of the Atlantic. The authors draw attention to the development of fossil gas (LNG) infrastructure projects and describe grassroots climate activists' efforts to join forces across the Atlantic to prevent the build-out in Germany, Ireland, and the United States.

In Chapter 31, "Climate Activism and the Political-Economic Landscape in Nigeria," Joy Egbe describes how growing up with climate and poverty-induced hardship in rural Nigeria motivated her to co-found a youth climate movement at her school and eventually to become one of several young "climatepreneurs" as the head of a company that provides renewable energy to local communities. Focusing on young green innovators like herself who respond to environmental challenges by founding companies to address the urgent challenges of energy poverty and climate change, the chapter underscores the importance of green skills empowerment and argues that prioritising climate action will shape a workforce with the skills to develop solutions for a just transition. Egbe concludes by identifying the barriers and challenges activists in Nigeria face compared to their counterparts in developed nations, which include the skills gap and the lack of access to infrastructure for technological enterprises.

In Chapter 32, "Eco-socialist militancy and Neo-rural experimentations in present Environmental Activism (Italy)," Elena Apostoli Cappello provides a historical perspective over several ethnographic studies of the radical leftist "antagonista" communities in contemporary Italy, focusing on the utopian roots of their environmentalism. She argues that this utopian way of thinking, together with "primitivist" projections and aspirations, seems to provide a central axis for considering social change and rethinking political outreach. She also presents utopian imaginaries as a form of narrative that produce shared perceptions and understandings of time among activist communities. With a socio-anthropological approach, the chapter explores the visions of societal transformation that are at play in activists' aspirations.

Chapter 33, "Promise Motivation: Films that Encourage Grassroots Climate Activism," takes a look at films that aim to encourage viewers to get actively involved in solving the climate crisis. After introducing a range of films from around the world that foreground local climate action, Sabine von Mering centres her analysis on the French documentary film *Demain* (2015). She also briefly discusses the Australian documentary *2040* (2020) as well as the American documentary film *Bidder 70* (2012) and the American feature film *How to Blow Up a Pipeline* (2022), inspired by Swedish scholar Andreas Malm's book of the same title. Von Mering shows that, while the most well-known climate-change films focus on "risk motivation," portraying the gloom and doom of climate disruption, the films discussed in this chapter prioritise what she calls "promise motivation," a form of visualising and storytelling that focuses on solutions to encourage viewers to become engaged.

In Chapter 34, "Insider Climate Activism: Slowing Down or Speeding Up to Decarbonize?," Annika Skoglund explores climate activism in relation to speed, detailing two emerging polarised forms of "insider activism," conceptualised as movements pursued in professional contexts. Both movements respond to the expressed "climate emergency," the first by propagating a fast movement and the second in favour of a slow movement. By first presenting a spectrum of velocities that have emerged in response to the climate emergency, this chapter replays these responses in two fictional polarisations of insider climate activism. The first illustration portrays employee activism within a CO_2-polluting corporation, to investigate how fast movements can develop within business. The second illustration, in contrast, portrays insider activism performed by public servants who work for a regional office, to explore how slow movements can prosper within state institutions. This method produces a polarisation, facilitating a reflective comparison of how activism pursued at work can take different transformative paces, sometimes successfully aligning with other organisational velocities and professional practices. While very diverse activist movements, slow as well as fast, could prosper within both the private and public sectors, these two polarised fictional examples provide insights into the largely unexplored desynchronised velocities that exist in climate activism.

Climate activists are typically portrayed as fighting *against* something – against polluting industries, against government inaction, against entrenched financial interests. But all climate activists are also simultaneously fighting *for* something; for a just transition, for a liveable climate, for a better future. The chapters in this section engage with how activists in different bioregional and political contexts probe the limits of grassroots climate activism by confronting governments, polluting corporations, and their financial backers, while helping to support sustainability and building resilience within and across communities. Increasingly, activists are doing so through efforts to connect with others in different countries for information sharing and support, often across long distances. The future, these analyses demonstrate, lies in a web of planetary civics and solidarity.

References

Abdallah, T. (2017/2023). *Sustainable Mass Transit: Challenges and Opportunities in Urban Public Transportation*. Amsterdam: Elsevier.

Ahmad, F. M. (2017). Flying high. International air travel and climate change. *International Law News*, 46, 1–6.

Bush, M.-J. (2020). *Climate Change and Renewable Energy: How to End the Climate Crisis*. London: Palgrave MacMillan.

De Lucia, V. (2014). The climate justice movement and the hegemonic discourse of technology. In M. Dietz and H. Garrelts (eds.) *Routledge Handbook of the Climate Change Movement*. London and New York: Routledge.

Felius, L. C., Brandser, A. and Smits, F. (2023). Practical solutions for building envelope retrofitting of historic buildings in cold climates. *Journal of Physics: Conference Series*, 2654(1), (December). http://doi.org/10.1088/1742-6596/2654/1/012005

Morris, C. and Jungjohann, A. (2016). *Energy Democracy. Germany's Energiewende to Renewables*. London: Palgrave Macmillan.

Sovacool, B. K., Baum, C. M. and Low, S. (2023). Risk–risk governance in a low-carbon future: Exploring institutional, technological, and behavioral tradeoffs in climate geoengineering pathways. *Risk Analysis*, 43(4), 838–859.

30

NO GREENWASHING OF FOSSIL GAS

Creating a Grassroots Transatlantic Climate Bridge Against False Solutions

Andy Gheorghiu and Sabine von Mering[1]

Introduction: Fossil Gas and the Climate

One of the areas of rapidly growing and internationally coordinated grassroots climate activism is the fight against fossil gas, especially on both sides of the Atlantic. Fossil fuels and money have crossed the Atlantic for decades. ConocoPhillips, Chevron, Exxon, Shell, Total, etc.: no matter where their headquarters are located, their products and services are available. Canadian energy giant Enbridge even calls its East Coast gas pipeline project Atlantic Bridge, suggesting an expanding fossil gas connection along – and (eventually) across – the Atlantic. As Naomi Oreskes (2010), co-author of *Merchants of Doubt* and others (Neujeffski and Goldenbaum, 2021) have shown, after climate scientists had begun to raise public awareness about global warming in the latter half of the twentieth century,[2] the fossil-fuel industry and their partners responded by spreading climate denial.[3] These strategies by the industry, which also meant scientists and activists had to spend enormous efforts defending the need to take action on climate, were sold across the Atlantic as well.[4] It is thus high time for more transatlantic climate activism.

Fossil gas consists primarily of methane. When fossil methane is released into the atmosphere, it is much more harmful to the climate than CO_2: up to 108 times more in the first 20 years, falling to up to 41 times more over 100 years (Moseman, 2024).[5] According to Robert W. Howarth's calculations (Howarth, 2019), the climate impact of fracked shale gas has been underestimated for a long time and may have contributed more than half of all of the increased emissions from fossil fuels and approximately one-third of the total increased emissions from all sources globally over the past decade. Yet, despite over a decade of scientific research suggesting the contrary,[6] fossil gas continues to be touted as a clean alternative, promoted by PR execs as "natural" gas (or *Erdgas* [earth gas] in German) to give it a clean and healthy image and distinguish it from "dirty" oil and coal. The fact that the European Union approved gas as "clean energy" in its 2022 taxonomy must be viewed as a major success by fossil gas lobbyists who were apparently able to convince policymakers to disregard the growing number of studies that have long confirmed that, far from being a bridge, gas/methane is a climate killer. On the other hand, due to the decade-long fight by the European anti-fracking movement, which then became an anti-gas movement, the European Union eventually had to acknowledge the significant global warming problem

DOI: 10.4324/9781003396567-36

related to methane emissions and in 2023 agreed on a regulation which aims at cutting these emissions in the energy sector (Council of the European Union, 2023). From 2027 onwards, this methane regulation will also apply to imports – a date climate activists warn is too late into the future, but admittedly it is better late than never.

Climate activists see the need for a proper analysis of the true climate impact of fossil gas confirmed by new studies, which underline the importance of the decade-long work of Prof. Robert Howarth (Cornell University, Ithaca, NY) and others that methane is being seriously undercounted in global warming calculations (Rathi, 2022). Yet, even governments such as the Scholz administration in Germany and the Biden administration in the United States, which prided themselves on their commitment to tackling the climate crisis and to addressing methane emissions with the Global Methane Pledge,[7] supported investments in the expansion of fossil gas infrastructure, particularly around Liquefied Natural Gas (LNG) export and import.[8] Climate scientists and climate activists had been increasingly vocal and successful at mounting resistance to these developments until Russia attacked Ukraine in February 2022. Until then, the most attention was paid to pipeline gas in Europe, most importantly the controversy surrounding the Russian North Stream 2 pipeline. However, the geopolitical dimension related to this controversial project and its connection to a growing LNG market with a strong transatlantic perspective was the elephant in the room long before Russia's attack on Ukraine (Gheorghiu, 2020). Yet, the exponential growth of the LNG industry and the growing protests against it received only muted attention, already before and especially after February 2022.

The tight relationship between profits and denial in the industry places those working to stop the climate crisis at a disadvantage. From the Kyoto Protocol through to the United Nations' Conference of the Parties (COP) process that includes the Paris Agreement, scientists and advocates determined to enact climate protection have had to fight an uphill battle against entrenched fossil-fuel interests. And so, as climate scientists improve their international collaborations and provide hard evidence of global warming and policymakers push forward at the UN level to implement and improve the Paris Agreement, grassroots climate activists have increasingly begun to form international and global alliances in order to counter fossil-fuelled resistance to the profound transformations that are urgently necessary.

In this chapter, we introduce how activists have collaborated to oppose fossil gas infrastructure projects in Germany, Ireland, and the United States in order to demonstrate the effectiveness of what we are calling a *Grassroots Transatlantic Climate Bridge* for climate protection. Motivated by the experiences of the anti-fracking/gas movement and aware of the fact that only a trans-border collaboration has the significant resistance power to oppose these developments, European activists aim to show solidarity with impacted communities in the US who are often forced to live in so-called sacrifice zones, and all hope to keep their countries on the 1.5-degrees Celsius path.

The State of Research

A number of studies have been published in recent years about grassroots climate activism against fossil gas in Europe. The report *Global Resistance to Fracking* was first presented at the COP in Paris in 2015 (Martin-Sosa Rodriguez, 2015).[9] In *The Anti-Fracking Movement in Ireland: Perspectives from the Media and* Activists, Tamara Steger and Ariel Drehobl (2017) look at articles published about fracking in Irish national and local newspapers by journalists, activists, and community members with a wide variety of backgrounds. Highlighting the ongoing struggle over activist credibility, economic interests, and the need for public trust, they find that anti-fracking activists are mentioned frequently, especially in local Irish newspapers and are often depicted as either violent or peaceful, hippies or serious citizens. Ella Muncie analysed protests against a proposal for onshore drilling for gas that would involve fracking in North England and the East Midlands in the summer of 2018. Through interviews conducted at three protests, she concludes in " 'Peaceful protesters' and 'dangerous criminals': the framing and reframing of anti-fracking activists in the UK" (Muncie, 2020) that activists felt that the reasons for arrests seemed to differ on a daily basis, meaning that "protesters did not know whether they would be seen in a criminal light from one day to the next" (Muncie, 2020). Similarly, in "The Anti-Fracking Movement and the Politics of Rural Marginalisation in Lithuania: Intersectionality in Environmental Justice" Diana Mincyte and Aiste Bartkiene study rural anti-fracking activists protesting against Chevron's plan to extract shale gas in Lithuania (Mincyte and Bartkiene, 2019). Mincyte and Bartkiene researched local media and the activists' representation in the broader discussion of fracking in the country and concluded that negative representation of the activists pulls from stereotypes of rural communities that are viewed as backward and unintellectual. The rural community framed their protests against Chevron as protecting rural livelihoods and landscapes in Lithuania, which was then misconstrued by the media as being anti-progress and anti-economic advancement. In "Fighting Science with Science: Counter-Expertise Production in Anti-Shale Gas Mobilizations in France and Poland," Roberto Cantoni looks at different movements within Poland and France and how they communicate scientific information between communities and NGOs to create smarter and more powerful activist campaigns (Cantoni, 2022). The author concludes that good communication is an important aspect of expanding the anti-fracking movement and that locals in both France and Poland turned to environmental NGOs and members of their respective country's green parties for help in supporting the anti-fracking cause. Others, like Evensen, Whitmarsh, and Devie-Wright, have shown that the growing understanding of climate change has impacted communities' resistance to fossil gas (Evensen et al., 2023). Hopke's work found that anti-fracking activists make sophisticated use of Twitter (Hopke, 2015). This chapter contributes to this body of work about grassroots activism against fossil gas by providing an overview of both the expansion of and the protests against transatlantic LNG. Such protests are not restricted to the transatlantic region, of course. They occur anywhere fracking appears – from Poland (Stasik, 2018) to Australia (Howey and Neale, 2023). After introducing the problem of LNG, we present examples of successful grassroots activism in Germany, the US, and Ireland, and highlight how the activists have benefitted from connecting across borders.

What Is LNG and Why Is It a Problem?

Liquefied natural gas (LNG) is fossil gas that has been cooled to a liquid state. The liquefaction process was developed in the nineteenth century and involves cooling the gas to about –260 degrees Fahrenheit. The volume of fossil gas in liquid form is about 600 times smaller than in gas form, making it fit for shipping overseas. At least 10% of the energy value of fossil gas is being lost during the production process of LNG. In 2023, the USA took the pole-position as the No. 1 global LNG exporter – made possible by large-scale shale gas extraction (Pospíšil et al., 2019; Nechyporenko, 2024, see also Figure 30.1).

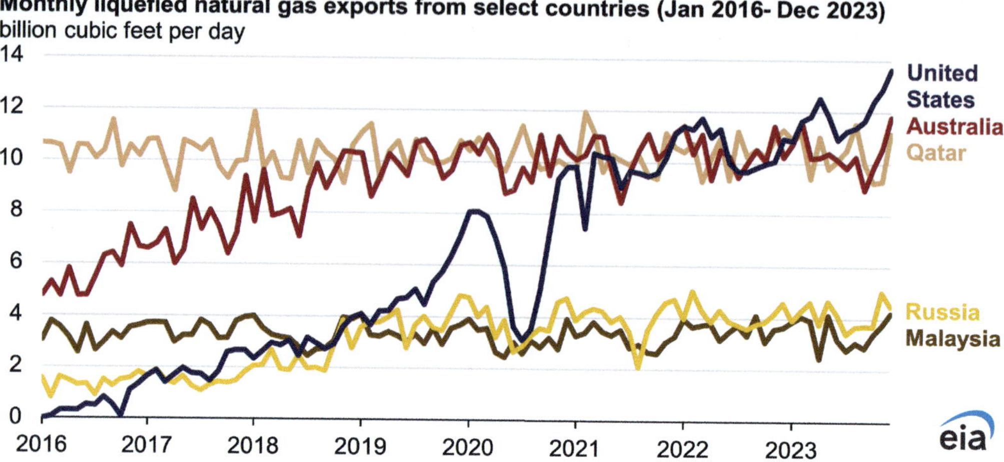

Figure 30.1   Monthly liquefied natural gas exports from select countries (Jan. 2016–Dec. 2023) in billion cubic feet per day.

Source: https://www.eia.gov/todayinenergy/detail.php?id=61683.

The so-called US shale gas boom was triggered by the controversial extraction technique hydraulic fracturing or fracking.[10] According to the *Compendium of Scientific, Medical, and Media Findings Demonstrating Risks and Harms of Fracking and Associated Gas and Oil Infrastructure* (the Compendium) – a fully referenced compilation of evidence, spanning many scientific and social scientific fields, documenting continuing and increasing impacts on health and environment: "no evidence [has been found] that fracking can be practised in a manner that does not threaten human health directly or without imperilling climate stability upon which human health depends."[11]

The shale gas or fracking boom in the US was enabled in 2005 by the "Halliburton Loophole," which exempts fracking from reporting requirements of federal regulation under the Safe Drinking Water Act (Marusic, 2023). Seven years later in 2012, Rex Tillerson (back then CEO of ExxonMobil) was famously quoted as saying "We're losing our shirts today," referring to the depression (and non-profitability) of gas prices due to the produced shale gas glut in a closed US market (DiColo, 2012). Back then, ExxonMobil and others were already talking about using exports of US LNG to escape the closed market and fallen prices. Existing restrictions for exports of US crude oil and LNG were lifted a few years later in 2015 under the Obama administration (Tsafos, 2018). The transatlantic LNG trade then received a huge boost under the Trump administration, which threatened to impose tariffs on European steel and automobile products. To avoid this, Jean-Claude Juncker, the EU commission president at the time, flew to Washington in July of 2018 to strike the so-called Trump-Juncker-Deal (Ibid). According to an EU Commission factsheet, US LNG exports grew by a staggering 2.418% between July 2018 and shortly before Russia's invasion of Ukraine in February 2022 (European Commission, 2022a). The US competed primarily with Qatar and Russia for the European LNG market (European Commission, 2022b). The US Energy Information Administration (US EIA) website shows the dramatic overall increase in shale gas production and, correspondingly, LNG exports in recent years, as well as the projected further increase in the coming decades:

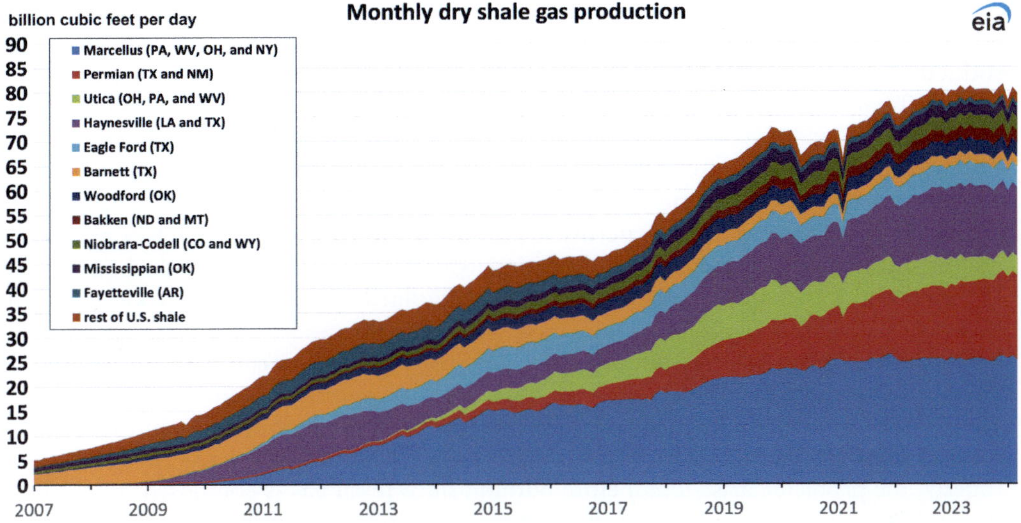

Figure 30.2   Monthly dry shale gas production in the United States.

Source: https://www.eia.gov/naturalgas/weekly/#tabs-storage-3.

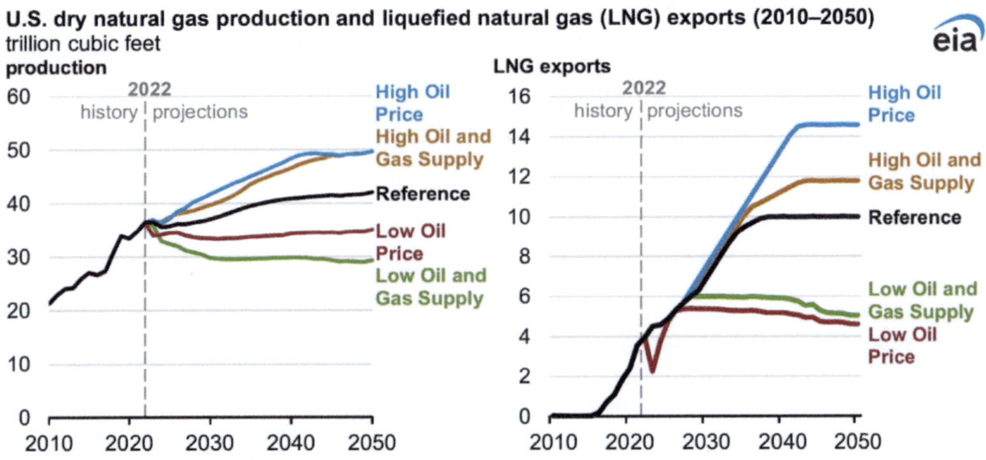

Figure 30.3   US dry natural gas production and liquified natural gas (LNG) exports (2010–2050).

Source: https://www.eia.gov/naturalgas/reports.php#/T202.

In 2023, the US became the -largest global LNG exporter leaving Australia, Qatar, and Russia behind (IGU, 2024). Of total US LNG exports in 2023, 67% went to Europe.[12] According to the US EIA, 80% of all fossil gas extracted in the USA was shale gas, which is fracked gas.[13] We can rightfully assume that nearly 100% of the exported LNG is fracked gas. Not accounting for the more than 100 peak-shaving facilities across the US (Mannan, 2012), the US currently has 9 LNG terminals that are operational, 6 that are under construction, and a total of 26 that are planned (Sierra Club, 2024).

A majority of the US exports to Europe comes from the Gulf Coast (mainly Texas and Louisiana). Due to their proximity, EIA also sees a direct link between shale gas production growth in the Haynesville and Permian basins and LNG exports from the states of Texas and Louisiana.[14] The Permian basin is considered to be a so-called climate bomb.[15]

Some still argue that fossil gas is beneficial for meeting climate goals as public subsidies/tax exemptions benefit companies on both sides of the Atlantic – despite the fact that construction of infrastructure is not in the public interest (Black et al., 2023) and the cost of renewables keeps dropping. Companies have also raised prices for US residential customers, exposing them to the volatility of the gas market (Cowan, 2024).

The other important assumption in the "bridge fuel" argument is that fossil gas is cleaner than coal and oil. While the more efficient gas power plants emit 50–60% less CO_2 than coal (Environmental Impacts of Natural Gas | Union of Concerned Scientists) and gas produces primarily carbon dioxide and water vapour ("Natural Gas Facts") when burned properly, the problems arise long before burning since fossil gas is composed primarily of the potent greenhouse gas methane (Fact Sheet: Natural Gas Greenhouse Gas Emissions). Therefore, the leakages and intentional venting and flaring of gas that take place before the gas is burned have to be taken into account. The emission of unburned fossil gas is particularly concerning because, as mentioned earlier, fossil methane is up to 108 times more potent than CO_2 over a 20-year period and up to 41 times over 100 years. The issue of unburned gas emissions has been under increased scrutiny, with new satellite data highlighting the high methane intensity of US fossil fuel production and consumption.[16] LNG can be particularly problematic because

> the greenhouse gas (GHG) emissions from the extraction, transport, liquefaction, and re-gasification of LNG can be almost equal to the emissions produced from the actual burning of the gas, effectively doubling the climate impact of each unit of energy created from gas transported overseas.
>
> *(NRDC)*

A 2014 report by the US Department of Energy already stated that "US LNG would likely be nearly as bad as coal when exported to Europe and worse than coal when exported to Asia when the climate impacts of methane leakage are measured over a 20-year timeframe" (Chesapeake Climate Action Network, n.d.). A new study by Prof. Howarth suggests that LNG might actually be 44% to more than two-fold worse than coal from a climate perspective (Howarth in Cain, 2023).

In 2021, the International Energy Agency said that there should be no investments in new gas, oil, or coal production (Muttitt). Indeed, it makes no sense to make any new investments in fossil fuels of any sort when we have less than a decade to dramatically reduce GHG emissions to avoid the worst of the climate crisis (IPCC, 2022). "We cannot save a burning planet with a fire hose of fossil fuels," UN Secretary General Antonio Guterres said in a speech to the COP28 summit in Dubai: "I urge governments to help industry make the right choice – by regulating, legislating, putting a fair price on carbon, ending fossil fuel subsidies, and adopting a windfall tax on profits," he said (Reuters, 2023b).

Towards a Transatlantic Climate Bridge: Resistance in North America and Europe

Protests Against LNG in Germany

In 2022 Germany was Europe's largest importer of fossil gas (Gross, 2022), but because of the opposition of the German Climate Alliance Against LNG[17] (founded 2018 by grassroots groups and the German anti-fracking and climate activist Andy Gheorghiu) and unprofitable projections coupled with binding climate targets, Germany had no LNG terminals until then. Three onshore LNG terminals had been proposed in northern Germany: one in Brunsbüttel, one in Wilhelmshaven, and one in Stade, and it was expected that "[t]ogether the three terminals would have accounted for ~ 30 billion cubic metres (bcm) of natural gas" (Brauers et al., 2021). In November 2020, less than a year and a half before Russia attacked Ukraine, the Wilhelmshaven project was cancelled. Uniper, the developer, claimed there wasn't enough market interest in LNG (and decided to switch its focus on hydrogen). No application had been put forward for Stade.

One of the reasons why the project at Wilhelmshaven was abandoned by Uniper was a small but strong-enough transatlantic opposition. Canadian activists and German climate activist Andy Gheorghiu fought together against the interconnected LNG export project Goldboro in Nova Scotia, Canada. Uniper's Canadian partner Pieridae even attempted to silence the activists by threatening them with legal action for informing the public about the climate-damaging impacts of the LNG project and in particular its need for public subsidies on both sides of the Atlantic (Urgewald, 2021).

Before Russia's fully invasion into Ukraine, the LNG terminal in Brunsbüttel had been the most advanced project in Germany. Plagued by delays caused by the opposition and market dynamics, none of the investors for the proposed terminals had been able to make a final investment decision before February 2022 – despite the fact that they had received full political support at the local, state-wide, and national level and were offered public subsidies. Of course, the war changed those dynamics dramatically.

The construction of German LNG import terminals and the export of US fracked gas directly to Germany had been a central part of the geopolitical debate in the years before Russia's invasion – mainly because of Germany's support for the North Stream pipelines (Gheorghiu, 2020). In 2020, when he was still Finance Minister under Chancellor Angela Merkel, her later successor Olaf Scholz even offered the Trump administration €1 billion (approximately $1.18 billion) in taxpayer funds to build two LNG import terminals in exchange for ending sanctions against the Nord Stream 2 pipeline (Deutsche Welle, 2021). After Russia's full invasion of Ukraine, the significance of the geo-political dimension grew and gave the industry a justification to push even harder for LNG infrastructure on both sides of the Atlantic. In 2022 Germany passed an LNG Acceleration Act, which basically declares fossil fuel terminals approvable until 2043 (Exner-Pirot, 2024). It also suspends environmental impact assessments for floating terminals and makes a mockery of public consultation, turning the permitting process into a farce. The political will is now basically sufficient to get the permit for an LNG terminal and related infrastructure (see also Ziehm, 2023). The German government will also use the state bank KfW to directly finance 50% of the onshore LNG terminal in Brunsbüttel and has chartered five FSRUs (Kullik, 2022). At the time of writing, 11 terminals (onshore and offshore) are listed by the LNG Acceleration Act along with connecting pipelines as in principle eligible for approval.[18] The import sites of Brunsbüttel, Wilhelmshaven, and Stade remain, and import sites, such as the ones off the Island of Rügen in the Baltic Sea (a pristine destination for tourism) have been added. German private companies and state-owned enterprises such as EnBW, RWE, Uniper, and SEFE (Securing Energy for Europe, formerly Gazprom Germania) have signed long-term contracts for or are buying spot-market available fracked gas from the US. German banks have also been investing in US LNG export terminals (Deutsche Umwelthilfe/Urgewald/Gheorghiu, 2023).

Transatlantic Climate Activist Connections Before LNG

US grassroots climate organisations like Rising Tide and 350.org had been active internationally several years before Greta Thunberg began her school strikes that launched the Fridays for Future youth climate movement in 2018. Rising Tide was active on both sides of the Atlantic as early as 2000, while 350.org, which was founded by author Bill McKibben and students at Middlebury College in 2008, had expanded to Europe by 2010. Both organisations helped create connections, shared expertise, and inspired new leadership. In 2019 Rising Tide America brought representatives of the German anti-coal movement Ende Gelände [Here and No Further][19] to the United States for a series of workshops at activist meetings. A visible sign of these connections was NoCoalNoGas' very effective use of Ende Gelände's signature white Tyvek suits at an action demanding the shutdown of a coal plant in Bow, New Hampshire in September 2019 (compare to Figure 21.1 in Chapter 21 in this volume).[20]

Figure 30.4 NoCoal NoGas activists at a protest in Bow, NH, USA in fall 2019 wearing Tyvek suits inspired by German Ende Gelände anti-coal activists.[21]

Photo credit: Johnny Sanchez, NoCoal NoGas campaign. Permission granted.

German climate activist groups, which represent the biggest climate movement in Europe's largest economy, organised with Ende Gelände [Here and No Further][22] and Fridays for Future, have been prioritising the fight against the country's dirtiest energy, lignite coal, while the climate movement in the US has been focused primarily on pushing back against coal and gas and oil pipeline projects – most famously KXL[23] and the Dakota Access Pipeline (Grable, 2023). Yet, resistance against fracked gas has also been mounting for some time.[24]

Long before Fridays For Future, Extinction Rebellion or *Letzte Generation* existed, and before the larger climate movement paid attention to the significant global warming potential of fossil gas, resistance against fracking and fracked gas spread throughout Europe and worldwide. The EU-wide anti-fracking movement, consisting mainly of grassroots groups or networks such as Gegen-Gasbohren[25] in Germany, was the very first opposition against fossil gas in Europe. It united on the basis of core demands formulated through the Korbach Resolution, which, since it was first published in May 2013, has been signed by NGOs, numerous associations, grassroots initiatives, activist networks, political parties/organisations, companies, cities, boroughs, and counties.[26] Before Russia attacked Ukraine, anti-LNG-activists in Germany had managed to make quite a bit of progress and delayed all projects significantly. All of that with no mass mobilisation or direct actions but through a process of confronting political decision-makers at the regional, state-wide, and national levels with facts and by informing the public about the waste of public money involved and the negative impacts of fracked gas imports. The German Climate Alliance Against LNG – initially launched by only a small group of local activists and long-time anti-fracking activist Andy Gheorghiu – was able to secure significant support from German environmental NGO *Deutsche Umwelthilfe* (Environmental Action Germany), as well as from the most well-known German grassroots climate action group *Ende Gelände* (known primarily for its fights against coal). In July 2021, they all joined forces for a first mass mobilisation on site at the proposed LNG terminal in Brunsbüttel. The action involved around 2,000 activists setting up a climate camp. At the time, activists saw this action as a turning point in the anti-LNG and anti-gas fight in Germany. They received a lot of press coverage, and more and more people became interested in getting involved in the activism against gas.

Figure 30.5 Ende Gelände camp and anti-LNG protest in Brunsbüttel, 2021.

Source: © Andy Gheorghiu (chapter author).

Since the German coalition government that came into power in early December 2021 involved the Green Party, and at the time North Stream 2 was close to being finished, many expected the issue of LNG terminals to be dead. However, all that changed quickly after Russia's full scale attack on Ukraine. With the Green Party leading the ministry for economy and climate protection and the ministry for environmental and nature protection, the German government started pushing for a massively oversized LNG import infrastructure which, according to analysts and climate experts, will either create stranded assets or critically undermine the energy transition and torpedo achieving crucial climate targets (Höhne et al., 2022). But instead of shying away from opposition, the German Climate Alliance Against LNG grew and worked towards creating a strong transatlantic and pan-European network. Apart from *Deutsche Umwelthilfe* and *Ende Gelände* there are local grassroots groups (such as the *Bürgerinitiativen* gegen CO2-Endlager or *Lebenswertes Rügen*) or Friends of the Earth Germany regional chapters active at almost all proposed LNG import sites. Despite the enormous legal hurdles created by the LNG Acceleration Act, all organisations/groups are nonetheless making use of all remaining public participation and legal tools to stop the expansion. *Ende Gelände* has started focusing strongly on the fight against LNG on the island of Rügen and has organised two camps plus additional demonstrations and actions in the spring and fall of 2023 – for public panel debates, impacted community members from Texas and Louisiana were zoomed in.[27]

Building a Transatlantic Anti-LNG Network

Based on information and research which provides a good overview concerning the direct connection between US LNG export terminals and German LNG import terminals, activists and groups on both sides of the Atlantic (including Texas Campaign for the Environment, Chispa Texas, Louisiana Bucket Brigade, Better Bayou, and Andy Gheorghiu) decided to establish the transatlantic anti-LNG network at the end of 2022. The network, which continues to expand, is focused on stopping the expansion of LNG infrastructure and trade on both sides of the Atlantic and on policy and regulatory changes for the permitting of LNG infrastructure. By connecting frontline communities in the Gulf South with organisers and decision-makers in Europe and by challenging European energy companies and banks, the network has raised awareness about the human rights risks, environmental damages, and significant contribution to global warming related to LNG and fracked gas. It has also managed to list LNG as a commodity with potential human rights risks in Germany, which will help limit the import of US fracked gas and maybe even force companies to terminate contracts. The network is also focusing on EU banks who are co-financing US LNG export terminals and is supporting the Irish climate movement in defending the unique policy of Ireland, which does not support the import of fracked gas. In addition, the transatlantic anti-LNG alliance has worked towards getting a strong EU methane regulation, which will make it harder for fracked gas imports to enter European markets, and its European members have reached out to the US Federal Energy Regulatory Commission (FERC) in an attempt to stop the permitting of new US LNG export terminals.

Several speakers tours have been organised on both sides of the Atlantic – with German activists and a lawmaker visiting US LNG export sites (Parker, 2023), and engaging with impacted communities and a US delegation of impacted women and mothers travelling to Europe to meet activists and European decision-makers (Siefken, 2023). Groups on both sides of the Atlantic participated in public consultations of permitting procedures on the other side of the pond and filed corresponding objections (Allaire and Malek-Wiley, 2023; Jackoby et al., 2023; Gheorghiu et al., 2024). A webinar, hosted by a German MEP with participation from Chispa Texas and Louisiana Bucket Brigade;[28] an EU parliamentary event hosted by a German and a French MEP with participation of the Port Arthur Community Action Network, Texas; and several transatlantic anti-LNG webinars[29] were also organised.[30] In January 2024 German civil society groups and Green MP Kathrin Henneberger addressed the US Federal Energy Regulatory Commission (FERC) in two separate letters – urging the approval authority not to grant a permit to US company Venture Global who plans an expansion of its operationally plagued LNG facility in Louisiana's wetlands (Henneberger, 2024; Morehouse, 2024).

In March 2024, another US Gulf anti-LNG delegation attended the People's Summit against the European Gas Conference in Vienna[31] – an event which was postponed due to the fear of protest. And in summer 2024, local activists John Beard Jr. and Elida Castillo from the US gulf coast travelled to Germany to help tell the story of the communities most heavily – and negatively – impacted by the LNG expansion. They were accompanied by the authors of this chapter.[32]

Based on the good documentation available on the negative effects on people and the environment in the US LNG export regions, a stakeholder dialog between German energy companies and NGOs has revealed that LNG – especially fracking gas – poses significant

Figure 30.6 Anti-LNG protest of European MEPs/NGOs/grassroots groups and John Beard Jr. (Port Arthur Community Action Network, Texas) in front of EU parliament. 29 November 2023.

Source: Food & Water Europe, Deutsche Umwelthilfe, Andy Gheorghiu. Permission granted.

human rights risks. Based on the German Supply Chain Act, the extraction, processing, and import of LNG and fracked gas are extremely problematic, and the structural human rights violations must be addressed (García and Grüning, 2023).

On 26 January, the Biden administration announced a "temporary pause" on gas export projects (The White House, 2024). In July 2024, a federal judge in Louisiana ruled against the pause of the Biden administration on new LNG export permits (Bikales, 2024), but in September 2024, US DOE insisted that the pause remain in place – after it authorized LNG exports for New Fortress's Energy Altamaria Mexico project (Landry, 2024).

The transatlantic anti-LNG collaboration is expanding and ongoing.

Between the US Gulf Coast and Germany: Texas and Louisiana, USA

What the wins by grassroots climate activists against LNG infrastructure in the US gulf region show is that communities must share knowledge, be well-organised, grow a network of allies, and remain constantly vigilant while pushing decision-makers at all levels.

The total of 26 export facilities that are currently planned (Sierra Club, 2024), with 6 LNG terminals already under construction, pose a serious threat to the health of vulnerable communities already severely impacted by the environmental racism of multiple toxic industrial complexes in the vicinity. At the same time, they threaten wetlands in a total area equal to about half of Washington, D.C. Wetlands are sites of high biodiversity and mitigate the effects of hurricanes. The most threatened wetlands are in Louisiana (Gibbons, 2023).

The fossil gas liquefaction facility *Calcasieu Pass 2* (Louisiana), the LNG export facilities *Plaquemines* and *Commonwealth* (Louisiana), the liquefaction facility in Port Arthur (Texas), and the expansion of the liquefaction facility in Corpus Christi and Sabine Pass (Texas) are all relevant in the transatlantic context.

Figure 30.7 View of Sabine Pass LNG Export Terminal in Texas on the border of Louisiana. *Photo: Andy Gheorghiu.*

Here is an overview of recent grassroots wins:

- in the Rio Grande Valley of Texas, frontline Indigenous leaders (Baddour, 2022) and their allies (Guevara, 2022) have been able to fend off three proposed LNG export terminals, preventing more than 13 million tons per year of domestic greenhouse gas emissions and preventing more than 30 million tons of LNG from being exported and burned every year. And they keep winning. In August of 2024, the US Court of Appeals for the DC Circuit revoked the authorisations given by the US Federal Energy Regulatory Commission (FERC) to the Rio Grande and the Texan LNG projects for not properly assessing environmental, health, and climate aspects (Cunningham, 2024).
- in Corpus Christi, Texas, advocates have successfully prevented five desalination plants from being built (Pskowski, 2023) to generate clean water for industrial expansion – holding back as much as six million tons per year of new oil, gas, and petrochemical pollution.
- in Port Arthur, Texas, community groups achieved a recent legal win vacating Port Arthur LNG's state air permit (Williams, 2023) and in turn delaying projected greenhouse gas emissions of up to seven million tons/year of domestic emissions and preventing up to 27 million tons of LNG from being exported annually.
- community groups in St. James Parish, Louisiana were successful in getting the South Louisiana methanol plant cancelled in 2022 (Trimble, 2022). This would have been the largest methanol production facility in North America, emitting 2 million tons per year of climate pollution, along with huge amounts of local toxins. Community groups in St. James Parish are also preventing the Formosa plastics plant from being built (Laughland, 2021). This plant would have emitted 13 million tons of climate pollution per year (Rosenberg, 2022).
- advocates in Lake Charles, Louisiana, have so far prevented Venture Global, the company behind the CP2 LNG export terminal, from making a final investment decision. CP2 would emit 17 million tons per year of greenhouse gases domestically and would export 28 million tons of LNG annually. Analysts regarded the announcement of the Biden administration for an LNG pause as having an impact on DOE permitting for CP2. However, US FERC approved the project in June[33] – a move that has been challenged by environmental advocates, landowners, and fishermen who filed two petitions with the US Court of Appeals for the D.C. Circuit in September 2024 (Sneath, 2024).

Groups on both sides of the Atlantic vow to continue the fight against the transatlantic LNG trade and build-out plans.

Protests Against LNG in Ireland

Another transatlantic alliance was formed between activists in Brownsville, Texas and activists in Ireland. The company NextDecade had proposed Cork LNG to be built in the Port of Cork in Ireland as an import terminal for gas from the Rio Grande LNG. In 2017, NextDecade and Port of Cork signed a Memorandum of Understanding

to advance a joint business development opportunity in Ireland for a new Floating Storage and Regasification Unit ("FSRU") and associated LNG import terminal infrastructure. Under the terms of the MOU, the potential development at the Port of Cork would receive LNG from NextDecade's planned Rio Grande LNG ("RGLNG") project in South Texas.

(businesswire, 2017)

Since this made the Port of Cork a major stakeholder, they had a large say in whether the project would be built. Not Here Not Anywhere, a volunteer group established initially in 2017 to stop offshore drilling in Ireland and now working on various anti-fossil-fuel campaigns, founded a coalition with Save RVG in Brownsville, Texas in 2018, and together they set up a petition against Cork LNG for the local county and city governments. Over 40 civil society organisations signed a letter against Cork LNG. In November 2019 a resolution against Cork LNG passed almost unanimously by Cork City and Cork County Council and the project was cancelled. The coalition with activists in Brownsville across the ocean allowed Irish activists in Not Here Not Anywhere to communicate the environmental justice crisis in Brownsville to politicians in Ireland. Thus, humanising the impacts of LNG in the US was extremely important for eroding its social licence in Ireland.

Another project activists have been fighting against in Ireland for over a decade is Shannon LNG. Shannon LNG was a proposed LNG plant and power station in County Kerry, Ireland. It would receive gas from Pennsylvania in the US. Local opposition groups in Ireland, established in 2007, turned the project into the subject of several court cases on environmental grounds. And in February 2019,

> the High Court referred the case to the European Courts, with the European Court of Justice ruling in September 2020 that the project should be subject to a new environmental assessment under the EU Habitats Directive. On that decision the High Court rejected Shannon LNG planning permission.
>
> *(SafetyBeforeLNG)*

Activists were also fighting Shannon LNG by raising awareness of impacts of fracked gas/climate emissions and by pushing to move Shannon LNG off of the European Union's list of "projects of common interest" for Ireland. In February 2020, Shannon LNG and its pipeline were included yet again in the fourth European Union Projects of Community Importance (PCI) list. Inclusion on the list entitles projects to the "most rapid treatment legally possible," consideration as being of public interest from an energy policy perspective, and, possibly, of "overriding public interest" (Carolan, 2020). In November 2023, Friends of the Irish Environment (FIE) succeeded in the High Court of Ireland which quashed the planning extension Shannon LNG received from the Irish authority An Bord Pleanála – forcing the company to apply again for a new planning permission (Elliott, 2020).

When Ireland's new coalition government was formed in June 2020, the coalition's "Programme for Government" included pledges to ban new gas exploration licences and to not support LNG terminals importing "fracked" US gas. They also committed to withdrawing the Shannon LNG terminal from the EU Projects of Common Interest (PCI) list in 2021. (Elliot). In the Irish updated policy statement of 2021, the government promised to work "with like-minded European States to promote and support changes to European energy laws . . . in order to allow the importation of fracked gas to be restricted" and to "work with international partners to promote the phasing out of fracking at an international level" (Department of Environment, Climate, and Communications, 2021). Following an appeal by FIE, the Irish court of appeal heard in February 2022 that Shannon LNG has "voluntarily" dropped its 4th PCI Project status of "overriding public interest" (Safety Before LNG,

2022). Over a year later, in September 2023, the agency in charge of planning, An Bord Pleanála refused to give Shannon LNG planning permission – citing the Irish Government's policy statement against fracked gas imports (Safety Before LNG, 2023).

Ireland banned fracking in 2017. It's a "dirty word" there, and it is thus politically not palatable to have anything to do with it. The third LNG project proposed in Ireland, which is the only one that would not receive gas from the US, had promised that they would not use fracked gas.[34]

Conclusion

The global Methane Pledge of 2021 notwithstanding, decisions by governments and the fossil gas industry to continue to expand fossil gas infrastructure pose a major threat to communities and the climate, but the many examples of successful coordinated intervention by grassroots climate activists on both sides of the Atlantic should encourage more people to stand up and say no. A combination of savvy mobilisation, sharing real stories of real people, combined with in-depth knowledge and vigilant observation of permitting processes, involvement in legal challenges, and the collaboration with others to fight against fossil gas infrastructure are key to activists' success. The alliances activists have formed across the Atlantic enable them not only to share information, tactics, and strategies but also to educate their respective legislators through connecting them with impacted locals on the other side. The transatlantic climate activist bridge has major advantages over corporate interests: the fearlessness, doggedness, and unlimited creativity of people rising up in defence of their livelihoods and joining forces with those on the other side of the pond appeal to people's sense of justice.

Notes

1 Special Thanks to Maya Kattler-Gold and Preston Merrill who helped with the research for early drafts of this chapter.
2 See for example https://www.nytimes.com/1988/06/24/us/global-warming-has-begun-expert-tells-senate.html.
3 https://www.climatefiles.com/exxon-knew/.
4 According to a 2023 column in the weekly magazine DER SPIEGEL by Christian Stöcker, professor for digital communication in Hamburg, for example, the German "Liberals," part of the governing coalition at the federal level, not only suffer true climate deniers in their midst, who adhere to the "merchants of doubt" ideology, but they also successfully employ their tactics. See https://www.spiegel.de/wissenschaft/mensch/klimaschutz-die-heimlichen-herrscher-der-fpd-kolumne-a-d0defee9–85ea-4cdb-adac-93e49e3539de.
5 See Global Methane Pledge (2021). https://www.ipcc.ch/report/ar6/wg1/downloads/report/IPCC_AR6_WGI_Chapter07.pdf (Table 7.15, page 1017 (pdf-page 95)
6 See https://www.research.howarthlab.org/pubs_all.php.
7 https://www.globalmethanepledge.org/.
8 See for example https://www.cleanenergywire.org/news/german-government-plans-extensive-lng-infrastructure-build-ensure-security-european-supply and https://www.theguardian.com/commentisfree/2023/dec/19/fossil-fuel-expansion-us-biden. See more in later mention.
9 Also in Spanish: https://www.ecologistasenaccion.org/30037/libro-resistencia-global-al-fracking/.
10 See https://energytransition.org/wp-content/uploads/2021/07/What-is-fracking.jpg.
11 See https://concernedhealthny.org/compendium/.
12 https://www.eia.gov/todayinenergy/detail.php?id=55920.

13 https://www.eia.gov/tools/faqs/faq.php?id=907&t=8. See also https://www.igu.org/resources/2024-world-lng-report/, graph p. 19 and Table 3.1 p. 30.

14 https://www.eia.gov/todayinenergy/detail.php?id=56320.

15 See https://www.permianclimatebomb.org/.

16 See https://www.kayrros.com/u-s-methane-emissions-from-fossil-fuels-at-risk-of-worsening-in-2022-extending-2021-trend/.

17 Opposition work consisted of open letters, media work, providing facts and informing people during public events (sometimes in panel debates with industry and politicians), and engaging with political decision-makers at the regional, state-wide, and national level.

18 See https://www.gesetze-im-internet.de/lngg/anlage.html.

19 https://www.ende-gelaende.org/.

20 See https://www.nocoalnogas.org/what-weve-done.

21 See for example in 2016: https://www.energiezukunft.eu/umweltschutz/ende-gelaende-augenzeugen-berichten/.

22 See https://www.ende-gelaende.org/en/.

23 See https://www.nrdc.org/stories/killing-kxl.

24 In the US, see for example https://www.americansagainstfracking.org/about-the-coalition/.

25 See https://www.gegen-gasbohren.de/.

26 https://www.gegen-gasbohren.de/aktionen-forderungen-und-ziele/korbacher-resolution. See also https://www.ecologistasenaccion.org/wp-content/uploads/adjuntos-spip/pdf/fracking_global_resistance.pdf.

27 See https://fruehling-auf-ruegen.de/zeitplan-camptage/ and https://www.ende-gelaende.org/en/news/no-lng-on-ruegen-no-lng-worldwide-together-against-the-dirty-lie-of-clean-gas-from-22-24-september-2023/.

28 See https://www.youtube.com/watch?v=WxN51HNdcnI&list=PLbk18iV_zcoypM9h_BnIun6m KWMxZU-eF&index=6.

29 See https://www.youtube.com/watch?v=KiE82n6HV8w and https://www.youtube.com/watch?v=4JN4h0txlS4.

30 See Ernst (2023). https://x.com/ErnstCornelia/status/1729525367428317352?s=20 [Accessed: 24 March 2024].

31 https://www.peoplessummit.at/en/.

32 See for example the event in Kassel on July 11, 2024: https://www.uni-kassel.de/uni/en/nachhaltigkeit/nachhaltigkeitsforschung/kassel-institute-for-sustainability/news-events/event/2024/07/11/podiumsdiskussion-importiertes-leid-und-klimachaos-die-dunkle-seite-von-fracking-lng-1?cHash=ea9fef01c8cdd2f001bbce00319f1577.

33 See https://lngprime.com/americas/ferc-approves-venture-globals-cp2-lng-project/115769/

34 See (http://www.safetybeforelng.ie/).

References

Allaire, J. and Malek-Wiley, D. (2023). *Objection Against the Permitting/Construction of the LNG (Liquefied Natural Gas) Import Terminal Stade (Hanseatic Energy Hub)*, 26 June. https://energytransition.org/wp-content/uploads/2023/08/US-objection-against-permitting-of-LNG-import-terminal-Stade_HanseaticEnergyHub.pdf [Accessed 24 March 2024].

Baddour, D. (2022). Indigenous leaders in texas target global banks to keep LNG export off sacred land at the port of brownsville. *Inside Climate News*, 18 October. https://insideclimatenews.org/news/18102022/indigenous-leaders-in-texas-target-global-banks-to-keep-lng-export-off-of-sacred-land-at-the-port-of-brownsville/

Bikales, J. (2024). Federal judge blocks Biden's pause on LNG Export permits. *Politico*, 1 July. https://www.politico.com/news/2024/07/01/judge-blocks-biden-lng-pause-00166157

Black, S., Parry, I. and Vernon, N. (2023). *Fossil Fuel Subsidies Surged to Record $7 Trillion. IMF.* https://www.imf.org/en/Blogs/Articles/2023/08/24/fossil-fuel-subsidies-surged-to-record-7-trillion [Accessed 3 February 2024].

Brauers, H., Braunger, I. and Jewell, J. (2021). Liquefied natural gas expansion plans in Germany: The risk of gas lock-in under energy transitions. *Energy Research & Social Science*, 76, 102059. https://doi.org/10.1016/j.erss.2021.102059

businesswire (2017). *NextDecade Signs MOU with Port of Cork to Develop FSRU and LNG Import Terminal Infrastructure*, 19 July. https://www.businesswire.com/news/home/20170719006183/en/NextDecade-Signs-MOU-Port-Cork-Develop-FSRU [Accessed 24 March 2024].

Cain, Ch. (2023). New study shows liquified natural gas might be worse for climate change than coal. *NBC Bay Area*, 29 November. https://www.nbcbayarea.com/news/national-international/new-study-shows-liquefied-natural-gas-might-be-worse-for-climate-change-than-coal/3384705/

Cantoni, R. (2022). Fighting science with science: Counter-expertise production in anti-shale gas mobilizations in France and Poland. *NTM*, 30, 345–375. https://doi.org/10.1007/s00048-022-00342-x

Carolan, M. (2020). Legal challenge to shannon LNG project has EU implications. *Irish Times*, 27 February. https://www.irishtimes.com/news/environment/legal-challenge-to-shannon-lng-project-has-eu-implications-1.4187146 [Accessed 24 March 2024].

Chesapeake Climate Action Network (n.d.). *US Department of Energy Report Confirms: U.S. LNG Exports to Asia Would Likely Be WORSE Than Coal for the Atmosphere for Decades to Come.* https://fossil.energy.gov/app/DocketIndex/docket/DownloadFile/206 [Accessed 24 March 2024].

Climate Files. *Exxon Knew.* https://www.climatefiles.com/exxon-knew [Accessed 23 March 2024].

Copernicus (2024). 2023 is the hottest year on record, with global temperatures close to 1.5 degree limit. *ECMWF*, 9 January. https://climate.copernicus.eu/copernicus-2023-hottest-year-record

Council of the European Union (2023). Climate action: Council and Parliament reach deal on new rules to cut methane emissions in the energy sector. *Press Release.* https://www.consilium.europa.eu/en/press/press-releases/2023/11/15/climate-action-council-and-parliament-reach-deal-on-new-rules-to-cut-methane-emissions-in-the-energy-sector/ [Accessed 22 January 2024].

Cowan, T. (2024). US residential gas customers bear brunt of LNG exports. *Institute for Energy Economics and Financial Analysis.* https://ieefa.org/resources/us-residential-gas-consumers-bear-brunt-lng-exports [Accessed 29 April 2024].

Cunningham, N. (2024). Rio Grande LNG construction authorization revoked. In *Gas Outlook*, 7 September. https://gasoutlook.com/analysis/rio-grande-lng-construction-authorisation-revoked-by-federal-court/ [Accessed 7 Sept 2024].

Department of Environment, Climate, and Communications (2021). Policy statement on the importation of fracked gas. *gov.ie*, 18 May. https://www.gov.ie/en/publication/f3774-policy-statement-on-the-importation-of-fracked-gas/

Deutsche Umwelthilfe/Urgewald/Gheorghiu (2023). *Investing in Climate Chaos. How German Banks and Companies Enable Fracking LNG Projects*, 19 April. https://www.duh.de/fileadmin/user_upload/download/Pressemitteilungen/Energie/LNG/US_LNG_terminals_EN.pdf [Accessed 24 March 2024].

Deutsche Welle. (2021). *Germany Offered US 'Dirty Deal' Over Nord Stream 2 Sanctions*, 2 October. https://www.dw.com/en/germany-offered-us-dirty-deal-to-drop-nord-stream-2-sanctions/a-56517249

DiColo, J. A. (2012). *Exxon CEO: 'Losing Our Shirts' on Low Natural Gas Prices*, 27 June. https://www.rigzone.com/news/oil_gas/a/118907/exxon_ceo_losing_our_shirts_on_low_natural_gas_prices/ [Accessed 24 March 2024].

Elliott, S. (2020). New blow for US LNG in Europe as Irish Court quashes Shannon LNG consents. *S&P Global*, 10 November. https://www.spglobal.com/commodityinsights/en/market-insights/latest-news/natural-gas/111020-new-blow-for-us-lng-in-europe-as-irish-court-quashes-shannon-lng-consents.

Ende Gelände. (n.d.). *No LNG on Rügen. No LNG Worldwide. Together Against the Dirty Lie of Clean Gas from 22–24 September 2023.* https://www.ende-gelaende.org/en/news/no-lng-on-ruegen-no-lng-worldwide-together-against-the-dirty-lie-of-clean-gas-from-22-24-september-2023/

Ernst, C. (2023). *Liquified Natural Gas: Asset or Threat? European Parliament, Brussels, Belgium*, 29 November. https://x.com/ErnstCornelia/status/1729525367428317352?s=20.

European Commission (2022a). *EU-US LNG Trade.* https://energy.ec.europa.eu/system/files/2022-02/EU-US_LNG_2022_2.pdf [Accessed 24 March 2024].

European Commission (2022b). *Quarterly Report*. https://energy.ec.europa.eu/system/files/2022-04/Quarterly%20report%20on%20European%20ga%20markets_Q4%202021.pdf [Accessed 24 March 2024].

Evensen, D., Whitmarsh, L., Devine-Wright, P., et al. (2023). Growing importance of climate change beliefs for attitudes towards gas. *Nature Climate Change*, 13, 240–243. https://doi.org/10.1038/s41558-023-01622-7

Exner-Pirot, H. (2024). From emergency to miracle: Germany's LNG acceleration law shows that Western states can still build when they need to. *MLI*, 15 January. https://macdonaldlaurier.ca/from-emergency-to-miracle/

García, B. and Grüning, C. (2023). Potential human rights risks along the supply and value chains. In Federal Ministry of Labour and Social Affairs Division VI b 3 "CSR – Corporate Social Responsibility." https://www.csr-in-deutschland.de/EN/Business-Human-Rights/Implementation-support/Sector-dialogues/Energy-sector-dialogue/publication-potential-human-rights-risks-along-supply-and-value-chains.html

Gardner, T. and Williams, C. (2024). Biden administration taking heat from all sides over Louisiana LNG project. *Reuters*, 19 January. https://www.reuters.com/business/energy/biden-administration-taking-heat-all-sides-over-louisiana-lng-project-2024-01-19/

Gheorghiu, A. (2020). US fracked LNG and Russian gas via Nord Stream 2: Germany's double game heats up both geopolitics and the climate. *Energy Transition*, 14 September. https://energytransition.org/2020/09/us-fracked-lng-and-russian-gas-via-nord-stream-2-germanys-double-game-heats-up-both-geopolitics-and-the-climate/ [Accessed 22 January 2024].

Gheorghiu, A. (2023). Global South issues in the Global North? A fossil toxic tour through Texas and Louisiana – Part 1. *Energy Transition. The Global Energiewende*, 3 August. https://energytransition.org/2023/08/global-south-issues-in-the-global-north-a-fossil-toxic-tour-through-texas-and-louisiana-part-1/

Gheorghiu, A., et al. (2024). Deny permit for Venture Global's Calcasieu Pass 2 (CP2) fracked gas export terminal. *Comments, Federal Regulatory Commission eLibrary*, 17 January, https://elibrary.ferc.gov/eLibrary/filelist?accession_number=20240117-5018

Gibbons, B. (2023). Proposed LNG export terminals threaten 22,000 acres of wetlands, many in Louisiana. *OilandGasWatch.org*, 1 March. https://news.oilandgaswatch.org/post/proposed-lng-export-terminals-threaten-22-000-acres-of-wetlands-many-in-louisiana [Accessed 23 March 2024].

Global Methane Pledge (2021). https://www.globalmethanepledge.org [Accessed 22 January 2024].

Global Resistance to Fracking (2015). https://www.ecologistasenaccion.org/wp-content/uploads/adjuntos-spip/pdf/fracking_global_resistance.pdf [Accessed 22 January 2024], available in Spanish at https://www.ecologistasenaccion.org/30037/libro-resistencia-global-al-fracking/

Grable, J. (2023). A new hope for shutting down the Dakota access pipeline. *Sierra*, 14 November. https://www.sierraclub.org/sierra/new-hope-shutting-down-dakota-access-pipeline [Accessed 24 March 2024].

Gross, C. (2022). Germany is the largest importer of natural gas in Europe. *Climate Scorecard*, 3 June. https://www.climatescorecard.org/2022/06/germany-is-the-largest-importer-of-natural-gas-in-europe/ [Accessed 22 January 2024].

Guevara, E., et al. (2022). With history of community success, fight to stop LNG in the RGV continues. *Sierra Club Lone Star Chapter*, 26 September. https://www.sierraclub.org/texas/blog/2022/09/history-community-success-fight-stop-lng-rgv-continues-0

Henneberger, K. (2024). Comments of Kathrin Henneberger, German MP re the Expansion of CP2 express pipeline project and calcasieu Pass LNG facilities under CP22–21, et al. *ELibrary*, January, Federal Energy Regulatory Commission (FERC). https://elibrary.ferc.gov/eLibrary/filelist?accession_num=20240116-5255

Höhne, N., et al. (2022). Plans for German liquified natural gas terminals are massively oversized. *New Climate*, 13 December. https://newclimate.org/resources/publications/plans-for-german-liquefied-natural-gas-terminals-are-massively-oversized

Hopke, J. E. (2015). Hashtagging politics: Transnational anti-fracking movement Twitter practices. *Social Media + Society*, 1(2). https://doi.org/10.1177/2056305115605521

Howarth, R. W. (2019). Ideas and perspectives: Is shale gas a major driver of recent increase in global atmospheric methane? *Biogeosciences*, 16(15), BG, 16, 3033–3046. https://doi.org/10.5194/bg-16-3033-2019

Howarth, R. W. (2024). *The Greenhouse Gas Footprint of Liquified Natural Gas (LNG) Exported from the United States*. Energy Science and Engineering. Modelling and Analysis, 3 October, (online). doi: 10.1002/ese3.1934

Howarth-Marino Lab Group (n.d.). Publications. https://www.research.howarthlab.org/pubs_all.php [Accessed 24 March 2024].

Howey, K. and Neale, T. (2023). Divisible governance: Making gas-fired futures during climate collapse in Northern Australia. *Science, Technology, & Human Values*, 48(5), 1080–1109. https://doi.org/10.1177/01622439211072573

International Gas Union (IGU) (2023). *2023 World LNG Report*. https://safety4sea.com/wp-content/uploads/2023/07/IGU-World-LNG-Report_2023_07.pdf

IPCC (2022). *Working Group I Full Report*. https://www.ipcc.ch/report/ar6/wg1/downloads/report/IPCC_AR6_WGI_FullReport.pdf.

Jackoby, J., et al. (2023). *Statement/Objection against the Permitting/Construction of the LNG Import Terminal Brunsbüttel*, 10 July. https://energytransition.org/wp-content/uploads/2023/08/USobjectiontoGermanLNGimportterminalBrunsbuttel.pdf [Accessed 24 March 2024].

Kayrros (2022). *US Methane Emissions from Fossil Fuels at Risk of Worsening in 2022, Extending 2021 Trend*, 28 June. https://www.kayrros.com/de/u-s-methane-emissions-from-fossil-fuels-at-risk-of-worsening-in-2022-extending-2021-trend/ [Accessed 23 March 2024].

Kullik, O. (2022). Fast track to independence. *Kfw.de*, 1 July. https://www.kfw.de/stories/lng-terminals.html

Landry, C. (2024) DOE: Pause on US LNG export permits remains despite New Fortress authorization. *Oil and Gas Journal*, 5 September. https://www.ogj.com/general-interest/government/article/55138183/doe-pause-on-us-lng-export-permits-remains-despite-new-fortress-authorization

Laughland, O. (2021). Multibillion-dollar Louisiana plastics plant put on pause in a win for activists. *The Guardian*, 18 August. https://www.theguardian.com/us-news/2021/aug/18/louisiana-plastics-plant-toxic-emissions-cancer-alley

Mannan, S. (2012). Liquefied natural gas. *Lees' Loss Prevention in the Process Industries (Fourth Edition)*, Butterworth-Heinemann, *ScienceDirect*, 2492–2506. https://doi.org/10.1016/B978-0-12-397189-0.00039-2

Martin-Sosa Rodriguez, S. (2015). *Global Resistance to Fracking*. Ecologistas en Accion. Madrid, Spain. https://www.ecologistasenaccion.org/wp-content/uploads/adjuntos-spip/pdf/fracking_global_resistance.pdf

Marusic, K. (2023). How the Halliburton loophole lets fracking companies pollute water with no oversight. *Environmental Health News*, 18 May. https://www.ehn.org/halliburton-loophole-2659983182.html [Accessed 24 March 2024].

Mincyte, D. and Bartkiene, A. (2019). The anti-fracking movement and the politics of rural marginalization in Lithuania: Intersectionality in environmental justice. *Environmental Sociology*, 5, 177–187. https://doi.org/10.1080/23251042.2018.1544834

Morehouse, C. (2024). German environmentalists call on FERC to reject massive LNG Terminal. *Politico Pro*, 17 January. https://subscriber.politicopro.com/article/2024/01/german-environmentalists-call-on-ferc-to-reject-massive-lng-terminal-00136102

Moseman, A. (2024). Why do we compare methane to carbon dioxide over a 100 year timeframe? Are we underrating the importance of methane emissions? *Ask MIT*, 4 January. https://climate.mit.edu/ask-mit/why-do-we-compare-methane-carbon-dioxide-over-100-year-timeframe-are-we-underrating

Muncie, E. (2020). 'Peaceful protesters' and 'dangerous criminals': The framing and reframing of anti-fracking activists in the UK. *Social Movement Studies*, 19, 464–481. https://doi.org/10.1080/14742837.2019.1708309

Nechyporenko, D. (2024). How much energy is needed to liquefy and re-gasify natural gas for LNG? *LinkedIn*, 13 February. https://www.linkedin.com/pulse/how-much-energy-needed-liquefy-re-gasify-natural-gas-lng-dima-kv1qc/

Neujeffski, M. and Goldenbaum, M. (2021). *Klimawandelleugner*innen in Deutschland: Zwischen Verschwörungsideologien und Wohlstandsegoismus. Fachstelle Radikalisierungsprävention und Engagement im Naturschutz. Die extreme Rechte zwischen Klimawandelleugnung und Klimanationalismus.* https://www.nf-farn.de/system/files/documents/broschuere_farn_klimavonrechts_web.pdf [Accessed 22 January 2024].

Oreskes, N. and Conway, E. (2010). *Merchants of Doubt. How a Handful of Scientists Obscured the Truth on Issues from Tobacco Smoke to Global Warming.* London: Bloomsbury.

Parker, H. (2023). German lawmaker tours liquefied natural gas sites on Gulf Coast, opposes expansion. *New Orleans Public Radio*, 19 July. https://www.wwno.org/coastal-desk/2023-07-19/german-lawmaker-tours-liquefied-natural-gas-sites-on-gulf-coast-opposes-expansion

Permian Climate Bomb. https://www.permianclimatebomb.org/

Pospíšil, J., Charvát, P., Arsenyeva, O., Klimeš, L., Špiláček, M. and Klemeš, J. J. (2019). Energy demand of liquefaction and regasification of natural gas and the potential of LNG for operative thermal energy storage. *Renewable and Sustainable Energy Reviews*, 99, 1–15.

Pskowski, M. (2023). Texas eyes marine desalination, oilfield water reuse to sustain rapid growth. *Inside Climate News*, 15 April. https://insideclimatenews.org/news/15042023/texas-water-legislation-desalination-oilfield-wastewatereyes-marine-desalination-oilfield-water/

Reuters (2023b). *UN Chief Says Ending Fossil Fuel Use Is Only Way to Save Burning Planet*, 1 December. https://www.reuters.com/business/environment/un-chief-says-ending-fossil-fuel-use-is-only-way-save-burning-planet-2023-12-01/ [Accessed 24 March 2024].

Rosenberg, J. (2022). Fishermen, shrimpers stage boat convoy to protest methane refinery build-out in Lake Charles area. *The Lens*, 4 November. https://thelensnola.org/2022/11/04/fishermen-shrimpers-stage-boat-convoy-to-protest-methane-refinery-buildout-in-lake-charles-area/

Safety Before LNG (2022). *Press Release*, 5 February. http://www.safetybeforelng.ie/pressreleases/pressrelease20220205ShannonLNGDropsPCIAdvantagesInNewPlanningApplication.html

Safety Before LNG (2023). *Press Release*, 15 September. http://www.safetybeforelng.ie/pressreleases/pressrelease20230915-Shannon-LNG-Refused-Planning-Permission.html

Siefken, H. (2023). Netzwerk Energiedrehscheibe beklagt Umweltvergiftung in den USA durch LNG-Terminals. *NWZ Online*, 29 March. https://www.nwzonline.de/wilhelmshaven/wilhelmshavener-lng-kritiker-beklagen-umweltzerstoerung-in-usa_a_4,0,181201495.html [Accessed 24 March 2024].

Sierra Club (2024). US LNG export tracker. *Mapbox*. https://www.sierraclub.org/dirty-fuels/us-lng-export-tracker

Sneath, S. (2024). Massive CP2 LNG export facility faces new legal hurdle over FERC approval. *desmogblog*, 6 September, https://www.desmog.com/2024/09/06/venture-global-cp2-lng-export-louisiana-ferc-legal-challenge/

Stasik, A. (2018). Global controversies in local settings: Anti-fracking activism in the era of Web 2.0. *Journal of Risk Research*, 21(12), 1562–1578. http://doi.org/10.1080/13669877.2017.1313759

Steger, T. and Drehobl, A. (2017). The anti-fracking movement in Ireland: Perspectives from the media and activists. *Environmental Communication*, 12, 1–13. https://doi.org/10.1080/1752403 2.2017.1392333

Stöcker, C. (2023). Die heimlichen Einflüsterer der FDP. *DER SPIEGEL*, 25 June. https://www.spiegel.de/wissenschaft/mensch/klimaschutz-die-heimlichen-herrscher-der-fpd-kolumne-a-d0defee9-85ea-4cdb-adac-93e49e3539de [Accessed 22 January 2024].

The White House (2024). *Fact Sheet: Biden-Harris Administration Announces Temporary Pause on Pending Approvals of Liquified Natural Gas Exports*, 26 January. https://www.whitehouse.gov/briefing-room/statements-releases/2024/01/26/fact-sheet-biden-harris-administration-announces-temporary-pause-on-pending-approvals-of-liquefied-natural-gas-exports/

Third Act (n.d.). *Stop the Expansion of LNG Exports in the Gulf.* https://thirdact.org/blog/help-stop-the-massive-expansion-of-lng-exports/

Trimble, A. (2022). Louisiana residents celebrate stopping South Louisiana Methanol's petrochemical complex. *EarthJustice*, 9 September. https://earthjustice.org/press/2022/louisiana-residents-celebrate-stopping-south-louisiana-methanols-petrochemical-complex

Tsafos, N. (2018). US LNG into Europe after the Juncker-Trump agreement. *CSIS*, 9 August. https:// www.csis.org/analysis/us-lng-europe-after-trump-juncker-agreement [Accessed 22 January 2024].
Urgewald (2021). Fortum/Uniper's partner Pieridae attempts to silence activists. *Press Release*, 30 March. https://www.urgewald.org/en/medien/fortumunipers-partner-pieridae-attempts-silence-activists
Williams, C. (2023). US appeals court scraps Sempra's Port Arthur LNG emissions permit. *Reuters*, 15 November. https://www.reuters.com/business/energy/us-appeals-court-decision-scraps-sempras-port-arthur-lng-emissions-permit-2023-11-15/
Withers, P. (2023). Proposed Goldboro LNG plant officially abandoned after more than a decade. *CBC News*, 24 November. https://www.cbc.ca/news/canada/nova-scotia/proposed-goldboro-lng-plant-abandoned-after-11-years-1.7038936.
Ziehm, C. (2023). *Stellungnahme*, 29 June. https://www.bundestag.de/resource/blob/955986/12596d ce6a0d7e3fc208e911a845bc6c/Stellungnahme_RAin_Dr-_Ziehm-data.pdf

31
CLIMATE ACTIVISM AND THE POLITICAL-ECONOMIC LANDSCAPE IN NIGERIA

Joy Egbe

Introduction

In recent years, Nigeria has borne the brunt of escalating impacts of climate change, with severe consequences for its people and environment. According to the Intergovernmental Panel on Climate Change (IPCC), temperatures in Nigeria have risen by approximately 1.1 degrees Celcius since the pre-industrial era (IPCC, 2021).[1] This warming temperature trend is particularly pronounced in the northern region of the country, where temperatures are rising faster than the global average. Prolonged periods of extreme heat have become commonplace, impacting the agricultural sector, water resources, and human health, leading to social and environmental crises in the region. This has also played a role in security challenges. Scarce resources and land degradation have led to conflicts between communities, often pushing them to a breaking point (Obaze, 2022). Droughts leading to famine and unemployment are exacerbating existing poverty not only in the northern region but across different parts of the country. This has spurred people and organisations to start speaking up for climate action. According to *Daily Trust*, in 2022 alone Nigeria was one of the countries experiencing the most devastating effects of climate change. Flooding, which lasted for months across many states and regions of the country, displaced over 1.4 million people, claimed over 603 lives, and left 2,400 injured. It destroyed over 200 homes and 500 hectares of farmland, which meant the agricultural sector lost massive amounts of farm produce, resulting in food scarcity and price hikes the following year. This means that at the same time as thousands of leaders from around the world were gathering in Egypt for the COP27 climate conference, over 1.4 million people were being directly impacted by climate change further south. Similar events have happened across different countries in sub-Saharan Africa. In 2023, the world recorded the hottest year in history. And sub-Saharan Africa again recorded a growing number of environmental disasters.[2]

Experiencing the climate crisis and the extreme weather events first-hand has pushed young activists to pivot from cleaning beaches, organising protests, and raising awareness to a more holistic approach, where they are focusing on implementing solutions in their communities. The urgent need and desire to address the interconnected challenges have

DOI: 10.4324/9781003396567-37

spurred the emergence of a number of grassroots activist movements, NGOs, community-based organisations, and social enterprises dedicated to finding solutions.

The Rise and Evolution of Grassroots Climate Activism

For many activists and communities in developing countries, climate change is seen as a social challenge that will exacerbate existing poverty among vulnerable people. Many countries have low or no capacity to adapt or mitigate climate change, given that the region is already faced with socio-economic issues that are negatively affecting culture, development, and social wellbeing. Indeed, countries in sub-Saharan Africa (similar to Asian nations like India and Bangladesh) have long been living with economic hardship, which climate change is exacerbating. Activists in these regions therefore perceive their work as a means of addressing poverty and inequality by focusing on ensuring their communities have access to basic amenities and restoring resources affected by the climate crisis, such as water and land, as well as protecting the rights of Indigenous peoples and marginalised groups and demanding developed nations provide financial support for vulnerable countries to cope and adapt to climate change.

As global temperature keeps rising, many have been displaced from their homes, many children are out of school, becoming victims of child labour, and girls are faced with the risk of child marriage and sexual abuse. One of the major reasons the climate crisis keeps worsening existing poverty and economic hardship is that Africa depends mostly on agricultural activities as a source of livelihoods and economic development. (Diao, 2007). The IPCC assessment on vulnerability has identified the role that agriculture plays in many countries' socioeconomic developments and how the extent of these countries' vulnerability to climate change affects agricultural practices and in turn livelihoods (IPCC, 2021).

Climate activism in Nigeria has changed significantly over the last 20 years. Early environmentalists advocated above all for the preservation of nature and its resources and were primarily concerned with the immediate consequences of pollution on communities and ecosystems. They protested against oil extraction and exploitation, especially in the country's Niger Delta region (Amnesty International, 2015). Environmentalists like Nnimmo Bassey and Chima Williams were very vocal in mobilising grassroots environmental and legal movements against oil companies like Shell in the oil-producing region of Nigeria (Aljazeera, 2022). Their work and the work of past environmental heroes like Kenule Beeson, popularly known as "Ken" Saro-Wiwa, laid the foundation for environmental justice in Nigeria, especially for Indigenous communities (Gianotti, 2018).

However, the landscape of climate activism in Nigeria has evolved significantly over the past two decades, especially in response to the increasingly urgent and dire scenarios posed by climate change. As the impacts continue to intensify globally, they are also redefining the environmental and social challenges faced by Nigerians. The frequent increase in climate-related disasters in the country and across different parts of Africa has been a major contributor to the current surge in the number of young people advocating for climate justice and actions. This new climate movement is majorly championed by youth and youth-led organisations calling on leaders to enact measures to reduce the impact of climate change on vulnerable communities. Many of these activists at the frontline of climate advocacy have directly witnessed the climate crisis, and this experience has heightened their sense of urgency. Their personal connection to climate issues leads them to perceive climate change as a threat not only to their communities and ecosystems but also to their own lives.

The uniqueness of the youth movement is its ability to engage local communities in traditional and local initiatives to address climate issues at the local level and develop radical solutions to best meet the needs of people and communities directly affected. This current youth movement is radical, proactive, and does not like to conform to any political narrative. The grassroots environmental movement is led by civilians and organisations, including non-governmental organisations (NGOs), civil society organisations, and social enterprises. Activism of this nature works hand in hand with local residents to develop local initiatives for local problems, directly related to the needs of the community.

The goal of this social and environmental movement is to provide members of the local community with concrete actions to survive the catastrophic impacts of climate change. The insights presented in this chapter draw from a combination of my lived experience as a climate activist and social entrepreneur leading energy innovations to address climate change in Nigeria, the on-the-ground work of Newdigit Technologies in providing clean energy access, as well as an extensive review of relevant grey literature and academic sources. By combining grassroots perspectives with an extensive review of literature, reports, and stakeholder conversations, the aim is to holistically capture the unique nature and challenges of Nigeria's rising youth-led climate activism movement. Additionally, authoritative reports from organisations like the IPCC, academic journals, government policy documents, and media coverage were examined to contextualise the landscape within the broader socio-political dynamics of grassroot activism and climate change impacts facing Nigeria.

How Do Young Africans Engage in Activism?

To African citizens the impact of the climate crisis ripples through every facet of our existence, leaving no aspect of our society and our lives untouched. From our sustenance to our safety, from education access to employment, our health to our very means of livelihood, the climate crisis exerts its influence with unrelenting force. The most disheartening part of this crisis is the unequal burden it has on the most vulnerable communities, especially those with limited resources (Acevedo et al., 2017). These communities' very survival hinges on the natural resources that surround them. And as the climate crisis intensifies, these communities' little or no adaptive capacity means they continue to bear the brunt of its wrath.

Growing up as a farmer's daughter, I was initially unaware of how much climate change affected my daily life. I did not know why farming was not enough to provide for our family at times, and why we had to wake up at 4 am to travel miles to get water, which in most cases was not even clean enough for consumption. After the death of a family member caused by the dangerous fumes in their home, the grief I felt was profound and indescribable. I couldn't understand how a seemingly harmless household device, a small power generator, could silently become a lethal weapon, taking not just one life but potentially thousands. It was a stark and heartbreaking reminder of the invisible threats that climate change can bring into our lives. I took these questions with me to university during my undergraduate studies. This was when I was able to connect most of the issues I experienced growing up as a child, and it dawned on me that climate change was the invisible thread weaving through the various challenges my family and community faced.

When I shared my experiences, I discovered everyone around me was experiencing the same challenges, while many did not call it climate change or had any name for it. With a group of like-minded students, we started a climate advocacy program – an initiative we called Xicos. We started with about 7 students and eventually over 20 students joined the movement. We started going from one school to another, from one community to another, educating people and school pupils about climate issues, participating in community market clean-ups and working on university conservation expeditions. At this time our main focus was for people to be aware of climate change and how it affects them. And with this passion we started working with local communities. I was able to empathise with what they were going through on a daily basis, such as their lack of electricity, health challenges, and other social and economic issues. It brought back many childhood memories for me. I could relate to the feeling of not having clean drinking water or other issues they were facing.

When we started the initiative, the idea was to ensure people wouldn't have to experience the same climate impact we did, and by creating this awareness, communities would be able to live a better social life in harmony with nature. But after about a year of the program, I discovered it was not delivering that much impact. People were aware of climate change and what the resulting impacts would be, but they continued with their old way of living. This was when I understood that awareness alone is not enough to drive climate action. Especially in Africa and in a country like Nigeria, communities need alternative solutions to improve their lives. This changed my perspective about how to approach climate change. Yes, the climate is changing. However, the exact nature of the change and the exact experience for countries and regions vary greatly depending on the social and economic landscape. We may be on the same sea, but we are not in the same boat. I remembered a group of women asking me, "Joy, are you saying cooking with firewood is dangerous to the environment and general health, so how do we cook our food if you are advocating against firewood? How do we power our homes if we can't use hydrocarbon generators? This is the only way we get to cook and have electricity, we don't have access to the grid." This question broke my heart, and it felt like I was taking their livelihoods away in the name of climate advocacy. These questions made me and my team found NewDigit Technologies in order to be able to provide alternative energy solutions to households and communities.

This was the beginning of my transition from a climate activist to a social entrepreneur or climatepreneur (as I call myself). I am still an activist but with a different perspective. Now my goal is to provide solutions to climate and social issues. I realise the connection between energy poverty and most of the social issues that we face in Nigeria, from access to clean water to quality education, access to health facilities, security and safety of girls and children, to quality of life. The devastating impact of fossil-fuel generators on our communities, including the tragic loss of lives, and the global agenda of the hydrocarbon economy spurred us into action. If the world economy does not prioritise decarbonisation to fairly transform energy access and economy in Africa, we risk perpetuating existing inequalities (Bumpus and Liverman, 2008). Access to clean energy should be a priority for every country but especially for developing nations like mine. Energy access is not a luxury and should not be seen as a luxury in Africa. Our green hydrogen technology called "Just Add Water" is designed to be a game-changer in expanding access to reliable, affordable, and sustainable electricity in Africa. It is a compact, single-unit product that uses dirty water to generate electricity, cooking gas, and oxygen and clean drinking water as a by-product. We are on a mission to domesticate green hydrogen.

Entrepreneurship as Climate Activism

As one of the many climate activists in Nigeria, the focus for me is to provide solutions to the climate crisis at the local level. I ask questions such as: *what are the problems that climate change has caused in our communities? How does it impact people and exacerbate the existing poverty? How do we mitigate future crises?* According to the World Bank and the International Energy Agency (IEA) (World Bank, 2022), over 43% of Nigeria's population is living in energy poverty, that is about 90 million Nigerians who have no access to modern cooking means and electricity whatsoever. The remaining 57% who somehow can access energy either rely on individual hydrocarbon generators for electricity or the national grid, both of which face reliability issues, as the national grid supply is unstable and gets interrupted by frequent outages. The Nigerian government recently increased the energy tariff by about 200% (Vanguard, 2024), and the removal of subsidies means that people have to pay five times the initial price to fuel their generators. How will the poor survive when every economic activity in Nigeria revolves around oil? This will only exacerbate the existing poverty in the country and cause many to become jobless (Ozili and Obiora, 2023).

This is an important opportunity for green innovators and entrepreneurs. Although Nigeria is an oil-producing country, the energy divide has widened dramatically. One would think that as an oil producing nation, Nigeria should long have been fully electrified, but the reality shows the inequality of the hydrocarbon economy. The exploration and extraction of oil in the Niger Delta region of Nigeria have raised unemployment in many communities where these oils are being extracted, communities whose main occupations were farming and fishing The oil has polluted the water and killed many creatures. The smoke and soot in the air have caused diseases. Sometimes oil leaks have caused explosions and fires burning hectares of farmland and homes. Communities where oil is discovered and extracted face many more social and economic issues. Therefore, many activists from this region still have to focus on legal actions against big corporations and oil companies.

Beyond Awareness: An Entrepreneurial Approach to Climate Solutions

The relentless environmental challenges thrust upon communities in Nigeria have had an unexpected consequence – the birth of numerous green innovators. For those like me who have embarked on the path of climate entrepreneurship, the motivation goes far beyond mere business ventures. As a co-founder of NewDigit Technologies, I have had the privilege of witnessing first-hand the power of innovative solutions, and this transformative potential is the driving force of our organisation. One of the key motivations behind the founding of NewDigit Technologies is its commitment to addressing urgent challenges facing communities – energy poverty and climate change. For us, these challenges are not just abstract concepts; we have experienced the crippling effects of energy poverty first-hand and witnessed how it stifles economic growth, limits educational opportunities, and impacts Nigeria's healthcare system, employment opportunities, and overall economic activity. Tackling the interconnected issues of energy poverty and climate change requires moving beyond just raising awareness to delivering practical, localised, and scalable decarbonisation solutions. NewDigit's entrepreneurial approach combines technological innovations with dynamic business models, capacity building, and consistent community engagement to make this shift. Technology demonstrations to build trust can gain buy-in for green solutions.

Young innovators and climatepreneurs like myself are exploring the power of technology and innovation to address not only the issue of climate change or energy poverty in Nigeria but also to mitigate climate change by providing communities and homes with an alternative energy solution that will replace the hydrocarbon economy. Since we started in 2018, we have made a critical impact on communities in Nigeria. Our products are used to power a large number of homes and small businesses, including three community hospitals. In 2019 we launched the Light Up Nigeria Project, a campaign to electrify rural communities, especially those without access to the grid. With this project, we were able to power hundreds of homes and communities. The Light Up Nigeria Project spanned into the COVID-19 lockdown, where many people were locked down in their homes; the project helped many homes to have access to the world and information. 2023 has been one of the most exciting years for us, A partnership with King Abdullah University of Science and Technology (KAUST), one of the biggest technological and research institutes in Saudi Arabia, has helped to further the advancement and development of what we call "Just Add Water" energy technology that uses water and solar cells to provide gas for electricity and cooking. The core goal of this technology is to domesticate green hydrogen and create a portal for green energy solutions that will replace household fossil fuel generators. Our partnership with KAUST helps us to conduct research in their advanced facility to develop minimum viable products of the New-digit flagship line called Just Add Water and pilot it in local communities to ensure solutions are tailored for Nigerian contexts, providing electrification to these communities and bringing attention of governments and relevant stakeholders to the challenges of the communities. Outside of the innovation and energy technologies we provide for households and communities, one of our major initiatives is empowerment through green skills. We believe that if we prioritise climate action, the new generation of the workforce will be people who have the skills and acumen to build and develop solutions for a just transition. As a clean energy organisation in Nigeria, we are joining forces with other innovators and social organisations who are developing solutions and products to address the climate crisis and energy poverty in different parts of Nigeria and Africa as a whole.

Figure 31.1   Author and her team during energy deployment in a local community in Lagos, Nigeria.
Source: Author's own image.

Greater Focus on Community-based Adaptation

In response to the escalating effects of climate change, grassroots climate activists and organisations in the Global South have been emphasising community-based adaptation measures, equipping and empowering communities with resources and tools such as climate-resilient farming techniques, supplying seeds and seedlings for tree planting, water conservation, and the utilisation of clean energy sources for optimal agricultural farming in flood and drought regions. Besides empowering these communities involved with the skills necessary for them to build resilience and adapt to climate impacts, these types of activism also create a basis for these communities to see themselves as part of the solution to the climate crisis rather than as the victims, and this further creates a sense of climate solution-driven empowerment. Much of this type of activism is championed by Civil Society Organisations (CSOs) and Non-Governmental Organisations (NGOs).

Education and Awareness Outreach

It is important to acknowledge the crucial role of climate education and awareness in driving climate action. Many NGOs and activists in Nigeria have been at the forefront of sharing information to enhance public understanding about climate change. While climate awareness is crucial, personal experiences and the ongoing devastating impact of climate change on many communities show that education needs to be supplemented with tangible solutions that deliver socioeconomic impact at scale. This realisation catalysed my shift towards an entrepreneurial mindset focused on innovation and scaling decentralised climate solutions, which was what birthed Newdigit Technologies. NewDigit's approach is bridging this gap through education, innovation, and social impact. People should not only be aware but understand the need to change their lifestyle and embrace alternative solutions that both address their social needs and address climate change.

Climate education is critical for this to succeed, especially educational programs that incorporate skill training and empowerment. This understanding is shared today by many NGOs and climate activists in Nigeria who are working to incorporate initiatives that empower communities to drive and accelerate climate action and build resilience. Some activists work directly with educational agencies like the Nigerian Federal Ministry of Education, and the Ministry of Environment, especially the state-level ministries and parastatals, which are most accessible to many citizens. They lobby for policies that will include climate education in the school curriculum for primary and secondary education. Activists use different means, including workshops, public speaking, and social media initiatives, to increase public awareness and understanding of climate change, its impacts, and potential solutions. They hold educational activities in schools, educating and inspiring the next generation of leaders about climate change, but they also visit religious assemblies, as well as traditional leaders and local community organisations, to help them understand the concept and contribute to sustainability activities.

These activities are led by NGOs, CSOs, and individual activists. Organisations like The Global Environmental and Climate Conservation Initiative (GECCI), which is an independent global campaign organisation that aims to create leaders in environmental and climate conservation. GECCI operates in over 90 countries, with a membership of over 22,000 predominantly young individuals aged 18–45. The organisation's teams campaign for sustainable agriculture, biodiversity protection, and socially responsible farming. GECCI's operations are spread across different nations of the world across six regions: from Africa, Europe, West Asia, Asia Pacific to Latin America and the Caribbean, and North America, establishing partnerships with various governmental and non-governmental organisations worldwide, including the United Nations Information Center Nigeria, the African Union Great Green Wall Initiative, and the International Human Rights Commission. With a presence in all 36 states of Nigeria, the organisation has directors in each geopolitical zone and presidents representing each state, ensuring effective coordination and representation at the local level. Since its inception, GECCI has successfully executed over 200 high-impact projects worldwide, sensitising over 40,000 primary school pupils in Nigeria on environmental issues and establishing approximately 1,500 Environmental Conservation Clubs in selected primary and secondary schools in Nigeria.

The Recycling Scheme for Women and Youth Empowerment (RESWAYE) runs a buy-back program that empowers women and youth (above 16 years) through collection, aggregation, and recycling of plastic waste. This is a solution that aims at combating the menace of plastic pollution in coastal areas in Nigeria, as well as promoting financial independence among the participants. Their work helps communities to become more sustainable through waste management initiatives and advocacy, focused on recycling, upcycling, sustainable packaging, and litter reduction in the waterways. Waste products are used to make needed household materials, including school bags to help students get back to school.

The Limits of Grassroots Climate Activism in the Global South

Recognising the role of political-economic landscapes in shaping opportunities and challenges for grassroots activism, scholars have highlighted the impact of economic inequality and political instability on environmental activism in the Global South. The intricate connection between the political and economic realms assumes a pivotal role in shaping the opportunities and hurdles encountered by climate activism. A favourable political-economic landscape can furnish grassroots activists with the necessary resources and support to effectively pursue their activism. In nations where robust environmental laws and regulations are in place, in countries like Sweden, Japan, and other developed countries, grassroots activists find it comparatively easier to confront polluters and hold governments accountable for climate policies (Michanek and Zetterberg, 2005).

Sweden's stringent environmental regulations for example provide a legal framework that empowers activists to challenge and demand action from authorities through protest and legal means. With a strong tradition of civic engagement and a political climate conducive to environmental protection, this type of environment provides activists with public support for their initiatives. In countries where environmental laws and regulations are weak or susceptible to manipulation by powerful interests and elites who can twist laws and policies in their own interests, however, grassroots activists face formidable obstacles. In Nigeria, for instance, the oil industry holds strong economic and political power, and this hinders activists from being able to advocate for nature- and human-favourable environmental regulations and policies. The intricate connections between political elites and these powerful interests are also the main obstacle for the enforcement of environmental laws and policies, resulting in a culture of impunity for environmental violations. In many countries in the Global South, including Nigeria, the lack of access to financial resources poses a major challenge for grassroots activists who rely on donor funds for their activities.

Systemic Barriers to Environmental Justice

Limited access to legal and regulatory frameworks can also pose an additional obstacle for activists. In an inadequate environment that favours special interests on the one hand and pushes people to focus on their self-interest on the other, it becomes difficult for activists to advocate for a collective push for meaningful policy reforms. Moreover, marginalised communities and grassroots organisations often face barriers to accessing the legal system and participating in decision-making processes, further compounding the challenges they face in advocating for political change. Concrete examples are Indigenous communities in the Niger Delta region and rural communities in the north. These communities are mostly located in remote areas. They have very limited access to legal resources and also face language barriers due to lack of formal education. When these communities advocate for land rights or protection of their ancestral land against environmental degradation and oil pollution from big oil companies, they often lack the financial resources to hire legal representation to defend themselves and their communities. Hence, they struggle to navigate the legal complexities, leaving them at a significant disadvantage in the decision-making process. The corporations come to these communities presenting themselves as the messiah who will bring development to the communities and lift them out of poverty but are actually preying on the people's lack of education to degrade and pollute their land. Many have ended up deceived by the corporations through legal clauses that forced them to sell their land under value for oil extraction.

It is important to recognize that activists desiring policy change, particularly from politically and economically disadvantaged countries, often face challenges stemming from skills shortages and limited access to legal and regulatory frameworks. These factors can severely limit their ability to promote policy reform and shape legislative agendas. Effectively navigating the complex political landscape and understanding the intricacies of policymaking requires specialised skills that may not be available to them. Some of these activists may also not be able to understand the role of the legal framework. Though the devastating impact of the climate crisis on their livelihoods drives many to protest, without proper training and support, they often find it difficult to effectively articulate their concerns, engage with policymakers, advocate effectively for desired policy changes, or even get their voice and message out.

Power Imbalance and Colonialism

Cultural norms also play a big role in shaping the context for climate activism in Africa. Every community or society is shaped by unique cultural traditions and beliefs that can influence the way people engage in climate activism, especially the youth. A strict culture or tradition that discriminates based on gender, age, or ability can pose a problem driving change for certain age groups and genders. There are many communities in Nigeria where women are not allowed to own land or engage in public activities without the consent or permission of a male – a husband, father, brother or community leader. These gender expectations and societal norms can limit women and hinder their ability to fully participate in climate activism efforts, as they are forced to remain at home to take care of their families. These restrictions tend to limit grassroots engagement. When activists visit some of these communities, the goals are to empower the local residents to engage in environmental activities and build resilience against climate impact. Additionally, the colonial legacy has created persistent power imbalances that continue to impair African activists to mobilise and organise across borders.

Colonialism, which has fragmented African societies and created artificial borders, continues to be a hindrance to collective action against issues like climate change and socioeconomic and political crisis in the region. Many environmental problems in Africa, such as deforestation, land degradation, and resource depletion, have transnational dimensions that require collective action across national borders. However, this division and geopolitical complexities stemming from colonial legacies often hamper the ability of African activists to mobilise effectively and collectively address social issues imposed by climate change.

Barriers to African Climate Innovation

The persistent power imbalances created by colonial histories have not only fragmented African societies and hindered collective action, but they have also led to significant disparities in technological development, economic stability, and access to critical infrastructure. This creates an additional obstacle for climate activists, grassroots organisations, and green entrepreneurs working to deploy innovative solutions. One of the major challenges social enterprises and hardware start-ups, especially in the clean energy sector, face aside from access to financial resources has been accessing advanced technology infrastructure and facilities locally. In most cases, start-ups working on hardware are forced to outsource their technology prototype and manufacturing development abroad to a country like China to develop prototypes and designs. This drastically increases the price of these technologies and makes it more difficult for low-income communities to access them due to naira-to-dollar inflation and custom duties. This lack of domestic technological capacity not only impacts the price point and accessibility of sustainable solutions, but it also hampers skills development, technology transfer, and local job creation – crucial elements for building resilient, self-reliant communities.

For an energy company aiming to domesticate green hydrogen solutions as alternative energy solutions for communities, access to high-level research technological facilities is a major driver in deploying innovation. As a company, we were in this phase for years leading up to our recent partnership with King Abdullah University of Science and Technology (KAUST). Their key interest in green hydrogen and clean energy access gave us the platform to use their resources and laboratory to modify and develop our products to fit the market. With the help of their resources and knowledge, we were able to conduct intensive research and produce various versions of our innovation for commercialisation in Nigeria. This helped us conduct a live demonstration in one of the major vulnerable communities in Lagos. This type of partnership, resource sharing, and knowledge transfer also shows the importance of higher institutions of learning for contributing to helping the community in achieving a sustainable society.

Start-up acceleration and entrepreneurship programs have also been a major platform for social organisations to access funding opportunities and business resources, but many green entrepreneurs have not been able to access financial support as much as other business sectors in Nigeria, though there are government-established programs, the Nigeria Climate Innovation Center (NCIC), which has been resourceful supporting green innovators in the country – Newdigit was part of the first cohort. The program offers businesses and green innovators access to knowledge and expertise, access to a network of experts and mentors who can guide them in developing their green innovations, and business development support. Many Nigerians developing solutions to address climate issues desire access to institutions that can provide financial support or investment for green innovations and technology resources and facilities that can help accelerate the development and scaling of their ideas.

Global Solidarity for Grassroots Activism

Nigerian activists have found solidarity in the global movement for climate action, aligning with international initiatives like the Paris Agreement and many coalitions and initiatives of youth-led organisations and activists to help amplify our activism (Bulkeley et al., 2014). Youth climate movements like the African Youth Initiative on Climate Change (AYICC), the Loss and Damage Youth Coalition, and YOUNGO (an official UNFCCC youth constituency) are examples of global networks of climate activists that are very important, particularly for Africa. These coalitions have provided the platforms for many climate activists in Nigeria to network and engage like-minded activists from other parts of the world, sharing resources and opportunities. The youth networks have developed more formal structures and governance systems for effective mobilisation in recent years. For example, AYICC has national chapters in several African countries and hosts annual conferences to bring young people together to share knowledge and develop strategies, like the Local Conference of Youth (LCOY), which serves as the national youth conference proceeding before the Conference of Youth (COY) led and organised by YOUNGO, which usually starts a few days before the COP. Being part of this type of youth movement serves as a capacity-building opportunity for youth to understand and follow the UN and UNFCCC processes, equipping activists with the skills needed to contribute to climate policies and implement their projects.

Social media platforms like Twitter and Facebook have become digital soapboxes that help organisations, activists, and social entrepreneurs to amplify their voices and get their messages across to wider audiences. The use of hashtags, retweets, and shares has been instrumental for climate activists to turn a local issue into a global conversation and gather global support. Technological advancement plays an important role in facilitating real-time communication and collaboration across multiple regions.

YOUNGO and AYICC use social media platforms to connect with other youth-led climate organisations across the continent and beyond. This has enabled them to share knowledge, resources, and policies and has amplified their voices in the global climate movement. Events, programs, and policy development are communicated via email, capacity-building events are organised on virtual meeting platforms, while social media platforms like Twitter and Facebook are used to share campaigns and publicise events and share opportunities for young people. This global movement has helped many of us to follow and understand the international processes and policies around climate change.

Conclusions

Climate activism is very dynamic and varies depending on region, country, and local circumstances, and most importantly the social and economic state of every country depends on how they react or adapt to the climate crisis, and this also influences the activism of the region or locality. While many developed countries focus on reducing emissions and transitioning to sustainable energy sources, developing regions are more likely to focus on adaptation and building resilience by providing tools and resources to adapt to the immediate impacts of climate change on people's livelihoods. The IPPC 2020 report on regional climate-change vulnerability recognized the regional differences and experience of climate impacts and how climate change affects different natural resources in the world and the economic and social settings of different regions. These differences in the impacts of the climate crisis also mean that grassroots climate activism in different regions must reflect the political, social, economic, and environmental conditions (Dryzek et al., 2011). For a country like Nigeria where citizens struggle to make ends meet, either as a result of climate issues or other crises, climate activism is inherently dynamic, with grassroots movements adapting their approaches to reflect the specific social, economic, and environmental realities of their local contexts.

This chapter has examined the multifaceted nature of climate activism in Nigeria, spanning awareness-raising, legal advocacy, and entrepreneurial innovations. From educational campaigns that enhance climate literacy, to leveraging regulatory frameworks to hold polluters accountable, to the pioneering work of organisations like NewDigit Technologies in deploying community-centric sustainable solutions – these diverse modes of engagement are all vital components of the climate justice movement. As the co-founder of NewDigit Technologies, I see our work as embodying a multifaceted approach to climate activism that combines education, innovation, and grassroots empowerment. While traditional awareness-raising campaigns are crucial in enhancing climate literacy and galvanising community engagement, we recognize that knowledge alone is insufficient to catalyse the systemic changes required, and entrepreneurial approaches that address societal challenges imposed by either climate or economic crisis impacting local communities are the best solutions for addressing climate change in Africa. Africa is one of the regions most affected by the climate crisis and also a growing economy which has its social, economic, and political challenges. The approach of climate action in Africa will also determine the state of our planet in the next few decades. Africa must be empowered to grow its economy, justly transition, and have the capacity to meaningfully participate in the global economy.

Notes

1 See also https://climateknowledgeportal.worldbank.org/country/nigeria/climate-data-historical.
2 See for example https://www.carbonbrief.org/analysis-africas-extreme-weather-have-killed-at-least-15000-people-in-2023/.

References

Acevedo, S., Mrkaic, M., Pugacheva, E., Topalova, P. (2017). The Unequal Burden of Rising Temperatures: How Can Low-Income Countries Cope. IMF Blog, 27 December. https://www.imf.org/en/Blogs/Articles/2017/09/27/the-unequal-burden-of-rising-temperatures-how-can-low-income-countries-cope

African Youth Initiative on Climate Change. *About Us.* https://ayicc.org/about-us/

Aljazeera. (2022). *Oil-Rich Niger Delta Still 'Land of Misery': Goldman Prize Winner.* https://www.aljazeera.com/features/2022/5/25/oil-rich-niger-delta-now-land-of-misery-gold man-prize-winner

Amnesty International (2015). *Nigeria: A New Generation Fights for a Pollution-Free Future.*

Bulkeley, H., Andonova, L. Betsill, M. M., Compagnon, D., Hale, T., Hoffmann, M. J. and Rogers, A. (2014). Transnational climate governance. *Global Environmental Politics*, 14(4), 1–18.

Bumpus, A. G. and Liverman, D. M. (2008). Accumulation by decarbonization and the governance of carbon offsets. *Economic Geography*, 84(2), 127–155.

Daily Trust (2022). https://dailytrust.com/nigeria-lost-n3tr-during-2022-flood-fg/

Diao, X. et al (2007). The Role of Agriculture in Development. Implications for Sub-Saharan Africa. Research Report 153. International Food Policy Research Institute.

Dryzek, J. S., Norgaard, R. B. and Schlosberg, D. (2011). Climate justice and global governance: Introduction. In *The Oxford Handbook of Climate Change and Society* (pp. 21–34). Oxford: Oxford University Press.

Empowerment of environmental NGOs (2005). *Michanek, Gabriel, and Charlotta Zetterberg. "Den virtuella miljökampen."* (The Virtual Environmental Struggle). Statsvetenskapligtidskrift 107.1.

Gianotti, Z. (2018). A continuing struggle over oil. Markkula Center for Applied Ethic, Santa Clara University. https://www.scu.edu/environmental-ethics/environmental-activists-heroes-and-martyrs/ken-saro-wiwa.html

International Energy Agency (IEA) (2022). *Nigeria Energy Outlook.* https://www.iea.org/countries/nigeria

Intergovernmental Panel on Climate Change (IPCC) (2021). *Climate Change 2021: The Physical Science Basis. Contribution of Working Group I to the Sixth Assessment Report of the Intergovernmental Panel on Climate Change.* https://ww.ipcc.ch/report/ar6/wg1/

King Abdullah University of Science & Technology (KAUST) (2023). *Energizing Change: KAUST and NEwdigit Team Up to Support Energy Security in Developing Nations.* https://sustainability.kaust.edu.sa/news-center/start-up-newdigit-technologies-collaborating-at-kaust

Obaze, O. H. et al (2024). *Deconstructing Nigeria, Global Patriot Newspapers,* 04 September. https://globalpatriotnews.com/deconstructing-nigeria-by-oseloka-h-obaze/

Ozili, P. and Obiora, K. (2023). *Implications of Fuel Subsidy Removal on the Nigerian Economy. Public Policy's Role in Achieving Sustainable Development Goals.* https://papers.ssrn.com/sol3/papers.cfm?abstract_id=4535876

Vanguard (2024). *Electricity: Concerns as Tariff Hike Hits Consumers.* https://www.vanguardngr.com/2024/04/2m-consumers-affected-by-hike-in-electricity-tariff/

World Bank (2022). *Access to Electricity (% of Population) – Nigeria.* https://data.worldbank.org/indicator/EG.ELC.ACCS.ZS?locations=NG

32

GROUNDING THE GLOBAL CLIMATE MOVEMENT

Eco-socialist Militancy and Neo-rural Experimentations in Present Environmental Activism (Italy)

Elena Apostoli Cappello

Introduction

This study aims to investigate a case study rooted in a very specific and local declination of the movements against climate change from the perspective of political anthropology. I will focus on a regional dimension of a phenomenon that is actually continental and even reaches beyond Europe, exploring alternative lifestyles and social experimentation. I intend to show that, at least in these cases of local grassroots activism in Italy, instances of mobilisation against climate change are embodied in creative and cross-cutting ways in communities and actors whose visions we might in part ascribe to the classical categories of nostalgic aspirations (Herzfeld, 1997) and, simultaneously, to actors we can identify as eco-socialists (Apostoli Cappello, 2022). This family of rural field-interlocutors, new farmers claiming to act for environmental sustainability, move through the field in a basically harmonious movement, intercepting and embodying the ecological expectations (Lallement, 2019) of a variety of urban consumer-activists (Apostoli Cappello, 2019), promoters of an ecological movement (Carlsson and Manning, 2010) embodied primarily in the locavore concern. Those who populate it claim today to be operating on the front lines of a battle against climate change. Activists here believe that that of production, notably agribusiness, is a nodal field in which to experiment with alternatives to the capitalist production paradigm, itself held responsible for climate change.

The overall tension of this chapter is to argue for the assumption that a transnational movement such as the movement against climate change absorbs, engages, and coordinates local struggles whose specific purpose often in unexpected ways complements the claims of the "global" face of the movement, strategically exploiting its successes and watchwords.

I will describe a set of activist paths that I started to follow since 2006 in the framework of my long-term ethnographies of Italian antagonistic, eco-socialist cultures. In the last 15 years, I have carried out a number of ethno-anthropological studies of Italian social movements, especially eco-socialists. Based on the considerable amount of data collected in Italy and also in the Mexican Chiapas insurgent Zapatista movement

(Kaufman and Reyes, 2011; Baschet, 2020; Nash, 2001), I have observed a strong reconversion of activism and eco-socialist activism towards the rural as an object of concern. The purpose of this chapter is to trace a transformation in its processual dimension, describing the paths of activists who are, in different ways, part of the eco-socialist milieu, who find their *raison d'être* precisely in the search for and practice of alternative ways of living one's daily life and of projecting one's future. They are, globally, in search of alternative orientations toward ways of life that look at social and material resources in a non-extractive way.

This chapter positions itself within the anthropological debate on Utopias, (Wallerstein, 1998; Harvey, 2000; Graeber, 2004; Anderson, 2006; Khasnabish, 2008; Chodorkoff, 2014; Cooper, 2014; Krøijer, 2015; Hébert, 2016). In an age when it seems easier to imagine the end of the world than the collapse of capitalism (Jameson, 1994), the analysis of utopian imaginaries helps us to grasp contemporary social critiques and their associated social transformations (Maskens and Blanes, 2018, 2016; Piette and Torterat, 2016; Appadurai, 2013; Wallman, 1992; Schaer, 2000; Shukaïtis, 2007; Moore, 1990). By analysing the spatial and temporal dimensions of specific utopian imaginaries, I seek to understand both the creative potential (Cossette Trudel, 2010) and the social values and contradictions they transmit.

Methodology

As an ethnologist, I worked in the field observing activists over a long period of time, often participating first-hand in their daily activities, their militant trips to Zapatista Chiapas, and mobilisations on specific campaigns. I collected militant life stories, to analyse the changes of direction and the evolution of the political and existential priorities of my interlocutors. I have mostly worked by collecting little structural or informal conversations, in a perspective of hermeneutic circularity and constant and progressive co-construction of the interpretative framework with my interlocutors. Often, they themselves had strong reflexivity and common theoretical references with me, and this created very strong feedback loops and dynamics, which are characteristic of my research. I have always positioned myself explicitly as a researcher in these contexts, despite having the posture of a committed anthropologist, as Herzfeld describes it.

My analysis arises from criticism of the capitalist system, the identification of consumption as a field of political action finds roots in the practices, in the Italian radical left "antagonista" counter-culture, in the laboratories of trash-ware (re-elaboration of computer devices recovered in the trash) that I have studied inside the Occupied Social Centres, in Italian *Centri Sociali Occupati* (CSO), squats mainly devoted to cultural and political activities, mainly in urban and/or peri-urban sites) in different Italian cities already from the first years of the twenty-first century. This world has been mixed with that of the groups of collective solidarity purchase *Gruppi di Acquisto Solidale* (GAS). I will also consider a long-term period in which, in many different ways, I have followed the Italian antagonists and eco-socialists through ethnography.

My first fieldwork focused on the experience of Italian alter globalist and antagonist activists, which I observed in their respective national contexts and in their journeys to insurgent, peasant, and Indigenous Chiapas, in Southern Mexico. Then I observed forms of collective eco-socialist mobilisation in Italy and Europe (Apostoli Cappello, 2013, 2017a, 2017b, 2017c), the "for the commons" *momentum*, and their neo-rural character.

Mobilising for "the Commons" and a Non-extractive Relationship with the Environment

During the years 2010–2020, I followed the ecological side of the eco-socialist milieu and of the Italian radical left ethnographically in different cities of Italy. I conducted research among activists who had formed part of the alter globalist mobilisations in the 2000–2010 and who are today sometimes part of important associations and NGOs of the environmentalist world in Italy: *Legambiente, Greenpeace Italia, Terra! Altragricoltura*, and the movement for proximity agriculture in urban contexts *"Gruppi di Aquisto Solidale"* (GAS), solidarity purchasing groups, and *"Orti Urbani Occupati"* (squatted urban gardens), as well as in the less structured and very heterogeneous communities of neo-rural communities.

In particular, I conducted fieldwork during a big referendum campaign, in spring 2011. During that time, the mentioned associations, together with many representatives of Italian civil society, mobilised for the defence of the common property (i.e. municipal public institutions) of water, and against the plan of re-nuclearisation in Italy. The campaign watchword here was "the defence of the Commons" — *Acqua Bene Comune* (Dardot and Laval, 2014): which was a public appeal that highlights the overall process of privileged attention and intensive politicisation of everyday practices. Most of the observed activists translated their moral conceptions into practical commitment in the material sphere, while proposing alternative practices of relation with the capitalist market. The idea of a non-utilitarian and non-extractive relationship with the environment spread at this historical moment in the Italian movements described here. This idea will be the political basis of today's ecological claims.

I observed the mobilisations mainly from three vantage points:

1. activist groups of Italy's largest environmental NGO, *Legambiente*, with several branches also in all Italian regions and major cities. I participated in the campaign with this group from an engaged anthropology perspective (Herzfeld).
2. groups that referred to the Occupied Social Centres (CSO), squats of post-Marxist inspiration with branches throughout Italy. These have long been my main object of study, I followed them on militant trips to Chiapas of the Zapatista insurrection and documented how it was these trips that brought the environmental issue to the top of their agenda (Apostoli Cappello, 2017a, 2017b) and who today are mobilising primarily for environmental activism.
3. grassroots Catholic-inspired groups, which I observed from a distance.

This campaign, which originated in the city of Naples and within a few months brought together unorganised organisations and activists from all over Italy, was from a historical point of view the last great moment of collective mobilisation of ecological, anarchist, radical left, and Catholic movements. All these movements at the beginning of the 2000s had mobilised together, in an unprecedented way, under the watchword of "another world is possible," that is, under the aegis of the anti-capitalist anti-globalisation (Dalla Porta, 2006) movement that began with the "Battle of Seattle" in 1999 (Klein, 1999).

The one for "the Commons" was the last moment when all these movements moved together, first and foremost to openly reclaim issues of social justice and linking them explicitly to issues of environmental justice and combating climate change. At this moment, a political and existential shift in the mode of mobilisation took place in Italy. Subsequently, activists who had participated in it would mobilise on a more local scale, with at most a regional horizon of struggle and with primarily ecological demands. This, while maintaining a solid base of critique of the capitalist model, in which, as noted earlier, all the activists in question identify the culprit of the climate crisis and the social crises that follow.

Primitivist Utopia as a Vector for a New Political Imagination and Generational Justice: How the Environment Became a Central Political Issue

Through ethnographic data collected over a longer period, I explored the transnational circulation of political imaginaries of eco-socialist CSO activists between Italy and Mexico. In numerous urban and rural locations in Italy and in Chiapas, rural Zapatista communities in Mexico, I analysed the political imaginary and the circulation of practices associated with contemporary eco-socialist movements. My aim was to conduct an anthropological study of a historical aspect of radical left movements in contemporary Italy and their management of political violence. In this context, I studied the anthropological transformations in the so-called libertarian and antagonistic Italian movements. The term *antagonisti* refers to a contemporary, heterogeneous political subject, today eco-socialist and rooted in the so-called heretical Marxist movements of the 1970s. The biography of some militants includes a close experience with the armed struggle of the Red Brigades (*"Brigate Rosse"*) or the Workers' Autonomy (*"Autonomia Operaia"* organisation) described by Bianchi e Caminiti (2007) for example. Today, these activists use utopian registers from the Zapatista movement to distance themselves from older, violent practices and to present the current social conflict in non-violent terms. I argued that, more broadly, Zapatista inspiration and militant travel practices structure an entire primitivist utopia, ductile enough to provide unifying elements of coherence to a network of activists and political organisations that share a political imaginary and their own idea of democracy and its implication in the relationship with the environment.

Starting from 2010, my long-term interlocutors have converted their experiences of political participation by focusing on the environmental and notably food production sphere – in their view eminently political – of the creation of small militant peasant communities. We are witnessing the spread of a phenomenon in which actors are making significant existential investments by engaging in agriculture that is often self-certified, self-produced, and local, resistant to GMOs and the chemical industry – the reading of these choices being that we must get out of a capitalist system of exchange and production and towards a non-extractive relationship with "the land" (intended as a polysemic object, described in Apostoli Cappello, 2022) by questioning the ideal of economic growth.

It is from here, by observing these new producers and their match with critical and often urban consumers, that I observed that critical, supportive, and/or sustainable consumption is now a widespread place of construction and identification of the self as a responsible subject, which is part of a committed environmentalist political ecumene. The articulations of the devices linked to the consumption and to the "alternative", i.e. self-managed production seem to constitute therefore from this season of environmentalist mobilisations a fundamental framework to build a new shared ideological horizon, where the climatic changes become the new horizon of the political struggle.

It is possible to identify three important aspects accompanying this change, aspects whose genealogy I will attempt to trace here:

1. The tension toward a utopian communitarianism, imagined or practised, in which the dimensions are small enough to allow for collective self-management. This point is made

explicitly by activists, for most of whom politics is done at the territorial level, with a global "critical" community in mind.

2. My interlocutors constantly conjure up the imagery of a "return to the land." This shows, in my view, the search for a connection between generations that has arisen among people of my generation who were mostly born into urbanised nuclear families and who have never lived with their grandparents who often lived in the country. So they perceive the lack of an intergenerational transmission of these practices that are related to rural self-subsistence.

   I want to say that the new activists from the horizons of neo-ruralist practices express in the spatial dimension (imaginary of an "exit" from the city, rootedness in rural life) the need for temporal rootedness concerning the search to reinforce the relationship with their roots, imagined or not.

3. There is a deep and very concrete rationality in the orientation of these generations of activists toward agriculture: the occupational reason. In fact, I am talking about generations struggling to find a job (especially in Italy) and, consequently, a recognised place in the society in which they live. In this sense, converting one's existential investment into micro-enterprise activities embedded in exchange networks that allow, if not subsistence, at least a large economic aid to it, is an alternative choice of an occupational as well as a political nature.

The last two points are not, not always at least, explicitly claimed by my interlocutors as the main reasons for their actions. This is largely my own reading, based on a methodology that involves analysing the subjectification processes of people whose paths and existential reorientations I have observed in the course of long-term research. Especially outside the context of interviews, in informal exchanges and by participating in dozens of meetings in public and private settings, I could not avoid to give weight to discursive recursivities expressed *au passage* by these unemployed or underemployed young Italian converts to politically committed agriculture.

However, this is a terrain anchored in the Italian context, characterised since decades by problems of social mobility that considerably hinder the inter-generational changes in the country. It is a country with a very elderly ruling class, in which young people are considered young up to the age of 35, live for a long time at their parents' home and depend on them economically, and have little representation in society. In this sense, the paths that have led my activists to reconsider the "return to the countryside" correspond to a choice that responds to political questions that also involve generational injustice. This element strongly contributed to their agenda transformation, from a social justice priority to an environmental one, without forgetting the social dimension of climate change.

Libertarian and/or eco-socialist self-managed communities try to found living examples of alternative utopian development "at home," getting out or trying to get out of the economic system of capitalist production, while allowing a rooting in the territory and opening the possibility of rooting in the present, a precious possibility for a generation that, as I said, has difficulty in inserting itself in mechanisms of vertical social mobility.

Imagined Utopia and Concrete Utopias

Besides local action and precisely in this context, starting from the end of the Nineties CSO activists undertaking trips to "go to learn" about community cohesion in places that are supposed to be "naturally" democratic, that is, the rural and "primitive" environments of Chiapas, build into their narratives the exotic framing described by Edward Said in his concept of Orientalism (1978). Alongside the emancipatory dimension of their practices exists an attempt to establish an anchoring in time: political communities originated from a Marxist-heretical thought are today characterised by a libertarian primitivism (along the path that leads from Rousseau, 1775 and Thoreau to Zerzan) and an anti-capitalist pre-industrial romanticism that nostalgically (Herzfeld, 1997) opposes "nature" against "progress": the temporal dimension of this structural nostalgia politically grounds a whole political community. In this quest for anchoring, all my interlocutors repeatedly refer to the need to reconnect with past generations in a sometimes Edenic perspective (Apostoli Cappello, 2017b).

Through its circulation, over two decades (since 1999 with the Seattle Movement, recently discussed again by Adler, 2021), a primitivist utopia has brought about an environmental reorientation (Apostoli Cappello, 2017a, 2017b) through ideas anchored in militant "back to the land" practices and micro-practices throughout Europe. For example, today, every CSO, in Rome and beyond, organises a political event at least once a week focused on agriculture. This kind of practice has been widespread for years in many places such as the occupied social centres of Veneto and Emilia Romagna where I have conducted several ethnographies (Apostoli Cappello, 2013, 2017a, 2017b, 2019, 2022).

The activists I followed try to find meaning through spatially located actions, in particular through initiatory journeys and political pilgrimages (Apostoli Cappello, 2017b) to the source of their political inspiration (the southern Mexican region of Chiapas, where the Zapatista insurged in 1994). They seek rootedness in the land understood as territory but also claim that their time is a new, apocalyptic one, which is, the decay of an order (the individualist and capitalist order of the Western world). The search for concrete communitarian Utopias arises from this double imperative.

From a spatial point of view, the eco-socialist Utopia consists in lives and territories emancipated from the capitalistic logic. However, alongside the emancipatory dimension of their practices exists an attempt to establish an anchoring in time: political communities rooted in post-Marxist ideological origins are today characterised by a libertarian primitivism (along the path that leads from Rousseau, 1775, and Thoreau, 1854 to Zerzan, 1994) and an anti-capitalist pre-industrial romanticism that nostalgically (Herzfeld, 1997) opposes "nature" against "progress": the temporal dimension of this structural nostalgia politically grounds a whole political community. In this quest for anchoring, all my interlocutors repeatedly refer to the need to reconnect with past generations in a sometimes Edenic perspective (Apostoli Cappello, 2017b).

Activists moved in a constant search for new places, culturally built as inspiring Utopias, where they attempt to put down new roots. So, a new rural aspiration contains for activists the possibility of founding their own citizenship, rediscussing the parameters of "critical" political participation per se, by the small militant peasantry. In its entirety, the research I have done so far allows me to observe the red thread that leads from a traditional political situation in prevalence of protests and urban-industrial antagonism to the other, ruralist, less explicitly political.

As I said in the three points stated earlier, small groups coming from an urbanised and ultra-individualised society thus show a great nostalgia for peasant community life, taking up a subterranean flow that combines a political tradition of proletarian internationalism with movements of ecological inspiration, associated with exoticising projections in the search for small authentic communities that live in a supposedly virtuous symbiosis with nature. In this sense, the alterglobalisation movement has also covered a deep ecological matrix.

An Ethnographic Example

Some points that emerged from fieldwork I conducted recently in northern Italy may be helpful for understanding what the arguments outlined so far mean in concrete terms. I observed and followed in their daily work a group of young "new" farmers (under 35) from the industrial town of Mestre (Venice), whom I interviewed and followed in 2016. These young organic producers recently settled on the Venetian island of Sant Erasmo but were originally from the town of Mestre. They crossed paths some 10 years earlier, when they were studying at the university in Venice, through activism against the "MOSE" dike system ("No Mose" movement) and against the entry of large cruise ships into the city ("*No Grandi Navi*" movement). They regularly frequented the local antagonist squat (CSO *Rivolta*, Mestre). These militant producers (30–40 years old) are considered "young producers" by other native producers on the island. They have now become recognised organic farmers on the island of San Erasmo (by their colleagues but also in the city of Venice). My "new" farmer interlocutors are part of an openly anti-capitalist network inspired by Zapatista Chiapas, led by activists close to the CSO who have travelled and lived in Chiapas and who redefined their political and existential project on their return to Italy. For example, one of the two farmers interviewed, S., travelled to Chiapas in a period of difficult business management, to regain some enthusiasm.

As said, I ethnographically followed for a long time the eco-socialist network of pro-zapatista neo-ruralists in the Venice in-between-land and the local *Rivolta* CSO activist network, which also supported and travelled to insurgent Chiapas in previous years. Within this eco-socialist macro-group, structured around relationships of friendship and political sharing, my Sant Erasmo new peasant interlocutors have decided to set up a small militant agri-food business in the lagoon island. What makes this experiment militant is not only its political background, but the fact that activists see it as the most effective way to fight against the industrial agribusiness production system. This production system, in fact, is considered the prime agent of a capitalism that is destroying the environment in general and the climate balance. The loss of this balance is particularly evident in the Venice lagoon, a very fragile ecosystem into which river waters flow that are laden with chemicals from industrial agriculture in the Veneto hinterland. Here, the rising of the sea water is already ongoing, along with the dry summers that have been bending agricultural production for a few years now. Within this framework, then, the action of my interlocutors makes sense and is understood by them as militant.

Sometimes the rural horizon is understood as an exodus, an exit from a political space, especially a metropolitan one, regulated by productivist logic; therefore the only possible political action, intellectual and practical, is the mass defection from the system and rather the parallel construction of an alternative society. The activists I follow are driven by the propensity, sometimes through occupations of small pieces of land.

As for the CSO counter-cultures, there is here the aspiration towards the "Temporary Autonomous Zones" (as defined by an American anarcho-libertarian ideologue whose pseudonym is Hakim Bey, who coined the acronym T.A.Z in 1993), which are occupied and squatted, highly politicised (even antagonistic) social spaces that have been very important in Italian protest cultures since the 1980s (Balestrini and Moroni, 2005; Bravo, 2008;

Bracke, 2015; Ginsborg, 1998). In this kind of alternative space, at least ideally, people seek a primarily psychological liberation from the self-repressions they consider generated by the conditioning of the principles of authority and hierarchy and by the alienations produced by work. That recalls when Graeber (2004) highlighted that "all forms of systemic violence are (among other things) assaults on the role of the imagination as a political principle" and the role of imagination as "a kind of creative reservoir, too, of potential revolutionary change"(2004:31).

My interlocutors are divided into a network of very small cooperatives and unstructured activist farming experiments also present in the Venetian hinterland of Mestre. One of the aims of this network is, they say, "to share problems and feel less alone." J. continues, "we don't have any labels, we're trying to build a self-certification with other groups that have certain political characteristics." J. clarifies the political meaning of the operation:

> We are what we eat. This must be our political activity, to make people understand that we shouldn't expect strawberries or zucchini in December. That's what the food industry does, but what impact do these products have?
>
> *(J., SE, 10/16)*

The implicit nature of the certification that J. has in mind is made clear through the link to the grandparents' generation: "The rule is to only buy things whose ingredients would also be recognized by your grandmother." The symbolic link with previous generations and the validating power of this close past are also apparent in other anecdotes. S. says: "People in their eighties tell us they haven't had potatoes as good [as theirs] since they were children. And if a grandmother says so, you can trust her." This role of validating intergenerational relations has its seat in the plane of the symbolic, for a generation of activists who on the material level, on the other hand, believe that they suffer a set of injustices from previous generations. This ambiguous and seemingly contradictory relationship with past genera-tions – and with a pre-modern past in general – is at the heart of the meaning framework of the current ecological activists I have studied.

In a self-documentary about this organic enterprise, K., one of the many international woofers who work with the cooperative, explains:

> Young people are fed up with today's society, with the frenzy of the cities. They're running away. In agriculture, they find a slower, more human rhythm, far from com-puters and smartphones. We just want to go back to the past, when we didn't have all this. . . . We want to enjoy the land and not work like crazy.
>
> *(K. SE, 7/16)*

Woofing is a practice of voluntary work in agricultural exploitations where people find food and a change in work in the field. The objective of this international movement is to create a global community mainly oriented towards barter logics and sustainability.

Elsewhere (Apostoli Cappello, 2019) I have shown how unromantic and actually strenu-ous this kind of agricultural work, based in fact on forms of self-exploitation, is. This phenomenon finds its own internal explanation in the fact that many agro-activists are try-ing to escape the logic of capitalist comfort. However, we must consider that they talk to me about 15 hours of work a day as a stable work rhythm, and actually my interlocutors complain about it.

If we take a closer look at the reasons for the choice that K. mentions, an economic logic emerges, in which dis-continuity is expressed in terms of control over one's life. J. says: "Before, there was only economic insecurity," but he then adds, paradoxically, that what he misses most now is the continuity of the salary from his salaried job. J. has held various positions in the catering and cultural events sector in Venice, while S. has worked in the construction industry. Both complain of the marginality caused by the economic insecurity of the period. Today, they say they find it harder than before to meet their expenses. Nevertheless, they report a gain in terms of control over their quality of life, expressed in the pleasure they derive from enjoying the landscape and socialisation based on shared values within an agro-militant subculture.

They do not, however, neglect the economic aspect of their activity, J. introduces: "Through the rediscovery of the relationship with nature, we want to build something" and goes on to explain the choice to practise only direct sales: "it's to maximise earnings, it's our customers who support us and come and visit us in the garden . . . it's also a choice in terms of personal growth" (understood as development and subjectification).

The non-commercial relationship with the land is an important point in the case of J., S., and their comrades; it's about the material possibility of rediscussing the relationship to work, the environment, and the whole contemporaneity.

Conclusion

What I have tried to show is how environmental mobilisations in Italy today mix the participation of new generations with a diverse world of long-time activists. For the latter, at the heart of my analysis, the environmental issue has gained priority along the way and is the result of interactions among political cosmologies, imaginaries, and ideology of very different matrices.

A general observation is that the forms of management of an ideological orthodoxy typical of the forms of classic militancy did not include, in the narrowly political sphere, those aspects of the daily and personal life of the Foucauldian matrix and what the feminist movement (Apostoli Cappello, 2017c) has highlighted as places of powers and micro-powers. Instead, today it is exactly these daily-life spheres that are more significant for micro-political action, following my interlocutors' understanding.

The "return to the land" meets the search for something that responds to the social need for rootedness and control over one's existence, perceived as strongly hetero-controlled by a multiplicity of devices of the advanced capitalist system. Food and its production lie at the intersection of these two needs. Nutrition has already been highlighted as a place of power and micro-power.

Deeply aware of the biopower dimension (Foucault, 1976), today's activists are reflective subjects with ecological sensibilities as places of political contestation, the spaces and aspects ignored by the activists of yesteryear, that is, the direct and immediate implication between the dimension of politics and that of life understood in its strictly biological characterisation. The philosophical proposals of Foucauldian thought, simplified and vulgarised in the political rhetoric of the organic intellectuals of the heretical Marxist and eco-socialist movements during the 90s and 2000s, have assumed a notorious weight for their capacity to give shape to political convictions and in general to the cosmology of eco-socialist activists (Apostoli Cappello, 2017a, 2017b).

Finally, this militant return to the land can be read from the perspective of a resistance to modernity, expressed as a refusal of the fragmented and urbanised individual and as the search for a community dimension projected onto a sometimes idealised pre-industrial rural life, which is the ideal horizon of global struggles against climate change. But it is, at the same time, a strategy to be rooted in the Italian present and to respond to a generational injustice, which economically excludes young people while shifting the consequences of climate change onto them.

Activists who never imagined having a holistic worldview acquired it by travelling to Central America and ideologically hybridising with environmental NGOs and Catholic movements in Italy. Today, therefore, they mix aspirations for a productive paradigm shift not (only) for social justice reasons but also – and especially – for environmental reasons.

References

Adler, P. (2021). *No Globalization Without Representation: U.S. Activists and World Inequality.* University of Pennsylvania Press.

Anderson, B. (2006). "Transcending without transcendence": Utopianism and an ethos of hope. *Antipode*, 38(4), 691–710.

Appadurai, A. (2013). *The Future as Cultural Fact: Essays on the Global Condition.* London: Verso.

Apostoli Cappello, E. (2013). *Tutti Siamo Indigeni! Giochi di specchi tra Europa e Chiapas.* Padua: Cleup-Padua UP.

Apostoli Cappello, E. (2017a). Autochtonies contraires. Circulations d'idées et de pratiques politiques et de résistances indigènes transatlantiques. *Autrepart*, 2017/4(84), 177–195.

Apostoli Cappello, E. (2017b). Buissonnière et initiatique: la participation locale à travers des pèlerinages politiques. *Participation*, 2017/3(19), 73–95.

Apostoli Cappello, E. (2017c). Emotional logics in militancy: An ethnography of subjectivation processes in Italian radical movements. *Emotions, Space and Society*, 25, 136–143.

Apostoli Cappello, E. (2022). Polysémie, émancipation et nostalgie dans deux processus de politisation de la terre en italie du nord. *Ethnologie Francaise*, 52, 487–503.

Apostoli Cappello, E. (2019). Economies morales et politisation de l'agriculture dans des distributions à filière courte (Venise). *Social Anthropology*, 27, 671–688.

Balestrini, N. and Moroni, P. (2005). *L'Orda d'oro 1968–1977 – La grande ondata rivoluzionaria e creativa, politica ed esistenziale.* Milano: Feltrinelli.

Baschet, J. (2020). Autonomie et espaces libérés. A propos de l'expérience zapatiste et des mondes post-capitalistes en gestation. *Mouvements*, 1(101), 117–130.

Bianchi, S. and Caminiti, L. (2007). *Gli Autonomi, Vol.1: le teorie, le lotte, la storia.* Roma: DeriveApprodi.

Bracke, A. M. (2015). Our bodies, ourselves. The transnational connections of 1970s Italian and Roman Feminism. *Journal of Contemporary History*, 50(3), 560–580.

Bravo, A. (2008). *A colpi di cuore. Storie del Sessantotto.* Milano: Laterza.

Carlsson, C. and Manning, F. (2010). Nowtopia. Strategic exodus? *Antipode*, 42(4), 924–953.

Chodorkoff, D. (2014). *The Anthropology of Utopia.* Porsgrunn: New Compass Press.

Cooper, D. (2014). *Everyday Utopias. The Conceptual Life of Promising Spaces.* Durham, NC: Duke University Press.

Cossette-Trudel, M. A. (2010). La temporalité de l'utopie: entre création et réaction. *Temporalités*, [Online], 12 | 2010, Online since 15 December 2010, connection on 31 December 2020. https://journals.openedition.org/temporalites/1346

Dalla Porta, D. (2006). *The Global Justice Movement: Cross-National and Transnational Perspectives.* Boulder, CO: Paradigm Publishers.

Dardot, P. and Laval, C. (2014). *Commun.* Paris: La Découverte.

Foucault, M. (1976). *La volonté de savoir.* Paris: Gallimard.

Foucault, M. (1971). *L'ordre du discours.* Paris: Gallimard.

Ginsborg, P. (1998). *Storia d'Italia 1943–1996. Famiglia, società, Stato.* Torino: Einaudi.

Graeber, D. (2004). *Fragments for an Anarchist Anthropology.* Chicago: Prickly Paradigm Press.

Harvey, D. (2000). *Spaces of Hope.* Berkeley: California University Press.

Hébert, M. (2016). Words not yet in begin. Reconciling anthropology and utopianism. *Anthropology and Materialism*, 3, 1–19.

Herzfeld, M. (1997). *Cultural Intimacy. Social Politics in the Nation-State.* New York: Routledge.

Jameson, F. (1994). *The Seeds of Time.* New York: Columbia University Press.

Kaufman, M. and Reyes, A. (2011). Sovereignty, indigeneity, territory: Zapatista autonomy and the new practices of decolonization. *South Atlantic Quarterly,* 110(2), 505–525.

Khasnabish, A. (2008). *Zapatismo Beyond Borders. New Imaginations of Political Possibilities.* Toronto: Toronto University Press.

Klein, N. (1999). *No Logo.* London: Picador.

Krøijer, S. (2015). *Figurations of the Future. Forms and Temporalities of Left Radical Politics in Northern Europe.* Oxford: Berghahn.

Lallement, M. (2019). *Un désir d'égalité. Vivre et travailler dans des communautés utopiques.* Paris: Seuil.

Maskens, M. and Blanes, R. (eds.) (2016). Introduction [SI] Ethnography and the mutualizing Utopia. *Journal of the Anthropological Society of Oxford Online,* New Series, 8(2).

Maskens, M. and Blanes, R. (eds.) (2018). *Utopian Encounters. Anthropologies of Empirical Utopias.* Limerick: Peter Lang.

Moore, H. L. (1990). Vision of the good life. *The Cambridge Journal of Anthropology,* 14(3), 13–33.

Nash, J. (2001). *Mayan Visions: The Quest for Autonomy in an Age of Globalization.* New York. Routledge.

Piette, A. and Torterat, G. (2016). A hominist manifesto: Anthropology, utopia and ethics. *Journal of the Anthropological Society of Oxford-online,* 2016, in *Ethnography and the Mutualizing Utopia,* VIII (2), 230–250.

Rousseau, J. J. (1996/1755). *Discours sur l'origine et les fondements de l'inégalité parmi les hommes.* Paris: Le Livre de Poche.

Said, E. (1978). *Orientalism.* New York and London: Verso.

Shear, B. and Burke, B. (2013). Anthropology of and for Non-Capitalism. *Anthropology News,* communityeconomies.org

Shukaitis, S., Graeber D. and Biddle, E. (2007). *Constituent Imagination: Militant Investigations // Collective Theorization.* Oakland: AK Press.

Thoreau, H. D. (1921/1854). *Walden ou La vie dans les bois.* Paris: Gallimard.

Wallerstein, I. (1998). *Utopistics or, Historical Choices of the Twenty-First Century.* New York: Free Press.

Wallman, S. (1992). *Contemporary Future: Perspectives from Social Anthropology.* London: Routledge.

Zerzan, J. (2001/1994). *Futuro Primitivo.* Turin: Nautilus.

33

PROMISE MOTIVATION

Films that Encourage Grassroots Climate Activism

Sabine von Mering[1]

Mary Mazzio's documentary film *Bad River* (2024) chronicles the ongoing struggle of Indigenous activists of the Bad River Band of Lake Superior Chippewa against Canadian oil giant Enbridge and its Line 5 pipeline. Author and climate activist Bill McKibben highlights the film in his substack newsletter as an example of how the powerful climate defense of frontline activists like the Bad River tribal members should inspire a broad coalition of resistance similar to what was built against the Keystone XL pipeline and what is currently being organised against LNG infrastructure (see Chapter 30 in this handbook). The film, McKibben writes, can serve as "a powerful chronicle of some of the saddest chapters in American history, and a hopeful picture of the emerging possibilities for power in the crucial fights of our time" (McKibben, 2024). Although Mazzio does pull together that "hopeful picture," she does not ignore the community's struggle with the PTSD of centuries of colonial violence and exploitation that continue to this day with the opioid crisis and insufficient access to basic services, among many other challenges.[2]

By highlighting frontline communities' struggles, *Bad River* joins films like Briar March's *There Once Was an Island – Te Henua e Nnoho* (2010), Rasmus Avike and Lars Jeremiassen's *Thule, Tuvalu* (2014), Julia Dahr and Kisilu Musya's *Thank You for the Rain* (2017), Chiwetel Ejiofor's *The Boy Who Harnessed the Wind* (2019), Alejandro Loayza Grisi's *Utama* (2022), and Rich Felgate's *Finite: The Climate of Change* (2022) in drawing attention to the people mobilising for action through describing the impact of the climate crisis and fossil fuel infrastructure in their communities and on the ground. Other films chronicle the youth climate movement – from Slater Jewell-Kemker's *Youth Unstoppable* (2018) to Nathan Grossman's *I am Greta* (2020), Jim Rakete's *Now* (2020) and Phie Ambo's *70/30* (2021), or Jacqueline Van Meygaarden, Antia Khanna, and Rehad Desai's *Temperature Rising* (2023).[3] There are also many films dedicated to mobilising action on specific issues like the Dutch documentary *Meat the Truth* (2007) or *Cowspiracy* (2014) and *Seaspiracy* (2021) devoted to advocating against animal agriculture (partly) as a way to address the climate crisis and the need to adopt a plant-based diet – a recommendation supported by Project Drawdown, which lists "Reduce Food Waste" and "Plant-Based Diet" as the top two most cost-effective measures to address the climate crisis[4] (see also Chapter 29 in this handbook).

DOI: 10.4324/9781003396567-39

Showing what is already being done by people on the ground is one of the ways film can help spur grassroots climate action. This should not be a task left only to independent filmmakers, neither should it solely be an issue for documentary film, as Amanda Shendruk wrote in her OpEd for the *Washington Post*, "Audiences Want a Different Climate Message. Hollywood Should Deliver" (Shendruk, 2021).[5] In the following, I take a closer look at a number of films that set out specifically to mobilise viewers for climate action by highlighting solutions. After introducing a simple method for assessing how a film uses the promise of improvements to motivate viewers to take action on climate, I begin my analysis with the French documentary film *Demain* (2015), followed by a brief look at the Australian documentary *2040* (2020), the American documentary *Bidder 70* (2012), and the American feature film *How to Blow Up a Pipeline* (2023).

Climate Films and Climate Activism

Documentary films play an important role in the climate movement. From Al Gore's *An Inconvenient Truth* (2006), to the British documentary/drama *Age of Stupid* (2009), to Leonardo DiCaprio's *Before the Flood* (2016), they succeed at keeping viewers angry, informed, involved, and motivated. They help activists recruit and convince others and to sharpen their own arguments. They help nurture and grow the climate activist community and demonstrate the importance of storytelling for climate activism.

In contrast to *Demain*, the majority of documentary films about the climate crisis – and even more so the mostly dystopian feature films about climate, from *Day After Tomorrow* (2004) to the 2023 TV series *Extrapolations* – scare people into action by expounding upon the risks of inaction that would lead to climate chaos, with numerous climbing CO_2 and temperature graphs, footage of storms, floods, wildfires, and the resulting suffering of people and animals. I call this strategy *risk motivation*. However, some say what is needed more than the overwhelming images of gloom and doom is concrete examples of what is already being done that could be modelled and serve as a source of hope, inspiration, and encouragement, in other words a focus on the many benefits of climate action. This idea was perhaps best captured humorously by Joel Pett in his 2009 "Climate Hoax" cartoon.[6] I call this *promise motivation*. Indeed, a better world should be of interest to everyone, and *Demain* clearly prioritises *promise motivation*. *An Inconvenient Truth*, *Age of Stupid*, and *Before the Flood* include both *risk* and *promise motivation*. I was interested in seeing what the balance was. With the help of a timer, I determined how much screen time was devoted to presenting risks versus promises (see Table 33.1). Images, footage, and voice-overs that showed or mentioned greenhouse gas pollution, extreme weather events, scientific information about rising temperatures, and also climate denial and lobbyist activity against climate protection I counted as *risk motivation*. Images, footage, and voice-overs showing urban gardening, renewable energy, community engagement for climate protection, and so on I counted as *promise motivation*.[7]

As Table 33.1 shows, all the three most well-known films mentioned earlier include only a minimal amount of *promise motivation*, usually in the form of to-do lists attached to the end, and at least in some cases these lists include questionable recommendations of neoliberal consumer behavior more akin to greenwashing than a true transformation.

To be sure, research is also not conclusive as to whether positive messaging is indeed better able to motivate action. For one thing, bad news seems more likely to trigger a response

Table 33.1 Risk motivation and promise motivation in well-known climate documentaries.

	An Inconvenient Truth	*Age of Stupid*	*Before the Flood*
Year made	2006 (US)	2009 (UK)	2016 (US)
Director	Al Gore	Franny Armstrong	Leonardo DiCaprio
Imdb	https://www.imdb.com/title/tt0497116	https://www.imdb.com/title/tt1300563	https://www.imdb.com/title/tt5929776
Length of actual film	1 hour 36 minutes	1 hour 26 minutes	1 hour 30 minutes
Promise Motivation	7 minutes (=7%)	6 minutes (=6%)	9 minutes (=10%)
Risk Motivation	79 minutes (=82%)	80 minutes (=93%)	77.5 minutes (=86%)

Source: Original creation.

than good. As Roy Greenslade wrote in the *Guardian* in 2007, summarising a recent Pew research survey, "the regular calls for papers to publish 'good news' rather than bad is largely a waste of time. People are stimulated to read by the latter. They want to know what has gone wrong rather than what has gone right" (Greenslade, 2007). Neuropsychologists call this "negativity bias" and suggest that it is embedded in our human psyche – our survival instinct requires us to know about risks, and hence we pay more attention to negative information. (Moore, 2019). Similarly, at least one recent pedagogical study claims that the stick is indeed more effective than the carrot: in an interview, scholar Jan Kubanek explained that "negative feedback may be more effective than positive feedback at modifying behavior. . . . From an evolutionary perspective, people tend to avoid punishments or dangerous situations. Rewards, on the other hand, have less of a life-threatening impact" (Remerowski, 2015).

That said, a 2016 study published in *Nature: Climate Change*, led by Paul G. Bain, found that communicating co-benefits of climate action did have a positive impact on motivating people: "communicating co-benefits could motivate action on climate change where traditional approaches have stalled" (Bain, 2016). It may also be too simplistic to equate positive messaging with positive behavior change. Ann Kaplan, who discusses Michael Madsen's documentary *Into Eternity: A Film for the Future* (2010) and Edward Burtynsky's *Manufactured Landscapes* (2008) in her book about climate trauma, argues, for example, that understanding risk is a precondition for developing hope: "to have hope for change entails recognizing first that things are wrong." (Kaplan, 2016, pp. 118–119).

One reason for the lack of robust bipartisan political action is fossil-fuel interests still actively spreading doubt about climate science, as explained in the powerful 2014 documentary film *Merchants of Doubt*, based on Harvard historian Naomi Oreskes' research into the matter – a must-see for anyone who wishes to understand why more progress has not been made to date.[8] But in terms of promoting climate action, that is not the full story. Ashley Bieniek-Tobasco and her colleagues conducted qualitative interviews with viewers of season two of the documentary television series *Years of Living Dangerously*, to "evaluate the effects of a mass media climate change program on audiences' efficacy beliefs, outcome expectations, emotional responses, and motivations and intentions to address climate change" (Bieniek-Tobasco et al., 2019, p. 1) The researchers were optimistic:

Existing research demonstrates that film has a unique potential to promote both individual and collective climate action through a combination of imagery, narrative storytelling, and celebrity messengers in both cognitive and emotional appeals (Manzo, 2017). Studies of films such as *An Inconvenient Truth, Gasland,* and *The Age of Stupid* demonstrate this potential . . . although longer term impacts are unclear (Howell, 2014).

(Ibid, p. 2)

However, when looking to assess efficacy rather than potential, the researchers discovered that people did not necessarily feel motivated to take action after watching episodes of *Years of Living Dangerously,* since they did not feel they could make enough of a difference. No matter whether the film employed celebrities or ordinary people, scary graphs or innovative renewable energy projects, interviewees said it too rarely showed that climate action would lead to success. Bieniek-Tobasco et al. conclude that "low efficacy combined with high levels of concern or risk perceptions . . . can lead people to reduce fear through coping mechanisms as opposed to taking action to reduce the threat." We must, they write, "emphasize the importance of balancing threat with efficacy and hope in communicating about climate change and in efforts to stimulate participation in climate action" (Ibid, p. 4).

The French documentary film *Demain* [*Tomorrow*], directed by Cyril Dion and Mélanie Laurent, is one film that balances threat with efficacy and hope especially well – though I will discuss later that the concept of "success" employed by Bieniek-Tobasco et al. may need further reflection.

Demain *(2015)*[9]

The film *Demain* opens with a black screen and an ominously thundering humming sound. Credits in thin white lines fade in and out as we hear a series of five intercut male and female news anchor voices interrupted by crackling radio signals announcing that a new study predicts "the potential demise of the human race" as a consequence of mass extinctions. The humming sound gives way to a piano imitating the interval of a French emergency vehicle's siren. A woman's voice from off camera recounts how, when her friend shared the news about that study with her, she realised that she could be leaving her newborn child in a world of devastation, which motivated her to take action. The black screen gives way to a sunrise over a cityscape and then cuts to a sequence of brief scenes from mass climate protests as Cyril, the friend, chimes in: "I had never read such a terrifying report. Yet only a few thousand of us were demonstrating in the street, and politicians weren't reacting." The piano begins to play chords, and we are introduced to a silent recording of a group of the filmmaker's friends sitting around a table engaged in lively discussion. Still from off camera, the woman's voice explains: "We weren't green freaks or activists, but most of us had kids, and none of us could just stand by after hearing this terrifying news. We all felt we had to do something." Thus motivated, the group decide to put their filmmaking skills to the service of climate action. Together they embark on a journey in search of solutions to the climate crisis. The journey begins with a plane trip to Northern California and an interview with the Stanford biologist Liz Hadley and her husband, the UC Berkeley palaeontologist Tony Barnosky, the two scientists who had coordinated the aforementioned study published in *Nature* magazine in 2012 (Sanders,

2015). Already in the opening scenes, the filmmakers thus model the behavior they aim to inspire in their viewers:

1. educate yourselves about climate science!
2. talk about it!
3. join up with others!
4. seek out solutions!
5. get active!

Demain also employs a number of "winning" methods of documentary filmmaking to achieve its goals, beginning with celebrity spokespeople reminiscent of Al Gore in *An Inconvenient Truth* (2006) and Leonardo DiCaprio in *Before the Flood* (2016). After all, the two voices in *Demain* are not just anybody's voices: the woman is the celebrated French actor Mélanie Laurent, who has become well-known abroad with roles in feature films such as Quentin Tarantino's *Inglourious Basterds* (2009) and Louis Leterrier's *Now You See Me* (2013), and more recently *Le Bal des Folles* (2021). Laurent co-wrote, co-directed, and co-produced the film with Cyril Dion. Dion is a renowned French author, poet, and activist, whose 2022 film *Animal* won the Green Award at the Trento Film Festival.[10] Other familiar methods employed by the filmmakers are the inclusion of expert interviews, travel to different locations to underscore the global reach of the problem (carrying with it the same problematic "baggage" of multiple shots of high-emitting plane rides that Gore and DiCaprio were criticised for), and repeated visual reminders of the human responsibility for/consequences of climate change with shots of traffic, polluting power plants, asphalted streets, poverty, etc.

However, the filmmakers also made an unusual choice. Anticipating that viewers might be "fed up with catastrophes," the scene and sound shift to a more positive outlook less than six minutes into the movie. And as the soft voice of the female singer tells us "we knew that we could save us, but never really tried," we instead begin to see that, no, many people have actually been trying, and they have great ideas for a better future. This is the core message of the second interview with Rob Hopkins, the founder of the Transition Town network in Totnes, UK, a growing movement of active communities founded in 2005 with the goal of working towards a "low-carbon, socially just future with resilient communities" that has since spread around the globe.[11] Countering the prevalent message of gloom and doom that dominates the news about a climate-changed future (and most films about climate change), Hopkins has the best line in the film when he shouts with a mixture of exasperation and glee, "It could be fantastic!" The way the film immediately cuts from this to the song "What a Wonderful World" is symptomatic of its core message – there *is* good news behind all the bad. This is a film that wants to encourage, not scare its viewers into action. The filmmakers use Hopkins' encouragement as their marching orders to find more people like him, so that they "can inspire people to create a different world" – a goal that gets re-stated repeatedly throughout the film.

The film accomplishes this goal by providing a glimpse of solutions to the climate crisis from a host of different angles. Its five "chapters" address five areas where action is already being taken: agriculture, energy, economy, democracy, and education. To bring the action to the screen, the team travels through Europe and across to the United States and also to India to show how individual activists are making a difference. In each location they interview experts and ordinary people who have made efforts to address climate change. With its lush greens and bright colours, in combination with the stunning music composed specifically for the film by the Swedish singer/songwriter Fredrika Kahle, *Demain* is a cinematic feast. It has an effect on the viewer similar to that of a nature walk (Tidball and Krasny, 2014) – it creates a sense of healing and hope that garnered great praise when it first came out in France in 2015. According to the French newspaper *Libération*, many in France were moved by the film: "Cyril Dion achieved in one hour and 58 minutes what decades of environmental activism did not achieve: to lay the foundation for a new collective fiction" (Lisarelli, 2017).[12]

Demain was made during the years after the financial crisis, and the filmmakers provide us with voices of those who suffered through it – local gardeners, people who are unemployed, small-business owners. This film was not intended for the proverbial choir. And yet it is also clearly a project of the political left, deriding fossil capitalism. Indeed, the film presents many of the key features of a degrowth or post-growth economy – such as urban gardening, local currencies, a change in bank lending away from debt, and an educational system moving away from testing towards a holistic approach – albeit without ever using the term degrowth (for an excellent introduction to the concept of degrowth see Schmelzer et al., 2022).

A closer look at a few more scenes will illustrate how the film employs cinematic techniques to encourage action. Chapter 1, "Agriculture," begins with a visit to Detroit, MI, where, we are told, the population had dropped from 2 million to 700,000 since 1960. The camera first shows dilapidated buildings along a vast terrain of formerly inhabited streets. But then it zooms in on a central urban garden project that literally lights up the city center with greenery, and we meet the people who took it upon themselves to start growing food "by Detroiters for Detroiters." The message is clear: anyone can do what these people have done. All you have to do is pick up a shovel and get active, and all will benefit.

The scenes from the urban gardeners in Detroit are intercut with scenes from Todmorden, north of Manchester in the United Kingdom, where Pam Warhurst and Mary Clear of the local Incredible Edible homegrown food initiative[13] explain how they started their version of "guerrilla gardening": "I've heard people talk about it for a long time, but I've never heard anybody do anything about it, really, not ordinary folks, you know?" Pam says. And Mary adds, "We didn't start with 'Shall we save the planet?' 'cause that was too grand. We just started with where we are." Cheerfully, Mary quips, "We've created a reason to have a conversation over a plant. How crazy is that?" The local activists may have been motivated initially by their concern about the climate crisis. But what keeps them going is the many benefits they derived from deciding to do something about it. Not only homegrown food but new skills, new friends, new pride about one's accomplishments, and a new sense of community, all of which in turn have made the community more resilient. This models for the viewer how change can be affected and that it may require employing a combination of tactics that may include civil disobedience. The positive outcome allows the Mayor of Todmorden to comment with a shrug and a smile when asked how he reacted to vegetables being grown outside the police station, guerrilla style: "it took the dialogue that we had."

The film alternates the focus on community building with interviews with scientific experts that help show how local actions connect to the global climate crisis. One example is the interview with Nick Green, a PhD candidate in biochemistry, a specialist in agroecology, and Todmorden's head farmer. Green explains – while viewers are treated to shots of Todmordeners cheerfully growing vegetables together – how a few years after starting Incredible Edible[14] they decided to use the opportunity to train young people, and now they produce half a ton of produce a year from about 300 square meters. He goes on to contextualise their work:

> the bulk of the world's food comes from tiny farmers, and they . . . produce most of the world's food, and the big industrial farmers produce a tiny fraction. In terms of production, the industrial farmers are hopelessly inefficient. Where they're good is they produce money.

From this interview with Todmorden's community garden visible in the background, the film cuts to an interview with Olivier de Schütter, UN Special Envoy on the Right to Food, and his confirmation of Green's assessment: "70–75% of what we consume in the world comes from small farmers."

The combination of grassroots activism and scientific expertise, of community building and connecting the local to the global, and of beautiful shots of nature and a compelling musical track is how *Demain* brings its point across. The positive outcomes of climate action, the film shows, go far beyond decarbonisation. The true benefits, it turns out, address lots of other problems that have plagued us for decades – from the decline of rural communities to people's growing isolation, to lack of access to healthy foods, and so on. The authenticity of the local voices in this film is what lends it credibility – a key feature of a documentary film done right. But the film also benefits from the beauty of its cinematography, in combination with its core message of building resilience for a better tomorrow.

The most visually arresting scenes introduce the viewer to a permaculture project in Normandy, France, where farmers Charles-Hervé Gruyer and his wife Perrine show how "working on a tiny piece of land working by hand you can produce what a tractor can on 10 times more land." The couple aimed to work with as little fossil fuel as possible, reminding us that "Right now each calorie of food we eat requires 10–12% of fossil fuels to produce." Perrine describes what makes her love her work: "Planting your own food gives you a boost – I can survive!" and Charles explains, "You can grow over $100 of vegetables in one square yard." As the camera zooms in on the colourful display of plump fruits and vegetables and the lush greens of the forest that the two have grown from scratch, the film seems to make its point even more through its powerful visuals: This is how it could be, the images seem to say. This is doable and desirable. Why would we not want to try this?

Of 116 minutes, 96 minutes of *Demain* are devoted to hands-on solutions, many of which can easily be replicated; just 20 minutes (or 17%) of the film are devoted to portraying the risk of climate change. In other words, 82% of the film is devoted to *promise motivation*. (See Table 33.2)

Cyril Dion directed a sequel to *Demain*, called *Après-Demain* [*After Tomorrow*], in 2018. The 71-minute documentary chronicles people who implemented projects after seeing *Demain*, thus providing proof that, with its emphasis on *promise motivation, Demain* did indeed achieve its goal of activating people.

There are actions portrayed in the film that everyone can take individually – from joining a community garden to starting a transition town neighbourhood. Others serve to inspire innovative city planners or politicians. In Copenhagen, Denmark, Morton Kabell, the Mayor of Environmental Administration, shows how the city is working towards becoming the first carbon-neutral capital in the world. In Iceland and the French island La Reunion, people model how we can reduce energy consumption by 60–65%. The film also features

Table 33.2 Risk motivation and promise motivation in *Demain*.

	Demain/Tomorrow. Take concrete steps to a sustainable future
Year made	2015 (France)
Director	Cyril Dion, Mélanie Laurent
Imdb	https://www.imdb.com/title/tt4449576/
Website	https://www.tomorrow-documentary.com/
Length	116 minutes
Promise Motivation	96 minutes (=82%)
Risk Motivation	20 minutes (=17%)
Original Language	French

Source: Original creation.

the world-renowned Danish architect Jan Gehl, one of the brains behind many of the decarbonisation efforts in the city of Copenhagen, who reminds viewers,

> If we build more roads, we will have more traffic! If you create space for people to walk, more people will walk. We are building cities in ways that are very unhealthy – where people sit in cars. And it is cheap to make walking and biking infrastructure. For social inclusion it is important to meet face to face so we are not afraid of each other.

In the wake of the COVID-19 pandemic, there is of course renewed awareness of these matters.

The film's final chapter on education at first does not appear to deal with the climate crisis at all. The focus is on how we can best prepare young people for a different tomorrow. The filmmakers visit a school in Helsinki, Finland, where the main goal is to teach children independence. The content itself is less important in the curriculum than training the students to figure things out on their own. The decision to include a chapter on education in the film may have been influenced by Rob Hopkins as well. In addition to having co-founded the Transition Town Network, Hopkins's most recent book, *From What Is to What If,* based on interviews with over 100 climate activists, artists, and thinkers, is all about the importance of the power of the imagination for climate action, and of course imagining a better future is what this film is all about. Education is a key to this goal, Hopkins says, because "As we know from the field of positive psychology, imagining a certain outcome can increase the likelihood of its coming to pass" (Hopkins, 2019, p. 11). And crucial for that to happen, he says, is time: "we don't give imagination enough time" (*Ibid,* p. 13) He feels we must give children more free time to play again, that we must reduce or get rid of standardised testing and bring back more free, unstructured time, playfulness, and improvisation, because "imagination is the only thing we have that is – or could be – radical enough to get us through, provided it is accompanied, of course, by bravery, and by action" (*Ibid,* p. 14) and that "we need to be able to imagine positive, feasible, delightful versions of the future before we can create them"(*Ibid,* see also Chapter 12 and other Chapters in Part II of this handbook).

Dion, Laurent, and their friends were no doubt intrigued by the Finnish school's focus on teaching children to practice their problem-solving. As documentary filmmakers, they are storytellers first and foremost, and addressing the climate crisis, they seem to say, will require many more problem-solvers who become imaginative storytellers with a vision for a better future. The film concludes with a series of split screens that present initiatives from around the world while Laurent concludes,

> we can promise [our children] that there are solutions, that thousands of women and men are taking a stand every day. And if we give it everything we've got, if we all join our forces and hearts we can all start to change the world . . . tomorrow.

The focus on the children brings us full circle back to the beginning of the film. The split screen that keeps splitting to reveal additional screens provides proof that the small number of positive examples shown in *Demain* represent only a fraction of the total worldwide, providing one last and powerful instance of *promise motivation*. Indeed, listing initiatives for people to join has become standard in climate documentaries, many of which incorporate specific action campaigns. This is also true for films that minimised *promise motivation* on screen. Al Gore's *Climate Reality Project* training, for example, has trained tens of thousands of climate activists since *An Inconvenient Truth* first came out.[15]

2040 *(2020)*

The Australian filmmaker Damon Gameau's *2040* has a lot of similarities to *Demain*. Gameau, too, begins by saying he was motivated to make the film by the arrival of his daughter, Velvet. He, too, goes on a journey to many destinations around the world to showcase people who are already implementing solutions. He, too, zooms in on a number of aspects, from energy, transport, food, and oceans to girls' education. He, too, intersperses his travels with expert interviews, including Tony Seba from RethinkX, Paul Hawken from the Drawdown Project, economist Kate Raworth, and marine permaculture entrepreneur Brian von Herzen, among others. What distinguishes Gameau's approach from that of Dion/Laurent's is that he not only begins each segment with close-up shots of children from different parts of the world expressing their wishes for the future, but he also includes brief fictionalised sequences of that imagined future of 2040 to demonstrate what it might look like "if we just embraced the best that already exists." He calls it an "exercise in fact-based dreaming."

Like *Demain*, *2040* combines the depiction of degrowth ideas with shots of innovations such as solar-powered microgrids (in Bangladesh), marine permaculture, autonomous vehicles, and regenerative agriculture, ending with a segment on educating girls.[16] At the end of his trip Gameau comes to a conclusion about the possibility of addressing climate change that is pure *promise motivation*: "we have everything we need right now to make it happen" (see Table 33.3).

Demain and *2040* are based on the premise that decarbonisation (in conjunction with lifestyle changes) is worth striving for, that a transition to a green economy is still possible if only enough people adopt this mindset and get to work, and that people are more likely going to be motivated by the visualisation of the promise of what can be done than by viewing examples of the risks they are facing.

Table 33.3 Risk motivation and promise motivation in 2040.

	2040
Year made	2020
Director	Damon Gameau
Imdb	https://www.imdb.com/title/tt7150512/
Website	https://whatsyour2040.com/available-courses/ with curricula specifically for the US.
Length of actual film	91 minutes
Promise Motivation	68 minutes (=74%)
Risk Motivation	23 minutes (=25%)
Original Language	English (Australia)

Source: Original creation.

In the following I briefly discuss two films that suggest that these steps are not sufficient. Like other social movements, the climate movement has access to a large toolbox of tactics and strategies, from educational initiatives and general petitions to direct action and civil disobedience.[17] *Demain* and *2040* do not include scenes of disobedience. In films that are sympathetic, civil disobedience can also function as a form of *promise motivation*.[18]

Bidder 70 (2012)

The award-winning documentary film *Bidder 70*, produced and directed by Beth and George Gage, follows climate activist Tim DeChristopher as he poses as a wealthy bidder at a public oil and gas lease auction in Utah in order to prevent the sale of large swaths of public lands adjacent to a pristine conservation area for fossil fuel extraction. The film chronicles Tim's journey through the court system – he eventually ends up spending 21 months in federal prison.[19] Throughout, Tim expresses the hope that his example will empower others to take action. I can attest to its success. At the end of one screening of the film at a local church, for example, many in the audience lifted their fists in solidarity. Fellow activist Dillon Hase from the group Peaceful Uprising summarises its impact thus: "Tim's action made me realize that one person can make a difference, and this voice went off in my head – you're powerful. You can change the world – and why aren't you?"

The film presents participation in civil disobedience as a healthy way to confront the cognitive dissonance of living at a time threatened by climate chaos. In that sense, the visualisation of civil disobedience is a marker of *promise motivation* in this film, an example of how people can engage in radical climate protection. As Tim DeChristopher says, "It's really healing and peaceful to be engaged in that." I know many climate activists who were inspired by Tim to risk arrest as part of their climate activism, including myself.

Risk motivation is not visualised in this film through footage of extreme weather events and climate disasters, but one could argue that the portrayal of the enormous power of the fossil fuel companies that profit from polluting our environment and destroying the atmosphere – and the fact that our legal system protects them rather than those who take huge personal risks to hold them accountable – could equal *risk motivation*. *Promise motivation* in this film is visualised in the activists' campaign that poses a threat to that status quo.

The film focuses mainly on accompanying Tim and his team through the drawn-out legal process, which stretches over years, long after the arrest.

How to Blow Up a Pipeline *(2023)*

Up to now, the large majority of climate activists have continued to espouse nonviolence.[20] The increasing criminalisation of climate activists even in countries that accept the scientific consensus on climate change and pride themselves on upholding the value of individual freedom and free speech is therefore extremely worrisome. In several Global South countries, peaceful nonviolent environmental activists have paid with their lives over the decades, such as Ken Saro-Wiwa in Nigeria or Berta Cáceres in Honduras. The killing by police of Manuel "Tortuguita" Teran during protests against "Cop City" in Atlanta in January 2023 – and the raids against climate activists of the "Last Generation" movement in Germany in May of 2023 – send chilling messages to activists, as authorities are criminalising those who aim to protect the climate instead of going aggressively after the criminals who are destroying it. (see also chapter 23 in this handbook). As the window of time to decarbonise closes and the blatant disregard for human life by those who continue to emit for profit gets exposed, more climate activists will be moved to embrace direct action, including civil disobedience, and there will no doubt also be more aggressive acts against fossil fuel and other emitting infrastructure in the future.

Harvard political scientist Erica Chenoweth and Lund University scholar Andreas Malm highlight the two sides of the debate about nonviolence: whereas Chenoweth found that, historically, nonviolent movements have had the best chances of succeeding (Ranalli, 2019), Malm, in his provocatively titled *How to Blow Up a Pipeline: Learning to Fight in a World on Fire*, argues that these movements (such as the suffragettes, for example) succeeded only when their objectives were at least seconded by a more radical wing that threatened – and in some cases applied – the use of violence against objects, if not against people.

Malm's book inspired Daniel Goldhaber and his team to make the feature film *How to Blow Up a Pipeline* (2023). The film begins with an opening shot of the back of a young, hooded person holding a knife while walking down a street and then bending down to let the air out of the tires of an SUV. The person leaves a yellow flyer on the car's dashboard, of which the viewer is able to read only the title: Why I Sabotaged Your Property.[21] Though the film credits state twice that it is "based on" Malm's book – and there is a brief scene in the film where one of the protagonists actually reads the book – it is of course rather a dramatisation of Malm's theories. As the character Logan says when he sees Shawn hold Malm's book, "That's got some good shit. Doesn't teach you how to do it, though." However, Goldhaber has clearly read the book as well, as he shows great sensitivity to the core question of when/why violent acts of sabotage that involve property damage may be justified in times of climate emergency.

The film dramatises to great effect how a seemingly random collection of eight young people, all victimised in some way by fossil fuel companies (there is mention of cancer clusters, land grab through eminent domain, toxic pollution, etc.) conspire to successfully blow up two sections of a Texas oil pipeline in the hope that this will disrupt the global oil market and reveal the vulnerability of fossil-fuel infrastructure, thereby moving authorities to finally act on climate. The *risk motivation* of climate disasters is given only seconds of airtime in this film – in the form of a brief diegetic radio report about the flooding in Pakistan, background shots of power plants polluting the air, or direct references the young people make as to their motivation. As to *promise motivation* – unlike in the case of Tim DeChristopher, the young people in this film do embrace using violence against the pipeline in full awareness of the risks. Shawn, for example, reflects on the possibility that they "could end up killing somebody or causing an ecological disaster" and Alisha initially rejects the idea of participating in what they all seem to agree is "a form of terrorism," saying "This flashy shit is pure ego." But the film also shows the incredible lengths to which the group goes in order not to harm other people – while taking great personal risks and ending up with one person shot in the arm, one with a broken leg, and two in prison. One could argue that the fact that the group not only succeeds but also ends up sending a powerful message of unity carries a lot of *promise motivation*, but clearly this will depend on the viewer's attitude towards sabotage as a form of civil disobedience.

At first the young activists speak very little, and the camera is not intent on making us sympathise with them throughout the first half of the film. Indeed, at this point, the viewer is made to believe that they do not really know or like each other all that much. However, this impression changes dramatically over the course of the film, as the plot is interrupted by flashback scenes that fill in the backstory for each character, one after the other, culminating in the realisation that two of the girls had agreed from the start to give themselves up to the authorities, while the others would go free, as one of them had been forced to collaborate with the police to avoid a long prison sentence for participating in a previous civil disobedience action. In retrospect, what appeared to be the characters' lack of empathy turns out to be an expression of the deep commitment the eight have towards their collective goal of protecting human life on planet Earth and working together in solidarity.

By zooming in on young people, the majority of them people of colour, many from poor families, the film gives voice to the most vulnerable populations that have contributed the least and face the biggest threats from climate disaster, raising the viewer's empathy and in turn the chance that we accept the group's actions as an example of *promise motivation*.[22]

Conclusion

It would be foolish to assume that all we need to activate people is the right kind of climate films with the right amount of *promise motivation*. Indeed, viewers who are disappointed that there is a lack of efficacy presented to them in climate films will eventually face a rough awakening. Our societies will change by design or by disaster, and once certain tipping points are reached, there will be almost only *risk motivation* left. For sure, we will need more climate documentaries about climate adaptation and resilience to teach us how to survive climate chaos.

Still, when we demonstrate against fossil fuel infrastructure projects in the various climate activist groups I am a part of, we chant, "We are unstoppable. Another world is possible." In view of the rapid increase of extreme weather events we are witnessing all around us, this chant may sound naive or at least falsely optimistic. But we are not in denial; we are what Fridays for Future's Luisa Neubauer and Alexander Repenning call "possibilists" (Neubauer and Repenning, 2023). Our chant is also not a prophecy. It is an expression of commitment to do what must be done. More people need to be inspired to join in, and we need all kinds of climate films to help move this conversation forward – a conversation that is still not happening enough, as polls by Yale University have shown: in 2021 only 35% of Americans responded that they "discuss global warming at least occasionally" (Marlon et al., 2023).

We need the storytelling of good climate films that jolt people out of apathy, complacency, or even climate anxiety with *risk motivation* but also the kind that provide *promise motivation* that goes beyond greening our economy and instead focuses on modelling how we build the solidarity that is needed to make big changes fast. The climate movement is a mass social movement, yet films tend to focus on a single hero. *Day After Tomorrow, Don't Look Up, Extrapolation,* even Benedikt Erlingsson's hilarious Icelandic comedy *Woman At War* (2018) certainly do. Although even the films discussed in this chapter have elements of that, what unites the films presented here is that they show that activists find *promise motivation* in the activism itself – regardless of efficacy – in the experience of a shared purpose, in becoming part of a committed community based in empathy and solidarity. And that, as Margaret Mead reminds us, is what actually makes change happen.[23] In her Op-Ed, Shendruk quotes Anna Jane Joyner of Good Energy, a Hollywood climate consultancy devoted to producing different kinds of films about climate change: "A really important part of the role that television and film play is making us realise that we're not alone, which is a step towards action." The films discussed in this chapter highlight this message repeatedly and return to it in the final scenes. In *How to Blow Up a Pipeline* (2022) it is in the "reveal" that members of the group did not, after all, betray each other. In *2040* the emphasis is throughout on the power of community and our responsibility for the (imagined) future. In *Demain* it is hidden in the ever-growing kaleidoscope of images of thousands of activists across the globe that closes the film. In *Bidder 70* it is in a quotation from Tim DeChristopher's sentencing statement:

> At this point of unimaginable threats, this is what hope looks like. In these times of a morally bankrupt government that has sold out its principles, this is what patriotism looks like. With countless lives on the line, this is what love looks like. And it will only grow.

Climate activists, the films suggest, are not primarily concerned with efficacy, because that would lead to constant frustration in light of the overwhelming threat of climate change. Instead, the films, and the activists portrayed, model how we can support each other and accept being "in it for the long run," no matter the outcome. As Wen Stephenson put it, what we are fighting for now is each other (Stephenson, 2015). These films provide the *promise motivation* that there is no limit to what such shared commitment can potentially accomplish – even against all odds.

Notes

1 An earlier and more expansive version of this article appeared under the title "Promise Motivation: Good News About Climate Change" in *The Canadian Journal of Film Studies* 32:2, 2023, pp.35–60. I am grateful to University of Toronto Press for granting permission to reprint material from that article in this volume.
2 That climate activism itself is dangerous for members of frontline communities was brought home when Indigenous filmmaker and activist Myron Dewey, 49, who co-directed the documentary *Awake: A Dream from Standing Rock* (2017) about the Indigenous-led protests against the North Dakota Access pipeline (see also Chapter 2 in this handbook), was killed in 2021 by a head-on crash that the judge ruled was a case of "vehicular manslaughter" – not to mention the travesty of justice expressed in the 30-day jail sentence for the killer. (Thompson, 2023).
3 There are also plenty of short films on the subject of climate activism like *She Changes Climate* – https://www.youtube.com/watch?v=ydbRQYYJbOY (2022) about the role of gender equality.

4 https://drawdown.org/.

5 "Only 2.8% of the 37,453 scripted films and TV episodes released in the US between 2016 and 2020 mentioned climate terms like solar energy, fracking, sea level, or renewable energy, and less than 0.6% specifically mention 'climate change.'" (.Johnson, 2024, p. 200).

6 See https://granthaminstitute.com/2018/10/30/6-things-we-learned-from-the-authors-of-the-1-5c-report/.

7 In some cases, there were also brief scenes (biographical information about the protagonists for example) that provided neither *risk* nor *promise motivation,* which is why the total percentage may not add up to the full length of the film.

8 Amy Taubin summarises the argument in her review of *Merchants of Doubt*: "Kenner relies on Oreskes' and Conway's research to demonstrate that the techniques and even some of the cast of characters employed by the climate change deniers are the same as those used for 50 years by the tobacco industry to cast doubt on what secret in-house studies by cigarette companies had already proved in the 50's – that cigarettes are carcinogenic, and that nicotine is addictive" (Film/Comment May/April 2015, 51, 2, ProQuest, 69). Given that Marc Morano, James Taylor, and the like have not stopped promoting climate denial, Taubin ends with the recommendation "Kenner should consider a sequel." Unfortunately, the organisations behind the spread of climate denial, though often based in the United States (Milman and Harvey, 2019), have been taking their messaging around the world, as evidenced by the rise of right-wing parties with platforms of climate denial in Europe, Australia, and South America.

9 The film opened in North America in March 2017.

10 He also wrote a best-selling book about ecological resistance, *Petit Manuel de Resistance Contemporaine,* in 2018.

11 See https://transitionnetwork.org/.

12 Susan Hayward wrote that *Demain* was a box office hit that generated $11 million. (Hayward, 2021). It won the 2016 César Award for Best Documentary Film and was viewed by more than a million in theaters and shown in over 30 countries. The film now has a US website: https://www.tomorrow-documentary.com/.

13 https://www.incredible-edible-todmorden.co.uk/.

14 Incredible Edible was formed "to create kind, confident, and connected communities through the power of food" (https://www.incredibleedible.org.uk/).

15 See https://www.climaterealityproject.org/.

16 Paul Hawken's Project Drawdown has long maintained that educating girls (in conjunction with family planning) is one of the five most cost-effective climate actions we can take, for the simple reason that there are over 68 million girls in the world who do not currently receive an education, which makes them more likely to become mothers at an early age, whereas educated girls postpone having children, have fewer children, and will also see to educating their own daughters.

17 Gene Sharp provided a list of 198 methods in his seminal work *The Politics of Nonviolent Action* (Boston: Porter Sargent, 1973). Luisa Neubauer and Alexander Repenning include a slightly updated list of 34 methods specifically useful for climate activists in their chapter entitled "Organize!" in *Beginning to End the Climate Crisis: A History of Our Future* (2023).

18 The geographer Kate Manzo's criteria for judging the "usefulness" of documentaries are *teachability* and *integrity*. Useful films, Manzo writes, are "educative, truthful and trustworthy, in ways not always intended by filmmakers." Kate Manzo, "The Usefulness of Climate Change Films," *Geoforum*, No. 84 (August 2017), 88–94. I have found *Demain* to be a very teachable film. My students have welcomed the film for being upbeat and for offering practical solutions.

19 https://www.bidder70film.com/.

20 Only a few, such as the brave water protector and Catholic worker Jessica Reznicek, have chosen to expand the repertoire. Reznicek was recently sentenced to 8 years in federal prison for publicly admitting to a series of attacks on the dangerous Dakota Access Pipeline, and – thanks to the Biden administration – was charged as a terrorist, as was her partner, Ruby Montoya, sentenced to 6 years in prison, also as a terrorist – although the women did not harm anybody and had turned themselves in to the authorities. A 2021 study showed that delaying pipelines can also be counted as a win: the many years of nonviolently protesting Line 3 prevented billions of tons of CO_2 emissions. (Flatow, 2021).

21 While the film uses a knife for effect, the activists who employed this form of sabotage in Sweden and other countries actually only let the air out, which did not ruin the tires in the process.

22 Young people often hear they should simply "go vote," but of course their generation faces a double whammy: they are going to bear the consequences of our inaction, and in most countries in the Global North they will not have a majority at the ballot box for decades. To tell them to use their vote instead is therefore pure cynicism. Only if older generations begin to support them (for example by joining Bill McKibben's Third Act – https://thirdact.org/) do they have a chance.

23 Margaret Mead is said to have said (the quotation could not be verified, but is widely attributed to her): "never doubt that a small group of thoughtful, committed citizens can change the world. Indeed, it's the only thing that ever has."

References

Films

70/30 (2021). dir. Phie Ambo.
2040 (2019). dir. Damon Gameau.
The Age of Stupid (2009). dir. Franny Armstrong.
Awake: A Dream from Standing Rock (2017). dir. Myron Dewey, Josh Fox, James Spione.
Bad River (2024). dir. Mary Mazzio.
Before the Flood (2016). dir. Leonardo DiCaprio.
Bidder 70 (2012). dir. Beth and George Gage. https://www.bidder70film.com/.
The Boy Who Harnessed the Wind (2019). dir. Chiwetel Ejiofor's.
Climate Warriors (2017). dir. Carl-A. Fechner and Nicolai Niemann.
Cowspiracy (2014). dir. Kip Andersen and Keegan Kuhn.
Demain (2015). dir. Cyril Dion and Mélanie Laurent, https://www.tomorrow-documentary.com/.
Extrapolations, TV Series, created by Scott Z. Burns.
Finite: The Climate of Change (2022). dir. Rich Felgate.
How to Blow Up a Pipeline (2022). dir. Daniel Goldhaber.
I Am Greta (2020). dir. Nathan Grossman.
An Inconvenient Truth (2006). dir. Al Gore.
Meat the Truth (2007). dir. Karen Zoeters, Gertjan Zwanikken.
Now (2020). dir. Jim Rakete.
Seaspiracy (2021). dir. Ali Tabrizi.
She Changes Climate (2022). dir.Natalie Hodgkins. https://www.youtube.com/watch?v=ydbRQYYJbOY.
Temperature Rising (2023). dir. Jacqueline Van Meygaarden, Antia Khanna, and Rehad Desai.
Thank You for the Rain (2017). dir. Julia Dahr and Kisilu Musya.
There Once Was an Island – Te Henua e Nnoho (2010). dir. Briar March.
Thule, Tuvalu (2014). dir. Rasmus Avike and Lars Jeremiassen.
Utama (2022). dir. and Alejandro Loayza Grisi.
Woman At War (2018). dir. Benedikt Erklingsson.
Years of Living Dangerously. https://www.natgeotv.com/me/years-of-living-dangerously/about.
Youth Unstoppable (2018). Slater Jewell-Kemker.

Secondary Sources

Bain, P., et al. (2016). Co-benefits of addressing climate change can motivate Action around the world. *Nature: Climate Change*, 1(2), 154–157. http://doi.org/10.1038/nclimate2814.
Bieniek-Tobasco, A., et al. (2019). Communicating climate change through documentary film: Imagery, emotion, and efficacy. *Climate Change*, 154(2019), 1–18.
Dion, C. (2018). *Petit Manuel de Resistance Contemporaine*. Arles: ACTES SUD.
Flatow, I. (2021). Indigenous activists helped save almost a billion tons of carbon per year. *NPR Science Friday*, 15 October. https://www.sciencefriday.com/segments/indigenous-protests-fossil-fuel/#segment-transcript [Accessed 6 May 2024].
Greenslade, R. (2007). The good news about bad news – It sells. *The Guardian*, 4 September. https://www.theguardian.com/media/greenslade/2007/sep/04/thegoodnewsaboutbadnewsi.
Hayward, S. (2021). *Ecology Documentaries: Their Function and Value Seen Through the Lens of Doughnut Economics*. London: Routledge.

Hopkins, R. (2019). *From What Is to What If: Unleashing the Power of Imagination to Create the Future We Want*. London: Chelsea Green Publishing.

Howell, R. A. (2014). Investigating the long-term impacts of climate change communications on individuals' attitudes and behavior. *Environment and Behavior*, 46(1), 70–101. https://doi.org/10.1177/0013916512452428

Johnson, A.E. (2024) *What If We Get It Right? Visions of Climate Futures*. New York: One World.

Kaplan, A. (2016). Getting real: Traumatic climate documentaries into eternity and manufactured landscapes. In Ann Kaplan (ed.) *Climate Trauma: Foreseeing the Future in Dystopian Film and Fiction* (pp. 118–119). New Brunswick: Rutgers University Press.

Lisarelli, D. (2017). Le Monde Après Demain. *Liberation*, 12 April. https://www.liberation.fr/debats/2017/04/12/le-monde-apres-demain_1562297

Malm, A. (2019). *How to Blow Up a Pipeline: Learning to Fight in a World on Fire*. London and New York: Verso.

Manzo, K. (2017). The usefulness of climate change films. *Geoforum*. https://doi.org/10.1016/j.geoforum.2017.06.006

Marlon, J., et al. (2023). *Yale Public Opinion Maps*. https://climatecommunication.yale.edu/visualizations-data/ycom-us/ [Accessed 6 May 2024].

McKibben, B. (2024). *Good Trouble on the Bad River*. https://billmckibben.substack.com/p/good-trouble-on-the-bad-river [Accessed 7 May 2024].

Milman, O. and Harvey, F. (2019). US is hotbed of climate change denial. *The Guardian*, 8 May. https://www.theguardian.com/environment/2019/may/07/us-hotbed-climate-change-denial-international-poll.

Moore, C. (2019). What is negativity bias? *Positive Psychology*, 30 December. https://positivepsychology.com/3-steps-negativity-bias/

Neubauer, L. and Repenning, A. (2023). *Beginning to End the Climate Crisis. A History of Our Future*. Waltham: Brandeis University Press.

Ranalli, R. (2019). Erica Chenoweth illuminates the value of nonviolent resistance in societal conflicts. *Harvard Kennedy School*. https://www.hks.harvard.edu/faculty-research/policy-topics/advocacy-social-movements/paths-resistance-erica-chenoweths-research

Remerowski, G. (2015). Carrot or stick? Punishments may guide behavior more effectively than rewards. *Science Daily*, 6 May. https://www.sciencedaily.com/releases/2015/05/150506120525.htm

Sanders, R. (2015). Dire forecast inspires upbeat film about solutions for a warmer tomorrow. *University of California*, 3 December. https://www.universityofcalifornia.edu/news/dire-climate-forecast-inspires-upbeat-film-about-solutions-warmer-tomorrow

Schmelzer, M., et al. (2022). *The Future Is Degrowth: A Guide to a World Beyond Capitalism*. London and New York: Verso.

Sharp, G. (1973). *The Politics of Nonviolent Action*. Boston: Porter Sargent.

Shendruk, A. (2021). Audiences want a different climate message. Hollywood should deliver. *The Washington Post*, 17 May. https://www.washingtonpost.com/opinions/interactive/2023/climate-change-disaster-movies-narrative/ [Accessed 24 April 2024].

Stephenson, W. (2015). *What We're Fighting for Now Is Each Other: Dispatches from the Frontlines of Climate Justice*. Boston: Beacon Press.

Thompson, D. (2023), Nevada man sentenced to 30 days in jail for fatal car accident that killed Paiute filmmaker Myron Dewey. *Native News Online*, 8 June. https://nativenewsonline.net/currents/nevada-man-sentenced-to-30-days-in-jail-for-fatal-car-accident-that-killed-paiute-filmmaker-myron-dewey [Accessed 23 March 2024].

Tidball, K. and Krasny, M. E. (2014). *Greening in the Red Zone: Disaster, Resilience, and Community Greening*. Dordrecht: Springer.

Websites:

350.org. https://350.org.

Climate Reality Project. https://www.climaterealityproject.org/.

Incredible Edible. https://www.incredible-edible-todmorden.co.uk/.

Project Drawdown. https://drawdown.org/.

Third Act. https://thirdact.org/.

Transition Town Network. https://transitionnetwork.org/.

34

INSIDER CLIMATE ACTIVISM – SLOWING DOWN OR SPEEDING UP TO DECARBONIZE?

Annika Skoglund

Introduction

Activism and collective political action can be understood in relation to different "movements," such as social movements, feminist movements, online movements and slow movements. These movements are often spoken about as consisting of a group of people whose collectivism is located in civil society or the public sphere (Dixon, 2014). This sphere, which even has been conceptually expanded to a "global civil society," has thus been historically constructed to function as a contained place where from citizens could – or even should – raise their voice against an external force or enemy (Ehrenberg, 2017, p. 195). In stark contrast to this view, however, this chapter sets out to explore activist movements in relation to speed, with focus on two polarized temporalities in climate activism – the Slow movement and the Fast movement. By drawing on previous ethnographic research, these two contrasting movements will be fictionally polarized in the case of employee activism (activism in commercial organizations) (Skoglund and Böhm, 2020) and insider activism (activism in public organizations) (Hysing and Olsson, 2018).

There is a reason for turning to speed when seeking to understand activist movements that emerge through commercial and public organizations. Hoofd (2012) traces the historical confluence of early activist beliefs in transformative pace and developments of neoliberal capitalism. Activist and capitalist shared beliefs in energetic action, Hoofd (2012, p. 12) details, have deepened and spread over time and left us with a so-called speed culture. This is a culture that relies on unrelentless change and accelerated production, relayed by the globalized meritocratic "speed-elite subject," "an exponent of the Eurocentric humanist subject" who "is often prevalent in many forms of activism" (Hoofd, 2012, p. 12) – but not all. There is thus a temporal intimacy between capitalism and activism that is important to consider when activism is pursued through commercial and public organizations.

Different paces, rhythms and "movements" are intrinsic to both activism and capitalism. Few have taken this as seriously as the Accelerationists, a theoretically-driven movement that seeks to maximally utilize fastness for an anti-capitalist agenda (Noys, 2014; Chistyakov, 2022). By speeding things up hysterically, Accelerationists wish to take capitalism to its limits, in efforts to create an implosion from within. The Slow movement, on the other

DOI: 10.4324/9781003396567-40

hand, works with desynchronization between the fast and the slow. Their contrasting belief is that capitalism should be obstructed, perhaps even extinguished, by deceleration instead of acceleration. According to scholarship on speed and politics, however, slowness has been shown to be as intrinsic to capitalism as fastness. To thrive, capitalism is in need of constant limitations to its otherwise uncontrollable acceleration of circulation (Virilio, 2006; Glezos, 2012) – a condition the Slow movement accordingly provides.

In light of this short summary of how activism and capitalism reciprocally thrive on each other's movements, this chapter will disclose tensions that arise due to polarizations of fast employee activism in commercial organizations and slow insider activism in public organizations. Both of these are pursued covertly or overtly from an embedded professional "insider" position and have been found to cut across organizational hierarchies horizontally and lead to activist-business-state conglomerations (Skoglund and Böhm, 2022). Instead of tracing such activism by conventional analysis of civil society, a focus on speed makes it possible to discern activism as a boundaryless movement in professional contexts. Investigating employee/insider activism as velocities also makes it possible to explain how certain paces align, while others remain desynchronized and oppositional.

"Labour activists" (Marens, 2013, p. 459), "internal activists" (Wickert and Schaefer, 2015, p. 107) and "organizational activists" (Spicer et al., 2009, p. 552) have previously been analysed with focus on hierarchical struggles and resistance between workers and managers. Even if many of these activists are interested in transforming their employer from within, they can also be further interested in having an impact on other organizations (Briscoe and Gupta, 2016). To understand this phenomenon, some scholars have used social movement studies to emphasise how professional work contexts can become prosperous places for "the working out of political issues" that elicit broader changes of society (Zald and Berger, 1978, p. 825). This strand of research stretches from studies of enterprising activism that deploys aggressive investments and methods to challenge the "ethic of ever-increasing consumption" (Mirvis, 1994, p. 83), to piecemeal and destabilizing employee activism driven by different goals (Scully and Segal, 2002). Employee activism can, for example, be performed by minority groups who use "distinctive types of inside knowledge" at the same time as they are outsiders – able to detach and recognize problems others are blind to (Meyerson and Scully, 1995, p. 588).

Turning to employee climate activists as "insiders," these have been found to share inside knowledge about climate change horizontally, at the same time as they have had to defend their outsider position and connect with non-organizational members (Skoglund and Böhm, 2020). Employee climate activists can, for example, take action against their employer's slowness to decarbonize, especially if it is a large CO_2-polluting industrial corporation. Based on their will to speed up solutions to climate change, they can exercise excessive, covert or overt, climate actions at work and beyond (Skoglund and Böhm, 2020). Promoting de-growth, they may even demand a quick and acute closing down of large parts of the emitting units, which threatens the employment security of their colleagues (Skoglund and Böhm, 2022). This may grow into a Fast movement that can have very different effects on the employer as well as the activist employees themselves. The counter-actions may be utilized rhetorically and/or be harnessed in a productive way by the employer to mobilize a strategically endorsed transformation of core operations. Employee activists may thus "build bottom-up pressure for transformation internally and prevent or overcome window-dressing" (Girschik, 2020, p. 35). Employee activism may in other cases be suppressed, whereby it often goes undercover (Scott, 1990) but can lead to a slower transformation via fragmented actions (Scully and Segal, 2002).

Employees within the public sector and state authorities have in comparison to activists within commercial organizations rather been conceptualized as "insider activists" (Olsson and Hysing, 2012; Hysing and Olsson, 2018; Hochstetler and Keck, 2007; Abers, 2019). One reason for the conceptual distinction between insider and employee activism could be due to researchers' ethical considerations of how public officials are contractually obliged to be apolitical at work. It is indeed also ethically sensitive to research commercial employee activism, but such research does not raise the same legal reasons for dismissal, as in the case of public officials. Despite these research-ethical concerns, it is important to address the unstable ideal of neutral professionality among public officials.

Public officials may, for example, be as impatient as employees within business. Speeding things up, too much, is nonetheless untenable for public officials, since slowness per definition is built into democratic institutions (Saward, 2017). This makes Fast movements, and an obvious acceleration of actions, difficult for a public official insider activist. If wishing to be intentionally activist at work, despite their choice of apolitical profession, insider activists need to work more covertly and align with the pace of their (public) organization. As a study of personal political agency within the public sector clearly shows, such incremental actions can slowly build up momentum and finally lead to a fundamental transformation, without noticeable radical expressions (Olsson and Hysing, 2012; Hysing and Olsson, 2018). One of the activist movements that could fit quite well, hidden, within many democratic governmental authorities, is thus the Slow movement. Slowness would escape the radar, perhaps be thought of as proper democratic deliberation (Saward, 2017), at the same time as it could transform governmental authority, decelerate business and infuse change within society.

To be able to grasp the different movements that can emerge with employee and insider climate activism, it is necessary to first address a broader temporal spectrum that is present in relation to the expressed climate emergency. After exploring this spectrum of transformative paces, the chapter will exemplify how different velocities align or become desynchronized depending on the professional work context of the climate activism.

Climate Emergency and the Fast Movement

The early debate about human-induced global warming has been strengthened by an intensifying language, from a "climate emergency" (Climate Emergency UK, 2019) to an "accelerating climate disaster" (Goswami, 2022). Not only seeking to mirror the scenarios created about the biospheric conditions of our common Earth, this velocity-ingrained vocabulary testifies to a politics of speed in climate-change discourse. Due to climate change having been proven a problem of unrelenting acceleration, prioritized solutions have been cast in similar temporality. The UN secretary, António Guterres, made the threat of "runaway climate change" explicit in 2018, stating that it is "moving faster than we are – and its speed has provoked a sonic boom SOS across our world" (United Nations, 2018). Governments have been urged to take more drastic action more quickly, not by 2050 but already by 2040, since "humanity is on thin ice, and that ice is melting fast" (United Nations, 2023). By condensing the most recent Intergovernmental Panel on Climate Change (IPCC) report, which details the increasing rate of temperature change, Guterres also emphasized the need "to massively fast-track climate efforts by every country and every sector and on every timeframe," including the speeding up of climate justice (United Nations, 2023). The IPCC clarifies that the hurry to respond is due to "a rapidly closing window of opportunity to secure a liveable and sustainable future for all" (IPCC, 2023). Indeed, "deep, rapid and sustained mitigation and accelerated implementation of adaptation actions" would have an impact "now and for thousands of years (*high confidence*)" (IPCC, 2023). But it would also mean potentially "disruptive changes" why it is "accelerated and equitable action" that is needed, where "finance, technology and international cooperation are critical enablers for accelerated climate action" (IPCC, 2023). Humans should not be outpowered by the climate emergency but should instead out-speed it. The solutions need to be implemented faster than the problems hit.

Some national governments have taken the call for rapid action seriously. In Sweden, for example, the digital tool Panorama "visualises and accelerates Sweden's climate transition" (The Swedish Climate Policy Council, 2023). The Panorama tool comes across as useful for a balancing of potentially clashing needs within sustainability. A balancing that is to be pursued by "speeding up development," pushed by a European Union (EU) regulation that came into effect on 30 December 2022, promoting faster permissions for an expansion of renewable energy (The Swedish Climate Policy Council, 2023). The reason for addressing the slow permission processes in the EU member states was not only the threatening rate of climate change but the neglected principle of national and European energy security, which became paramount after the full-scale Russian invasion of Ukraine in February 2022 (Ibid). Two different emergencies, energy security and climate change, which both were in need of an acceleration of solutions, conveniently cohered to be mutually addressed. Renewable energy permissions were to be fast-tracked for 18 months (Panorama, 2023), "to further accelerate renewable energy projects and remove remaining administrative obstacles" (European Commission, 2023). Bureaucracy was limited to accelerate climate action.

As is well known, the private sector is often expressing their struggles with administrative obstacles and slow governmental bureaucracy, wishing it to be more efficient and less uncertain. The right speed is indeed important for monetary flows, where the operating conditions created for business should lure investments to infuse capital circulation that consistently generates a prosperous milieu supportive of competitive business. The organization Fossil Free Sweden, for example, seeks to "increase the pace of the climate transition" to make Sweden competitive as "one of the first fossil free nations in the world" (Fossil Free Sweden, 2023). The organization was spawned by the Swedish Government in 2015 and consists of private, civic and governmental actors. Their official idea is to bring these various actors together to find "common ways ahead so as to speed up the transition to a fossil free welfare nation" and "accelerate developments" (Fossil Free Sweden, 2023). The National Coordinator of Fossil Free Sweden, a former agronomist who previously held the position as Secretary-General of the Swedish Society for Nature Conservation, suggests that major industrial actors that are important to Sweden, "must go fossil free relatively quickly to cope with their international competition" (Fossil Free Sweden, 2023). For this reason, the task is to "speed up important reforms in the Parliament" and make sure the state secures conditions that will "enable the sector to speed up the sector's transition" and thereby "reduce emissions more rapidly" (Fossil Free Sweden, 2023). This is just one example of how the climate emergency – and urgent rapid action called for by the IPCC – is taken closer to where speeding up has been of historical interest – within business.

Speedy solutions to climate change have of late been affirmed by the corporate sector, exemplified by a US corporation: "the impacts of climate change are an urgent global threat. And at Intel, we're committed to accelerating progress against the world's critical challenges" (Intel, 2023b). In their Corporate Responsibility Report, Intel stresses the ambition to "accelerate improvements across" the industry, to "go faster and be more effective working together" (Intel, 2023a, p. 9). Their partnerships with others will thus "accelerate the sustainability of PCs, improve the energy efficiency of data centres, and accelerate handprint projects" (Intel, 2023a). The handprint is meant to balance out the carbon footprint of the industry by "harness[ing] digital technologies to accelerate environmental progress" (Intel, 2023a, p. 26). Overall, their Corporate Responsibility Report is full of quests for progress of different types, ultimately to be realized by a talented, purpose-driven and diverse workforce that "accelerate[s] Intel's growth" (Intel, 2023a, p. 20), perhaps based on the product category called "accelerator products" (Intel, 2023a, p. 14).

Abovementioned demands for immediate responses to the climate emergency elicit acceleration and the need for fastness within business, as well as among increasingly entrepreneurial states that seek to attract investments to their territory. However, a counter movement has arisen, which mainly is against the speed that feeds capitalism and its negative impacts on climate change. If capitalism thrives on fastness (Hoofd, 2012), then exchanging acceleration for slowness would perhaps be a better response to the climate emergency.

Climate Emergency and Deceleration

One of the early – but highly debated – Slow movements targets population growth. According to calculations and advancements of the so-called Kaya Identity (González-Torres et al., 2021), slowing down population growth would have a "negative" effect on economic progress due to less consumption per time unit. And by extension, such de-growth of people and decelerated consumption would generate fewer CO_2 emissions. Clarified by the UK-based campaigning charity called Population Matters, which was founded in 1991: "Every additional person increases carbon emissions – the rich far more than the poor – and increases the number of climate change victims – the poor far more than the rich" (Population Matters, 2023b). The Zero Population Growth (ZPG) and related movements, such as The Overpopulation Project (2023), also seek to make science-based population deceleration legitimate, stretching to proposals of enforced family planning or soft empowerment of women to promote smaller families (Population Matters, 2023a). As Slow movements, ZPG, Population Matters and The Overpopulation Project also go beyond climate changes to cut an environmentally healthy balance between births and deaths and stimulate a better householding with our shared, limited, natural resources. Similarly, other Slow movements also seek to limit CO_2 emissions by decelerating consumption, but instead of targeting reproductive desires they prefer to work towards a change of behaviours during a human lifetime.

Well-known Slow movements are the Slow Food and CittaSlow (Slow city), which have branched out over time. Among many others, less known movements are perhaps The Japanese Sloth Club, Slow Money, Slow Mail and Slow Technology. The Sloth Club sees fastness as the core problem and uses the evolution of the slow and sleepy three-toed sloth as a model for the solution (Kumar, 2015). For Slow Mail, a subversive App called Pony has been created to revive the old feel and function of slow post services (Bogost, 2022). Mailing slowly can also be attained by relating differently to digitized communication, sometimes facilitated in workplaces by initiating Email Light Time between certain dates (Hope-Hailey, 2023). This deceleration of emails sent hither and thither across the office corridor or the Earth has only occasionally been highlighted as decreasing CO_2 (Molloy, 2020). Slow Technology is similarly less interested in targeting CO_2 emissions and more interested in decreasing stress in life. It is either introduced to create a break in our increasingly mandatory harried digital connectedness, or, more profoundly, suggested to produce deceleration by designing for reflexivity and relaxing brainwork during use (Hallnas and Redstrom, 2001).

Slow technology builds on ideas of a more general form of deceleration that was popularized in a book by Carl Honoré in 2004. Drawing on various Slow movements, he wrote for a broad audience about his own struggles to counter everyday acceleration (Honoré, 2004). In the foreword to the fifteenth anniversary edition of *In Praise of Slow, How a Worldwide Movement is Challenging the Cult of Speed*, Honoré emphasizes the rising acceptance of slowness seen as key for the creation of an "economic system that is humane, fair and sustainable" (Honoré, 2004, p. xv). Within such a sustainable economic system, the Earth's resources would not be used up as quickly but be given time for renewal or recycling. Slowness would in addition secure quality, for example by resulting in fewer technical incidents and thereby higher productivity. The slowness created would not equal the affirmation of de-growth (Meissner, 2019) or Luddite rejection of innovation (Jones, 2013). It would rather enforce sensible progress through moderation between fast and slow, accelerating technologies and decelerating mindfulness (Honoré, 2004).

Means of taking yourself out of harried and speedy situations occasionally, or performing a constant enduring deceleration, can be found in a general turn to Slow Living. Slow Living elicits "a mindset whereby you curate a more meaningful and conscious lifestyle that's in line with what you value most in life" (Slow Living ldn, 2023). Pausing and dwelling is key to such everyday deceleration, facilitated by certain interior designs, or why not a Slow Journalism magazine (Delayed Gratification, 2023)? As a branch of deceleration, Slow Journalism has been proposed to counter fast media, crucial for providing and grasping in-depth explanations of such a complex issue as climate change (Gess, 2012). The overall ambition with Slow Journalism is also to better connect with the audience and build a deeper relationship between the writer and the reader community. Different Slow movements bring this generation of a heightened sensation of presence and connectedness to the fore. Before the birth of the Slow movement, this sensation existed without the need to be articulated as "slow" (Slow Movement, 2023).

Making the "slow" explicit became of increasing importance in the 1980s within a growing anti-capitalist activism. The Slow Food movement, for example, polarized itself in the wake of fast-food chains and pre-prepared mass-produced meals heated up by the accelerating technology of the microwave oven. Criticism was first raised against poor diets as well as disconnection from the craft of cooking, after which the Slow Food movement expanded to promote local produce and thoughtful cooking in relation to seasonal shifts and nature's "natural" resources (van Bommel and Spicer, 2011). Not only cooking slowly but eating slowly and training one's senses became crucial (Munro, 2014). To linger over the food in togetherness was a way to embrace a "relaxed pace of life, create conviviality and hospitality," which spilled over into the formation of the Slow City movement (Leonard, 2020).

Inspired by localism, the Slow City movement originated in Italy during the late 1990s to counter globalization and the spatial disengagement that the culture of speed now was understood to generate (Çiçek et al., 2019). The Slow City movement follows natural rhythms – such as the regular pace of animals and life cycles of organic material – and fittingly uses the snail as a model for its brand (Cittaslow.org, 2023). The aim is to generate place-specific pace, authenticity and correlating identities, which stand out from the mass-produced culture – and food – found elsewhere. In addition to the first anti-capitalist and anti-globalization pace found within the Slow Food and Slow City movements, these have of late been strengthened by both tourism and climate activism. They attract visitors to have unique experiences, critiquing existing massifying food supply chains, promoting vegetarian and vegan low-emitting options.

In comparison to the clear agency to decelerate in civil society Slow movements, as exemplified above, the movement "Slow Democracy" faces the difficult task of balancing between acceleration and deceleration within state agencies (Saward, 2017). According to Saward, "juggling urgency and caution" is not only difficult in practice, it is also problematic to mediate. Depending on the problem and situation, political leaders or public officials may claim that they are acting fast, but behind the scenes they are utterly precautionary or just uninterested. Or the opposite, rhetorically they can come across as taking their time, deeply contemplating different alternatives and acting calmly. In practice, however, they act rapidly. Saward (2017) therefore turns to the Slow Food and CittaSlow movements to academically counter social accelerationism via Slow Democracy. From a formal political perspective, deceleration could be achieved through, for example, radical decentralization with the aim to harness "autonomous local decision-making" (Saward, 2017, p. 11). Since local actors have more stake in relation to their close environment, the idea is to facilitate engagement with traditions and locally relevant innovations, which would limit blind change due to the unstoppable force of speed itself (Saward, 2017).

Fast Employee Activism Versus Slow Insider Activism

Based on this temporal spectrum and strikingly different paces of activism, this section will hypothetically explore potential tensions between two fictional cases, one of employee climate activism within a commercial organization and one of insider climate activism within a public organization. This established, yet underdeveloped method will work as a "benchmark device" (Morgan, 2006, p. 7) where the fictions allow an accentuation and synthesis to appear (Cutcher et al., 2020). Cutcher et al. (2020) use their fiction of three "ideal types" of academic positions within feminist theorizing in Organization Studies to accentuate these for further analysis and scrutinization. In contrast, this chapter draws on lived experience, to fictionally illustrate fast climate employee activism within a made-up CO_2 emitting corporation and slow climate insider activism within an imagined regional office. The method allows for a polarized situation to appear, with the aim to reflect on how desynchronized rhythms can unfold between climate employee/insider activism. Even if the opposite fiction could have been chosen, a focus on Fast movements and commerce as well as Slow movements and state bureaucracy follow an existing "general ideal" to be exaggerated and exposed to stimulate critical reflection (Cutcher et al., 2020; Morgan, 2006).

A Fictional Conversation Between
Three Employee Activists in a Corporation

"I have had enough," Sidney thought. "This corporation is crazy, why can't they listen to the science? It's crystal clear, the IPCC consensus. It is an EMERGENCY, for God's sake . . ."

Unable to keep these thoughts to themselves, Sidney burst out during a shared lunch at a nearby restaurant:

"Aren't your children worried about the climate emergency? Mine are absolutely devastated."

"Yes, mine too. All of them, actually. They gang up on me and accuse me of greenwashing, saying the food I bring to the table is filthy, tainted by dirty money made for this enormous hypocritical polluter," says Parker, gesticulating with the arms to imitate an emotional righteous teenager. "They even feel filthy, being connected to me, they said yesterday. The issue is just getting worse every day, and I have no clue how to defend myself. I mean, I agree. So, what can I say? Like, we are trying hard but things are more difficult than you think. Innovative technology takes time."

People were now glancing at them across the room in the convivial vegan restaurant, perhaps wondering what big ugly polluter Sidney and Parker represented, when a third colleague, Charlie, joins the distraught conversation.

"No, it will not be enough. New technology is just not enough. It's too slow. Far too slow. The climate emergency can hit us any day, and not just future generations. We have to do more. Much more. And faster."

"What do you mean?" asks Sidney.

"Well, let's start with the problems with all the permits we need for transitioning to renewable energy." Charlie raises their eyebrows and continues with a raised voice. "What are those public officials doing, really, thwarting our decarbonization? The EU was clear on the need for speeding things up. So, how can we take action to get around their stupid scepticism? It is devastating for the climate, their absurd reluctance to process the permits."

"Yeah, someone really has to take it up with the politicians," Sidney proposes. "Seriously, they just can't let it be that slow. They have to understand that the climate emergency is already here, and it needs faster bureaucracy, much faster."

"Yeah, well, that's true," Parker continues, "but, lobbying will also take a very long time."

"I know," adds Charlie, even more distressed than before. "So, why not at least implement decarbonization across the entire organisation, in all those corners of our activities we know best, and where we actually are in control, and no one else?"

"Do you think that would really work?" Parker asks cynically.

"I don't care, we just have to try," Charlie continues.

"Perhaps we can secure the workflow of these climate actions somehow," Sidney contemplates loudly.

"We also need to brand the actions in some way that works," Charlie announces with retained hope.

"Okay, well, that's my managerial expertise," Parker assures, "let's see. . . . What about Low-carb ASAP? Or, Accelerating Decarbonization Now?"

"I like the Now," Sidney confirms. "Then it is not about the future, as everyone else is claiming, but about what we are actually doing, in this moment, to speed up. I guess even your friends in the Board of Directors will buy into this, they probably know it has to happen at some point anyway, to stay competitive, zero carbon."

The three of them happily return to the office, invigorated by their collective decision to take climate actions further and faster, now. Leaving no stone unturned, finding all the dirty carbon in every corner of the operations, they continue to intensify the movement within their CO_2- polluting corporation, opening up for others to speed up actions, receiving support from all levels to elicit transformation internally but also among their users and suppliers. In the end, the corporation manages to decarbonize by behavioural changes; nevertheless, the implementation of new decarbonizing technology remains slow on the uptake.

A Fictional Conversation Between Three Insider Activists in a Regional Office

"Have you heard what that crazy CO_2-emitting corporation is up to now?" Rowan asks Robin and Ash. "First they demanded to get all those environmental permits processed in no time, and now they are taking every opportunity they can to whine about slow bureaucracy to the politicians. As if we were thwarting their operations intentionally. I mean, this is democracy after all! It is supposed to be slow."

"Yes, we definitely have to involve all the relevant stakeholders, and gosh, they take their time. Especially the universities, when they are asked to be expert commentators. They are definitely the slowest," Robin continues.

"Well, from my perspective as a biologist," Ash adds, "there is so much both the corporation and the stakeholders just cannot see. They are all so very enthusiastic about decarbonization, but they totally neglect other environmental aspects which play into the equation of carbon dioxide equivalents, such as biodiversity. From my point of view, everything should be done with utter precaution."

"Well, hasn't the precautionary principle been dead for decades now?" wonders Robin.

"Yeah, well, guess so, guess it's just too slow for business," Ash admits with resignation.

"Perhaps nothing a corporation does to speed up decarbonization actually leads to sustainability? I mean, speeding things up is part and parcel of capitalism, to keep the wheels spinning," Robin reveals.

"Yeah, that's certainly how it is, we all know that, of course, good to be reminded that acceleration is the core problem of human-induced climate change," Rowan adds.

"Which means . . ." Robin gradually reveals, "a slowing down would be much better for everyone in the long run."

"Yeah, exactly," Ash confirms. "The EU certainly missed that one."

"Perhaps we should think harder about that, slowing things down." Robin suggests. "Perhaps invite that expert who already wanted our application to the CittaSlow community to meet the climate emergency with the correct tools. Traditional knowledge. A slow, enjoyable, pace of life. I mean, what would that take?"

"Not much, I reckon, I definitely think we are very good at slowing things down already," Rowan asserts with a big smile.

"Yeah, let's do it!" Ash and Robin retort in chorus, already feeling surprisingly empowered.

They do not return to their respective computers but make it an early day and leave for a slow-cooked meal in a close-by Italian restaurant. Over the Slow food they continue to contemplate the prospect of slowing down the whole city and life in general. They feel strengthened in their new, well-paced togetherness and decide to meet the next day outside the office, to go for a long meandering morning walk. The relaxed start of the day is then followed by their own sanctioning of overt Email Light Time. During the following month they only respond sporadically and shortly to mail, while anchoring the application to the CittaSlow organization with the rest of the regional office. Being very convincing about the merits of going slow, they successfully finalize the application within one and a half years. Their city is now officially using a brand based on the insect Phasmatodea (walking stick), devoted to a slow beautiful life and stopping industrious activities within digitalization, plastics, mining, cement and steel.

Conclusion

The methodological choice to polarize fast and slow activist movements to generate analytical accentuation (Cutcher et al., 2020) and position them in different professional organizational contexts illustrates employee/insider climate activism through a perspective on speed (Glezos, 2012). The fictional dialogue of an emerging employee Fast movement points to the "ideal type" of business to generate energetic action (Hoofd, 2012), while the fictional dialogue of an emerging insider Slow movement aligns with the "ideal type" of democratic action (Saward, 2017). Instead of analysing employee/insider activism as a potential alignment (Girschik, 2020) or misalignment (Scully and Segal, 2002) with existing organizational values and strategies, this shows how employee/insider activism elicits contrasting professional paces depending on their organizational context. Slow and fast movements were intentionally orchestrated by using two fictional illustrations, to further reflect on how climate activism pursued in the private and public sectors can become oppositional. That is, despite their mutual ambition to respond to the climate emergency, temporal desynchronization emerged among the climate activists portrayed.

The fictional examples clarify that employees within corporations would find it difficult to gather overt collective support for a Slow movement, just as public officials in established democracies would find it difficult to successfully mobilize a Fast movement. In comparison to fast movements cultivated through commerce and capitalism (Hoofd, 2012), less is known about the expected strategy – to pursue slowness – via insider activism in the public sector. The main contribution in this chapter thus follows up on previous studies of insider activism in public organizations, known for being underpinned by weaker directions from political parties and increased individual ideational flexibility and mobilization of key responsible individuals, such as technical experts (Olsson and Hysing, 2012). To act on and sustain this flexibility, insider activism can appropriate and slightly twist formalized Slow movements, inspired by the CittaSlow, Slow mail and Slow food movements. Insider activism could disappear successfully among already slow work tasks in the public sector (Saward, 2017), by configuring unique, context-specific, Slow movements. Slow movements would let public officials calibrate a legitimate pace for their climate activism, to accomplish a transformation with focus on other target organizations, such as commercial organizations with affluent fast movements furthered by employee climate activists.

References

Abers, N. R. (2019). Bureaucratic activism: Pursuing environmentalism inside the Brazilian State. *Latin American Politics and Society*, 61(2), 21–44. http://doi.org/10.1017/lap.2018.75

Bogost, I. (2022). The subversive genius of extremely slow email. *The Atlantic*. https://www.theatlantic.com/technology/archive/2022/01/slow-internet-email/621232/ [Accessed 3 February 2023].

Briscoe, F. and Gupta, A. (2016). Social activism in and around organizations. *Academy of Management Annals*, 10(1), 671–727. http://doi.org/10.1080/19416520.2016.1153261

Chistyakov, D. I. (2022). Philosophy of accelerationism: A new way of comprehending the present social reality (in Nick Land's Context). *RUDN Journal of Philosophy*, 26(3), 687–696. http://doi.org/10.22363/2313-2302-2022-26-3-687-696

Çiçek, M., Ulu, S. and Uslay, C. (2019). The impact of the slow city movement on place authenticity, entrepreneurial opportunity, and economic development. *Journal of Macromarketing*, 39(4), 400–414. http://doi.org/10.1177/0276146719882767

Cittaslow.org. (2023). *International Network of Cities Where Living Is Good*. https://www.cittaslow.org/ [Accessed 25 April 2023].

Climate Emergency UK. (2019). *Climate Emergency Declarations with a Target Date by 2030*. https://climateemergency.uk [Accessed 30 January 2023].

Cutcher, L., Hardy, C., Riach, K. and Thomas, R. (2020). Reflections on reflexive theorizing: The need for a little more conversation. *Organization Theory*, 1(3), 263178772094418. http://doi.org/10.1177/2631787720944183

Delayed Gratification. (2023). *The World's First Slow Journalism Magazine*. https://www.slow-journalism.com/ [Accessed 25 April 2023].

Dixon, C. A. (2014). *Another Politics – Talking Across Today's Transformative Movements*. Oakland: University of California Press.

Ehrenberg, J. R. (2017). *Civil Society, Second Edition The Critical History of an Idea* (2nd ed.). New York: New York University Press.

European Commission. (2023). *Enabling Framework for Renewables*. https://energy.ec.europa.eu/topics/renewable-energy/enabling-framework-renewables_en [Accessed 30 March 2023].

Fossil Free Sweden. (2023). *About Fossil Free Sweden*. https://fossilfrittsverige.se/en/about-us/. [Accessed 30 March 2023].

Gess, H. (2012). Climate change and the possibility of 'slow journalism'. *Ecquid Novi*, 33(1), 54–65. http://doi.org/10.1080/02560054.2011.636828

Girschik, V. (2020). Shared responsibility for societal problems: The role of internal activists in reframing corporate responsibility. *Business & Society*, 59(1), 34–66. http://doi.org/10.1177/0007650318789867

Glezos, S. (2012). *The Politics of Speed: Capitalism, the State, and War in an Accelerating World, Interventions*. London: Routledge.

González-Torres, M., Pérez-Lombard, L., Coronel, J. F. and Maestre, I. R. (2021). Revisiting Kaya identity to define an emissions indicators Pyramid. *Journal of Cleaner Production*, 317, 128328. http://doi.org/10.1016/j.jclepro.2021.128328

Goswami, U. (2022). UN warns of 'accelerating climate disaster'. *The Economic Times*. https://economictimes.indiatimes.com/news/india/un-warns-of-accelerating-climate-disaster/articleshow/95132014.cms

Hallnas, L. and Redstrom, J. (2001). Slow technology – designing for reflection. *Personal and Ubiquitous Computing*, 5(3), 201–212. http://doi.org/10.1007/PL00000019

Hochstetler, K. and Keck, E. M. (2007). *Greening Brazil: Environmental Activism in State and Society*. Durham: Duke University Press.

Honoré, C. (2004). *In Praise of Slow, How a Worldwide Movement Is Challenging the Cult of Speed*. London: Seven Dials.

Hoofd, M. I. (2012). *Ambiguities of Activism: Alter-Globalism and the Imperatives of Speed, Routledge Research in Cultural and Media Studies* (p. 43). New York: Routledge.

Hope-Hailey, V. (2023). *Email Light Time*, 3 April.

Hysing, E. and Olsson, J. (2018). *Green Inside Activism for Sustainable Development: Political Agency and Institutional Change*. Cham: Springer.

Intel. (2023a). *Corporate Responsibility Report*. https://csrreportbuilder.intel.com/pdfbuilder/pdfs/CSR-2021-22-Full-Report.pdf#page=77 [Accessed 30 March 2023].

Intel. (2023b). *Our Commitment to Sustainability*. https://www.intel.com/content/www/us/en/environment/intel-and-the-environment.html [Accessed 30 March 2023].

IPCC. (2023). *AR6 Synthesis Report, Headline Statements*. IPCC. https://www.ipcc.ch/report/ar6/syr/resources/spm-headline-statements [accessed 30 March 2023].

Jones, E. S. (2013). *Against Technology: From the Luddites to Neo-Luddism*. New York and London: Routledge.

Kumar, S. (2015). Slow, small, simple. *Resurgence and Ecologist*. https://www.resurgence.org/magazine/article4485-slow-small-and-simple.html

Leonard, B. (2020). *Citta Lente – Slow Cities*. https://finiteeartheconomy.com/citta-lente-slow-cities/ [Accessed 25 April 2023].

Marens, R. (2013). What comes around: The early 20th century American roots of legitimating corporate social responsibility. *Organization*, 20(3), 454–476. http://doi.org/10.1177/1350508413478309

Meissner, M. (2019). Against accumulation: Lifestyle minimalism, de-growth and the present post-ecological condition. *Journal of Cultural Economy*, 12(3), pp. 185–200. http://doi.org/10.1080/1 7530350.2019.1570962

Meyerson, E. D. and Scully, A. M. (1995). Tempered radicalism and the politics of ambivalence and change. *Organization Science*, 6(5), 585–600. http://doi.org/10.1007/978-1-137-16511-4_15

Mirvis, H. P. (1994). Environmentalism in progressive businesses. *Journal of Organizational Change Management*, 7(4), 82–100. http://doi.org/10.1108/09534819410061397

Molloy, D. (2020). *Climate Change: Can Sending Fewer Emails Really Save the Planet?* https://www. bbc.com/news/technology-55002423 [Accessed 26 April 2023]

Morgan, M. S. (2006). Economic man as model man: Ideal types, idealization and caricatures. *Journal of the History of Economic Thought*, 28(1), 1–27. http://doi.org/10.1080/10427710500509763

Munro, I. (2014). Organizational ethics and Foucault's 'art of living': Lessons from social movement organizations. *Organization Studies*, 35(8), 1127–1148. http://doi.org/10.1177/0170840614530915

Noys, B. (2014). *Malign Velocities: Accelerationism and Capitalism*. John Hunt Publishing.

Olsson, J. and Hysing, E. (2012). Theorizing inside activism: Understanding policymaking and policy change from below. *Planning Theory and Practice*, 13(2), 257–273. http://doi.org/10.1080/1464 9357.2012.677123

Panorama. (2023). *278 Styrmedel och åtaganden för Sverige*. https://app.climateview.global/v3/ public/board/ec2d0cdf-e70e-43fb-85cb-ed6b31ee1e09?action=e1301de4-117b-41fe-9ad7-37800c3b531e [Accessed 30 March 2023].

Population Matters. (2023a). *Choosing a Smaller Family*. https://populationmatters.org/choosing-a-small-family/ [Accessed 25 April 2023].

Population Matters. (2023b). *Our Vision*. https://populationmatters.org/ [Accessed 26 April 2023].

Saward, M. (2017). Agency, design and 'slow democracy'. *Time & Society*, 26(3), 362–383. http:// doi.org/10.1177/0961463X15584254

Scott, C. J. (1990). *Domination and the Arts of Resistance: Hidden Transcripts*. New Haven: Yale University Press.

Scully, M. and Segal, A. (2002). Passion with an umbrella: Grassroots activism in the work-place. *Social Structure and Organizations Revisited*, 19, 125–168. http://doi.org/10.1016/ S0733-558X(02)19004-5

Skoglund, A. and Böhm, S. (2020). Prefigurative partaking: Employees' environmental activism in an energy utility. *Organization Studies*, 41(9), 1257–1283. https://doi.org/10.1177/0170840619847716

Skoglund, A. and Böhm, S. (2022). *Climate Activism – How Communities Take Renewable Energy Actions Across Business and Society*. Cambridge: Cambridge University Press.

Slow Living ldn. (2023). *What Is Slow Living?* https://slowlivingldn.com/what-is-slow-living/ [Accessed 25 April 2023].

Slow movement. (2023). https://www.slowmovement.com/ [Accessed 25 April 2023].

Spicer, A., Alvesson, M. and Kärreman, D. (2009). Critical performativity: The unfin-ished business of critical management studies. *Human Relations*, 62, 537–560. http://doi. org/10.1177/0018726708101984

The Overpopulation Project. (2023). *Family Planning Success Stories*. https://overpopulation-project. com/family-planning-success-stories/ [Accessed 30 March 2023].

The Swedish Climate Policy Council. (2023). *The Swedish Climate Policy Council*. https://www.kli-matpolitiskaradet.se/summary-in-english/ [Accessed 30 March 2023].

United Nations. (2018). *Secretary-General's Address to the General Assembly*. https://www.un.org/ sg/en/content/sg/statement/2018-09-25/secretary-generals-address-general-assembly-delivered-tri-lingual [Accessed 4 April 2023].

United Nations. (2023). *Guterres Comments on the IPCC Report*. UN Audiovisual Library.

van Bommel, K. and Spicer, A. (2011). Hail the Snail: Hegemonic struggles in the slow food move-ment. *Organization Studies*, 32(12), 1717–1744. http://doi.org/10.1177/0170840611425722

Virilio, P. (2006). *Speed and Politics: An Essay on Dromology*. Translated by Mark Polizzotti, Semiotext(e) foreign agents series. South Pasadena, CA: Semiotexte.
Wickert, C. and Schaefer, M. S. (2015). Towards a progressive understanding of performativity in critical management studies. *Human Relations*, 68(1), 107–130. http://doi.org/10.1177/0018726713519279
Zald, N. M. and Berger, A. M. (1978). Social movements in organizations: Coup d'Etat, insurgency, and mass movements. *American Journal of Sociology*, 83(4), 823–861.

PART VII

Epistemological Questions and New Praxis

Introduction to Part VII: Epistemological Questions and New Praxis

The final part of the handbook returns to questions of public pedagogy and praxis. It considers how practitioners and scholar-activists can and must work: how do we best research and teach grassroots climate activism in the twenty-first century? And can we support activist platforms that serve to empower resistance, amplify marginalised voices, and nurture new critical publics (see Ruth Machen, 2018)? As such, it reflects a key aim of this handbook: to help connect and disseminate some of the powerful, urgent work of the many activist practitioners and researchers working with communities to address the climate emergency.

As a civic agenda, grassroots climate activism offers alternatives to the dominant model of neoliberal research and education, modelling insights into what a re-imagined platform for civil society to co-produce place and common goods might look like. The vision of an activist commons builds on a rich tradition of feminist, queer, anti-racist, and working-class struggles. As Esra Erdem (2020) describes, this is about "grassroots spaces created by a community for the sharing of knowledge in which knowledge and ideas can be freely shared among equals . . . [where] space is not given: it has to be established and occupied" (p. 316). Erdem highlights four key principles/themes that underpin climate activism as community commons:

- *access*: research and education should be socially inclusive and foster the sharing of knowledge and practices by committing to co-created learning. This involves re-shaping spaces of learning to include community settings such as parks, libraries, churches, trade union halls, community centres, cafés, bookstores, galleries, etc.
- *commoning practices*: recognition of the social labour undertaken by the climate activist community to produce and sustain collective resources, acknowledging in particular the diversity of experiences and knowledges that are involved and the multiplicity of skills in everyday activist practices (including that involved in administrative, coordinating, and logistical work).

DOI: 10.4324/9781003396567-41

- *collective self-management*: the processes of collective decision-making that enable more participatory and inclusive forms of decision-making in research and education. Informality, autonomy and response-ability are key shared characteristics across diverse decision-making, practices of scholar activism, and public pedagogy.
- *community*: the development of alternative power and knowledge relations in and through research and education at the grassroots activist level that help rebuild a sense of community as commons. Community-building in this sense includes nurturing a sense of learning, belonging and commitment, "being-in-common."

This section begins by discussing how scholars (can) engage with grassroots climate activism and how the accelerating climate crisis impacts what methods and methodologies we (should) use as researchers in the field and in the classroom. How is knowledge produced under these conditions? What barriers do teachers and researchers face? How can these be overcome? As the threat (and the cost) of major climate disruptions increases, so does hostility towards the drastic (and in some cases costly) measures that must be taken. This increasingly volatile environment impacts scholars and teachers today and will likely become more volatile in the years ahead. The chapters in this section demonstrate that scholarship and teaching are themselves important forms of grassroots climate activism.

In Chapter 35, "Knowledge in the Frame: The Epistemology of Extinction Rebellion," Georgina Treloar views the global climate movement at its heart as an epistemic project. It is largely concerned with the transmission and interpretation of knowledges to mobilise publics. When Extinction Rebellion (XR) emerged as a campaign of mass nonviolent civil disobedience in the UK in October 2018, it amplified a master frame of a climate and ecological emergency, whilst a climate justice master frame was eschewed at a strategic level. Drawing on primary data from interviews with XR co-founders, central team members, and the co-founders and "rank-and-file" members of a local UK group, this chapter discusses how knowledges factored into the frames of XR, the contentions that arose around certain framing decisions, and concepts that might aid a better grasp of how knowledges fit into the effective framing of climate and ecological collapse, in terms of motivating participation in the climate movement.

Chapter 36, "Convening Climate Activism in Canada: Conflicting Expertise and the Production of the Problem," by Adam Fleischman, focuses on the role expert-activists in the climate movement in Canada play in aligning disparate interests and moderating the conflicts that can emerge in the encounter of plural voices. This chapter examines the Climate Action Network-Réseau action climat (CAN-Rac), a key organisation that bridges science and grassroots politics in the climate debate. The Canadian context situates a climate and energy discourse that is entrenched with technological nationalism and rooted in settler-colonial relations. The author explores the challenges of mediating dissent and conflict within the movement and the complexities of engaging with Indigenous activists and white settler-colonial environmentalists. This chapter highlights how climate-change politics is contentious with respect to norms, concepts, and institutions within Western science – which often is enmeshed in dissent and conflict – and makes the case for the role of mid-level experts in moderating diverse agendas across grassroots climate activism in Canada.

Chapter 37, "Climate Activism in K-12 Formal Education Across North America," is a collaboration between scholars and activists across several places in North America, including Ontario and Alberta in Canada and Oregon in the US. The authors are all involved in various aspects of climate activism within and surrounding the K-12 formal education system, and they discuss their perspectives on climate-change education and activism in North America through a series of vignettes. Field et al. suggest that, while many youth around the globe have led climate activism efforts and centred "justice" as integral for pathways towards liveable futures, formal K-12 institutions in North America have tended to lag in their attention to climate. To investigate tensions and possibilities within the formal education system in Canada and the US, this chapter explores several sites of climate activism within the educational arena, from individual schools to the district, state, and federal level.

Chapter 38, "People-powered Climate Justice: Challenges and Possibilities in Engaging Activists in Strategic Capacity Building," by Rumbidzai Irene Mphalo, Joseph Worthy Jr., and Alexander Repenning, showcases strategies to improve the effectiveness of frontline climate justice activism across experiences in Africa and Europe. Drawing on Gene Sharp's nonviolent action principles and a comprehensive assessment of the challenges faced by the climate movement, the authors describe a training program that aims to enable activists to achieve a more successful implementation of strategies. The authors identify the need for coalition building, strengthening North-South alliances, and engagement with intersecting movements such as human rights and land rights as essential areas of action. The chapter also discusses the importance of fostering diversity in the climate movement and providing hands-on training to activists. Challenges related to political and economic dependencies, funding inaccessibility, and related asymmetries are explored, with a focus on developing guiding principles for supporting and amplifying grassroots climate activism.

In conversation with Treloar's discussion of XR, Hannah Fitchett suggests in Chapter 39, "Building Regenerative Cultures in Extinction Rebellion Activism and Academic Research," that the most distinctive characteristic of the Extinction Rebellion (XR) movement's approach to mitigating climate breakdown is their aim to build "regenerative cultures." Drawing on 11 months of engaged anthropological fieldwork with XR London, this chapter explores the practice of engaged anthropology when working with XR activists: the challenges this involves and some tools to help researchers navigate them. By bringing together academic theory from engaged anthropology with knowledge about regenerative cultures gained through fieldwork, Fitchett explores what a "regenerative" approach to research with grassroots climate activists may look like. Taking three core principles of regenerative cultures – prioritising care, recognising co-construction, and balancing reflexivity and action – the author details how each relates to XR activism and academic theory and informs climate research praxis. These three principles can help guide approaches to studying climate activism which are grounded in an ethics of climate justice.

As the concluding section of this handbook, these chapters speak to the possibilities and the challenges faced by scholar-activists. They invite readers to consider the relationships between grassroots activism and the academy or the classroom. In their own ways, each of these chapters begs the question: what are the relationships between the people teaching and researching climate change and grassroots climate activism? What role can scholarship play in illuminating the often hard-to-spot dynamics of this kind of collective action? And how can scholars collaborate *with* activists to tell these stories?

There are no simple answers to these complex questions, but the chapters offer an interesting starting point for considering the work of the scholar activist in times of multiple crises. As part of a greater whole, this section mirrors the rest of the handbook by zooming in on the grassroots, with the help of a diverse set of voices who present current developments in grassroots climate activism in different parts of the world, from a large variety of disciplinary angles and a diversity of perspectives. Collectively, the chapters demonstrate alternatives to how public research and education can be organised around the principles of commons, shared by a community who themselves define the modes of use and production, distribution, and circulation of resources as an integral part of grassroots climate activism as a form of critical praxis.

References

Erdem, E. (2020). Free universities as commons. In J. K. Gibson-Graham and Kelly Dombroski (eds.) *The Handbook of Diverse Economies* (pp. 316–322). Edward Elgar Publishing.

Machen, R. (2018). Towards a critical politics of translation: (Re)-producing hegemonic climate governance. *Environment and Planning E – Nature and Space*, 1(4), 494–515.

35
KNOWLEDGE IN THE FRAME
The Epistemology of Extinction Rebellion

Georgina Treloar

Introduction

Grassroots climate activists draw on a variety of knowledges, including climate science, to generate public understanding and mobilise mass participation, with "listen to the science" being a common catch cry. When Extinction Rebellion (XR) mobilised in late 2018, it leant heavily on scientific knowledge (Karnik Hinks and Rödder, 2023; Thierry, 2023) within an emergency master frame. Knowledge of local and regional climatic events also helped to propel this framing into the political sphere; local councils started declaring climate emergencies from November 2018, and the UK parliament declared a climate and environment emergency in May 2019. But, despite XR's contentious decision to suspend justice frames and eschew a climate justice master frame at a broad strategic level, XR activists still sought to represent Indigenous knowledges within a global climate justice master frame. An often-overlooked aspect of XR is the Internationalist Solidarity Movement.

In the following chapter, I explore how knowledge factored into the frames that XR UK employed during its nascency. Data was collected from in-depth, semi-structured interviews, participant observation, and textual analysis of key documents. I conducted interviews with XR co-founders, key activists from the central UK teams, and the co-founders and "rank-and-file" members of XR Canterbury in 2020. As a participant observer, I drew on my experience volunteering with the UK media and messaging team from late 2018 into 2019, organising locally, and taking part in the two main London "Rebellions" in April and October 2019, including being arrested in October 2019 for obstructing the highway on Whitehall. I also drew on my experience as an elected local politician and parliamentary candidate for the Green Party, including proposing the motion for my council to declare a climate and ecological emergency in July 2019. Textual data included early versions of XR's Heading for Extinction talk, early iterations of XR's "demands", and the "Declaration of Rebellion", which was read out at XR's public launch in Parliament Square on 31 October 2018. It also included additional information, such as that supplied to me by Cllr Carla Denyer, of Bristol City Council (now co-leader of the Green Party of England and Wales and Member of Parliament for Bristol Central), who was the first person to propose a climate emergency motion in local government in the UK. I also accessed other XR-related content including talks and social media posts.

DOI: 10.4324/9781003396567-42

Framing as an Entry Point for Epistemic Inquiry

Social movements can be understood from a range of perspectives. These include: the mobilisation of resources (McCarthy and Zald, 1977; Gamson, 1975), political processes and opportunity structures (Tilly, 1978; McAdam, 1999; Kitschelt, 1986), framing (Benford, 1997; Snow et al., 1986; Snow et al., 2019), and cultural perspectives including concerns to do with collective identity (Hunt and Benford, 2004; Polletta and Jasper, 2001; Melucci, 1989) and emotions (Jasper, 2018; Goodwin et al., 2001; Gould, 2004). New social movement theories aimed to explain the wave of social movement activity that emerged in the 1960s as distinct from activity that had gone before (Buechler, 1995). Social movements have also been viewed as epistemic struggles (Icaza and Vázquez, 2013), as cognitive "praxis" and responsible for the social shaping of knowledge (Eyerman and Jamison, 1991), and as producers of knowledge (Cox and Flesher Fominaya, 2009; Chesters, 2016). Indeed, the environmental movement more broadly has been previously described as a largely epistemic project (Yearley, 1992).

I suggest that because of its inherent cognitive emphasis, alongside its concern with how resonance is generated amongst target publics, the framing perspective is best equipped to deal with epistemic questions relating to social movement mobilisation and the resonance required to generate participation. Whilst some literature to date has addressed – and even questioned – XR's use of knowledge (Berglund and Schmidt, 2020; Karnik Hinks and Rödder, 2023; Matthews, 2020; Thierry, 2023) and pedagogical processes (Zantvoort, 2023), there have been limited attempts to investigate how knowledge was implicated in the mobilising frames of Extinction Rebellion.

A frame analysis has the potential to elucidate how climate activists negotiate different knowledges, including the contentions and contradictions that arise from the framing of knowledge. Ultimately, an interrogation of these questions may help foster *epistemic justice* within activist communities and across global activist networks, and within the academic community. Kidd et al. (2017, p. 1) define epistemic *injustice* as:

> those forms of unfair treatment that relate to issues of knowledge, understanding, and participation in communicative practices. These issues include a wide range of topics concerning wrongful treatment and unjust structures in meaning-making and knowledge producing practices.

I therefore take *epistemic justice* as the amelioration of such *epistemic injustices*. Further, since Anderson (2012) suggests that epistemic justice is "a virtue of social systems" (p. 163), it is also a virtue of social movements.

An Overview of Extinction Rebellion's Framing

Extinction Rebellion (XR) emerged as a campaign of mass nonviolent civil disobedience in the UK in October 2018. It pushed for radical political change in government policy on climate and the environment. Co-founders were associated with the Rising Up activist network and plans for a "rebellion" started taking shape in April 2018. Campaign strategy was an evolution of experiments in nonviolent direct action by the Stop Killing Londoners and Vote No Heathrow campaigns, as well as university divestment campaigns in the UK, with which members of the Rising Up network were involved. Issues of framing were deliberated by the nascent group throughout 2018, and a strategy was devised with a view of attracting a support base for a mass "rebellion" in the autumn. It was decided amongst the nascent group to use a different set of frames which the climate justice movement had previously attempted to mobilise on.

According to Benford and Snow (2000), *collective action frames* are "action-oriented sets of beliefs and meanings that inspire and legitimate the activities and campaigns of a social movement organisation" (p. 614). Collective action frames, according to the authors, consist of three elements: the *diagnostic frame*, which identifies the problem at hand; the *prognostic frame*, which proposes what needs to happen to address the problem; and *motivational framings*, which serve to motivate people to take collective action.

XR's collective action frame during its nascency operated on multiple levels. On a *macro* level, the diagnostic frame (the problem) was that there is a climate and ecological emergency; the corresponding prognostic frame was that the UK had to decarbonise by 2025 and establish a Citizens' Assembly to define the specific prognostics on how to get there. On a *meso* level, the diagnostic frame was an implication that the UK government had not been telling the truth about the climate and ecological emergency; the corresponding prognostic frame here was that a campaign of mass nonviolent civil disobedience was needed to get the government to "tell the truth" and "act now". And on a *micro* level, diagnostic framing involved a personal interrogation of the truth about the climate and ecological emergency; my data indicates that participants did not necessarily simply accept any knowledge XR presented but felt compelled to do their own research before participating; the prognostic frame on this personal level being that what was needed, as a response to this truth, was personal involvement in nonviolent civil disobedience or other ways to support the movement. My data suggests that, for some participants, including central UK team members and decision makers, "the truth" was not necessarily just centred on scientific knowledge but was indeed associated with a sense of global climate injustice. Indeed, some participants in my study had previously been very involved with social and global justice activism, including with the Occupy movement.

When XR launched in late 2018, it did so with four dominant motivational frames. These included a framing of "it's worse than you think and time's running out", which was largely based on the premise that reporting from the Intergovernmental Panel on Climate Change was conservative and the knowledge it was based on was prone to scientific reticence (see also Thierry, 2023). A "failure of previous movements" frame asserted that the tactics of the climate movement had so far failed to make headway and that a campaign of mass civil disobedience was now the only option; this also related to the contention that climate justice frames had failed to gain sufficient traction in the UK. This was aided by a motivational framing best described as "the efficacy and propriety of nonviolent civil disobedience", and this was supported by social science knowledge about mass mobilisation. Finally, there was a "moral imperative and virtue ethics" framing which posited that taking part in the campaign was the right thing to do.

A *master frame* is defined as a "kind of master algorithm that colours and constrains the orientations and activities" of social movements (Benford and Snow, 2000, p. 618); as such, it serves as an umbrella frame under which other movement frames operate. Master frames are often transposable between social movements (Benford and Snow, 2000; Benford, 2013) and can serve as common ground between different social movements (Carroll and Ratner, 1996). Rights frames, choice frames, injustice frames, and environmental justice are some such master frames (Benford and Snow, 2000).

Instead of a justice master frame, XR operated within an emergency master frame, both contributing to and benefiting from the elevation of emergency framing to master frame status in 2018 through 2019. There was, however, pushback from within XR – and from within the wider social movement landscape – to adopt a climate justice master frame. XR, at the time of writing, now includes a climate justice master frame alongside an emergency master frame, as evidenced in the various iterations of its demands.

The Intergovernmental Panel on Climate Change Special Report on 1.5°C Global Warming (IPCC, 2018) was highly compelling for participants. However, crucially, it is important to note that this report was published in early October 2018: months after the nascent XR group began organising in April 2018. John, co-founder of XR Canterbury, highlights that in 2018 there was a fortuitous "confluence" of events that bolstered the case that XR was making about the state of the planet; as well as the IPCC report; this included publication, shortly after, of the World Wildlife Fund's (WWF) *Living Planet Report* on 22 October (Grooten and Almond, 2018).

XR also placed epistemic matters at the heart of its first demand: the UK government should *tell the truth* about the climate and ecological emergency by communicating how dire a situation the climate science was projecting. Climate science was also the key form of knowledge referenced in XR's Declaration of Rebellion. The Declaration states: "the science is clear: we are in the sixth mass extinction event and we will face catastrophe if we do not act swiftly and robustly". But, whilst these key strategic communications emphasised scientific knowledge, XR invited a deeper interrogation of climate science via a framing of "it's worse than you think and time's running out". In this sense, XR encouraged a critical attitude towards science. There was also space for the amplification of epistemologies other than climate science in XR.

Global Climate Justice Frames in XR

Whilst there was an emphasis on scientific knowledge – and although justice frames were suspended – there was an important tract of XR that worked to amplify global climate justice frames, which included local and Indigenous international knowledges. This is often overlooked in the burgeoning literature on XR and by its critics. The XR Internationalist Solidarity Network (XR ISN) was formed just after the London bridges occupation on 17 November 2018 and has been active in connecting XR UK with a global network of frontline activists and decolonial perspectives since, often having a presence at the main London "rebellions" (XR ISN, 2024b).

The XR ISN arose from the Global South Witness Group: a band of around eight activists including of Ghanian, Mongolian, West Papuan, Indian, Bangladeshi, and Jamaican heritage, who set about touring several of the bridges on the day of the bridges' occupation to share the knowledge and experiences of those on the frontline of climate breakdown in the regions to which they were connected. Members of the Global South Witness Group and the subsequent XR ISN have strong links with the international social movement for African reparations in the UK, including the "Stop the Maangamizi: We Charge Genocide/Ecocide!" campaign, which demands that the UK government establish an "all-party parliamentary commission for Truth and Reparatory Justice" and consider the "various proposals for reparations in accordance with the United Nations guidelines and principles on a right to remedy and reparation" (Stop the Maangamizi, 2024).

This does not just mean financial and political repairs for the oppression and genocide of African people under colonialism but also for many other types of repair, including social and cultural repairs for African people and the African diaspora in the UK. Further, the international social movement for African reparations sits within the wider global movement for reparations including the struggles of Indigenous peoples all around the world (Stanford-Xosei, 2024). Representatives of the XR ISN have been present at the major London Rebellions, have appeared on panels and given talks within XR, and the XR ISN co-hosted the Global Justice Rebellion at Millbank during the 2019 October Rebellion in London. It is led by "voices and experiences of frontline Global South and diaspora activists" and is:

> shaped by the generations of resistance that are our history and stand tall in our power – a power that is rising to meet the climatic, ecological and civilisational crises with true transformative justice.

> *(XR ISN, 2024a)*

An excellent introduction to reparations and how it factors in the struggle for climate justice is provided by XR ISN co-founder, reparations legal expert, and doctoral researcher, Esther Stanford-Xosei, in a presentation given in the London XR office in May 2019, which was recorded and posted by Extinction Rebellion UK on its YouTube channel (Stanford-Xosei, 2024). In the talk, Stanford-Xosei explains how:

> The peoples that we're working with by way of internationalist solidarity and the knowledge foundations that they base their liberation struggles and their resistance on is key.
>
> *(Stanford-Xosei, 2024, 01:02)*

She recognises that "there can be no reparatory justice without global cognitive justice" and that there is "the right of multiple forms of knowledge to exist" (Stanford-Xosei, 2024, 07:32). And she quotes her XR ISN co-founder, Kofi Mawuli Klu, when she says cognitive justice is the "repair of knowledge to do justice to every form of knowledge that exists; knowledge does not exist only in its European forms" (08:02).

The XR ISN thus provided a channel through which Indigenous and local epistemologies of those on the frontline of climate and ecological breakdown and affected by historical and current social injustices caused by colonialism could be amplified through XR – and these were carried by a climate justice master frame. The dominant master frame that Extinction Rebellion adopted, however, was the climate and ecological emergency master frame.

Knowledge/s in the Emergency Master Frame

Master frames of the climate movement have previously been identified as *climate change*, which was associated with more benign messaging historically adopted by NGO climate organisations, and *climate justice*, which was associated with more "radical" groups attempting to highlight the disproportionate impact of anthropogenic climate change on those who are least responsible for it (della Porta and Parks, 2014). In 2018 a climate and ecological emergency master frame rose to prominence, becoming a dominant master frame of the global climate movement.

Extinction Rebellion certainly was not the first to frame climate change as an emergency; prominent politicians, commentators, activists, and scientists have used emergency framing for climate change at least since the mid 2000s (Hodder and Martin, 2009). Prior to the climate movement, emergency framing was employed by the anti-nuclear movement in the 1980s (Hodder and Martin, 2009) albeit not necessarily explicitly, and it has been associated with social movement activity on protest rights (della Porta, 1998). In the climate movement, emergency framing emerged alongside advocacy for a World War Two-style mobilisation as a response to the problem. In 2011, environmentalists and advocates including Lester Brown and Bill McKibben called for such a mobilisation, petitioning President Obama in an open letter (Brown et al., 2011). The emergency frame was cemented in the climate movement in 2014 with the formation of The Climate Mobilization in the United States, who later, in 2016, helped organise a "Climate Emergency Caucus" to tie in with the Iowa Presidential Caucuses, shifting the emergency frame further into the political sphere. McHugh et al. (2021) suggest it was in 2016 that the term "climate emergency" was adopted in the mainstream media and that the emergency frame replaced one of future risk. That same year, the first local authority in the world declared a climate emergency: Darebin Council, in the northern suburbs of Melbourne, Australia.

The first local council in the UK to declare a climate and ecological emergency was Bristol City Council on 13 November 2018. The motion was proposed by Cllr Carla Denyer of the Green Party, who later became co-leader of the Green Party of England and Wales. Interestingly, the motion was already in the pipeline before XR's first major actions, although it had been suggested to Denyer by two local Rising Up activists who went on to join XR. Denyer, who kindly offered me insights into her motion, says:

It produced a massive response, because by then the first XR mobilisation had happened, the IPCC 1.5 degree report had just come out and there had been a summer/autumn of extreme weather events, so climate was very high on the political agenda (pers. comms).

In other words, scientific knowledge from the IPCC report underpinned the emergency framing of her motion, but knowledge of local or regional climatic events also contributed to a sense of emergency. Further, XR had just made its first mass public outing in London with a flash of colour and a big red banner stating, "THIS IS AN EMERGENCY", giving Denyer's emergency motion additional leverage.

Extinction Rebellion participants and co-founders that I interviewed also recognise how intrinsic the release of the IPCC Special Report 1.5°C was to the amplification of the emergency master frame. Ronan, central UK media and messaging team coordinator, says, "I love the emergency messaging . . . and with the IPCC and all of that, it all just collided at the right time". An early member of XR UK's political circle and founding member of the co-liberation and visioning circles, Mothiur, reflects:

the key argument, the key thing for me [was that] we had to bring the idea of an emergency into the climate debate . . . and that came through from the IPCC report.

Denyer mentions how extreme weather events contributed to the success of her emergency motion. Extreme weather also was intrinsic to how I framed my emergency motion, when, in July 2019, as an elected local politician in the Southeast of England, I proposed it to my council. Whilst I drew heavily on knowledge from the IPCC Special Report 1.5°C to evidence my motion, it felt more impactful to reference extreme local weather events; on the day my motion was passed by full council, a wildfire had ripped through a semi-rural area in my district: a highly unusual event for where I live. This rare occurrence gave empirical credibility to my argument for climate action; this was the climate emergency at proximity. Canterbury XR member Mike reflects on how extreme weather events bolster the emergency master frame. He says:

> These emergencies are a climate emergency. . . . We've seen them in Australia, we've seen them in California. We've also seen flooding locally. So, it is very apparent what is happening and I think that reinforces the urgency.

Canterbury group co-founder, Julie, however, indicates that people in the UK are perhaps too removed from the most severe impacts of climate change to feel motivated to act:

> Even in the face of wildfires in Australia, floods across the globe, wildfires in California . . . we're still in a situation where people here don't really get the climate emergency. They're detached from it still. . . . So, I guess my great hope is in the Australians because they've had the shit kicked out of them with the fires.

Canterbury XR member Henry says:

> It's obviously clear in Australia . . . they're experiencing temperatures which are hotter than any temperatures that they've experienced before . . . I think because we're a temperate island, we haven't had our toes singed.

Framing Contentions and Their Epistemic Implications

A key contention in XR's framing was that, whilst scientific knowledge was at the core of the diagnostic framing, social diagnostics about the root social-economic causes of climate change were overtly suspended, at least initially. This tied in with the broader strategic decision to suspend social justice frames and eschew a climate justice master frame. My interviews with co-founders, key members of the UK central teams, and the co-founders and members of a local group provide further insight into the contentions arising from the suspension of social justice frames and the eschewing of a climate justice master frame. In the following section I will explore what some of my interviewees said about these contentions.

The primary problem, as XR overtly framed it at the outset, was that there was an emergency because too much greenhouse gas is being emitted into the atmosphere and this threatens the planet's life systems and civilisation. Rather than explicitly name the social-economic systems that might be causing this, XR's strategy was to call for a Citizens' Assembly, through which such social diagnostics could be made. A Citizens' Assembly would also decide the best path forward, including solutions to reduce emissions; in other words, XR also deferred prognostic framing about what needed to happen to a Citizens' Assembly. As press team co-ordinator Andrew suggests, deferring to a Citizens' Assembly means:

> we don't have to say whether we support wind turbines or veganism or making reparations to African nations a key part of what we're all about . . . we don't have to

have a position on anything . . . as soon as you start having a position on stuff, you're immediately cutting away people who could be a part of it.

Further, the rationale for having a Citizens' Assembly was to bypass the problem of the press and the media frames that interfere with people's understanding of the climate and ecological crisis; in a Citizens' Assembly, a representative selection of citizens can be directly presented with scientific and social evidence to better identify the problem and the way forward. Andrew says:

All of that really important stuff can be part of the conversation, but let's put off having it be part of the conversation until we've actually created the forum in which the conversation can be had.

He also highlights the contention that this frame suspension caused:

that's largely why friends of mine dropped out of XR who had been key to setting it up in the first place . . . it was over these disagreements with other co-founders.

The ongoing contention over the suspension of social diagnostics and social justice frames often centred around the inclusion of a fourth demand that named these things. Some local groups in the UK, such as XR London, did include a fourth demand about "global inequalities caused by centuries of colonial, gender and class exploitation" (Extinction Rebellion London, 2022), which they added to the original three: Tell the Truth, Act Now, and for a Citizens' Assembly. Groups elsewhere in the world, such as XR USA and XR Canada, also added a fourth demand along these lines.

Co-founder Simon explains some of the background to this contention amongst members of the Rising Up network and other activists, as they set about designing the campaign of mass nonviolent civil disobedience, which was to become Extinction Rebellion:

We had our first talk about deeper framings, with the guys from Reclaim the Power and also Wretched of the Earth, and we were all talking about the meta framings involved . . . it was very much along what you might call the climate justice framings that the left has developed over the past thirty years. Some of us were very very happy with that kind of framing and wanted to continue with that. Some of us weren't. There was a lot of push-back [from a particular person]. . . around using those kinds of frames, saying that they didn't work, that it wouldn't get people involved.

Simon also explains how some of the first graphics that the nascent group produced for the campaign stated that "the climate crisis is a racist crisis". He recalls, however, that at some point a decision was made to not explicitly use justice framings. He also expresses regret about that decision and indicates that perhaps, strategically, XR would be in a much stronger position if more careful thought had been given around justice frames. Further, on emergency framing, Simon adds:

many of us don't believe that we're in an emergency, as it were . . . we're within an inevitability of a system that's become catabolic, cannibalistic and is ultimately devouring itself.

Early member of the political circle and founding member of the co-liberation and visioning circles, Mothiur, supported the idea of emergency framing and indicates that he was aligned with the suspension of justice frames. He says:

> There were six of us at the beginning in that political circle . . . and it was felt, and I agreed, that we needed to create a different narrative to what's been going on in terms of what the left have been talking about in terms of social justice.

But, for Mothiur, XR has been a journey, with the Black Lives Matter movement being a catalyst for a development in his perspective. He says that prior to Black Lives Matter:

> I intellectually knew, or I'd heard the arguments on climate justice, that the climate crisis is sourced in coloniality and oppression . . . and I kind of knew it but I didn't really give it much attention.

Mothiur recalls reading criticism directed towards XR: that XR leaned too much on certain scientific knowledge, such as that which evidences the melting of the icecaps today, when there is also scientific evidence that shows the historical changes in climate resulting from the colonisation of the Americas and the subsequent extinction of races and cultures centuries ago (see Koch et al., 2019). However, there is an indication in Mothiur's account which suggests that a burgeoning awareness of global climate justice issues did not negate his support for emergency framing.

Co-founder Robin highlights, however, that whilst XR did suspend overt justice frames during its initial mobilisation, the early versions of the Heading for Extinction talk script did emphasise the disproportionate impacts of climate change on the regions that are least responsible for it. The talk script states:

> to knowingly do so in benefit of our society's wealthiest few while hundreds of thousands of people, mainly from the Global South, die as a result of climate change is criminal.

Injustice was therefore implicit, although the explicit framing in this particular example is one of criminality. Robin says, however, in relation to overt justice frames: "people rightly wanted us to talk in more detail about that and how we have a responsibility in the UK to be acting on that". He also recognises that the Black Lives Matter movement focussed attention on how climate and racial crises intersect: "we need to make that analysis, because if we don't, we'll be siloed and not fighting together on this".

For Sam, member of the central UK media team and political strategy team, justice framings were of critical importance. He cites Naomi Klein's book *This Changes Everything* (Klein, 2015) as his wakeup call on global climate injustice. Going to COP24 in Poland cemented his sense of justice:

> Seeing how the elite who operate in the corridors of power and the total disregard that they do have to human life, and how black and white an issue this actually is and how existential it actually is, really hit home and really scared me.

Member of XR Canterbury Meg says, "we didn't get everything right" but that in her experience, from listening to co-founders and others give talks on behalf of XR, climate justice *was* the underlying narrative. Meg says there was a clear message that:

> we have to be talking about the Global South but not in the sense that we're somehow these sorts of saviours trying to save these people who have been fighting for this for years, we've got to follow their lead.

Whilst this message, for Meg, was at the heart of Extinction Rebellion, she says, "that somehow got lost on the ground . . . and from the perspective of others there were things that were really clunky and insensitive". Co-founder Frieda says:

> The question is how are we going to decrease this massive injustice . . . the answer is to create a mass movement . . . and how we're going to create a mass movement is we're going to talk in a language that people understand.

In other words, explicit justice frames are considered, by some, to be potentially obstructive to mass public participation in climate activism in the UK, and perhaps the very act of climate activism, no matter what frames are amplified, is in essence an act in the name of global climate justice. Co-founder Simon, however, suggests that social justice issues can gain traction in climate activism and he uses collective action around the Dakota access pipeline as an example:

> The idea that they do not gain traction has been found to be quite false over in America, where you've had all these groups of diverse activists who've used this social justice framing, this climate justice framing, and have had quite significant wins.

Sophie, from the UK media and messaging team, says:

> If our top line was "climate justice now", I would be asking to change it, because it's not going to achieve what we need it to, which is climate justice.

She also suggests that "a lot of people who don't understand it feel patronised by activists". But she also indicates that within XR, especially in XR Youth, there was a sense that suspending justice frames leaves no space to talk about the Global South. But, for people to know and understand about the experiences of people in the Global South, Sophie suggests:

> We need to get them to care first, and you will not get them to care first by essentially trying to make them feel guilty about all these people across the world that they don't connect with emotionally.

Nevertheless, Sophie also recalls that when XR tried to amplify voices from the Global South in media appearances, there appeared to be resistance from UK media outlets. She recalls how representatives of the XR ISN would be poised to give media interviews, but:

> so many times, black or brown people were just dropped and then it would be white people who'd still be platformed . . . we would get a call saying "oh sorry, the

interview's off" and we were like oh, that's funny, because it was [a member of the ISN] we were putting forwards, you know, instead of a young white girl.

The mass media is a major site of contest over meaning and social movements (Gamson, 2004). The media can influence public perception and the trajectory of movements (Gitlin, 2003), and this includes a subtle and unconscious influence on the frames which are amplified through the media (Gamson, 2004). Equally, the media plays a role in helping the public identify which actors represent a social movement (Klandermans and Goslinga, 1996). It is apparent that, whether consciously or unconsciously, media outlets acted as gatekeepers in terms of who represented XR in the mediasphere. And, whilst largely participation in XR reflects that of previous environmental activism (Hayes et al., 2020), media attention that homed in on XR as being primarily white and middle class (Bell and Bevan, 2021) was perhaps reinforced through media bias.

Yet, XR UK activist Charlie (pseudonym) highlights that in the early XR UK political circle there were two white people, three people of colour, two of them Muslim, two of them women, and XR ISN representatives were in the "anchor group", these two groups being decision-making bodies since the beginning of XR's public emergence. Emergency framing and the overt suspension of justice frames and eschewing of a climate justice master frame apparently did not necessarily preclude participation in strategic decision-making based on race – or class, for that matter. Charlie points to Gail Bradbrook and Roger Hallam's working class backgrounds and suggests that this had, in fact, played a part in why the "exclusive" established activist world rejected their ideas.

XR, however, later came under criticism for its perceived lack of awareness of working-class people when an action was organised by a small group that caused disruption in a predominantly working-class neighbourhood in Canning Town, London in October 2019. And one participant in my study, Mark, of XR Canterbury, who identifies as working class, reports that he found it "much harder to be listened to" because of his working-class background. Yet Mark did not indicate a particular affinity with emergency or justice framing: his motivation appears to arise primarily from the deep sadness he feels about the destruction of nature.

It is interesting to note that academic criticism of XR which claims that XR is by design white and privileged in terms of its demographic, tactics, and discourses (such as Berglund and Schmidt, 2020) completely ignores the work of the XR Internationalist Solidarity Network. A Google Scholar search reveals little to no mention of the XR ISN in the academic literature. Thus, epistemic injustice prevails in scholarly critiques of XR, even in research that is concerned with representation.

A similar dynamic appears to play out regarding gender representation – in the mediasphere and in the academic literature. Co-founder Clare offers insight into the perceived and projected power of social movement founders and the tendency for men to be perceived as leaders, especially by the media. She describes how at the time of XR's rapid scaleshift in late 2018, "everybody was talking about Roger as the founder". Co-founder Gail Bradbrook, according to Clare, subsequently highlighted that this tendency to focus on Roger was "how women get written out of history" and "it's not ok for Roger to be the founder . . . we have to do something about that". Clare indicates how "the press want to recognise some sort of hierarchy or box they can put you in". And so it was agreed that a group of them would all be called co-founders, and there are ten, perhaps twelve people who can be recognised as such. "And that was just done", Clare says, "for that purpose to prevent it from being Roger's movement".

An unconscious bias is evident in the academic literature, too. A Google Scholar search for literature on XR with Roger's name versus Gail Bradbrook generates vastly more results for Roger. Scholarly critiques of XR which focus on race and class appear to have a particularly strong tendency to elevate Roger's voice, even if it is to criticise it. There are even fewer mentions of Clare (Farrell) in the literature, who was hugely influential as one of the creative minds behind XR's visual identity and was, and still is, at the time of writing, a prominent media spokesperson, including during XR's major rebellions in 2019.

Contention around justice frames also extended to the social media sphere. In a movement that grappled with decentralisation in theory and practice, what frames were amplified across social media often came down to who had the passwords to XR's social media accounts. For instance, a member of the media and messaging team told me, "when I had control of the Twitter account, that meant that we did talk about capitalism, when someone else did it, it meant that we didn't". This person also suggests that, whilst there had been many conversations about social justice issues within XR, including the effects of capitalism and colonialism and their relationship to the climate emergency, most members across the UK were not engaged with those topics. A perceived lack of engagement with notions of global climate justice was the central concern raised by activist network Wretched of the Earth (2019) in a public letter to XR in May 2019.

However, Charlie, who was instrumental in the formation of the XR ISN, recalls one of the ISN co-founders saying to him about XR and the platform it gave them:

> We've been waiting for this . . . we've been let down consistently by the established climate activist world, who either use us tokenistically or ignore us completely, but always try to keep control.

Charlie, who has also been previously heavily involved in social justice activism, also says:

> I'm sure there's lots of other groups who would say they've been doing the same thing for years which is fine and all respect to them . . . but for the Majority World activists who were involved in this they said they've never had the opportunity to do this before.[1]

Some concepts from the social movement framing literature can help to tease out some of these contentions and how epistemic matters factor in.

Frame Resonance and Its Component Parts

Whilst social movement scholars have offered interpretations of Extinction Rebellion's collective action frame (de Moor et al., 2018; Smiles and Edwards, 2021; Buzogány and Scherhaufer, 2022; Hayes et al., 2020; de Moor et al., 2021), and I have offered a brief outline of my interpretation, a deeper investigation of processes to do with frame resonance may further an understanding of how knowledge factored into the frames of Extinction Rebellion.

The concept of frame resonance helps to explain the mobilising potency of social movement frames (Benford and Snow, 2000). Benford and Snow (2000) describe how frame resonance hinges on two sets of interacting factors: the credibility of a frame and a frame's salience. Frame credibility, according to the authors, is determined by three factors: the consistency between frames and the actions of a social movement, the empirical credibility of frames, and the credibility of frame articulators. Salience is established in three respects: if a frame is central to the beliefs, values, and ideas of a target public; if it is commensurable with the daily lived experience of people (experiential commensurability); and if the frame resonates on a cultural level. Three of these aspects of frame resonance are particularly relevant to epistemic concerns: empirical credibility, the credibility of frame articulators, and experiential commensurability.

Empirical credibility is established not in terms of whether diagnostic and prognostic claims are factual or valid but whether claims can be empirically verified (Benford and Snow, 2000). This explains why local extreme climatic events served to elevate and propel emergency framing into the political sphere. Knowledge of local climatic events could be used as evidence of the claims embedded in XR's emergency framing. This indicates the mobilising potency of such events and suggests it is important to continue to draw on such events in climate movement activity and future mobilisation. But it is less clear whether such local "emergencies" offer empirical credibility to global climate justice framings in the UK.

This leads to the credibility of frame articulators. To be sure, within the emergency master frame, which has rested largely on scientific reporting, scientists have served to lend credibility to XR's scientific claims; Scientists for Extinction Rebellion is a subgroup of XR which not only aims to help verify scientific claims and provide scientific credibility to XR but to shift scientific reporting away from a culture of reticence towards emergency framing (Scientists for Extinction Rebellion, 2024). In terms of the global climate justice frames which the XR ISN amplified, members of the XR ISN provided knowledge and experience of regions which are directly unjustly affected by climate change and the economic and colonial systems which have given rise to it; this includes scholarship and expertise on the international social movement for African reparations in the UK, which is the research focus of XR ISN co-founder and doctoral researcher Esther Stanford-Xosei.

Also related to the credibility of frame articulators, co-founder Roger, in our interview, suggests that "normal" people don't have time for articulators of justice frames who have not experienced injustice, and "what's the point of talking about justice when you're not in jail, basically". This also leads to the question: who *is* the arbiter of justice? Do articulators of justice frames risk assuming the role of arbiter, and is this problematic if they themselves do not have experience of injustice?

Finally, the concept of experiential commensurability describes how the salience of a frame – and consequentially a frame's resonance – is in part contingent on the frame's relevance to a person's life and everyday experiences. It follows that, if a person has experienced the impact of climate change or ecosystem collapse and if this is perceived to be an emergency, then there may be more support for an emergency frame. Though it would also follow that justice frames do not gain as much traction where climate injustices are less felt or visible. An apparent contradiction to this is the global school strikes movement which appeared to mobilise on global climate justice and emergency frames; though it is conceivable that mobilisation would not have occurred on justice frames alone.

External cultural factors related to salience are also relevant in understanding how frames factored into XR's mobilisation. Cultural resonance, that is, how "culture out there" imposes on social movement framing activity, is of great import (Benford and Snow, 2000). When XR rose to prominence, the United Kingdom was in the grips of a deep political and symbolic cultural battle around the UK's decision, via a referendum in 2016, to leave the European Union. It is conceivable that, for a nation bound up as it was so tightly in a solipsistic quandary over its own identity, issues of global justice may not have had the broad cultural resonance required for mass mobilisation or penetration into the media and political spheres. This, perhaps, explains XR's pursuit of a "moral imperative and virtue ethics" motivational frame over an injustice framing. Indeed, some of the language XR initially used, including in its Declaration of Rebellion, has patriotic overtones. An experiment in liberal and conservative framings also included the invocation of the political philosophy of Thomas Hobbes and John Locke in early versions of the Heading for Extinction talk.

Conclusion

The frames and framing processes which XR mobilised on are complex; unpicking them, including the contentions that arose, provides insight into how knowledge factored into personal participation and mass mobilisation. Controversially, XR initially suspended overt social justice frames in its collective action frame and eschewed a climate justice master frame, and it deferred social diagnostics to a Citizens' Assembly. Instead, it adopted a climate and ecological emergency master frame and, in place of injustice frames, employed motivational framings including a moral imperative and virtue ethics frame, to galvanise support.

Despite XR's contentious decisions around justice framing, XR's ISN was – and still is – at the time of writing, an important tract of XR's work, yet it is largely overlooked in the academic literature on XR. Academic criticism of XR which focuses on XR's ostensible lack of engagement with social justice issues still tends to amplify the voices of white male co-founders and key activists.

In her XR talk, Esther Stanford-Xosei (2024) expresses some of the resonances between concepts found in ancient African knowledge and some of the cultural frames expressed by many people within XR to do with fostering a re-connection to the earth; Stanford-Xosei highlights that Gail Bradbrook, for example, was an articulator of such frames. Stanford-Xosei also highlights how shared dialogue between the reparations and environmental movements have been occurring for some time, and she highlights a shared focus on the legal notion of ecocide and reflects on a conversation she had with lawyer Polly Higgins on a radio show Stanford-Xosei hosted on Voice of Africa Radio, years previously.

Issues of epistemic justice certainly do arise in the decisions social movements make about the frames they mobilise on and the knowledge included in frames. Whilst XR leant heavily on scientific knowledge via an emergency master frame, it did not exclude Indigenous knowledges. Epistemic justice also relates to the representation of knowledge and frame articulators in the media – and it is a challenge of the climate movement to puncture media bias. But it also relates to the representation of knowledge and frame articulators in the academic literature.

Finally, the framing perspective offers useful concepts to aid an understanding of social movement participation and mobilisation, including the epistemic aspects of frame resonance. But there are indeed other aspects to personal motivation and mass mobilisation. Cultural, emotional, organisational, and wider structural political analyses are amongst other useful perspectives that can help build knowledge about – and contribute to – climate and ecological activism.

Note

1 To note: this view was relayed, not directly from an XR ISN activist; I was unable to secure interviews with ISN activists during my period of data collection.

References

Anderson, E. (2012). Epistemic justice as a virtue of social institutions. *Social Epistemology*, 26(2), 163–173.

Bell, K. and Bevan, G. (2021). Beyond inclusion? Perceptions of the extent to which Extinction Rebellion speaks to, and for, Black, Asian and Minority Ethnic (BAME) and working-class communities. *Local Environment*, 26(10), 1205–1220.

Benford, R. D. (1997). An insider's critique of the social movement framing perspective. *Sociological Inquiry*, 67(4), 409–430.

Benford, R. D. (2013). Master frame. In D. A. Snow, D. della Porta, B. Klandermans and D. McAdam (eds.) *The Wiley-Blackwell Encyclopedia of Social and Political Movements* (pp. 1–2). Oxford: Blackwell Publishing Ltd. https://doi.org/10.1002/9780470674871.wbespm126 [Accessed 20 February 2024].

Benford, R. D. and Snow, D. A. (2000). Framing processes and social movements: An overview and assessment. *Annual Review of Sociology*, 26(1), 611–639.

Berglund, O. and Schmidt, D. (2020). *Extinction Rebellion and Climate Change Activism: Breaking the Law to Change the World*. London: Palgrave Macmillan.

Brown, L., et al. (2011). *Open Letter to President Barack Obama and President Hu Jintao*. https://350.org/getting-message-where-it-counts/ [Accessed 4 October 2023].

Buechler, S. M. (1995). New social movement theories. *The Sociological Quarterly*, 36(3), 441–464.

Buzogány, A. and Scherhaufer, P. (2022). Framing different energy futures? Comparing Fridays for Future and Extinction Rebellion in Germany. *Futures*, 137, 102904. https://doi.org/10.1016/j.futures.2022.102904 [Accessed 20 February 2024].

Carroll, W. K. and Ratner, R. S. (1996). Master framing and cross-movement networking in contemporary social movements. *The Sociological Quarterly*, 37(4), 601–625.

Chesters, G. (2016). Social movements and the ethics of knowledge production. In K. Gillan and J. Pickerill (eds.) *Research Ethics and Social Movements* (pp. 11–26). London: Routledge.

Cox, L. and Flesher Fominaya, C. (2009). Movement knowledge: What do we know, how do we create knowledge and what do we do with it? *Interface: A Journal for and About Social Movements*, 1(1), 1–20.

de Moor, J., Doherty, B. and Hayes, G. (2018). The 'new' climate politics of Extinction Rebellion? *Open Democracy*. https://www.opendemocracy.net/en/new-climate-politics-of-extinction-rebellion/ [Accessed 20 February 2024].

de Moor, J., et al. (2021). New kids on the block: Taking stock of the recent cycle of climate activism. *Social Movement Studies*, 20(5), 619–625.

della Porta, D. (1998). Protests, protestors, and protest policing: Public discourses in Italy and Germany from the 1960s to the 1980s. In M. Giugni, D. McAdam and C. Tilly (eds.) *How Social Movements Matter* (pp. 66–96). Minneapolis: University of Minnesota Press.

della Porta, D. and Parks, L. (2014). Framing processes in the climate movement: From climate change to climate justice. In M. Dietz and H. Garrelts (eds.) *Routledge Handbook of the Climate Change Movement* (pp. 19–30). Abingdon: Routledge.

Extinction Rebellion London (2022). [Twitter], 23 February. https://twitter.com/XRLondon/status/1496523453066850321 [Accessed 23 February 2024].

Eyerman, R. and Jamison, A. (1991). *Social Movements: A Cognitive Approach*. Cambridge: Polity Press.

Gamson, W. (1975). *The Strategy of Social Protest*. Homewood, IL: The Dorsey Press.

Gamson, W. (2004). Bystanders, public opinion, and the media. In D. A. Snow, S. A. Soule and H. Kriesi (eds.) *The Blackwell Companion to Social Movements* (pp. 242–261). Oxford: Blackwell Publishing.

Gitlin, T. (2003). *The Whole World Is Watching: Mass Media in the Making and Unmaking of the New Left*. Berkeley and Los Angeles, CA: University of California Press.

Goodwin, J., Jasper, J. M. and Polletta, F. (eds.) (2001). *Passionate Politics: Emotions and Social Movements*. Chicago: University of Chicago Press.

Gould, D. B. (2004). Passionate political processes: Bringing emotions back into the study of social movements. In J. Goodwin and J. M. Jasper (eds.) *Rethinking Social Movements: Structure, Meaning, and Emotion* (pp. 155–176). Oxford: Rowman & Littlefield.

Grooten, M. and Almond, R. E. A. (2018). *Living Planet Report – 2018: Aiming Higher*. Gland, Switzerland: World Wide Fund for Nature.

Hayes, G., Doherty, B. and Saunders, C. (2020). *A New Climate Movement?: Extinction Rebellion's Activists in Profile*. Aston University School of Social Sciences and Humanities Working Paper. https://publications.aston.ac.uk/id/eprint/41725/ [Accessed 20 February 2024].

Hodder, P. and Martin, B. (2009). Climate crisis? The politics of emergency framing. *Economic and Political Weekly*, 44(36), 53–60.

Hunt, S. A. and Benford, R. D. (2004). Collective identity, solidarity, and commitment. In D. A. Snow, S. A. Soule and H. Kriesi (eds.) *The Blackwell Companion to Social Movements* (pp. 433–457). Oxford: Blackwell Publishing.

Icaza, R. and Vázquez, R. (2013). Social struggles as epistemic struggles. *Development and Change*, 44(3), 683–704.

IPCC (2018). *Global Warming of 1.5°C: An IPCC Special Report on the Impacts of Global Warming of 1.5°C Above Pre-Industrial Levels and Related Global Greenhouse Gas Emission Pathways, in the Context of Strengthening the Global Response to the Threat of Climate Change, Sustainable Development, and Efforts to Eradicate Poverty*. Edited by V. Masson-Delmotte, P. Zhai, H. O. Pörtner, D. Roberts, J. Skea, P. R. Shukla, A. Pirani, W. Moufouma-Okia, C. Péan, R. Pidcock, S. Connors, J. B. R. Matthews, Y. Chen, X. Zhou, M. I. Gomis, E. Lonnoy, T. Maycock, M. Tignor and T. Waterfield. https://www.ipcc.ch/sr15/ [Accessed 25 November 2022].

Jasper, J. M. (2018). *The Emotions of Protest*. Chicago: The University of Chicago Press.

Karnik Hinks, E. and Rödder, S. (2023). The role of scientific knowledge in Extinction Rebellion's communication of climate futures. *Frontiers in Communication*, 8. http://doi.org/10.3389/fcomm.2023.1007543 [Accessed 20 February 2024].

Kidd, I. J., Medina, J. and Pohlhaus Jr., G. (2017). Introduction to the Routledge handbook of epistemic injustice. In I. J. Kidd, J. Medina and G. Pohlhaus, Jr. (eds.) *The Routledge Handbook of Epistemic Injustice* (pp. 1–9). London: Routledge.

Kitschelt, H. P. (1986). Political opportunity structures and political protest: Anti-nuclear movements in four democracies. *British Journal of Political Science*, 16, 57–85.

Klandermans, B. and Goslinga, S. (1996). Media discourse, movement publicity, and the generation of collective action frames: Theoretical and empirical exercises in meaning construction. In D. McAdam, J. D. McCarthy and M. N. Zald (eds.) *Comparative Perspectives on Social Movements* (pp. 312–337). Cambridge: Cambridge University Press.

Klein, N. (2015). *This Changes Everything: Capitalism vs. the Climate*. New York: Simon and Schuster.

Koch, A., et al. (2019). Earth system impacts of the European arrival and Great Dying in the Americas after 1492. *Quaternary Science Reviews*, 207, 13–36.

Matthews, K. R. (2020). Social movements and the (mis) use of research: Extinction Rebellion and the 3.5% rule. *Interface: A Journal for and About Social Movements*, 12(1), 591–615.

McAdam, D. (1999). *Political Process and the Development of Black Insurgency, 1930–1970*. Chicago: University of Chicago Press.

McCarthy, J. D. and Zald, M. N. (1977). Resource mobilization and social movements: A partial theory. *American Journal of Sociology*, 82(6), 1212–1241.

McHugh, L. H., Lemos, M. C. and Morrison, T. H. (2021). Risk? Crisis? Emergency? Implications of the new climate emergency framing for governance and policy. *Wiley Interdisciplinary Reviews: Climate Change*, 12(6), 1–15. https://doi.org/10.1002/wcc.736 [Accessed 20 February 2024].

Melucci, A. (1989). *Nomads of the Present*. London: Hutchinson Radius.

Polletta, F. and Jasper, J. M. (2001). Collective identity and social movements. *Annual Review of Sociology*, 27, 283–305.

Scientists for Extinction Rebellion (2024). *Scientists for Extinction Rebellion Homepage*. https://www.scientistsforxr.earth [Accessed 25 February 2024].

Smiles, T. and Edwards, G. A. S. (2021). How does Extinction Rebellion engage with climate justice? A case study of XR Norwich. *Local Environment*, 26(12), 1445–1460.

Snow, D. A., et al. (1986). Frame alignment processes, micromobilization, and movement participation. *American Sociological Review*, 51(4), 464–481.

Snow, D. A., Vliegenthart, R. and Ketelaars, P. (2019). The framing perspective on social movements: Its conceptual roots and architecture. In D. A. Snow, S. A. Soule, H. Kriesi and H. J. McCammon (eds.) *The Wiley Blackwell Companion to Social Movements* (pp. 392–410). Oxford: John Wiley & Sons Ltd.

Stanford-Xosei, E. (2024). *XR Talks – Esther Stanford-Xosei – An Introduction to Reparations – Extinction Rebellion'*. https://www.youtube.com/watch?v=KI5WsvwVIYA [Accessed 23 February 2024].

Stop the Maangamizi (2024). *Stop the Maangamizi: We Charge Genocide/Ecocide Campaign Homepage*. https://stopthemaangamizi.com [Accessed 23 February 2024].

Thierry, A. (2023). Heading for extinction: The representation of scientific knowledge in Extinction Rebellion's recruitment talks. *Frontiers in Communication*, 8. https://doi.org/10.3389/fcomm.2023.1237700 [Accessed 20 February 2024].

Tilly, C. (1978). *From Mobilization to Revolution*. New York: Random House.

Wretched of the Earth (2019). *An Open Letter to Extinction Rebellion*. https://www.redpepper.org.uk/environment-climate/climate-change/open-letter-to-extinction-rebellion/ [Accessed 24 February 2024].

XR ISN (2024a). *Extinction Rebellion Internationalist Solidarity Network about Page*. https://www.xrisn.earth/about [Accessed 23 February 2024].

XR ISN (2024b). *Extinction Rebellion Internationalist Solidarity Network Homepage*. https://www.xrisn.earth [Accessed 23 February 2024].

Yearley, S. (1992). *The Green Case: A Sociology of Environmental Issues, Arguments and Politics*. Abingdon: Routledge.

Zantvoort, F. (2023). Movement pedagogies in pandemic times: Extinction Rebellion Netherlands and (un)learning from the margins. *Globalizations*, 20(2), 278–291.

36

CONVENING CLIMATE ACTIVISM IN CANADA

Conflicting Expertise and the Production of the Problem

Adam Fleischmann

Introduction

Early Saturday morning, 6 October 2018, push notifications lit up phones across the eastern half of North America just as the rising sun hit the weekend coast. Messages were coming in from a time zone half a world and more than half a day away – from Incheon, South Korea. The 48th session of the Intergovernmental Panel on Climate Change (IPCC) had just come to a close. North American climate civil society organisations – never a cohort accused of respecting normal business hours – were writing home in exhausted celebration. The victory being celebrated? The approval of the IPCC's Special Report on the impacts of 1.5°C (or 2.7°F) of global warming.

Heading that delegation of observer civil society organisations sending daybreak dispatches was Catherine Abreu, then-Executive Director of *Climate Action Network-Réseau action climat* or CAN-Rac (Climate Action Network-Réseau action climat Canada, 2018).[1] They were not celebrating the results of the IPCC's research, per se. In fact, the victory for civil society groups was their successful effort to meaningfully include a powerful and honest description of the impacts of 1.5°C in the report's Summary for Policymakers (Masson-Delmotte et al., 2018). Hard-won was the inclusion of the very real human and non-human suffering, ecosystem devastation and biodiversity loss due by around 2040 if we as a species – some groups with more impact than others – continue living together as we currently do. In her role at CAN-Rac, Abreu was described to me as, among other things, "a wizard at taking research and evidence and translating it into good policy."[2] Abreu and CAN-Rac played the important role in Incheon of bringing diverse voices to the table and civil society to scientists, who would then write the much-cited Summary for Policymakers.

This chapter focuses on the Canadian climate movement and the role of expert-activists in convening the movement, coordinating disparate groups and interests and mediating disputes – sometimes about the very nature of the problem of climate change. Colloquially pronounced "can-rack," CAN-Rac is a network of more than 130 non-governmental and local non-profit organisations across Canada, part of the international CAN system. Founded in the 1980s, CAN calls itself the largest climate network in the world (Climate Action Network, 2023), composed of nearly 2,000 non-industry, non-governmental and

DOI: 10.4324/9781003396567-43

community non-profit organisations with regional and national nodes of the network, including CAN-Rac. The latter is a coalition of more than one-hundred Canadian groups: activist organisations from local grassroots groups to national NGOs, faith-based, humanitarian and physicians groups, First Nations assemblies, labour unions and more. Individual members of member organisations are as diverse as students, professionalised activists, Indigenous activists, retired religious community members and more. CAN-Rac staff act as network conveners and policy coordinators in Canada and beyond. They also act as a bridge between large international events and what needs to happen on the ground on a national and sub-national scale. Despite having a large network of members and "being a powerhouse in this realm," as one staff member described the organisation,[3] they accomplished this work with a staff of only four people at the time of my research.

CAN-Rac is one group of grassroots organisations within a broader, informal network of non-governmental organisations (NGOs) in the United States and Canada. This broader, informal network is the focus of the larger anthropological study on which this chapter is based. The broader study, based on 17 months of ethnographic field research in 2017–2019, examines how climate change experts in NGOs in the US and Canada approached climate change at the zone of encounter between politics and science. In order to study this network of NGOs, I employed qualitative fieldwork, combining in-person and digital ethnographic methods (such as participant observation and in-depth, open-ended interviews) with fieldwork in the non-geographically defined spaces of complex global phenomena and knowledge systems (Fleischmann, 2022; Knox, 2020). The organisations in this network are *not* the formal institutions of the "science-policy interface," in the sense that the term is often used in governance spheres, in Science and Technology Studies (STS) and in other social sciences studying climate change or environmental governance more broadly (e.g. Lahsen, 2008; United Nations Committee of Experts on Public Administration, 2021; van den Hove, 2007). Rather, my broader research emphasises the non-state spaces of action among what I call "mid-level" climate change experts: climate actors with expertise in climate science or politics and on-the-ground experience in climate organising spaces (Fleischmann, 2022). The broader research shows how, in these spaces of science communication and education, policy and data analysis, coordination and convening, both climate change science and politics are negotiated, problematised and made intervenable.

The work of mid-level experts in this context organises itself into an informal network of organisations who operate between climate change science and climate change politics. The organisations exist along a continuum: some organisations more than others work more closely with the data, modelling and dynamics of climate science; others perform their work closer to the advocacy, policy work and activist-organising of climate politics. CAN-Rac is situated closer to the political action end of the continuum. Their role as conveners and policy analysts situates CAN-Rac in the unique position of simultaneously supporting and convening the activist work of its member organisations, as well as acting as a key civil society player in the realm of domestic and international climate politics (though the latter will not be a focus of this chapter). The organisations in this broader network, like CAN-Rac, are commonly staffed by telecommuters, working remotely over conference calls and digital communication technology. They periodically meet in person, bringing together organisations and broader networks in the realm of climate action. These reunions often occur at the diplomatic and organising summits that are the outcome of months of work.

Ethnography and Settler-Indigenous Environmentalist Conflict

This chapter ethnographically analyses one part of one such event: CAN-Rac's ClimaCon 2018 conference and a conflict that occurred at it, which speak to larger questions in the study of grassroots climate activism. Grounded in anthropological perspectives and ethnographic methods, this chapter employs an in-depth qualitative research methodology that follows anthropological and ethnographic approaches to studying events, institutions and organisations (e.g. Mauksch, 2020; Gellner and Hirsch, 2001; Smith, 1987). These approaches understand events as reflexive, processual spaces where power relations are performed and ethical evaluations and relations are enacted. They understand work and world as enmeshed; work as articulated to larger structures via social relations. An ethnographic approach in these settings is committed to the "exploration, description and analysis of such a complex of relations" through precisely the everyday work and worlds they organise and in which they play out (Smith, 1987, p. 160).

As part of an ethnographic approach to studying grassroots climate activism, this chapter must account for the larger context in which these events play out. Climate and energy discourse in the Canadian context are actively entrenched in an extractive industry that is at the centre of a divisive technological nationalism (Barney, 2017). This technological nationalism, the state and attendant extractive industries are, of course, predicated on the existence of the settler-colonial relation. Settler-colonialism is understood, following Yellowknives Dene theorist Glen Coulthard, as a condition of political action in this context: "the inherited background field within which market, racist, patriarchal, and state relations *converge* to facilitate a certain power effect" in the present (Coulthard, 2014, p. 14). That power effect is one that, often through projects of extractivism, continues to "facilitate the *dispossession* of Indigenous peoples of their lands and self-determining authority" (Coulthard, 2014, p. 7, emphasis original). These settler-colonial politics are intimately enmeshed in environmental politics in Canada, including the politics of climate change.

Canadians have had consistently positive attitudes toward and media coverage of climate change (Callison and Tindall, 2017, p. 3). Despite this, settler environmental groups and Indigenous groups have often had discrete goals and understandings of climate change and other environmental issues. Characterising climate change and the environmental movement in Canada, Callison and Tindall note that "[i]n some cases, environmental groups have worked with Indigenous groups, but they have largely separate goals and structures" (2017, pp. 29–30). The authors use the example of oil pipelines crossing Indigenous lands not covered by treaties in British Columbia. While both groups have protested pipelines, they have often done so for different reasons: First Nations groups have protested the lack of Indigenous consultation and consent required by law, while the primary motivation of (settler) environmentalists has been limiting greenhouse gas emissions by keeping oil in the ground. The two groups have at times faced conflict.

Anjali Helferty is a fellow researcher of Canadian climate politics and was a fellow member of the Steering Committee for CAN-Rac's ClimaCon 2018. As she put it in a 2020 study of settler anti-oil pipeline activists and their collaborations in solidarity with Indigenous Peoples, Canadian climate activism is immersed in an ongoing history of violence: "Living just below the surface of interactions between settler environmentalists and Indigenous peoples today is the extended history of colonialism perpetuated by settler environmentalists in the name of conservation, wilderness protection, or animal rights" (Helferty, 2020, p. 16). She cites recent attention to the damages to Inuit lives and livelihoods caused by anti-seal hunt campaigns since the 1960s thanks to the 2016 documentary *Angry Inuk* (Arnaquq-Baril, 2016). Yet even within protest campaigns deemed legendarily successful in the Canadian environmentalist imaginary today, conflict emerged between Indigenous and settler activists. During the Clayoquot Sound protests of the 1980s and 1990s to stop logging on Vancouver Island, the Nuu-chah-nulth peoples and settler protestors had different understandings of the problem and different goals. These differing articulations of the problem led to conflict, to the extent that "the framing of the Clayoquot Sound protests as an unreserved victory becomes complicated and contested" (Helferty, 2020, p. 17). Nearby, the violent, racially charged encounters of the Makah whaling conflict provide a more unambiguous example of settler-Indigenous environmentalist conflict. As formerly endangered grey whale populations boomed in the mid-1990s, Nuu-cha-nulth relations in the nearby Makah Tribe of Northwest Washington State looked to re-establish their Treaty right to whale after an 80-year voluntary moratorium. Settler environmentalists, led in the media by Greenpeace, acted with the preservationist goal of a complete prohibition on whaling; traditional Makah whaling societies worked to secure their Treaty rights in accordance with their cultural and spiritual systems.[4]

Within this broad context, the work of CAN-Rac and ClimaCon proved eventful. The events of the conference revealed something greater about the Canadian climate change movement at the time: about its place in the history of the institution of white settler-colonial environmentalisms in Canada; the state of the diversity of the movement in the late 2010s; and the diverse visions, ethics and epistemologies of climate change and its futures at work in the movement. Alongside the history of Indigenous-settler conflict in grassroots environmental movements, the events described later speak to climate change's challenges to the kinds of ethical and political relations made constitutive among epistemology – knowledge, its history and how we know what we know – expertise and political action. As I explore later, in doing climate activist work together, diverse members of the CAN-Rac climate coalition sometimes enact competing forms of advocacy and visions of the future, entailing differing articulations of ethical reasoning and relations, knowledge and expertise. These are constitutive of contesting articulations of the problem of climate change.

This chapter therefore addresses larger questions of longstanding interdisciplinary interest, about the individual's relations to larger collectives, global systems, settler colonialism, difference and ethics. In addition, this chapter responds to key comparative questions in the study of grassroots climate activism – questions that guide the research plotted by this handbook – while exploring their limits. How do diverse climate activists understand their work in their own terms? What visions of societal transformation are climate activists articulating in different contexts and what forms of politics, socio-economic futures, and alternative ways of living and working do such visions generate? How do different activists engage differently with the moral dimensions of the climate crisis, such as responsibility and justice?

After introducing ClimaCon, I next employ ethnographic narrative and its exploration, description and analysis of the everyday at the conference, before turning specifically to a central conflict that occurred, its disruption and relative resolution. Through an analysis of these events, I enter into conversation with queer and feminist studies scholar Sara Ahmed and Indigenous media, STS and anthropology scholar Candis Callison to better understand the legacies of ongoing violent histories, the active work being done to combat the inheritance of these legacies in grassroots activist spaces and the differing epistemologies, ethics and articulations of the problem of climate change that exist simultaneously in the same activist space.

CAN-Rac ClimaCon 2018

ClimaCon 2018 was a two-day conference convened by Climate Action Network-Réseau action climat Canada on 10–11 October 2018 at Toronto, Canada's York University, an annual gathering of Canadian climate groups in its third year. As a researcher and participant-observer immersed in CAN-Rac's work and world, I served on the conference's online organising Steering Committee and as a volunteer on the ground during the conference. Conference organisers targeted numbers of 200–250 attendees. Open to the public but focused on network members, the conference was organised around priorities garnered from a poll of members early in the planning process in spring 2018: members indicated that they wanted opportunities for more strategic planning, relationship building and enhanced engagement within the network. The program subsequently involved few plenary and panel sessions and instead opted for smaller workshops and groups sessions that reflected these inward-looking priorities.

Sessions were planned as targeted relation-building events, meant to allow participants to reflect on some harder questions, such as who is in the room, physically, that day but also "in the room" of Canadian climate politics more broadly. Primarily, however, the conference revolved around ten two-day issues-based, small-group breakout sessions, with the goal of members leaving the conference with some strategic takeaways by the end of the second day of these sessions. Breakout session themes included just transition and labour, international climate policy, legal tools for climate action, Indigenous rights and the climate movement, "Canadian climate politics + elections" and more. For each breakout session, speakers and a facilitator or two with expertise in the session theme were recruited from among CAN-Rac staff and member organisations: staff from the environmental law and First Nations legal defence groups, Ecojustice and RAVEN Trust; staff and volunteers from Indigenous Climate Action and Red Rising Magazine; organisers from the Canadian Labour Congress and the Green Economy Network; CAN-Rac's International Policy Coordinator for the session focused on that topic and more. With the narrative tone and tools of ethnographic description, I next turn to some of the events of the conference and what they can tell us about the Canadian climate movement and beyond.

The Opening Ceremony

The conference began with an opening ceremony, putting Indigenous perspectives up front. It was initiated with a territorial welcome conducted by Anishinaabekwe (Anishinaabe woman in Anishinaabemowin – or Ojibwe language) elder Kim Wheatley. Seated at stage left at one of the fifteen or so round, ten-person tables in the conference hall, I listened and wrote in my field notes as Wheatley then stood at the narrow podium on the short stage and framed the events of the days ahead. "This is all our land," she began. "We've been welcoming for five-hundred years and I want to continue to be that." She mentions the Mississauga Treaty of 1805, the 13th of the Numbered Treaties, sometimes called the Toronto Purchase Treaty, which included surrounding areas of what are now Etobicoke, Toronto, North York, York and Vaughan – and which, in 2010, was subject to a $145 million dollar settlement between the government of Canada and the Mississaugas. Swept up in Wheatley's words, I wrote in my notebook later how wonderfully articulate, passionate and patient a welcoming speech it was, explaining the basics of Indigenous sovereignty and stewardship on those lands. "Mother Earth never makes mistakes," Wheatley declared, continuing, "the Western science module is only one. There are other ways of knowing." For thousands of years, she notes, "We [Indigenous people] did a really good job at stewardship. There were not species at risk." In 200 years, this has all been undone, she indicates. Perhaps there are other, older ways of knowing and governing to which we can return. From the start, ClimaCon organisers attempted to centre Indigenous perspectives and presences.

After Wheatley's opening ceremony, CAN-Rac Executive Director Cat Abreu emphasises that what Wheatley spoke of is an important part of the context of ClimaCon 2018. Before the ClimaCon can continue, a CAN-Rac member, Mitchell Beer, hops on stage to give a gift to Cat in congratulations of her achievements at the 48th session of IPCC in South Korea, which ended just four days prior. "Eddy, get up here!" she calls to CAN-Rac International Policy Coordinator, as Beer congratulates their efforts in leading civil society groups at the meeting "to make sure the IPCC released a *real* 1.5°C report." Setting the stage for the global stakes of the rest of the conference, "the difference between 1.5° and 2° is the difference between the paradise we live in now and climate catastrophe," Cat emphasises. With the group of conference goers welcomed onto the territory by Wheatley and the stakes articulated for the work laid out for us, the conference began.

Breakout Sessions: Conflict and Conditions of Experience

After the opening ceremony, an icebreaker activity and a facilitated exercise ensued, which outlined the challenges and possibilities of living and doing climate activist work in diverse coalition settings, across differences: settler and Indigenous, economic, cultural, racial and gender. Before the full group of conference participants split up into breakout groups, a CAN-Rac staff member went over a series of principles, projected on three of the four large screens of the conference room, high up near the tall ceilings. The slide read: "Group Agreements." Not unlike ground rules, guidelines or group norms I had established with students before leading undergraduate seminar discussions – or, more closely, calls for attention to who is making and taking space in other activist settings – these agreements were meant to induce reflection in participants and establish welcoming spaces for open discussion in milieux with people of diverse backgrounds in terms of gender, class, race and other positionalities. The Group Agreements laid out here in two columns on the large projection screens included "Challenge the concept, not the person" and "Engage tension, don't indulge drama." Others included "Assume best intent – attend to impact" or "W.A.I.T. why am I talking" to encourage reflection before speaking, especially for those for whom our society often gives time and a platform to speak. The emphasis was on community-building and creating caring, open spaces of conversation, despite the differences in the room.

As it turns out, while I was busy taking notes for our relatively straightforward third breakout session on international climate politics, other conference-goers in other breakout groups were having more complicated conversations. My ClimaCon field notes began a new page with a pointed question: "what can I say about the afternoon?"

The question follows a lacuna, a gap in time and notes filled with pointed exchanges and a reluctant intervention. During the afternoon report-back period back in the main conference room, one rapporteur from each breakout group was supposed to present an "ah-ha" moment, one thing our group wanted everyone to know from our breakout and one action point, to later write on a post-it note and stick onto a large easel paper pad with all the rest. The groups went around and eventually it was the turn for the Indigenous Rights and the Climate Movement breakout room. Two women stood up and shared not takeaways but an intervention, a call out and a call in, revealing of the presence, here and now, of the challenges intimated in Group Agreements and dramatised in the morning's facilitated exercise.[5]

In ethnographic research, the partiality of the researcher is assumed. Knowledge and knowledge production are understood to be situated, interpretative, coming from a particular standpoint. As throughout much of the interpretive social sciences, claims to objectivity and devotions to positivism are met with ambivalence, understood as "particularly bad guides to how scientific knowledge is actually made" (Haraway, 1988, p. 576). This is the case for ethnography of circumstances directly observed as much as for those for whom the researcher was absent, the repercussions of which nonetheless reverberate through the present as a condition of experience. One cannot be everywhere at once. But the ethnographer can attempt to render the conditions of other people's experience as the ethnographer comes to know them through fieldwork. The researcher can then ask what these conditions tell us of the constellations of power and knowledge in and through which people live.[6] Much ethnographic research in contexts of war, political violence or post-conflict settings, for example, deals with consequential data beyond the temporality of the ethnographic moment, beyond direct participant observation (e.g.: Das, 2007; Gamarra and Pimentel Mauricci, 2023). This is important to bear in mind when accounting for how settler-colonialism forms the background of settler-Indigenous environmentalist conflict.

Faced with significant events beyond the temporality of my direct observations at ClimaCon, I focus here not on a so-called objective accounting of events but on the after-effects of actions that proved objects of critique within broader ethical evaluations of the event and articulations of the problem of climate change and its attendant futures. After the breakout sessions at ClimaCon, I pieced together what happened from various sources: over the course of the initial short intervention during report backs; in multiple hushed conversations at the bar during the Pecha Kucha event that evening with CAN-Rac staff, Indigenous breakout room leaders and some of the more seasoned, professional organisers working for organisations like 350.org and Greenpeace; on the car ride home as conference facilitator Amara graciously drove me back to my friends' apartment where I was staying that evening; in the next morning's early meeting and during the subsequently reworked agenda of the second day of the conference, guided by CAN-Rac Director, Cat. Through these first- second- and third-hand accounts, I learn that, in at least two breakout sessions, including the Indigenous Rights and the Climate Movement group, the Group Agreements laid out before we broke out into our breakout groups had not been respected. I am later told that during the breakout session one Indigenous woman session leader, whom I would later learn was an activist I had befriended during the conference, had a "very circular, poetic 'auntie' way of speaking" not familiar to some of the white participants.

Back at the afternoon's report-back sessions, the two leaders of the Indigenous Rights and the Climate Movement breakout room shared their accounts of the events. Notetakers in the Indigenous Rights and the Climate Movement room had complained that conversation was not "linear." Some of the Indigenous people of colour session leaders and participants felt their experiences, expertise and ways of speaking were not being respected, the two speakers say; they felt that, despite the Group Agreements, space was being taken up by others who perhaps thought that they knew better. In one session, an unfortunate "ugly dynamic" emerged, where white women conference participants were challenging women of colour presenters, who themselves were experienced professional and volunteer organisers with Indigenous Climate Action, *Red Rising* Magazine and Idle No More, for example. In that same session or perhaps in another, participants contributed to conversation with some "super problematic statements," such as asking "why are we even centering Indigenous perspectives here?" I wrote these quotes in my notebook as the two leaders spoke in the main conference hall.

The fact that comportment in the breakout sessions was an object of ethical and epistemological evaluation – for Indigenous volunteer group leaders and, as we will see, CAN-Rac staff – is illuminating. It was as if the challenges identified earlier in the day had come to life in this very space. It was as if the potential difficulties portended in the Group Agreements had not been fully absorbed by some white participants – or worse, had fallen on ears not yet ready to hear them. Although the conference organisers and facilitators – and, I imagine, at least some of the conference participants – were on the same page as the volunteer group leaders, many of the conference participants, it seemed, were not. I learned that the facilitator Amara, CAN-Rac staff and several breakout leaders had re-worked the conference agenda for the next day to address this conflict. Cat would open the conference the following day with "a bit about the context in which we're working" and some framing for how to move forward, and then another breakout leader whom I got to know over dinner at the bar that evening would further address conference participants.

Reaction, Resolution, Redress

"Time for some real talk," Cat opened the next day's conference. Rather than a standard daily welcome of introductions, announcements and housekeeping, the second day of the conference opened with the executive director sitting in a chair, alone on stage, out front and next to the podium, to address the conflict of the previous day. During her intervention, Cat outlined the fundamentals of a political and moral philosophy and a set of operating principles. The primary underlying cause of climate change is not fossil fuels, she insisted to the enrapt, silent audience. Rather, it is fundamentally the exploitation of non-human worlds. This is an exploitation and violence toward non-human animals, broader nature and other humans, she says, falling short of naming capitalism and settler colonialism but explaining them in simpler terms. At CAN-Rac, she continues, they *always* seek to involve conversations about these underlying issues. In fact, it is "an organisational priority." She emphasises these last words with a more emphatic tone of voice and then a pause. However, yesterday, she explained to the group, some Indigenous people and people of colour, participants and presenters, felt their experiences and expertise were not being respected, that space was being taken up and taken over. She continued: when Indigenous people, people of colour, gender non-conforming people, people whose first language is not English come into spaces such as these, they are coming in with vulnerability and, in participating in difficult conversations about these underlying issues, perhaps traumas. After concluding that we as activists need to be aware of these dynamics, she called on Brendan Campbell, whom I had met the previous night, to help her open up a broader conversation on these topics. He took the stage, at ease with the microphone in hand.

Brendan was the leader from a different breakout group, Legal Tools for Climate Action, a young Cree and Métis man representing the RAVEN Trust, a Canadian non-profit NGO working on Indigenous rights and legal defence. Speaking from the perspective of a gay and Indigenous man, he said that "a lot of us adults don't admit we can make mistakes and our learning is, in fact, just beginning." In situations like this, when one has been called out, told they have made a mistake, accountability can be a guiding framework: "accountability to yourself, accountability to those you've harmed (intentionally or not) and accountability to communities." If your first reaction upon being told you've caused some harm is to be defensive, Brendan continued, you have some work to do; self-conscious work on yourself. Instead of feeling defensive, one should recognise that one is being granted a degree of trust: the BIPOC[7] person who calls you out is trusting you to learn, offering you an education, he said. However, not all Native or queer folks are the same; "so how do white or straight people learn?" Brendan asked. "Same as everyone! By active reading and listening." He concluded, noting that when intervening or calling out one of your own group, for example in conference participants' member organisations back home, one should be careful not to criticise or ostracise but to teach. Moreover, that moment is an opportunity to check in with your people, see how they are doing, their mental health, as organising work can be stressful, anxiety-inducing, heart-breaking and joyous when the stakes are so high.

After Cat and Brendan's opening for the morning, another conference organiser announced that four white folks, breakout leaders including the activist from 350.org Canada I had met the night before and Christian from my breakout group, had volunteered to field questions from other conference participants, questions folks felt like they'd never had a chance to ask. This was later described to me as "off gassing" or as a pressure valve for the BIPOC conference participants and breakout leaders, to relieve some of the pressure of having to educate or answer a barrage of questions from the majority white or straight or settler conference participants. Although I did not hear how this experiment in off gassing went, I did see a group of people gathered around Christian and others before the second day's breakout room activities.

The second day of the breakout groups went about in more humbled, hushed tones, occupying less than half the time in the morning's agenda than the previous day's sessions. Soon after, we headed back to the main room. Near the end of the report-backs, the two Indigenous women who had first intervened with a complaint, as well as the Indigenous activist who had been characterised as speaking in the "circular, poetic, 'auntie' way," stood to address the group again. They passed the microphone between each other and spoke in an intentional gesture at resolution, toward moving forward in a good way. My friend with the "auntie" style of rhetoric emphasised again that their ways of speaking were not respected, that their ways of speaking were tied to their ways of being in the world. They concluded by reminding the group that the issues at hand at this conference are all connected. Structures of power, histories of violence and trauma connect with one another such that they were not only talking about Indigenous rights and climate change but the issue of what is deemed in Canada as Missing and Murdered Indigenous Women, about land and language and settler colonialism. And with this, these women not only articulate a distinct problematisation of climate change but also emphasise the labour of complaint to prevent the inertial force of the reproduction of an institutional legacy. I next briefly turn to the work of Ahmed in order to better understand the labour of complaint and its relation to the broader Canadian grassroots and NGO-based climate movement. I then look to Callison on global climate change, its politics and epistemologies, exploring what the concept of articulation can tell us of the conflict at ClimaCon and the work toward climate-safe futures.

Complaint, Climate Change, Knowledge and Ethics
Anticipatory of a Future Yet Unwritten

Complaint, Its Inheritance and Disclosure

In Sara Ahmed's phenomenology of complaint and ethnography of (academic) bureaucracy, she characterises complaint as "non-reproductive labour: the labour of trying to intervene in the reproduction of a problem" (Ahmed, 2021).[8] By studying complaint from a queer feminist lens, she is shining a spotlight on the ongoing histories of violence that do the labour of reproduction – of themselves, of violence, of types of people and possibilities. It is a reproductive labour that has momentum, against which complaint works; institutions labour to reproduce themselves. As a non-reproductive labour, complaint is the active work to *not* reproduce what Ahmed calls an inheritance. "If you can become a complainer by virtue of not reproducing an institutional legacy," Ahmed writes, "not reproducing an institutional legacy could be described as *the work of complaint*" (Ahmed, 2021, emphasis original). Here, the institutional legacy under scrutiny is not that of CAN-Rac Canada as an organisation but that of the institution of environmental and climate activism in Canada, of which CAN-Rac is only a part. This is an institution of activism in Canada that itself is a product of, an inheritance of, the "unended and ongoing histories" of settler colonialism (Ahmed, 2019).

In intervening after the breakout sessions, the queer and women Indigenous activists at ClimaCon were doing the work of complaint to stop the inertial reproduction of an inheritance. In intervening to change the second day's agenda and open a space of conversation and redress in response to the complaint, CAN-Rac conference organisers committed themselves to the labour of complaint as "an effort to stop something from happening" (Ahmed, 2021). Following Cat, Brendan, Amara and others' interventions on Day 2 of the conference, I argue that another formulation might name this the work not only of "not reproducing an institutional legacy" but also the work of producing a different articulation of the problem of climate change, its causes and the relations needed to fully address it. These differing articulations were based in different kinds of expertise with different knowledge bases; in this case, activists with a white settler background and mentality simply lacked the understanding of the Indigenous expertise.

On the one hand, the leaders who spoke up demonstrated through their intervention their ability to transform their commitment and resolve into a solution to what was seen as a poor or undesired articulation: an inheritance of the institutional legacy of settler colonialism. Rather than a problem, their intervention was central to articulating the collective vision shared with CAN-Rac staff and volunteers. On the other hand, conference organisers were, clearly, prepared and skilled to handle the issue. Indeed, the Indigenous experts who spoke up may not have chosen to intervene in the articulating form of a complaint had they not felt that, with CAN-Rac, the conditions were worth it. As I will explore later, CAN-Rac staff and volunteers running the conference were able to co-articulate the problem of climate change with them. To conclude my analysis of these events, I turn more explicitly to this idea in order to better understand the work of CAN-Rac, the conflict at ClimaCon and the visions of climate change and its futures they produce.

Climate Change: Articulations and Relations, Advocacy and Epistemology

In her 2014 monograph, Candis Callison (2014) investigates how climate change comes to matter for diverse publics in North America, outlining the advocacy and activist work they do to encourage people to care about climate change. *How Climate Change Comes to Matter* examines how five groups – Inuit at the Inuit Circumpolar Council, the media and science journalists, the evangelical Christian group Creation Care, scientists and researchers themselves and the corporate social responsibility nonprofit Ceres – navigate the political and scientific realms surrounding climate change. In doing so, she lays out not only the vernaculars and imaginaries through which people articulate and understand their worlds but also how they produce climate change – shifty and unstable as it is – "as object, issue, cause, experience, and body of scientific research, evidence and predictions" (Callison, 2014, p. 11). Of course, the complex global phenomenon that is climate change is more than the global knowledge infrastructure through which we know it. Climate change is produced – and as many things, multifariously.

The conflict at ClimaCon arose out of differing sets of vernaculars, ethics, theories of change and epistemologies surrounding climate change. As the complaint and its resolution made clear, competing forms of advocacy for the future and visions of shared history were afoot. What can their instantiations here tell us? What are the lessons for our understanding of grassroots climate activism to take away from these proceedings?

Problematisation

Put in conversation with Callison and Ahmed, they first teach us how "differently configured and articulated notion[s] of the problem of climate change" came into contestation at ClimaCon (Callison, 2014, p. 6). On the one hand, objections from certain white conference participants that conversation in breakout groups was off topic or was "non-linear" indicate an understanding of what does and does not count as part of the issue of climate change. Questions about why it was at all important to focus on Indigenous perspectives are revealing of assumptions about whose knowledge and expertise matter (most), what and whose histories should come to bear on the present. Here, Indigenous women's oratory norms (see: Kianga Judge, 2022) become representative for the epistemologies and worldviews they express. This dynamic is not unique to ClimaCon but is representative of larger dynamics within settler-colonial contexts. As Sium and Ritskes note, the nonlinearity of Indigenous people's styles of narration can function as a form of knowledge that works "against the colonial epistemic frame to subvert and recreate possibilities and spaces for resistance" (Sium and Ritskes, 2013, p. iii).[9] They are "threatening because they position the teller outside the realm of 'objective' commentary, and inside one of subjective action" (Sium and Ritskes, 2013, p. iv).

On the other hand, in their interventions the Indigenous breakout leaders and the executive director, Cat, insisted on an understanding of climate change as enmeshed in an overlapping series of issues. It is a problem whose underlying cause is not, to recall, fossil fuels, Cat emphasised in the morning intervention, but a history of exploitation and of violence toward particular human and non-human worlds. Rather than technical and science-centred, it is an issue of knowledge and relations, ethical and political. White conference participants' questioning of tone and topic, whether intentional or not, revealed a different instantiation of the problem of climate change and an understanding that their instantiation was the more valuable and correct. Importantly, these are articulations of the problem of climate change that enlist differing assemblages of knowledge and relations, past events and political philosophies.

Articulation, Ethical Reasoning and Relations

Second, these competing forms of advocacy and visions of the future show us that contesting articulations of the problem of climate change reveal the existence of "different modes of ethical reasoning" (Callison, 2014, p. 5) and different relations at play among conference-goers and members of the Canadian climate movement more broadly. Put another way, the conflict represented different kinds of co-articulations of the problem of climate change. Following Callison in drawing on James Clifford (2001) and Stuart Hall (1986), with *articulation* I mean to bring attention here to not only manners of understanding and expressing the problem of climate change but also pragmatic, indeterminate political coalition- and subject-making. An articulation is "a linkage which is not necessary, determined, absolute and essential for all time" (Hall, 1986, p. 53). Articulations are "actively produced and potentially challenged" with "messy and pragmatic" politics (Clifford, 2001, p. 481, 483). They are never inevitable and therefore able to change. Articulation, then, becomes about new forms of political expression and organisation; as much a voicing of epistemological relations as a recombinant political coalition, made up of concrete connections that can be hooked and unhooked, made and unmade and remade differently. It is in this sense that articulations are ethical (relational, evaluative, subject making) and conditional (situated, partial, impermanent). It is in this sense that the history of settler-colonialism – and the history of conflict between settler and Indigenous environmental activists – comes to bear on this case. "You have to ask," writes Hall, "under what circumstances can a connection be forged or made?" (Hall, 1986, p. 53).

Following arguments elsewhere (Fleischmann, 2020, 2022), I argue with Callison that a key condition of global anthropogenic climate change's conceptual force in the world is a challenging imposition of global relations into various articulations: "climate change challenges people to see themselves as part of global environmental, industrial, and capital systems, and in many ways it demands co-articulation of how to locate oneself in a larger collective" (Callison, 2014, p. 23). This larger collective may include (articulation with) groups of people and perspectives with which one is not familiar. While these demands force people to reckon with where they fit into a problem and its solutions that are about more than individual consumers, they are not easy demands and they are not taken up in the same ways everywhere. Despite (or, more cynically, because of) their ostensible involvement in the Canadian climate movement, for some of the participants at ClimaCon, the conference may have been their first co-articulation with a larger collective that includes diverse Indigenous perspectives and CAN-Rac's articulation of the problem of climate change, which attempts to include those perspectives as an organisational priority.

This insight further indicates that, although global climate change *may* present a challenge to dominant ways of thinking and being in what is deemed a positive light, it need not inherently do so. As white conference participants' questioning of Indigenous conference leaders' expertise and belonging – along with the latter's intervention of a complaint, the labour to not reproduce a history and institution of settler-colonial environmentalisms – makes clear, global climate change can challenge people to see themselves as part of global systems, while nonetheless allowing them to reproduce institutions and histories of domination and oppression.

Knowledge, Epistemology, Advocacy, Futures

Third and last, this problematisation of climate change, this particular co-articulation, this reckoning with locating oneself in a larger collective in new ways, has as much to do with knowledge as it does with relations. As Callison puts it, "how we learn to make 'best judgements' and recognise facts as problems is part of an epistemological and collective process" (2014, p. 166). What is more, epistemology – how we know what we know, the history and genealogy of that knowledge and how we express it (Callison, 2014, p. 46) – matters as much as who is communicating that knowledge. ClimaCon's Indigenous speakers were different kinds of experts with different epistemologies than those to which the older white members or, even, the largely white student base of non-CAN-Rac-member conference participants were perhaps habituated. This was *not* the epistemic community of experts by which some of the climate activists present were perhaps used to being addressed. In other words, these were *not* science experts practising "advisory science" to play the delicate balance of "near-advocacy" as in the case of Callison's climate scientists and researchers (2014, cf. Chapter 4); neither was this scientists of diverse expertise wielding the "epistemic agency" of "charismatic data" to translate it for policymakers or an educated public, as in the case of Jessica O'Reilly's 2017 ethnography of Antarctic technoscientific governance via what she calls epistemic technocracy (O'Reilly, 2017, cf. Chapter 7, etc.). Instead, for some (settler, white) ClimaCon participants, the conference involved listening to and centring the leadership of experts whose authority did originate in the scientific realms with which they were likely used to dealing in the realm of climate change politics.[10]

Much of the work of CAN-Rac's climate politics is indeed advocacy, in the sense that staff, network-member activists and others are looking to advance a cause and petition decision-makers who are deemed capable of delivering solutions at scales beyond the individual, at the level of an industry, a nation, the global. In her tour de force *Advocacy After Bhopal*, anthropologist Kim Fortun defines advocacy as the performance of ethics in anticipation of *a* – or *the* – future (Fortun, 2001). If we understand advocacy as the enactment of an anticipatory ethics, anticipatory of a future yet unwritten, both the futures one anticipates and the ethics by and with which one conducts oneself differ greatly according to how one defines the problem of climate change, the forms of ethical reasoning and relations articulated and the knowledge and expertise, through which one knows and lives it. At play in the conflict at ClimaCon were competing articulations of ethics and politics, anticipatory of competing visions of a climate-safe future.

Conclusion

The closing panel of the ClimaCon 2018 conference is a "fireside chat" between Cat – CAN-Rac's executive director – and Ellen Gabriel, a Mohawk elder, activist and artist from the Kanahsatà:ke Nation. Gabriel came to public prominence in 1990 as the official spokesperson of the Mohawk side during Kanahsatà:ke Resistance, otherwise known as the Oka Crisis. Since then, she has worked as an artist and teacher and has remained in the public eye through her political activist work and public speaking, including at the international institutions of the United Nations. Gabriel speaks about Indigenous people coming into the fight for the climate with ongoing struggles for land and land use rights, for their ways of life and relations with the natural world, all in the face of genocide and trauma. She, like Brendan, emphasises that not all Indigenous people are the same. She highlights the importance of language, relating that "language is an extremely important part of the environmental movement for Indigenous peoples." It is both the protocol and medium for relations with the land, she emphasises, relating the importance of both Indigenous languages and ways of speaking for human and nonhuman ethics. She concludes with three issues or points of priority for the Canadian climate movement leading up to the 2019 federal election.

By a few weeks after the conference, CAN-Rac staff and ClimaCon organisers deemed the conference a success, despite some difficulties. They conducted debriefs with the presenters on what was deemed by CAN-Rac staff as a demonstration of lateral violence (violence directed at members of marginalised groups) and uncompassionate behaviour, and they were content with how they were able to convert these challenges into teachable moments of intervention, brought on by the intervention of complaint. In the end, they felt they had successfully worked to create a powerful space of healing.[11]

Climate Action Network-Réseau action climat Canada's work as network conveners and coordinators places them in a unique position to navigate their members' – and the broader Canadian climate movement's – diverse articulations, relations, visions of advocacy and epistemology of climate change. As an instantiation of the CAN-Rac network, ClimaCon 2018 was a nodal convergence of the production of climate change and its politics through the network. Understood as the work to not reproduce an inheritance of settler-colonial environmental relations, an active reinforcement of a shared vision, the Indigenous women's intervention with a complaint at ClimaCon spoke to the different articulations of the problem of climate change and its politics and knowledge at hand. Yet, how much do these differences matter for the future of climate action in Canada and more broadly? For the co-articulation and collaboration that it will indubitably necessitate? As Callison expresses the question, "what does collaboration mean when goals related to climate risks are differently configured? How much do epistemological differences matter? Configured as differences in epistemology, 'speaking up for the facts' might require as much listening as it does speaking" (Callison, 2014, p. 245). Moving forward, these are questions with which practitioners of climate politics will need to grapple.

Notes

1 Climate Action Network-Réseau action climat's name is steadfastly bilingual due to Canada's two official languages.
2 Teika Newton, interview, 20 September 2018.
3 Teika Newton, interview, 20 September 2018.
4 For more on the cultural, religious, subsistence and economic stakes of Makah whaling, their exclusive right to seal and whale and the often violent, racist encounters of this conflict, see Bowechop, 2004; Brand, 2009; Beldo, 2019; NOAA Fisheries, 2022.
5 Note that, given the sensitivity of topics and the vulnerability with which they intervene (see later mention), and considering I did not have their explicit permission, I do not here explicitly name these two women or the friend who, later, will be described as presenting with a "poetic, 'auntie' way of speaking."
6 My thanks to Alonso Gamarra for his thoughtful dialogue on these matters. See Gamarra and Pimentel Mauricci on post-civil war Peru: a narration of past events gives the present form; memories of the disappearance of a brother render political violence and war as part of the background of the ethnographic moment (2023, p. 651).
7 Black, Indigenous, person of colour
8 The following citations of Ahmed (2021) are from a non-paginated EPUB file e-book and therefore do not have stable page numbers. The passages cited here are from Chapter 4 "Occupied" in section titled "Nonreproductive labour," as well as the conclusion of that chapter.
9 The other side of that crooked coin, Linda Tuhiwai Smith notes, is that often Indigenous activists (and scholars) also get criticised when they speak with the language and experience of Western education: "One of the many criticisms that gets levelled at indigenous intellectuals or activists is that our Western education precludes us from writing or speaking from a 'real' and authentic

indigenous position. Of course, those who do speak from a more 'traditional' indigenous point of view are criticised because they do not make sense" (Smith, 1999, pp. 13–14).

10 The most facile contemporary example of the power of discursive recourse to scientific authority in the realm of climate change politics and activism is, of course, the slogan popularised by teenage Swedish climate activist Greta Thunberg: "Listen to the science!"

11 As of the time of writing, ClimaCon 2018 was the latest ClimaCon to happen. CAN-Rac staff had wanted to take a year off the labour-intensive planning process to refocus after the lessons learned in 2018. However, when the novel coronavirus pandemic first touched North American shores in early 2020, it was soon clear that the conference would not occur in 2020. Subsequently, as the pandemic raged on in 2021 and 2022, the conference did not occur the next years, either.

References

Ahmed, S. (2019). *Complaint as a Queer Method* [Lecture]. Montreal: McGill University, 4 October.

Ahmed, S. (2021). *Complaint!* [EPUB Kobo eBook]. Durham: Duke University Press.

Arnaquq-Baril, A. (dir.) (2016). *Angry Inuk*. https://www.nfb.ca/film/angry_inuk/ [Accessed 31 August 2020].

Barney, D. (2017). Who we are and what we do: Canada as a pipeline natiom. In S. Wilson, A. Carlson, and I. Szeman (eds.) *Petrocultures: Oil, Politics, Culture* (pp. 91–132). Montreal and Chicago: McGill-Queen's University Press.

Beldo, L. (2019). Stock morality: Whalers, activists, and the power of the state in the Makah whaling conflict. *American Ethnologist*, 46(1), 47–60. https://doi.org/10.1111/amet.12733.

Bowechop, J. (2004). Contemporary Makah whaling. In M. Mauzä, M. E. Harkin and S. Kan (eds.) *Coming to Shore: Northwest Coast Ethnology, Traditions, and Visions* (pp. 407–419). Lincoln: University of Nebraska Press.

Brand, E. (2009). The struggle to exercise a treaty right: An analysis of the Makah Tribe's path to whale. *Environs: Environmental Law & Policy Journal*, 32(2), 287–319.

Callison, C. (2014). *How Climate Change Comes to Matter: The Communal Life of Facts*. Durham: Duke University Press.

Callison, C. and Tindall, D. B. (2017). Climate change communication in Canada. In *Oxford Research Encyclopedia of Climate Science*. Oxford: Oxford University Press. https://doi.org/10.1093/acrefore/9780190228620.013.477.

Clifford, J. (2001). Indigenous articulations. *The Contemporary Pacific*, 13(2), 468–490.

Climate Action Network (2023). *About CAN*. https://climatenetwork.org/overview/

Climate Action Network-Réseau Action Climat Canada (2018). *Climate Action Network Canada Responds to the Release of the 'Special Report on Global Warming of 1.5°C' from the Intergovernmental Panel on Climate Change*. https://climateactionnetwork.ca/2018/10/16/climate-action-network-canada-responds-to-the-release-of-the-special-report-on-global-warming-of-1-5c-from-the-intergovernmental-panel-on-climate-change/. [Accessed 19 May 2022].

Coulthard, G. S. (2014). *Red Skin, White Masks: Rejecting the Colonial Politics of Recognition*. Minneapolis: University of Minnesota press.

Das, V. (2007). *Life and Words: Violence and the Descent into the Ordinary*. Berkeley, Los Angeles, London: University of California Press.

Fleischmann, A. (2020). *How to Make Climate Change Feel Real*. SAPIENS. https://www.sapiens.org/technology/climate-interactive/ [Accessed 27 August 2020].

Fleischmann, A. (2022). *Possibility in an Era of Climate Change: Anthropology, Knowledge, Politics*. PhD Dissertation. McGill University.

Fortun, K. (2001). *Advocacy after Bhopal: Environmentalism, Disaster, New Global Orders*. Chicago: University of Chicago Press.

Gamarra, A. and Pimentel Mauricci, E. (2023). That's the negative moment of the dialectic . . . *American Anthropologist*, 125(3), 648–652. https://doi.org/10.1111/aman.13892.

Gellner, D. and Hirsch, E. (eds.) (2001). *Inside Organizations: Anthropologists at Work*. Oxford: Berg Publishers.

Hall, S. (1986). Gramsci's relevance for the study of race and ethnicity. *Journal of Communication Inquiry*, 10(2), 5–27. https://doi.org/10.1177/019685998601000202.

Haraway, D. (1988). Situated knowledges: The science question in feminism and the privilege of partial perspective. *Feminist Studies*, 14(3), 575–599. https://doi.org/10.2307/3178066

Helferty, A. T. (2020). *"We're Really Trying, and I Know It's Not Enough": Settler Anti-Pipeline Activists and the Turn to Frontline Solidarity with Indigenous Peoples*. Thesis. https://tspace.library.utoronto.ca/handle/1807/103742 [Accessed 13 January 2024].

Kianga Judge, S. (2022). *Starting Where You Are: First Nations Non-Linear Storytelling, The Australian Museum*. https://australian.museum/learn/first-nations/burra/non-linear-stories/australian.museum/learn/first-nations/burra/non-linear-stories/ [Accessed 13 January 2024].

Knox, H. (2020). *Thinking Like a Climate: Governing a City in Times of Environmental Change*. Durham: Duke University Press.

Lahsen, M. (2008). Commentary on "Gone the bull of winter? Grappling with the cultural implications of and anthropology's role(s) in glocal climate change" by Susan A. Crate. *Current Anthropology*, 49, 587–588.

Masson-Delmotte, V., et al. (eds.) (2018). IPCC, 2018: Summary for policymakers. In *Global Warming of 1.5°C. An IPCC Special Report on the Impacts of Global Warming of 1.5°C Above Pre-Industrial Levels and Related Global Greenhouse Gas Emission Pathways, in the Context of Strengthening the Global Response to the Threat of Climate Change, Sustainable Development, and Efforts to Eradicate Poverty*. World Meteorological Organization (p. 32). Geneva, Switzerland: IPCC.

Mauksch, S. (2020). Five ways of seeing events (in anthropology and organization studies). In R. Mir and A.-L. Fayard (eds.) *The Routledge Companion to Anthropology and Business*. London and New York: Routledge.

NOAA Fisheries (2022). *Makah Tribal Whale Hunt Chronology | NOAA Fisheries, NOAA*. https://www.fisheries.noaa.gov/west-coast/marine-mammal-protection/makah-tribal-whale-hunt-chronology [Accessed 7 September 2023].

O'Reilly, J. (2017). *The Technocratic Antarctic: An Ethnography of Scientific Expertise and Environmental Governance* (1st ed.). Ithaca: Cornell University Press.

Sium, A. and Ritskes, E. (2013). Speaking truth to power: Indigenous storytelling as an act of living resistance. *Decolonization: Indigeneity, Education & Society*, 2(1). https://jps.library.utoronto.ca/index.php/des/article/view/19626 [Accessed 30 May 2023].

Smith, D. E. (1987). *The Everyday World as Problematic: A Feminist Sociology*. UPNE.

Smith, L. T. (1999). *Decolonizing Methodologies: Research and Indigenous Peoples*. Zed Books.

United Nations Committee of Experts on Public Administration (2021). *CEPA Strategy Guidance Note on Science-Policy Interface*. UN Strategy Guidance Note, p. 26. https://publicadministration.desa.un.org/publications/cepa-strategy-guidance-note-science-policy-interface [Accessed 25 April 2024].

van den Hove, S. (2007). A rationale for science–policy interfaces. *Futures*, 39(7), 807–826. https://doi.org/10.1016/j.futures.2006.12.004

37

CLIMATE ACTIVISM IN K-12 EDUCATION ACROSS NORTH AMERICA

Ellen Field, Sarah Stapleton, Gregory Lowan-Trudeau, Colin Harris, Chloe Nguyen, and Aishwarya Puttur

Introduction

The global climate crisis is the greatest social challenge of the twenty-first century. The Paris Climate Agreement outlined required actions for limiting warming to 1.5 degrees Celsius (UN, 2015). Keeping heating below 1.5 degrees Celsius requires rapid mitigation efforts to reduce greenhouse gas emissions across society as well as adaptation measures to make communities and infrastructure as resilient as possible to both existing and predicted climate consequences. Within Intergovernmental Panel on Climate Change (IPCC) and United Nations Framework Convention on Climate Change (UNFCCC) policies, education is considered an essential lever for transitioning societies and economies to become carbon neutral and resilient (see Article 12 of the Paris Agreement or Action for Climate Empowerment stemming from UNFCCC). It is a key for both individual behaviour change and generating public support for systems-level climate policy.

However, the integration of climate change content into curriculum policy seems to be doggedly slow in comparison to how important it is deemed by the IPCC and UNFCCC. The slow adoption and lack of leadership has caused us to wonder about the socio-political factors that hinder climate action among school leaders and educational policymakers. According to UNESCO, 53% of national curriculum frameworks analysed (100 in total) included climate change at least once; however, the depth of inclusion was usually minimal. The remaining 47% of national curriculum frameworks made no mention of climate change (UNESCO, 2021a).

According to the Monitoring and Evaluating Climate Communication and Education (MECCE) project, of 50 countries reviewed, only 39% of countries have a national level law, policy or strategy specifically focused on climate change education, and only 27% have publicly available budgets for climate change education (Benavot and McKenzie, 2022). The slow integration of climate change education into curricula has been echoed in the streets by youth climate activists in countries around the world (Verlie and Flynn, 2022), leading to the formation of some youth-led organisations focused on advocating for curricular reform, such as Teach the Future (2021) across Europe or Climate Education Reform British Columbia (CERBC) in Canada, as well as private member bills to increase regional climate change education through legislation.

DOI: 10.4324/9781003396567-44

Benchmarking research on the state of climate change education in formal education systems in Canada and the US has shown many gaps and opportunities for improving climate change education policy and practice. This includes a lack of explicit expectations within formal curriculum (Field et al., 2023; NAAEE, 2022) and minimal climate change coverage during instructional time: 35% of Canadian educators indicated that they do not teach climate change in any subject and 41% indicated they teach climate change for 1–10 hours of instruction per year or semester (Field et al., 2019; LSF, 2022); and in the US, seven in ten public school administrators say climate change is being taught in their school or district but fewer than four in ten teachers say they are personally teaching it (NAAEE, 2022).

In both contexts, educators report needing professional development and few teachers are very confident or feel prepared to teach the topic (Field et al., 2019; LSF, 2022; NAAEE, 2022). Concern over pushback from parents or members of the general public is noted as a concern for pre-service teachers (Berger et al., 2015), and administrators are more likely to worry about the politics of parents than teachers are (NAAEE, 2022). However, there is a fairly high level of concern across society on climate impacts, which challenges the perceived risk of parental pushback.

The lack of policy leadership on climate change education in Canada and the US has created a relative policy vacuum, in which the implementation of climate change education seems to fall on the shoulders of teachers, students, and concerned citizens who are not in positions to make educational policy decisions. This chapter presents four vignettes which explore climate activism at the nexus of climate change education and the sociopolitical challenges of change-making within formal education systems. The vignettes are based on our research and experiences as scholars working in climate change education (Ellen, Sarah, Greg), as a professional working within a non-profit to support climate change education (Colin), and as youth climate activists, volunteering to change education systems (Aishwarya and Chloe).

The first vignette explores activism in Oregon where educators and high school students took it upon themselves to introduce state-wide legislation requiring climate education. The second vignette depicts the tensions and hesitations that Albertan educators experience about teaching climate change content while working in a province whose politics and culture – including curriculum policy – have been shaped by a reliance on oil and gas production. The third vignette outlines the powerful mobilisation of the group Climate Education Reform British Columbia, composed of frustrated high school students and recent high school graduates, who put together a comprehensive framework on intersectional, interdisciplinary, and action-oriented climate education that policymakers would be wise to consider and adopt. The fourth vignette outlines the powerful work of youth activists in Canada who are mobilising and educating through peer-workshops for financially just climate futures through advocating for divestment away from banks that continue to invest in fossil fuel projects.

While there is a lack of climate change education policy leadership within North American education systems, these vignettes are quick glimpses into select sites, providing a partial view of climate change action and activism that is bubbling away on the fringes in school cafeterias, in social media messages, and in conversations within parent/community groups.

Vignette 1: Educator and Youth Activism Toward Required K–12 Climate Education in Oregon

In the US, the formal K–12 education sector has typically been slow to recognise its responsibility regarding climate change (e.g., Henderson and Long, 2023). Youth have repeatedly led the way to push the education sector to do better in preparing them for the changing climate. A major part of doing better by them is including robust climate education in schools within the formal, required curriculum.

Across the US, the Next Generation Science Standards (adopted in 20 states) introduce climate change-related standards explicitly starting in middle school. Beyond these standards, several other states in the US have taken specific measures to expand climate education (Cho, 2023, Feb 9). Based on a legislative proviso requested by the governor, since 2018 Washington State has incorporated extensive teacher support and professional development for climate education into K–12 science education through its Climetime program. Thanks to a governor mandate, New Jersey is the foremost state to include climate education across all grades and disciplines. This was accomplished through direct input of climate change into the state standards (Madden, 2022, Feb). In 2022, Connecticut passed a state law requiring climate change be taught within science classes, with fifth, eighth, and eleventh graders taking state tests that include climate change concepts (Cho, 2023, Feb 9). California is working to connect university systems to teacher training around climate education as well.

In Oregon, Oregon Educators for Climate Education, a small group of dedicated educators, including Sarah, in collaboration with high school students, has organised since 2020 to crowd-source and introduce state-wide legislation that aims to require climate education in all grades (K–12) across all disciplines. The impetus for creating the Oregon climate education bill began when several teachers hosted a climate change summit in their district for teachers and students to share their ideas and work in climate education. High school students attending the event remarked at the inequities across schools and classes in learning about climate issues. They noted that coverage of climate change across grades and disciplines depended upon individual teacher interest and commitment. These students implored that climate be taught across all disciplines and all grades.

A number of discussions ensued in which a group of us (concerned educators and students) wondered how to create widespread change. Inspired by recent Oregon legislative measures that require teaching of Indigenous perspectives (Senate Bill 13), Holocaust recognition (Senate Bill 664), and Ethnic Studies (House Bill, 2845), we decided to focus on crafting state-wide legislation on climate education. Initially, this goal seemed lofty, particularly for our small group of educators, inexperienced in legislative procedures. But, what once seemed formidable, over months, began to feel essential.

Our core group of six to eight educators consists of three teachers on special assignment (e.g., subject area leaders), one high school social studies/Spanish teacher, one retired high school campus supervisor, a special education teacher, and me (Sarah), a teacher education professor. We have met at least once a week for nearly 2 years. We also have a student advisory committee, convened by another non-profit, that meets regularly with several of our steering committee members.

Our team has taken great efforts to make our climate education legislation non-partisan. A major intention has been to show that climate education can be responsive to individuals and local communities, as a pathway toward resilience. Indeed, our rural communities are on the frontlines of major climate impacts in our state, namely wildfires and droughts. Though Oregon is typically viewed as a liberal state and currently has a supermajority of liberal legislators as well as a liberal governor, the rural areas of Oregon (not unlike the rest of the US) tend to be very conservative. Partisan divides are so significant in Oregon that conservative state legislators have repeatedly walked out of the Capitol building to stymy progressive legislative attempts, including climate change bills. In that way, potential opposition forces have loomed large.

Despite a very positive Senate education committee hearing, our bill did not move forward out of committee in the 2023 session. We were told of hesitation by legislators who had a difficult time understanding what climate education looked like in our context and who worried whether teachers were prepared to do it. We were also told that conservative senators on the committee were the most concerned.

Despite our bill not moving forward in the 2023 legislative session, several state senators have vowed to continue working with it in a future session. For us as activists, this feels too far into the future given the urgency of the climate crisis. In the meantime, we are working to establish groundwork for a stronger case for the next legislative session. Partnering with another nonprofit, we have created an open-source climate education hub for educators in the state to access. We are also creating professional development opportunities for teachers in the state around climate education.

In just under 2 years, our small team has:

- met with numerous stakeholders across the state, including teachers, students, other climate groups, education organisations, non-profits, watershed councils etc.
- garnered support of the state's largest teachers' union
- crowd-sourced, drafted, and continually edited a bill and climate learning principles
- partnered with another nonprofit to host and work with a state-wide Student Advisory Council
- communicated with journalists
- met with state legislators and gained two state senator sponsors for our bill
- applied and awarded two foundation grants
- created a partnership with another nonprofit to collaboratively build an online open-access climate education resource hub
- organised two summer teacher climate lesson creation institutes

The work is slower than we feel is necessary to address the climate crisis, but in the absence of federal climate education legislation, we are left to work in state government contexts. As a small team of dedicated volunteers, we are extremely committed but would benefit from systemic change across primary and secondary education systems in the US and globally.

Vignette 2: Towards a Just Transition: Climate Change Education in Alberta, Canada

The western Canadian province of Alberta is a national and global centre for oil and gas extraction, refinement, and financial management. The oil/tar sands in the northern portion of the province are particularly infamous contributors to climate change and environmental degradation in general (Katz-Rosene, 2017). Amidst Alberta's predominantly pro-oil and gas and conservative sociopolitical context (Uechi, 2015), educators in a range of settings often experience conflict, tension, and a general chill regarding environmentally oriented initiatives (Chambers, 2008) and climate change-related endeavours specifically (Henderson et al., 2017). As a university professor in a school of education (Greg) and executive director of an outdoor and environmentally focused educational non-profit (Colin) in Alberta, we observe, experience, and navigate these trends on a regular basis.

Alberta's social, political, economic, and environmental dynamics are often best illustrated by news media. For example, a recent controversy emerged regarding the Alberta government's vigorous objection to the federal government's promotion of a "just transition" as inherently anti-oil and gas (Ibbitson, 2023; Thurton, 2023). The concept of a just transition was introduced internationally as part of the 2015 Paris Agreement (United Nations, 2015) that emerged from the twenty-first Conference of Parties (COP21). It echoes the theory of just sustainabilities developed by Agyeman (2013) to denote the imperative of simultaneously addressing social and environmental justice concerns while transitioning towards lower carbon societies. In objecting to the general use of the term "just transition" and its underlying premise, Alberta's current conservative energy minister Sonya Savage (Thurton, 2023), premier Danielle Smith (Ibbitson, 2023), and even the leader of the social democrat opposition, Rachel Notley (Cryderman, 2023), have demonstrated a lack of care in keeping with what Arenas (2021) describes as the capitalocene – the current era which is still primarily driven by a capitalist economic paradigm at the expense of social and environmental concerns.

Another recent event made headlines in educational contexts specifically when a teacher in Blackfalds, a small central Alberta community, was threatened by local parents simply for facilitating a climate change-themed activity and discussion in her fourth-grade class. The tensions and threats escalated to the point that the school leadership decided to cancel an upcoming dance (Parsons, 2019).

Such events and dynamics undoubtedly create a chilling effect upon educators in Alberta. This is further reinforced through legislative action such as the introduction of a revised provincial curriculum which erroneously de-emphasises humanity's culpability for contemporary climate change (Lowan-Trudeau, 2022). In explicitly striving to present a falsely "balanced" understanding of climate change while still recognising the vested interests of many Albertans in the oil and gas industry, the new curriculum does both teachers and students a serious disservice.

Pro-oil and gas and anti-climate change rhetoric in Alberta often intersect with other social concerns. For example, in anticipation of youth climate activist Greta Thunberg's visit to our province amidst the Youth Strikes for Climate in 2020 (Verlie and Flynn, 2022), overtly sexual and misogynistic imagery was produced and displayed by an Alberta oil and gas service company (Smellie, 2020). Such intersectional dynamics are also exemplified by the disproportionate prevalence of violence against Indigenous women near oil and gas "man camps" (Awasis, 2014).

In the non-profit realm, organisations such as Take Me Outside (TMO), of which Colin is the executive director, also engage with environmental topics in Alberta and across Canada. TMO's mandate is to encourage and support teachers in taking their students outside. As such, they often engage in conversations and activities related to climate change; however, as an outdoor-focused organisation, much of their work involves fostering a broad appreciation of the natural world. This common but admittedly somewhat contested (e.g., Gough, 1999; Payne, 1999; Thomashow, 1996) approach is grounded in the notion that students will develop and carry forward an appreciation of the environment that they will apply to specific topics of concern such as climate change. TMO also attempts to introduce intersectional consideration of, for example, Indigenous knowledge, through a variety of events and engagements with knowledge keepers. These can be challenging endeavours at times in the non-profit sector, which relies on funding and in-kind support from a range of government and industry stakeholders.

From a post-secondary perspective, Greg's work primarily involves engaging with graduate students, many of whom are practising teachers and administrators, in both research and environmentally themed courses. His own research also connects him with K–12 schools, non-profits, and associated curricula. Despite assurances of academic freedom, similar chilling dynamics exist in post-secondary contexts in Alberta to those encountered by K–12 teachers. As such, conversations and endeavours with colleagues and students alike can prove challenging but also encouraging at times.

Faced with such existential and sociopolitical challenges, we also look to promising and inspiring events and trends such as the presence of youth voices advocating for increased climate education (Derworiz, 2020; Lawrynuik, 2019), the continued emergence of a range of initiatives within and beyond schools (Worobec et al., 2021), and scholarly evidence that Alberta youth value climate science and corporate environmental responsibility (Laurie et al., 2021).

One inspiring example is Alberta Youth Leaders for Environmental Education (AYLEE), a group of junior and senior high school students with a strong focus on climate action and education. AYLEE's activities range from collaborating with other climate and environmentally oriented organisations to review Alberta's recently revised curriculum (Worobec et al., 2021), to the initiation of a survey-based study which demonstrated that Alberta students want more climate, energy, and environmentally related education (AYLEE, 2021; CBC News, 2021).

In the midst of the dynamics, tensions, initiatives, and experiences described earlier, we continue our work in climate and environmental education and related areas in Alberta. We also work to provide teachers and students alike with the knowledge and confidence to critically and effectively engage with climate and environmentally related teaching and learning.

Vignette 3: Canadian Youth-mobilising for Transformative Shifts in Formal Education

While youth have been motivated by multiple injustices (intergenerational, geo-political, racial, socio-economic, colonial, etc.), the formal education system and its integration – or lack thereof– of climate change education has not been motivated by these same climate injustices: "education systems lag behind, preoccupied with the 'what' and 'how' of climate change, rather than engaging it as a social issue in which students themselves are implicated" (Karsgaard and Davidson, 2021, p. 1).

In Canada, the education system has been slow to respond and integrate climate change education in a holistic, comprehensive, or systematic way (Field et al., 2023). Canada, as a signatory to the Paris Agreement in 2016, agreed to Article 12, which requires signatories to enhance "climate change education"; however, education is only mentioned in Canada's Nationally Determined Contributions in terms of job-training for new skills in growing sectors (Canada, 2021). In Canada, K–12 formal education and postsecondary education is under the jurisdiction of provinces and territories, meaning there are 13 distinct education systems in Canada.

Recent analysis of the climate change education curriculum policy within provincial and territorial jurisdictions of Canada shows fractured and uneven inclusion of climate change expectations across subjects and grades – for instance, in British Columbia, there were two climate change expectations in mandatory science courses and 17 climate change expectations in elective courses, whereas there were 27 climate change expectations in mandatory social studies and zero in social studies electives (Field et al., 2023). The study also found that the majority of climate change expectations occur in elective courses at the secondary level (seven of the 13 provinces) (Field et al., 2023). As a climate change education scholar (Ellen), I have been perplexed by the lack of leadership and progress within Canadian schools. When I met youth (including Chloe) from the Climate Education Reform British Columbia (CERBC, 2021) group, I was so impressed with their conceptualisation of what climate change education could be that I tried to support their efforts however I could.

While there are calls for mandatory climate change education (UNESCO, 2021b), there is a glaring gap for policymakers to work from, that is, there is no agreed upon framework of what climate change education is or anticipated learning outcomes (Reid, 2019) that would guide integration into curriculum. After a recent review of proposed frameworks and conceptualisations, there is one framework that is holistic and comprehensive and centres justice. The framework was developed by a group of high school students in British Columbia, Canada, who got together after their shared experiences of the lack of climate change education in their high school classes. The group, Climate Education Reform British Columbia, have brought proposals to the British Columbian Minister of Education, Jennifer Whiteside. The youth are asking the Ministry of Education to declare a climate emergency, create specialised committees at regional and district levels to support teachers to integrate climate change pedagogy, create a youth advisory committee to work alongside the BC Ministry of Education and district-level committees, and create a comprehensive framework of intersectional, interdisciplinary, and action-oriented climate education across subjects with the following learning outcomes (abbreviated for publication – full text available on website):

- understand the physical climate system and the causes and effects of climate change
- understand the urgency of the climate crisis and the need to act now
- understand the political, economic, and sociological aspects of the climate emergency
- understand the relationship of social justice issues with climate change and climate solutions
- critically engage in politics
- feel a stronger connection to the environment through further out-of-classroom learning
- contribute to climate solutions by understanding the types of changes needed and the many ways to be involved to enact those changes
- envision a better world and feel empowered and energised (CERBC, 2021)

The work of CERBC highlights the mostly untapped potential of students in schools and how their time and energy is (mostly not) directed towards climate action with impact. The work of the CERBC group, who organise in their spare time between classes and jobs, is compelling not only in their planning for improving climate change education within British Columbia but also in their comprehensive framing of climate change education and required learning outcomes.

There is a strong need for drastic change at the educational policy level to address climate change and climate justice more robustly. In particular, there is a need for both top-down policy change and bottom-up cultural shifts by placing significance on youth engaged in systems change making while also highlighting the agency-limiting practices common in today's formal educational settings (Trott, 2021).

Vignette 4: Youth-mobilising for Financially-just Climate Futures

Signatories to the Paris Agreement recognised the need for specific climate financing, which would ensure "finance flows consistent with a pathway towards low greenhouse gas emissions and climate-resilient development" (UN, 2015, Article 2.1c). Some companies and major investors are adopting sustainable business plans that are compatible with 1.5 degrees of warming; however, there are many major pension funds and investment firms portfolios that are still aligned with 3+ degrees of warming (Global Commission on Economy and Climate, 2018). Many investors, as well as banks and companies, continue to underestimate the risks of climate change and are still making short-sighted decisions to expand investment into carbon-intensive assets (Global Commission on Economy and Climate, 2018).

Regarding climate education, there is an inherent disconnect between youth and finances. It is seen as an idea that one must understand only upon going off to university; even then it's a bumpy learning curve. Aishwarya Puttur, member of Banking on a Better Future, describes this disconnect as:

> a veiled secret hidden beyond the common person's reach – the money behind the climate crisis. I see youth out on the streets screaming, crying, pleading for justice. For a world in which they will be able to live in. They point fingers at the assumed culprits; fossil fuel companies and it is by no doubt that they are destroying our future by simply continuing to exist and expand. Similar to any other culprit; however, exists underlying supporters; those who do the work behind the scenes to ensure the successes of these crimes. One such pillar of support are financial institutions.

In colonially known Canada, those financial institutions include the big five banks of Canada: Canadian Imperial Bank of Commerce (CIBC), Toronto Dominion (TD), Royal Bank of Canada (RBC), Scotia Bank, and Bank of Montreal (BMO). They are among the top global funders of fossil fuels, providing $1.1 trillion since 2016 to fossil fuel companies (Merleaux, 2023).

Public understanding of the role financial institutions have in lending and investing in fossil fuel projects is not well established and linked. Arguably, 2022 has brought forward these interconnections due to increased activism led by young people and advocacy groups. If young people are relying on formal education to learn about how divesting is a powerful climate action, they will not find these lessons in their classrooms. The Ontario curriculum financial literacy scope and sequences, which is a guide with teacher prompts and curriculum-linked examples, considers the financial costs of climate impacts and costs of mitigation and adaptation. It does not suggest analysis of how financial flows of lending and investing impact greenhouse gas emissions or how to research carbon-disclosure by financial institutions or companies.

The organisation, Banking on a Better Future, is a youth-led group that recognised this important gap. Aishwarya and other members developed workshops aimed to help high school students think critically about personal financial decisions, the finance industry, and the economic system and their interconnectedness to climate change. The workshop discusses how systemic climate solutions seem out of reach for many young people and that these feelings induce climate anxiety. Banking on a Better Future shows how young people can make financial decisions about financial institutions' climate policies and lending through comparing major banks in Canada's climate commitments and actions. They offer a pledge for young people to commit to, and by taking this pledge, the person then lets the branch manager know that they are closing their account as the bank has not made meaningful progress on climate commitments. They list alternative banking options, such as credit unions, besides the big five banks that continue to finance fossil fuel projects.

The Royal Bank of Canada in particular has increasingly come under scrutiny in Canada as they lend funding for the Coastal Gas Link pipeline, which is planned to run through Wet'su'wet'en land and is unceded territory. The Dinï ze' and Ts'akë ze' (chiefs) oppose this development and have launched two separate legal actions to defend their rights and title from unwanted industrial activity and to hold the BC provincial and Canadian government to its climate commitments (Proctor, 2022). The Royal Bank of Canada is also being investigated for greenwashing by Canada's competition bureau, and recent data shows that it financed $42 billion in 2022 towards fossil fuel projects (Merleaux, 2023). Banking on a Better Future has also successfully mobilised university students across Canada through their campaign to remove fossil fuel banks from campuses. Banking on a Better Future, through their campaigning and workshops, have garnered national media attention which has further catalysed discussion among university students. Recent research in the US has shown how more than 1,500 lobbyists are working on behalf of fossil-fuel companies, while at the same time representing or working for hundreds of liberal-run universities, technology companies, and environmental groups that say that they are addressing the climate crisis (Milman, 2023). This research raises the alarm of the influence of fossil fuel lobbyists on climate change education.

Both Banking on a Better Future and Climate Education Reform British Columbia are examples of youth-developed and youth-driven projects that are responsive to collective change-making and have the potential to aid deep decarbonisation through providing quality climate education or accountable climate financing. However, for the youth involved, their work is coordinated and facilitated between classes and part-time jobs. Their work is not supported by a teacher or has not been adopted by a larger organisation with additional support or funding. This moment in time requires educational leaders to find ways to connect with youth-developed activism and engage deeply in change-making processes – not to shy away from them.

Conclusion

This chapter has outlined four sites of climate activism that illustrate how these efforts often occur based on the enthusiasm, conviction, and talents of educators and youth (Eames, 2017). This is what happens in a relative policy vacuum. Quite often, these individuals and groups find themselves confronting the task single-handedly or in small groups, exposing a clear gap where climate activists could step in to exert pressure on formal education systems. While some youth activists focused on improving climate change education in K-12 systems, there has been relatively little activist focus on demanding better climate change education in schools. Additionally, this research has highlighted the chilling effect of political dynamics on educators, which highlight the critical need for clearer guidelines, targeted strategies, and consistent policies to foster a cohesive and effective approach to integrating climate change education into formal education – even in places where climate change has been weaponised or ideologically challenged. In petrostates, like Alberta, is teaching climate science and climate impacts an act of resistance or activism?

Recently, Peter Kalmus, a climate scientist, tweeted "I don't know how more parents aren't climate activists" (2 June 2023). Parents are important stakeholders in the education system and could ask principals, superintendents, and school board trustees to improve climate change education in similar ways that parents are increasingly requesting particular topics to not be taught in schools in some regions. While there are some parent-advocacy climate groups in Canada, these are relatively small groups engaging only pockets of parents. We would love to see groups like Parents for Future (parentsforfuture.org) engage Canadian adults. When we look to health advocacy groups, it is often parents who are the strongest advocates for children's treatment.

The absence of legislated guidelines providing a framework for integrating climate change education into formal education systems forces us to be strategic in how we organise within the sector. This has led some of us to engage in direct democracy by drafting legislation. However, through this, professional norms and teacher beliefs about activism revealed interesting patterns.

Educators often resort to using personal emails or meeting outside of work hours to discuss or coordinate climate-action-related activities. This behaviour illustrates the blurred boundary between their professional obligations and their personal commitment to enact change. This may be due to teachers' perceptions that they need to maintain a "neutral" position. However, in this moment in time, given the risk and uncertainty that society faces, should we not all be activists?

We hope this chapter will prompt further discussion and instigate action among educators, decision-makers, activists, and students alike, to ensure that climate change education is recognised as an integral part of formal education systems, thereby enabling children and youth to be better prepared for climate-altered futures and hopefully transform climate trajectories.

References

Agyeman, J. (2013). *Introducing Just Sustainabilities: Policy, Planning, and Practice*. London and New York: Zed Books.

Arenas, A. (2021). Pandemics, capitalism, and an ecosocialist pedagogy. *The Journal of Environmental Education*, 52(6), 371–383. https://doi.org/10.1080/00958964.2021.1999197

Awasis, S. (2014). Pipelines and resistance across Turtle Island. In T. Black, S. D'Arcy, T. Weis and J. Kahn Russell (eds.) *A Line in the Tar Sands: Struggles for Environmental Justice* (pp. 253–266). Between the Lines.

AYLEE (2021). *AYLEE Survey Report: The Youth Led Environmental Education Poll 2021*. Alberta Youth Leaders for Environmental. Education. https://www.abcee.org/sites/default/files/AYLEE SurveyReport.pdf

Benavot, A. and McKenzie, M. (2022). Indicators interactive data platform – Primary and secondary education. *Monitoring and Evaluating Climate Communication and Education Project*. https://mecce.ca/data-platform/idp/?_sft_idp_cce_approach=1-1-primary-secondary-education-2

Berger, P., Gerum, N. and Moon, M. (2015). "Roll-up your sleeves and get at it!" Climate change education in teacher education. *Canadian Journal of Environmental Education*, 20, 154–172.

Canada (2021). *Canada's 2021 Nationally Determined Contribution Under the Paris Agreement*. UNFCCC. https://unfccc.int/sites/default/files/NDC/2022-06/Canada%27s%20Enhanced%20 NDC%20Submission1_FINAL%20EN.pdf

CBC News (2021). Youth led survey suggests teenage Albertans want more environmental education. *CBC News*. https://www.cbc.ca/news/canada/edmonton/youth-led-survey-teenage-albertans-want-climate-education-1.6294508

Chambers, J. M. (2008). Human/nature discourse in environmental science education resources. *Canadian Journal of Environmental Education*, 13(1), 107–121.

Cho, R. (2023). Climate education in the U.S.: Where it stands, and why it matters. State of the Planet, Columbia Climate School. *News Climate*, 9 February. https://news.climate.columbia.edu/2023/02/09/climate-education-in-the-u-s-where-it-stands-and-why-it-matters/ [Accessed 8 June 2023].

Climate Education Reform British Columbia (2021). *Reform to Transform Campaign*. https://www.climateeducationreformbc.ca/

Cryderman, K. (2023). Dispute over Ottawa's just transition to become the top issue in Alberta election. *Globe and Mail*, 19 January. https://www.theglobeandmail.com/opinion/article-dispute-over-ottawas-just-transition-become-the-top-issue-in-alberta/

Derworiz, C. (2020). Report suggests Alberta students want more education on climate change. *National Post*, 15 November. https://nationalpost.com/pmn/news-pmn/canada-news-pmn/report-suggests-alberta-students-want-more-education-on-climate-change

Eames, C. (2017). Climate change education in New Zealand. *Curriculum Perspectives*, 37(1), 99–102. http://doi.org/10.1007/s41297-017-0017-7

Field, E., Schwartzberg, P. and Berger, P. (2019). *Canada, Climate Change and Education: Opportunities for Public and Formal Education* [Technical Report]. https://www.researchgate.net/publication/337111645_Canada_Climate_Change_and_Education_Opportunities_for_Public_and_Formal_Education

Field, E., Spiropolous, G., Nguyen, A. and Grewal, R. (2023). Climate change education within Canada's regional curricula: A systematic review of gaps and opportunities. *Canadian Journal of Educational Administration and Policy*, 202(2023), 155–184. https://journalhosting.ucalgary.ca/index.php/cjeap/article/view/74980

Global Commission on Economy and Climate. (2018). *Unlocking the Inclusive Growth Story of the 21st Century: Accelerating Climate Action in Urgent Times*. Global Commission on the Economy and Climate. https://newclimateeconomy.report/2018/wp-content/uploads/sites/6/2018/09/NCE_2018_FULL-REPORT.pdf

Gough, N. (1999). Surpassing our own histories: Autobiographical methods for environmental education research. *Environmental Education Research*, 5(4), 407–418. https://doi.org/10.1080/1350462990050406

Henderson, J. and Long, D. (2023). Climate change as superordinate curriculum? *Research in Education*. https://doi.org/10.1177/00345237231160080

Henderson, J., Long, D., Berger, P., Russell, C. and Drewes, A. (2017). Expanding the foundation: Climate change and opportunities for educational research. *Educational Studies*, 53(4), 412–425. https://doi.org/10.1080/00131946.2017.1335640

Karsgaard, C. and Davidson, D. (2021). Must we wait for youth to speak out before we listen? International youth perspectives and climate change education. *Educational Review*, 75(1), 1–19. https://doi.org/10.1080/00131911.2021.1905611

Katz-Rosene, R. M. (2017). From narrative of promise to rhetoric of sustainability: A genealogy of oil sands. *Environmental Communication*, 11(3), 401–414. https://doi.org/10.1080/17524032.2016.1253597

Lakind, A. (2022). Reading the youth climate movement: Social media, literary creation, and allyship. In R. Young (ed.) *Climate Change Education: Reimagining the Future with Alternative Forms of Storytelling* (pp. 9–28). Lanham: Rowman & Littlefield.

Laurie, R., Atkinson, K., Ohm, J. and Bishop, M. (2021). Communicating climate change to Alberta's youth: Lessons learned from the Alberta narratives project. In F. Weder, L. Krainer and M. Karmasin (eds.) *The Sustainability Communication Reader: A Reflective Compendium* (pp. 419–435). Wiesbaden: Springer VS.

Lawrynuik, S. (2019). 'It's kind of frightening': Students worry climate change education lacking in Alberta classrooms. *The Narwhal*, 10 July. https://thenarwhal.ca/its-kind-of-frightening-students-worry-climate-change-education-lacking-in-alberta-classrooms/

Learning for a Sustainable Future (2022). *Canadians' Perspectives on Climate Change Education: 2022*. Report by Leger for Learning for a Sustainable Future. https://lsf-lst.ca/wp-content/uploads/2023/03/Canadians-Perspectives-on-Climate-Change-and-Education-2022-s.pdf

Lowan-Trudeau, G. (2022). Climate change curricula in Alberta, Canada: An intersectional framing analysis. *Northwest Journal of Teacher Education*, 17(3), 10. https://doi.org/10.15760/nwjte.2022.17.3.10

Madden, L. (2022). *Report on K-12 Climate Education Needs in New Jersey*. https://www.njsba.org/wp-content/uploads/2022/02/climate-change-ed-online-2-2.pdf [Accessed 8 June 2023].

Merleaux, A., et al., (2023). *Banking on Climate Chaos 2023*. https://www.bankingonclimatechaos.org/wp-content/uploads/2023/04/BOCC_2023_vFinal-1.pdf

Milman, O. (2023). 'Double Agents': Fossil-fuel lobbyist work for US groups trying to fight climate crisis. *The Guardian*, 5 July. https://www.theguardian.com/us-news/2023/jul/05/double-agent-fossil-fuel-lobbyists

NAAEE, Edge Research. (2022). *The State of Climate Change Education: Findings from a National Survey of Educators*. https://naaee.org/sites/default/files/2023-02/NAAEE_State%20of%20Climate%20Change%20Education%20Report_SUBMITTED%2012_12_22%5B1%5D.pdf

Parsons, P. (2019). Social media threats force cancellation of Alberta school's holiday dance. *CBC News*, 10 December. https://www.cbc.ca/news/canada/edmonton/social-media-threats- force-cancellation-of-blackfalds-elementary-school-holiday-dance-1.5390831

Payne, P. (1999). The significance of experience in SLE research. *Environmental Education Research*, 5(4), 365–381. https://doi.org/10.1080/1350462990050403

Piispa, M. and Kiilakoski, T. (2022). Towards climate justice? Young climate activists in Finland on fairness and moderation. *Journal of Youth Studies*, 25(7), 897–912. https://doi.org/10.1080/1367 6261.2021.1923677

Proctor, J. (2022). Wet'suwet'en members sue RCMP and Coastal GasLink for alleged harassment and intimidation. *CBC News*, 22 June. https://www.cbc.ca/news/canada/british-columbia/wet-suwet-en-pipeline-rcmp-cgl-lawsuit-1.6498022

Reid, A. (2019). Climate change education and research: Possibilities and potentials versus problems and perils. *Environmental Education Research*, 25(6), 767–790. https://doi.org/10.1080/135046 22.2019.1664075

Samuel, K. (2021). *Youth as Coalitional Possibility in the Youth Climate Movement: An Analysis of the Climate Justice Rhetoric of Hilda Nakabuye, Autumn Peltier, & Greta Thunberg*. Masters Thesis. James Madison University. JMU Scholarly Commons.

Smellie, S. (2020). The horrible Greta Thunberg sticker highlights Alberta's toxic oil culture. *Vice*, 10 March. https://www.vice.com/en/article/wxe7yw/the-horrible-greta-thunberg-sticker-highlights-abertas-toxic-oil-culture

Teach The Future. (2021). *Prepare Students for Tomorrow, Teach the Future Today*. https://www.teachthefuture.org/

Thomashow, M. (1996). *Ecological Identity*. Cambridge, MA: MIT Press.

Thurton, D. (2023). Ottawa must scrap polarizing term 'just transition': Alberta environment minister. *CBC News*, 8 January. https://www.cbc.ca/news/politics/ottawa-just-transition-alberta-environment-1.6704486

Trott, C. D. (2021). What difference does it make? Exploring the transformative potential of everyday climate crisis activism by children and youth. *Children's Geographies*, 19(3), 300–308. https://doi.org/10.1080/14733285.2020.1870663

Uechi, J. (2015). NDP crushes 44-year conservative dynasty in Alberta. *National Observer*, 5 May. http://www.theglobeandmail.com/news/alberta/albertas-political-dynasties/article24255480/

UNESCO (2021b). *UNESCO Urges Making Environmental Education a Core Curriculum Component in All Countries by 2025*. UNESCO. https://en.unesco.org/news/unesco-urges-makingenvironmental-education-core-curriculum-component-all-countries-2025

UNESCO. (2021a). *Getting Every School Climate-Ready: How Countries Are Integrating Climate Change Issues in Education*. UNESCO. https://unesdoc.unesco.org/ark:/48223/pf0000379591

UNESCO. (2022). *Youth Demands for Quality Climate Change Education*. UNESCO. https://unesdoc.unesco.org/ark:/48223/pf0000383615

United Nations (2015). *Paris Agreement*. United Nations. https://unfccc.int/files/essential_background/convention/application/pdf/english_paris_agreement.pdf

United Nations. (n.d.). *Action for Climate Empowerment*. UNFCCC. https://unfccc.int/ace.

Verlie, B. and Flynn, A. (2022). School strike for climate: A reckoning for education. *Australian Journal of Environmental Education*, 38(1), 1–12. https://doi.org/10.1017/aee.2022.5

Worobec, K., Janzen, J. and MacKinnon, S. (2021). Opinion: Alberta's draft curriculum needs to equip children for climate change. *Edmonton Journal*, 23 October. https://edmontonjournal.com/opinion/columnists/opinion-albertas-draft-curriculum-needs-to-equip-students-for-climate-change

38

PEOPLE-POWERED CLIMATE JUSTICE. CHALLENGES AND POSSIBILITIES IN ENGAGING ACTIVISTS IN STRATEGIC CAPACITY BUILDING

Rumbidzai Irene Mpahlo, Joseph Worthy Jr., and Alexander Repenning

Introduction

The climate crisis is worsening despite the climate movement's sustained efforts. In this chapter, we will reflect on an effort to engage the climate justice movement in improving its strategic capacity and increasing its effectiveness on the frontlines. Our reflections are grounded in a 15-month fellowship program for climate justice activists. The program, carried out by the Albert Einstein Institution (AEI), an organisation founded by nonviolent action scholar Gene Sharp in 1980 and funded by the non-profit organisation Right Livelihood, is designed to make frontline activists more effective. It engages at the intersection of theory and practise with an emphasis on the development of the individual activists and increasing their capabilities, skill sets, and ways of thinking. The 31 activists selected for the fellowship were drawn from different regions on the European and African continents, constituting a diverse group in terms of age (between 20 and 45 years old), gender, language, years of experience in the movements, and areas of focus on climate justice. They brought together rich knowledge, perceptions, attitudes, and experience and ways of dealing with their climate justice struggles.

The text will make the case of why strengthening strategic understanding and capacity, rooted in the insights of Gene Sharp's work, is critical for addressing some of the movements' challenges. Sharp's publications on strategic nonviolent action have been a source of inspiration and insight to numerous global struggles against authoritarianism, invasion, occupation, and other violations, and the decades of AEI's experience in providing this support hold valuable lessons for the climate justice movement. We will share learnings from the evaluation of the first cohort during 2022–2023 with climate justice activists based in Europe, and observations from the second cohort during 2023–2024 with climate justice activists based in Africa. The chapter focuses on how the fellowship has effectively engaged frontline climate justice activists in learning and how it has impacted them individually and the work in their movements and organisations.

DOI: 10.4324/9781003396567-45

The next section discusses the internal dynamics which confront climate justice movements in Europe, Africa, and elsewhere, followed by outlining how Gene Sharp's theories connect with the climate justice movement's work. Needs of the movement will then be examined from the information that was provided by activists from Europe and Africa. Part two introduces the fellowship's design principles, developed to ensure learning and behaviour change in the activists. Part three presents the structure and content of the fellowship, followed by the identified improvements and challenges and a conclusion.

Dynamics Within Climate Justice Movements in Europe, Africa and Beyond

The climate justice struggle continues to grow and evolve as different ideologies, principles, strategies, tactics, and contexts shape the way activists articulate and implement their work. The specific issues that people and communities experience inform the nature and form of their programmes and actions, as they strive to build solidarity and agency. Bond (2012) provides what he calls the climate justice political traditions in chronological order from the 1990s to 2010. He shows the emergence and build-up of environmental and climate justice actions and movements locally, regionally, and globally with various demands targeting local actors, the COP process, and the World Trade Organisation (WTO) among others. Some movements have targeted their attention towards international and regional processes which seek to fight the effects of climate change and the involvement of their governments in the implementation of the resolutions in their countries. Top-down approaches, such as the United Nations negotiations, have been criticised for being elitist and disconnected from the people's realities on the ground (Bond, 2012). The climate mitigation strategies include carbon capture projects, carbon trading, and carbon markets, which have limited chances of success (Bond, 2012). Tramel (2020, p. 76) discusses how climate and environmental justice movements are converging " 'from below' in resistance and as an act of building political power". This is against the climate mitigation strategies which promote resource grabbing and thus dispossession, expulsion, and exploitation of Indigenous and local communities, which rely on land and water for their survival.

Various organisations and movements have recorded some successes at the local level in their specific struggles. Therefore, they have extended their reach from local to regional and international through building and joining transnational movements (Tramel, 2020). In the Global North, efforts of the climate movement recently shifted from the mass demonstrations that peaked in 2019 to more confrontational tactics, e.g. blocking traffic and throwing cans of soup on paintings (Fisher et al., 2023). Both approaches have been largely ineffective so far. Protests caused a spike in public concern about climate change, but political action of decision-makers falls far behind what's needed to reach the 1.5 degrees Celcius target, and "governments seem more concerned with criminalising non-violent protesters than acting on climate" (Fisher et al., 2023, p. 912).

The climate struggle in Africa is grounded in the fact that Africa has contributed the least to the climate crisis, yet it bears the brunt of its effects while lacking tangible redress from the Global North. The crisis weighs on various groups disproportionately, and the response of social movements is shaped by an attempt to address their needs. Most of the grassroots climate activism in Africa is closely tied to its countries' appetite for economic development, because the dominant economic model in use in African economies has worsened the situation for African people. Economic policies and strategies are promoting multinational companies that engage in extractive projects that lead to increased climate vulnerability.

The projects force the displacement of people and their relocation in environments that make them prone to climate hazards, pollution, food insecurity, and limited access to most basic rights. Limited national climate adaptation and mitigation leave communities at the mercy of climate disasters. This means that the struggle for community climate justice must be informed by a strong national and transnational climate action framework.

The climate justice movement is mobilising against the neo-colonial, capitalist, and extractivist nature which pushes communities in Africa further into poverty and vulnerability (Allen, 2015). Climate justice activists in Africa also engage in campaigns and protests against their government policies and companies that worsen their plight. Goldblatt and Hassim (2023) illustrate feminist struggles in South Africa as women push for climate justice through litigation against a titanium mine that has threatened to destroy their food production and the land they reside on. Certain movements rally their struggle from a feminist approach because societal and governance systems do not promote women's inclusion in climate action, yet they suffer most from the effects. In African countries like South Africa, young people feel excluded from decision-making within the climate justice movement, which also has implications on their effectiveness (Nkrumah, 2021).

Relevance of Gene Sharp's Work to the Climate Justice Movement

The fellowship emphasises the use of strategic nonviolent action within climate justice struggles. Nonviolent action is seen as "a technique of struggle involving the use of psychological, social, economic, and political power" (Engler, 2013). While acknowledging that nonviolent actions are widely used in climate justice movements, there is a need to increase effectiveness in how they are planned and conducted. Gene Sharp's theories offer a comprehensive framework to develop their strategic direction. He defines strategic nonviolent struggle as "a nonviolent struggle that is conducted in accordance with a previously prepared strategy" (Sharp, 2012, p. 285).

Sharp was a renowned theorist in the field of strategic nonviolent action who significantly contributed to the development of this field as a social science through his publications. His most significant work is the three-volume set titled "The Politics of Nonviolent Action" (Sharp, 1973), while his most well-known work is "From Dictatorship to Democracy" (Sharp, 2002), which has been credited with inspiring and informing the Arab Spring uprisings. However, his most relevant contribution is a self-liberation toolkit – a collection of core readings published on AEI's website – that includes extracts from various publications by Gene Sharp and other scholars, guiding individuals in planning a grand strategy to achieve freedom and dignity through strategic nonviolent action. These readings serve as the foundation for applying strategic nonviolent action principles.

The 15-month fellowship discussed in this chapter and its sessions are derived from Gene Sharp's work, which suggests that the power of corporations, governments, and international institutions is based on the assistance, cooperation, and obedience of multitudes of people. Without that, they can no longer function. Therefore, struggles need to engage in processes that increase their power while reducing that of their opponents. To achieve this, a struggle must engage in planning and organising and, in most cases, endure a cost to shift and re-establish the power dynamic.

In the climate justice movement, the same dynamic exists: those in power rely on the cooperation of the masses. The movement must find a way to induce large-scale noncooperation targeting key institutions, fuelled by the people most affected and the creation of alternative institutions, norms, and systems that replace the entities causing climate harm. The next section outlines current roadblocks identified within the movement that are hindering its success.

Mapping Key Needs Emerging from the Climate Justice Movement

Before the start of the fellowship, we conducted an analysis of current needs and trends in the climate justice movement. We were interested in further understanding the work that climate justice movements engage in, to review how effective it has been and whether the process of devising their activities can be enhanced by employing nonviolent strategies. A qualitative approach was used to gather the information. The analysis was based on interviews and conversations with 23 climate activists and experts from civil society organisations based in Europe (Germany, Sweden), North America (USA, Canada), and Africa (Nigeria, Ivory Coast). The interviews were conducted in 2021. Ten of the interviewees have a national or regional focus in their work and activism, and 13 work with a focus on international and transnational movement building, covering dozens of countries in Europe, Africa, North America, and Latin America.

In the following sections, we share some of the insights from the analysis. The summary is enriched with insights from the applications of the European and African Fellowship cohort from 2022 and 2023. Applicants answered a range of questions regarding their role in the climate justice movement, their motivation, their political understanding and experience, and their analysis of the current state of the movement. A total of 70 applications[1] were reviewed for this purpose. The needs expressed in the interviews and applications ranged from building different kinds of alliances; to training, education, and capacity building; to vision and strategy development for the movements.

Building Alliances and Increasing Diversity

A dominant topic was the need for coalition building and coalition management, especially between the Global North and the Global South, as well as alliances with other movements, especially workers and groups dedicated to defending human rights, land rights, and food sovereignty. One applicant lamented the lack of synergy between movements and lack of building on collective achievements. Overall, an impression remained from the applications that the climate justice movement is fragmented and, as several applicants stressed, in need of consolidation.

On a similar note, several interviewees and applicants in Europe pointed out the lack of diversity in the European movement and brought up the fact that many black people, people of colour, and Indigenous people are tired of the climate movement, as their voices are not being heard or they are often put to the front to show diversity ("tokenism"). More importance should be given to feminist perspectives, to women of colour, and to voices from the Global South, to Indigenous and rural actors. These groups in Europe lack support compared to, for example, the US where the Climate Justice Alliance is led by frontline communities.

Training, Capacity Building, Education, and Direction

Many interviewees and applicants expressed a great need for training and capacity building for a variety of skills. One of the organisations emphasised that there are now plenty of offers for training and capacity building for the climate movement but that most lack hands-on training. This includes learning how to make decisions in big groups, how to improve communication skills, spread knowledge about successful approaches, and encourage their replication.[2] Others highlighted the importance of bringing communication and logistical capacity to movements. Some mentioned the importance of assistance with conflict mediation and that support has to be quick in order to cause as little harm as possible.

Another organisation talked about the importance of educational material for activists in the Global North informing about false climate solutions and other crucial perspectives often missing when speaking about Climate Justice. This includes linking land rights to the climate issue, as land that is controlled by the people who depend on it for food production is less likely to be destroyed.

The need expressed most strongly by applicants from Africa is for more training and capacity building for a variety of target groups. Some emphasised the need for an increased engagement and education of decision-makers through workshops and training; others emphasised the importance of educating personnel such as judges and legislators on the perils of climate change and existing and potential legislative means to address them. Affected communities and climate advocates were also mentioned as target groups in need of further training and empowerment.

Related to this issue is a stronger focus in the movement on national, regional, and local grassroots organising to work on changing things closer to home, as many civil society groups have lost hope in COP as a mechanism for change since COP15 in Copenhagen, one interviewee stated. Many applicants in Africa emphasised a need for engaging the local and rural communities, the importance of supporting local initiatives and community resilience and the engagement of community-based organisations. Another important issue is language. Several applicants stated the need for overcoming language barriers, using simple language or, for example, by providing resources in French. In addition to the need for strong organising work, applications and interviewees painted a picture of a movement in need of strategic direction. Some said that there is no clear vision for the way ahead, so the movement needs to dare to support a variety of ideas to then learn what worked and what did not. Another pointed out a need to generate new utopian ideas.

Funding, Mental Health, and Other Concerns

Another issue brought up repeatedly was the lack or inaccessibility of financial resources to sustain and grow the movement in a healthy way. In Europe, interviewees mentioned a lot of offers of funding for climate activism today, but very often the application, the criteria, and the reporting are too complicated to be useful for grassroots activists or are simply inaccessible. In other cases, when funding has been provided, it often led to changing power dynamics in the groups and created a special "class" of paid activists vs. non-paid activists, which in turn caused additional problems.

A major concern in the European context was mental health, as many activists showed symptoms of burnout – a phenomenon that the facilitator and co-author, Joseph Worthy Jr., has observed repeatedly in the over 15 years of experience building and working with movements. Other issues raised were a need for visible wins, for broadening the base of support, for identifying more tangible campaign targets, or for mass mobilisations. Interviewees also mentioned a trend of disillusioned activists moving towards more violent action and property destruction.

Overall, applicants and interviewees painted a picture of a movement in need of strategic direction, for more diversity, mental health support, and the tools to build coalitions with other social justice struggles in the Global South and the Global North. Activists focused on the involvement of certain target groups (local communities, governments, workers, climate advocates), increasing certain capacities (building alliances, decision-making, language), and certain methods (mass mobilisation, teaching).

Many of these issues are not unique to the climate justice movement. Other movements throughout history have faced and overcome similar challenges, and we think there are lessons to be learned from these historical examples. Gene Sharp's systematisation of these learnings (Sharp, 1973) holds valuable lessons to increase the climate justice movement's capacity to use effective nonviolent action. To ensure learning and behaviour change in the activists, we coupled Sharp's theories with design principles detailed in the next section.

Fellowship Design Principles

To increase the strategic capacity and effectiveness of frontline activists, a deep and operational understanding of Gene Sharp's theories is required in order to apply them within their context. This is the only way the theories become effective. They are not grid-like applications to every problem on the planet. To foster this, we sought to create the conditions for learning grounded in the activists' experience, worldview, and opinions yet firmly committed to underlying theories that guide the development of strategic nonviolent action practitioners.

The critical question becomes how to create conditions grounded in the frontline activists' worlds, experiences, and opinions and still hold Gene Sharp's theories up for examination and implementation. This has been a long-held struggle in movement-building spaces, where one is either sacrificed for the other or at least watered down to benefit the other. For effective learning to occur, conditions should allow both the full presence of the activists and theories to enter into the space unaltered in order to move a group of people towards actionable results in their craft and their movement space.

To make this space for learning happen, we've developed design principles for the fellowship based on personal experience of practitioners, long-solidified and viable approaches to learning, and some new experimental features. These have shown some promising results in this particular space and could be applied to other spaces in regard to learning how to develop movements. We developed this experience based on four design principles. Affirming presence, connection to materials, creating safe space for theory exploration, and responsive coaching and increasing leadership capacity. The following sections expand on each of these four principles.

Affirming Presence

Affirming the group's presence ensures that each participant recognises they are an integral part of the learning space and that their experience in this field can contribute to knowledge development in our time together (Hooks, 2014). The facilitator is careful not to overstate or overburden the activists with his/her experience, relevant or not, that threatens to create conditions that stifle the opinions and thoughts of the frontline practitioners. This approach seeks to remove a barrier between facilitator/participant. It creates conditions to deepen the sharing of experiences that contribute to the richness of the learning space and connects the theories to the participants' lived experiences.

This is done by having the activists share who they are in writing and through presentations, which enables the facilitators to understand the activists and their work better. Additionally, in our kick-off meeting, activists and facilitators brought a memento that represented the deepest parts of who they were and what they brought into the space other than their need for knowledge. This created conditions conducive for learning and knowledge sharing.

Building Connections to the Materials

Connection to the material is paramount to successful learning for anyone but especially for frontline activists who, in some cases, are withdrawing from conflict to engage in learning. This withdrawal has to be justified and seen as critical to the success of their movement. Movements can be resistant to this process as it's time-consuming and can feel like inertia in the face of oppression happening in real-time.

Because of this reality, facilitators must create a connection to the material that illuminates the validity and the necessity of these processes. To do this, we allow activists to engage in activities that connect aspects of the theory to fictional scenarios to help make those connections. These scenarios detail an issue a community is facing, and activists decide on a course of action. Over a timespan we reveal new "conditions" and allow them to change, improve, or keep their courses of action. As the groups unveil and discuss their approaches, they also discover that those approaches were not issue-specific, rather they were specific to the conditions within which the issues were existing. This is where the conditions for learning grounded in the activists' experience, worldview, and opinions come to life and make them effective. These types of connections are sought throughout the fellowship. Wherever possible, we try to take real-life experiences of the activists and connect them to the materials to show the relevance of the theories in question.

Gamifying the theories is another approach we took to facilitate connection with the material. While games and exercises will never be as effective as "on the ground" praxis, if designed well, they can offer meaningful growth to the practitioners. To do this, we take the key theories we're engaging with and try to create a "game" or exercise that may be coupled with guiding questions to integrate the learners' thinking better.

We've also introduced the Feynman Method, a process of reducing complex ideas down to their most simple, understandable components (Feynman et al., 2011), into our work for key theories. While we do engage in learning assessments for the fellowship, we want to ensure that the learners aren't simply regurgitating knowledge but truly understanding it.

Safe Spaces for Theory Exploration

Successful application of strategic nonviolent action needs safe spaces in which activists can concretely apply the theories within their own context and gauge what is possible for their situations. We also understand that activists, especially – yet not exclusively – in the Global South, risk great personal harm in pursuit of the dignity they seek. One of the activists has already been on trial for "terrorism". Others have faced arrest or have been the cause of arrest of others. To be clear, these aren't the same arrests that people face in the Global North; they can involve the absence of due process and certainty of release.

In the face of these realities, how do we create space for learning that satisfies the pedagogical approach of "praxis" – reflection against the backdrop of actions (Freire, 1993) – but limits the consequences of mistakes and miscalculations? Engaging in strategic nonviolent action is not safe and yet we know that, historically, people engaged in struggles have usually "learned by doing", which is through trial and error and experimenting with carrying out action and gauging outcomes. Through study of the technique and requirements for effectiveness, we have learned that dangers can be mitigated through careful planning. Still, when using fictional scenarios, there is space for engaging in concrete practice without the risks of action on the ground.

This may be a crucial step in closing the loop between the different design principles for frontline learning. It can create opportunities to fill any of the holes from connection to practice and offer opportunities to identify and address challenges with theories.

Responsive Coaching and Increasing Leadership

Even the most strategically sound activist can have an issue implementing new processes into their movement. The material alone is insufficient to ensure a well-thought-out strategic nonviolent movement. Questions of leadership and structure remain salient to the strategy issue in the climate justice movement. Our coaching process is designed to address this reality so that the new skills can see the light of day within the climate justice movement. During the fellowship, we have observed some activists grappling with leadership and structural concerns as they engage in the learning process. Some have expressed doubt that other movement leaders will be receptive to new discussions, particularly when a woman leads the conversation about internal power dynamics and direction. These challenges can impact the successful implementation of the knowledge gained from the sessions. We should be clear that this issue persists in the Global North and in the Global South. That said, the coaching process allows participants to unpack the issues in their movements and, through thought partnership, develop interventions that increase the likelihood of success.

Fellowship Structure, Formats, and Main Components

To achieve our design, we structured the fellowship around four points of contact: theory grounded sessions, practice-problem sessions, individualised coaching, and teach-backs.

Theory grounded sessions are whole-group meetings that include an overview of theoretical concepts, strategic principles, best practices, case studies, trends, and phenomena shaping the "battlefield", as well as current developments in the field of nonviolent action and climate. Practice-problem sessions are brainstorming meetings with the whole group where activists present their most pressing problems, and facilitators and other activists collaborate to help find creative solutions.

In addition to the group sessions, activists engaged in one-on-one coaching sessions with the facilitator to address topics that they feel need the most attention. Teach-backs are a three-step process for activists to gain a better understanding of the content: after each theory-grounded session, activists are encouraged to report back on their learning via a short video or a one-to-two-page essay. Then, facilitators review and suggest short readings to fill gaps in the fellows' learning. Finally, the learnings are reviewed in one-to-one coaching sessions.

Main Components of the Theory Sessions

We developed theory sessions based on six components: the strategic estimate, operational planning, understanding power, nonviolent methods and tactical sequencing, maintaining nonviolent discipline, and mental health. These components cover Sharp's theoretical framework, complemented by a session on mental health. They prepare fellows to plan effectively, deploy resources with intention, and maintain their persistence by taking care of themselves and their people.

Strategic Estimate

The Strategic Estimate is a core instrument in the work of AEI and a critical tool for strategic planners. Based on a military planning tool adapted by Robert Helvey for a nonviolent context, it provides a systematic approach to developing the best course of action to accomplish a mission. It does this by identifying and analysing essential factors such as the environment (physical, political) and the capabilities of those expected to be participants (both friendly forces and the opponent, as well as third parties who could become involved in the struggle). Then it compares strengths and vulnerabilities to develop courses of action. Based on this analysis, the best course of action is selected. Since the implemented actions are based on information in a strategic estimate, the quality and quantity of information analysed significantly influence the chances for success (Sharp, 2005).

A carefully formulated strategic estimate will allow for the activists' application of nonviolent strategy to be focused and relevant to the conditions their movement finds itself in. In this group of exercises, they go through everything they should consider before progressing in the strategic process. The steps are:

1. Mission statement (campaign statement) – this big-picture statement should show the "vision for tomorrow". This allows new leadership to step up and pick up where someone else left off in the campaign
2. The situation and course of action – contains all the information about the situation in which the mission will be conducted, including but not limited to transportation, political climate, communications, climate, weather, power, etc. This should be done for the offensive group and the opposing group
3. Analysis of opposing courses of action – determine and write out the effect each opposing group's actions could have on the campaign's course of action
4. Comparison of own courses of action – the advantages and disadvantages of each course of action are weighed
5. Decision – the course of action is decided upon and formulated into an organising statement.

Operational Planning

Each strategic objective needs to be carefully thought out and planned to ensure success. Operational planning provides a framework to think through each individual operation. For example, if recruitment is a high priority, there should be a clear plan as to who, how many, and when you will recruit people to carry out a specific action. If a movement identifies a hole in their power capacity, they would use operational planning to ensure that they are taking the necessary steps to fill that hole to enable them to win their campaign.

Understanding Power

Understanding power helps activists engage with those who hold power and recognise the power inherent in their communities. We dispel misconceptions about power and discuss how power structures are fluid and that compliance (of the people) is necessary to maintain the status quo. During these discussions, we introduce the activists to concepts like the pillars of support, demonstrating that centralised power is only as potent as the pillars that uphold it. We use tools like the relative power matrix to illustrate the symbiotic relationship between a movement's power and its opponent's power and how this dynamic can shift based on different conditions and the approach taken.

Nonviolent Action Methods and Tactical Sequencing

During our meetings, we spend a considerable amount of time exploring nonviolent action methods and their relevance to the previous components. These methods are the most visible aspect of any movement, and they are what the public notices and responds to. However, if a movement selects its methods based on assumptions rather than a sound strategic foundation, it becomes less effective, conveying the message to the public that strategic nonviolent action is ineffective as a means of struggle. To avoid this, we incorporate methods into our discussions on strategic thinking and power, allowing activists to practise thinking critically about which method of strategic nonviolent action makes sense with the information they have.

Maintaining Nonviolent Discipline

There are several moments in a movement's life cycle where the validity of strategic nonviolent action comes into question. For this component, we want to ensure the activists are engaged in a way that strengthens their struggles for nonviolent discipline through proven practices and safeguards their movements against violent flanks and other ineffective behaviour.

Maintaining Mental Health

Maintaining mental health is critical in strategic nonviolent action struggles. Mental health affects everything from the ability to think strategically and maintain direct action for the necessary period. We discuss with activists the best ways to maintain their mental health and that of their movements.

What Improvements and Challenges Are We Seeing?

We have seen the fellowship allow movements to advance and introduce new concepts, stabilise the structure of their campaigns, and make plans to deepen the capacity of their movements. Moreover, we have seen an improvement in the effectiveness of some of the activists' activities.

An example from one of the international movements we have coached illustrates this observation. This group was looking for a way to increase their power internationally. They had sought an endorsement from Pope Francis, arguably the world's most powerful religious figure. However, the internal struggle was whether the Catholic Church aligned with their values. The group deliberated and decided to seek the endorsement and got it. While the movement's founders are salient political thinkers, they sought counsel for language and the perspective to convince their group that this particular effort was in their favour beyond certain philosophical disagreements. In the coaching there was no endorsement by the facilitator in one way or the other. It was a simple explanation of power in its purest form and how it could be leveraged or not. Since then, the group has been able to move forward, leveraging the endorsement of Pope Francis in their favour.

Another promising development is that one group of activists has begun to integrate both constructive and obstructive methods into their movement work, revitalising bare land and building the capacities of villages to resist or seek reparations for stolen land. As a direct consequence of implementing the learnings of the fellowship, the group saw an increase of their power in less than 6 months.

A third example to highlight is a group fighting a dam project in their country. We took a session to do a group strategic estimate and use the activists' situation to fine-tune everyone's understanding of their strategic estimates. During the session, they found a novel approach to addressing their problem because the company was French, which opens avenues to hold the corporation accountable to business standards in France. The activists connected with an interested student in France who began to research possible interventions.

The improvements we are seeing are because activists have taken the time to look at their situations more closely and have been given the space and time to consider their courses of action against the backdrop of Sharp's theories.

Throughout the two first cohorts we could also observe the benefits of engaging with the material as a team. On the rare occasion where we have had two or more activists from the same groups within the fellowship, we have seen them be able to build each other's understanding of the theories in relation to their movements. In addition to the importance of decision-making power on strategy mentioned earlier, this experience holds relevant learnings for future cohorts and similar programs aiming to build strategic capacity.

Another challenge particularly relevant to activism on the frontlines is the tension between the need for careful planning and the urge to react to new situations. Most activists we have collaborated with have demonstrated high dedication, engagement, and willingness to apply their knowledge and skills in real time during the fellowship. Since their communities are often the ones at risk, new developments can add urgency to their campaigns, while the strategic planning process does not lend itself to expediency. While this tension cannot be entirely resolved, real-time coaching with people whose expertise is rooted in experience may help to address this issue.

While the online format makes continuous participation possible for activists from different geographical locations, even in remote areas, the reliance on a stable internet connection poses a challenge in engaging with frontline activists. Communities at the frontlines of climate justice struggles may lack stable access to the internet and face power shortages. This has presented limitations with some of the interactive online tools introduced to create a more enriching learning environment.

In the future, we're considering how we can be more creative when engaging with frontline activists, for example by working with hard-copy materials and taking pictures of their plans and progress. However, this could add more questions regarding security and organisation of knowledge.

Outlook and Conclusion

Our preliminary findings show that the fellowship has enabled the activists to reflect and improve the way they confront their struggles and actions. At an individual level, some activists have expressed a better understanding of how to build and assess power and how to strategise considering a variety of methods and tactical considerations.

At an organisational level, several activists have indicated that they are applying non-violent strategies in a systematic manner which has assisted them in achieving some wins. However, others claim that though they have applied the knowledge gained, they are still facing stiff resistance to produce any tangible results though they have made some progress.

On a broader societal level, we will continue to examine the effects of our engagement and its impact on overall perception of the climate justice movement. Overall, the programme is relevant to improve the impact of climate justice activists; however, the analysis has shown that certain conditions should be in place. The skills can be effectively applied if there is supportive organisational leadership and a cooperative team who are willing to make adjustments to their usual ways of doing things. This is supported by our observation that movements of activists in leading positions benefit more from the fellowship than those of staffers or campaigners engaging in it. As the program focuses on building strategic understanding and capacity, it is best suited for activists who are part of the group making strategic plans and decisions in the movement. In the opposite case, we have seen practitioners who struggle with communicating their newfound insights to the leadership. The result is frustration from the activist with no avenue to implement learnings. The goals of particular campaigns need to be strategically shared and understood to get buy-in and to keep activists participating despite the harsh political environments, especially in Africa.

What is currently emerging amongst activists in Africa is that the fellowship has filled knowledge and skills gaps. The fellowship has enhanced the activists' ability for strategic thinking and interrogation of their projects, thus improving their strategies, methods, and tactics and potentially increasing their chances of success in their specific struggles. The acquired knowledge and skills are not only applicable during the fellowship but are meant to influence the culture of working in the organisations and movements that the activists belong to.

The fellowship has also generated interest among other partner organisations that stand to benefit from the knowledge shared by the activists. The fellowship has opened up space for different activists to build synergies according to language, regions, and thematic focus. The activists have the capacity to strengthen the network they have established to prioritise important aspects and train other activists, partner organisations, and movements within their spheres of influence, thus enhancing the broader movement to increase their "wins".

While we acknowledge that the fellowship is not the panacea to the challenges that climate justice movements are facing, the fellowship has proven to be a valuable puzzle piece in the much-needed strategic development of the climate justice movement. It broadened the thinking of activists to assess their ever-changing environments, the actors, and their dynamic contexts and to identify what works and conditions which enable better results. As activists from both continents mentioned, more synergies need to be forged and sustained between activists in the Global North and those in the Global South. The fellowship is a stone in the mosaic of creating a platform for activists to find common ground across regions and engage in strategies that amplify their fight.

Notes

1 Thirty-one applications from people based in ten European countries and 39 applicants from 14 African countries.
2 As argued later, we are convinced that methods and tactics are only effective when applied as part of a broader strategy based on the situation analysis. Hence, the replication we speak of is limited to certain processes that lead to deeper strategic analysis and action.

References

Allen, F. (2015). The state of the climate justice movement in South Africa. *Capitalism Nature Socialism*, 26(2), 46–57.

Bond, P. (2012). *Politics of Climate Justice: Paralysis Above, Movement Below*. Cape Town: University of Kwa Zulu Natal Press.

Engler, M. (2013). The Machiavelli of nonviolence: Gene Sharp and the battle against corporate rule. *Dissent*, 60(4), 59–65.

Feynman, R. P., Leighton, R. B. and Sands, M. (2011). *Six Easy Pieces Essentials of Physics Explained by Its Most Brilliant Teacher*. New York: Basic.

Fisher, D. R., Berglund, O. and Davis, C. J. (2023). How effective are climate protests at swaying policy – and what could make a difference? *Nature* 623 (7989), 910–13. https://doi.org/10.1038/d41586-023-03721-z

Freire, P. (1993). *Pedagogy of the Oppressed*. Translated by M. B. Ramos. New York: Continuum.

Goldblatt, B. and Hassim, S. (2023). Grass in the cracks': Gender, social reproduction and climate justice in the Xolobeni struggle. In C. Albertyn (eds.) *Feminist Frontiers in Climate Justice* (pp. 246–267). Cheltenham and Northampton: Edward Elgar Publishing.

Hooks, B. (2014). Introduction: Teaching to transgress. In *Teaching to Transgress. Education as the Practice of Freedom* (pp. 8–12). New York: Routledge.

Nkrumah, B. (2021). Eco-activism: Youth and climate justice in South Africa. *Environmental Claims Journal*, 33(4), 328–350.

Sharp, G. (1973). *The Politics of Nonviolent Action*. Boston: Porter Sargent Publisher.

Sharp, G. (2002). *From Dictatorship to Democracy. A Conceptual Framework for Liberation*. Boston: The Albert Einstein Institution.

Sharp, G. (2005). *Waging Nonviolent Struggle*. Boston: Porter Sargent Publishers.

Sharp, G. (2012). *Sharp's Dictionary of Power and Struggle: Language of Civil Resistance in Conflicts* (Kindle ed.). Oxford: Oxford University Press.

Tramel, S, (2020). Convergence as political strategy: Social justice movements, natural resources and climate change. In T. Moreda et al. (eds). *Converging Social Justice Issues and Movements* (pp. 64–81). London and New York: Routledge.

39

BUILDING REGENERATIVE CULTURES IN EXTINCTION REBELLION ACTIVISM AND ACADEMIC RESEARCH

Hannah Fitchett

Introduction

Extinction Rebellion (XR) emerged in November 2018, just weeks after the peak of the transnational Fridays for Future school strikes inspired by Greta Thunberg (Malm, 2021, pp. 12–13). The rapid growth of both movements occurred in the wake of the 2018 Intergovernmental Panel on Climate Change (IPCC) report, which stressed that limiting global warming to 1.5 degrees Celsius required "rapid and far-reaching" systemic changes that were "unprecedented in terms of scale" (IPCC, 2018, p. 15). XR's mass civil disobedience in April 2018 positioned the movement "at the crest of the twenty-first-century cycle" of climate activism (Malm, 2021, p. 13). For two weeks, activists occupied key roads across central London, sitting, gluing, and locking-on to infrastructure until, one-by-one, they were arrested and carried away. Almost 1,100 activists were arrested, making this the biggest act of non-violent civil disobedience in the UK for decades (Malm, 2021, p. 13). The XR movement spread rapidly across the globe – resulting in 1,010 XR groups across 88 countries in early 2024 (XRGlobal, 2024) – and inspired the emergence of many similar climate groups in the Global North, including Just Stop Oil, Letzte Generation, and Scientist Rebellion.

XR aims "to halt mass extinction and minimise the risk of [resulting] social collapse" using non-violent civil disobedience to pressure governments to meet its three demands (XRUK, 2024). These can be summarised as: (1)"All institutions must communicate the danger we are in" due to climate breakdown, (2) society must reach net zero greenhouse gas emissions by 2025, and (3) "the Government must create and be led by a Citizens' Assembly on Climate and Ecological Justice" (XRUK, 2024). Many XR groups also demand that climate justice is prioritised in how these demands are achieved. While civil disobedience is the most outwardly visible aspect of XR's approach, it is the movement's less outwardly visible efforts to build "regenerative cultures" that are most distinctive within the broader climate movement (Harms, 2022, p. 519).

Fundamental to XR's efforts to build "regenerative cultures" is an ethic of practising "care" across "all levels, including individuals, communities, our soil, water and air", with the aim to "heal and improve" (XRGlobal, 2024). The wellbeing of these "levels" is

DOI: 10.4324/9781003396567-46

considered interdependent, meaning that caring for any one requires caring for all (XRUK, 2024). Within XR, as in alternative economics and farming (Harms, 2022, p. 517), the term "regenerative" therefore refers to a form of relationality, based on a perception of all life as inherently interconnected, that attempts to enact care in the place of extraction (XRUK, 2024). Regenerative cultures are understood to take many forms: rather than a fixed definition, XRUK's (2024) website offers "a framework based on natural principles" which people can use to develop their own approaches to "practising a regenerative way of being". Practices that my interlocutors considered "regenerative" included sharing "grief" for what climate breakdown has destroyed, waiting outside police stations to greet and assist activists when they were released, and "regen reminders" during meetings (which typically involve reading empowering poetic statements about interconnectedness and the resulting importance of care).

This chapter treats XR's notion of regenerative cultures as a provocation to examine what a "regenerative" approach to researching grassroots climate activism could look like and how this approach can increase the ability of such research to benefit the activism studied. As this chapter was written mid-fieldwork, I primarily explore this approach in relation to conducting ethnographic fieldwork. By thinking through some of the principles and theoretical perspectives that guided my ethnographic research with XR in London between November 2021 and September 2022, I bring my interlocutors' ideas about regenerative cultures into conversation with scholarship on engaged ethnography. The chapter focuses on three of the core ideals within XR's regenerative cultures as understood by the activists I worked with, which I summarise as: prioritising care, recognising co-construction, and balancing reflexivity and action. I illustrate how academic knowledge and the knowledge I gained from my interlocutors can be woven together to form the regenerative research approach that I propose.

The chapter is underpinned by the propositions that engaged ethnography is well suited to research on climate movements, and that the ethical and methodological insights it contributes can benefit such research across other disciplines. While there is no formula to practising engaged ethnography, since the suitability of methods depends on one's relationship to the ever-shifting fieldwork context, I argue that these three principles of XR's regenerative cultures can provide important guidance for academic research on grassroots climate activism.

The chapter begins by describing what is meant by engaged anthropology, what it entailed in my fieldwork, and why I chose this approach, which leads to an explanation of why I propose a "regenerative" research approach. I then explain the relevance of this approach for other disciplines, for the study of other activist movements, and in particular how it relates to efforts to break down hierarchies of knowledge and the broader inequalities they underpin. Against this background the chapter then explores in turn three key principles of regenerative cultures. Starting with prioritising care, I describe how I attempted to enact care during my fieldwork by relating to my interlocutors as a "comrade" in shared political struggle (Scheper-Hughes,1995, p. 430) and how this allowed me to engage in deep yet constructive critiques. Second, I describe how my interlocutors' recognition of the co-constructed nature of social and environmental worlds both informs my efforts to break down hierarchies related to the researcher/researched dichotomy and can facilitate research that "stands with" (TallBear, 2014) the climate movement. Third, I explore the principle of balancing reflexivity and action and how this can guide efforts to produce research that is valued by grassroots climate movements despite often-limiting academic conventions.

Practising "Regenerative" Engaged Research

My interest in XR as a social anthropologist emerged through my engagement in XR as an activist since April 2019; this political participation in XR has shaped my research from the beginning. I practise engaged anthropology when working with XR, which involves being "personally engaged and politically committed" in one's ethnographic fieldwork (Scheper-Hughes, 1995, p. 419), as well as in one's written work (Hale, 2006, p. 98; Kirsch, 2018). During my fieldwork with XRLondon, I participated as an activist as well as an anthropologist, planning and facilitating meetings, events, and actions; participating in outreach; and committing to a role in XR for 7 months.

When researching activists, supporting their political struggles is often a precondition for access to the group (Kirsch, 2018, p. 5). Indeed, my active participation helped me gain my interlocutors' trust and access to non-public aspects of the movement. However, my engagement as a researcher was – and continues to be – motivated by my view that during fieldwork anthropologists maintain the "human responsibility to take an ethical (and even a political) stand" on historical events they witness (Scheper-Hughes, 1995, p. 411). Indeed, I believe that anthropologists can – and often should – contribute more than just texts to addressing injustice (Kirsch, 2018, p. 230). Through practising engaged anthropology while studying XR, I therefore aim to support the movement's efforts to mitigate climate breakdown and deeply intertwined social injustices. It should be noted that engaged anthropology as I have described it may also be referred to as militant anthropology (Scheper-Hughes, 1995, Juris, 2008; Russell, 2015) or activist anthropology (Hale, 2006). I prefer the term engaged anthropology, since activist and militant anthropology place the focus on the activism of the *anthropologist* rather than of the people they are researching, and consequently mask the inherently collaborative nature of ethnographic fieldwork (Kirsch, 2018, pp. 17–18). Engagement, in contrast, better describes my activism as a collaboration with that of my interlocutors, highlighting the relational nature of this activism. Moreover, "engaged anthropology" avoids restrictive terms for political struggle that may not align with a community's understanding of their activities.

During my research, I recognised several parallels between XR's concept of regenerative cultures and the aims and values motivating my engaged approach. Building on these parallels, this chapter outlines an approach for what I propose could be considered "regenerative research", since like XR's regenerative cultures it concerns a form of relationality that prioritises care, recognises co-construction, and aims to balance action and reflection. Indeed, like XR's regenerative cultures, it aims to propose an alternative to the forms of relationality and resulting extractive ethics within neocolonialism and capitalism, in order to facilitate the development of different "ways of being" and "doing". In other words, it attempts to guide an approach to studying climate activism that is rooted in the ethics of grassroots climate justice activism, rather than research ethics shaped by the neoliberal logic increasingly present within academia or epistemological hierarchies constructed through colonialism.

I contend that such engagement with the concept of regenerative cultures can also assist the study of the shift occurring within the climate movement towards creating new forms of relationality (Harms, 2021, p. 239), in which this concept has played a central role. As I will illustrate through examples from my fieldwork, the principles within regenerative cultures provided useful guidance through many of the challenges that the ethnographic study of such activism involves.

Although I draw on the concept of regenerative cultures in relation to XR, this approach is also relevant to research of other activist groups. Indeed, regenerative cultures have become part of the Western climate movement zeitgeist due to XR's rapid expansion into a transnational movement. The concept has been adopted by subsequent activist groups including Letzte Generation in Germany and Futuro Vegetal in Spain. Moreover, it encapsulates many ideas that have been popular in Western environmentalism since the 1970s (Taylor, 2009, p. 14), and can be traced back to Rousseau (*ibid*, p. 10). Core characteristics of XR's concept of regenerative cultures – such as veneration of the biosphere understood as comprising interdependent parts, using natural processes as models for how to organise sustainably, and scepticism of the supernatural – fit neatly into the worldview that the scholar of environmental ethics Bron Taylor (2009, p. 16) termed "Naturalistic Gaian Earth Religion", before the existence of XR. Furthermore, efforts to protect activists' well-being through self-care have long been integral to feminist and anti-racist activism (Harms, 2022, p. 517).

The regenerative research approach I propose is therefore relevant to the research of many other activist groups. Moreover, concepts from psychology, climatology, and agronomy intersect with XR's concept of regenerative cultures (Harms, 2022, p. 518), and the argument that effective climate breakdown mitigation must recognise interconnection has been made across many academic disciplines (Haraway, 2016; Latour, 2017; Mies and Shiva, 2014; Morton, 2007). Therefor the research framework I outline also has transdisciplinary relevance.

Treating Interlocutor Ideas as Forms of Knowledge

Before examining what regenerative research may involve, it is important to lay out the theoretical basis for using an emic concept belonging to my interlocutors to shape my academic approach. Here, ecofeminist scholar Donna Haraway's (1988) concept of "feminist objectivity" is foundational. In contrast to dominant notions of scientific objectivity that offer a "conquering gaze from nowhere", feminist objectivity aims to produce "situated knowledges" that are embodied and partial (p. 581). Haraway argues that knowledge is inherently partial and situated, and that recognising this is the only way to achieve some form of objective, rational knowledge. Situated knowledge makes no pretence at disengagement (p. 590). Feminist objectivity breaks down the subject/object binary and situates the researcher, thus rendering them accountable for the knowledge that they produce (p. 583). This approach aims to transform knowledge systems and ways of seeing by privileging "deconstruction, passionate construction, [and] webbed connections" (Haraway, 1988, pp. 584–585).

My approach diverges somewhat from Haraway when she argues that transforming knowledge systems in this way necessitates focusing on the perspectives of the subjugated, since they are "least likely to allow denial of the critical and interpretive core of all knowledge" (p. 584). Despite recognising the risk of appropriating and romanticising the vision of the subjugated, she argues their perspective offers more "objective, transforming accounts of the world" (p. 584). Although XR activists' perspectives are marginalised in the UK by most politicians, press, and powerful corporations, XR London activists are predominantly middle class, white, and university educated (Doherty et al., 2020). Most of my interlocutors therefore hold relatively privileged positionalities within the Global North and are consequently some of the most privileged climate activists globally. There is, however, also significant potential for "transforming knowledge systems" through anthropological research on groups with relative privilege and power. Indeed, the importance of studying the powerful has been recognised in social anthropology since the 1970s when the anthropologist Laura Nader (1974, p. 302) claimed that a correction to the systematic lack of anthropology of the powerful was urgent "by virtue of [their] world influence". As she argued, the lives of the oppressed cannot be fully understood without also studying the powerful and vice versa, since they are inherently interrelated.

Turning the research lens towards those more commonly positioned as *researchers* rather than *the researched* is part of my effort to challenge knowledge hierarchies (as I explain later). I therefore pair feminist objectivity with an awareness that my interlocutors and I, as part of largely the same demographic, simultaneously hold subjugated and subjugating positionalities, and I apply careful critique to both their and my own knowledge and the assumptions implicit within it. Additionally, I pay particular attention to the perspectives of my interlocutors with identities that are under-represented within XR London. By sharing their experiences and concerns with me, many of these people have helped me challenge my assumptions and recognise issues in XR which I now try to address in my activism and academic work.

The researcher/researched (or subject/object) binary is characteristically permeable within the ethnographic research that characterises social anthropology. Anthropological knowledge production is an inherently collaborative process between the anthropologist and their interlocutors (Ingold, 2021) and has been increasingly recognised as such thanks to contributions from feminist and postmodern anthropology (Lassiter, 2005). Anthropologists such as Kim TallBear (2014) and María Casas-Cortés (2009) aim to further deconstruct the subject/object binary by treating their interlocutors' knowledge as "knowledge in its own right", rather than as an object of study that requires "translation" by an anthropologist to be taken seriously (TallBear, 2014, p. 2). During her research among feminist anti-precarity activists in Madrid, for example, Casas-Cortés (2009, p. 86) read the articles the activists produced as "knowledge in its own right" before analysing them anthropologically. Similarly, in exploring how my interlocutors' understandings of regenerative cultures could guide research, I attempt to engage with my interlocutors' knowledge on this topic in the same way I engage with academic knowledge to shape my research approach, while also analysing it as anthropological data. This more democratic approach to knowledge production helps to challenge knowledge hierarchies and the inequalities they justify – within and beyond ethnographic research – and consequently helps facilitate research that effectively supports one's interlocutors' political struggles (TallBear, 2014, p. 2). I argue that this research approach is increasingly vital in an escalating climate crisis largely resulting from the domination of a Modernist form of knowledge that depicts "nature" – and people portrayed as "closer" to it – as objects to be exploited (Bookchin, 1987; De Castro, 2019; Haraway, 2016; Knox, 2015; Latour, 2017, 2018; MacCormack and Strathern, 1980; Macy and Johnstone, 2012; Mies and Shiva, 2014; Morton, 2007; Strang, 2016).

This approach to knowledge production informs my efforts to propose a research approach that draws on a combination of academic scholarship and three key principles that summarise *ideals* that my interlocutors *generally* considered central to regenerative cultures as forms of knowledge in their own right. These are: prioritising care, recognising co-construction (in relation to both social change and the conditions for life), and balancing reflexivity and action. I present these principles as ongoing actions to emphasise that enacting them is a continual process, within both XR and research. I will now explore each in turn.

Prioritising Care

The activities that my interlocutors considered "regenerative" were generally those they regarded as expressions of *care* for the self, others, communities, and/or the Earth: for instance, walking in nature, expressing gratitude during meetings, sharing food with strangers, and gardening. The importance they attributed to such acts of care was largely informed by a perception that all living beings are interconnected, therefore all actions have rippling impacts. Care based on perceived interconnection is also central to engaged anthropology. Indeed, Nancy Scheper-Hughes (1995, p. 419), who advocates for "militant anthropology", describes it as based on an "ethic of care and responsibility" stemming from the conviction that anthropologists are not separate from the social worlds they study. She argues that anthropologists must therefore participate in their interlocutors' lives as "comrades", engaging with the associated "demands and responsibilities" (p. 430).

In relation to my research, I understand these to involve commitment to a shared political struggle with my interlocutors and a responsibility to reflect critically on how to practise this at each stage of my research. My engaged approach can therefore be understood as "a combination of thought and action orientated towards understanding and changing collective praxis" (Russell, 2015, p. 223). My alignment with XR's intentions to mitigate climate breakdown via direct-action and efforts to build "regenerative cultures" is of course essential to this approach. Crucially, such political and emotional investment in XR does not preclude my ability to reflect critically on the movement. Indeed, the ethic of care that political engagement introduces to the research encounter can aid critique and often renders it more constructive (Schuurman and Pratt, 2002, p. 291). As Haraway (1998, p. 589) argues, enacting "loving care" within research generates complexity, yet is necessary to overcome the challenges involved in understanding another's partial, embodied, situated knowledge.

My active engagement in my interlocutors' political struggles during fieldwork rendered internal conflicts and tactical and ethical issues more visible; complicating the picture offered by political theory, media articles, and XR's external-facing messaging. For instance, the role I took on in XR required me to engage in a tense debate about whether to adopt a fourth demand that climate justice must be prioritised in how XR's demands are met. A process had been set up to respond to calls to add this demand, but most of my interlocutors felt it had failed. Consequently, many London groups had autonomously adopted climate justice demands.

During one of my first London-wide XR meetings, the adoption of a London-wide climate justice demand was proposed, with the idea of encouraging other XR groups to do the same to eventually achieve its adoption by XR UK. This sparked weeks of debate over questions such as whether adopting this demand was tokenistic, whether the wording was inclusive, and what decision-making process should be used. Participating in these discussions complicated my preference to push for UK-wide adoption of this demand, as I realised this risked groups adopting it in order to end, rather than begin, conversations about climate justice. While my political investment therefore enabled me to develop a more critical yet constructive perspective, a tension nevertheless remained between the care that motivated my engagement and the academic necessity for critique. My care for the activists pushing for a UK-wide demand – and for the values motivating them – made me reluctant to critique their efforts. Furthermore, my critiques led me to question many of my fundamental assumptions, including at points that XR had an effective approach for mitigating climate breakdown and thus was a movement I wanted to engage in, raising ethical dilemmas regarding my engaged approach.

Engaging in care and critique simultaneously was therefore often uncomfortable and complicated to navigate, however, the complexity produced by such situated approaches helps make them accountable (Haraway, 1998, p. 589). Nuanced understanding is essential for both good activism and good anthropological research (Shah, 2017, pp. 56–57). Additionally, constructive critique driven by care was also practised by my interlocutors, all of whom perceived various issues within XR which they tried to address. Engaging with one's interlocutors as a caring comrade can therefore take the researcher beyond instrumentally simplified demands and slogans, facilitate deep yet constructive critique, and help researchers to understand tensions experienced by their interlocutors.

Recognising Co-construction

The second important principle of XR's regenerative cultures – recognising co-construction – refers to the idea that, due to the interconnection of all life, ecological conditions and social change are co-constructed. This perspective is used to challenge hierarchies and is illustrated by XR's depiction of the relationships between the self, others, our communities, and the Earth using concentric circles (see Figure 39.1), as well as in my interlocutors' descriptions of the climate movement as an "ecosystem": placing value on a diversity of

Figure 39.1　Diagram showing intersecting forms of care within XR's regenerative cultures.

Source: XRUK Newsletter 16: The Rebellion Returns, 1 September, 2020.

approaches and depicting agency as dispersed. Additionally, the intentionally loose definition of regenerative cultures justifies multiple possible practices and forms of knowledge without placing these in a hierarchy. While this open-ended definition makes it difficult to know exactly what regenerative approaches to activism – and indeed research – involve, it offers valuable guidance for my efforts to break down hierarchy within the researcher/researched dichotomy.

As the anthropologist Tim Ingold (2021, p. 289) describes, "the world and its inhabitants, human and nonhuman, are [anthropologists'] teachers, mentors and interlocutors"; they shape the anthropologist's way of thinking and being, meaning that anthropology is always a study *with* humans. Consequently, ethnography is an unavoidably "improvisational practice" (Malkki, 2007, p. 179) requiring anthropologists to adapt their research aims, methods, and outputs in response to interactions during fieldwork (Adams et al., 2017; Chua, 2015; Malkki, 2007; Shah, 2017). Nevertheless, anthropologists do not always fully recognise and value their interlocutors' role in anthropological knowledge production (Merry, 2005) or their interlocutors' knowledge as knowledge in its own right (Casas-Cortés, 2009; TallBear, 2014).

Many anthropologists practising the engaged approach I have described strive to treat their interlocutors' knowledge as knowledge in its own right by inviting their interlocutors to shape their research directly, from co-designing the research to co-authoring publications (Cepek, 2020; Hale, 2006; Casas-Cortés, 2009; Merry, 2005; TallBear, 2014). As part of my effort to break down the researcher/researched dichotomy, I invited my interlocutors to co-design my research at the start of fieldwork, waiting till this moment to ensure the research could be shaped by those I worked with directly, in a manner relevant to this fast-changing movement during my fieldwork. Since XR strives to organise in a non-hierarchical manner, there was no authority to direct this invitation to. When introducing myself and my research and asking for consent from each group I worked with, I explained that my initial research focus on regenerative cultures was open to adaptation in response to ideas people had for focuses they felt could produce useful research. I briefly explained my research methodology to help illustrate the kind of insights my research could provide; however, I received little feedback.

Since pushing to co-design the research when my interlocutors appeared disinterested would conflict with my aim to practice non-hierarchical relationships with my interlocutors, I attempted to allow my interlocutors' knowledge to influence my research design in ways which demanded less of their time. As Bonilla (2015 cited in Weiss, 2021, p. 951) argues, all people are "perpetually analysing their world and making arguments about it"; consequently anthropologists should engage with their interlocutors' arguments, rather than merely describe them. Therefore, I used the analyses and arguments the activists made during fieldwork to select a research focus that I hoped could assist my interlocutors with the tensions they experienced in their activism.

I drew inspiration from TallBear's (2014, pp. 2–3) call for a researcher who "stands with" the community they work with by striving to develop shared goals, "shared conceptual ground and shared stakes". This approach relies on the researchers' willingness to allow themselves and their "stakes in the knowledge to be produced" to be "altered" (p. 2). While all anthropologists expect to be altered through relationships and experiences with their interlocutors, "standing with" one's interlocutors involves embracing this transformation as a tool for more democratic knowledge production (*ibid*) and involves greater recognition and acceptance of one's interlocutors' agency in this process of alteration. Indeed, building on TallBear, the anthropologist Michael Cepek (2020, p. 5) encourages anthropologists to invest "political, scholarly, and emotional energies" into social worlds shared with one's interlocutors and to allow this to "transform" their lives to the extent that they cannot claim to "unilaterally control [their] work or even [their] intellectual, political, and personal positions".

My engagement in my interlocutor's activism and lives undoubtedly shaped my thinking and consequently my research approach. Trying to actively encourage this process, when I noticed that my interlocutors often lamented a loss of regenerative cultures – and attributed many of the movement's issues to this loss – I suggested to them that I focus my research on this "loss". Although the reception was generally positive, again I received little feedback. As Cepek (2020, p. 20) states, "anthropologists can never be certain that they are standing with their collaborators". Indeed, although I cannot claim unilateral control over my research focus, this does not guarantee that my interlocutors feel it reflects their interests. Initially I had hoped a direct exchange with my interlocutors could allow their knowledge to shape the research and avoid perpetuating knowledge hierarchies, yet I realised that instead my research must continually attempt to encompass and respond to the knowledge my interlocutors shared with me throughout fieldwork. Such continual adaptation is also essential since one's interlocutors are continually changing (Cepek, 2020) – particularly within climate activism where novelty and relevance to the current moment are vital. Of course, academic requirements to provide fixed research outcomes limit continual adaptation (Cepek, 2020, p. 21), yet as I later explain, this problem can be mitigated through balancing action and reflexivity.

The regenerative research approach this chapter lays out results from my attempt to "stand with" my interlocutors by enabling their knowledge to alter my perspectives, methods, and research aims, and thus to co-construct my research approach. I will continue to adapt this approach in response to the knowledge I encounter as my fieldwork progresses. I consider the ideas within this regenerative approach methodological tools in an "open, flexible, highly context-dependent, and time-sensitive repertory" (Malkki, 2007, p. 180), which I draw on in response to the knowledge I gain from my interlocutors. Engaging with my interlocutors' concept of regenerative cultures in this way has created shared stakes in exploring what regenerative cultures may constitute and how to build them, which I hope may also assist my interlocutors' activism.

Balancing Reflexivity and Action

The third core principle of regenerative cultures that informs my research approach is balancing action and reflection. XR's "regenerative action cycle" encourages activists to divide their time between civil disobedience and reflective activities, such as grieving the casualties of climate breakdown, debriefing actions, and sharing gratitude. Many of my interlocutors who advocated for regenerative cultures considered this balance vital, arguing that activism without sufficient reflexivity could unintentionally perpetuate the systemic exploitation they considered responsible for climate breakdown. This tension among my interlocutors paralleled debates about balancing intellectual and material decolonisation (Fanon, 1963; Moosavi, 2020; Tuck and Yang, 2012). Similarly, both critical academic reflexivity and practical fieldwork are foundational to anthropological research. However, within anthropology, reflexive cultural critique is often prioritised over engaged fieldwork methodologies as a tool to support political struggles, despite increasing recognition of knowledge as situated and collaboratively produced (Hale, 2006, p. 100). Unfortunately, academic cultural critique is often neither accessible nor necessarily valued by the people involved in the struggles the anthropologists are attempting to support.

This was highlighted for me two months into fieldwork, when I reminded activists in a London-wide meeting (which I was helping to facilitate) about my research and attempted to ask the many new members if they consented to being involved. One activist immediately asserted that they resented discussing this in the meeting as they considered it a distraction from activism. I explained that I would be as brief as possible but that I needed to ask for consent and preferred to do so openly in a meeting so that people could hear each other's questions or concerns. Another activist commented that since "XR is based on research and that's integral to the movement", he considered my research valuable. He was concerned, however, that by the time my thesis was written "we may not have PhDs and XR may not exist" due to climate-breakdown-induced societal collapse. I explained that I also planned to share accessible summaries of my research while writing it up. A third activist commented that they felt my research could help "see things that we cannot see about group dynamics", while another added that it could help to recognise and embed "the learning that's happening" in the group. A fifth activist politely declined being interviewed, explaining she did not object to my research but was tired of social scientists misunderstanding her.

The opposition to my request for consent during the meeting surprised me. Although I had felt reluctant to raise my research in the meetings since we were always short of time and it was not directly XR activism, I believed that my research could assist XR. I therefore considered it part of my activism (although sometimes in conflict with it) and had assumed my interlocutors would, too. Additionally, the comments about how my research could aid XR also concerned me, as I could not guarantee the insights they hoped for. Indeed, exploration of contentious topics and efforts to co-design and co-author often encounter complications that render engaged anthropology less likely to provide the academic outputs it proposes (Kirsch, 2018, p. 9). Such research therefore "has the potential of betrayal as well as support" (Merry, 2005, p. 255).

This fieldwork experience highlighted to me that my engagement in activism during fieldwork was the most reliable means through which my interlocutors may feel my research benefited them. Indeed, the geographer Laura Pulido (2008, p. 350) argues that, for activists working with anthropologists, the extent of the anthropologist's political engagement in their activism during research often primarily determines whether they feel the research benefited them. Investing time and energy into one's interlocutor's activism, Pulido argues, is therefore a vital means of reciprocating the knowledge gained during fieldwork. Framing such exchanges as "giving back", however, risks reinforcing the researcher/researched dichotomy and associated power inequalities (TallBear, 2014). Framing the activist labour of engaged anthropologists as "community service" devalues its importance as a means of knowledge production (Scheper-Hughes, 2009, p. 3). Indeed, I took on additional responsibilities in XR during fieldwork as an effort to reciprocate my interlocutors *and* to practise activism I consider necessary *and* to further my understanding of XR.

I continue to engage in XR activism while writing up my research as this helps ground my analysis in practical action and to contribute insights from my research directly without barriers such as inaccessible academic language or slow publishing processes. Balancing critical academic analysis with practical engagement in activism in this way allows the knowledge gained from each to feed into the other and, as TallBear (2014, p. 6) argues, is likely to achieve deep change more quickly. While all XR activists contribute insights and reflections to the movement, much of which are grounded in more experience than I have, observations and analysis developed through engaged academic research can nevertheless offer unique perspectives and insights on the climate movement. Indeed, as both the anthropologist Sally Merry (2005, p. 243) and one of my interlocutors in the meeting described earlier have pointed out, activism often draws directly from academic research. The principle within XR's concept of regenerative cultures that balancing reflexivity and action is necessary to achieve sufficiently radical change therefore provides valuable guidance for efforts to research climate activism in a manner that assists such activism.

Conclusion: The Need for Regenerative Research

This chapter has woven together knowledge about regenerative cultures gained from my interlocutors during fieldwork, with scholarship in engaged anthropology, to suggest a framework for a "regenerative" approach to research with grassroots climate activists. This approach was structured around three principles that I consider central to my interlocutors' understanding of regenerative cultures: prioritising care, recognising co-construction, and balancing reflexivity and action. I began by illustrating how an ethic of care, informed by a perception of interconnection, underpins XR's concept of regenerative cultures and also engaged anthropology's aim to relate to one's interlocutors as "comrades" in a shared struggle. Through examples from my fieldwork, I demonstrated how such care can complicate the researcher's relationship with their interlocutors, challenge their political involvement in ways that are constructive academically and for activism, and help researchers of climate activism to understand the tensions their interlocutors experience.

I then explained the second principle, recognising co-construction, which describes the idea within XR's regenerative cultures that both social and environmental conditions are co-constructed due to the inevitable interconnection between all life. I combined this idea with scholarship from feminist and engaged anthropology that, based on a rejection of knowledge hierarchies, encourages building shared conceptual ground and aims with one's interlocutors during research to allow the anthropologist's thinking and the research to be more democratically co-constructed. I argued that recognition of co-construction is increasingly crucial in the study of climate activism, as it enables research to advance shared goals with the activists we study.

Turning to the third principle, balancing reflexivity and action, I then explained how it relates to my interlocutors' perception that achieving the change they desired required a careful balance between reflexive activities and direct action. I illustrated how this idea overlaps with anthropological arguments to combine intellectual efforts to support the struggles of the communities we study with practical engagement in these struggles. Drawing on both these ideas, I proposed that regenerative research of climate activism requires balancing practical political engagement with academic critique so that insight from each may feed easily into the other, increasing the power of both to assist struggles for climate justice. Although I have separated these three principles for the sake of clear explanation, each informs and motivates the others, therefore all three must be engaged with to produce regenerative research of the climate movement.

A degree of ethical alignment with one's interlocutors is of course a prerequisite for this regenerative approach. However, since much research on grassroots climate activism is motivated by a desire to mitigate climate breakdown, this approach is well suited to the study of climate activism. While I have demonstrated the value of these three principles in relation to my anthropological research with XR, the principles of care, recognising knowledge as co-constructed, and balancing action and reflexivity can arguably be applied to the study of climate activism from any discipline, regardless of whether ethnographic fieldwork is involved. Moreover, the points where these principles jar with one's research approach – or the climate activism one studies – may provide some of the most productive areas for critical reflection.

As I have illustrated, attempting to practise a "regenerative" research approach raises many ethical and practical dilemmas about who the research benefits, how to reconcile conflicting perspectives, and where one's responsibilities lie. The resulting complication of my perception on – and relationship with – XR has aided both my academic critique and my activism. There is no roadmap for navigating the dilemmas of engaged research, as each dilemma is highly personal and context specific. However, I have demonstrated that while these three principles of regenerative cultures generate productive complications, they simultaneously offer valuable guidance for navigating the resulting complexities.

Escalating climate breakdown demands approaches to researching grassroots climate activism that share the aims of this activism, that engage with the concepts these activists develop to meet those aims as forms of knowledge, and that are attuned to adapt to emerging forms of climate activism. The regenerative research approach I have proposed is an attempt to respond to this need. This chapter has demonstrated how engaging with one's interlocutors' perspectives as a form of knowledge as well as research data can open the mind "to working in non-standard ways" (TallBear, 2014, p. 6). Navigating the complexities involved in "non-standard" approaches and relationships with one's interlocutors helps increase recognition of one's biases and the systemic inequalities underpinning them. Such approaches are vital if our research is to share goals with the Western climate movement, which is itself increasingly striving to oppose systemic forms of exploitation through generating alternative forms of relationality.

References

Adams, T. E., Ellis, C. and Jones, S. H. (2017). Autoethnography. *The International Encyclopedia of Communication Research Methods* (pp. 1–11). Hoboken, NJ: John Wiley & Sons.

Bonilla, Y. (2020). *Non-Sovereign Futures: French Caribbean Politics in the Wake of Disenchantment*. Chicago: University of Chicago Press.

Bookchin, M. (1987). Social ecology versus deep ecology: A challenge for the ecology Movement. In *Green Perspectives: Newsletter of the Green Program Project*. The Anarchist Library. https://theanarchistlibrary.org/library/murray-bookchin-social-ecology-versus-deep-ecology-a-challenge-for-the-ecology-movement [Accessed 18 August 2023].

Casas-Cortés, M. I. (2009). *Social Movements as Sites of Knowledge Production: Precarious Work, the Fate of Care and Activist Research in a Globalizing Spain*. Doctoral Dissertation. The University of North Carolina at Chapel Hill.

Cepek, M. (2020). Engaging, standing, and stepping: Evaluating anthropological activism on an Amazonian petro-frontier. *Commoning Ethnography*, 3(1), 5–24.

Chua, L. (2015). Troubled landscapes, troubling anthropology: Co-presence, necessity, and the making of ethnographic knowledge. *Journal of the Royal Anthropological Institute*, 21(3), 641–659.

De Castro, V. E. (2019). On models and examples: Engineers and bricoleurs in the Anthropocene. *Current Anthropology*, 60(20), 296–308.

Doherty, B., Hayes, G. and Saunders, C. (2020). A new climate movement? Extinction Rebellion's activists in profile. In *CUSP Working Paper No 25*. Guildford: Centre for the Understanding of Sustainable Prosperity. https://www.cusp.ac.uk/themes/p/xr-study/#1579557747983-bfc4b6a4-fd4ccb6c-f99e1cd0-f327 [Accessed 30 March 2023].

Fanon. F. (1963). *The Wretched of the Earth*. New York: Grove Press.

Hale, C. R. (2006). Activist research v. cultural critique: Indigenous land rights and the contradictions of politically engaged anthropology. *Cultural Anthropology*, 21(1), 96–120.

Haraway, D. (1988). Situated knowledges: The science question in feminism and the privilege of partial perspective. *Feminist Studies*, 14(3), 575–599.

Haraway, D. J. (2016). *Staying with the Trouble: Making Kin in the Chthulucene*. Durham: Duke University Press.

Harms, A. (2021). What kinds of activism do regenerative cultures fuel and how might we research them. *Social Anthropology*, 29(1), 238–240.

Harms, A. (2022). Beyond dystopia: Regenerative cultures and ethics among European climate activists. *American Anthropologist*, 124(3), 515–524.

Ingold, T. (2021). *Being Alive: Essays on Movement, Knowledge and Description*. Abingdon and New York: Routledge.

IPCC. (2018). Summary for policymakers. In V. Masson-Delmotte, P. Zhai, H.-O. Pörtner, D. Roberts, J. Skea, P. R. Shukla, A. Pirani, W. Moufouma-Okia, C. Péan, R. Pidcock, S. Connors, J. B. R. Matthews, Y. Chen, X. Zhou, M. I. Gomis, E. Lonnoy, T. Maycock, M. Tignor and T. Waterfield (eds.) *Global Warming of 1.5°C. An IPCC Special Report on the Impacts of Global Warming of 1.5°C Above Pre-Industrial Levels and Related Global Greenhouse Gas Emission Pathways, in the Context of Strengthening the Global Response to the Threat of Climate Change, Sustainable Development, and Efforts to Eradicate Poverty* (pp. 3–24). Cambridge: Cambridge University Press.

Juris, J. S. (2008). Performing politics: Image, embodiment, and affective solidarity during anti-corporate globalization protests. *Ethnography*, 9(1), 61–97.

Kirsch, S. (2018). *Engaged Anthropology: Politics Beyond the Text*. Berkeley: University of California Press.

Knox, H. (2015). Thinking like a climate. *Distinktion: Scandinavian Journal of Social Theory*, 16(1), 91–109.

Lassiter, L. E. (2005). Collaborative ethnography and public anthropology. *Current Anthropology*, 46(1), 83–106.

Latour, B. (2017). *Facing Gaia: Eight Lectures on the New Climatic Regime*. Cambridge: John Wiley and Sons.

Latour, B. (2018). *Down to Earth: Politics in the New Climatic Regime*. Cambridge: John Wiley and Sons.

MacCormack, C. and Strathern, M. (eds.) (1980). *Nature, Culture and Gender*. Cambridge: Cambridge University Press.

Macy, J. and Johnstone, C. (2012). *Active Hope: How to Face the Mess We're in Without Going Crazy*. Novato, CA: New World Library.

Malkki, L. H. (2007). Tradition and improvisation in ethnographic field research. In A. Cerwonka and L. H. Malkki (eds.) *Improvising Theory: Process and Temporality in Ethnographic Fieldwork* (pp. 162–188). Chicago: University of Chicago Press.

Malm, A. (2021). *How to Blow Up a Pipeline*. London: Verso Books.

Merry, S. E. (2005). Anthropology and activism: Researching human rights across porous boundaries. *PoLAR: Political and Legal Anthropology Review*, 28(2), 240–257.

Mies, M. and Shiva, V. (2014). *Ecofeminism*. London: Zed Books Ltd.

Moosavi, L. (2020). The decolonial bandwagon and the dangers of intellectual decolonisation. *International Review of Sociology*, 30(2), 332–354.

Morton, T. (2007). *Ecology Without Nature: Rethinking Environmental Aesthetics*. Cambridge, MA: Harvard University Press.

Nader, L. (1974). Up the anthropologist: Perspectives gained from studying up. In D. Hymes (ed.) *Reinventing Anthropology* (pp. 285–311). Pantheon Books.

Pulido, L. (2008). FAQs: Frequently (Un)asked questions about being a scholar activist chapter author(s). In C. R. Hale (ed.) *Engaging Contradictions: Theory, Politics, and Methods of Activist Scholarship* (pp. 341–366). Berkeley: University of California Press.

Russell, B. (2015). Beyond activism/academia: Militant research and the radical climate and climate justice movement(s). *Area*, 47(3), 222–229. https://doi.org/10.1111/area.12086.

Scheper-Hughes, N. (1995). The primacy of the ethical: Propositions for a militant anthropology. *Current Anthropology*, 36(3), 409–440.

Scheper-Hughes, N. (2009). Making anthropology public. *Anthropology Today*, 25(4), 1–3.

Schuurman, N. and Pratt, G. (2002). Care of the subject: Feminism and critiques of GIS. *Gender, Place & Culture*, 9(3), 291–299.

Shah, A. (2017). Ethnography? Participant observation, a potentially revolutionary praxis. *HAU: Journal of Ethnographic Theory*, 7(1), 45–59.

Strang, V. (2016). Justice for All: Inconvenient truths and reconciliation in human-non-human relations. In H. Kopnina and E. Shoreman-Ouimet (eds.) *Routledge Handbook of Environmental Anthropology* (pp. 259–273). Abingdon and New York: Routledge.

Tallbear, K. (2014). Standing with and speaking as faith: A feminist-indigenous approach to inquiry [Research note]. *Journal of Research Practice*, 10(2), Article N17. http://jrp.icaap.org/index.php/jrp/article/view/405/371.

Taylor, B. (2009). *Dark Green Religion: Nature Spirituality and the Planetary Future*. Berkeley: University of California Press.

Tuck, E. and Yang, K. W. (2012). Decolonization is not a metaphor. *Decolonization, Indigeneity, Education & Society*, 1(1), 1–40.

Weiss, M. (2021). The interlocutor slot: Citing, crediting, cotheorizing, and the problem of ethnographic expertise. *American Anthropologist*, 123(4), 948–953.

XRGlobal (2024). *Wellbeing*. https://rebellion.global/about-us/ [Accessed 18 January 2024].

XRUK (2024). *Wellbeing*. https://extinctionrebellion.uk/ [Accessed 18 January 2024].

XRUK Newsletter (2020). *Newsletter 16: The Rebellion Returns*, 1 September. https://extinctionrebellion.uk/2020/07/03/uk-newsletter-16-the-rebellion-returns-1-september/ [Accessed 30 August 2023].

40

A FUTURE THROUGH GRASSROOTS CLIMATE ACTIVISM

An Outlook

Sabine von Mering, Wendy Steele, and Alexandre da Silva Faustino

As one hottest year on record chases another and extreme weather events become the norm rather than the exception, more and more people are motivated to take action – whether they join an existing organisation, become active in their neighbourhood, or engage through the many forms of activism described in this volume. The more the impacts of climate disruption are felt not only in countries of the Global Majority – where entire regions are threatened to become uninhabitable within this century[1] – but also in the Global North, where too many are still viewing themselves as immune to the crisis (Ballew et al., 2024), the more climate change will act as the "threat multiplier" that it is (*Age of Consequences*, 2016). This will increasingly lead to disruptions of the global economy and its globalised supply chains as witnessed during the COVID-19 pandemic and increased migration from South to North and from coastal areas to interior regions, which often go hand-in-hand with a rise in xenophobia, racism, and right-wing extremism that threaten even relatively stable democratic societies (Gonzalez, 2020). While the origins of the COVID-19 pandemic remain unresolved, scientists are convinced that planetary-scale heating will continue to lead to the spread of new pathogens and raise the threat of pandemics.[2]

This handbook has specifically focused on grassroots climate activism. In doing so, it draws together a wide range of climate activism undertaken in different parts of the world to demonstrate the challenges activists are facing and the variety of tactics and strategies they employ to meet them. Two important meta-themes emerge as critical for the future of climate activism as a grassroots phenomenon. *First*, we see this as part of a decentralised and multi-actor model of transformative change. Working at the grassroots level emphasises the rhizomatic nature of activist agency and a co-creative approach to building the capabilities of multiple actors – the more-than-human. The challenge and opportunity for decentralised activism is how best to harness the unfurling and adjacent grassroots possibilities already in emergence or in varying stages of development, in the time needed to address the impacts of climate change. Part of this entails demonstrating alternatives towards a future of mutual thriving and finding ways to share insights openly for collective learning-led climate action.

DOI: 10.4324/9781003396567-47

A *second* key meta-theme to emerge from the handbook centres around what John Goddard (2018, p. 363) characterises as civic groups and institutions being place-based and grounded in ecological and cultural contexts, which are "not just in the place but of the place." Characteristics that distinguish this grassroots activist orientation that connects to the need for greater solidarity and action on climate change include: active engagement with both the local community and the planetary; taking a holistic approach to engagement so that it is not just confined to specific individuals or teams; having a strong sense of place and acknowledging extent to which place-based activism helps to form identity and purpose; holding a clear activist understanding not just about what activism is good at, but what it is good for; being willing to invest in people and resources in order to have impact; and a commitment to accountability to community, to the public, and to dwelling well in the world at large.

Together, the chapters in the handbook highlight the innovative capacity, expansive scope, and relentless viability of grassroots climate activism. From Indigenous communities with their millennial history of caring with Country and successfully resisting destructive forces on this planet and their fighting for the right to protect their land, communities, and non-human kins; to artists, writers, and filmmakers who are finding new ways to make the climate crisis visible, audible, and feel-able; to young and old climate strikers who are developing new modes of protest; to those fighting climate adaptation and climate litigation within traditional institutions; to the scholars and teachers whose job it is to translate what they are studying so that it becomes understandable for everyone else. All are using the tools and creative methods available to them and developing new ones all the time.

The chapters demonstrate these and other characteristics that are critical to the future of grassroots climate activism: pluralism and different forms of knowing; a porousness of boundaries particularly across and between disciplines and activist practices; a holism of research, education, scholarship, and learning that brings together the practical, technical, abstract, aesthetic, and spiritual; cooperativism and an ethos of collaboration over competition; compassion and the recognition of the importance of an ethics of care in all aspects of life; and collectivism and the meaningful learning within and across a web of relationships between people and with the non-human world (Brown and McGowan, 2018, Steele and Rickards, 2021).

This volume cannot possibly do justice to the millions of people around the world who identify as grassroots climate activists. There are chapters we hoped to include that did not materialise, and there are stories that we wanted to feature, but we were of course limited by our own connections, within both academic and activist circles. We would have loved to include chapters that explore the unique contributions of other groups, for example labour activists who are allying themselves with the climate movement or psychologists who are addressing the growing epidemic of climate anxiety and depression through their expertise (Stanley et al., 2021; Clery et al., 2022). We would have loved to have included more elaborate chapters on the Degrowth movement, the fight against climate change from within the public health field (Sasser, 2018), the Climate Justice Alliance and the Just Transition Framework in the US (see Chapter 1), and more on the ongoing fights against extractive industries and for a just transition (see Chapter 2 and 21). Luckily, much has been written about these elsewhere (see for example Estes, 2019; Fairchild and Weinrub, 2017; Saito, 2024; Rodriguez et al., 2023).

What all the research confirms and portrays is that, in response to the exacerbating crisis, climate activists are developing new initiatives all the time. Climate gaming is one area we did not cover in this volume since it tends to be driven by game inventors and companies rather than grassroots activists. Future grassroots organisations will no doubt utilise new forms of awareness raising, organising, and mobilising with the help of gaming, social media, and artificial intelligence as well, while also continuing to probe the pitfalls of technological solutions.

In this volume we focus on nonviolent grassroots climate activism, because we firmly believe in the power of the people to effect change, and it is our fervent hope that our readers will be inspired to get involved in one of the many nonviolent climate organisations studied in these chapters – or start forming one of their own. There is no question that violence is already part of climate activism – violence perpetrated not only by a verified mendacious and criminally ruthless fossil-fuel industry and its enablers but also by the systems that criminalise those who are challenging their dominance in efforts to protect a liveable planet. Whether one agrees with Andreas Malm – whose 2021 book *How to Blow up a Pipeline* inspired the 2022 feature film of the same title directed by Daniel Goldhaber – or not, frustration with government inaction may well lead more people to adopt more violent responses, whether it be targeted acts of sabotage similar to US Catholic worker Jessica Reznicek who paid for her act of civil disobedience with a terrorism conviction and 8 years in prison (Grubb, 2021) or acts of self-immolation similar to David Buckel in 2018 and Wynn Bruce in 2022 (Cameron, 2022).

The critiques in Part V also remind us that activists and those who support them must remain open to criticism and be willing to change course. Rather than viewing those who expose blind spots on the part of the climate movement as betraying the cause or the harmony of unity, confronting uncomfortable truths and allowing for critical voices to emerge is essential to ensuring that the movement remains vital and agile and able to embrace needed course corrections. Inherent to nurturing collective processes, there are many productive outcomes that emerge from acknowledging, listening to, and moderating differences. They are also important reminders that the conflicts – either internal or external – fuelling grassroots activism must not be overlooked, as this risks homogenising activist narratives that are plural by essence and not always will converge in priorities and strategies (Kakenmaster, 2019). Climate activism can be enacted in many different ways, coloured by the granularity of the always growing diversity of positionalities that come together under organised climate actions (Bond et al., 2020).

Just as with other activist organisations in the past, climate activists' methods will continue to evolve, and people's association with different groups will remain fluid. New forms of climate activism are bound to arise. While we don't have a crystal ball, we can foresee that grassroots climate activists will continue to be creative in their targeting, messaging, and organising and also become better connected across borders. At the same time, climate disruptions will continue to shift the focus to adaptation measures and to mitigating conflicts arising from displacement.

As Naomi Klein and others have repeatedly warned, threats to democracy coincide with escalating climate threats for very obvious reasons, as taking action to address climate change is also a "chance to build a fairer and more democratic economy" (Klein, 2019). Since an increase in fairness threatens the entrenched inequities of the neoliberal order, those who benefit from the status quo have shown great resistance to embracing the change that is needed, whether they propagate climate denial and support outright anti-democratic, right-wing populism or hide behind a democratic mantle only to weaken and slow climate action (Lu, 2020). Efforts at realising a Green New Deal (in the US) and concretising the European Green Deal continue to face strong headwinds, adding yet another dimension to the complex list of challenges grassroots climate activists face. Effective coalition-building across the socio-economic spectrum and the fight for climate justice must embrace "more democracy, not less" to succeed (Willis, 2020, p. 5). The European Fridays for Future movement has been particularly successful at organising for climate justice *and* democracy in the context of European Parliament elections (Pollex and Berker, 2024). At the same time, the wars in Ukraine and the Middle East have caused activists to pay attention not only to the heart-breaking loss of life and blind misguidedness of the "eye for an eye and tooth for a tooth" mentality but also the excesses of largely unaccounted-for military carbon emissions and devastating environmental destruction. Above all, this has laid bare the brutal intersections between the climate crisis, land-based violence, deep human and non-human community suffering, the fragility of democracy, and our inability to protect both people and planet.

Finally, a note to the disaffected: headline-grabbing claims that "the climate movement is dead" have been around for years (Slowcode, 2016). We hope this volume instead helps amplify the abundance of grassroots climate activism that is not only alive and well on six continents but is growing by leaps and bounds in creative innovation, sophistication, and organisational reach. Those who suggest that "the climate movement has failed" are not only wrongfully shifting the blame from those who are perpetrating the crisis to those who are doing everything they can to move humanity forward into a better direction. They are also inadvertently suggesting that there is a definite endpoint at which we should all throw up our arms and give up, centring themselves and their own frustrations instead of com-munal needs. The many grassroots climate activists portrayed in this volume instead devote their time and energy to those – in the present and in the future – who are most directly impacted by the climate crisis.

While the scholarship of Grassroots Climate Activism will continue to chronicle, ana-lyse, and explain these developments, scholars will also continue to be activists for more equitable and regenerative futures. And we will have to do so in broadly transdiscipli-nary collaborations, which carry their own sets of challenges. However, as scholars we can model the solidarity across differences that this urgent work requires. In solidarity, we wel-come the future of climate activism, recognising that its roots are deeply embedded in the actually existing activist practices already underway in local community and place-based contexts focused at a planetary scale.

Notes

1 Based on scientific models, the German paper *Berliner Morgenpost* visualised these regions pow-erfully in 2022: https://interaktiv.morgenpost.de/klimawandel-hitze-meeresspiegel-wassermangel-stuerme-unbewohnbar/.
2 See https://www.hsph.harvard.edu/c-change/subtopics/coronavirus-and-climate-change/.

References

Age of Consequences. USA (2016). Jared P. Scott, 80 min. Kelly Nyks, Jared P. Scott, Sophie Robin-son. https://filmsfortheearth.org/en/film/the-age-of-consequences/
Ballew, E., Carman, J., Verner, M., Rosenthal, S., Maibach, E., Kotcher, J. and Leiserowitz, A. (2024). The attitude-behavior gap on climate action: How can it be bridged? *Yale Climate Change Communication. Climate Note,* 21 March. https://climatecommunication.yale.edu/publications/attitude-behavior-gap/
Bond, S., Thomas, A. and Diprose, G. (2020). Making and unmaking political subjectivities: Climate justice, activism, and care. *Transactions of the Institute of British Geographers,* 45(4), 750–762. https://doi.org/10.1111/tran.12382

Brown, E. and McGowan, T. (2018). Buen vivir – Reimagining education and shifting paradigms. *Compare*, 48(2), 317–323. https://www.pbs.org/wnet/peril-and-promise/2023/01/extinction-rebellion-shifts-away-from-public-disruption/

Cameron, C. (2022). Climate activist dies after setting himself on fire. *New York Times*, 22 April. https://www.nytimes.com/2022/04/24/us/politics/climate-activist-self-immolation-supreme-court.html [Accessed 28 April 2024].

Clery, P., Embliss, L., Cussans, A., Cooke, E., Shukla, K. and Li, C. (2022). Protesting for public health: A case for medical activism during the climate crisis. *International Review of Psychiatry*, 34(5), 553–562. https://doi.org/10.1080/09540261.2022.2093627

Estes, N. (2019). *Our History is the Future. Standing Rock Versus the Dakota Access Pipeline, and the Long Tradition of Indigenous Resistance.* London: Verso.

Fairchild, D. and Weinrub, A. (2017). Energy democracy. In D. Lerch (ed.) *The Community Resilience Reader* (pp. 195–206). Washington, DC: Island Press. https://doi.org/10.5822/978-1-61091-861-9_12

Goddard, J. (2018). The civic university and the city. In *Geographies of the University* (pp. 355–373, 363). Cham: Springer.

Gonzalez, C. (2020). Climate change, race, and migration. *Journal of Law and Political Economy*, 1(1). https://doi.org/10.5070/LP61146501

Grubb, C. (2021). A tale of two damages: Double standard for Jessica Reznicek and energy transfer partners. *Common Dreams*, 25 October. https://www.commondreams.org/views/2021/10/25/tale-two-damages-double-standard-jessica-reznicek-and-energy-transfer-partners.

Kakenmaster, W. (2019). Articulating resistance: Agonism, radical democracy and climate change activism. *Millennium: Journal of International Studies*, 47(3), 373–397. https://doi.org/10.1177/0305829819839862

Klein, N. (2019). *On Fire: The (Burning) Case for a Green New Deal.* New York: Simon & Schuster.

Lu, C.-Y. (2020). *Surviving Democracy: Mitigating Climate Change in a Neoliberalized World.* London: Routledge.

Malm, A. (2021). *How to Blow Up a Pipeline.* London: Verso.

Pollex, J. and Berker, L. E. (2024). The European parliament and Fridays for Future: Analysing reactions to a new environmental movement by Europe's climate policy champion. *Journal of European Integration*, 1–18. https://doi.org/10.1080/07036337.2024.2334079.

Rodriguez, I., Walter, M. and Temper, L. (eds.) (2023). *Just Transformations. Grassroots Struggles for Alternative Futures.* London: Pluto Press.

Saito, K. (2024). *Slow Down. The Degrowth Manifesto.* Astra House.

Sasser, J. S. (2018). Public health in the anthropocene: Exploring population fears and climate threats. In C. O'Manique and P. Fourie (ed.) *Global Health and Security* (pp. 166–178). London: Routledge.

Slowcode, D. (2016). The climate movement is dead. A reportback from the COP21 in Paris. *Earth First! The Radical Environmental Journal*, 36(1), 42–44.

Stanley, S. K., Hogg, T. L., Leviston, Z. and Walker, I. (2021). From anger to action: Differential impacts of eco-anxiety, eco-depression, and eco-anger on climate action and wellbeing. *The Journal of Climate Change and Health*, 1, 1–5. https://doi.org/10.1016/j.joclim.2021.100003

Steele, W. and Rickards, L. (2021). *The Sustainable Development Goals in Higher Education: A Transformative Agenda?* London: Palgrave.

Willis, R. (2020). *Too Hot to Handle? The Democratic Challenge of Climate Change.* Bristol University Press.

INDEX

Note: page numbers in **bold** indicate a figure and page numbers in *italics* indicate a table on the corresponding page.

Gage, George 517
Galindo, Regina José 163–164
Gamarra, Alonso 571n6
Gameau, Damon 516–517, **517**
Garcia, Alixa 159
GARN *see* Global Alliance for the Rights of Nature (GARN)
Gaskell, Elizabeth 182
Gates, Bill 179
GECCI *see* Global Environmental and Climate Conservation Initiative (GECCI)
Geddes, Stuart 196
Gee, Maggie 177
Gehl, Jan 515
generational tensions 285–286
Gerhardt, Christina 178
Germany 322–334, 356–357, **411, 413**, 452–453, 469–472
Gheorghiu, Andy 471
GHNRE *see* Global Network for Human Rights and the Environment (GNHRE)
Ghosh, Amitav 180
Giddens, Anthony 170n7
Giugni, Marco 279
Global Alliance for the Rights of Nature (GARN) 167
Global Climate Strike 7
Global Environmental and Climate Conservation Initiative (GECCI) 490–491
Global Network for Human Rights and the Environment (GNHRE) 356
Global North 3, 5, 44, 163, 170n2, 223, 231–232, 241, 276, 315, 323, 386, 435, 599
Global South 3, 163, 170n2, 223, 231, 345, 383, 518
global warming 1, 3, 6, 45, 156, 280
Goade, Michaela 176–177
Goddard, John 613
Gold, Heloise 137
Goldhaber, Daniel 518, 615
Goldtooth, Tom 156
Gore, Al 185n13, 509, **510**, 512, 516
Gramsci, Antonio 182
grassroots climate activism(s): collaborative 73; community-centric 73; fiction as 173–184; future through 613–616; from the ground up 1–3; mobilisation of 53–55; nonfiction as 173–184; in North Africa 381–392; Pacific youth and 64–74; rise and evolution of 485–486; in unequal waterscapes 44–59; as urgent 1–3
Grassroots Organizing Cools the Planet: Letter From the Grassroots to 1Sky 4
Green, Nick 513
GreenFaith 297, 301–302
green hydrogen 384, 487, 489, 493

Green New Deal 5–6, 101, 103, 615
Greenpeace 145, 258, 259, 357, 448, 451, 564
Greenslade, Roy 510
greenwashing 447–454, 463–477
Greta and the Giants (Tucker) 177
Grist 173–174
ground-up 1–3
growthism 169n2
Guardians of the Forest 39
Guddling About 168
Gulf Coast 468, 474
Gunditjmara First Nations People's Ocean Defenders 114
Guterres, António 109, 526

habitus 65–66, 73–74
Haikola, Pirjo 199
Hall, Stuart 152
Hallam, Roger 310
Hambach Forest 324, 328–330, *330*
Hammersley, Martyn 372n1
Haque, Kamil 164–165
Haraway, Donna 602
Hardt, Michael 312
Harris, Rutha Mae 133
hashtag 22, 256–262
Hawken, Paul 179, 516, 521n16
Haydon, Kirsten 199
Hayward, Susan 521n11
Helguera, Pablo 127, 132
Helvey, Robert 595
heritage, activism as connection to 72–73
High, Mette 439
Hodgson, Nikki 149
Hogan, Linda 129
holistic 90, 506, 513, 579–580
Honduras **410**
Honoré, Carl 529
honour 296
hooks, bell 133
hope 190–203
Hopepunk 173–174
Hopkins, Rob 179–182, 512, 515
Howard, John 367
Howarth, Robert W. 463
Howland, Daniel 24
How to Blow Up a Pipeline (film) 518–520
Huang, Zechen 199, 201–202
Hulbert, Tammy 199
human rights 28, 52, 54, 92, 348–350, 352–358
human-water entanglement 110, 114, 199
Hurricane Katrina 164–165
hydrogen, green 384, 487, 489, 493
hyperobjects 182–183